Scalable Parallel Computing

McGraw-Hill Series in Computer Engineering

McGraw-Hill Series in Computer Science

Senior Consulting Editor
C.L. Liu, *University of Illinois at Urbana-Champaign*

Consulting Editor
Allen B. Tucker, *Bowdoin College*

Fundamentals of Computing and Programming
Computer Organization and Architecture
Computers in Society/Ethics
Systems and Languages
Theoretical Foundations
Software Engineering and Database
Artificial Intelligence
Networks, Parallel and Distributed Computing
Graphics and Visualization
The MIT Electrical and Computer Science Series

McGraw-Hill Series in Networks, Parallel and Distributed Computing

Ahuja: *Design and Analysis of Computer Communication Networks*
Filman and Friedman: *Coordinated Computing: Tools and Techniques for Distributed Software*
Forouzan: *Introduction to Data: Communications and Networking*
Hwang: *Advanced Computer Architecture: Parallelism, Scalability, Programmability*
Hwang and Xu: *Scalable Parallel Computing: Technology, Architecture, Programming*
Hwang and DeGroot (eds.): *Parallel Processing for Supercomputers and Artificial Intelligence*
Keiser: *Local Area Networks*
Kershenbaum: *Telecommunications Network Design Algorithms*
Laksmivarahan and Dhall: *Analysis and Design of Parallel Algorithms*
Quinn: *Designing Efficient Algorithms for Parallel Computers*
Siegel: *Interconnection Networks for Large-Scale Parallel Processing*
Walrand: *Communication Networks*

Scalable Parallel Computing

Technology, Architecture, Programming

Kai Hwang
University of Hong Kong
University of Southern California

Zhiwei Xu
Chinese Academy of Sciences
University of Hong Kong

WCB McGraw-Hill

Boston   Burr Ridge, IL   Dubuque, IA   Madison, WI   New York   San Francisco   St. Louis
Bangkok   Bogotá   Caracas   Lisbon   London   Madrid
Mexico City   Milan   New Delhi   Seoul   Singapore   Sydney   Taipei   Toronto

WCB/McGraw-Hill
A Division of The McGraw-Hill Companies

Scalable Parallel Computing
Technology, Architecture, Programming

This book is printed on acid-free paper.

1 2 3 4 5 6 7 8 9 0 DOC DOC 9 0 0 9 8

ISBN 0-07-031798-4

Editorial Director: Kevin Kane
Publisher: Tom Casson
Sponsoring editor: Lynn Cox
Marketing manager: John Wannemacher
Project manager: Richard DeVitto
Production supervisor: Richard DeVitto
Cover designer: Francis Owens
Editorial Assistant: Nina Kreiden
Printer: Quebecor Printing Dubuque

Library of Congress Cataloging-in-Publication Data
Hwang, Kai.
    Scalable parallel computing : technology, architecture,
  Programming / Kai Hwang, Zhiwei Xu.
        p.   cm.
   Includes index.
   ISBN 0-07-031798-4
   1.   Parallel processing (Electronic computers)   2.   Computer
architecture. 3. Computer networks. I. Xu, Zhiwei. II. Title.
QA76.58H85 1997
004'.358—dc21   97-41663
                                   CIP

http://www.mhhe.com

Trademark Notice

ANSI Standards X3T11, X3T9, and X3H5 are trademarks of the American National Standards Institute.

ALLICACHE and KSR-1 are trademarks of Kendall Square Research.

X Window, Alewife, J-Machine, and *T are trademarks of Massachusetts Institute of Technology.

CM-2, CM-5, C*, *Lisp, CM Fortran, and CMOST are trademarks of Thinking Machines Corporation.

HP/Convex Exemplar is a trademarks of Hewlett-Packard Company and Convex Computer Systems.

Cray 1, Cray 2, Y-MP, C90, CRAFT, T3D, T3E, UNICOS are trademarks of Cray Research, Inc.

Dash and SPLASH are trademarks of Stanford University.

Ethernet is a trademark of Xerox Corporation.

Gigabit Ethernet is trademark of the Gigabit Ethernet Alliance.

SCI, POSIX Threads, and Futurebus+ are trademarks of the Institute of Electrical and Electronic Engineering, Inc.

Illiac IV, Cedar, and Parafrase are trademarks of University of Illinois.

Intel 80x86, i960, i860, SHV, Paragon, Intel Hypercube, Pentium, Pentium Pro, and Pentium-II are trademarks of Intel Corporation.

MIPS R2000, R4000, and R10000 are trademarks of SGI/MIPS Technology, Inc.

Mach/OS, C.mmp, and Cm* are trademarks of Carnegie-Mellon University.

NYU Ultracomputer is a trademark of New York University.

SX/4 is a trademark of NEC Information Systems.

Symmetry and NUMA-Q are trademarks of Sequent Computers.

Gigaplane, Gigaplane-XB, S-Bus, TSO, PSO, NFS, SPARC, Ultra Enterprise, SPARCcluster, SPARCstation, and Solaris MC are trademarks of Sun Microsystems, Inc.

Tera is a trademark of Tera Computer Systems.

Cosmic Cube and Mosaic C are trademarks of California Institute of Technology.

UNIX 4.3 BSD, NOW, GLUnix, IRAM, xFS, and 4.4BSD are trademarks of the University of California at Berkeley.

DEC, DECsafe, VAX, VAXCluster, VMS, Digital Unix, TruCluster, Digital NT Cluster, Alpha 21164, and Alpha AXP are trademarks of Digital Equipment Corporation.

Apollo Domain, HP, Series 9000, PA-RISC, and UX are trademarks of the Hewlett-Packard Company.

IBM PC, AIX, ESSL, 801, RP3, EUI, HACMP, IBM, LoadLeveller, NetView, POWER, PowerPC, RS/6000, S/370, S/390, SP2, SP, and Sysplex are trademarks of International Business Machines Corporation.

Lifekeeper, NCR and Teradata are trademarks of National Cash Register.

M680x0 is trademark of Motorola Corporation.

OSF/1, DCE, and DFS are trademarks of the Open Software Foundation, Inc.

Oracle Parallel Server is a trademark of Oracle Corporation.

Express is a trademark of Parasoft Corporation.

Load Sharing Facility is a trademark of Platform Computing Corporation.

Origin 2000, Power C, POWER CHALLENGEarray, and Cellular IRIX are trademarks of Silicon Graphics Computer Systems.

Sybase is a trademark of Sybase, Inc.

UNIX and SVR5 are trademarks of X/Open Company, Ltd.

Microsoft, Wolfpack, and Windows NT are trademarks of Microsoft Corporation.

OpenMP is a trademark of OpenMP Standards Evaluation Board.

Myrinet is a trademark of Myricom, Inc.

About the Authors

Kai Hwang presently holds a Chair Professor of Computer Engineering at the University of Hong Kong (HKU), while taking a research leave from the University of Southern California (USC). He has been engaged in higher education and computer research for 26 years, after earning the Ph.D. in Electrical Engineering and Computer Science from the University of California at Berkeley. His work on this book started at the USC and was mostly completed at the HKU.

An IEEE Fellow, he has published extensively in the areas of computer architecture, digital arithmetic, parallel processing, and distributed computing. He is the founding Editor-in-Chief of the *Journal of Parallel and Distributed Computing*. He has chaired the international conferences: ICPP86, ARITH-7, IPPS96, ICAPP 96, and HPCA-4 in 1998. He has received several achievement awards for outstanding research and academic contributions to the field of parallel computing.

He has lectured worldwide and performed consulting and advisory work for US National Academy of Sciences, MIT Lincoln Laboratory, IBM Fishkill, TriTech in Singapore, Fujitsu and ETL in Japan, GMD in Germany, CERN School of Computing, and Academia Sinica in China. Presently, he leads a research group at HKU in developing an ATM-based multicomputer cluster for high-performance computing and distributed multimedia, Intranet, and Internet applications.

Zhiwei Xu is a Professor and Chief Architect at the National Center for Intelligent Computing Systems (NCIC), Chinese Academy of Sciences, Beijing, China. He is also a Honorary Research Fellow of HKU. He received a Ph.D. in Computer Engineering from the University of Southern California. He participated in the STAP/MPP benchmark projects led by Dr. Hwang at USC and HKU during the past 5 years.

He has taught at the Rutgers University and New York Polytechnic University. He has published in the areas of parallel languages, pipelined vector processing, and benchmark evaluation of massively parallel processors. Presently, he leads a design group at NCIC in building a series of cluster-based superservers in China. His current research interest lies mainly in network-based cluster computing and the software environments for parallel programming.

Dedication

To my family, for their love and support -- k.h.

To my high-school teacher,
Ms. Zhang Dahua, who taught me the
serenity of science at a tumultuous time -- z.x.

Table of Contents

About the Authors   viii

Foreword   xix

Preface   xx

Guide to Instructors/Readers   xxiii

Part I   Scalability and Clustering   1

Chapter 1   Scalable Computer Platforms and Models   3

   1.1   Evolution of Computer Architecture   5
      1.1.1   Computer Generations   5
      1.1.2   Scalable Computer Architectures   6
      1.1.3   Converging System Architectures   8

   1.2   Dimensions of Scalability   9
      1.2.1   Resource Scalability   9
      1.2.2   Application Scalability   11
      1.2.3   Technology Scalability   12

   1.3   Parallel Computer Models   13
      1.3.1   Semantic Attributes   14
      1.3.2   Performance Attributes   17
      1.3.3   Abstract Machine Models   18
      1.3.4   Physical Machine Models   26

   1.4   Basic Concepts of Clustering   30
      1.4.1   Cluster Characteristics   30
      1.4.2   Architectural Comparisons   31
      1.4.3   Benefits and Difficulties of Clusters   32

   1.5   Scalable Design Principles   37
      1.5.1   Principle of Independence   37
      1.5.2   Principle of Balanced Design   39
      1.5.3   Design for Scalability   44

   1.6   Bibliographic Notes and Problems   47

Chapter 2   Basics of Parallel Programming   51

   2.1   Parallel Programming Overview   51
      2.1.1   Why Is Parallel Programming Difficult?   52
      2.1.2   Parallel Programming Environments   55
      2.1.3   Parallel Programming Approaches   56

   2.2   Processes, Tasks, and Threads   59

    2.2.1   Definitions of an Abstract Process   59
    2.2.2   Execution Mode   62
    2.2.3   Address Space   63
    2.2.4   Process Context   65
    2.2.5   Process Descriptor   66
    2.2.6   Process Control   67
    2.2.7   Variations of Process   70

2.3 Parallelism Issues   71
    2.3.1   Homogeneity in Processes   72
    2.3.2   Static versus Dynamic Parallelism   74
    2.3.3   Process Grouping   75
    2.3.4   Allocation Issues   76

2.4 Interaction/Communication Issues   77
    2.4.1   Interaction Operations   77
    2.4.2   Interaction Modes   80
    2.4.3   Interaction Patterns   82
    2.4.4   Cooperative versus Competitive Interactions   84

2.5 Semantic Issues in Parallel Programs   85
    2.5.1   Program Termination   85
    2.5.2   Determinacy of Programs   86

2.6 Bibliographic Notes and Problems   87

Chapter 3   Performance Metrics and Benchmarks   91

3.1 System and Application Benchmarks   91
    3.1.1   Micro Benchmarks   92
    3.1.2   Parallel Computing Benchmarks   96
    3.1.3   Business and TPC Benchmarks   98
    3.1.4   SPEC Benchmark Family   100

3.2 Performance versus Cost   102
    3.2.1   Execution Time and Throughput   103
    3.2.2   Utilization and Cost-Effectiveness   104

3.3 Basic Performance Metrics   108
    3.3.1   Workload and Speed Metrics   108
    3.3.2   Caveats in Sequential Performance   111

3.4 Performance of Parallel Computers   113
    3.4.1   Computational Characteristics   113
    3.4.2   Parallelism and Interaction Overheads   115
    3.4.3   Overhead Quantification   118

3.5 Performance of Parallel Programs   126
    3.5.1   Performance Metrics   126
    3.5.2   Available Parallelism in Benchmarks   131

3.6 Scalability and Speedup Analysis   134
    3.6.1   Amdahl's Law: Fixed Problem Size   134
    3.6.2   Gustafson's Law: Fixed Time   136
    3.6.3   Sun and Ni's Law: Memory Bounding   139
    3.6.4   Isoperformance Models   144

3.7 Bibliographic Notes and Problems   148

Part II   Enabling Technologies   153

Chapter 4   Microprocessors as Building Blocks   155

4.1 System Development Trends   155
    4.1.1 Advances in Hardware   156
    4.1.2 Advances in Software   159
    4.1.3 Advances in Applications   160

4.2 Principles of Processor Design   164
    4.2.1 Basics of Instruction Pipeline   164
    4.2.2 From CISC to RISC and Beyond   169
    4.2.3 Architectural Enhancement Approaches   172

4.3 Microprocessor Architecture Families   174
    4.3.1 Major Architecture Families   174
    4.3.2 Superscalar versus Superpipelined Processors   175
    4.3.3 Embedded Microprocessors   180

4.4 Case Studies of Microprocessors   182
    4.4.1 Digital's Alpha 21164 Microprocessor   182
    4.4.2 Intel Pentium Pro Processor   186

4.5 Post-RISC, Multimedia, and VLIW   191
    4.5.1 Post-RISC Processor Features   191
    4.5.2 Multimedia Extensions   195
    4.5.3 The VLIW Architecture   199

4.6 The Future of Microprocessors   201
    4.6.1 Hardware Trends and Physical Limits   201
    4.6.2 Future Workloads and Challenges   203
    4.6.3 Future Microprocessor Architectures   204

4.7 Bibliographic Notes and Problems   206

Chapter 5   Distributed Memory and Latency Tolerance   211

5.1 Hierarchical Memory Technology   211
    5.1.1 Characteristics of Storage Devices   211
    5.1.2 Memory Hierarchy Properties   214
    5.1.3 Memory Capacity Planning   217

5.2 Cache Coherence Protocols   220
    5.2.1 Cache Coherency Problem   220
    5.2.2 Snoopy Coherency Protocols   222
    5.2.3 The MESI Snoopy Protocol   224

5.3 Shared-Memory Consistency   228
    5.3.1 Memory Event Ordering   228
    5.3.2 Memory Consistency Models   231
    5.3.3 Relaxed Memory Models   234

5.4 Distributed Cache/Memory Architecture   237
    5.4.1 NORMA, NUMA, COMA, and DSM Models   237
    5.4.2 Directory-Based Coherency Protocol   243
    5.4.3 The Stanford Dash Multiprocessor   245
    5.4.4 Directory-Based Protocol in Dash   248

5.5 **Latency Tolerance Techniques 250**
  5.5.1 Latency Avoidance, Reduction, and Hiding 250
  5.5.2 Distributed Coherent Caches 253
  5.5.3 Data Prefetching Strategies 255
  5.5.4 Effects of Relaxed Memory Consistency 257
5.6 **Multithreaded Latency Hiding 257**
  5.6.1 Multithreaded Processor Model 258
  5.6.2 Context-Switching Policies 260
  5.6.3 Combining Latency Hiding Mechanisms 265
5.7 **Bibliographic Notes and Problems 266**

Chapter 6 System Interconnects and Gigabit Networks 273
6.1 **Basics of Interconnection Network 273**
  6.1.1 Interconnection Environments 273
  6.1.2 Network Components 276
  6.1.3 Network Characteristics 277
  6.1.4 Network Performance Metrics 280
6.2 **Network Topologies and Properties 281**
  6.2.1 Topological and Functional Properties 281
  6.2.2 Routing Schemes and Functions 283
  6.2.3 Networking Topologies 286
6.3 **Buses, Crossbar, and Multistage Switches 294**
  6.3.1 Multiprocessor Buses 294
  6.3.2 Crossbar Switches 298
  6.3.3 Multistage Interconnection Networks 301
  6.3.4 Comparison of Switched Interconnects 305
6.4 **Gigabit Network Technologies 307**
  6.4.1 Fiber Channel and FDDI Rings 307
  6.4.2 Fast Ethernet and Gigabit Ethernet 310
  6.4.3 Myrinet for SAN/LAN Construction 313
  6.4.4 HiPPI and SuperHiPPI 314
6.5 **ATM Switches and Networks 318**
  6.5.1 ATM Technology 318
  6.5.2 ATM Network Interfaces 320
  6.5.3 Four Layers of ATM Architecture 321
  6.5.4 ATM Internetwork Connectivity 324
6.6 **Scalable Coherence Interface 326**
  6.6.1 SCI Interconnects 327
  6.6.2 Implementation Issues 329
  6.6.3 SCI Coherence Protocol 332
6.7 **Comparison of Network Technologies 334**
  6.7.1 Standard Networks and Perspectives 334
  6.7.2 Network Performance and Applications 335
6.8 **Bibliographic Notes and Problems 337**

Chapter 7 Threading, Synchronization, and Communication 343
7.1 **Software Multithreading 343**

7.1.1 The Thread Concept   344
7.1.2 Threads Management   346
7.1.3 Thread Synchronization   348

7.2   Synchronization Mechanisms   349
7.2.1 Atomicity versus Mutual Exclusion   349
7.2.2 High-Level Synchronization Constructs   355
7.2.3 Low-Level Synchronization Primitives   360
7.2.4 Fast Locking Mechanisms   364

7.3   The TCP/IP Communication Protocol Suite   366
7.3.1 Features of The TCP/IP Suite   367
7.3.2 UDP, TCP, and IP   371
7.3.3 The Sockets Interface   375

7.4   Fast and Efficient Communication   376
7.4.1 Key Problems in Communication   377
7.4.2 The Log P Communication Model   384
7.4.3 Low-Level Communications Support   386
7.4.4 Communication Algorithms   396

7.5   Bibliographic Notes and Problems   398

Part III   Systems Architecture   403

Chapter 8   Symmetric and CC-NUMA Multiprocessors   407

8.1   SMP and CC-NUMA Technology   407
8.1.1 Multiprocessor Architecture   407
8.1.2 Commercial SMP Servers   412
8.1.3 The Intel SHV Server Board   413

8.2   Sun Ultra Enterprise 10000 System   416
8.2.1 The Ultra E-10000 Architecture   416
8.2.2 System Board Architecture   418
8.2.3 Scalability and Availability Support   418
8.2.4 Dynamic Domains and Performance   420

8.3   HP/Convex Exemplar X-Class   421
8.3.1 The Exemplar X System Architecture   421
8.3.2 Exemplar Software Environment   424

8.4   The Sequent NUMA-Q 2000   425
8.4.1 The NUMA-Q 2000 Architecture   426
8.4.2 Software Environment of NUMA-Q   430
8.4.3 Performance of the NUMA-Q   431

8.5   The SGI/Cray Origin 2000 Superserver   434
8.5.1 Design Goals of Origin 2000 Series   434
8.5.2 The Origin 2000 Architecture   435
8.5.3 The Cellular IRIX Environment   443
8.5.4 Performance of the Origin 2000   447

8.6   Comparison of CC-NUMA Architectures   447

8.7   Bibliographic Notes and Problems   451

Chapter 9   Support of Clustering and Availability   453

9.1   Challenges in Clustering   453
  9.1.1   Classification of Clusters   453
  9.1.2   Cluster Architectures   456
  9.1.3   Cluster Design Issues   457

9.2   Availability Support for Clustering   459
  9.2.1   The Availability Concept   460
  9.2.2   Availability Techniques   463
  9.2.3   Checkpointing and Failure Recovery   468

9.3   Support for Single System Image   473
  9.3.1   Single System Image Layers   473
  9.3.2   Single Entry and Single File Hierarchy   475
  9.3.3   Single I/O, Networking, and Memory Space   479

9.4   Single System Image in Solaris MC   482
  9.4.1   Global File System   482
  9.4.2   Global Process Management   484
  9.4.3   Single I/O System Image   485

9.5   Job Management in Clusters   486
  9.5.1   Job Management System   486
  9.5.2   Survey of Job Management Systems   492
  9.5.3   Load-Sharing Facility (LSF)   494

9.6   Bibliographic Notes and Problems   501

Chapter 10   Clusters of Servers and Workstations   505

10.1   Cluster Products and Research Projects   505
  10.1.1   Supporting Trend of Cluster Products   506
  10.1.2   Cluster of SMP Servers   508
  10.1.3   Cluster Research Projects   509

10.2   Microsoft Wolfpack for NT Clusters   511
  10.2.1   Microsoft Wolfpack Configurations   512
  10.2.2   Hot Standby Multiserver Clusters   513
  10.2.3   Active Availability Clusters   514
  10.2.4   Fault-Tolerant Multiserver Cluster   516

10.3   The IBM SP System   518
  10.3.1   Design Goals and Strategies   518
  10.3.2   The SP2 System Architecture   521
  10.3.3   I/O and Internetworking.   523
  10.3.4   The SP System Software   526
  10.3.5   The SP2 and Beyond   530

10.4   The Digital TruCluster   531
  10.4.1   The TruCluster Architecture   531
  10.4.2   The Memory Channel Interconnect   534
  10.4.3   Programming the TruCluster   537
  10.4.4   The TruCluster System Software   540

10.5   The Berkeley NOW Project   541
  10.5.1   Active Messages for Fast Communication   541

10.5.2 GLUnix for Global Resource Management   547
10.5.3 The xFS Serverless Network File System   549
10.6 TreadMarks: A Software-Implemented DSM Cluster   556
10.6.1 Boundary Conditions   556
10.6.2 User Interface for DSM   557
10.6.3 Implementation Issues   559
10.7 Bibliographic Notes and Problems   561

Chapter 11   MPP Architecture and Performance   565
11.1 An Overview of MPP Technology   565
11.1.1 MPP Characteristics and Issues   565
11.1.2 MPP Systems – An Overview   569
11.2 The Cray T3E System   570
11.2.1 The System Architecture of T3E   571
11.2.2 The System Software in T3E   573
11.3 New Generation of ASCI/MPPs   574
11.3.1 ASCI Scalable Design Strategy   574
11.3.2 Hardware and Software Requirements   576
11.3.3 Contracted ASCI/MPP Platforms   577
11.4 Intel/Sandia ASCI Option Red   579
11.4.1 The Option Red Architecture   579
11.4.2 Option Red System Software   582
11.5 Parallel NAS Benchmark Results   584
11.5.1 The NAS Parallel Benchmarks   585
11.5.2 Superstep Structure and Granularity   586
11.5.3 Memory, I/O, and Communications   587
11.6 MPI and STAP Benchmark Results   590
11.6.1 MPI Performance Measurements   590
11.6.2 MPI Latency and Aggregate Bandwidth   592
11.6.3 STAP Benchmark Evaluation of MPPs   594
11.6.4 MPP Architectural Implications   600
11.7 Bibliographic Notes and Problems   603

Part IV   Parallel Programming   607

Chapter 12   Parallel Paradigms and Programming Models   609
12.1 Paradigms and Programmability   609
12.1.1 Algorithmic Paradigms   609
12.1.2 Programmability Issues   612
12.1.3 Parallel Programming Examples   614
12.2 Parallel Programming Models   617
12.2.1 Implicit Parallelism   617
12.2.2 Explicit Parallel Models   621
12.2.3 Comparison of Four Models   624
12.2.4 Other Parallel Programming Models   627

12.3   Shared-Memory Programming   629
   12.3.1   The ANSI X3H5 Shared-Memory Model   629
   12.3.2   The POSIX Threads (Pthreads) Model   634
   12.3.3   The OpenMP Standard   636
   12.3.4   The SGI Power C Model   640
   12.3.5   C//: A Structured Parallel C Language   643
12.4   Bibliographic Notes and Problems   649

Chapter 13   Message-Passing Programming   653

13.1   The Message-Passing Paradigm   653
   13.1.1   Message-Passing Libraries   653
   13.1.2   Message-Passing Modes   655
13.2   Message-Passing Interface (MPI)   658
   13.2.1   MPI Messages   661
   13.2.2   Message Envelope in MPI   668
   13.2.3   Point-to-Point Communications   674
   13.2.4   Collective MPI Communications   678
   13.2.5   The MPI-2 Extensions   682
13.3   Parallel Virtual Machine (PVM)   686
   13.3.1   Virtual Machine Construction   687
   13.3.2   Process Management in PVM   689
   13.3.3   Communication with PVM   693
13.4   Bibliographic Notes and Problems   699

Chapter 14   Data-Parallel Programming   705

14.1   The Data-Parallel Model   705
14.2   The Fortran 90 Approach   706
   14.2.1   Parallel Array Operations   706
   14.2.2   Intrinsic Functions in Fortran 90   708
14.3   High-Performance Fortran   711
   14.3.1   Support for Data Parallelism   712
   14.3.2   Data Mapping in HPF   715
   14.3.3   Summary of Fortran 90 and HPF   721
14.4   Other Data-Parallel Approaches   725
   14.4.1   Fortran 95 and Fortran 2001   725
   14.4.2   The pC++ and Nesl Approaches   728
14.5   Bibliographic Notes and Problems   733

Bibliography   737

Web Resources List   765

Subject Index   787

Author Index   799

Foreword

Michael J. Flynn, Stanford University

Parallel processors are the future of computer design. The replicability of technology has proven an ever-increasing impetus towards parallel processor implementations. The demand for increasing performance in important applications is another persuasive argument. However, building efficient, scalable parallel processors has proven to be a very difficult problem.

For decades, it seemed that large-scale ($n > 100$ processors) parallel processors could be realized efficiently with just a little bit more effort and research. This has not proven so. It has been difficult to find the prerequisite parallelism in programs, and when this parallelism was found, it did not translate into a speedup of program execution proportional to the number of processing elements involved. This has been especially true when n exceeds 100.

One important problem in the realization of efficient parallel processors is the computational model that underlies the program model and language used to represent applications. This model was developed for and efficiently suits the single processor. Because our linguistic notions of sequential specification of actions are familiar, they are readily mapped onto a serial computational and programming model. Attempts to remap these representations onto the parallel model have proved so far quite inefficient.

There are three legs that support the understanding of efficient parallel processing: *computational models, underlying alternatives,* and the *programming paradigms.* This book is uniquely concerned about all three issues. Metrics for performance provide a quantitative basis for understanding basic models of computation.

Professors Hwang and Xu comprehensively analyze hardware models of processors, interconnection networks, and serial wide area networks giving a complete picture of the hardware state of the art. Their coverage extends from the hardware ILP (instructional level parallelism) to forms of parallelism achieved by NOW (networks of workstations). Their systems viewpoint integrates the computational model, the hardware, and the programming model in a review of major parallel processor implementation efforts that are currently underway.

It is in this integration of hardware and software that this book is most valuable. It is only by understanding the mapping from application onto system and into processor configuration that we can expect the consistent progress in building efficient scalable parallel processor systems.

Preface

Oh, A Digital World!

This book covers scalable architecture and parallel programming of multiprocessors, multicomputers, and network-based cluster platforms. Digital technology has made the computer industry. Now, digital technology is making another wave of fundamental impact to telecommunications and information industries. Converting everything to digital is the key to future success in a highly automated society.

The crossbreeding of technologies demands a new generation of computers that can adapt to scalable, parallel, and distributed computing. These changes in computer and information technologies has prompted computer professionals to study the material presented in this text. The ultimate goal is to become ready for new challenges in the 21st Century.

A Glance at the Book

The book consists of 14 chapters presented in four parts. We provide a balanced coverage of four aspects: *principles, technology, architecture,* and *programming*:

- In Part I, three chapters cover scalable computer platforms and models, basics of parallel programming, and parallel performance metrics.

- Part II assesses commodity microprocessors, distributed cache and memory architecture, switched interconnects, Gigabit networks, and communications.

- Part III covers *symmetric multiprocessors* (SMP) and *cache-coherent, nonuniform memory-access* (CC-NUMA) machines, *clusters of workstations* (COW), and *massively parallel processors* (MPP).

- Part IV presents parallel languages, programming models with emphasis on Unix programming environments, message passing, data parallelism, and the use of PVM, MPI, Fortran 90, and HPF on scalable computers.

A Trilogy on Computer Systems

Over a period of 15 years, Hwang and his associates have produced a trilogy on computer systems, all published by McGraw-Hill Book Company.

- ***Computer Architecture and Parallel Processing***,
  by K. Hwang and F. A. Briggs (1983)

- ***Advanced Computer Architecture***:
  Parallelism, Scalability, Programmability, by K. Hwang (1993)

- ***Scalable Parallel Computing***:
  Technology, Architecture, Programming, by K. Hwang and Z. Xu (1998)

More than 90% of the topics treated in this book are based on new technological advances and research development within the past 5 years. This book is newly written, not a revision of Hwang's previous books. Unique features are highlighted below:

Hot Chips and Interconnects

We assess commodity microprocessors and hot chips for building scalable multiprocessors and multicomputer clusters. Distributed cache/memory and Gigabit networks are studied along with latency hiding mechanisms. In particular, we study multiprocessor buses and crossbar switches, SAN (*System Area Network*), and LAN (*Local Area Network*) such as Gigabit Ethernet, SCI (*Scalable Coherence Interface*), and ATM (*Asynchronous Transfer Mode*) networks.

Scalable Platforms and Clusters

We focus on scalable architectures, fast messaging mechanisms, latency hiding, distributed shared memory, cache coherence protocols, and memory consistency models. We cover software extensions for higher availability, single system image, failure recovery systems, and job management in clusters of computers.

Case studies include the HP/Convex Exemplar, Cray T3D/T3E, IBM SP2, Digital TruCluster, Microsoft Wolfpack, Sun Ultra Enterprise 10000, SGI Origin 2000, Sequent NUMA-Q, Intel/Sandia ASCI Option Red. We discuss the lessons learned from Stanford Dash, Berkeley NOW, Princeton SHRIMP, and Rice TreadMarks.

Parallel Software Environments

This book devotes more than half of the material to software tools and parallel programming systems. In the shared-memory approach, we study the ANSI X3H5, Pthreads, SGI Power C; OpenMP, and C// language. We study Solaris MC and LSF (*Load Sharing Facility*) for availability, single system image, and cluster job management.

For parallel programming, we study data-parallel, message-passing, shared-memory, and implicit paradigms. We study MPI (*Message-Passing Interface*), PVM (*Parallel Virtual Machine*), Fortran 90, and HPF (*High Performance Fortran*) for explicit parallelism; and languages and compliers for implicit parallelism.

Benchmark-based Evaluation

This book benefits from our benchmarking experience with six scalable computers: namely, the SP2 at Maui High Performance Computer Center, the T3D/T3E and Paragon at the San Diego Supercomputer Center, the T3D at Cray Eagan Data Center, the SP2, SGI server, and Pearl cluster at the University of Hong Kong.

Collective MPI communications in various machine platforms are evaluated with firsthand benchmark results. We reveal architectural implications from NASA parallel NAS and USC/HKU STAP benchmark results. These benchmark performance results are evaluated along with scalability analysis over machine sizes and problem sizes.

Web Resources

In a rapidly changing world, any computer book becomes obsolete in a few years. We have strived to prolong the life cycle of this book by selecting practical topics and discussing fundamental issues that can last over generations of computer systems. Examples and quantitative data are drawn from real designs or benchmark experiments.

We have complied at the end of the book an extensive *Web Resources List*, linking to thousands of home pages of computer companies, research projects, information technology centers, and major application groups across academic, business, and government sectors. An on-line home site of this list is also maintained at HKU. See *Guide for Instructors/Students* to access our home pages.

Acknowledgments

We thank the professional reviews of our draft manuscript by six leading experts in this field. Their suggestions are very useful in revising the manuscript to its present form and contents. We appreciate Choming Wang for preparing the indices and the help from Dr. Cho-Li Wang in maintaining the book's Web site at HKU.

Intellectual exchanges with Dharma Agrawal, Jean-Loup Baer, Gordon Bell, David Culler, Jack Dongarra, Michael Flynn, Ian Foster, Jeffrey Fox, Mark Franklin, Wolfgang Giloi, Allan Gottlieb, Anoop Gupta, John Hennessy, Ken Kennedy, Duncan Lawrie, Charles Leiserson, Kai Li, Guojie Li, Lionel Ni, David Patterson, Gregory Pfister, John Rice, Sartaj Sahni, Chuck Seitz, Bruce Shriver, H. J. Siegel, Burton Smith, Daniel Tabak, H. C. Torng, and Ben Wah are always inspiring and appreciated.

We appreciate the sponsorship from McGraw-Hill editors, Eric Munson, Lynn Cox, and Betsy Jones. The production work from Richard DeVitto, Francis Owen, and Nina Kreiden and assistance from Polly Leung of HKU are gratefully acknowledged.

During the courses of writing this book, research funding supports from MIT Lincoln Laboratory, Hong Kong Research Grants Council, and the University of Hong Kong are appreciated. In particular, the excellent facilities and environment provided by HKU make the writing of this book a very pleasant undertaking.

Points of Feedback

For all technical contacts, suggestions, corrections, or exchanges of information, readers and university instructors are advised to contact either author via Email:

kaihwang@cs.hku.hk   zxu@apple.ncic.ac.cn

We appreciate your feedback and hope you enjoy reading the book.

Kai Hwang
Zhiwei Xu

November 15, 1997
Hong Kong

Guide to Instructors/Readers

This book is designed as a standard text for classroom adoption in Computer Science or Computer Engineering curriculum at college/university levels. Suitable courses include: Computer Architecture, Parallel Processing, Distributed Computing, Concurrent Programming, Network-based Computing, Computer Engineering, etc.

Flowchart for Reading   The flowchart shows the logical flow of the 14 chapters in this book. The four parts are indicated on the side labels. There are two chapters in theory and modeling (slashed boxes), six unshaded boxes for hardware and architecture chapters, and six shaded boxes for software and programming chapters.

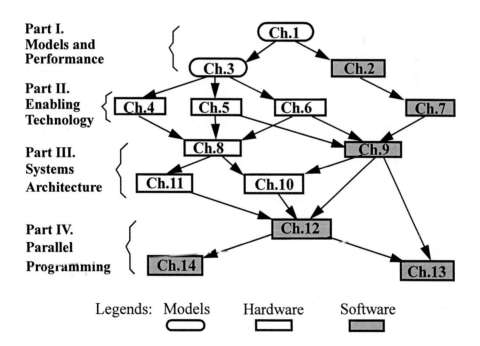

Course Offerings

Suggested below are five possible course offerings in adopting this book for use in an one-semester course with 45 hours of lectures:

- **Computer Architecture**: For hardware-oriented students in Electrical and Computer Engineering programs, cover Chapters 1, 3-6, and 8-11.

- **Parallel Programming:** For programming and software-oriented students in Computer Science programs, Chapters 1, 2, 7-10, and 12-14 are suitable.

- **Parallel Processing:** For mixed students from Computer Science/Engineering and Electrical Engineering programs, cover Chapters 1, 2, 4-6, 8-10, and 12.

- **Distributed Computing:** For mixed CS and EE students, Chapters 1, 2, 5-7, 9-10, and 13-14 are suitable for one-semester use.

- **Computer Engineering:** An advanced course in computer technology and software system design. Chapters 1, 2, 4-7, and 8-10 should be covered.

All readers should start with the material in chapter 1. Engineering students may read more of the technology and architecture chapters on the left side and software-oriented students on the right subtree of the flowchart. Logically, the reading of this book should flow from the top to the bottom chapters shown by the arrows.

The six hardware and architecture chapters form the core of the course in *Computer Architecture* or in *Computer Engineering.* The six chapters in software and programming form the core for the *Parallel Programming* course. The *Parallel Processing* course covers all scalable parallel systems with more emphasis on shared-memory multiprocessing. The *Distributed Computing* course emphasizes message-passing systems and network-based cluster computing.

Instructor's Manual

An *Instructor's Manual* is available to proven instructors without charge. The Manual provides solutions to selected homework problems, viewing-graph masters, some sample tests, and topics suggested for term projects.

Instructors can request the Manual by writing to: Lynn Cox, McGraw-Hill College Division, 55 Francisco Street, Suite 200, San Franscico, CA. 94133-2117, U.S.A. or submit a written request to Fax No. 415-989-7702.

Web Site Access

As an updated reference, the book is also intended for self-study use by system designers, academic researchers, application programmers, system analysts, resource managers, solution providers, and computer professionals in general. To avoid becoming obsolete, readers are invited to visit our Web site for updated WWW links.

http://www.cs.hku.hk/~kaihwang/book98.html

This Web site is updated dynamically. If you want your organization or your projects to be added into the list. Contact Dr. C. L. Wang of the University of Hong Kong by Email: **clwang@cs.hku.hk.**

Scalable Parallel Computing

Part I

Scalability and Clustering

I.1 Objectives

This part introduces the principles behind the scalability and programmability of parallel and distributed computer systems. Its purpose is to set up the necessary background for readers to study subsequent chapters.

Chapter 1 Scalable computer platforms are modeled in Chapter 1. The concept of scalability is defined along three orthogonal dimensions: resources, applications, and technology. Then we characterize three abstract machine models: the PRAM, BSP, and phase parallel models; and five physical machine models: the PVP, SMP, MPP, COW, and MPP systems.

Design principles for scalability are introduced with examples. The purpose is to achieve system designs independent of the technology, architecture, algorithm, language, application, and network environment used. The basic concepts of balanced design, overdesign, and backward compatibility are described.

Chapter 2 This chapter is specially tailored for programming scalable parallel computers. Besides basic concepts on processes, tasks, threads, and environments, we cover all the important issues in parallelism management, process interactions, program semantics, algorithmic paradigms, and software portability.

We introduce the four parallel programming models listed below. Details of the models are postponed until Part IV.

- The *parallelizing compiler model*
- The *data-parallel model*
- The *message-passing model*
- The *shared-memory model*

Chapter 3 This chapter covers basic performance benchmarks and metrics. The purpose is to identify attributes toward scalable performance. We start with a comprehensive introduction of parallel benchmark suites. Then we elaborate on the tradeoffs between performance and costs. The caveats of sequential program execution are identified.

Overheads in parallelism management and software interactions are analyzed with a quantitative approach. Granularity, available parallelism, parallel performance metrics, Amdahl's law, Gustafson's law, Sun and Ni's law, and various isoperformance models are quantitatively analyzed with illustrative benchmark results.

I.2 Notes to Readers

Chapter 1 must be read ahead of all remaining chapters. It is required for all four possible course offerings suggested in the Preface.

Chapter 2 must be read before those software-oriented Chapters 7, 9, 12, 13, and 14. For hardware-oriented readers, these chapters can be skipped in the first reading.

Chapter 3 will be helpful to understand the performance-sensitive material presented in Chapters 4, 5, 6, 8, 10, and 11.

For an introductory course taken by mixed students from Computer Science and Electrical Engineering majors, Chapter 3 can be skipped in the first reading.

However, research-oriented students may find Chapter 3 extremely useful, as long as the research topic chosen is related to system performance.

Chapter 1

Scalable Computer Platforms and Models

This chapter presents basic models of parallel and cluster computers. Fundamental design issues and operational principles of scalable computer platforms are introduced. We review the computer technology over the last 50 years. Scalable and cluster computer systems are modeled with key architectural distinctions. Scalability will be introduced in three orthogonal dimensions: *resource*, *application*, and *technology*.

Abstract and physical machine models are specified in Section 1.3. In Section 1.4, we introduce basic concepts of multicomputer clustering. The differences among symmetric multiprocessors, clusters of computers, and distributed computer systems are clarified. Three basic principles are studied in Section 1.5 to guide the design and application of scalable parallel computers.

Bits, Bytes, and Words   The following units are widely used in the computer field, but sometimes were wrongly used with confusing notations and ambiguous meanings. To cope with this problem, we present below a set of notations that will be used throughout the book. In particular, readers should not be confused with the shorthand notations for basic units of *time*, *byte*, and *bit* respectively.

The basic unit in time is *second*, abbreviated as s. The two basic information units are *byte* and *bit*. One byte (1 B) is 8 bits (8 b). Byte is always abbreviated as B and bit as b. Other information units are *word* (16 b or 2 B), *doubleword* (32 b or 4 B), and *quadword* (64 b or 8 B). This is based on convention used by Intel, Motorola, and Digital Equipment.

Mainframe vendors consider a word to have 32 b. Some supercomputer designers consider 64 b in a word. A frequently used workload unit is the number of *floating-point operations*, abbreviated as *flop*. A unit for computing speed is the number of *floating-point operations per second* (flop/s). A unit for information transfer rate is the number of *bytes per second* (B/s). The execution rate of a processor is often measured as *million instructions per second* (MIPS), which is equivalent to the notation Mi/s used in Europe.

Notations and Conventions We will use the conventions and notations specified in Table 1.1. Note that decimal and binary interpretations differ to some extent in the high-order units. Readers should be aware of their differences.

For instance, a *kilo* could mean either $10^3 = 1000$ or $2^{10} = 1024$. In computer science, K is usually used to mean 1024; in other scientific disciplines, k denotes 1000. This ambiguity must be clarified, even the relative difference is small. In general, the binary interpretation is applied to information units such as bits, bytes, or words. The decimal interpretation is often used in reference to other units such as execution time, power, weight, pin count, and computational workload, etc.

The convention for plurals is as follows: The unit will be pluralized when the whole word is used, and not pluralized when an abbreviation is used. For instance, we say 2 KB, 2 kilobytes, or even 2 Kbytes. They all mean the same thing: 2048 bytes. But we do not use 2 KBs or 2 Kbyte. Note that 2 Kb means a completely different thing: 2048 bits, or 256 bytes.

Similarly, 2 Mflop/s is a proper notation, meaning a speed of executing 2 million floating-point operations per second. We should avoid using 2 Mflops/s, 2 MFLOPS, or 2 Mflops, even some people use them. Throughout this book, the logarithmic function is always base 2. That is, $\log n$ means $\log_2 n$, unless otherwise specified.

Table 1.1 Notations and Conventions for Numbers

Prefix	Abbreviation	Meaning	Numeric Value
milli	m	One thousandth	10^{-3}
micro	μ	One millionth	10^{-6}
nano	n	One billionth	10^{-9}
pico	p	One trillionth	10^{-12}
femto	f	One quadrillionth	10^{-15}
atta	a	One quintillionth	10^{-18}
kilo	K (or k)	Thousand	10^3 (or 2^{10})
mega	M	Million	10^6 or 2^{20}
giga	G	Billion	10^9 or 2^{30}
tera	T	Trillion	10^{12} or 2^{40}
peta	P	Quadrillion	10^{15} or 2^{50}
exa	E	Quintillion	10^{18} or 2^{60}

1.1 Evolution of Computer Architecture

Computers have gone through 50 years of development. We summarize below five generations of development, approximately 10 to 15 years per generation, since the 1940s. A formal definition of a scalable computer is given below. Then we describe the converging architecture for future parallel computers.

1.1.1 Computer Generations

Over the past 50 years, digital electronic computers have gone through five generations of development. Table 1.2 provides a summary of the technology, architecture, software, application, and representative systems in each generation of computers. Each generation has improved from the previous generations in hardware and software technologies used and in application levels.

Table 1.2 Computer Generations in 50 Years

Generation Period	Technology and Architecture	Software and Operating System	Representative Systems
First 1946-1956	Vacuum tubes and relay memory, single-bit CPU with accumulator-based instruction set	Machine/assembly languages, programs without subroutines	ENIAC, IBM 701, Princeton IAS
Second 1956-1967	Discrete transistors, core memory, floating-point accelerator, I/O channels	Algol and Fortran with compilers, batch processing OS	IBM 7030, CDC 1604, Univac LARC
Third 1967-1978	Integrated circuits, pipelined CPU, microprogrammed control unit	C language, multiprogramming, timesharing OS	PDP-11 IBM 360/370, CDC 6600
Fourth 1978-1989	VLSI microprocessors, solid-state memory, multiprocessors, vector supercomputers	Symmetric multiprocessing, parallelizing compilers, message-passing libraries	IBM PC, VAX 9000, Cray X/MP
Fifth 1990- present	ULSI circuits, scalable parallel computers, workstation clusters, Intranet, Internet	Java, microkernels, Multithreading, distributed OS, World-Wide Web	IBM SP2, SGI Origin 2000, Digital TruCluster

In hardware technology, the first generation used vacuum tubes and relay memory. The second generation was marked by the use of discrete transistors and core memory.

The third generation began to use *small-scale integrated* (SSI) circuits. The fourth generation began with the use of *very large-scale integrated* (VLSI) microprocessors. The fifth generation uses *ultra large-scale integrated* (ULSI) circuits. For example, in 1997, 10M-transistor microprocessors and 256M-transistor DRAMs (*dynamic random-access memories*) are in use.

In the software area, the first generation used machine/assembly languages. The second generation began to use *high-level languages* (HLLs), such as Algol and Fortran. The third generation was marked by the use of C language, multiprogramming and time-sharing operating systems. The fourth generation emphasized parallel and vector processing. The fifth generation emphasizes scalable and clustered computing as covered in this book. As we cross into the 21st century, future generations of computers will evolve with even higher capability and user friendliness.

1.1.2 Scalable Computer Architectures

A common feature of modern computers is parallelism. Even within a single processor, parallelism has been exploited in many ways. This will be discussed in detail in Chapter 4. More recently, the concept of *scalability* has been reemphasized, which includes parallelism. The scalability concept is central to this book and will be elaborated on later. Given below is a short definition of scalable systems.

Definition 1.1 A computer system, including all its hardware and software resources, is called *scalable* if it can *scale up* (i.e., improve its resources) to accommodate ever-increasing performance and functionality demand and/or *scale down* (i.e., decrease its resources) to reduce cost.

∎

Although we will focus on the scaling-up aspect most of the time, scalability is not equivalent to being big. Scalability involves the ability to scale down. Scaling issues arise even on a uniprocessor system. More specifically, saying that a system is scalable implies the following:

- *Functionality and Performance*: The scaled-up system should provide more functionality or better performance. The total computing power of the system should increase proportionally to the increase in resources. Ideally, the users would like to see computing power increase close to n times when the system resource is improved n times.
- *Scaling in Cost*: The cost paid for scaling up must be reasonable. A rule of thumb is that scaling up n times should incur a cost of no more than n or $n \log n$ times.
- *Compatibility*: The same components, including hardware, system software, and application software, should still be usable with little change. It is unreasonable to

expect the users to pay for a completely new operating system and to redevelop their application codes. Often scaling involves only a part of a system, e.g., adding more processors or upgrading a processor to the next-generation. The improvement should be compatible with the rest of the system. That is, the old memory, disk, interconnect, and peripherals should still be usable.

The Computer Pyramid   The main motivation for a scalable system is to provide a flexible, cost-effective information processing tool. As illustrated in Fig. 1.1, the computer classes form a pyramid with respect to sales volume and performance and cost. At the bottom of the pyramid is the class of personal computers (PCs), which have a quite affordable price and a huge annual sales volume. Their performance is also the lowest.

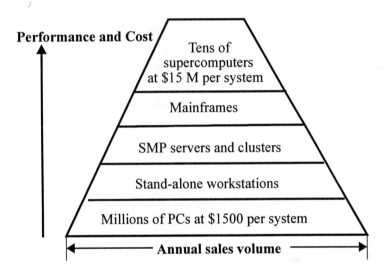

Figure 1.1   The pyramid of computer classes.

At the top is the class of supercomputers. Their sales volume is very low because of the expensive price. However, a supercomputer integrates many resources into a single system, employs cutting-edge technology, and has the highest performance.

The scalability concept advocates a common scalable architecture across all computer classes. This has several advantages:

- It satisfies different performance and cost requirements. For instance, a user could first purchase a low-end system. When his performance requirement increases, he could scale up the system, knowing that the original software and hardware components can still be used in the new system, and his investment is protected.

- High-end machines may use low-end components to cut cost. For example, with high volume, PCs have low-cost, off-the-shelf components. With a scalable architecture, supercomputers can use such components to reduce system cost. In fact, using commodity components (processors, memory chips, disks, I/O controllers, etc.) has become a trend in high-performance systems development.

- The cutting-edge technology developed for high-end systems may eventually migrate down the pyramid to improve the performance of low-end systems, if cost-effectiveness can be increased with improved production technology.

1.1.3   Converging System Architectures

Scalable parallel computers are converging toward three architectures, which share much commonality with increasing levels of resources sharing as shown in Fig. 1.2.

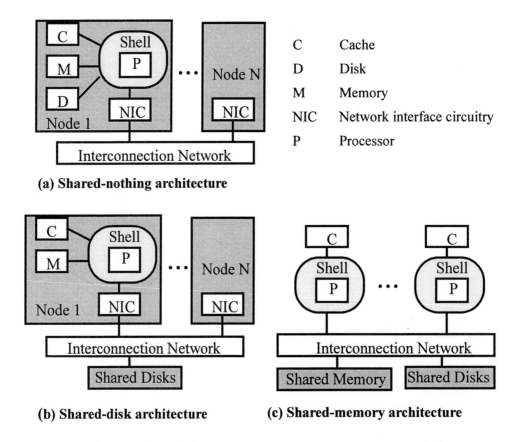

(a) Shared-nothing architecture

C	Cache
D	Disk
M	Memory
NIC	Network interface circuitry
P	Processor

(b) Shared-disk architecture   (c) Shared-memory architecture

Figure 1.2   Common architecture for scalable parallel computers.

The *shared-nothing* architecture consists of a number of *nodes* connected by an *interconnection network*. The node usually follows a *shell* architecture [571], where a custom-designed circuitry (called the *shell*) interfaces a commodity microprocessor to the rest of the node, including a (board-level) cache, a local memory, a *network interface circuitry* (NIC), and a disk. There may be more than one processor in a node.

The *shared-disk* architecture differs from the shared-nothing architecture in that the disk modules are moved out of the nodes to be shared among the nodes. In the *shared-memory* architecture, even the main memory becomes shared.

Macro-Architecture vs. Micro-Architecture The overall structure of a computer system (e.g., as shown in Fig. 1.2) is called the *macro-architecture*, while the processor and the shell surrounding it are called the *microarchitecture*. An advantage of the shell architecture is that when the processor is upgraded to the next generation or changed to a different architecture, only the shell (or the microarchitecture) needs to be changed.

All of the above architectural concepts are simplifications of real parallel computers. For instance, shared memory or shared disks can be attached to the interconnect of a shared-nothing architecture [Fig. 1.2(a)]. This will lead to the architectures with both local and shared disk and/or shared memory as shown in Fig. 1.2(b-c). More clarification will be given in later chapters.

1.2 Dimensions of Scalability

There are different ways of scaling (improving or reducing system resources), which leads to different aspects and definitions of scalability [69, 513]. In the following, we will focus on scaling up. We first need to define several basic terms:

A *processor* refers to a CPU, including one or more instruction pipelines, floating-point units, and caches, among other things. A processor may be implemented on a single chip (die), on multiple dies mounted on a single multichip module, or as multiple, separate chips. A *node* may have one or more processors, a local memory, and other circuitry. For simplicity, we will assume single-processor nodes in the models shown in Fig. 1.2, unless otherwise specified.

1.2.1 Resource Scalability

Resource scalability refers to gaining higher performance or functionality by increasing the *machine size* (i.e., the number of processors), investing in more storage (cache, main memory, disks), improving the software, etc.

Size Scalability The most obvious way to scale up a computer system is to increase the *machine size*, i.e., the number of processors. Size scalability measures the maximum number of processors a system can accommodate. Not all parallel computers are equally size-scalable. For instance, as of 1997, a *symmetric multiprocessor* (SMP) system can scale up to about 64 processors, while an IBM SP2 can scale up to 512 processors.

In current parallel computers, size scaling is often not just a simple matter of adding processors. The communication subsystem, including interconnect, interface, and communication software, may also need to be improved. How to effectively utilize the greater hardware parallelism, i.e., how to program the scaled-up system, is another concern.

Two main factors limiting the scalability of parallel systems are *programming* and *communication*. We will study these issues extensively in later chapters and learn how current commercial and research systems address them effectively.

Scaling Up in Resources While adding processors is probably the most obvious way to scale up a system, it is not the only way. One can keep the same number of processors, but invest in more memory, bigger off-chip caches, bigger disks, etc.

Example 1.1 Memory requirement in the IBM SP2

When the Maui High-Performance Computing Center (MHPCC) decided to upgrade its 400-node SP2 system, it opted to increase the memory and disk capacity, instead of adding more nodes. The scaled-up storage capacity is summarized in Table 1.3. This table says that, for instance, there are eight wide nodes in group 1, each having a 1-GB main memory and a 4.5-GB local disk. Also 1000 MB is used as the paging space for swapping, and another 2000 MB is used as a local scratch space.

Table 1.3 Memory and Disk Capacity of the IBM SP2 at MHPCC

Group	Node Type	Nodes	Memory (MB)	Disk (GB)	Paging Space (MB)	Scratch Space (MB)
1	Wide	8	1024	4.5	1000	2000
2	Wide	56	256	4.5	500	2000
3	Wide	16	256	2.0	500	1000
4	Thin	160	128	2.0	256	1000
5	Thin	160	64	1.0	180	250

Scaling up the memory capacity is more than just buying more memory chips. The system must be designed to allow for this expanded memory capacity. Real systems always have an upper limit on the maximum memory capacity. For instance, the IBM SP2 can accommodate at most 2-GB memory per node, and the Cray T3D only 64 MB. These capacities are mainly limited by cost and technology tradeoffs. ∎

Software Scalability The software of a scalable computer system can be improved in many ways, as suggested below:

- A newer version of the operating system, which has more functionalities such as multithreading, supports for more user processes and larger address space, more efficient kernel functions, etc.
- A better compiler with more efficient optimizations
- More efficient mathematical and engineering libraries
- More efficient and easy-to-use applications software
- More user-friendly programming environment

1.2.2 Application Scalability

To fully exploit the power of scalable parallel computers, the application programs must also be scalable. This means that the same program should run with proportionally better performance on a scaled-up system. Two most interesting measures are scalability in *machine size* and that in *problem size*.

Scalability in Machine Size This indicates how well the performance will improve with additional processors. For instance, suppose an *n*-processor computer system is used as a database server. It holds the U.S. population database, is normally queried by 100 U.S. scientists, and gives a performance of 1000 *transactions-per-second* (TPS).

Now, if we double the number of processors to $2n$, how much speed improvement can we expect? If the system's speed can increase to close to 2000 TPS, we are justified in saying that the system scales well over machine size. Note that the resources increased are most frequently processors, but they could also be memory capacity and I/O capability.

Scalability in Problem Size This indicates how well the system can handle larger problems with larger data size and workload. Consider the database server example again. How well will the server behave if the server now holds the population database of China? Note that the size of the database is increased 5 times. What will happen if the number of users is increased to 200 (100 U.S. and 100 Chinese scientists in a joint study)?

There are three points worth noting when studying application scalability:

- Many practical parallel applications have built-in limitations regarding machine size and problem size. For instance, a parallel radar signal processing program may use at most 256 processors and handle at most 100 radar channels. These limitations cannot be exceeded by simply increasing machine resources. The program has to be significantly modified to handle more processors and radar channels.
- We should consider an application on a specific machine as a combination. This application/machine pair is sometimes referred to as a *system*.

- Application scalability does not depend on just machine size and problem size. It also depends on the memory capacity, I/O capability, and communication capability of the machine. All these factors jointly affect the scalability. We will treat these issues in Chapter 3.

1.2.3 Technology Scalability

Technology scalability applies to a scalable system that can adapt to changes in technology. It can be further divided into three categories: *generation scalability*, *space scalability*, and *heterogeneity scalability*.

Generation (Time) Scalability A system can be scaled up using the next-generation components, such as a faster processor, a faster memory, a newer version of operating system, a more powerful compiler, etc. Again, the computing power should increase when the system migrates to the next generation. Furthermore, the rest of the system should be usable with as few modifications as possible. It is unreasonable to require an overhaul of the entire system when the user just wants to upgrade the processor or the operating system.

The fastest evolving component of a computer system is the processor, which is expected to double its performance approximately every 18 months to 2 years. The slowest evolving is probably programming languages (Fortran 77 is still widely used), although more powerful compilers come out every few years.

A performance-conscious user may want to upgrade the processor of the computer system every 2 years and upgrade other components at much slower rates. Unfortunately, this is not what has happened with parallel computers in the past, where every generation often required new investment in system software and redevelopment of user applications.

Example 1.2 Generation scalability of IBM personal computers

The most generation-scalable computers are probably the IBM PCs. Many PC users are frequently upgrading their system components such as processor, RAM capacity, and multimedia extensions, knowing that the computing power will increase and the rest of the system can stay the same. The rest system include the monitor, the hard drive, and the printer, etc.

Most importantly, they know that the *binary code* of their existing system and application programs (DOS, Windows, database, spreadsheet, word processing software, etc.) can run faster on the upgraded system without change. This is because the PC system (from processor to motherboard, I/O cards, and software) is designed to be generation-scalable.

■

Space Scalability This phrase was coined by Gorden Bell to refer to the ability of a system to scale from multiple processors in a box, in a room, or in a building to multiple buildings and geographic regions. In this sense, SMPs and MPPs have limited space scalability, while the Internet has excellent space scalability.

Heterogeneity Scalability This property refers to how well a system can scale up by integrating hardware and software components supplied from different designers or vendors. This calls for using components with a standard, open architecture and interface. In the software area, this is called *portability*.

Example 1.3 Software portability in scalable parallel computers

The IBM *Parallel Operating Environment* (POE) is scalable over any size of RS6000 system. With POE, a parallel program can run without change on any network of RS6000 nodes, where each node can be a low-end PowerPC workstation or a high-end SP2 wide node.

These nodes can be connected by any popular interconnect, from the slow Ethernet to the *high-performance switch* (HPS) of SP2. The nodes can even reside at geographically distanced locations, one in the University of Southern California and the rest in Maui High-Performance Computing Center.

Another example is the *parallel virtual machine* (PVM), which is also heterogeneously scalable: It allows a parallel program to run on a network of nodes from different vendors.

■

1.3 Parallel Computer Models

A *parallel machine model* (also known as *programming model*, *type architecture*, *conceptual model*, or *idealized model*) is an abstract parallel computer from the programmer's viewpoint, analogous to the von Neumann model for sequential computing. According to a Purdue Report [556], such a model must characterize those capabilities of a parallel computer that are fundamental to parallel computation. All but the most specialized parallel computers can be expected to provide them.

The abstraction need not imply any structural information, such as the number of processors and interprocessor communication structure, but it should capture implicitly the relative costs of parallel computation. Such a model should be precise enough about performance without being too explicit about the implementation details. Such an abstract model provides many benefits to computer architects, software developers, programmers, and algorithm designers. Every parallel computer has a *native model* that closely reflects its own architecture.

1.3.1 Semantic Attributes

A parallel machine model can be characterized by five *semantic attributes* and several *performance attributes*. These attributes can be understood by first looking at a theoretical machine model. A *parallel random-access machine* (PRAM) of size *n*, as shown in Fig. 1.3, consists of *n* processors, all accessing a shared memory.

A parallel program on a PRAM consists of *n* processes, where the *i*-th process resides in the *i*-th processor and is composed of a sequence of instructions. Each processor executes one instruction at each basic time step (called a *cycle*). The instructions include data transfer, arithmetic/logic, control flow, and I/O instructions, as can be found in a typical sequential computer.

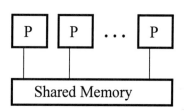

- MIMD
- Fine grain
- Tightly synchronous
- Zero overhead
- Shared variable

Figure 1.3   The PRAM (parallel random-access machine) **model.**

Homogeneity This attribute characterizes how alike the processors of a parallel computer behave when executing a parallel program. A PRAM of size 1 degenerates to the traditional *random access machine* [16], where the program executes a single instruction stream and accesses a single data stream. Based on Flynn's classification [244], such a machine is a *single-instruction* (stream) *single-data* (stream) (SISD) machine, which models conventional sequential computers.

When there is more than one processor, a PRAM will access multiple data streams and, in general, may execute multiple instruction streams. Thus a PRAM is a *multiple-instruction* (streams) *multiple-data* (streams) (MIMD) machine. However, if at each cycle all processors must execute the same instruction, there is only one instruction stream, and we have a *single-instruction* (stream) *multiple-data* (streams) (SIMD) machine.

A special case of MIMD is the *single-program multiple-data* (SPMD) computation [363], where all the processes execute the same program, parameterized by the process index. A difference between SIMD and SPMD is that in an SPMD computation, different instructions can be executed at the same cycle.

Synchrony This attribute states how tightly synchronized the processes are. A central feature of the PRAM model is that it is *synchronous*. At each cycle, all memory read operations from all the *n* instructions must be performed before any processor can perform a memory write or a branch.

We say that the PRAM is *synchronous at instruction level*. Such strong synchrony is absent in most real parallel computers, except SIMD machines. Rather, real MIMD parallel computers are *asynchronous*: Each process executes at its own pace, independent of the speed of other processes. If a process has to wait for other processes to enforce correct semantics, additional *synchronization operations* must be executed.

There are other synchrony schemes between these two extremes. An interesting alternative is suggested by Valiant in his *bulk synchronous parallel* (BSP) model [623]. Instead of synchronizing at every instruction, the processes synchronize with one another only at the end of every *superstep*, where a superstep can be considered as a bulk of L instructions. Within a superstep, the processors execute their instructions asynchronously. Again, within a superstep, all memory read operations from all processors must be performed before any processor can perform a memory write operation.

Anther *loosely synchronous* scheme [253, 254] is very similar to the BSP model. A parallel program is divided into a sequence of *phases*. A phase could be as simple as a single assignment statement, or as complex as a sequence of arbitrary statements. Multiple processes are executed within each phase asynchronously. They synchronize at the end of each phase.

Interaction Mechanism This attribute characterizes how parallel processes may *interact* to affect the behavior of one another. In the PRAM model, processes interact through *shared variables* (or shared memory). Another important interaction mechanism is *message passing*. With this scheme, processes do not have shared variables (i.e., all variables accessible by a process are invisible to other processes).

Instead, processes interact by sending and receiving messages. An asynchronous MIMD machine interacting through shared variables is often called a *multiprocessor*. An asynchronous MIMD machine with message passing is often called a *multicomputer*. These interaction mechanisms may be combined in various ways. Multiprocessors and multicomputers will be further distinguished in Section 5.4.1 and Section 8.1.1.

Address Space The *address space* of a process is the set of memory locations accessible by the process. In some machines (e.g., the PRAM model), all memory locations reside in a *single address space*, from the programmer's viewpoint. Not all parallel computers have a single address space. For instance, in a multicomputer, each processor has its own separate address space. Such machines are said to have *multiple address spaces*. It is for this reason that processors of a multicomputer communicate not through shared variables, but by passing messages.

One important feature of the theoretical PRAM model is that all processes have equal access time to all memory locations. We say that such a machine has *uniform memory access* (UMA). On the other hand, if it takes different amounts of time for processes to access different word locations, we say the machine has *nonuniform memory access* (NUMA). These memory models will be studied in Chapters 5 and 8 in details.

ι

For instance, in a *distributed shared memory* (DSM) multiprocessor, the programmer sees a logically shared memory, which is physically distributed to all processors, called *local memories*. One processor's local memory may be another's *remote memory*. Some computers also have a hierarchical memory organization. Besides local memories, the processors have access to a global memory. The time to access a local word is usually less than that of accessing a remote or global word, sometimes by 1 or 2 orders of magnitude. Thus for NUMA machines, it is important to exploit both *program-* and *data locality*.

Memory Model   This attribute specifies how the machine model handles shared-memory access conflicts. Generally, a number of *consistency rules* are used to resolve such conflicts. For the PRAM model, one can use the *exclusive read exclusive write* (EREW) rule, by which a memory cell can be read or written by at most one processor per cycle.

The *concurrent read exclusive write* (CREW) rule says that at each cycle, a memory cell can be read by multiple processors, but can be written by at most one processor. With the *concurrent read concurrent write* (CRCW) rule, at each cycle, multiple processors can both read from and write to the same memory location. In case of conflicts, concurrent writes are resolved by some prespecified access policy.

A shared memory in a real multiprocessor is a complex system that incorporates many performance-enhancing techniques. As a consequence, the order of memory accesses in a parallel program may not directly match what is supported by a multiprocessor. The behavior of a shared memory is characterized by various memory *consistency rules*, commonly known as the *memory consistency model*, to be treated in Chapter 5.

Definition 1.2   An *atomic operation* is one that is

(1) *Indivisible*: Once it starts, it cannot be interrupted in the middle, meaning other processes cannot see an intermediate state.

(2) *Finite*: Once it starts, it will finish in a finite amount of time.

Definition 1.3   A stronger definition of an atomic operation requires it to satisfy the following four properties. Such an atomic operation is known as a *transaction*.

- *Atomicity*: A transaction satisfies Definition 1.2. Putting it in another way, either all or none of the suboperations of the transaction are performed.
- *Consistency*: A transaction always transfers a program from one consistent state to another. Exactly what is consistent depends on program semantics. For instance, a write-a-word operation should not write just half of a word.
- *Isolation*: The effect (result) of a transaction is not revealed to other transactions until it is *committed*.
- *Durability*: Once committed, a transaction's effect persists when system fails.

■

1.3.2 Performance Attributes

The semantic attributes are important in that a programmer must understand them in order to write correct parallel programs. These attributes are relatively long-lasting, usually staying the same from one generation of a parallel computer to the next. There are other attributes of a parallel computer, which a programmer should know in order to develop *efficient* parallel programs. These *performance attributes* are highly platform-dependent and are usually improved with each new generation.

Basic performance attributes are summarized in Table 1.4. Machine size n is the number of processors in a parallel computer. Workload W is the number of computation operations in a program. P_{peak} is the peak speed of a processor.

Table 1.4 Performance Attributes of Parallel Systems

Terminology	Notation	Unit
Machine size	n	Dimensionless
Clock rate	f	MHz
Workload	W	Mflop
Sequential execution time	T_1	s
Parallel execution time	T_n	s
Speed	$P_n = W / T_n$	Mflop/s
Speedup	$S_n = T_1 / T_n$	Dimensionless
Efficiency	$E_n = S_n / n$	Dimensionless
Utilization	$U_n = P_n / (nP_{peak})$	Dimensionless
Startup time	t_0	μs
Asymptotic bandwidth	r_∞	MB/s

Definition 1.4 There are three types of operations in a parallel program: *Computation* operations include arithmetic/logic, data transfer, and control flow operations that can be found in a traditional sequential program; *parallelism* operations are needed to manage processes, such as creation and termination, context switching, and grouping; *interaction* operations are needed to communicate and to synchronize processes.

∎

The parallelism and interaction operations can be either *explicit* or *implicit*. An explicit operation is one that appears in the parallel program text. For instance, a process can be created by explicitly calling the UNIX function *fork*( ). An implicit operation does not appear in the program, but is done silently by the system.

For example, when the time slice of a process runs out, the operating system may perform a *process switch* to put the current process into a waiting queue and schedule another process to run. Such a process switch is not explicitly specified in the parallel program. Both explicit and implicit operations take time to execute.

Definition 1.5   The parallelism and the interaction operations are the sources of *overhead*, in that they need extra time to carry out in addition to the time needed to execute pure computational workload. More specifically, these overheads can be divided into the following four types:

- *Parallelism overhead* caused by process management
- *Communication overhead* caused by processors exchanging information
- *Synchronization overhead* in executing synchronization operations
- *Load imbalance overhead* incurred, when some processors are idle while the others are busy

                                                                        ∎

If there were no overheads, the speedup of using n processors would always be n. These overheads will be treated in detail in Chapter 3. For now, we briefly define two overhead metrics (first introduced by Hockney [313]) for one node to communicate a message to another node. Such a communication is called *point-to-point*. There is another type, called *collective communication*, where more than one messages are communicated among multiple nodes simultaneously.

The *start-up time*, denoted by t_0, is the time (in μs) to communicate a 0-Byte or a short (e.g., one-word) message. The start-up time is also known as the *communication latency*, or simply *latency*. The *asymptotic bandwidth*, denoted by r_∞, is the rate (in MB/s) for communicating a long message. These terms will be used in estimating the total communication time in point-to-point and collective message passing operations.

1.3.3 Abstract Machine Models

An abstract machine model is mainly used in the design and analysis of parallel algorithms without worrying about the details of physical machines. We show three abstract models below. Each has many variations under different assumptions.

The PRAM Model   We have already described the semantic attributes of the PRAM model in Section 1.3.1. Its performance attributes are summarized below:

- The machine size n can be arbitrarily large.
- The basic time step is called a *cycle*.
- Within a cycle, each processor executes exactly one instruction. The instruction could be a *null* instruction (i.e., do nothing), in which case we say the processor is idle at that cycle.
- All processors implicitly synchronize at each cycle, and the synchronization overhead is assumed to be zero. Communication is done through reading and writing of shared variables. The communication overhead is ignored. The parallelism overhead is also ignored. Thus the only overhead accounted for in a PRAM program is the load imbalance overhead.
- An instruction can be any *random-access machine* instruction [16]. For instance, an instruction could perform the following three operations in one cycle: (1) fetch one or two words from the memory as operands, (2) perform an arithmetic/logic operation, and (3) store the result back in memory.

Example 1.4 parallel execution on the PRAM machine

To compute the inner product s of two N-dimensional vectors A and B on an n-processor EREW PRAM computer, one can assign each processor to perform $2N/n$ additions and multiplications to generate a local result in $2N/n$ cycles, and then one can add all the n local sums to a final sum s by a treelike reduction method in $\log n$ cycles. The total execution time is $2N/n + \log n$ cycles. Compared with the sequential algorithm which takes $2N$ cycles, the parallel PRAM algorithm has a speedup of $n/\{1 + [n/(2N)]\log n\} \to n$, when $N \gg n$.

■

Note that the time complexity of the sequential algorithm is a function of just one parameter: the problem size N. The parallel PRAM algorithm introduces another parameter: the machine size n. The time complexity of most PRAM algorithms is expressed as a function of N and n. The PRAM model does account for the load imbalance overhead. For instance, in the last step of the parallel summation algorithm above, only one processor is busy computing the final sum, and all other processors are idle.

Due to its simplicity and clean semantics, the PRAM model has been favored by many computer scientists. The majority of theoretical parallel algorithms are specified with the PRAM model or a variant thereof. This model is also widely used in analyzing the complexity of parallel algorithms. The main shortcoming of the PRAM model lies in its unrealistic assumptions of zero communication overhead and instruction-level synchrony.

Another problem with the PRAM model is that the time complexity of a PRAM algorithm is often expressed in the big-O notation, which is often misleading, because the machine size *n* is usually small in existing parallel computers. One should not rely the big-*O* complexity in estimating real machine performance at all.

Example 1.5 Computational Complexity in PRAM steps

Suppose three PRAM algorithms *A*, *B*, and *C* have time complexities of $7n$, $(n \log n)/4$, and $n \log \log n$, respectively, when executing on an *n*-processor PRAM computer. According to the big-*O* notation, algorithm *A* is the fastest ($O(n)$), followed by *C* ($O(n \log \log n)$), with *B* being the slowest ($O(n \log n)$).

In reality, on machines with no more than 1024 processors, we have $\log n \le \log 1024 = 10$ and $\log \log n \le \log \log 1024 < 4$. Thus the fastest algorithm is really *B*, followed by *C*, while *A* is the slowest for systems with less than 1024 processors.

■

A parallel sorting algorithm based on a PRAM model can sort $O(n)$ items in $O(\log n)$ time [18]. However, sorting on real parallel computers does not use such an "optimal" algorithm, because the big-*O* notation hides a big constant factor. Furthermore, even when this $O(\log n)$ algorithm is improved, it may still not be suitable for practical use, since it ignored all the overheads.

Despite the inaccuracy, the PRAM model abstracts away many details, and the resulting simplicity makes it an exellent model for high-level parallel algorithm development. Many parallel algorithms, developed with the use of the PRAM model, are practically good algorithms. Sometimes an unrealistic PRAM algorithm can be improved to generate a practical parallel algorithm.

With unrealistic assumptions, the PRAM model has not been used as a machine model for *real-life* parallel computers. Other abstract models were proposed to amend this inaccuracy. They all introduce additional parameters to capture the overheads and are more complex than the PRAM model. A common feature of all these abstract models is that they are independent of the communication network topology.

The BSP Model The *bulk synchronous parallel* (BSP) model was proposed by Leslie Valiant of Harvard University to overcome the shortcomings of the PRAM model, while keeping its simplicity. A *BSP computer* consists of a set of *n* processor/memory pairs (*nodes*) that are interconnected by a communication network as shown in Fig. 1.4.

A BSP program has *n* processes, each residing on a node. The basic time unit is a *cycle* (or *time step*). The program executes as a strict sequence of *supersteps*. In each superstep, a process executes the computation operations in at most *w* cycles, a communication operation that takes *gh* cycles, and a *barrier synchronization* that takes *l* cycles. The barrier forces the processes to wait so that all processes have to finish the current superstep before any of them can begin the next superstep.

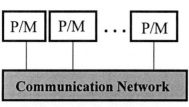

w: Maximum computation time
   within each superstep
l: barrier synchronization overhead
g: h relation coefficient

- MIMD
- Superstep:
    Computation
    Communication
    Barrier
- Variable grain
- Loosely synchronous
- Nonzero overhead
- Message passing or
  shared variable

Figure 1.4   The bulk synchronous parallel (BSP) model.

The BSP computer is an MIMD system, because the processes can execute different instructions simultaneously. It is loosely synchronous at the superstep level, compared to the instruction-level tight synchrony in the PRAM model. Within a superstep, different processes execute asynchronously at their own pace. The BSP model does not require any specific interaction mechanism: It could be either shared variable or message passing. There is a single address space, and a processor can access not only its local memory but also any remote memory in another node.

Within a superstep, each computation operation uses only data in its local memory. These data are put into the local memory either at the program start-up time or by the communication operations of previous supersteps. Therefore, the computation operations of a process are independent of other processes. A communication is always realized in a point-to-point manner. Thus it is not allowed for multiple processes to read or write the same memory location in the same cycle.

Because of the barrier synchronization, all memory and communication operations in a superstep must completely finish before any operation of the next superstep begins. All of these imply that a BSP computer has a sequential consistency memory model.

In a real parallel computer, there are many component communication operations demanding different communication patterns among them. For simplicity, the BSP model abstracts the communication operations in a BSP superstep by the *h* relation concept. This greatly simplifies performance prediction and algorithm complexity analysis.

Definition 1.6 An *h relation* is an abstraction of any communication operation, where each node sends at most *h* words to various nodes and each node receives at most *h* words. On a BSP computer, the time to realize any *h* relation is no more than *gh* cycles, where g is a constant decided by the machine platform.

■

The BSP model is more realistic than the PRAM model because it accounts for all overheads except the *parallelism overhead* for process management. The execution time of a superstep is determined as follows:

- To account for load imbalance, the computation time w is the maximum number of cycles spent on computation operations by any processor.
- The synchronization overhead is l, which has a lower bound of the communication network latency (i.e., the time for a word to propagate through the physical network) and is always greater than zero.
- The communication overhead is gh cycles, where g is the proportional coefficient for realizing an h relation. The value of g is platform-dependent, but independent of the communication pattern. In other words, gh is the time to execute the most time-consuming h relation. The parameter g has a small value on a computer with more efficient communication support.
- The time for a superstep is estimated by the sum $w + gh + l$.

The BSP model does not disallows the overlapping of the computation, the communication, and the synchronization operations within a superstep. If all three types of operations are fully overlapped, the time for a superstep becomes $\max(w, gh, l)$. However, we will use only the more conservative $w + gh + l$.

Example 1.6 Parallel execution using the BSP machine model

Consider the inner-product problem in Example 1.4. We can solve this problem on an 8-processor BSP computer in 4 supersteps:

Superstep 1
 Computation: Each processor computes its local sum in $w = 2N/8$ cycles.
 Communication: Processors 0, 2, 4, 6 send their local sums to
      processors 1, 3, 5, 7. Apply 1 relation here.
 Barrier synchronization.
Superstep 2
 Computation: Processors 1, 3, 5, 7 each perform one addition ($w = 1$).
 Communication: Processors 1 and 5 send their intermediate results
      to processors 3 and 7. A 1 relation is applied here.
 Barrier synchronization.
Superstep 3
 Computation: Processors 3 and 7 each perform one addition. ($w = 1$).
 Communication: Processor 3 sends its intermediate result to
      processor 7. Apply 1 relation here.
 Barrier synchronization.

Superstep 4

Computation: Processor 7 performs one addition ($w = 1$) to generate the final sum.

No more communication or synchronization is needed.

The total execution time is $2N/8 + 3g + 3l + 3$ cycles. In general, the execution time is $2N/n + \log n\,(g + l + 1)$ cycles on an n-processor BSP. This is in contrast to the $2N/n + \log n$ time on a PRAM computer. The two extra terms, $g\log n$ and $l\log n$, correspond to *communication* and *synchronization overheads*, respectively.

∎

Phase Parallel Model The authors have proposed a *phase parallel model* [657] for parallel computation, that is further refined from the above two abstract machine models. This model is similar to the BSP model with the following distinctions:

- A parallel program is executed as a sequence of phases. The next phase cannot begin until all operations in the current phase have finished. There are three types of phases as defined below:

  (1) **Parallelism phase**: This refers to the overhead work involved in process management, such as process creation and grouping for parallel processing.

  (2) **Computation phase**: One or more processors execute a number of local computation operations. Here local means that all data needed by a node are available in its local memory.

  (3) **Interaction phase**: This refers to tasks needed to perform interaction operation: a communication, a synchronization, or an aggregation (e.g., *reduction* and *scan*) operation.

- For a given a computation phase, denote the workload (e.g., in Mflop) by w and t_f is the average time (in µs) to execute a computation operation. Different computation phases may execute different workloads at different speeds, corresponding to different w and t_f values. This is in contrast to the previous two models, where every computation operation is assumed to take a uniform *cycle*.

- Different interaction operations may take different times. However, there is a general form for the time to execute an interaction operation:

$$T_{interact}\,(m, n) \; = \; t_0\,(n) + \frac{m}{r_\infty\,(n)} \; = \; t_0\,(n) + m \cdot t_c\,(n) \qquad\qquad (1.1)$$

Here m is called the *message length* in bytes, the *start-up time* $t_0(n)$ and the *asymptotic bandwidth* $r_\infty\,(n)$ are functions of machine size n. The parameter

$t_c(n) = 1/r_\infty(n)$ is called the *per-byte messaging time*. We explain below how to sequence parallelism, computation, and interaction phases. Analogous to the BSP model, we call such a sequence a *superstep*. We focus on the case when the interaction is a communication with a message length of m B. Note that some, but not all, phases may be missing in a superstep. For example, a superstep may contain just a computation phase, while another superstep may contain just an interaction phase.

∎

Example 1.7 Parallel execution using the phase parallel model

The inner-product problem in Example 1.4 is solved by the following three-phase parallel algorithm:

Parallelism phase:	**parfor** $(i=0; i<n; i++)$
	$\{$
Computation phase:	$(\text{localSum})[i] = \sum\limits_{j=in}^{(i+1)n} A[j] \cdot B[j]$
Interaction phase:	$s = \text{sum\_reduction}(\text{LocalSum}[i]);$
	$\}$

Figure 1.5   Parallel inner-product computation using the phase parallel model

The first phase creates n processes. In the second phase, process i computes LocalSum$[i]$ for all $0 \le i < n$. In the third phase, all processes participate in a reduction operation to generate the final result s.

In the computation phase, each process computes a workload of $2N/n$ flop at a speed of $1/t_f$, where t_f is the average time to execute a flop by a processor, and is equal to the reciprocal of the single-processor speed. That is, $t_f = 1/P$.

Assume that the parallelism overhead is $t_p(n)$ and the interaction overhead for the reduction operation is $t_0(n)$ (the reduction operation has a constant message length $m = 4$ B). Then the total execution time is $2Nt_f/n + t_p(n) + t_0(n)$.

∎

Suppose the total computational workload performed by the n processors in a superstep is $W = nw$, which can be divided into n independent computations. The execution times of the n independent computations are assumed to be random variables, with a mean value of wt_f and a standard deviation of σt_f. Therefore, the parameters w and σ represent the mean and the standard deviation, respectively, of the per-processor workload. Intuitively, w indicates *granularity* and σ indicates *load imbalance*, respectively. A small σ indicates a nearly balanced parallel computation phase. When $\sigma = 0$, the total workload W is assumed evenly divided among the processors.

Since the n processors perform independent computations, the execution time of the computation phase, denoted by T_{comp}, should be the maximum of the n individual execution times of the processors. Kruskal and Weiss [379] have shown that this maximum value can be approximated by

$$T_{comp} = (w + \sigma \sqrt{2 \log n}) \, t_f \qquad (1.2)$$

Definition 1.7 Assume the message length is $m = \alpha w$ B in the interaction phase of a superstep. The parameter α, called the *communication-to-computation ratio* (CCR) of the superstep, measures how many bytes need to be communicated per flop computation. ∎

From Eq.(1.1), the interaction overhead is expressed by

$$T_{interact} = t_0(n) + m/r_\infty(n) = t_0(n) + \alpha \cdot w \cdot t_c(n) \qquad (1.3)$$

The total execution time of the superstep on n processors is expressed by:

$$\begin{aligned} T_n &= T_{comp} + T_{interact} + T_{par} \\ &= (w + \sigma \sqrt{2 \log n}) \, t_f + t_0(n) + \alpha \cdot w \cdot t_c(n) + t_p(n) \end{aligned} \qquad (1.4)$$

Improved from the PRAM and the BSP models, the phase parallel model is closer to covering real machine/ program behavior. All types of overheads are accounted for, as shown in Eq.(1.4): the load imbalance overhead (the σ term), the interaction overhead (the t_0 and the t_c terms), and the parallelism overhead (the t_p term). Different expressions are used for different interactions.

These expressions can be obtained by measuring the interaction operations on real machines. Once obtained, they can be used many times by different applications. This point will be treated again in Chapter 3. The parameters used in the phase parallel model are summarized in Table 1.5.

Table 1.5 Parameters of the Phase Parallel Model

Parameter	Unit	Meaning	Depending on
t_0	μs	Start-up time	System
t_c	μs/B	Time to communicate a byte	System
t_f	μs	Time to execute a flop	System and application
t_p	μs	Parallelism overhead	System
w	Mflop	Mean workload	Application
σ	Mflop	Standard deviation of workload	Application
α	B/flop	Communication-to-computation ratio	Application

1.3.4 Physical Machine Models

Large-scale computer systems are generally classified into six practical machine models: the *single-instruction multiple-data* (SIMD) machines, the *parallel vector processor* (PVP), the *symmetric multiprocessor* (SMP), the *massively parallel processor* (MPP), the cluster of workstations (COW), and the *distributed shared memory* (DSM) multiprocessors. SIMD computers are used mostly for special-purpose applications. The remaining models are all MIMD machines, as summarized in Table 1.6.

Table 1.6 Semantic Attributes of Parallel Machine Models

Attributes	PRAM	PVP/SMP	DSM	MPP/COW
Homogeneity	MIMD	MIMD	MIMD	MIMD
Synchrony	Instruction-level synchronous	Asynchronous or loosely synchronous	Asynchronous or loosely synchronous	Asynchronous or loosely synchronous
Interaction mechanism	Shared variable	Shared variable	Shared variable	Message passing
Address space	Single	Single	Single	Multiple
Access cost	UMA	UMA	NUMA	NORMA
Memory model	EREW, CREW or CRCW	Sequential consistency	Weak ordering is widely used	N/A
Example machines	Theoretical model	IBM R50, Cray T-90	Stanford DASH, SGI Origin 2000	Cray T3E, Berkeley NOW

Five MIMD parallel machine models are shown in Fig. 1.6. Most modern parallel computers are built with commercially available, off-the-shelf commodity hardware and software components. The only exception is the PVP machines, where many building blocks are custom-made.

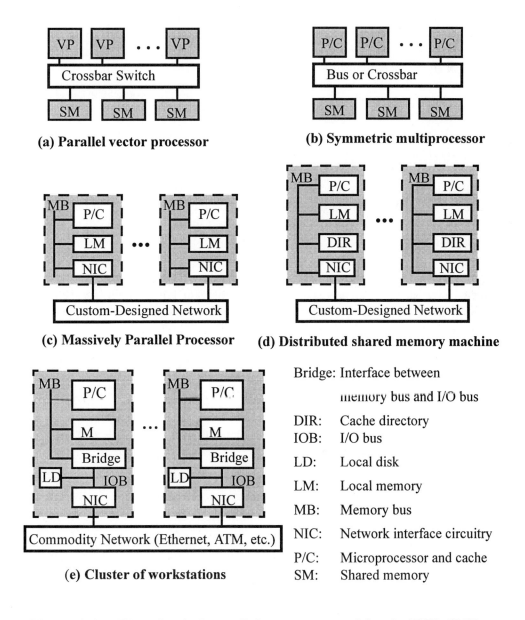

(a) Parallel vector processor

(b) Symmetric multiprocessor

(c) Massively Parallel Processor

(d) Distributed shared memory machine

(e) Cluster of workstations

Bridge: Interface between memory bus and I/O bus

DIR: Cache directory
IOB: I/O bus
LD: Local disk
LM: Local memory
MB: Memory bus
NIC: Network interface circuitry
P/C: Microprocessor and cache
SM: Shared memory

Figure 1.6   Five physical parallel computer models: the PVP, SMP, MPP, DSM, and COW.

Parallel Vector Processors The structure of a typical PVP is shown in Fig. 1.6(a). Examples of PVPs include the Cray C-90, Cray T-90, and NEC SX-4. Such systems contain a small number of powerful custom-designed *vector processors* (VPs), each capable of at least 1 Gflop/s performance.

A custom-designed, high-bandwidth crossbar switch network connects these vector processors to a number of *shared-memory* (SM) modules, which provide high-speed data access. For instance, in the T-90, the shared memory can pump data to a processor at 14 GB/s. Such machines normally do not use caches, rather they use a large number of vector registers and an instruction buffer.

Symmetric Multiprocessors The SMP architecture is shown in Fig. 1.6(b). Examples include the IBM R50, the SGI Power Challenge, and the DEC Alpha server 8400. Unlike a PVP, an SMP system uses commodity microprocessors with on-chip and off-chip caches. These processors are connected to a shared memory through a high-speed snoopy bus. On some SMPs, a crossbar switch is used in addition to the bus.

SMP systems are heavily used in commercial applications, such as databases, on-line transaction systems, and data warehouses. It is important for the system to be *symmetric*, in that every processor has equal access to the shared memory, the I/O devices, and the operating system services.

Being symmetric, a higher degree of parallelism can be released, which is not possible in an *asymmetric* (or *master-slave*) multiprocessor system. In 1997, most PVPs and SMPs have at most 64 processors, such as the Sun Ultra Enterprise 10000. The limitation is mainly caused by using a centralized shared memory and a bus or crossbar system interconnect, which are both difficult to scale once built.

Massively Parallel Processors To take advantage of higher parallelism available in applications such as scientific computing, engineering simulation, signal processing, and data warehousing, we need to use even higher scalability computer platforms by exploiting the distributed memory architectures, such as MPPs, DSMs, and COWs.

The term MPP (*massively parallel processor*) generally refers to a very large-scale computer system having the following features:

- It uses commodity microprocessors in processing nodes.
- It uses physically distributed memory over processing nodes.
- It uses an interconnect with high communication bandwidth and low latency.
- It can be scaled up to hundreds or even thousands of processors.
- Like the multiprocessor model, it is an asynchronous MIMD machine. However, processes are synchronized through blocking message-passing operations, not shared-variable synchronization operations.
- The program consists of multiple processes, each having its private address space. Processes interact by passing messages.

The MPP modeled in Fig. 1.6(c) is more restricted, representing machines such as the Intel Paragon and TFLOP. Such a machine consists a number of *processing nodes*, each containing one or more microprocessors interconnected by a high-speed memory bus to a local memory and a *network interface circuitry* (NIC). The nodes are interconnected by a high-speed, proprietary communication network. The nodes are said to be *tightly coupled*.

Distributed Shared-Memory (DSM) Machines The DSM machines are modeled in Fig. 1.6(d), based on the Stanford DASH architecture [405]. *Cache directory* (DIR) is used to support distributed coherent caches. The Cray T3D is also a DSM machine [3]. But it does not use the DIR to implement coherent caches. Instead, the T3D relies on special hardware and software extensions to achieve the DSM at arbitrary block-size level, ranging from words to large pages of shared data.

The main difference between DSM machines and SMP is that the memory is physically distributed among different nodes. However, the system hardware and software create an illusion of a single address space to application users. A DSM machine can be also implemented with software extensions on a network of workstations such as the TreadMarks, to be studied in Chapter 10.

Clusters of Computers The cluster concept is shown in Fig. 1.6(e). Examples include Digital's TruCluster [197], IBM SP2 [14], and the Berkeley NOW [178]. Clusters are the low-cost variation of MPPs in some cases. Important cluster distinctions are listed below:

- Each node of a COW is a complete workstation, minus some peripherals (e.g., monitor, keyboard, mouse, etc.). Such a node is sometimes called a "headless workstation". A node may also be an SMP or a PC.

- The nodes are connected through a low-cost commodity network, such as Ethernet, FDDI, Fiber-Channel, and ATM switch, although proprietary networks are also used in some commercial clusters.

- The network interface is *loosely coupled* to the I/O bus in a node. This is in contrast to the *tightly coupled* network interface of an MPP [Fig. 1.6(c)], which is connected to the memory bus of a processing node.

- There is always a local disk, which may be absent in an MPP node.

- A complete operating system resides on each node, whereas in some MPPs only a microkernel exists. The operating system of a COW is the same workstation UNIX, plus an add-on software layer to support single system image, availability, parallelism, communication, and load balancing.

The boundary between MPPs and COWs is becoming fuzzy these days. The IBM SP2 is considered an MPP. But it has a cluster architecture, except that a proprietary *high-performance switch* is used as the communication network. Clusters have many cost/performance advantages over the MPPs. Clustering is becoming a trend in developing scalable parallel computers as highlighted in the next section.

1.4   Basic Concepts of Clustering

In this section, we elaborate further on basic concepts concerning clusters of computers. Multicomputer clusters are distinguished below from SMPs, MPPs, and distributed systems. Benefits and challenges are identified in building clusters. Detailed network technologies will be treated in Chapter 6. Principles of clustering are treated in Chapter 9, and example cluster systems are studied in Chapter 10.

1.4.1   Cluster Characteristics

A *cluster* is a collection of complete computers (nodes), that are physically interconnected by a high-performance network or a *local-area network* (LAN). Typically, each computer node is an SMP server, or a workstation, or a personal computer. More importantly, all cluster nodes must be able to work together collectively as a single, integrated computing resource; in addition to filling the conventional role of using each node by interactive users individually.

Five architectural concepts (indicated in *italic*) are merged into a cluster as an *interconnected* set of whole computers (*nodes*) that work collectively as a *single system* to provide uninterrupted (*availability*) and efficient (*performance*) services. A conceptual architecture of a cluster is shown in Fig. 1.7.

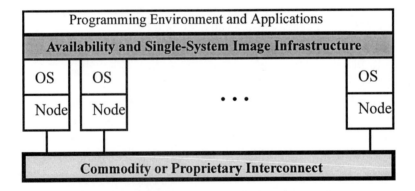

Figure 1.7   Typical architecture of a cluster of multiple computers.

The five architectural concepts of clusters are characterized below. More can be found in Chapters 9 and 10.

- *Cluster Nodes*: Each node is a complete computer. This implies that each node has its processor(s), cache, memory, disk, and some I/O adapters. Furthermore, a complete, standard operating system resides on each node. A node can have more than one processor, but only one copy of the OS image. Chapter 10 will describe the details of some cluster nodes.

- *Single-System Image*: A cluster is a single computing resource. This is in contrast to a *distributed system* (e.g., a local computer network), where the nodes are used only as individual resources. A cluster realizes this single-resource concept by a number of *single-system image* (SSI) techniques. SSI makes a cluster easier to use and manage. Most commercially available clusters have not achieved the status of full SSI operations.

- *Internode Connection*: The nodes of a cluster are usually connected through a commodity network, such as Ethernet, FDDI, Fiber-Channel, and ATM switch. However, standard protocols are used to smooth internode communication. Different networks for cluster construction are treated in Chapter 6.

- *Enhanced Availability*: Clustering offers a cost-effective way to enhance the availability of a system, which is the percentage of time a system remains available to the user. The availability techniques will be discussed in Section 9.2.

- *Better Performance*: A cluster should offer higher performance in a number of service areas. One area is to treat a cluster as a *superserver*. If each node of an *n*-node cluster can serve *m* clients, the cluster as a whole can serve *mn* clients simultaneously. Another area is to use a cluster to minimize the time to execute a single large job by distributed parallel processing.

1.4.2 Architectural Comparisons

Clusters, SMPs, MPPs, and distributed systems are four overlapped architectural concepts, as shown in Fig. 1.8. An good example of a distributed system is a LAN. A node in a LAN can be a PC, a workstation, or an SMP server. The *node complexity* refers to the hardware and software capability. A cluster node is more complex than an MPP node because the former has a disk and a full-fledged operating system, while the latter may not have a disk and may use only a microkernel.

An SMP server is more complex than a cluster node because it has more peripherals, such as terminals, printers, external *redundant array of inexpensive disks* (RAID), and tape units. Some of these peripherals may be absent in cluster nodes. We project the cluster boundary will be continuously expanded to overlap even more with the MPPs and SMPs in the future.

The horizontal axis shows the degree of SSI at various abstraction levels: ranging from a single application, to subsystem, runtime system, OS kernel, and hardware (memory and/or I/O) levels. In other words, the SSI is a relative concept, depending on the boundary of the SSI from a user's perspective. The SMP always offers SSI at all levels. The MPP supports SSI only in some application and system levels. The clusters offer SSI at a reduced level from SMPs to a comparable level of MPPs.

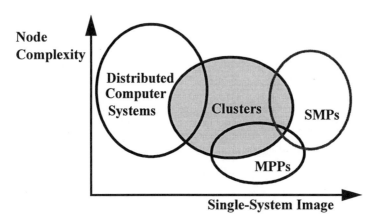

Figure 1.8   Overlapped design space of clusters, MPPs, SMPs, and distributed computer systems.

A cluster or an MPP can be used as a single resource (e.g., one giant workstation), while a distributed system always produces multiple system images due to the autonomy of individual computers in the network domain. We compare other characteristics of all four system types in Table 1.7. Current MPPs and distributed systems have a maximum size of thousands of nodes. Most clusters have tens of nodes, only as a few clusters have exceeded 100 nodes.

MPP nodes often run a microkernel, while nodes in other architectures use a full-size OS. The operating systems on a distributed system are usually heterogenous, but homogenous OS are preferred for the others. Internode communication is often done by message passing in an MPP or cluster; whereas shared files in the server are often used in a distributed system. The processors in an SMP communicate through the shared memory. Only SMPs maintain a single address space. The MPP has a single address space only if DSM is supported by hardware.

SMP has a single run queue for all processors. Multiple run queues are used in cluster and MPP nodes, but they are coordinated to balance the workload. Multiple queues in a distributed system are mostly independent. Distributed systems must use standard communication networks and protocols to accommodate heterogeneous node platforms. A distributed system often spans multiple organizations, thus requires internode security measures. These are not required for MPPs and clusters.

1.4.3 Benefits and Difficulties of Clusters

The cluster concept brings many benefits as well as challenges. Among them the most important ones are *usability, availability, scalability, available utilization,* and

Table 1.7 Comparison of Cluster, MPP, SMP, and Distributed Systems

System Characteristic	MPP	SMP	Cluster	Distributed System
Number of nodes (N)	$O(100)$-$O(1000)$	$O(10)$ or less	$O(100)$ or less	$O(10)$-$O(1000)$
Node complexity	Fine or medium grain	Medium or coarse grain	Medium grain	Wide range
Internode communication	Message passing or shared variables for DSM	Shared memory	Message passing	Shared files, RPC, message passing
Job scheduling	Single run queue at host	Single run queue	Multiple queues but coordinated	Independent multiple queues
SSI support	Partially	Always	Desired	No
Node OS copies and Type	N (microkernel) and 1 host OS (monolithic)	One (monolithic)	N (homogeneous desired)	N (heterogenous)
Address space	Multiple (single if DSM)	Single	Multiple	Multiple
Internode security	Unnecessary	Unnecessary	Required if exposed	Required
Ownership	One organization	One organization	One or more organizations	Many organizations
Network protocol	Nonstandard	Nonstandard	Standard or nonstandard	Standard
System availability	Low to medium	Often low	Highly available or fault-tolerant	Medium
Performance metric	Throughput and turnaround time	Turnaround time	Throughput and turnaround time	Response time
SSI: Single-system image, RPC: Remote procedure call, DSM: Distributed shared memory				

performance/cost ratio. Simply connecting a bunch of workstations through an Ethernet and dubbing them a cluster will hardly lead to a high-availability and high-performance system. A number of techniques (mostly software) must be in place to realize these potentials. In Chapter 5, 9, and 10, we will discuss these technological challenges in detail. Only basic concepts are given below for the time being.

Usability Since individual nodes of a cluster are a traditional platform, users can develop and run their applications in a familiar, mature environment. The platform provides all the powerful workstation programming environment tools and allows the thousands of existing (sequential) applications to run without change. Thus a cluster can be viewed as a huge workstation, providing much increased throughput and reduced response time for multiple sequential user jobs.

For parallel applications, a cluster is no more difficult to program than a message-passing MPP. In addition, many database vendors have modified their products to run on clusters, such as the IBM DB2 Parallel Edition, Oracle, and Sybase.

Availability The term *availability* refers to the percentage of time a system is available for productive use. Traditional monolithic systems, such as mainframes and fault-tolerant systems, achieve high availability through expensive, customized design. Instead of using custom components, clusters utilize inexpensive commodity components to provide higher availability, with multitudes of redundancy:

- *Processors and Memories*: A cluster has multiple memory and processor components. When one fails, the others can still be used to keep the cluster going. In contrast, when the shared memory of an SMP machine fails, the entire system will be brought down.

- *Disk Arrays*: A cluster has multiple local disks. Thus, the failure of one does not crash the entire system. In fact, a node with a faulty disk can still function, by using remote disks.

- *Operating System*: A cluster has multiple OS images, each residing in a separate node. When one system image crashes, other nodes can still work. This is in contrast to SMP, which has a single OS image residing in the shared memory. The failure of this image crashes the entire system.

The key to realizing the availability potential of clusters is a number of software techniques. Techniques are also needed to make shared components (e.g., the interconnect) highly available; otherwise they may become a *single point of failure*. These techniques will be discussed in Chapter 9.

Scalable Performance A cluster's computing power can increase with added nodes. Again, clusters' scalability is a multitude scalability. This can be best seen by comparing clusters to SMPs. SMPs are processor-scalable systems, while clusters scale in many components, including processor, memory, disks, and even I/O devices. Being loosely coupled, clusters can scale to hundreds of nodes, while it is extremely difficult to build an SMP of more than tens of processors.

In an SMP, the shared memory (and the memory bus) is a bottleneck. When several sequential programs are running on an SMP, they have to compete for the memory, even though these programs are totally independent. When the same set of programs is executed in a cluster, there is no memory bottleneck. Each program can be executed in one node, utilizing the local memory.

For such applications, clusters can provide much higher aggregate memory bandwidth and reduced memory latency. The local disks of a cluster also aggregate to large disk space, which can easily surpass that of a centralized RAID disk. The enhanced processing, storage, and I/O capability enables a cluster to solve large-scale problems by using some of the well-developed parallel software packages such as PVM or MPI, to be discussed in Chapter 13.

Clusters are compared with UP (uniprocessor), SMP, MPP, and fault-tolerant systems in Fig. 1.9, with respect to scalable performance and availability. SMPs are not highly scalable, due to the use of contention bus and centralized shared memory. The single operating system image and shared memory are potentially two single points of failure, which reduces the availability of an SMP.

Fault-tolerant systems have extremely high availability, but it is expensive to scale them up. MPPs are just the opposite. They can be easily scaled up and yet preserve a single-system image. Clusters are currently in the middle ground, moving toward higher performance and higher availability in future upgrades.

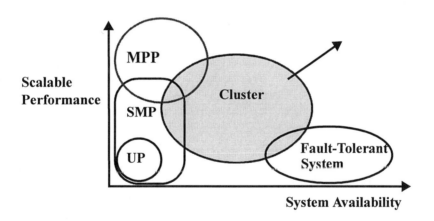

Figure 1.9   Comparison of uniprocessor, SMP, MPP, cluster, and fault-tolerant systems in system availability.

Performance/Cost Ratio   Clusters can achieve the above benefits cost-effectively. Traditional PVP supercomputers and MPPs can easily cost tens of millions of dollars. On the hand, the price of a cluster with the same peak performance is one or two orders of magnitude lower. Clusters are largely made of commodity components, and their performance and price follow Moore's law, which makes the performance/cost ratio of clusters grow faster than those of PVPs and MPPs.

Example 1.8 Performance/cost ratios of parallel computers and clusters

The Berkeley NOW group has compared the performance/cost ratios of two parallel computers and four NOW configurations, as reported in Table 1.8. The six configurations are a 16-processor Cray C90 PVP, a 256-node Intel Paragon MPP, and four versions of a network of 256 RS6000 workstations, when running an atmospheric chemistry application called Gator.

The program has three main steps. The computation step needs to perform 36 Gflop to solve *ordinary differential equations* (ODEs) in parallel. The transport step is communication-intensive. The I/O step is dominated by inputting 3.9 GB of data.

The execution times for C90 and Paragon are measured data, while those for the four versions of NOW are predicted by the Berkeley group. Just connecting 256 workstations using Ethernet and using PVM over TCP/IP for communication lead to a dreadful system, which is 1000 times slower than C90 and whose performance/cost ratio is 138 times lower. The main degrading factor is the communication overhead, which is 23,340 s versus 4 s on the C90.

Table 1.8 Performance/Cost Ratios of C90, Paragon, and Four NOW Systems (Courtesy: Anderson et al. *IEEE Micro,* [33], 1995)

System Configuration	ODE (s)	Transport (s)	Input/ Output (s)	Total (s)	Cost ($M)	Performance/ Cost Ratio (Mflop/s per $M)
Cray C90	7	4	16	27	30	44
Intel Paragon	12	24	10	46	10	78
NOW	4	23,340	4030	27,347	4	0.32
NOW+ATM	4	192	2015	2211	5	3.3
NOW+ATM+PIO	4	192	10	205	5	35
NOW+ATM+ PIO+AM	4	8	10	21	5	342

Replacing the Ethernet with a high-bandwidth ATM switch improved the performance 10 times, adding 25% cost. Implementing parallel I/O capability improves the performance by another order. Finally, using active message to speed up communication further reduces the execution time by an order of magnitude. With all improvements in place, the cluster is faster than C90, and the performance/cost ratio becomes 7 times better than that of the C90. In Chapters 9 and 10, we will learn how to develop SSI clusters with all the above hardware and software improvements. ∎

1.5 Scalable Design Principles

Designing a scalable high-performance computer is a complex engineering process. Over the past two decades, four scalable design principles have emerged, based on the experiences and lessons learned from both successful and failed system designs:

- Principle of independence
- Principle of balanced design
- Principle of design for scalability
- Principle of latency hiding

These principles come into play again and again in subsequent chapters. In this section, we explain the first three principles. Latency hiding issues will be discussed in Chapter 5.

1.5.1 Principle of Independence

This principle says that we should try to make the components of a system independent of one another. If complete independence cannot be achieved, we should make the dependence as small and clear as possible. Here the components include both hardware and software components.

One obvious benefit is the possibility of *independent scaling*: We can scale up a system along one dimension by improving one component, independent of others. The user is not required to simultaneously upgrade all components. This is also called *incremental scaling*. For instance, if the user wants to scale the size of a parallel computer system by adding nodes, the user should not be required to upgrade the operating system or the programming environment.

When the processor is upgraded to the next generation, the system can operate at higher performance without upgrading the remaining components. Another benefit is to enable heterogeneity scalability. Since a component is not tied to a particular architecture or system, it can be used in building many other systems. This can result in cost cutting. Ideally, the component becomes a commodity, with the following features:

- It has an open architecture with a standard interface to the rest of the system.
- It is an off-the-shelf product; or better yet, it is in the public domain.
- It has multiple vendors in the open market with a large volume.
- It is relatively mature, as it is used by many people for a sufficiently long time period and is reasonably debugged.
- Because of all the above, the commodity component has low cost and high availability and reliability.

We elaborate below the *independence principle* by listing some specific applications along with some concrete examples.

- The algorithm should be independent of the architecture.
- The application should be independent of the platform.
- The programming language should be independent of the machine.
- The language should be modular and have *orthogonal* features.
- The node should be independent of the network, and the network interface should be independent of the network topology.

Example 1.9 **Independence spirit in developing the Internet and IBM SP2.**

The success of the Internet is a perfect example of the benefits of the independence principle. The Internet is independent of the type of hosts, interconnection hardware, and application software. Different types of hosts are connected, ranging from PCs to supercomputers from different vendors. The interconnect hardware can be Ethernet, FDDI, etc. Users can use different software to browse the Web.

The IBM SP2 design incorporates the independence principle. The node architecture is designed to allow the use of different communication architectures (e.g., Ethernet or HPS). The communication protocol is independent of the communication hardware: on either an Ethernet or HPS, either the standard IP protocol or an IBM proprietary *user space* protocol can be used.

■

Example 1.10 **The MPI story and the legend of hypercube computers**

Message passing interface (MPI) is a good example of using a few independent (orthogonal) language features. MPI is based on four main concepts that are orthogonal to one another: *data type*, *communication operations*, *communicator*, and *virtual topology*. Any combination of the four is valid. This orthogonal independence brings a *multiplicative effect*: A few simple concepts, when combined together, can offer a lot of functionalities.

Many parallel algorithms developed for earlier hypercube computers explicitly used the hypercube topology of the interconnect, which may not perform well on a mesh-collected system. Learning from this lesson, current MPPs employ communication algorithms that are independent of the interconnect topology. A good example is the collective communication library in IBM SP system[59].

■

Two common techniques are employed to realize the independence principle: separating architecture from implementation, and using standard components.

Architectures vs. Implementations The concept of architecture should be separated from that of implementation. An *architecture* is a precise model (or specification) of the common behavior or functionalities of a family of computer systems or system components. An *implementation* is a specific realization of the model. The architecture model is used by both users and designers. An architecture can have many different implementations, which may have different performances, but all realize the same functionalities.

The architecture concept is an old one, which was first proposed and successfully applied in the design of the IBM 360 computers. Closely related to the architecture concept is the *family* concept. All members of a family share the same architecture. But different members may use different implementations and configurations to suit the different performance/cost requirements of users.

Another related concept that has gained popularity is the *open architecture* (or *open system*) concept, which means that the owner of the architecture (usually the vendor) allows users or third parties to see the architecture so that they can make components that are compatible with the architecture, or even modify or reengineer it. The success of IBM PCs testifies that this is indeed a technically powerful and commercially viable concept.

Using Standard Components   The above discussion about architecture already alludes to the importance of using standard components. There are two types of standard:

- The first is an *industry standard* (also called a *de facto* standard), which is usually pioneered by a company, then becomes widely used by the end users and accepted by the majority of the industry.
- The second type of standard is established by a national or international standards organization, such as the *International Standards Organization* (ISO), the *American National Standards Institute* (ANSI), and the IEEE Standard Committee.

Caveats   • We must be aware of two caveats when applying the independence principle:

- First, parallel computers are at the cutting-edge of computing technology. In any such system, there is usually some key component/technique that is novel, thus not quite a standard yet.
- Second, we cannot build an efficient system by simply scaling up one or a few component alone.

For example, using faster processors but slow memory modules or slow communication subsystems will not work. We must design a balanced system among all subsystems, as discussed in the next subsection.

1.5.2 Principle of Balanced Design

This principle states that we should try to minimize any performance bottleneck. We should avoid an unbalanced system design, where a slow component degrades the performance of the entire system, even though the remaining components are fast.

Furthermore, we should avoid *single point of failure*; i.e., the failure of one component breaks down the entire system. There are several theoretical and experimental results that can help achieve a balanced design.

Amdahl's Law   Suppose an application program is grouped into two types of computational structure: part X and part Y. The two parts take $X\%$ and $Y\%$ of the total execution time, respectively:

If the part X is improved to run n times as fast, the speedup S is defined

$$S = \frac{\text{Original time}}{\text{Improved time}} = \frac{1}{(X/n)\,\% + Y\%} \rightarrow \frac{1}{Y} \qquad n \rightarrow \infty \qquad (1.5)$$

This equation is known as *Amdahl's law*, which has the following implications:

- We should optimize the larger part X, i.e., speed up the common cases.
- The best speedup is upper bounded by the value $1/Y$.
- The slow part Y is called the *bottleneck*. We should minimize Y, i.e., make Y as small as possible.

Amdahl's Rule   This is also known as *Amdahl's other law*. It states that the processing speed should be balanced against the memory capacity and the I/O speed. More specifically, a rule of thumb is that a 1 *million instructions per second* (MIPS) computational speed should be balanced with 1 MB of memory capacity and 1 Mb/s of I/O rate.

Amdahl's rule is a rule of thumb. As such it is not necessarily always observed as far as the *peak* speed is concerned. Current processors have a peak speed in the range of 400-1200 Mi/s, but not many machines have installed with more than 400 MB local memory per processor. Amdahl's rule is more observed in recent systems when the sustained computational speed (e.g., sustained Mflop/s) is used, instead of the peak speed.

Example 1.11 The PetaFLOPS Project

A recent prediction from the PetaFLOPS Project [587] indicates that for a wide range of scientific/engineering simulation problems, the memory requirement (in GB) and the speed demand (in Gflop/s) have the following relationship:

$$\text{Memory} \sim \text{Speed}^{3/4} \qquad\qquad (1.6)$$

Thus a memory size of about 30 TB is appropriate for a Pflop/s machine. Note that 1 Pflop/s = 1,000,000 Gflop/s.

∎

Example 1.12 I/O and checkpoint problems

To understand the I/O speed requirement, consider the *checkpoint* problem: The system needs to periodically dump the memory contents to the disk so that in case of a system crash, users can restart their work from the last checkpoint, instead of from the very beginning. Assume we want the dumping to be finished in 90 s. Then for a 1-MB memory, we need a disk bandwidth of 1/90 MB/s = 0.9 Mb/s, which is close to Amdahl's rule.

For larger systems, the checkpoint time is even longer. Assume a dumping time of 900 s. Then for a machine with 1-GB memory, we need a disk bandwidth of 1000/900 = 1.1 MB/s. For 100-GB memory, the disk I/O requirement increases to more than 100 MB/s!

∎

The 50% Rule   From Eq.(1.4), the performance of a parallel program can be degraded by *load imbalance, parallelism overhead, communication start-up overhead, and per-byte communication overhead.* Another rule of thumb is that a parallel system is balanced if each of the four overhead factor may degrade the performance by a factor of no more than 50%. Using this rule, we can estimated desired bounds on various overhead factors. This will be treated in Chapter 3.

For instance, Table 1.9 lists the desired values of the *communication start-up overhead* t_0 under different granularity and speed conditions. Current message-passing computers have too large a start-up overhead to support fine-grained or even medium-grained parallel applications. The exception is some Cray vector supercomputers (PVPs), which have a small t_0 of only 2 μs for point-to-point communication.

When the message size $\alpha \cdot w$ or the granularity w is large, the bandwidth $r_\infty = 1/t_c$ becomes a more important factor. Table 1.10 lists the desired values of r_∞ under different assumptions of speed P_1 and *communication-to-computation* ratio α.

Example 1.13 Grid lattice in PDE solvers

Many numerical parallel *partial differential equation* (PDE) solvers for two-dimensional (2D) problems use the following scheme: The data domain is a 2D grid with $N \times N$ data points. Each data point needs X B of memory, and the total memory

Table 1.9 Desired Communication Startup Overhead Values

Grain Size w	Single-Node Speed P_1	t_0 should be less than
100 flop (Fine Grain)	4 Mflop/s	25 μs
	20 Mflop/s	5 μs
	100 Mflop/s	1 μs
1000 flop (Medium Grain)	10 Mflop/s	100 μs
	50 Mflop/s	20 μs
	250 Mflop/s	4 μs
10000 flop (Coarse Grain)	10 Mflop/s	1000 μs
	100 Mflop/s	100 μs
	1000 Mflop/s	10 μs

Table 1.10 Desired Communication Bandwidth Values

Communication-to-Computation Ratio α	Single-Node Speed P_1	r_∞ should be greater than
0.01 B/flop	4 Mflop/s	0.04 MB/s
	20 Mflop/s	0.2 MB/s
	100 Mflop/s	1 MB/s
0.1 B/flop	10 Mflop/s	1 MB/s
	50 Mflop/s	5 MB/s
	250 Mflop/s	25 MB/s
1 B/flop	10 Mflop/s	10 MB/s
	100 Mflop/s	100 MB/s
	1000 Mflop/s	1000 MB/s

requirement is $N^2 X$ B. The algorithm performs a number (e.g., 10,000) of time steps. In each step, a grid point needs Y-flop computation and access to its four neighbor points. The execution time using a single processor is

$$T_1 = YN^2 t_f \qquad (1.7)$$

On an n-processor MPP, the data domain can be decomposed into n square regions, each an $(N/\sqrt{n}) \times (N/\sqrt{n})$ subdomain containing N^2/n grid points. In each time step, the computational workload per processor is YN^2/n. Each processor needs to get $XN/\sqrt{n}$ B from each of the four neighbors. Thus on an n-processor machine, the execution time is approximately

$$T_n = \frac{YN^2 t_f}{n} + 8\left(t_0 + \frac{NXt_c}{\sqrt{n}}\right) \tag{1.8}$$

The constant factor 8 comes from the fact that each processor needs to send a message to and receive one from each of the four neighbors. The *speedup factor* is

$$S = \frac{T_1}{T_n} = \frac{n}{1 + \dfrac{8t_0}{YN^2 t_f} + \dfrac{8Xt_c}{YNt_f\sqrt{n}}} \tag{1.9}$$

In Fig. 1.10, we draw the speedup curves for four different parallel computers:

(1) The first machine has 50-Mflop/s processors connected by a slow communication network with $t_0 = 550$ μs and $r_m = 1$ MB/s. Such a network is comparable to Ethernet for point-to-point communication. When the problem size $N = 102$, machine A shows a poor speedup (denoted by the square curve at the lower part of Fig. 1.10), less than 10 over 128 processors.

(2) However, the speedup improves drastically when the problem size N increases 8 times to 8K (denoted by the diamond curve in Fig. 1.10). This increases granularity w and reduces communication-to-computation ratio α.

(3) The third machine (denoted by the triangle curve in Fig. 1.10) has 100-Mflop/s processors (twice as fast as the first machine) connected by the same slow network. Due to this unbalanced design, the speedup drops back for $N = 809$.

(4) The last machine (denoted by the circled curve at the top of Fig. 1.10) is more balanced than machine B, with 100-Mflop/s processors connected by a faster network similar to that of IBM SP2, with $t_0 = 46$ μs and $r_\infty = 35$ MB/s. This balanced design results in the best performance among the four parallel computers.

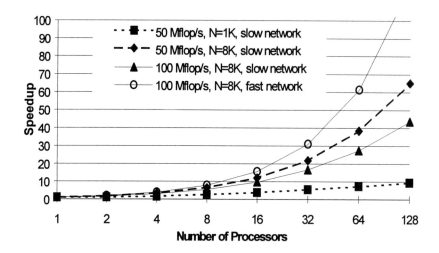

Figure 1.10   Speedup of a 2D PDE solver on four parallel computers.

■

1.5.3 Design for Scalability

This principle states that when designing a scalable system, one should consider scalability as a main objective right from the start, instead of as an afterthought. Provisions must be made so that the system can either scale up to provide higher performance or scale down to allow affordability or greater cost-effectiveness. Two popular approaches to designing for scalability are *overdesign* and *backward compatibility.*

Overdesign   Using the overdesign technique, a system is designed not to barely satisfy the minimal requirements of a current-generation system. The design must include additional features that anticipate improvement from scaled-up systems in the future. These features might be wasted on the current system, but they allow smooth migration into future, improved systems.

Example 1.14 Address space in modern processor design

One of the most important considerations in designing a processor is the *size of its address space,* i.e., the number of byte locations the processor is able to access directly. To quote Gordon Bell, a small address space is the only architecture design mistake that cannot be easily corrected later.

Modern processors support a 64-bit address space, or $2^{64} = 11.8 \times 10^{19}$ B. This huge address space may not be fully utilized by Unix supporting only a 32-b (4-GB) address space. This overdesign in address space will create a much easier transition when the operating system is scaled up to a 64-b Unix.

■

As a counterexample, the processor used in the original IBM PCs, the Intel 8086/8088 microprocessor, has a limited 20-bit or 1-MB address space. The DOS puts a limit of 640-KB for kernel and user software. This 640-KB limit caused many problems for compiler-writers and application software developers.

Since no DOS program (including Windows) can exceed this 640-KB limit, software designers had to invent elaborate techniques (e.g., high memory, expanded memory, and extended memory) to take advantage of the larger address spaces provided by later generations of processors (Intel 286, 386, 486, Pentium, and Pentium Pro).

Example 1.15 Overdesign in the IBM RS6000 SMP servers

The IBM RS6000 SMP is overdesigned for generation scalability. For instance, the first generation of SMPs is based on the PowerPC 601 processor. But the rest of system components, from memory, I/O, power, fan, and the clocking circuitry, are designed to accommodate the next two generations of PowerPC 604 and 620 processors.

Each processor port of the switch has a bandwidth of 600 MB/s, larger than needed by the 601. These overdesigned features make it easy to scale up in a future generation of SMP system by simply upgrading the processors.

■

Backward Compatibility   The overdesign technique says that one must consider requirements for scaled-*up* systems when designing a current system. A complementary technique, called *backward compatibility*, says that in designing a hardware or software component, one must consider the requirements of *scaled-down* systems. The scaled-up component should be usable in the old or in a scaled-down system. For instance:

- A new processor should be able to execute binary codes for old processors.
- A parallel program designed to run on n nodes should be able to run on a single node ($n = 1$), maybe with a reduced input data set.
- A program for a supercomputer should be able to run on a workstation. Again, the input data set must fit the limited main memory of a workstation.
- A new version of the operating system should preserve all old functionalities, except obsolete ones, which must be clearly documented.

Example 1.16 Backward compatibility of a data bus width in processors

The CPU bus of modern processors has a 64-b data bus. Some processors support backward compatibility by allowing a 64-b processor to be used in a motherboard that has a smaller data bus width (32, 16, or 8 b).

This is achieved by setting a few control pins of the processor. Suppose the processor needs two bus cycles for a memory *read* (one cycle to send the address and another to read the data from memory). When the *read* is realized in an 8-b motherboard, the processor setting will automatically read the 1 B data from memory in 9 bus cycles.

∎

Example 1.17 The parallel operating environment in IBM SP systems

The IBM *parallel operating environment* (POE) was initially designed for IBM ScalableParallel computers (SP1 and SP2), where a user application executes on multiple SP nodes. The POE supports backward compatibility as follows:

- The POE can be executed completely on one node, which allows multiple user tasks to run on one node.
- The node does not have to be an SP node. It can be any RS6000 workstation.
- The POE is implemented on top of the IBM Unix (AIX). Thus existing Unix programs can run in the POE.

∎

Caveats Backward compatibility does not mean all features of an old system should be kept. We should weed out obsolete features. In applying the overdesign technique, we should consider the cost incurred and not anticipate into the future too much. Sometimes, though, the overdesign can actually decrease the total development and production costs. This is explained by the following example from Intel experience.

Example 1.18 The Intel 486 DX and 486 SX

When Intel announced the 80486 microprocessor chips, two models were provided: the 486 DX and the 486 SX. Internally, both chips are the same. The only difference is that the *floating-point unit* (FPU) of the DX chip was disabled in the SX chip, which was marketed for users who did not need floating-point capability.

Although the chip area for the FPU is wasted on the SX chip, this overdesign saved Intel from designing two chips. The combined sales of the two chips showed much reduced design and production costs and thus increased profit.

∎

1.6 Bibliographic Notes and Problems

Parallel processors are also treated in Flynn [245] and Hennessy and Patterson [305]. Hwang [327], Lenoski and Weber [406], and Culler, et al. [181] have covered scalable multiprocessors, but with different scopes and emphases in their treatment of the subject matters. Discussions of system scalability are given by Hill [308], Rattner [513] and Smith [569]. Ni [464] has given a layered classification of parallel computers. Almasi and Gottlieb [26] cover a broad range of highly parallel computing systems up to 1994.

Basic concepts of clusters of computers are introduced in Pfister [497]. Two special issues on network-based cluster computing appeared in the *Journal of Parallel and Distributed Computing* in February and June of 1997. Recent advances in parallel and cluster computing are also discussed in Bell and Gray [71] and Davy and Dew [190]. Scalable design principles are treated in the articles by Hwang and Xu [333] and Harris and Topham [297].

The PRAM model was proposed by Fortune and Wyllie [246]. The BSP model is described in Valiant [623]. The phase parallel model was presented in Xu and Hwang [657]. Amdahl's law was originally presented in [29].

Gordon Bell has assessed supercomputers and MPPs against modern technologies [67] to [71]. Scalable computer platforms are reported by IBM [335] to [341], DEC [191] to [198], Cray Research [3] and [174], SGI [547] to [550], Sun Microsystems [598] to [600], Microsoft's Wolfpack [573], and Convex/HP [165] and [166], etc. These systems will be studied in subsequent chapters.

Journals and Conferences We recommend the following archival *journals* and *conference proceedings* for additional information on parallel processing, distributed and cluster computing, computer architecture, and scalable computing:

(1) *IEEE Concurrency: Parallel, Distributed and Mobile Computing*, a dedicated magazine specialized in applications of parallel computers, published first as *IEEE Parallel and Distributed Technology* since 1993.

(2) *IEEE Computer Magazine*, a general magazine for all subgroups of the IEEE Computer Society, published since 1967. This magazine often carries special issues that are related to parallelism or networking.

(3) *Communication of The ACM*, a general magazine for all subgroups of the Association for Computing Machinery, published since 1957.

(4) *Journal of Parallel and Distributed Computing* (JPDC), the very first journal devoted to this area, published by Academic Press, Inc. since 1983.

(5) *IEEE Transactions on Parallel and Distributed Systems*, a research journal published by IEEE Computer Society since 1990.

(6) *Parallel Computing*, an international journal on parallel computers and their programming and applications, published by Elsevier Science since 1984.

(7) *International Conference on Parallel Processing* (ICPP), the oldest conference in this field since 1972, often held in the Midwest United States.

(8) *International Parallel Processing Symposium* (IPPS), an annual symposium sponsored by the IEEE/Computer Society Technical Committee on Parallel Processing since 1986.

(9) *International Symposium on Computer Architecture* (ISCA), an annual symposium jointly sponsored by the IEEE/CS Technical Committee on Computer Architecture and by the ACM SIGARCH since 1972.

(10) *International Symposium on Architectural Support for Programming Languages and Operating Systems*, ACM annual conference since 1975.

(11) *Scalable High-Performance Computing Conference*, a conference sponsored by the IEEE/CS Tech. Committee on Supercomputer Applications since 1985.

(12) *International Symposium on High-Performance Computer Architecture* (HPCA), sponsored by IEEE Computer Society since 1995.

(13) *International Conference on Algorithms and Architectures for Parallel Processing*, an Asia-based annual conference sponsored by IEEE Singapore and Australia sections since 1995.

(14) The *COMPAR*, *VAPP*, and *EUROPAR*, three Europe-based conference series on parallel processing. Conference proceedings have been published by North-Holland or Springer-Verlag since 1981.

(15) *The ACM Symposium on Parallel Algorithms and Architecture*, an annual symposium sponsored by the ACM since 1989.

Homework Problems

Problem 1.1 Define and distinguish among the following terms on scalability:

(a)   Scalability over machine size

(b)   Scalability over problem size

(c)   Resource scalability

(d)   Generation scalability

(e)   Heterogeneity scalability

Problem 1.2 Compare the following abstract parallel computer models. Comment on their differences, relative merits, and limitations in modeling real-life parallel computers and applications:

(a) PRAM model

(b) BSP model

(c) Phase parallel model

Problem 1.3 Describe two unique features of each of the following five parallel architecture models. A unique feature means that it is not found in any of the other models.

(a) Parallel vector processor (PVP)

(b) Symmetric multiprocessor (SMP)

(c) Massively parallel processor (MPP)

(d) Cluster of workstations (COW).

(e) Distributed shared memory (DSM) machine.

Problem 1.4 Answer the following questions on parallel execution modes on existing parallel computers:

(a) What are the advantages and the disadvantages of SIMD, MIMD, SPMD and MPMD execution modes?

(b) How do you realize an MIMD parallel application on an SIMD computer? Use a concrete example to explain your scheme.

(c) How do you realize an MPMD parallel application on a computer that supports only the SPMD mode? Use a concrete example to explain your scheme.

Problem 1.5 With regard to the principle of independence:

(a) List as many independent pairs of architectural features as possible in a fully scalable computer system.

(b) Use examples to explain how to separate the concept of architecture from multiple implementations of the same architecture.

Problem 1.6 Define the following performance measures associated with scalable parallel and cluster computing:

(a) Parallel execution time

(b) Workload versus sustained speed

(c) Parallelism versus interaction overheads

(d) Point-to-point and collective communications

Problem 1.7 With regard to multiprocessors, multicomputers, and cluster of computers:

(a)   Distinguish between multiprocessors and multicomputers based on their structures, resource sharing, and interprocessor communications.

(b)   Explain the differences among UMA, NUMA, COMA, DSM and NORMA memory models.

(c)   What are the additional functional features of a cluster that are not found in a conventional network of autonomous computers?

(d)   What are the advantages of a clustered system over a traditional SMP server?

Problem 1.8 Use the PRAM to model existing parallel computers:

(a)   Which PRAM variant can best model an SIMD machine? Explain how.

(b)   Which PRAM variant can best model a shared-memory MIMD machine?

(c)   Criticize the inadequacy of the PRAM to model most parallel computers that have been built today.

Problem 1.9 Consider the speedup performance of parallel computers.

(a)   Formulate Amdahl's law and outline its implications.

(b)   What is Amdahl's rule? Explain how to apply it in parallel system design.

(c)   Check two computer systems you most frequently use and see if Amdahl's rule is followed.

Problem 1.10   Take the following design approaches to building scalable or cluster multiprocessor and multicomputer systems.

(a)   Use two example designs to explain how overdesign can enhance scalability.

(b)   Why is backward compatibility important in building scalable computers?

(c)   What is the checkpoint capability in a highly available cluster system?

(d)   Compare the differences between multicomputer clusters and distributed computer systems in node complexity, node operating system, internode communications, job scheduling, support for SSI, system availability, and security control.

Chapter 2

Basics of Parallel Programming

Parallel programming is the activity of constructing a parallel program from a given algorithm. It demands extensive interaction between algorithm designers and computer system architects. This was emphasized by H. J. Siegel et al. [556]:

> *A widespread misconception is that the two most important parts of the high-performance computing field are architecture and algorithm. However, the interface between them is a crucial issue as well.*

In this chapter, we review the state of parallel programming and examine the representation of parallelism, process interactions, and basic properties of parallel programs. In all example codes, the * refers to *multiply sign*, unless otherwise noted. The ++ means *increment by* 1, *x* % 7 means *x modulo 7*, and (x == y) refers to the test of the condition: (*x equal to y*) in C notation. The notation x:= 5 means the assignment of 5 to a variable x.

2.1 Parallel Programming Overview

It is generally agreeable that parallel programming is in a sorry state:

- Parallel software development has lagged far behind the advances of parallel hardware. The lack of adequate parallel software is the main hurdle to the acceptance of parallel computing by the mainstream user community.
- Compared to their sequential counterparts, today's parallel system software and application software are few in quantity and primitive in functionality.

Unfortunately, these trends are likely to continue for several reasons, which will be discussed shortly. However, there is light at the end of the tunnel. We will point out recent advances which may narrow the gap between parallel hardware and parallel software.

2.1.1 Why Is Parallel Programming Difficult?

The main reason that parallel software has been lagging behind is that parallel programming is a more complex intellectual process than sequential programming. First, parallel programming subsumes sequential programming. It involves all issues in sequential programming, plus many more issues that are intellectually more challenging.

Second, while there is only one basic model (the von Neumann model) for sequential programming, there are many different parallel programming models. Third, software environment tools such as compiler, debugger, and profiler are much more advanced for sequential programs development.

Fourth, more people have been practicing sequential programming than parallel programming and for a much longer time. In other words, sequential programming is more mature, with a huge base of accumulated knowledge, including mistakes discovered and lessons learned in the past. Let us compare parallel programming to sequential programming from a user's perspective as illustrated in Fig. 2.1.

Sequential Programming Suppose a user needs an application program. For a sequential machine, it is very likely that such an application already exists. Better still, the source code (e.g., in C or Fortran) is probably available. The user needs only to recompile the code and run it on the target machine. If an application needs to be developed, the user is likely to find an existing algorithm that can be adapted to suit the purpose. Even if the algorithm does not exist, the user is aided by some programming tools.

First, there are a number of algorithmic paradigms that have been long established [16] and can guide the user in algorithm design. Second, there is just one machine model, the von Neumann model [271], that universally supports all algorithmic paradigms, programming languages, and computer platforms. Finally, although many sequential languages have been proposed, there are just a few standard languages: Fortran dominates scientific computing; Cobol and fourth-generation languages (4GLs) are widely used in data processing; C is used everywhere.

Sequential programming tools are *general*. For instance, the C language can support all algorithmic paradigms and specify all computable functions. A C program, once developed, can run on any sequential computer. In other words, sequential programming is *heterogeneity scalable* (Section 1.2.3). Sequential programming tools are *stable* over many generations of computer evolution (i.e., they are *generation scalable*).

Fortran 77 was standardized in 1977 and is still widely used today. The C language used today has seen little change compared to the original version invented two decades ago [368]. The von Neumann model has been used for more than 40 years without much modification. Due to their generality and stability, these tools are mature. They have been tested for many years in different applications on different computer platforms. People learned how to best use them and avoid pitfalls.

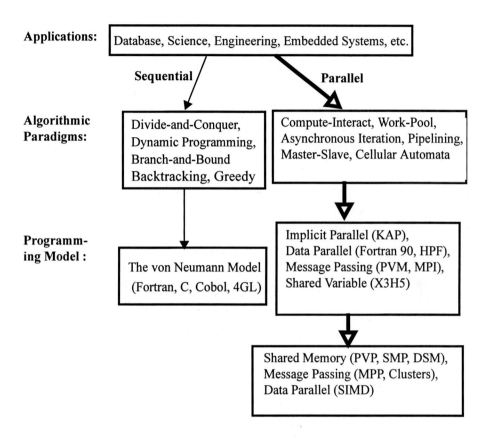

Figure 2.1   Comparison of parallel and sequential programming.

Parallel Programming   Parallel programming is inferior in all aspects. It is unlikely that a parallel code exists for a desired application. Even if it does, it often cannot be used for the user's parallel computer (e.g., an MPP), because the parallel code was originally written for a different architecture (e.g., an SMP server).

Parallel programming does not have the support of mature, general, and stable tools either. The parallel algorithmic paradigms are still not well understood or widely accepted. Instead of a single, general machine model, there are two levels of models, with many different models at each level.

A *programming model* is what the programmer sees and uses when developing a parallel program. A *native model* is the lowest-level user-visible programming model provided by a specific parallel computer platform. Other programming models can be implemented on top of this native model. For instance, on an SGI Power Challenge, which is an SMP, the native model is a shared-variable model (e.g., SGI Power C). But data parallel (e.g., HPF) and message-passing (e.g., MPI) can be implemented on top of it.

Most parallel languages used on current systems are some extension of Fortran or C. Such parallel languages at both programming and native levels are far less generation scalable and heterogeneity scalable than sequential ones. Consequently, a parallel program developed on one parallel platform is less likely to be portable to other and future parallel computers.

Advances in Parallel Programming   Despite the above pessimistic review, there has been much progress in the parallel programming field:

- Many parallel algorithms have been developed. Although most of them are based on the unrealistic PRAM model, some can be adapted for practical use.
- A small set of simple parallel algorithmic paradigms has emerged and is becoming accepted. We will discuss these paradigms in Section 12.1.1.
- The native models are converging toward two models: the single-address space, shared-variable model for PVPs, SMPs, and DSMs, and the multiple-address space, message-passing model for MPPs and clusters. The SIMD model has faded from mainstream, general purpose computing. However, it is still useful for special purpose, embedded applications such as signal, image, and multimedia processing.
- The high-level parallel programming models are converging toward three standard models: data parallel (e.g., HPF), message-passing (e.g., PVM and MPI), and shared-variable (e.g., ANSI X3H5). There is also a model where the user writes sequential programs, and the implicit parallelism is extracted by a parallelizing compiler (e.g., Kap). These models will be briefly reviewed in Section 12.2.1 and discussed in detail in Part IV.

Throughput Processing   Even with accelerated progress in parallel software development, parallel applications will be very few compared to sequential applications in the foreseeable future. A parallel computer system (including both hardware and software) should efficiently support sequential programs to increase system throughput of multiple, independent sequential jobs, in addition to decreasing response time of a single parallel application.

It is desirable that a high-performance parallel computer be viewed as a huge workstation with *single system image*, so that many users can take advantage of the enhanced processing and storage capability to run multiple sequential jobs. Gregory Pfister calls this mode the *serial program parallel system* (SPPS) model [497], also known as *throughput processing*.

2.1.2 Parallel Programming Environments

From a user's viewpoint, a typical parallel processing system has a structure as depicted in Fig. 2.2. An algorithm is first developed to solve an application problem. Then a user (programmer) needs to implement the algorithm in a high-level language (the source code). Neither the algorithm nor the source code needs to be explicitly parallel.

A compiler then translates the source code into the native code to execute on a *parallel platform*, which includes the operating system and the underlying parallel computer hardware. We will use the term *compiler* to include all the software involved in translating a source code, such as assemblers and linkers. A *preprocessor* is a source-to-source translator. An example is the familiar C preprocessor CPP.

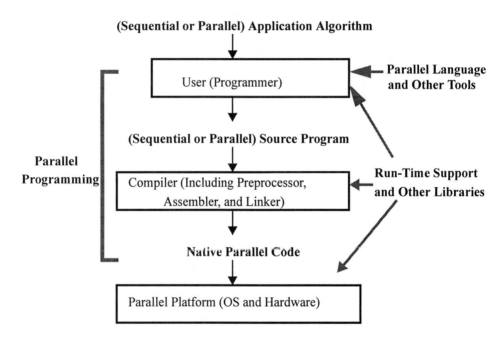

Figure 2.2   Components of a parallel processing system.

Any programming language has a *runtime support* system, which is a set of subroutines loaded with the user code to provide some necessary functionality such as initialization when a user code starts to execute, clean-up after the user code finishes, and the allocation and deallocation of data objects. Part of the runtime support is provided by *libraries*. A library is a set of frequently used subroutines compiled according to some special rules, to provide a clean, high-level interface to computer resources in realizing some needed functionalities.

A library can come with a language, e.g., the C "stdio" library that takes care of standard I/O. It can also come with an operating system, e.g, the thread library of Solaris, which provides multithreading. It could also be language- and platform-independent, such as MPI, which supports the message-passing model of parallel processing. The library subroutines can be linked to the user code before run time or dynamically at run time

Environment Tools   The term *tools* has been used to mean different things. In the broadest sense, a tool refers to any hardware and software utility that aids the development and execution of user applications. The set of all such tools is collectively referred to as a *programming environment* (or simply *environment*). Examples of tools include operating system utilities, programming languages, compilers, runtime libraries, etc.

Environment tools are the set of tools normally not associated with an operating system or a programming language. Environment tools include the following types:

- *Job management tools* are used to schedule system resources and manage user jobs. Examples include the *Network Queueing System* (NQS) and the *Load Sharing Facility* (LSF). Such tools may also provide for *load balancing* and *checkpointing* (i.e., periodically backing up user applications).
- *Debugging tools* are used to detect and locate semantic errors in parallel and sequential applications.
- *Performance tools* are used to monitor user applications to identify performance bottlenecks, which is also known as *performance debugging*.

2.1.3 Parallel Programming Approaches

There are currently four models of parallel programming that are widely used on real parallel computers: *implicit*, *data parallel*, *message-passing*, and *shared variable*. These models will be discussed briefly in Section 12.2 and in detail in Part IV. These parallel programming models are realized in real parallel programming systems, mostly by extending Fortran or C. There are three means of extension: *library subroutines*, *new language constructs*, and *compiler directives*.

- *Library Subroutines*: In addition to the standard libraries available to the sequential language, a set of new library functions is added to support parallelism and interaction operations. Examples of such libraries include the MPI message passing library and the POSIX Pthreads multithreading library.
- *New Constructs*: The programming language is extended with some new constructs to support parallelism and interaction. An example is the aggregated array operations in Fortran 90.
- *Compiler Directives*: The programming language stays the same, but formatted comments, called *compiler directives* (or *pragmas*), are added.

These approaches are illustrated in Fig. 2.3 with a simple code fragment. All three parallel programs perform the same computation as the sequential C code in Fig. 2.3a. The library approach is demonstrated in Fig. 2.3b, where the two loops are distributed to *p* processes to execute in parallel. The two library functions, my_process_id() and number_of_processes(), exploit parallelism. The barrier() function ensures that all processes are synchronized after the first loop, so that the second loop will use the correct values of the array A updated in the first loop.

```
for ( i = 0 ; i < N ; i ++ )  A[i] = b[i] * b[i+1] ;
for ( i = 0 ; i < N ; i ++ )  c[i] = A[i]  + A[i+1] ;
```

(a) A sequential code fragment

```
id = my_process_id () ;
p = number_of_processes () ;
for ( i = id ; i < N ; i = i+p )  A[i] = b[i] * b[i+1] ;
barrier () ;
for ( i = id ; i < N ; i = i+p )  c[i] = A[i]  + A[i+1] ;
```

(b) Equivalent parallel code using library routines

my_process_id(), number_of_processes (), and barrier()

```
A(0:N-1) = b(0:N-1) * b(1:N)
c = A(0:N-1) + A(1:N)
```

(c) Equivalent code in Fortran 90 using array operations

```
#pragma parallel
#pragma shared ( A, b, c )
#pragma local ( i )
{
    #pragma pfor iterate (i=0; N ; 1)
        for ( i = 0 ; i < N ; i ++ )  A[i] = b[i] * b[i+1] ;
    #pragma synchronize
    #pragma pfor iterate (i=0; N ; 1)
        for ( i = 0 ; i < N ; i ++ )  c[i] = A[i] + A[i+1] ;
}
```

(d) Equivalent code using pragmas in SGI Power C

Figure 2.3   Three parallelization approaches.

These parallel operations can be made much simpler by using new language constructs, which is illustrated in Fig. 2.3c. The Fortran 90 array assignment construct A(0:N-1) = b(0:N-1) * b(1:N) performs N element multiplications and assignments in one assignment statement. There is no need for an explicit synchronization between the two array assignments, because Fortran 90 statements are *loosely synchronous*: all operations of one statement are finished before the next statement can begin.

The compiler directive approach is exemplified in Fig. 2.3d. The parallel pragma indicates that the next statement (which happens to be a block) should be executed in parallel. The shared pragma indicates that the three array variables are shared by the parallel processes, while the local pragma indicates that each process has a local i variable. The SGI *pfor* pragma instructs the system to execute the next loop in parallel. The synchronize pragma generates a barrier synchronization.

The relative merits and shortcomings of the three approaches are summarized in Table 2.1. The library approach is currently the most widely used one, due to ease of implementation. All parallelism and interaction functionalities are realized by a set of library subroutines to supplement a sequential C or Fortran. As a consequence, there is no need to implement a new compiler. However, without the compiler's support, the user is deprived of compile-time analysis, error checking, and optimization.

Table 2.1 Three Approaches to Implementing Parallel Programming System

Approach	Example	Advantages	Disadvantages
Library	MPI, PVM, Cray Craft	Easy to implement, do not need a new compiler	No compiler check, analysis, optimization
New constructs	Fortran 90, Cray Craft	Allow compiler check, analysis, optimization	Difficult to implement, need a new compiler
Directives	HPF, Cray Craft	Between library and new constructs can be ignored on a sequential platform	

The new construct approach is just the opposite of the library approach. A new compiler is a must, which makes this approach more difficult to implement. However, the new compiler can perform extensive analysis and optimizations, which can be used to catch errors and to generate more compact and efficient code.

The compiler directive approach can be viewed as a compromise between the other two approaches. It has one additional advantage: A parallel program is just a sequential program plus some formatted comments. Therefore, it can be compiled and executed on any sequential platform, as long as the compiler is able to ignore all directives. Look at Fig. 2.3d. When all the pragmas are ignored (deleted), it is just the same old sequential C program in Fig. 2.3a.

The three approaches can be used to realize any programming model. The approaches and the programming models can all be combined in various ways on any parallel platform. For instance, the Cray MPP programming model utilizes all three approaches (new constructs, library functions, and compiler directives) to realize data parallel (Fortran 90), shared-variable (work-sharing) and message-passing (PVM) programming models in one integrated programming tool, called Cray Craft.

2.2 Processes, Tasks, and Threads

On a parallel computer, a user application is executed as *processes*, *tasks*, or *threads*. We will discuss their commonalities and then their differences. The traditional definition of *process* is *a program in execution*. This simple definition needs to be clarified.

2.2.1 Definitions of an Abstract Process

It is useful to consider the process concept at two levels. The abstract view, which will be discussed in this section, is simple and well suited for users of parallel computers. In the next section, we will present a more detailed view of process, showing how it is implemented. Conceptually, a process (thread, task) can be viewed as a dynamic entity with four components.

Definition 2.1 A *process P* is a 4-tuple $P = (P, C, D, S)$, where P is the *program* (or the *code*), C the *control state*, D the *data state*, and S the *status* of the process P. A process is *dynamic* in that it is not just the code text or/and the data, but includes the idea of *being in execution*. Exactly what this means is explained below.

∎

Program (Code) Any process is associated with a program. As a concrete example, consider the following C code:

```
main() {
    int i = 0;
    fork(); fork();
    printf("Hello!\n");
}
```

We can compile this program (assuming it is in file hello.c) :$ cc -o Hi hello.c and execute it as a simple command: $ Hi.

The operating system will create a process to say hello four times. In addition to the obvious code provided by the user (the source is in hello.c), which we will call the *user code*, the process uses *runtime support*, *library routines*, and *system functions*, as shown in Fig. 2.. The above process calls printf(), which is a routine in the standard I/O library of the C programming language, and fork(), which is a Unix system call.

Control and Data States Most programs are based on an *imperative* machine model, where a central concept of is *state updating*. An imperative program can be considered as a state machine (or an *automaton*), which maps the program from an initial state to one or more final states. Let us first define some terminology.

Definition 2.2 A program uses two sets of variables: *Data variables* are variables declared by the programmer to hold data values. *Control variables* are variables holding control flow information, which are not explicitly declared. In other words, control variables hold information regarding which operations should be executed next. The union of these two sets forms the set of *program variables*.

■

Definition 2.3 At any time, each data or control variable of a program is paired with a value, which could be a special value *undefined*. The set of all (data variable, data value) pairs at time *t* defines the *data state* of the program at time *t*. Similarly, The set of all (control variable, control value) pairs at time *t* defines the *control state* of the program at time *t*. The *program state* at time *t* is then the sum of the data state and the control state at *t*. Alternatively, we can say that the program state at *t* is the set of all (program variable, value) pairs.

■

Definition 2.4 A program starts with an initial state. When an *atomic operation* (cf. Definition 1.2) of the program is executed, the program is transformed from the current state to a next state. The program keeps executing atomic operations to update its state, until it terminates. Then the program is in a final state.

An imperative program can thus be viewed as producing one or more sequences of atomic operations which transform the state machine from an initial state to the final state(s). Of course, a program may not terminate. When a program enters a state such that it is guaranteed not to terminate, we say the program enters a divergent state.

■

For simplicity, we can regard control variables as program counters. For a process with a single *thread of control*, there is just one control variable: the program counter. For a process that has multiple threads of control, there are multiple control variables, one for each thread, holding the program counter value of that thread. For multiple subprocesses, the whole process can have several control variables, each for a component process.

The data variables may be declared either implicitly or explicitly in a user program. For instance, the data variables of a Unix process include hidden file pointers *stdin*, *stdout*, and *stderr* for standard input, standard output, and standard error, respectively, which may not be explicitly declared.

Process Status A process has a certain *status* at any time. (In some literature, what we mean by status is called *state*.) Some important statuses and their transitions are shown in Fig. 2.4.

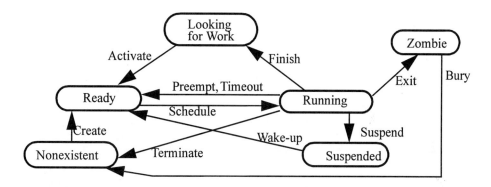

Figure 2.4   The process state transition diagram.

- At the beginning, a process does not exist (it has a *nonexistent* status). It comes into being through a *process creation* operation by its creator, the *parent process*. A newly created child process is *ready* to execute, but it actually begins *running* (executing its code) only after being scheduled. Several things can happen while a process is running:
- It could be *suspended* to enter a *suspended* status because it is not able to continue executing, due to a page fault or some other circumstance.
- Later on when the circumstance changes so that it is able to run again, it will be *woken up* to have the *ready* status.
- A running process could also be *preempted* by another process with higher priority, or it could exhaust the CPU time slice allocated to it (*timeout*). In both cases, the process itself is able (ready) to continue executing, but is forced to give up the CPU resource and change its status to *ready*.
- Finally, a running process could *terminate* itself, either normally after it finishes executing its code, or abnormally (aborts), and cease to exist.

Another frequently used operation is *process switching*, which refers to transferring a running process to either a suspended or ready status and scheduling the next ready process to run. The diagram in Fig. 2.4 is meant to cover major process states.

In real operating systems and parallel programming environments, the situation is much more complicated with additional statuses. For instance, the ANSI X3H5 standard for shared-memory parallel programming [39] allows a new status: A process, upon finishing execution, can enter a *looking-for-work* status. Such an unemployed process can be *activated* when it is assigned some new work to do. This scheme could reduce the overhead resulting from a process's being terminated and then recreated to do a new work.

The process concept presented above is suitable for *users*, e.g., someone who just needs to use a parallel language. For people who want to implement the process concept, e.g., implementers of a parallel language or a message-passing library, a more detailed view of process is needed. Implementation of processes must consider the following aspects: *execution mode, address space, context, process descriptor*, and *process control.*

2.2.2 Execution Mode

An operating system normally includes the following components:

- The **kernel** is the essential, indispensable program in an operating system, which directly manages system resources, handles exceptions, and controls processes. Only one kernel can reside in a computer at a time.
- A **shell**, also known as a *command interpreter*, is a user interface to the operating system. An example is the C shell on Unix.
- **Utilities** are additional operating system software that provide frequently used functions, such as compilers, editors, debuggers, etc.

A computer provides two *execution modes* to execute programs. Which mode is in effect is indicated by one or more bits in some special register. The kernel executes in the *kernel mode*, also known as the *supervisor mode*, the *system mode*, or the *privileged mode*. In addition, *kernel processes* execute in kernel mode. These are processes created by the kernel to help manage system resources. An example is the *swapper* process in Unix.

Other programs are executed as a process in the *user mode*. Such a process is called a *user process*. Note that although a shell or a utility (e.g., a compiler) may be considered part of an operating system, they are executed as a user process. When a user process is executing, the kernel also resides in memory. A program executing in the user mode is also called a *user-level* program. Otherwise it is a *kernel-level* program.

Mode Switch The execution mode can be switched back and forth between user and kernel modes. A computer starts up in the kernel mode. After initializing the system and creating a few kernel processes, the kernel eventually passes control to a shell, which is a user process that can create additional user processes.

When a user process executes, the execution mode can be switched to the kernel mode under two circumstances:

- When the user process generates a synchronous exception, including making a *system call* to request services from the kernel
- When an asynchronous exception (e.g., a timer or disk interrupt) occurs

After the kernel finishes the requested services, it can switch the execution mode back to the user mode. The services performed by the kernel can be divided into the following two classes: The first class is not performed for any particular user process. This includes process scheduling and handling asynchronous exceptions. The second class is performed for a user process when the process generates a synchronous exception.

For instance, when the user process generates a division-by-zero exception, the kernel will execute an exception handler and eventually will terminate the user process. Also, when the user process makes a system call trap, the kernel will execute the desired system service (i.e., an I/O function) and then resume the user process.

2.2.3 Address Space

Computers use *addresses* (or *locations*) to refer to any object, be it data or code. The address specified in an instruction is called the *virtual address*, while the address a processor puts on its (physical) address bus is called the *physical address*. In most computer systems, each address, be it physical or virtual, holds 1 byte of information. Addresses normally start at 0 and grow consecutively up to a maximum.

Address Space of a Processor For any processor, the set of all possible virtual (physical) addresses is called the *virtual (physical) address space* of that processor. The term *address space* is often used when it is not important to differentiate between physical and virtual address spaces.

The *size of an address space* is the total number of addresses A. Since each address holds a byte, we often say an address space of A bytes. Given an address space of A bytes, each address can be uniquely identified by $\log A$ bits. The size of a processor's virtual address space is given by its instruction set architecture, while the size of its physical address space is indicated by the width of its address bus.

For instance, the Intel Pentium processor has 32-bit physical addresses, and a process uses 46-bit virtual addresses. From a process's viewpoint, the processor has a $2^{32} = 4$ GB physical address space and a $2^{46} = 64$ TB virtual address space. It is easy to imagine why a 64-TB virtual address space is needed. Just think about multimedia.

Suppose one wishes to create an on-line movie library. Assume one movie frame has 1024×1024 pixels, and each pixel needs 4 bytes. That is 4 MB per frame. There are 30 frames in one second of movie. Without compression, a two-hour movie would need $2 \times 3600 \times 30 \times 4$ MB $= 0.864$ TB! It is not so obvious why one would ever need a 4-GB physical address space, partially due to a misconception that physical addresses are used only to access physical address space.

In the literature, the term *memory address space* (or simply *memory*) is often used in the place of *address space. For example,* Intel processor has a separate *I/O address space.* It should be emphasized that an address, even in the memory address space, can be used to refer to any devices, including memory, disk, and other I/O devices. A good analogy is to consider an address as a telephone number, which can be used for a telephone, a fax machine, or a modem, etc.

Even when we reserve all of the physical address space for main memory, 4 GB is not that large. When the IBM PC came out, it had only about 0.5-MB main memory. In 1997, it is common for an IBM PC to have 16- or 32-MB main memory. Following this trend, we would foresee 4-GB main memory for *personal computers* in 15 to 20 years. For current supercomputers, a 4-GB or greater physical address space is needed for shared memory, whether it is distributed or centralized.

Example 2.1 Address space in the Alpha 21064 processor

A case in point is the DEC Alpha 21064 processor, which has 34-bit physical addresses. When Cray Research used it to build the T3D MPP, it found the physical address space too small and had to add a new piece of hardware for each processor to extend the address space.

This is because the Cray T3D is designed to be scalable to 2048 processors, and each processor can have 64 MB local memory, which is remote to other processors. Being a distributed shared-memory machine, T3D allows each processor to access all local as well as remote memories.

The implication is that each processor needs a physical address space of 2048×64 MB = 128 GB, or 37-bit physical addresses.

∎

Table 2.2 shows the address space sizes of popular processors. As can be seen, later models have larger address spaces. For instance, the DEC Alpha architecture actually allows 64 bits for both virtual and physical addresses. Although only 43-bit virtual addresses are implemented in the 21064 processor, the remaining 21 bits are reserved, so that programs developed for 21064 can run on later processors, which may use part or all of the 21 bits.

Virtual Memory Layout A simplified view of the virtual address space layout of a process is shown in Fig. 2.5, as found in popular operating systems such as Unix. Part of the process address space can be accessed only in the kernel mode, which is called the *kernel space.* The other part is the *user space,* which is divided into three segments. The *text* segment contains the code of the process, which has fixed size, is not writable, and can be shared by other processes.

Table 2.2 Address Space Sizes of Popular Processors

Architecture	Model	Size of Physical Address Space	Size of Virtual Address Space
Intel 80x86	8086	1 MB (2^{20} B)	1 MB (2^{20} B)
	Pentium	4 GB (2^{32} B)	64 TB (2^{46} B)
PowerPC	601	4 GB (2^{32} B)	4 PB (2^{52} B)
	620	16 EB (2^{64} B)	16 EB (2^{64} B)
DEC Alpha	21064	16 GB (2^{34} B)	8 TB (2^{43} B)
	21164	1 TB (2^{40} B)	8 TB (2^{43} B)
MIPS	R4000	64 GB (2^{36} B)	1 TB (2^{40} B)
	R10000	1 TB (2^{40} B)	16 TB (2^{44} B)

The *data* segment holds both static and dynamic data objects of the process. The dynamic data are stored in the *heap* area of the data segment, which can grow (or shrink) at run time. The *stack* segment is used to store the activation records of procedure calls made by the process, among other things. The stack can also grow or shrink at run time, usually in the opposite direction of the heap.

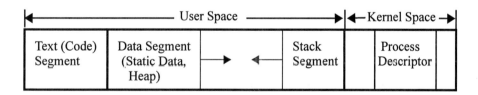

Figure 2.5   Address space layout of a process.

2.2.4 Process Context

The *context* of a process is that part of the program state that is stored in the processor registers. Not all processor registers are part of the process context. For instance, floating-point registers are not in the context of a process if it does not use them.

A *context switch* is the action to save the current process context and to load a new context. A context switch is needed when there is a mode switch. For instance, when a keyboard interrupt occurs while a user process is executing, the processor will pass control to the kernel to execute an *interrupt handler* (*interrupt service routine*).

However, before the handler is executed, the context of the user process must be saved to memory to free the registers for the handler to use. After the interrupt is handled, the kernel restores the context of the user process and passes control back to let the user process resume execution.

A context switch may also be needed when a *process switch* or *thread switch* takes place. A process (thread) switch occurs when a running process (thread) changes its status to ready or suspended and a new ready process (thread) is scheduled to execute. A thread switch needs to change at least the stack pointers, the program counter, and the status registers. A process switch may change all registers, as the address space is changed.

2.2.5 Process Descriptor

Additional information of the process is stored in some data structures in the kernel space. The most important is the *process descriptor*, which contains information for the kernel to manage processes, such as the following:

- *Process credentials*, such as the process identifier, the parent process identifier, the user identifier, and the group identifier, etc.
- *Process status*, such as ready, running, suspended, etc.
- *Context*, an area holding the context of the process. This area is used, for instance, to save the context when the process is switched from running to suspended.
- *Memory map*, such as the sizes and access rights of various memory segments, pointers to segment and page tables.
- *Other per-process Information*, such as opened files, received signals, etc.
- *Global data structures*, such as pointers to queues and tables managed by the kernel for all processes. For instance, all ready processes are waiting for execution in a ready queue.
- *Process control information*, which is described in Section 2.2.6.

Per-Process Information The process descriptor is also called *task descriptor* or *process control block*. The kernel maintains a table or list of descriptors for all processes. It is not necessary that all process information reside in the main memory at all times.

Operating systems divide the process descriptor into two parts: a *per-process* part that can be swapped out with the process's user space, and the other part that has to stay in the main memory at all times, even when the process is swapped out. In some implementations, the per-process part is allocated in the user space, but accessed only by the kernel.

An example of the per-process part is the *kernel stack*. When the kernel executes an exception handler (e.g., a system call or a division-by-zero handler) on behalf of a user process, the handler may need to make procedure calls and thus need a stack. Some early implementations use just one stack for all kernel functions, which makes the kernel stack a bottleneck especially in a multiprocessor system. Many modern implementations let each process keep both a kernel stack and a user stack in its user space.

2.2.6 Process Control

Process control refers to the functions performed by the kernel to actually manage the processes. The process control functions include the following:

Process Descriptors The kernel uses the process descriptors to create, suspend, wake up, and terminate processes, etc. It manages the global data structures (e.g., the queue for ready processes and the list of process descriptors) and all per-process data structures.

Protection The kernel performs extensive checks to ensure that a process only accesses the resources it has the right to. The process descriptor includes various privilege and rights information of the process. Normally, a user process can only access its user space (Fig. 2.5). It cannot access the kernel space or another process's user space. Two processes communicate through special *interprocess communication* (IPC) mechanisms, such as messages and pipes in Unix. Some systems allow an additional *shared-memory* area to enable processes to communicate through shared address space.

Note that in any computer system, all memory and I/O devices are referred to as addresses. By protecting the address space, the kernel ensures the protection of all memory, I/O devices, program and data files, etc.

Scheduling This refers to assignment of resources (processors, memory and I/O) to the processes by a kernel program called the *scheduler*. A scheduler should be *fair* and *efficient*. The process descriptor contains the process *priority* information. A scheduler is fair if it allows processes with equal priority to have equal access to system resources and a higher-priority process to have better access. Important types of schedulers are illustrated in Fig. 2.6.

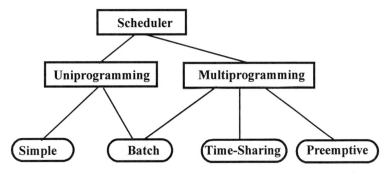

Figure 2.6   Various types of schedulers in the kernel.

For instance, a process should not be allowed to *starve*, i.e., to wait forever for a free processor to run. A scheduler is efficient if it incurs low overhead. The simplest systems allow only a single user process to be active at a time. In other words, only one user is allowed to use the system at a time. Furthermore, this single user can run one process at a time. When the user process is waiting for an I/O function to complete, no other processes can use the free processor resource.

The resources are *dedicated* to the single user. Such a system is called a *single-user single-tasking* (*uniprogramming*) system. An example is the Microsoft DOS. More sophisticated systems allow multiple processes from a single user to utilize the resource simultaneously.

An example of such single-user multitasking systems is the Microsoft Windows. Unix and Windows NT are examples of *multi-user multitasking* (or *multiprogramming*) systems, which allow many users processes to share the system resources simultaneously.

Resource sharing takes several forms.

- In the *dedicated* mode, a user program has exclusive use of the needed resources. It does not give up the resource until the program is finished. In other words, the resources are not shared.

- In the *batch* mode, a user process, once scheduled to run, executes to completion. However, in a multiprogramming batch system, the process may voluntarily give up the processor resource when it is waiting for an event, such as the completion of an I/O operation.

- In a *time-sharing* system, multiple user processes are simultaneously executed on one processor in an interleaved fashion. The processor time is divided into short *time slices* (or *quantums*). Each process executes for one quantum and then yields the processor to another process.

- In a *preemptive* system, a high-priority process can take away the processor from a low-priority process, even when the latter has not exhausted its current time quantum.

Example 2.2 Time-sharing versus batch processing

The time-sharing technique is especially useful for *interactive* applications, which need I/O operations from a terminal and value fast response time. Suppose 10 processes are using a processor. One interactive process needs 5 s of processor time to execute. The other processes each need 1000 s.

In a batch system, the user may have to wait for 9005 s after submitting the interactive process to see the final result, if the batch scheduler schedules the interactive process after the other nine processes. However, in a time-sharing system, the user only needs to wait for 50 s. The details of this example is left as an exercise. ∎

Example 2.3 Preemptive scheduler in real-time systems

Preemptive schedulers are valuable for *real-time systems*, which require that certain operations be guaranteed to finish within certain time constraints. Suppose a real-time process needs 5 s of processor time and is required to finish in 10 s. As shown in Example 2.2, this requirement can not be satisfied with simple time-sharing if more than nine other processes exist. However, with preemptive scheduling, the real-time process can preempt other processes and finish in 5 s, ignoring overhead.

■

Example 2.4 Multiprogrammed batch processing

Suppose a parallel application is run on a two-processor multicomputer in batch mode, and the execution time is observed to be 4 times longer than that when the application is run in dedicated mode. Why?

One answer is that the batch mode does not imply dedicated use of resources. When a process performs a communication operation through a system call, it may give up the processor to another process, which happens to require a long processor time. This is illustrated in Fig. 2.7.

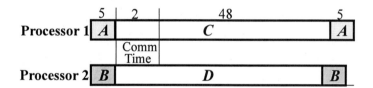

Figure 2.7   Multiprogrammed batch processing.

The parallel application has two processes *A* and *B*, each assigned to a processor and needing 10 s for computing. There is also a 2-s communication overhead. In dedicated mode, the program takes a total of 12 s over two processors. The processors are idle while the communication is in progress.

In the multiprogrammed batch mode, processor idling can be alleviated by overlapping communication with computation. When processes *A* and *B* make the communication call, they yield the processors to processes *C* and *D*. However, *A* and *B* cannot get the processors back until *C* and *D* finished their computation and yield the processors. The elapsed time (or wall clock time) increases to 60 s.

■

On parallel computers, the following terminology has been used for scheduling. *Space sharing* refers to assigning one process to a processor, and different applications are assigned to different processor sets (often called *partitions*). *Time sharing* refers to the multiprogramming mode, where multiple processes from different applications can be assigned to the same processor. The processes are scheduled by batch, time-sharing (time-slicing), or preemptive modes in Fig. 2.6.

A scheduler of parallel computers must take into account that a parallel application has *interacting* processes. Scheduling these processes as individual, independent processes may significantly degrade performance.

Example 2.5 A worst-case scenario in scheduling parallel processes

Consider Fig. 2.6 again. Suppose the communication is *synchronous*, i.e., it needs both processes A and B to participate. When process A begins executing the communication code, it finds that B is not running any more (e.g., timed-out or preempted). Worse yet, B is swapped out. Thus A has to wait for a long time until B is running again.

But at this time, A has exhausted its processor time quantum and is timed out. Thus B has to wait. This thrashing could go on for a long time. A technique for solving this thrashing problem is *gang scheduling*, also known as *group scheduling*. The basic idea is that the interacting processes should be scheduled as one entity (group, or gang). When a group member is running, all members are running. It is not allowed to schedule or deschedule just one member. These and other job-scheduling issues will be discussed in detail in Section 9.5.

■

2.2.7 Variations of Process

Traditional operating system processes have separate address spaces. This concept makes sense for security and protection in a multiuser, multitasking environment. Imagine what chaos would result if all user and system processes resided in the same address space. However, this separate-address space concept makes process management time-consuming. For instance, when a Unix process executes a fork() system call to create a child process, a new address space for the child must be created. This means that memory must be allocated, the data segment and the descriptor of the parent must be copied, and a runtime stack must be set up for the child. For this performance reason, Unix processes are said to be *heavy-weighted* processes.

The high overhead of process creation and switching could have a detrimental effect on parallel processing. As will be seen in Chapter 3, creating a Unix process could take hundreds of thousands of CPU clock cycles, and much more when real processes with large data sets are created on distributed memory computers. Thus heavy-weighted parallel processes are not suitable for scalable parallel computers, unless the parallel processes have coarse computation grains. To exploit finer-grained parallelism, it is necessary to have *light-weighted* processes.

The concept of light-weighted processes (also called *threads*) has been proposed and implemented in a number of operating systems, *thread libraries*, and parallel programming languages. The main difference between OS processes and threads is that within a heavy-weighted OS process, multiple threads can exist that share the same address space (including the process descriptor) of the process. When a (heavy-weighted) process is created, it normally has a single thread, called the *base* thread. Any thread can create additional threads by executing a thread creation operation, which has a typical form as shown below:

thread-create (*foo*, argument1, ..., argument$n$) ;

Such an operation is very similar to a procedure call. It creates a thread to execute function *foo* with the given n arguments. A separate runtime stack needs to be set up for the new child thread, which shares the rest of address space (the code segment, the data segment, and the process descriptor) with the parent thread.

Because much less memory allocation and copying is required, creating a thread is much faster than creating a heavy-weighted process. Furthermore, a user-level thread can be created without involving the kernel, which saves the mode-switching overhead.

A thread needs more than just a per-thread stack. Other information needs to be maintained, such as the following: the context, i.e., the processor registers, including the program counter; execution status, such as running, suspended, or ready; a per-thread data area holding thread private variables. Furthermore, thread management (creation, termination, scheduling, etc.) must be done either by the kernel or by a user-level thread library.

In this book, we use the term *process* to refer to either a heavy-weighted process or a thread, when the context is clear or when the weight is irrelevant. Otherwise, we will be specific with the following conventions:

task = process = heavy-weighted process = OS process
light-weighted process = thread

2.3 Parallelism Issues

Parallel programming is more complex than sequential programming. Many additional issues arise. In particular, we discuss issues arising from the specification of parallelism in user programs.

2.3.1 Homogeneity in Processes

This refers to the similarity of component processes in a parallel program. There are three basic possibilities:

- **SPMD**: The component processes in a *single-program-multiple-data* (SPMD) program are *homogeneous*, in that the same code is executed by multiple processes on different data domains.
- **MPMD**: The component processes in a *multiple-program-multiple-data* (MPMD) program are *heterogeneous*, in that multiple processes may execute different codes.
- **SIMD**: Both SPMD and MPMD programs are MIMD, in that different instructions can be executed by different processes at the same time. An SIMD program is more restrictive than SPMD, in that not only must multiple processes execute the same code, but also they must all execute the same instruction at the same time. In other words, SIMD programs is a special case of SPMD programs.

In this book, we focus on MIMD programs, which are the predominant parallel programs in real applications. This is not surprising as most parallel computers are MIMD machines. MPMD programs are usually specified by the *parallel block* construct or the *multiple-code* approach, while SPMD programs are usually specified by the *parallel loop* construct, the *data-parallel* construct, or the *single-code* approach discussed below.

Two additional terms are often found in the literature. A *data-parallel* program refers to an SPMD program in general and a program that uses only the data-parallel constructs (such as those in Fortran 90) in particular. It emphasizes exploiting parallelism in the data domain or the data structure. A *functional-parallel* (also known as *task-parallel* or *control-parallel*) program is usually a synonym for an MPMD program. The MPMD (functional-parallel) and SPMD (data-parallel) styles can be mixed in a parallel program.

Parallel Block A natural way to express MPMD programs is to use the *parbegin* and *parend* constructs. These structured constructs were originally proposed by Dijkstra and are also known as *cobegin* and *coend*. They are always used as a pair as follows:

$$\textbf{parbegin}\ S_1\ S_2\ ...\ S_n\ \textbf{parend}$$

is called a *parallel block*, where $S_1\ S_2\ ...\ S_n$ are its component processes, which could contain different codes. The parallel block is so named in contrast to a *sequential block* found in traditional sequential languages: **begin** $S_1\ S_2\ ...\ S_n$ **end**.

The n component processes $S_1\ S_2\ ...\ S_n$ of a parallel block start execution simultaneously when the parallel block is executed. They execute independent of one another, each at its own pace. Special interaction operations need to be executed to coordinate these processes. The parallel block terminates when all n component processes terminate.

Parallel Loop   When all processes in a parallel block share the same code, we can denote the parallel block with a shorthand notation called a *parallel loop* as follows:

parbegin Process(1) . . . Process(n) **parend**

can be simplified to the following parallel loop :

parfor (i:=1; i<=n; i++) { Process(i) }

Parallel loop is often used in specifying SPMD parallel programs.

Processes in a system program, such as an operating system or a networking system, are usually heterogeneous. Many parallel computing programs, especially for massively parallel computers, are highly homogeneous. It would be difficult to write a completely heterogeneous parallel program for a 1000-processor computer, because then we need to write a different program for each processor.

It is desirable for a parallel programming environment to provide for both homogeneous and heterogeneous processes. However, supporting SPMD is often enough for scalable parallel computers. When the number of different codes is small, one can fake MPMD by using an SPMD program. For instance, the MPMD code **parbegin** A; B; C; **parend** can be expressed equivalently as an SPMD parallel loop

```
parfor (i:=0; i<3; i++) {
    if (i=0) A;
    if (i=1) B;
    if (i=2) C;
}
```

Multi-Code versus Single-Code   Many programming languages used on current parallel computers, especially MPPs and COWs, do not provide parallel block or parallel loop constructs. On such systems, MPMD parallelism is specified by the *multi-code* approach. For instance, to specify **parbegin** A; B; C; **parend**, a user needs to write three programs and compile them to generate three executable programs *A*, *B*, and *C*, which are loaded to three processing nodes by a shell script:

```
run A on node 1
run B on node 2
run C on node 3
```

The programs *A*, *B*, and *C* are just sequential programs plus library calls for interaction.

An SPMD program can be specified using the *single-code* approach. For instance, to specify the parallel loop **parfor** (i := 0; i < N; i++) { foo(i) }, the user needs to write only one program such as the following:

```
pid = my_process_id();
numproc = number_of_processes() ;
for ( i := pid; i < N; i = i + numproc) foo(i) ;
```

This program is compiled into one executable program *A*, which is loaded to *n* nodes by executing a command: run A -numnodes n. Note that the single-code approach is easier for the user, and it allows the same program to be executed on different numbers of nodes repeatedly in SPMD mode.

Data-Parallel Constructs   In data-parallel languages (e.g., Fortran 90 and HPF), SPMD parallelism can be specified using data parallel constructs. For instance, to specify the parallel loop:

$\qquad$ **parfor** (i := 1; i <= N; i++) {C[i]=A[i]+B[i];}

the user can use an array assignment: C = A + B or the following loop:

$\qquad$ **forall** (i = 1, N) C[i]=A[i]+B[i].

2.3.2 Static versus Dynamic Parallelism

The *structure* of a program refers to the way it is composed from its components. For instance, the code **if** (C) S1 **else** S2 has a structure "**if** (...) ... **else**..." and three components: C, S1, and S2. A program exhibits *static* parallelism if its structure and the number of component processes can be determined before run time (e.g., at compile time, link time, or load time). Otherwise, we say the program exhibits *dynamic* parallelism, which implies that processes can be created and terminated at run time.

Static parallel programs tend to have less runtime overhead and thus are more efficient than dynamic parallel programs. This is because many initialization operations (such as memory allocation, stack allocation, system table initialization, etc.) can be avoided. Dynamic parallel programs tend to be more flexible.

Static parallelism can be expressed by the constructs in Section 2.3.1 (e.g., parallel blocks and parallel loops). Dynamic parallelism is usually expressed through some kind of *fork* and *join* operations. They can also be specified using the single-code or the multi-code approach.

For instance, a parallel block : **parbegin** P, Q, R **parend** is static if P, Q, R are. However, the code : **while** (C>0) **begin fork**(foo(C)); C:=boo(C); **end** exploits dynamic parallelism, because it can be known only at run time how many times the fork operations will be executed to create new processes.

Fork/Join   There have been different versions of fork/join. We will use an example to demonstrate the general idea. The following program has three processes, where *A* is the main process, which is created automatically when the program begins execution:

```
Process A:              Process B:              Process C:
begin                   begin                   begin
    Z := 1;                 fork(C);                Y := boo(Z);
    fork(B);                X := foo(Z);         end
    T := foo(3);            join(C);
end                         output(X+Y);
                        end
```

Process A executes a **fork**(B) operation to create a new process B, which begins to execute in parallel with A. In other words, while B is being executed, A can proceed to execute its subsequent statement [i.e., the assignment T:=foo(3)]. We say that A is a *parent* process of B and that B is a *child* process of A. The two processes execute asynchronously. It is possible that A finishes (i.e., it reaches its **end** statement) before B does. In that case, A cannot terminate until B has terminated. In general, a parent process cannot terminate until all its children have done so.

Sometimes we want a parent to wait for a child before the parent has reached the end of its code. This can be accomplished with the use of a **join** statement. In the above codes, process B in turn creates another process C. Now since the output statement in B needs the value of variable Y computed by C, a **join**(C) statement is inserted before the output statement. The **join**(C) statement forces B to wait until C terminates, before it can proceed to execute the output statement.

Fork and join are very flexible constructs. However, they are unstructured. They are like goto in a sequential language and must be used carefully. Many parallel computing programs can be specified using only parallel blocks and parallel loops.

2.3.3 Process Grouping

Processes in a parallel application are usually not independent. They need to interact with one another. We have already seen in Section 2.2.6 that it is desirable to schedule interacting processes as a group. In Section 2.4, we will see that the group concept has other uses, such as supporting *collective* interactions.

A *process group* is an ordered set of processes. The number of member processes is called the *group size*. Each group has a *group ID*, which uniquely identifies the group in the parallel program. Each member process has a *rank* in the group, which is normally an integer ranging from 0 to $n-1$, where n is the group size. A member process can be uniquely identified by its rank and the group ID.

To support the group concept, a parallel programming language needs to provide functions to manage groups, such as creation and destruction of groups, inquiring about the group ID, the group membership, and the members ranks, etc.

2.3.4 Allocation Issues

Any parallel program must perform certain computations (workload) on certain data objects. *Allocation* refers to *partitioning* of the data and the workload into the processes and *mapping* the processes to the nodes (processors). According to Section 1.3.3 and Eq.(1.4), a good allocation scheme should balance the workload (i.e., reduce σ/w) and minimize overhead (i.e., minimize $(t_p + t_0) / (wt_f) + (\alpha \cdot t_c) / t_f)$, so that the system is busy computing most of the time, instead of wasting time idling or interacting. This should be done without sacrificing parallelism.

Parallelism   An important task of partitioning is to choose the proper parallelism and granularity. The *degree of parallelism* (DOP) of a parallel program is usually defined as the number of component processes that can be executed simultaneously. A quantitative metric, known as *average parallelism*, which will be defined in Section 3.5.1, is often used to measure the overall DOP of a program.

Granularity   A related concept is the *granularity* (or *grain size*), which is defined to be the computation workload executed between two parallelism or interaction operations. For the phase parallel model (Section 1.3.3), granularity can be more precisely defined to be the computational workload w executed by one processor in a superstep.

The unit of granularity is number of instructions, number of floating-point operations, or seconds. Grain sizes are often described as *fine* (*small*), *medium*, or *coarse* (*large*). Although there is no exact definition, a rough classification is to consider grain size to be small for less than 200, medium for 20 to 2000, and large for thousands or more computation operations.

The term *granularity* is sometimes used to refer to process size, i.e., the total number of operations in a component process (not just those in a superstep). Along this line, we use the term *operation-level* (or *instruction-level*) *parallelism* when the component processes in a parallel program are as small as just one or a few computation operations or instructions. The term *block-level* refers to the case when the size of individual processes is a block of assignment statements. A special case of block-level parallelism is *loop-level*, where a loop creates a number of processes, each executing an iteration consisting of a number of assignment statements. When the component processes each consist of a procedure call, we have *procedure-level* parallelism, sometimes also called *task-level* parallelism.

The degree of parallelism and the grain size are often reciprocal, in that, other things being equal, increasing grain size tends to decrease the degree of parallelism, while decreasing grain size tends to increase the degree of parallelism. On the other hand, in real parallel computers the degree of parallelism and the overhead of communication and synchronization usually have a proportional relationship. Other things being equal, increasing the degree of parallelism often increases the overhead, while decreasing the degree of parallelism tends to reduce the overhead.

Implicit versus Explicit Allocation With *explicit allocation*, the user needs to explicitly specify how to allocate data and workload. With implicit allocation, this task is done by the compiler and the runtime system. Various combinations are possible. For instance, in symmetric multiprocessors, a common approach is to let the data reside in a central, shared memory so that all processes can access. The workload is distributed to the processes either statically or dynamically. When a process is allocated a piece of the workload, it gets the data needed from the shared memory.

In distributed memory systems, a popular method is the *owner-compute* rule: The data are first distributed among the processes. When process P is allocated with variable x, P is called the *owner* of x. Computations associated with x will be performed by the owner process P. Data and workload allocations have a large impact on performance. An important research topic is how to allocate data so that most of time, data needed by a process are nearby. This is called *exploiting data locality*, which can reduce cache and memory misses and communication overhead.

2.4 Interaction/Communication Issues

An *interaction* is an activity or operation that processes perform to affect the behavior of one another. Some authors simply call interaction operations as *communications.* We prefer to use the term interaction, because it is embedded with more implications. There are several important issues related to interaction: the interaction operations which need to be supported in a parallel system; the interaction modes and patterns; competitive versus cooperative interactions.

2.4.1 Interaction Operations

We briefly introduce the three most frequently used types of interactions: *communication, synchronization,* and *aggregation* operations. In the literature, these operations are often universally called *communication.* However, it is important to distinguish among them, as they have different requirements for architecture and programming supports.

Communication A *communication* operation passes data values among two or more processes. In a shared-memory program, one process can compute a value of a shared variable and store it in the shared memory. Later on another process can get this value by referencing the variable. This is called communication through *shared variable.*

A multiprocessor program using procedure-level parallelism can also allow a parent process to create a child process by forking a procedure, e.g., executing fork(foo(x)). Data values can be passed as parameters between the child process and the parent. This is called communication through *parameter passing.* Finally, in a multicomputer model, processes can communicate through *message passing.*

Synchronization   A *synchronization* operation causes processes to wait for one another, or allows processes that are waiting to resume execution. There are different types of synchronization operation:

- *Atomicity*: A process often needs to perform a sequence of operations as a single *atomic operation* (cf. Definition 1.2), as illustrated below:

  parfor (i:=1; i<n; i++) {
      atomic { x= x+1; y= y-1; }
  }

  The keyword **atomic** indicates that each of the n processes must execute the two assignments as one atomic operation. Implicit synchronizations are performed by the parallel system to enforce atomicity.

- *Control Synchronization*: A process executing a control synchronization operation will wait until the program execution reaches certain *control states*. A common example of control synchronization is the *barrier synchronization*, as shown in the following code:

  parfor (i:=1; i<n; i++) {
      P_i
      barrier
      Q_i
  }

  Here we have n processes. The ith process executing P_i is followed by a **barrier** followed by Q_i. When it finishes P_i and reaches the **barrier** statement, it has to wait until all other processes also reach their **barrier**. Another control synchronization construct is the critical region, as illustrated below:

  parfor (i:=1; i<n; i++) {
      critical { x= x+1; y= y-1;}
  }

  Note that a critical region is *mutually exclusive*, in that only one process is allowed to execute the two assignments at a time. In contrast, multiple processes can execute their atomic regions, as long as the atomicity is enforced.

- *Data Synchronization*: A process executing a data synchronization operation will wait until the program execution reaches certain *data states* (cf. Section 2.2.1). For instance, a process executing a **wait**(*x > 0*) statement will be delayed until variable x becomes positive. Examples of data synchronization operations include locks, conditional critical regions, monitors, and events. In most current systems, atomicity is realized through data synchronization, such as the following:

parfor (i:=1; i<n; i++) {lock(*S*); x= x+1; y= y−1; unlock(*S*); }

where the lock synchronization depends on the data state in semaphore *S*.

Control synchronization depends on only the program's control state, and is not affected by the program's data state. Control synchronization is generally easier to understand than data synchronization, although the later is more flexible. We will discuss synchronization in detail in Chapters 7 and 14.

Aggregation An *aggregation* operation is used to merge partial results computed by the component processes of a parallel program to generate a complete result. It can be realized as a sequence of supersteps, and each superstep consists of a short computation and a simple communication and/or synchronization.

As an example, consider the following program which computes the inner product of two vectors A and B. Here **aggregate_sum** merges the partial results x[0], x[1],..., x[n−1] to generate a final result inner_product = x[0] + x[1] + ... + x[n−1].

```
parfor (i:=0; i <n; i++) {
    x[i]:= A[i] * B[i];
    inner_product := aggregate_sum(x[i]);
}
```

This summation operation is called a *reduction*, because it reduces a number of values into a single value. Other types of aggregations include *scan* (also known as *parallel prefix*), *descend algorithms*, *ascending algorithms* [86-87,508], etc.

Example 2.6 A recursive doubling reduction operation

Suppose there are *n* processes *P*(0), *P*(1),..., *P*(n−1). An array element a[i] is initially distributed to process *P*(i). The reduction Sum=a[0]+...+a[n−1] can be described by the following single-code program for process *P*(i), where *i*=0 to *n*−1:

```
Sum = a[i];                   // Each process has a local variable Sum
for (j=1; j<n; j=j*2) {       // there are log(n) supersteps
    if ( i % j = 0) {
        get Sum of process P(i+j) into a local variable tmp;
        Sum = Sum + tmp;
    }
}
```

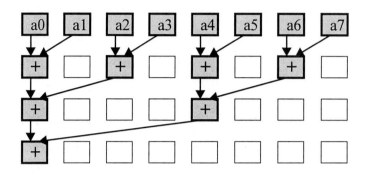

Figure 2.8   The recursive doubling operation.

The summation is realized in a treelike fashion in $\log n$ supersteps, as shown in Fig. 2.8 for $n = 8$. In each superstep, the processes get a scalar from another process and perform a scalar addition to compute the intermediate partial sum. In the end, process P(0) contains the final result.

This is the main characteristic of an aggregation: It consists of a sequence of supersteps, and within each superstep, each process performs a small grain of computation and then communicates a small message.

■

2.4.2 Interaction Modes

Consider Fig. 2.9, where n processes $P_1,..., P_n$ interact by executing interaction code C. The n processes are called the *parties* (or *participants*) of the interaction. We call the interaction *synchronous* if code C cannot be executed until all participants have reached C. We call the interaction *asynchronous* if when a process reaches C, it can proceed to execute C without having to wait for other processes.

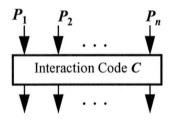

Figure 2.9   Interaction among n processes.

An interaction is called a *two-party* interaction if $n = 2$, and a *multiparty* interaction if n can be greater than 2. A two-party communication is called a *point-to-point* communication, when one process sends a message to another process. A multiparty interaction is often called a *collective* interaction.

Parallel programs with only synchronous interactions are easier to understand, because when the interaction code C is executed, we know all participant processes have reached C. In other words, when C starts to execute, the program is in one control state. With asynchronous interaction, when a process starts to execute the interaction code, the program can be in different control states. Any of the other processes may have not reached C, may have just reached C, or may have passed C. However, asynchronous interactions tend to be more flexible. Many parallel programs need only multiparty synchronous interactions.

With different ways to set the entry and the exit conditions, many interaction modes are possible when a process P encounters an interaction code C. The following status of a process P can be defined with respect to the interaction code C:

- *In*: Process P is in code C, including just entered, just finished its portion of C, or still executing C.
- *Out*: Process P is not in C, including not yet arrived or exited.
- *Arrived*: Process P just arrived at C, but has not entered yet.
- *Finished*: Process P just finished its portion of code C but has not exited yet.

For a two-party interaction, there can be $4^4 = 256$ combinations. Only a few are used in practice. Some examples are shown in Table 2.3.

Table 2.3   Combinations of Entry and Exit Conditions

Entry Condition		Exit Condition		Examples
Self	Other	Self	Other	
Arrive	X	In	X	Nonblocking send/receive
Arrive	X	Finish	X	Blocking send
Arrive	X	Finish	Finish	Blocking receive
Arrive	Arrive	Finish	Finish	Synchronous send/receive (CSP)
Arrive	Arrive	Finish	In or finish	Barrier
Arrive	Out	Finish	Out	Critical region

Three interaction modes are widely used:

- *Synchronous*: In this mode, all participants must arrive before the interaction begins. A participant process can exit the interaction and continue to execute the subsequent operation, only when all participants have finished the interaction. An example is the send/receive operations in CSP [312]. A simple way to tell if an interaction is synchronous is the *two-barrier test*: Place one barrier before and one after the interaction code. If the resultant parallel program has exactly the same semantics as the original one, the interaction is synchronous. Otherwise it is *asynchronous*.

- *Blocking*: A participant can enter an interaction as soon as it arrives, regardless of where the other parties are. It can exit the interaction when it has finished its portion of the interaction. Note that the other parties may have not finished (or even entered) the interaction code. An example is a blocking send, whose completion implies that the message is sent out, but has not necessarily been received in the destination.

- *Nonblocking*: A participant can enter an interaction as soon as it arrives. It can exit the interaction before it has finished its portion of the interaction. An example is a nonblocking send, whose completion implies only that the process has requested to send. The message is not necessarily sent out.

2.4.3 Interaction Patterns

The *pattern* of an interaction refers to which participants affect which other participants. We call an interaction *static* if its pattern can be determined at compile time. Otherwise it is a *dynamic* interaction. A special case of dynamic interaction occurs when the group of participants can change dynamically. For instance, processes may leave or join a group at run time, known as *dynamic grouping* [261].

In an *n*-party interaction, if the pattern can be specified as a simple function of the process index, we say the interaction has a *regular* pattern An interaction has a *regular* pattern, if there is an algorithm which, given only two index numbers i and j, can determine in constant time and space whether the ith process affects the jth process.

Example 2.7 Point-to-point and collective communication patterns

Representative regular communications are shown in Fig. 2.10, which can be divided into four classes, according to how many senders and receivers are involved in the communication:

- *One-to-One*: This is also known as *point-to-point* communication. There is one sender and one receiver, as illustrated in Fig. 2.10(a).
- *One-to-Many*: This includes *broadcast* [Fig. 2.10(b)], where one process (called the *root*) sends the same message to all processes, itself included. A *scatter* operation [Fig. 2.10(c)] is a generalization of a broadcast, in that the root sends a distinct (or *personalized*) message to a different process.

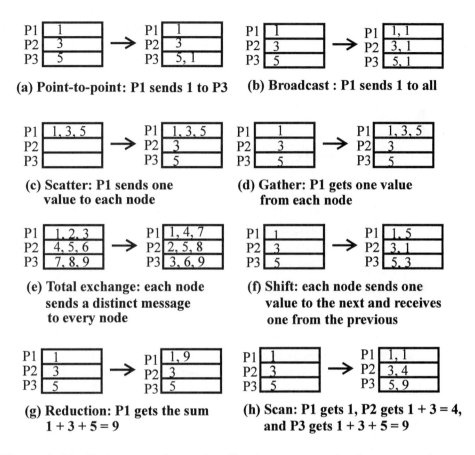

Figure 2.10 Point-to-point and collective communication operations

- *Many-to-One*: This includes *gather* [Fig. 2.10(d)], where the root process receives a distinct message from each process. In total, the root receives *n* messages, where *n* is the group size. Another example is *reduction* [Fig. 2.10(g)], which aggregates *n* local values, one from every process, into a final value in the root process.

- *Many-to-Many*: The simplest form of many-to-many communication is a *permutation*, where no process sends to and receives from more than one process. An example of permutation is the *circular shift* in Fig. 2.10(f). Another example is *scan* [Fig. 2.10(h)], which generalizes a reduction by aggregating *n* local values into *n* final values. In a *total exchange* [Fig. 2.10(e)], every process sends a personalized message to each of the *n* processes, itself included. ∎

While the user has some freedom in choosing the interaction mode (i.e., synchronous or asynchronous), the patterns of interaction are usually application-dependent. A parallel language often provides for both dynamic and irregular interaction patterns.

A class of irregular communications is the *h-relation* (cf. Definition 1.6) when the exact communication pattern is unknown. Then all one knows is that each node sends at most *h* words to various nodes and each node receives at most *h* words. Without knowing the pattern, the communication algorithm has to prepare for the worst case. That is why most *h*-relation algorithms are based on total exchanges.

2.4.4 Cooperative versus Competitive Interactions

When talking about parallel programs in this book, we are primarily concerned with developing parallel *application* programs, i.e., applications written by a user (programmer) to be run on a parallel computer. Such programs are different from system programs (e.g., concurrent operating system).

One main difference is that processes interact in different ways. In a concurrent operating system, processes interact mainly by competing for shared resources. We call such interaction *competitive interaction*. On the other hand, it is often the case that processes in a parallel application program cooperate in computing a function. When these processes interact, often their objective is not to safely share a resource.

Rather, they want to communicate information to one another, to synchronize with one another, or to generate combined values from partially computed values. Such interactions are called *cooperative*. Competitive interactions usually require the use of synchronization operations that depend on a program's data state, while cooperative synchronization operations often only depend on a program's control state.

A user generally expect her parallel computing program to be both *terminative* and *determinate* (i.e., the program should eventually stop and the result should be unique). In contrast, a concurrent operating system code is usually nonterminative and indeterminate. Some important differences of these two types of programs are summarized in Table 2.4.

Historically, competitive interactions have received much more attention, and many constructs have been proposed, such as locks, semaphores, events, conditional critical regions, and monitors [37]. Only a few years ago, programming languages used in many multiprocessors provided only constructs for competitive interactions.

With such a language, cooperative interactions need to be first converted (by the programmer) into resource sharing interactions, and then implemented through competitive interaction primitives. This roundabout way of implementation not only can be inefficient, but also forces user with unnecessarily coding. The developed program is often not structured, and is difficult to understand and debug. A main objective of this book is to study constructs, such as barrier and coherent region to be discussed in Part IV, that are specially designed for cooperative interactions.

Table 2.4 Competitive Versus Cooperative Parallel Programs

Attribute	Competitive Programs	Cooperative Programs
Example Applications	Operating Systems Networking Tools	Partial Differential Equations, Discrete Event Simulation
Granularity	A Few Large-Grain Heterogeneous Processes	Many Small-Grain Homogeneous Processes
Example Interactions	Critical Section, Producer-Consumer, Reader-Writer	Barrier Synchronization, Broadcasting, Permutation, Reduction, Scan, Descend, Work Pool or Queue
Constructs for Interaction	Lock, Semaphore, Event, Conditional Critical Region, Monitor	Barrier, Fetch-and-Add, Coherent Region
Properties	Non-Terminative, Indeterminate	Terminative, Determinate

2.5 Semantic Issues in Parallel Programs

Semantic issues refer to various behavioral properties of a parallel program. We will discuss only two issues: *termination* and *determinacy*. These issues are important because it is desirable for parallel programs to always terminate and to be determinate. In fact, one of the major applications of computing is to model the physical world, as exemplified by computational physics, computational chemistry, computational biology, etc.

To make parallel computing effective in scientific experiments, it should have two basic computational properties: *terminative* and *determinate*. The purposes are to guarantee a finite amount of time to complete and to ensure that the computational experiment can be repeated by other users.

2.5.1 Program Termination

The termination issue deals with whether a program will terminate normally, and if not, the possible reasons for nonhalting. As will be seen, there are three types of non-termination: *infinite looping*, *livelock*, and *deadlock*.

Definition 2.5 A program is *terminative* if it always finishes execution and enters a final state, from any initial state. A program is *divergent* if from some initial state, it can enter a state of continuously executing without halting. A program *deadlocks*, if it is stuck in the middle of execution without being able to continue execution and is not in a final state. ■

Nontermination manifests itself differently in sequential and parallel programs. Sequential programs do not deadlock. Thus if a sequential program is not terminative, it is divergent. An example sequential program that diverges is the infinite loop **"while** $(x > 0)$ x: = x + 1". In parallel programs, there is a second type of divergence, called *livelock*, where several processes are trapped in infinite loops due to mutual interference. A simple example of livelock is the following code:

```
par {
    while (y > 0) x := x + 1 ;
    while (x > 0) y := y + 1 ;
}
```

A parallel program can be non-terminative due to deadlocks. The compiler or the runtime system should detect all nonhalting situation. Unfortunately, this problem is equivalent to the Turing machine halting problem, which is undecidable.

Thus all we can hope for is that the parallel language is designed to either prevent deadlock and livelock or allow the compiler and the runtime system to detect them. In this respect, programming systems exploiting implicit parallelism are superior, since the language guarantees that deadlocks or livelock can not occur.

2.5.2 Determinacy of Programs

We first need to differentiate between two concepts: *determinism* and *determinacy*. Determinism means a unique computation (i.e., a unique sequence of atomic operations), while determinacy means a unique final result. Furthermore, the definition of determinism implies that if a program is run multiple times with the same initial state, the same unique computation sequence occurs in all these different runs.

The determinacy definition implies that the final result is the same from the same initial state for different runs. Conventional sequential programs are most deterministic, and thus determinate. However, even in sequential programs, indeterminacy is not an unknown phenomenon. A sequential program with the same initial data state may produce different results in different runs, when a variable is assigned a value through random number generation.

However, this type of indeterminacy is often desired by users in Monte Carlo simulations or in genetic algorithms. In other words, these applications are indeterminate in nature. Many application functions to be computed are mathematically determinate. However, the parallel programs realizing the functions may become indeterminate as a result of parallel execution.

It is this type of indeterminate anomaly that we must detect and deal with. Again, programming systems exploiting implicit parallelism are superior, since the language guarantees that all programs are determinate.

2.6 Bibliographic Notes and Problems

The importance and the sorry state of parallel programming have been pointed out by many experts. References [443, 556, 587] summarize some of their opinions and consensus in this field. Pancake and associates analyzed why parallel programming is difficult [481, 482]. Considering the difficulties, Pfister advocated the *sequential-program, parallel-processing* (SPPP) style (e.g., throughput processing) [497].

Although everyone agrees that the current parallel programming environments need a major improvement, there is no consensus as to what is the best way, or if there is a best way. Nevertheless, the parallel community, including most users, vendors, and researchers, have converged toward four imperative programming models, which will be discussed in Part IV.

Other researchers are still strongly advocating unimperative styles, such as the dataflow or functional programming models [123]. Recent progress in parallel object-oriented programming is discussed in [58,139,144]. Unimperative programming is beyond the scope of this book.

The concepts of process and thread have been discussed in [581,621] from the operating system's viewpoint. The IEEE Pthreads standard is published in [345]. Most current operating systems support threads, as described in [340] for AIX, in [132,601] for Solaris, and in [193] for Digital Unix.

Various language notations for specifying parallelism have been reviewed by Andrews [37]. The single-code and multi-code schemes have been used in both public-domain systems (e.g., PVM [261]) and commercial systems (e.g., IBM SP2 [337]). The data-parallel approach will be further discussed in Chapter 14. The partition and scheduling issues have been reviewed in [230].

Parallel programs have much more complex semantic issues than sequential programs. Robert Keller's seminal paper [366] first proposed the *labeled transition system* concept, which is the foundation of most operational semantic models. Apt and Olderog's book [41] is a good introduction to structured operational semantics. Algebraic semantics are discussed by Baeten and Weijland [53], and by Milner [449].

Homework Problems

Problem 2.1 Suppose a process P has two threads t and u. We know that t and u share the address space of P but each of them has a private stack. List three other data and control structures, in which t and u have private copies.

Problem 2.2 List the data variables and the control variables in Fig. 2.3(a). Do the same for Fig. 2.3(b) and Fig. 2.3(c). Comment on their differences in program variables.

Problem 2.3 Use Fig. 2.3 to explain Table 2.1.

(a)   What are the advantages of Fig. 2.3(b) over Fig. 2.3(c)?

(a)   What are the advantages of Fig. 2.3(c) over Fig. 2.3(b)?

(a)   Why is that Fig. 2.3(d) is a compromise between Fig. 2.3(b) and Fig. 2.3(c)?

Problem 2.4 Differentiate the following terms related to the process concept:

(a)   Mode switch, context switch, process switch

(b)   Process state, program state, and process status

(c)   Data state and control state

(d)   Process, task, and thread

Problem 2.5 Recently, some researchers have advocated *single-address-space operating system*. The idea is that all processes and the kernel reside in the same address space. Suppose an HTTP daemon, a DNS daemon, and the kernel runs in a Web server computer.

(a)   Explain a main advantage of the single-address-space approach over Unix.

(b)   Explain a main challenge to the single-address-space approach, and propose a solution.

Problem 2.6 Differentiate the following terms related to the process scheduling:

(a)   Simple uniprogramming processing

(b)   Uniprogrammed batch processing

(c)   Multiprogrammed batch processing

(d)   Time sharing processing

(e)   Preemptive multiprogrammed processing

Problem 2.7 Consider Example 2.2 and answer the following questions. The *elapsed time* is the wall-clock time from the submission of a process for execution to the completion of the execution.

(a)   How is the 9005 seconds elapsed time derived for the interactive process?

(b)   How is the 50 seconds elapsed time derived for the time-sharing case?

(c)   What is the elapsed time if the interactive process has very high priority and runs in a system with preemptive scheduling?

Problem 2.8 Differentiate and exemplify the following terms of parallelism issues:

(a)   SPMD and MPMD

(b)   Implicit versus explicit parallelism

(c)   Static versus dynamic parallelism

(d)   Data parallelism versus control parallelism

Problem 2.9 Consider the three parallel codes in Fig. 2.3. For each of them, answer the following questions.

(a)   Is the code SPMD or MPMD?

(b)   Does the code exploit implicit parallelism or explicit parallelism

(c)   Does the code exploit static parallelism or dynamic parallelism

(d)   Does the code exploit data parallelism or control parallelism

Problem 2.10 Differentiate and exemplify the following terms of interaction issues:

(e)   Data synchronization and control synchronization

(f)   Competitive and cooperative interactions

(g)   Static versus dynamic interactions

(h)   Point-to-point versus collective communications

(i)   Regular versus irregular communications

Problem 2.11 Consider Fig. 2.10(a) to Fig. 2.10(f). For each of them, explain what h-relation is the communication (e.g., is it a 1-relation, 5-relation? why?).

Problem 2.12 A scan is a generalization of reduction (see Example 2.6). Suppose there are n processes, and process $P(i)$ contains a value denoted as $a[i]$. Then after a scan, process $P(i)$ will get the result $a[i] + \ldots + a[n-1]$.

(a)   Show a short single-code program to compute the scan. The scan operation is realized in $\log n$ supersteps.

(b)   Show a figure to illustrate how your program computes the scan for $n = 8$. Follow the style in Fig. 2.8.

Problem 2.13 Consider the following code to compute the sum of an integer array A:

```
sum = 0;
for (i=0; i<N; i++) sum = sum + A[i];
```

(a)   Use the **parfor** and **atomic** region constructs to write a determinate parallel program to compute the sum function. Assume the arrays size is N, the number of processors is n, and N divides n.

(b)   Repeat part (a) but using the **parfor** and **aggregation** constructs.

(c)   Compare these two programs in terms of simplicity and performance.

(d)   Is there any circumstance such that your determinate code will *not* be determinate when executing on a *real* parallel computer?

Problem 2.14 Write a simple, SPMD parallel program to compute the sum of an integer array (Problem 2.13) using each of the following approaches. You can use any pseudo-code notation that is clearly defined.

(a)   The library approach

(b)   The new construct approach

(c)   The compiler directive approach

Chapter 3

Performance Metrics
and Benchmarks

Scalable systems are built to satisfy the ever-increasing demand of higher performance in computers. To achieve this goal, both computer users and designers must understand the performance issues addressed by the following fundamental questions:

- How should one characterize the performance of applications and systems? A popular method is *benchmarking*, discussed in Section 3.1.

- What are user's requirements in performance and cost? This issue is treated in Section 3.2. Tradeoff between these two requirements is discussed in terms of execution time, sustained speed, cost-effectiveness, etc.

- How should one measure the performance of application programs? What kind of performance metrics should be used? These questions are answered in Section 3.3 for sequential programs, and in Section 3.5 for parallel programs.

- How should one characterize the system performance when a parallel program is executed on a parallel computer? What are the factors (parameters) affecting the performance? What are the typical parameter values in current systems? These issues are addressed in Section 3.4.

- How should one quantify and analyze system scalability? How can one determine the scalability of a parallel computer executing a given application? These issues are discussed as basic performance laws in Section 3.6.

3.1 System and Application Benchmarks

A *benchmark* is a performance testing program that supposedly captures processing and data movement characteristics of a class of applications. Benchmarks are used to measure and to predict the performance of computer systems, and to reveal their architectural weakness and strong points. A *benchmark suite* is a set of benchmark programs together with a set of specific rules governing the test conditions and procedures, including the tested platform environment, the input data, the output results, and the performance metrics. A *benchmark family* is a set of benchmark suites.

Benchmarks Classification Benchmarks can be classified according to application classes, such as scientific computing, commercial applications, network services, multimedia applications, signal processing, etc. Benchmarks can also be divided into *macro benchmarks* and *micro benchmarks*.

A macro benchmark measures the performance of a computer system as a whole. It compares different systems with respect to an application class, and is useful for the system buyer. However, macro benchmarks do not reveal why a system performs well or badly. Micro benchmarks measure a specific aspect of a computer system, such as CPU speed, memory speed, I/O speed, operating system performance, networking, etc.

Benchmarks can be full-fledged applications or just *kernels*, which are much smaller and simpler programs extracted from the applications while maintaining the main characteristics. A benchmark can be a program that does real work or a *synthetic* program specifically designed for benchmarking. Micro benchmarks tend to be synthetic kernels. Over 100 benchmark suites have been proposed and are in use. Several popular benchmark suites are listed in Table 3.1.

Table 3.1 Representative Micro and Macro Benchmark Suites

Type	Name	Measuring
Micro benchmark	LINPACK	Numerical computing (linear algebra)
	LMBENCH	System calls and data movement operations in Unix
	STREAM	Memory bandwidth
Macro benchmark	NAS	Parallel computing (CFD)
	PARKBENCH	Parallel computing
	SPEC	A mixed benchmark family
	Splash	Parallel computing
	STAP	Signal processing
	TPC	Commercial applications

3.1.1 Micro Benchmarks

Three micro benchmark suites are briefly introduced below. Detail can be found in the individual benchmark documentation released by relevant organizations.

The LINPACK   The LINPACK benchmark was created and is maintained by Jack Dongarra at the University of Tennessee. Although we call it a benchmark, LINPACK is widely used for real computation work. It is a collection of Fortran subroutines that analyze and solve linear equations and linear least-squares problems, where matrices can be general, banded, symmetric indefinite, symmetric positive definite, triangular, and tridiagonal square.

It is simple and easy to use, yet a good indicator of the numerical computing capability of a system. Closely tied to LINPACK is a periodically published list of various systems' LINPACK performance numbers, including a Top-500 list, which lists the 500 most powerful computers in the world. LINPACK has been modified to ScaLAPACK for distributed-memory parallel computers.

A sample of LINPACK performance from the top 8 most powerful computers is shown in Table 3.2, where R_{max} is the sustained maximal speed achieved; N_{max} is the problem size when R_{max} is achieved, i.e., the order of the data matrix; $N_{1/2}$ is the problem size when half of R_{max} is achieved; and R_{peak} is the peak speed of the system measured.

Table 3.2 Sample of December 1996 LINPACK Report
(Courtesy: Dongarra, 1996 [211])

Computer	No. of Processors	R_{max} (Gflop/s)	N_{max} (order)	$N_{1/2}$ (order)	R_{peak} (Gflop/s)
Intel ASCI Option Red	7264	1068	215,000	53,400	1453
CP-PACS	2048	368.2	103,680	30,720	614
Intel Paragon XP/S MP	6768	281.1	128,600	25,700	338
Numerical Wind Tunnel	167	229.7	66,132	18,018	281
Fujitsu VPP500/153	153	200.6	62,730	17,000	245
Cray T3D 1024	1024	100.5	81,920	10,224	152
IBM SP2-T2	512	88.4	73,500	20,150	136
NEC SX-4/32	32	66.53	15,360	1792	64

As of December 1996, the fastest computer ever built was the 7264-processor Intel ASCI Option Red, which has a peak speed of 1453 Gflop/s. It achieved a sustained speed of 1.068 Tflop/s in solving a problem characterized by a $215{,}000 \times 215{,}000$ matrix.

The LMBENCH   The LMBENCH benchmark suite is maintained by Larry McVoy of SGI. It is a portable benchmark used to measure the operating system overheads and the capability of data transfer between processor, cache, memory, network, and disk on various Unix platforms. It is a simple, yet very useful tool for identifying performance bottlenecks and for the design of a system. Some LMBENCH results for three systems are shown in Table 3.3. The three systems are an IBM 990 server, a Sun Ultra workstation, and an Intel Adler PC system..

Table 3.3 Bandwidth, Latency, and System Overheads Measured by LMBENCH (Courtesy: McVoy and Staelin, *Proc. of USENIX Tech. Conf.* 1996 **[440]**)

Attribute		Intel Alder	Sun Ultra	IBM 990
Bandwidth (MB/s)	Memory copy	52	85	242
	File read	52	85	187
	Pipe	38	61	84
	TCP	20	51	10
Latency (μs)	Memory read	0.28	0.27	0.26
	File create	23,809	18,181	13,333
	Pipe	101	62	91
	TCP	305	162	332
System overhead (μs)	Null system call	7	5	16
	Process creation	4500	3700	1200
	Context switching	36	14	13

The STREAM   The STREAM benchmark is a simple synthetic benchmark maintained by John McCalpin of SGI. It measures sustainable memory bandwidth (in MB/s) and the corresponding computation rate. The motivation for developing the STREAM benchmark is that processors are getting faster more quickly than memory, and more programs will be limited in performance by the memory bandwidth, rather than by the processor speed.

This benchmark performs the four operations shown in Table 3.4 for a number of iterations, with unit-stride data accesses (i.e., the index variable i increments by 1 for each iteration). Both memory reads and writes are included in computing the bandwidth. The benchmark is designed to work with data sets much larger than the available cache. The a, b, c vectors in Table 3.4 are 2-million-element arrays, where one element is an 8-B word.

McCalpin proposed a *machine balance* metric, defined by:

$$\text{Machine balance} = \frac{\text{Peak floating-point speed (flop/s)}}{\text{Sustained TRIAD memory bandwidth (words/s)}} \qquad (3.1)$$

Table 3.4 The Four Operations in the STREAM Benchmark

(Courtesy J. D. McCalpin, at http://www.cs.virginia.edu/stream/ [437])

Name	Code	Bytes/Iteration	Flop/Iteration
COPY	a(i) = b(i)	16	0
SCALE	a(i) = q × b(i)	16	1
SUM	a(i) = b(i) + c(i)	24	1
TRIAD	a(i) = b(i) + q × c(i)	24	2

The machine balance metric can be interpreted as the number of flop that can be executed in the time period to read/write a word. As illustrated in Table 3.5, the machine balance values of many systems have been increasing over the years, implying that memory bandwidth lags more and more behind the processor speed.

An exception is IBM RS/6000 servers, which have always paid great attention to the memory system design. In recent years, other companies have also tried to improve the memory system performance (e.g., DEC 8400 and SGI Origin 2000).

Table 3.5 Historical Trends in Machine Balance Values

(Courtesy J. D. McCalpin, at http://www.cs.virginia.edu/stream/ [437])

Year	System	Memory Bandwidth (MB/s)	Peak Speed (Mflop/s)	Machine Balance
1978	DEC VAX 11/780	4	0.4	0.8
1991	DEC 5000/200	28	10	2.9
1993	DEC 3000/500	100	150	12
1995	DEC 600-5/300	169	600	28.4
1995	DEC 8400/350	234	700	24
1980	IBM PC 8088/87	2	0.1	0.2
1992	IBM PC 486/DX2-66	33	10	2.4
1994	IBM PC Pentium-100	85	66.7	6.3
1990	IBM RS/6000-320	60	40	5.3
1993	IBM RS/6000-580	240	126	4.2
1994	IBM RS/6000-590	654	262	3.2
1995	IBM RS/6000-591	800	310	3.1
1989	SGI 4D/25	13	8	5
1992	SGI Crimson	62	50	6.5
1993	SGI Challenge	57	75	10.5
1994	SGI Power Challenge	135	300	17.8
1996	SGI Origin 2000	317	388	9.8

3.1.2 Parallel Computing Benchmarks

A dozen or so parallel computing benchmark suites have been in use. For instance, Splash and Splash-2 are numerical computing benchmarks developed at Stanford University, which have been widely used to study distributed shared-memory machines [649].

The Perfect benchmark suite developed at the university of Illinois has been used to evaluate various parallelizing compiler systems and techniques [76,91,176].

We describe below three parallel benchmark suites: NPB, PARKBENCH, and STAP that represent major application groups in scientific computations.

The NPB Suite The *NAS Parallel Benchmarks* (NPB) is developed by the Numerical Aerodynamic Simulation (NAS) program at NASA Ames Research Center for the performance evaluation of parallel supercomputers. The NPB mimics the computation and data movement characteristics of large-scale *computational fluid dynamics* (CFD) applications.

NPB provides a "pencil and paper" specification as well as MPI-based Fortran source-code implementations. The benchmarking results are verified by NAS and published in a periodic NAS report. Because of these features, NPB has gained wide acceptance among parallel computer vendors, users, and researchers.

The NPB suite consists of five kernels (EP, MG, CG, FT, IS) and three simulated applications (LU, SP, BT) programs. The bulk of computation is integer arithmetic in IS, while the other benchmarks are floating-point computation-intensive. The EP (*Embarrassingly Parallel*) benchmark is aptly named since it can run on any number of processors with little communication. It estimates the upper achievable limits for floating point performance of a parallel computer. The IS (*Integer Sorting*) benchmark is a parallel sorting program based on bucket sort. It requires a lot of total exchange communication.

The MG (*MultiGrid* method) benchmark solves a 3D scalar Poisson equation. It performs both short- and long-range communications that are highly structured. The CG (*Conjugate Gradient* method) benchmark computes the smallest eigenvalue of a symmetric positive definite matrix. It features unstructured grid computations, requiring irregular long-range communications.

The FT benchmark solves a 3D partial differential equation using an FFT-based spectral method, also requiring long-range communication. The BT (*Block Tri-diagonal*), LU (*block lower triangular, block upper triangular*), and SP (*Scalar Penta-diagonal*) benchmarks apply different methods to solve the Navier-Stokes equations. LU performs a large number of small communications (five words each), while the other two use coarse-grain communications.

The PARKBENCH The PARKBENCH (*PARallel Kernels and BENCHmarks*) committee was founded at Supercomputing '92 by a group of people interested in parallel computer benchmarking. A contribution of the group is the establishment of a consistent set of performance metrics and notations, which we have adopted in this book.

The current benchmarks are for distributed-memory multicomputers, coded with Fortran77 plus PVM or MPI for message passing. Work is in progress to develop Fortran90 and HPF versions and benchmarks for shared-memory architectures. The group has put forth four classes of benchmarks so far:

- *Low-level benchmarks*: These micro benchmarks measure basic architectural parameters such as timer resolution, arithmetic operations speed, cache and memory speed, communication start-up time and bandwidth, and synchronization overhead.

- *Kernel benchmarks*: These measure subroutines frequently used in scientific computing, such as matrix operations, FFT, solving partial differential equations, the NPB kernels, etc.

- *Compact application benchmarks*: Currently, only parallel spectral transform shallow-water model application and the three NPB simulated applications are included.

- *HPF compiler benchmarks*: These are several simple, synthetic applications used to measure performance of HPF compilers, focusing on parallel implementation of explicitly parallel HPF constructs.

The Parallel STAP Suite   The *Space-Time Adaptive Processing* (STAP) benchmark suite is a set of real-time, radar signal processing benchmark programs, originally developed at MIT Lincoln Laboratory [101]. The sequential STAP from MIT was recently converted to a Parallel STAP at the University of Southern California to evaluate various MPPs, as reported in Hwang et al. [334, 634].

The STAP benchmark programs are computation-intensive, requiring one to perform $O(10^{10}—10^{14})$ floating-point operations over $O(10^2—10^4)$ MB of data in a fraction of a second. This translates to a speed demand of $O(10 — 10^4)$ Gflop/s in real time.

The STAP benchmark suite consists of five programs: *Adaptive Processing Testbed* (APT), *High-Order Post-Doppler* (HO-PD), *Beam Space PRI-Staggered Post Doppler* (BM-Stag), *Element Space PRI-Staggered Post Doppler* (EL-Stag), and *General* (GEN). The GEN program consists of four independent component programs to perform sorting (SORT), *fast Fourier transform* (FFT), vector multiplication (VEC), and *linear algebra* (LA), which represent kernel routines often used in radar signal processing applications.

The other four benchmark programs all start with a *Doppler Processing* (DP) step, in which the program performs a large number of one-dimensional FFT computations. All four programs end with a *target detection* (TD) step.

The APT performs a *Householder Transform* (HT) to generate a triangular learning matrix, which is used in a later *beamforming* (BF) step to null the jammers and the clutter; whereas in the HO-PD program, the two adaptive beamforming steps are combined into one step. The BM-Stag program and the EL-Stag program are similar to HO-PD, but use a staggered interference training algorithm in the beam space and in the element space, respectively.

Example 3.1 The APT benchmark program

With much simplification, the APT program in the STAP benchmark suite can be described as follows, where variable N is a problem parameter. The notation [.] indicates that all array elements along that dimension may be accessed. The variable *house* is a matrix containing about 80 KB of information, independent of N.

```
for ( j = 0 ; j < N ; j++)                /* The DP step */
    for ( k = 0 ; k < 32 ; k++)
        fft( data[.][j][k] ) ;

ht ( data[1][.][.], house ) ;             /* The HT step */
for ( i = 0 ; i < N ; i++)                /* The BF step */
    bf (data[i][.][.], house, detect[i][.] ) ;

for ( j = 0 ; j < N ; j++)                /* The TD step */
    for ( i = 0 ; i < N ; i++)
        td ( detect[i][j], target_report ) ;
```

∎

3.1.3 Business and TPC Benchmarks

The most popular benchmark suite for commercial applications is the TPC benchmarks developed by the *Transaction Processing Performance Council* (hence the name TPC), which is a nonprofit organization founded to develop transaction processing and database benchmarks. TPC provides open standard specifications of its benchmarks, which can be implemented by any tester, and the results are audited and reviewed by the TPC-authorized auditors before being published. So far, TPC has released four benchmarks. TPC-A and TPC-B became obsolete as of June 1995.

TPC-C is a data entry benchmark to measure the performance and price/performance of transaction processing systems, while TPC-D measures those of decision support systems. TPC is working on a new benchmark, called TPC-E (Enterprise), to quantify the ability of a given system to support the computing environment appropriate to large business enterprises.

Currently, TPC-C is the most widely used commercial application benchmark. TPC-C is an *on-line transaction processing* (OLTP) benchmark. It simulates a complete wholesale company environment where terminal operators execute transactions against a database. The company operates out of *N* warehouses, where each warehouse supplies 10 sales districts and each district serves 3000 customers. Each warehouse has 10 terminals, one for each district.

At any time, an operator can select one of the five transactions shown in Table 3.6, for creating an order, making a payment to the customer's database, examining the status of an order, the delivery of an order, and examining the current stock level. However, TPC requires that of all transactions executed, at least 43% must be Payment and 4% each for Order-Status, Delivery, and Stock-Level.

Table 3.6   Five Types of Transactions in TPC-C Benchmark

Transaction Type	Database Access	Transaction Weight	Execution Frequency	90% Percentile Response Time
New-Order	Read/write	Medium	High	< 5 s
Payment	Read/write	Light	At least 43%	< 5 s
Order-Status	Read only	Medium	At least 4%	< 5 s
Delivery	Read/write	Heavy	At least 4%	< 5 s
Stock-Level	Read only	Heavy	At least 4%	< 20 s

To publish any TPC-C results, a tester must provide a full-disclosure report to show that all the requirements stated in the TPC-C specification are met. The report discloses detailed system configuration information and performance and cost metrics. The *total system cost* covers all hardware and software, including on-line data storage needed for 180 days' operation and 5-year maintenance cost.

Two TPC-C results are most frequently used: the performance result *tpmC* and the price/performance result *$/tpmC*. The *TPC-C throughput*, or *TPC-C transactions per minute* (denoted by tpmC), measures the number of New-Orders that can be processed per minute, while the system is also executing the other four types of transactions according to the workload mix determined by the TPC-C benchmark specification. The *price/performance* is defined to be the total system cost divided by the throughput. As a concrete example, the IBM PC Server 704 has a total system cost of about $588K, a TPC throughput of 6679.50 tpmC, and a price-performance of $88/tpmC.

TPC-C allows the measured system to scale, but the number of terminals and the size of the database must also scale proportionally. Since a warehouse cannot hold the entire stock of the company, a percentage of transactions must go to other warehouses. This prevents TPC-C from becoming an embarrassingly parallel benchmark. The measured system must provide the ACID properties: *atomicity, consistency, isolation*, and *durability*.

3.1.4 SPEC Benchmark Family

The SPEC benchmark family is developed by a nonprofit corporation called the *Standard Performance Evaluation Corporation*. SPEC emphasizes developing real applications benchmarks that closely reflect the actual workload, instead of small, synthetic kernels. For each benchmark and benchmark suite, SPEC defines a few (two in many cases) metrics that measure the overall performance of the entire system.

SPEC started with benchmarks that measure CPU performance, but has extended to client/server computing, commercial applications, I/O subsystems, etc. Currently, SPEC has released the following benchmark suites:

- *SPEC95*, which measures the performance of CPU, memory system, and compiler code generation.
- *SPEChpc96*, which measures the performance of high-performance computing systems running industrial-style applications. It currently contains two benchmarks: the seismic processing benchmark SPECseis96 and the computational chemistry benchmark SPECchem96.
- *SPECweb96*, which measures Web server performance based on workload derived from logs of real-world WWW servers.
- *SFS*, for *System-level File Server*, which includes the LADDIS benchmark and measures the response time versus the throughput of NFS file servers for various load levels.
- *SDM*, for *System Development Multitasking*, measures how a system handles an environment with a large number of users issuing typical Unix software development commands (e.g., make, cp, grep, spell, etc.). It includes two multiuser Unix command benchmarks made of Unix shell scripts.
- *GPC*, for *Graphics Performance Characterization*, measures graphics performance. It includes three on-going projects. The *Picture Level Benchmark* (PLB) measures "vectors per second" and "polygons per second" graphics capabilities. The *X Performance Characterization* (XPC) project developed the Xmark93 benchmarking tool to measure X Window performance. The *OpenGL Performance Characterization* (OPC) project developed the Viewperf benchmark to measure OpenGL performance.

The SPEC95 The SPEC95 CPU benchmarks are the most famous SPEC benchmarks widely used by vendors and users. These are CPU benchmarks measuring the CPU speed, the cache/memory system, and the compiler as a whole. The time spent in operating system and I/O functions is negligible. SPEC95 consists of CINT95, a set of 8 integer programs; and CFP95, a set of 10 floating-point programs; both CPU-intensive applications.

SPEC periodically publishes performance results of various machines, both in hardcopy and on the Web. Table 3.7 shows the SPEC95 results of Digital AlphaStation 500/500 which uses a 500-MHz Alpha 21164 processor with 8-MB cache and 128-MB memory. All SPEC95 results of a given system are expressed as ratios compared to a Sun SPARC station 10/40, the reference machine. For instance, a value of 5 means that the measured machine achieves a performance that is 5 times that of the Sun SPARCstation 10/40, or 4 times faster.

Each metric is aggregated over all benchmarks of a suite by taking the geometric mean of the ratios of the individual benchmarks. The speed metrics measure the ratios to execute a single copy of the benchmark, while the throughput (rate) metrics measure the ratios to execute multiple copies of the benchmark. Results are obtained with conservative or aggressive optimizations and are reported as "baseline" and "peak" performance numbers. The peak speeds (boldface in Table 3.7) are frequently quoted SPEC95 numbers.

Table 3.7 SPEC95 Performance of an AlphaStation
(500-MHz 21164 Processor with 8-MB
External Cache and 128-MB Memory)

Metric	Speed		Throughput	
	95	_base95	_rate95	_rate_base95
SPECint	**15**	12.6	135	113
SPECfp	**20.4**	18.3	183	165

Example 3.2 Interpretation of SPECint and SPECfp results

The Digital AlphaStation 500/500 has SPECint_rate_base95 = 113 and SPECfp95 = 20.4. What do these numbers mean? The SPECint_rate_base95 number is obtained by taking the geometric mean of the rates of the eight benchmarks of the CINT95 suite, where each benchmark is compiled with a low optimization (e.g., -O).

The rate of each benchmark is measured by running multiple copies of the benchmark for a week, and the execution time is normalized with respect to Sun SPARCstation 10/40. The number 113 means that the Digital machine executes 112 times more copies of CINT95 than Sun SPARCstation 10/40 in a week.

The SPECfp95 number is obtained taking the geometric mean of the ratios of the 10 benchmarks of the CFP95 suite, where each benchmark is compiled with aggressive optimization (e.g., -O4). The benchmark is measured by running a single copy with execution time normalized with respect to a Sun SPARCstation 10/40. The number 20.4 means that the Digital machine is 19.4 times faster than Sun SPARCstation 10/40 for executing a single copy of CFP95.   ■

3.2   Performance versus Cost

How do we measure the performance of a computer system? Many people believe that *execution time* is the only reliable metric to measure a computer's performance. The approach is to run the user's application on the target machine and measure the wall clock time elapsed. Intuitively, this statement is sound. But this approach is sometimes difficult to apply, and it could permit misleading interpretations.

For instance, only a limited number of application programs are portable across different computer platforms. To run a given application on a newly introduced parallel computer requires many worker-months effort in modifying, debugging, and testing the application code, not to mention the high cost of paying the programmers and CPU time.

On current parallel computers, tens to hundreds of dollars are charged for a CPU-hour. Assume the rate is $20 per CPU-hour. An application will be charged more than $5000 for running on 256 processors for just a hour. Expensive as it may seem, the CPU-hour cost may be small compared to the labor cost in parallel code development. This is why many users opt for automatic tools such as parallelizing compilers, especially when maximal performance is not crucial or when the application is an evolving one, not a final production version.

Example 3.3   Pitfalls of using execution time as a performance measure.

Suppose after many worker-months of hard work, the user finally ran his code successfully on a parallel computer X and measured an execution time of 1000 s. This execution time figure alone does not tell him much.

Assuming the code ran on another machine Y with a 500-s execution time, can he conclude that machine X is slower than machine Y? The answer is "not necessarily". It could be that machine X is indeed slower. But it also could be that the algorithm is not implemented optimally for a particular system, or that wrong data structures were used in the machine X code.

It could be that machine X actually has excellent computing power, but the user's code is too fine-grained, which causes excessive communication overhead. The point is that execution time alone does not give the user much clue to a true performance of the machine.

■

Execution time is not the only performance requirement, either. Described below are six types of performance requirement that are frequently posed by users: *execution time, speed, throughput, utilization, cost-effectiveness,* and *performance/cost ratio.* These requirements could lead to quite different conclusions for the same application on the same computer platform.

3.2.1 Execution Time and Throughput

Table 3.8 lists the values of several performance metrics for three programs in the STAP benchmark suite when they are executed on a 256-node SP2.

Table 3.8 Measured STAP Performance on a 256-Node SP2

Program	Execution Time	Speed	Speedup	Utilization
APT	0.16 s	9 Gflop/s	90	13%
HO-PD	0.56 s	23 Gflop/s	233	34%
GEN	1.40 s	3.8 Gflop/s	86	6%

Execution time is critical to some applications. For instance, in a real-time application, the user cares about whether the job is guaranteed to finish within a time limit. Suppose it is required that a STAP benchmark be executed within 0.5 s. The measured performance data in Table 3.8 shows that only the APT program satisfies this 0.5-s requirement on a 256-node SP2. Both HO-PD and GEN fail to meet the requirement.

Processing Speed For many applications, the users may be interested in achieving a certain processing speed, rather than an execution time limit. The application may process different data inputs with different workloads, thus taking different execution times. However, the speed requirement should be maintained.

For instance, the STAP programs are *benchmarks*, in that they are not the real production programs. Rather, they capture the characteristics of the real codes. What the user really wants is a STAP system that can deliver a certain processing speed, say 10 Gflop/s. From Table 3.8, we see that on a 256-node SP2, only the HO-PD program meets this speed requirement.

System Throughput Another speed-related requirement is the throughput requirement, where *throughput* is defined to be the number of jobs processed in a unit time. If only one job is executed at a time, throughput is just the reciprocal of the execution time. For instance, from Table 3.8, the throughput for the APT program on a 256-node SP2 is one APT per 0.16 s, or 6.25 APTs per second.

The throughput is usually used when multiple jobs are executed simultaneously. In many cases, the system throughput can be increased by the following two methods:

- One method is by *pipelining*, where successive jobs overlap their executions among several pipeline stages. The throughput becomes the reciprocal of the execution time at the longest pipeline stage.
- Another method is by assigning a separate job to each node. Thus *n* jobs are processed on n nodes, simultaneously.

Example 3.4 **Calculation of throughput and speed in the parallel APT**

From Table 3.8, we can derive that the workload of the APT program is 9*0.16 = 1.44 Gflop. We can construct a two-stage pipeline for the APT program on a 256-node SP2, by using 128 nodes per pipeline stage.

One stage takes 0.14 s, while the other 0.11 s. Thus the total execution time increases from 0.16 to 0.25 s. However, the throughput increases from 1/0.16 to 1/0.14 = 7.14 APTs per second, or equivalently, 10.34 Gflop/s.

The APT program takes about 14 s to execute on one SP2 node. We can assign a number of APT runs to each of the 256 nodes. The system throughput then becomes 256/14 = 18.29 APTs per second, or equivalently, 25.6 Gflop/s.

■

Fourteen seconds may be too long an elapsed time for real-time radar signal processing. However, it is acceptable for applications such as on-line transaction processing. In many production environments, it is common practice to maximize throughput with the constraint of a set execution time limit.

For instance, the user may specify a requirement for an on-line transaction system to process 1000 transactions and that the maximal response time for each individual transaction is no more than 10 s.

3.2.2 Utilization and Cost-Effectiveness

Instead of searching for the shortest execution time, the user may want to run his application more cost-effectively. He may want a high percentage of the CPU-hours to be used for useful computing, instead of being wasted on load imbalance, communication overheads, etc.

A good indicator of cost-effectiveness is the *utilization factor* (or simply *utilization*), which is the ratio of the achieved speed to the peak speed of a given computer. The following example gives the physical meaning of this term.

Example 3.5 **Utilization of the IBM SP2 for APT execution**

Each SP2 node has a peak speed of 266 Mflop/s. Assume each CPU-hour is charged at $10. From Table 3.8, calculate the utilizations for running the APT program on 256 nodes and on 1 node. Which is more cost-effective?

A 256-node SP2 has a peak speed of $266 \times 256 = 68$ Gflop/s. The APT program achieved 9 Gflop/s on 256 nodes and 100 Mflop/s on 1 node, translating to a utilization of 13.3% and 37.6%, respectively. These utilization values indicate that running on one node is more cost-effective.

In fact, it takes 0.16 s to execute one APT run on 256 nodes. The total cost for one APT execution is $10 \times 256 \times 0.16/3600 = \$0.11$. The corresponding cost-effectiveness is (9 Gflop/s) / \$0.11 = 79 Gflop/s per dollar. On one node, the execution time is about 14.4 s. The total cost is only $10 \times 14.4/3600 = \$0.04$, and the cost-effectiveness is (9 Gflop/s) / \$0.04 = 225 Gflop/s per dollar.

∎

This example demonstrates that higher utilization corresponds to higher Gflop/s per dollar, which is always true if CPU-hours are charged at a fixed rate. The rate charged per CPU-hour could be higher or lower for different types of jobs, on different machines.

We find that execution time, speed, and utilization are the most important metrics. Special attention should be paid to the utilization metric, which is often overlooked, but more informative than execution time and speed. A low utilization always indicates a poor program or compiler. In contrast, a good program could have a long execution time due to a large workload, or a low speed due to a slow machine.

But what utilization values are considered low? This of course depends on the applications and platforms. Our experience and interaction with other MPP users give the following estimation: A sequential *application* executing on a single MPP processor has a utilization ranging from 5% to 40%, typically 8% to 25%. Some individual *subroutines* can be made faster to reach 75% or more.

However, when such subroutines are incorporated into a real application, they do not necessarily retain their high utilization. A *parallel application* executing on multiple processors has a utilization ranging from 1% to 35%, typically 4% to 20%.

A widespread misconception is that single-node, or sequential, computation always has the highest utilization, as parallel computing has extra communication and idling overheads. This is not always true. For instance, the parallel APT program executing on Intel Paragon achieved the highest utilization over four nodes, not on one node.

We give below utilization values measured on three MPPs (Intel Paragon, IBM SP2, and Cray T3D) using the NAS and the STAP benchmarks. In Fig. 3.1, we show the measured utilization and speed of two parallel STAP programs.

The utilization ranges from 5% to 38%. Generally, the utilization drops as more nodes are used. However, this is not always true. By processing the NAS parallel benchmark results [525], we found the utilization values shown in Table 3.9 and Fig. 3.2.

For the three MPPs, the utilization ranges from 2% to 60%, with a harmonic mean of 12%. Note that these results were generated from the vendors' benchmark programs, which are often highly optimized. An ordinary user application of similar type is unlikely to attain the same performance

The cost-effectiveness measure should not be confused with the *performance/cost ratio* of a computer system, which is defined as the ratio of the speed to the *purchasing price*. This ratio is depicted in Fig. 3.3 for various computers running the NAS benchmark.

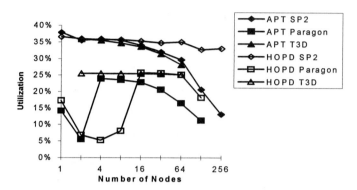

Figure 3.1   Measured utilization of the parallel APT and HO-PD benchmark programs on three MPPs.

Table 3.9 NAS Benchmark Utilization of Six Computers

Type	Machine Model	Max	Min	Harmonic Mean
PVP	Cray C90	73	28	54
MPP	Cray T3D	49	2	10
	IBM SP2	60	3	18
	Intel Paragon	36	3	9
SMP	SGI Power Challenge	44	10	25
	Convex Exemplar	35	1	8

Although there is a wide variation of the performance/cost ratio measured in peak Gflop/s per million dollars, the sustained performance/cost ratios are much more concentrated, around 1 Gflop/s per million dollars in 1995.

Example 3.6 The misleading peak performance/cost ratio

Using the peak performance/cost ratio to compare system is often misleading. In Fig. 3.3, the peak performance/cost ratio of the Cray J916 is much lower than those of the Convex SPP 1000, the Cray T3D, and the SGI Power Challenge. But its sustained performance-cost ratio is actually higher.

Often, users need to use more than one metric in comparing different systems. If we just use the cost-effectiveness or the performance/cost ratio alone, the current IBM PC could easily beat more powerful systems, due to the much lower price of PCs compared with mainframes or supercomputers.

■

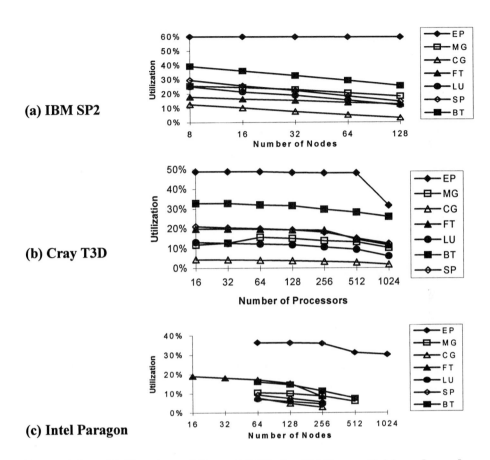

(a) IBM SP2

(b) Cray T3D

(c) Intel Paragon

Figure 3.2   Utilization of three MPPs for NAS parallel benchmarks
(Some points missing for unknown data)

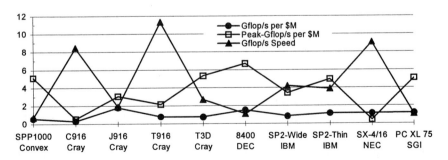

Figure 3.3   Performance/cost ratios of ten 1995 parallel computers

To summarize, the execution time is just one of several performance requirements a user may want to impose. The others include speed, throughput, utilization, cost-effectiveness, performance/cost ratio, and scalability. Usually, a set of requirements may be imposed. For instance, the user may want to require a certain level of throughput or utilization, subject to a limit on the execution time. She may want to select the system with the best performance/cost ratio above a minimal speed.

3.3   Basic Performance Metrics

This section studies the performance metrics used for uniprocessor systems, as well as parallel computers. The terminology is consistent with that proposed by the Parkbench group [317], which follows the conventions used in other scientific fields, such as physics.

3.3.1 Workload and Speed Metrics

Three metrics are frequently used to measure the computational workload of a program C: *the execution time, the number of instructions executed*, and *the number of floating-point operations executed*. The first metric is tied to a specific computer system. It may change when the program C is executed on a different machine.

The second metric, tied to an *instruction set architecture*(ISA), stays unchanged when the program is executed on different machines with the same ISA. The third metric is often architecture-independent. Table 3.10 summarizes all three performance metrics.

Table 3.10 Workload and Speed Metrics

Workload Type	Workload Unit	Speed Unit
Execution time	Seconds (s), CPU clocks	Application per second
Instruction count	Million instructions or billion instructions	MIPS or BIPS
Floating-point operation (flop) count	Flop, Million flop (Mflop), billion flop (Gflop)	Mflop/s Gflop/s

Instruction Count   For an arbitrary program C, we can use the *instruction count* as a measure of its workload, which is the number of instructions executed. Such a workload has the unit of *millions of instructions*. The corresponding speed metric is *millions of instructions per second* or MIPS. Several cautions must be taken in using MIPS rate:

- The workload is the instructions that the machine executed (i.e., *dynamic instruction count*), not just the number of instructions in the assembly program text (called *static program count*). These two counts are the same only for straight-line programs.

- Instruction count may depend on the input data values. For instance, a sorting program may execute 100,000 instructions for a given input, but only 2000 for another. For such an *input-dependent* program, the workload is defined as the instruction count for the worst-case input or a specific baseline input data set.

- Even with a fixed input, the instructions executed could be different on different machines. For instance, a RISC processor often executes 50% to 150% more instructions than a CISC processor, given the same high-level language program.

- Even with a fixed input, the instruction count of a program could be different on the same machine when different compilers or optimization options are used.

- Finally, a larger instruction count does not necessarily mean the program needs more time to execute.

Execution Time For a given program on a specific computer system, one can define the workload as the total time taken to execute the program. This execution time should be measured in terms of *wallclock time*, also known as the *elapsed time*, obtained by, for instance, the Unix function gettimeofday(). It should not be just the *CPU time*, which can be measured by the Unix function times(). The basic unit of workload is seconds, although one can use minutes, hours, milliseconds (ms), or microseconds (μs) as well.

The execution time depends on many factors. Listed below are some of the most important ones:

- *Algorithm*: The algorithm used has a great impact on execution time. For instance, to sort N items, a bubble sorting algorithm has an execution time of $O(N^2)$, while a merge sorting has only $O(\log N)$.

- *Data Structure*: How the data are structured also impacts performance.

- *Input Data*: The execution time of many applications does not depend on the input data values. For instance, an N-point FFT program will take $O(N \log N)$ time, no matter what the input FFT vector is.

    Other programs, such as sorting and searching, could take different times on different input data, because the control flow of the program could change with different inputs. When using execution time as the workload for such *input-dependent* programs, we need to use either the *worst-case time* or the time for a baseline input data set.

- *Platform*: Obviously, the machine hardware and operating system affect performance. What is not so obvious is that performance is affected by not just the processor. One can see quite different timing results on two platforms that use the same processor at the same clock rate. Other factors include the memory hierarchy (cache, main memory, disk), operating system versions, and whether the application has dedicated use of the computer or has to timeshare the resource with other applications.

- *Language*: Even when the same algorithm and data structure are used on the same platform, different execution times will be seen if the languages used to code the application are different. Furthermore, the compilers and the compiler/linker options used play an important role. Using binary-level library functions can also reduce the execution time.

Floating-Point Count   For scientific/engineering computing and signal processing applications where numerical calculation dominates, a natural metric is the number of *floating-point operations* that need to be executed. To use this metric meaningfully, we must follow some rules in counting the flop. Table 3.11 illustrates somerules used in practice.

Table 3.11 Rules for Counting Floating-Point Operations

Operation	Flop Count	Comments on Rules
A[2*i] = B[j–1]+1.5*C–2;	3	Add, subtract, or multiply each count as 1 flop Index arithmetic not counted Assignment not separately counted
X = Y;	1	An isolated assignment is counted as 1 flop
if (X>Y) Max=2.0*X;	2	A comparison is counted as 1 flop
X = (float) i + 3.0 ;	2	A type conversion is counted as 1 flop
X = Y / 3.0 + sqrt(Z) ;	9	A division or square root is counted as 4 flop
X = sin(Y) – exp(Z) ;	17	A sine, exponential, etc. is counted as 8 flop

When the application program C is simple and its workload is not input-dependent, the workload of C can be determined by code inspection. When the application code is complex, or when the workload varies with different input data (e.g., sorting and searching programs), one can measure the workload by running the application on a specific machine. This specific run will generate a report of the number of flop or instructions actually executed. This approach has been used in the NAS benchmark, where the flop count is determined by an execution run on a Cray Y-MP or on a Cray C90.

Example 3.7 Utilization rate in FFT processing

Consider Example 3.10 again. Each N-point FFT has a workload of $W = 5N\log N$ flop. The total workload for a Doppler processing step is $2048\times(5\times8192\times\log8192)$ flop, or about 1.09 Gflop.

With 50-s execution time on machine X, the speed is about 22 Mflop/s. Suppose machine X has a peak speed of 266 Mflop/s. The Doppler processing achieved an utilization rate of $22/266 = 8.27\%$, a rather low rate from the peak performance.

■

In theoretical performance analysis, it is often assumed that every instruction or flop takes the same amount of time. This *uniform-speed* assumption does not hold in real systems. For instance, on a single IBM SP2 node, the speed is observed to vary from 5 to 250 Mflop/s. Thus sequential execution time is also used widely to complement flop or instruction count in measuring workload.

3.3.2 Caveats in Sequential Performance

Using execution time or instruction count as the workload metric leads to a strange phenomenon: The workload changes when the same application is run on different systems. Furthermore, the workload can be determined only by executing the program.

The flop count metric is more stable. For instance, an FFT has the same workload of $W = 5N\log N$ flop, no matter how it is executed. In addition, the flop count (and, to a less degree, the instruction count) can be determined from code inspection, without execution.

The flop count and the instruction count have another advantage: Their corresponding speed and utilization measures give some clue as to how well the application is implemented. Current high-level language programs often achieve a utilization of 5% to 40% on sequential computers and 1% to 25% on parallel computers.

Highly tuned programs could achieve more. For instance, Agarwal et al. [13] show that by writing Fortran programs to take advantage of the special architecture features of the POWER2 processor, it is possible to achieve 90% utilization in some numerical subroutines.

Example 3.8 Sequential performance of the IBM SP2

Table 3.12 shows the sequential performance of the STAP benchmark programs on a single SP2 node, which has a 266-Mflop/s peak speed.

The workload values are obtained by inspecting the source STAP programs. The execution time values are from actual measurement of each of the component algorithms.

For example, the APT program is divided into four component algorithms: *Doppler Processing* (DP), *Householder Transform* (HT), *Beamforming* (BF), and *Target Detection* (TD).

Table 3.12 Sequential STAP Programs Performance on SP2

Program	Component Algorithms	Workload (Mflop)	Execution Time (s)	Speed (Mflop/s)	Utilization (%)
APT	Total	1447	14.37	100	37.85
	DP	84	4.12	20	7.65
	HT	2.88	0.04	72	27.07
	BF	1314	9.64	136	51.22
	TD	46	0.57	75	28.01
HO-PD	Total	12,853	130.61	98	37.00
	DP	220	11.62	19	7.12
	BF	12,618	118.82	106	39.92
	TD	14	0.17	82	30.96
GEN	Total	5326	121.05	44	16.54
	SORT	1183	22.80	52	19.51
	FFT	1909	79.14	24	9.06
	VEC	604	19.11	32	11.88
	LIN	1630	20.23	82	31.00

Each component algorithm performs different amounts of workload with different execution times, thus resulting in different speed and utilization values. The utilization values in Table 3.12 tell us immediately that the FFT program is too slow, not optimized for the POWER2 architecture of the SP2. On the other hand, the beamforming program is quite good. These facts cannot be seen by just looking at the execution times.

∎

Summary of Performance Metrics To sum up, all three metrics are useful, but especially the flop count and the execution time. It is important that a single workload be used consistently in predicting and measuring the performance of an application, even though on parallel machines, extra operations may need to be performed to solve the same problem.

In practice, the flop count workload is often determined by code inspection, following the rules in Table 3.11. The execution time is often measured on a specific machine (e.g., SPARCstation 10 or Cray Y-MP), under a set of specific testing conditions including hardware platform, compiler options, and input data set, etc.

3.4 Performance of Parallel Computers

We discuss performance attributes of parallel computer systems in this section and those for parallel applications in the next section. As already mentioned in Section 1.3.2, a parallel system has computational and overhead characteristics.

3.4.1 Computational Characteristics

Table 3.13 shows the historical values of performance parameters for three commercial parallel computer families, where the Cray family represents PVP machines, the Intel family MPPs, and the SGI family SMPs.

Table 3.13 Computational Parameters of Three Computer Families

Computer	Year	Clock Rate (MHz)	Memory Capacity	Machine Size n	P_{peak} (Mflop/s)	$n \cdot P_{peak}$ (Gflop/s)
Cray 1	1979	80	1 MB	1	160	0.16
Cray X-MP	1983	105	4 MB	2	210	0.42
Cray Y-MP	1987	166	1 GB	8	333	2.66
Cray C90	1991	238	2 GB	16	1000	16
Cray T90	1995	454	8 GB	32	2000	60
Intel iPSC	1985	10	64 MB	128	0.04	0.005
Intel iPSC/860	1989	40	1 GB	128	60	7.68
Intel Paragon	1992	50	64 GB	2048	75	153
Intel ASCI	1996	200	297 GB	9216	200	1800
SGI Power	1988	16	256 MB	2	1	0.002
SGI Challenge	1993	150	16 GB	36	75	2.7
SGI Power Challenge XL	1994	75	16 GB	18	300	5.4

The Memory Hierarchy The performance of a modern computer depends on how fast the system can move data between processors and memories. The memory subsystem hierarchyis s shown in Fig. 3.4. For each layer, we are interested in three parameters:

- *Capacity C:* How many bytes of data can be held by the device?
- *Latency L:* How much time is needed to fetch one word from the device?
- *Bandwidth B:* For moving a large amount of data, how many bytes can be transferred from the device in a second?

Typical values of the three parameters on current computers are shown in Fig. 3.4. The faster and smaller devices are closer to the processor. The devices closest to the processor are the registers, which are in fact a part of the processor chip. Normally, no extra cycles are needed to fetch a word from a register to a functional unit (e.g., an ALU). Thus the latency is 0. But registers have very limited capacity.

For instance, the Intel 80x86 processors have eight data registers, capable of holding only 32 bytes. Other processors (e.g., RISC processors and vector processors) have more registers, but the capacity rarely exceeds 2 KB. The bandwidth can be estimated in the following way: An add operation fetches two words from and stores one word to the registers. Assume a 100-MHz clock rate and a 64-bit word length. The bandwidth is $3 \times 8 \times 100 \times 10^6 = 2.4$ GB/s. Higher bandwidth is possible due to faster clock rates and parallel operations within the processor.

The level-1 cache is usually on the processor chip, while the level-2 cache is off the chip. The main memory includes the local memory in a node, and the global memory for machines with a centralized shared memory, such as PVPs and SMPs. The remote memory refers to all local memories of other nodes.

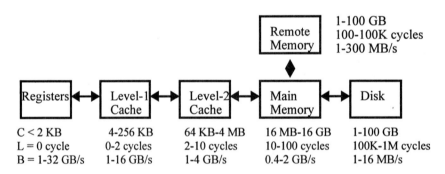

Figure 3.4   Performance parameters of a typical memory hierarchy.

Example 3.9 **Performance of Digital's TruCluster memory hierarchy**

The DEC TruCluster consists of eight 8400 SMP nodes interconnected by an eight-port Memory Channel interconnect. Each SMP node can have up to 12 processors or up to 14-GB shared main memory. Each processor has a 16-KB level-1 cache and a 96-KB level-2 cache, both on chip. There is an off-chip, 4-MB level-3 cache shared by two processors. Remote memories of other nodes are accessed through the memory channel interconnect.

Table 3.14 shows the parameters of the TruCluster memory hierarchy, assuming a processor clock of 300 MHz. These are best numbers (hardware peak) which cannot be exceeded. The best numbers attainable by a user process are shown in brackets. Note that the 300-MB/s main memory bandwidth refers to one processor accessing

Table 3.14 Performance of the TruCluster Memory Hierarchy

Location	Capacity	Bandwidth	Latency
Register	512 B	34 GB/s	3 ns (1 clock)
Level-1 cache	16 KB	9.6 GB/s	3 ns (1 clock)
Level-2 cache	96 KB	4.8 GB/s	10 ns (3 clocks)
Level-3 cache	4 MB	1.2 GB/s	20 ns (6 clocks)
Main memory	14 GB	1.6 GB/s [300 MB/s]	70 ns (21 clocks) [240 ns (80 clocks)]
Remote memory	98 GB	100 MB/s [61 MB/s]	5 μs (1500 clocks)

the memory. The bandwidth could be as high as 982 MB/s when multiple processors access the memory simultaneously.

∎

3.4.2 Parallelism and Interaction Overheads

In the theoretical PRAM model, parallelism and interaction overheads are ignored. For instance, the instruction level synchrony implicitly places a barrier after each instruction, at no cost. It takes no time to read/write any memory word. Such assumptions are false in real parallel computers. Assuming no overlapping, the time to execute a parallel program is

$$T = T_{comp} + T_{par} + T_{interact} \tag{3.2}$$

where T_{comp}, T_{par}, and $T_{interact}$ are the times needed to execute the computation, the parallelism, and the interaction operations, respectively.

Recall that in Section 1.3.2 and Section 1.3.3, the overheads in a parallel program can be divided into three classes: the load-imbalance overhead, the parallelism overhead, and the interaction overhead (including synchronization, communication, and aggregation). How to achieve load balance is often a parallel algorithm design problem. In this section, we focus on the other two classes of overhead.

From the discussion in Section 2.3, there are three types of parallelism operations, which are sources of *parallelism overhead*:

- *Process management*, such as creation, termination, context switching, etc.
- *Grouping* operations, such as creation or destruction of a process group.

- *Process inquiry* operations, such as asking for process identification, rank, group identification, group size, etc.

From the discussion in Section 2.4.1, there are three types of interaction operations, which are sources of *interaction overhead*:

- *Synchronization*, such as barrier, locks, critical regions, and events
- *Aggregation*, such as reduction and scan
- *Communication*, such as point-to-point and collective communication and reading/writing of shared variables

Knowing these overheads can help a programmer decide how to best develop a parallel program. For instance, if parallelism overheads are small, the programmer can afford to write *dynamic* parallel programs, in which processes are frequently created, destroyed, and context-switched.

If the overheads are large, *static* parallel programs should be used instead, where processes are created only at the beginning and are terminated only at the very end. If barriers are efficiently supported in a machine, one can use a synchronous iterative algorithm to solve a problem. However, if barriers are time-consuming, one may be better off with an asynchronous algorithm.

Large Overheads When studying modern computers, one can immediately notice two things about the parallelism and interaction overheads: They are often large compared to the basic computation time, and they vary a great deal from system to system.

We use three metrics to measure the overhead incurred. The basic metric is *execution time* (in μs). Sometimes, the user may want to use the *number of clock cycles*, or the *number of flop* that can be executed by a single node in the same amount of time.

For instance, the SP2 uses 66-MHz, 266-Mflop/s POWER2 processors. A *send/ receive* in SP2 takes at least 39 μs, which is equivalent to 2601 CPU clock cycles, or 10,374 peak flop. Table 3.15 shows the overhead of process management and communication on some uniprocessor Unix systems, obtained from McVoy's lmbench program [440]. The Process Creation column shows the time needed to fork an empty child process that does nothing but immediately exits.

The Context Switching column shows the time needed to switch between two processes. The Pipe Latency column shows the round-trip time for two Unix processes to pass a small token back and forth through a pipe. The Pipe bandwidth column shows what can be achieved with a 50-MB message between two processes through a pipe.

Table 3.16 shows the overhead for process creation and synchronization in a Sun workstation running the Solaris operating system [601]. Solaris supports three types of processes: the heavy-weighted *Unix processes*, the *light-weighted processes* (LWPs), and user-level *threads*, as discussed in Chapter 2, 7, and 12. The overhead of Unix processes are one to two orders of magnitude higher than those of user-level threads.

Table 3.15 Parallelism and Communication Overheads in Unix Systems
(Courtesy: McVoy and Staelin, *Proc. of USENIX Tech. Conf.* 1996 [440])

Processor	Operating System	Process Creation (μs)	Context Switching (μs)	Pipe Latency (μs)	Pipe Bandwidth (MB/s)
POWER2	AIX 3	1.4K	21	138	N/A
POWER	AIX 2	2.0K	20	143	34
Pentium	Linux 1.1	3.3K	66	157	13
Alpha	OSF1 V2.1	4.8K	25	185	32

Table 3.16 Parallelism and Interaction Overheads in Solaris

Operation	Overhead
Fork a Unix process	3,057 μs
Create a light-weighted process	422 μs
Create a user-level thread	101 μs
Process-level lock	105 μs
Thread-level lock	1.8 μs

These data demonstrate that the parallelism and interaction overheads could be very large. For instance, a POWER2 processor can execute 4 floating-point operations per clock cycle (15 ns). From Table 3.15, the time to create a Unix process (1.4K μs) is long enough to execute 372,000 flops!

The overhead values also vary greatly from one system to another, even when the same processor is used. Users cannot extrapolate from their past experience with a "similar" system to guess what the overhead will be. It is important for the user to know the overhead values so as to avoid using expensive parallelism and interaction operations.

Often the overhead is mainly caused by OS kernel or system software cost. Consequently, the overhead could be quite different from one system to another, even when all they use the same processor architecture.

The hardware latency of the switch in SP2 is less than 1 μs. But the latency jumps to several hundred μs if it uses the kernel-level IP protocol. IBM has implemented a user-space protocol to bypass the kernel, which reduces the latency to as little as 39 μs.

3.4.3 Overhead Quantification

Since parallelism and interaction overheads are significant, they should be quantified. The quantized overheads are still lacking for many of the real computers. Parallel computer companies provide computational performance data in various forms, such as the peak performance (MIPS or Mflop/s) per node, or per an n-node system.

In contrast, they rarely provide overhead data, except start-up and bandwidth numbers in point-to-point communications. It is very difficult to answer the question, How long does it take to create a process, to partition a process group, to do a barrier synchronization, or to perform a collective reduction on any existing parallel computer?. Sometimes, even the vendors themselves do not have the answers. The users often have to measure the overhead themselves, which is a rather costly exercise.

Although numerous benchmarks and metrics for parallel computers have been proposed, few provide estimation of overhead. Existing benchmarks that measure overhead include the PARKBENCH benchmark suite [317], which measure point-to-point communication and barrier synchronization for distributed memory MPPs.

Well-known overhead metrics are the parameters r_∞, $m_{1/2}$, t_0, and π_0, proposed by Hockney [313] for measuring point-to-point communication [314]. Parallel computer vendors should provide a set of closed-form expressions to quantify overheads, as vendors are in the best position to perform controlled experiments to measure and derive the fundamental metrics (i.e., the coefficients in the overhead expressions).

Ideally, these overhead expressions should quantify all operations in such standards as PVM and MPI. At least, they should quantify all interaction operations. General guidelines for this step are suggested below. We also show how these guidelines are used to generate the overhead expressions for the IBM SP2.

Overhead Measurement Conditions The exact conditions under which one conducts measurement experiments must be clearly spelled out. The following is a partial list:

- The data structures used.
- The programming language , library, and compiler option used.
- In general, overhead measurements should be done in the batch mode, because this is the mode in which most production runs are executed.
- The communication hardware and protocol used.
- Whether to measure the wall clock time or the CPU time. In general, the wall clock time is more useful.

Example 3.10 Communication overhead of a 400-node IBM SP2

We have measured in [655] the communication overheads of the IBM SP2 MPP for up to 256 nodes. The following measurement conditions were followed:

- The data structures used are always made small enough to fit in the node memory, so that there will not be extensive page faults.

- The test program is written in standard C. The only library functions used are the standard I/O and timing functions, and the MPI (or MPL) communication functions. No other library functions or assembly codes are used.

- The best compiler options are always used: mpcc -O3 -qarch=pwr2 program.c. The compiler "mpcc" is actually a script file that calls the IBM xlc compiler and links the MPI and MPL libraries.

- The system resource is used in as dedicated a manner as possible, so that interference from the operating system and other user processes can be minimized. Each of the 256 nodes is set to be solely used by one process, by specifying in the host.list file: hostname   dedicated unique

- The SP2 allows communication through either an Ethernet or proprietary High-Performance Switch (HPS). For each communication medium, one can realize communication through either the standard Internet Protocol (the IP protocol) or an IBM-developed *user space protocol* (the US protocol). To achieve the best communication performance, one can set the following environment variables for using the HPS through the US protocol:

      setenv EUIDEVICE css0   /* use HPS (instead of Ethernet) */
      setenv EUILIB us   /* use the US protocol (instead of IP) */

- Both the wall clock time and the CPU time (both the system time and the user time) are measured to gain confidence in the testing results. The Unix times() function is used to measure the CPU time, and the gettimeofday() function is used to measure the wall clock time.

■

Overhead Measurement Methods Although measuring overhead looks deceptively simple, to obtain accurate measurement results is a challenging undertaking. This is due to three reasons:First, most computer systems provide coarse timer resolutions, on the order of microseconds or even milliseconds. Second, the processors in a parallel computer, especially MPPs and clusters, often operate asynchronously, not following the beat of a common clock.

This causes the *time synchronization problem* [1]; i.e., it is very difficult to force the processors to start an operation at the same time. Third, the measurement results could vary significantly, even for the same communication operation.A popular method for measuring a point-to-point communication (e.g., between node 0 and node 1) is the *ping-pong* scheme: Node 0 executes a send operation to send a message of *m* bytes to node 1, which executes a corresponding receive to get the message.

Then node 1 immediately sends the same message back to node 0. The total time for this ping-pong operation is divided by 2 to get the point-to-point communication time, i.e., the time to execute a single send or receive operation. This procedure is better explained by the following example.

Example 3.11 The ping-pong scheme to measure latency

The ping-pong scheme is described below with analysis:

```
for (i=0; i < Runs; i++)
    if (my_node_id ==0) {                          /* sender */
        tmp = Second();
        start_time = Second();
        send an m-byte message to node 1;
        receive an m-byte message from node 1;
        end_time = Second();
        timer_overhead = start_time - tmp ;
        total_time = end_time - start_time - timer_overhead ;
        communication_time[i] = total_time / 2 ;
    } else if (my_node_id ==1) {                   /* receiver */
        receive an m-byte message from node 0;
        send an m-byte message to node 0;
    }
}
```

The *Runs* loop is used to obtain a number of communication_times, from which the maximum, the minimum, and the mean values can be calculated. The ping-pong scheme is not affected by the time synchronization problem, because the timer function Second() is executed on only a single processor (node 0).

■

A generalization of the ping-pong scheme is the *hot-potato* (also known as the *fire-brigade*) method. Instead of two nodes (the sender and the receiver), n nodes are involved. Node 0 sends a message of m bytes to node 1, which immediately sends the same message to node 2, and so on.

Finally, node $n-1$ sends the message back to node 0. The total time is divided by n to get the point-to-point communication time. For collective communications, the situation is more complex.

Example 3.12 **Measuring collective communication performance**

Consider the following SPMD program that is executed by each of n nodes of a distributed-memory multicomputer. Barrier is applied to synchronize the asynchronous operations in the measuring process.

```
for (i=0; i < Runs; i++) {
    barrier synchronization;
    tmp = Second();
    start_time = Second();
    for (j=0; j< Iterations; j++)
        the_collective_routine_being_measured;
    end_time = Second();
    timer_overhead = start_time - tmp;
    total_time = end_time - start_time - timer_overhead;
    local_time = total_time / Iterations;
    communication_time[i] = maximum of all n local_time values;
}
```

The barrier synchronization tries to time-synchronize the nodes, which may not be realized. A barrier is a logical synchronization operation, not a time synchronization one. As a consequence, the nodes do not necessarily start to execute the tmp = Second() statement at the same time. A scheme follows which generalizes the ping-pong method for collective operations:

```
for ( i=0 ; i < Runs ; i++) {
    if (my_node_id ==0) {
        tmp = Second();
        start_time = Second();
    }
    node 0 broadcasts an empty message to all n nodes ;
    for ( j=0 ; j< Iterations ; j++)
        the_collective_routine_being_measured ;
    all nodes perform an empty reduction to node 0 ;
    if (my_node_id ==0) {
        end_time = Second();
        timer_overhead = start_time - tmp ;
        communication_time[i] =end_time - start_time - timer_overhead;
    }
}
```

This method avoids the time synchronization problem by counting time only on node 0. The broadcast operation starts all nodes, while the reduction operation ends all nodes. Before the broadcast operation, no node can start to execute the *Iterations* loop. After the reduction operation, all nodes have finished the *Iterations* loop. If the value of *Iterations* is large enough, the disturbance caused by the broadcast and the reduction operations should be negligible.

■

Overhead Expressions After we obtain the overhead data from measurement, how do we interpret them? Three interpretation methods are suggested:

Method 1 Present the data in a table. For instance, Table 3.17 shows the timing results for point-to-point communication on SP2 by running the proprietary MPL communication library. It shows the overhead (in μs) for node 0 to send an *m*-byte message to node $n-1$. The overhead is relatively independent of *n*. This method presents the most accurate and detailed information. However, it is not very convenient for the user.

Table 3.17 One-Way Point-to-Point Communication Times on SP2

Message Size	Machine Size *n*			
m	2	8	32	128
4 B	46	47	48	48
1 KB	101	120	120	133
64 KB	1969	1948	1978	2215
4 MB	1.2M	1.2M	1.2M	1.2M

Method 2 Present the data as a curve, as shown in Fig. 3.5. A benefit is that curves can show the trend of communication overhead growth.

Metod 3 Ideally, the user prefers to have a simple, closed-form expression that can be used to estimate the overhead for various message lengths and machine sizes, not just those that have been measured. For instance, using least-square fitting of the measured timing data, we can express the point-to-point communication overhead on SP2 as a linear function of the message length: $t = 46 + 0.035m$ μs. If done carefully, the inaccuracy caused by curve fitting is small, as demonstrated in Fig. 3.5.

Parallelism Overhead In many parallel processing applications, the same parallel program is executed many times to process raw data. There is a one-time overhead for creating all processes and groups, which can be used by subsequent computations. In other words, the parallelism overhead is amortized.

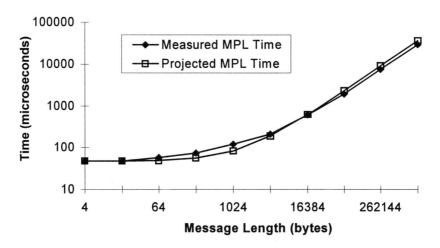

Figure 3.5   Overhead of point-to-point communication on SP2.

Creating a process or a group is expensive on parallel computers, especially MPPs and clusters. Our measurement [655] shows that creating a process on SP2 takes about 10000 μs or more, equivalent to millions of flop. A group creation takes about 1 ms. A program that needs to create many processe groups must have a large grain size.

Point-to-Point Communication Hockney [315] has proposed a model to characterize the communication time (in μs) for a point-to-point operation as follows, where the *communication overhead, t(m),* is a linear function of the message length m (in bytes):

$$t\,(m)\;=\;t_0 + m\,/\,r_\infty \tag{3.3}$$

Here t_0 is the *start-up time* in μs, and r_∞ is the *asymptotic bandwidth* in MB/s. Two additional parameters were introduced by Hockney. The *half-peak length,* denoted by $m_{1/2}$ bytes, is the message length required to achieve half of the asymptotic bandwidth. The *specific performance,* denoted by π_0 MB/s, indicates the bandwidth for short messages. Only two of the four parameters $r_\infty, t_0, m_{1/2}, \pi_0$ are independent. The other two can be derived using the following relations: $t_0 \;=\; m_{1/2}/r_\infty \;=\; 1/\pi_0$. The parameter $m_{1/2}$ is a good indicator of how well a system supports short messages communications.

For SP2, $t\,(m) \;=\; 46 + 0.035m$. In other words, the start-up overhead is $t_o = 46$ μs, the asymptotic bandwidth is $r_\infty = 1/0.035 = 28.57$ MB/s, and the half-peak message length is $m_{1/2} \;=\; t_0 \times r_\infty \;=\; 1314$ bytes.

Collective Communication We have measured the following collective communications: In a *broadcast* operation, processor 0 sends an *m*-byte message to all n processors.

In a *gather* operation, processor 0 receives an m-byte message from each of the n processors, so in the end, mn bytes are received by processor 0. In a *scatter* operation, processor 0 sends a distinct m-byte message to each of the n processors, so in the end, mn bytes are sent by processor 0.

In a *total exchange* operation, every processor sends a distinct m-byte message to each of the n processors, so in the end, mn^2 bytes are communicated. In a *circular-shift* operation, processor i sends an m-byte message to processor $i + 1$, and processor $n - 1$ sends m bytes back to processor 0. Note that in current message-passing systems, a collective communication always requires a process to send a message to itself. That is why mn^2 bytes are communicated in a total exchange, and not $mn(n-1)$ bytes.

We have extended Hockney's expression in Eq.(3.3), to as follows: The *communication overhead, $T(m, n)$,* is now a function of both m and n. However, the *start-up latency* depends only on n. The *asymptotic bandwidth* $r_\infty(n)$ also varies with the machine size n., no loger a constant as in Eq. (3.3).

$$T(m, n) = t_0(n) + m/r_\infty(n) \tag{3.4}$$

After fitting the measured timing data to different forms of $t_0(n)$ and $r_\infty(n)$, we derived the formulas for the five collective operations, as shown in Table 3.18.

Table 3.18 Collective Communication Overhead Expressions for SP2

Operation	Timing Formula
Broadcast	$(52\log n) + (0.029\log n)m$
Gather/Scatter	$(17\log n + 15) + (0.025n - 0.02)m$
Total Exchange	$80\log n + (0.03n^{1.29})m$
Circular Shift	$(6\log n + 60) + (0.003\log n + 0.04)\,m$

Collective Computations We measured three representative collective computation operations: *barrier, reduction,* and *scan*. The curve-fitted overhead expressions are shown in Table 3.19. Note that over 256 processors, the barrier overhead is 768 µs, equivalent to the time to execute as many as $768 \times 266 = 202{,}692$ flop. This answers the question, Should I use a synchronous algorithm? The answer is only when the grain size is large.

This method [656] requires one to quantify each operation individually to achieve accuracy. However, the overhead expressions need to be measured and derived only once for a given computer platform, and they can be used by the user community many times.

The method is compared in Fig. 3.6 for predicting the overhead of a total exchange for both short and long messages. In part (a), we show the relative errors of the two methods when $mn^2 = 16$ MB (e.g., $m = 1024$ bytes and $n = 128$). The traditional method

Table 3.19 Collective Computation Overhead for SP2

Operation	Time Expression
Barrier	$94 \log n + 10$
Reduction	$20 \log n + 23$
Scan	$60 \log n - 25$

underestimates the overhead, especially for large numbers of processors. In part (b), we compare the measured overhead with those projected by the two methods, when $mn^2 = 64$ KB (e.g., $m = 4$ bytes when $n = 128$). While our method is close to the measured result, the traditional method overestimates the overhead significantly for large machines.

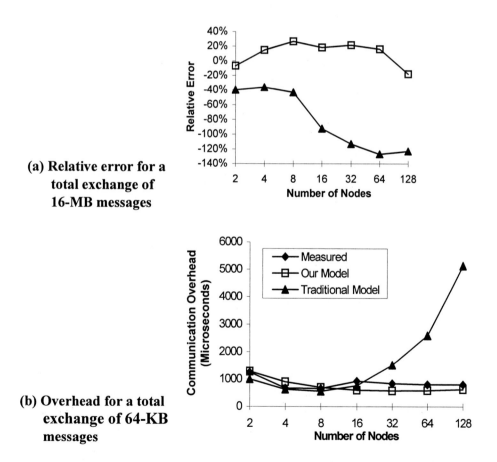

(a) Relative error for a total exchange of 16-MB messages

(b) Overhead for a total exchange of 64-KB messages

Figure 3.6   Two methods for predicting the total exchange overhead.

3.5 Performance of Parallel Programs

In this section, we discuss a number of performance issues and metrics for parallel applications. Our analysis is based on using the phase-parallel model for illustrative purposes. All the performance matrices presented apply to other types of parallel programs as well.

3.5.1 Performance Metrics

Consider a sequential program C consisting of a sequence of k major component computational phases $C_1, C_2, ..., C_k$. We want to develop an efficient parallel program from C by exploiting each phase C_i with a *degree of parallelism* DOP_i. Parallelism and interaction operations both introduce the overhead. A phase-parallel program is depicted in Fig. 3.7. We assume that there is a lumped parallelism overhead at the start of the parallel program.

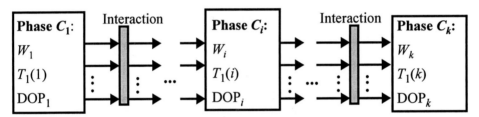

Figure 3.7   The phase parallel model of an application algorithm

Basic Metrics Semantically, all operations of one phase (step) must finish before the next step can begin. Step C_i has a computational *workload* of W_i million floating-point operations (Mflop) and takes $T_1(i)$ seconds to execute on one processor. It has a degree of parallelism of DOP_i.

In other words, when executing on n processors with $1 \leq n \leq DOP_i$, the parallel execution time for step C_i is $T_n(i) = T_1(i)/n$. The total parallel execution time over n nodes becomes

$$T_n = \sum_{1 \leq i \leq k} \frac{T_1(i)}{\min(DOP_i, n)} + T_{par} + T_{interact} \tag{3.5}$$

where T_{par} and $T_{interact}$ denote all parallelism and interaction overheads, respectively.

Extreme-Value Metrics There are several metrics of extreme values which give lower and upper bounds for T_n, P_n, and S_n. Let T_∞ be the length of the *critical path*, which is equal to the time to execute an application using an unrestricted number of nodes and excluding all parallelism and interaction overhead. From Eq.(3.5), we have

$$T_\infty = \sum_{1 \le i \le k} \frac{T_1(i)}{DOP_i}$$ (3.6)

The smallest n to achieve $T_n = T_\infty$ is called the *maximum parallelism,* denoted by N_{max}. This is the maximal number of nodes that can be used to reduce the execution time. This metric can be computed by $N_{max} = max_{1 \le i \le k}(DOP_i)$. The sustained speedup P_n has as its upper bound the *maximum performance* $P_\infty = W/T_\infty$. The n-node execution time T_n has as its lower bound T_1/n and T_∞. That is,

$$T_n \ge max\,(T_1/n,\, T_\infty)\ .$$ (3.7)

The *average parallelism* T_1/T_∞ provides an upper bound on the speedup. That is, $S_n \le T_1/T_\infty$. Brent has proved [109] that, excluding all parallelism and interaction overhead, T_n is bounded by the following inequalities:

$$T_1/n \le T_n < T_1/n + T_\infty$$ (3.8)

Incorporating Eq.(3.7), we have

$$max\,(T_1/n,\, T_\infty) \le T_n < T_1/n + T_\infty$$ (3.9)

These inequalities are useful to estimate the parallel execution time.

Other Metrics The *average granularity* is defined as W/T_o, where W is the total workload and $T_o = T_{interact} + T_{par}$ is the total overhead, including all parallelism and interaction overheads. A reciprocal metric, called *average overhead*, is defined as T_o/W. In many cases, the total overhead is dominated by the communication overhead. These metrics are summarized in Table 3.20.

Table 3.20   Performance Metrics Based on The Phase-Parallel Model

Notation	Terminology	Definition
T_1	Sequential time	$T_1 = \sum_{1 \le i \le k} T_1(i)$
T_n	Parallel time, n-node time	$T_n = \sum_{1 \le i \le k} \dfrac{T_1(i)}{\min(DOP_i, n)} + T_{par} + T_{interact}$
T_∞	Critical path	$T_\infty = \sum_{1 \le i \le k} \dfrac{T_1(i)}{DOP_i}$
P_n	n-node speed	$P_n = W/T_n$
S_n	n-node speedup	$S_n = T_1/T_n$
E_n	n-node efficiency	$E_n = S_n/n = T_1/(nT_n)$
U_n	n-node utilization	$U_n = P_n/(nP_{peak})$
T_o	Total overhead	$T_o = T_{par} + T_{interact}$
	Average parallelism	T_1/T_∞
	Average overhead	T_o/W
	Average granularity	W/T_o

Example 3.13   Performance metrics of the parallel APT benchmark program

A coarse-grain phase-parallel algorithm for the APT program is shown in Fig. 3.8. The parameter $N = 256$. The DP step can be distributed to up to 8192 nodes. The *total exchange* step is executed by all nodes, with an aggregated message length of 17 MB. The HT step is sequentially executed on a single node (DOP = 1).

The *broadcast* operation sends an 80-KB message to all nodes. The BF step performs beamforming using 256 nodes. The TD step needs to track the targets on 256 nodes locally. Then all local target reports are merged through the final collective *reduction* operation. The workload of each computation step is drawn from Table

3.12, shown in both Mflop and execution time on one SP2 node. Parallelism overhead is negligible, because static tasking is used on the SP2.

We predict the extreme values of performance metrics, ignoring all communication overheads. This is called *zero-overhead prediction*. From Fig. 3.8, the maximal parallelism is max(8192, 1, 256, 256) = 8192, the total workload is W = 1447 Mflop, the sequential execution time is T_1 = 14.37 seconds, and the critical path is

$$T_\infty = \frac{4.12}{8192} + \frac{0.04}{1} + \frac{9.64}{256} + \frac{0.57}{256} = 0.08 \text{ s}$$

Thus the maximum performance is $P_\infty = W/T_\infty = 1447/0.08 = 18087$ Mflop/s, and the average parallelism is $T_1/T_\infty = 14.37/0.08 = 180$.

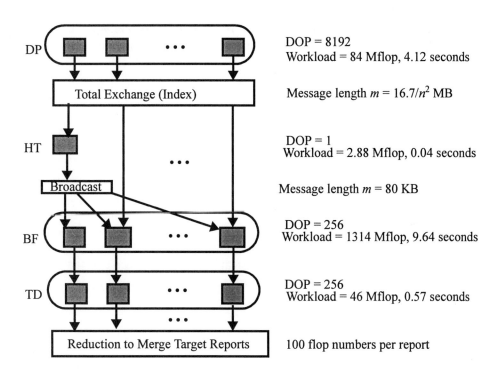

Figure 3.8   Structure of the parallel APT algorithm in STAP.

Example 3.14 Estimation of interaction overhead in the APT benchmark

We can use the overhead expressions in Table 3.18 and Table 3.19 to estimate the interaction overhead $T_{interact}$ for running the parallel APT program on SP2.

The interaction overhead is the sum of the three communications $T_{interact} = T_{index} + T_{bcast} + T_{reduce}$. From Table 3.18 and Fig. 3.8, the overhead of total exchange of 16.7 MB is

$$T_{index} = 80 \log n + 0.03 n^{1.29} m \; \mu s = 0.00008 \log n + 0.5 n^{-0.71} \quad \text{seconds} \qquad (3.10)$$

The broadcast overhead is espressed by

$$T_{bcast} = 52 \log n + (0.029 \log n) m \; \mu s = 0.00237 \log n \quad \text{seconds}$$

From Table 3.19, the time to reduce n flop numbers, one from each of n nodes, is $20 \log n + 23$ μs. In the reduction step of Fig. 3.8, we need to combine n target reports, each having 100 flop numbers. We conservatively estimate the reduction overhead to be:

$$T_{reduce} = 100 \, (20 \log n + 23) \; \mu s = 0.002 \log n + 0.0023 \quad \text{seconds} \qquad (3.11)$$

The total interaction overhead is then

$$T_o = T_{interact} = 0.5 n^{-0.71} + 0.00445 \log n + 0.0023 \qquad (3.12)$$

A popular belief in parallel processing is that communication overhead increases when more nodes are used. As the above example shows, this belief may be false. We see that in the APT program, the communication overhead has a constant component (0.0023), an increasing part ($0.00445 \log n$), and a decreasing part ($0.5 n^{-0.71}$). The net effect is that the total communication overhead *decreases* as the machine size increases, when no more than 256 nodes are used.

∎

Example 3.15 Expected execution time of the APT benchmark

We predict the execution time T_n of the parallel APT algorithm in Fig. 3.8 on an n-node SP2, where $n \le 256$. Also we compute the average granularity when 256 nodes are used. The detailed derivation is left as an exercise (Problem 3.6). The total execution time using n nodes is:

$$T_n = T_{comp} + T_{par} + T_{interact}$$

$$= \frac{14.33}{n} + 0.5n^{-0.71} + 0.00445\log n + 0.0423$$

The total workload is $W = 1447$ Mflop or 14.37 seconds on an SP2 node. The average granularity is

$$\frac{W}{T_o} = \frac{1447}{0.0479} = 30209 \qquad (3.13)$$

On the average, the 256 nodes perform 30,209 flop for each microsecond of communication, or 30209/256=118 flop per node for each μs of communication. The average overhead is $1/30209 = 33$ μs of communication for every Mflop of computation.

We can also use execution time as the workload. Then the average granularity becomes

$$\frac{W}{T_o} = \frac{14.37}{0.0479} = 300$$

Thus on the average, the 256 nodes perform 300 seconds of computation for each second of communication, or 300/256=1.17 second of computation per node for each second of communication.

■

3.5.2 Available Parallelism in Benchmarks

There is a wide range of potential parallelism in application programs. Engineering and scientific codes exhibit a high DOP due to data parallelism. Kumar (1988) has reported that computation-intensive codes may execute 500 to 3500 arithmetic operations concurrently in each clock cycle in an idealized environment.

However, instruction-level parallelism is much lower. Wall [632] indicated that limit of instruction-level parallelism is around 5, rarely exceeding 7. Bulter et al. (1991) reported that when all constraints are removed, the ILP in some scientific programs may exceed 17 instructions per cycle.

Some trace results show that, if the architecture and compiler work perfectly, one can expect an ILP of 2.0 to 5.8 instructions per cycle on a superscalar processor that is reasonably designed. Average parallelism of the PERFECT benchmark suite is given in Table 3.21 for each of the 12 programs in the suite [92]. .

Table 3.21 Average Parallelism of the PERFECT Benchmark Programs
(Courtesy: Blume et al., *IEEE Tarallel and Distributed Technology*, 1994 [92])

Program	Parallelism	Program	Parallelism	Program	Parallelism
ADM	10.8	FLO52	206.9	QCD	2.4
ARC2D	336.0	MDG	5.3	SPEC77	13.8
BDNA	139.5	MG3D	1.3	TRACK	38.7
DYFESM	17.9	OCEAN	272.4	TRFD	87.9

Example 3.16 Loop parallelism in symbolic and numerical programs

Larus [392] traced the average parallelism in six real programs, as shown in Table 3.22. Only parallel execution of loops is considered, assuming infinitely many processors are available and ignoring all parallelism and interaction overheads. The gcc, xlisp, and espresso are symbolic applications, the sgefa and dcgc are numerical applications, while the costScale is a combination of the two classes.

The Workload column shows the number of million clock cycles need to execute a program sequentially. The All column shows the average parallelism when all loop-carried dependencies need to be synchronized: an instruction must wait until its operands are all computed. The Standard column shows the average parallelism, ignoring those dependencies that can be easily eliminated by compiler optimization. The None column ignores all dependencies, which is the idealized case..

Table 3.22 Loop Parallelism in Six Programs (Courtesy: J. R. Larus,
IEEE Trans. Parallel and Distributed Systems, 1993 [392])

Program Name	Program Description	No. of Loops	Workload (M cycles)	Average Speedup		
				All	Standard	None
gcc	GNU C compiler	1328	23.3	1.3	2.3	205.6
xlisp	Lisp interpreter	204	14.9	2.5	3.0	6.4
espresso	PLA minimization	845	28.8	1.5	2.7	18.6
sgefa	Gauss elimination	154	26.6	1.1	68.7	1661.5
dcgc	Conjugate gradient	165	16.0	25.1	245.8	622.7
costScale	Network flow	135	86.8	1.3	1.7	4883.4

■

Example 3.17 Performance of three STAP benchmark programs

Table 3.23 shows some performance metrics of three programs in the STAP benchmark suite, using minimal, maximal, and nominal data sets. The input data size and the workload are given by the STAP benchmark specification [101].

The *maximum parallelism* is computed simply by finding the maximum of DOPs of individual steps. The *critical path* is the execution time when potentially infinitely many nodes are used.

For simplicity, we assume that every flop takes the same amount of time. Then each step's contribution to the critical path is its workload divided by its DOP. The *average parallelism* is the ratio of workload to critical path.

Table 3.23 Performance of Three STAP Benchmark Programs

Program	Input Data Size (MB)	Workload (Mflop)	Maximal Parallelism	Average Parallelism	Critical Path (Mflop)
APT (min)	0.23	5	1800	10	0.51
APT (norm)	18.35	1446	8192	177	8.19
APT (max)	3276.80	12,100,000	400,000	1005	12,036.05
HO (min)	0.15	21	1176	17	1.24
HO (norm)	50.33	12,852	49,152	261	49.35
HO (max)	3276.80	33,263,288	200,000	65,839	505.22
GEN (min)	0.26	6	2048	103	0.05
GEN (norm)	100	5326	196,608	108	49.27
GEN (max)	33,957	4,604,011	16,580,608	4332	1062.81

For instance, from Table 3.12, we know that the HO-PD program has three steps: Doppler Processing, Beamforming, and Target Detection. For nominal data set, these steps have a DOP of 49152, 256, and 256, respectively. The maximum parallelism there is computed to be max(49152, 256, 256) = 49152.

The critical path of the parallel HO-PD program is computed with the value: 220/49152 + (12618 + 14)/256 = 49.35 Mflop. The average parallelism is calculated by the ratio 12852/49.35 = 261.

∎

The above measures of available parallelism show that non-numeric computation has relatively little parallelism even when basic block boundaries are ignored. A *basic block* is a sequence or block of instructions that has a single entry and a single exit points.

While compiler optimization and algorithm redesign may increase the available parallelism in an application, limiting parallelism extraction to a basic block limits the potential ILP to a factor of about 2 to 5 in ordinary programs. However, the DOP may be pushed to thousands in scientific codes when multiple processors are used simultaneously to exploit parallelism beyond the boundary of basic blocks.

3.6 Scalability and Speedup Analysis

In this section, we discuss how to use the metrics developed in the previous sections to analyze and predict the scalability of parallel computer-program combinations. We introduce three performance models based on three speedup metrics. Then we discuss three approaches to scalability analysis, based on maintaining a constant efficiency, a constant speed, and a constant utilization, respectively.

3.6.1 Amdahl's Law: Fixed Problem Size

In many practical applications that demand a real-time response, the computational workload W is often fixed. As the number of processors increases in a parallel computer, the fixed workload is distributed to more processors for parallel execution. Therefore, the main objective is to produce the results as soon as possible. Consider a problem with a fixed workload W. Assume the workload W can be divided into two parts: $W = \alpha W + (1 - \alpha) W$, where α percent of W must be executed sequentially, and the remaining $1-\alpha$ percent can be executed by n nodes simultaneously. Assuming all overheads are ignored, a *fixed-load speedup* is defined by:

$$S_n = \frac{W}{\alpha W + (1 - \alpha) \ (W/n)} = \frac{n}{1 + (n - 1) \alpha} \to \frac{1}{\alpha} \quad \text{as } n \to \infty \qquad (3.14)$$

This equation is called *Amdahl's law*, which is one of the most fundamental laws in studying parallel systems. Amdahl's law has several implications:

(1) For a given workload, the maximal speedup has an upper-bound of $1/\alpha$. In other words, the sequential component of the program is a bottleneck. This α is known as the *sequential bottleneck* of a program. When α increases, the speedup decreases proportionally.

(2) To achieve good speedup, it is important to make the sequential bottleneck α as small as possible.

(3) When a problem consists of the above two portions, we should make the larger portion executed faster. In other words, one should optimize the common case.

Now let us incorporate into Amdahl's law with all the aforementioned overheads. The fixed-load speedup S_n in Eq.(3.18) becomes

$$S_n = \frac{W}{\alpha W + (1-\alpha)\ (W/n) + T_o}$$

$$= \frac{n}{1 + (n-1)\alpha + \dfrac{nT_o}{W}} \rightarrow \frac{1}{\alpha + \dfrac{T_o}{W}} \quad \text{as } n \rightarrow \infty \tag{3.15}$$

This extended Amdahl's law says that we must not only reduce the sequential bottleneck, but also increase the average granularity in order to reduce the adverse impact of the overhead. In other words, the performance of a parallel program is limited not only by the sequential bottleneck, but also by the average overhead.

Example 3.18 Upper bound on the speedup of parallel APT execution

Consider the parallel APT program in Fig. 3.8. Estimate the Amdahl upper bound on speedup by (1) ignoring all overheads; and (2) considering all overhead with the assumption that $T_o\,(\infty) \approx T_o\,(256)$.

(1) When all overhead is ignored, the workload equals $W = 14.37$ s. The sequential component is the HT step, which accounts for $\alpha = 0.04/14.37 = 0.278\%$. By Amdahl's law, the speedup has an upper-bounded by $1/\alpha = 359$. Note that the average parallelism of 178 in Table 3.23 gives a tighter upper bound than Amdahl's Law. These hard limits on speedup cannot be exceeded by increasing the number of nodes.

(2) From Eq.(3.12), $T_o(256)=0.0479$ s. The average overhead is $T_o(\infty)/W = 0.0479/14.37 = 0.00333$. By Eq.(3.15), the speedup has an upper-bound of a tighter value $1/(0.00278 + 0.00333) = 163$. ∎

The fixed-workload speedup drops as the sequential bottleneck α and the average overhead increase. The problem of a sequential bottleneck cannot be solved just by increasing the number of processors in a system. This property has imposed a very pessimistic view on parallel processing over the past two decades.

A major assumption in Amdahl's law is that the problem size (workload) is fixed and cannot scale to match the available computing power as the machine size increases. This often leads to a diminishing return when a larger system is employed to solve a small problem. The next section presents a solution to this problem.

3.6.2 Gustafson's Law: Fixed Time

The sequential bottleneck can be alleviated by removing the restriction of a fixed problem size John Gustafson (1988) proposed a *fixed-time* concept that achieves an improved speedup by scaling the problem size with the increase in machine size.

Scaling for Higher Accuracy There are many applications that emphasize accuracy more than minimum turnaround time. As the machine size is upgraded to obtain more computing power, we may want to increase the problem size in order to create a greater workload, producing more accurate solution and yet keeping the execution time unchanged.

Representative examples include the use of finite-element method to perform structural analysis or the use of finite-difference method to solve computational fluid dynamics problems in weather forecasting. Coarse grids require fewer computations, but finer grids require many more computations, yielding greater accuracy.

The weather forecasting simulation often demands the solution of four-dimensional PDEs. If one reduces the grid spacing in each physical dimension (X, Y, and Z) by a factor of 10 and increases the time steps by the same magnitude, then we are talking about an increase of 10,000 times more grid points. The workload thus increases to at least 10,000 times greater. With such a problem scaling, of course, we demand more computing power to yield the same execution time.

The main motivation lies not in saving time but in producing much more accurate weather forecasting. This problem scaling for accuracy has motivated Gustafson to develop a fixed-time speedup model. The scaled problem keeps all the increased resources busy, resulting in a better system utilization.

Fixed-Time Speedup In accuracy-critical applications, we wish to solve the largest problem possible on a larger machine with about the same execution time as for solving a smaller problem on a smaller machine. As the machine size increases, we also increase the workload. Let the original problem have a workload of W, of which α percent is sequential and $1-\alpha$ percent can be executed in parallel.

On a single node machine, the execution time is W. On an n-node machine, we scale the workload to $W' = \alpha W + (1-\alpha)nW$. The parallel execution time on n nodes is still W. However, the sequential time for executing the scaled-up workload is $W' = \alpha W + (1-\alpha)nW$. The *fixed-time speedup* with scaled workload is defined as

$$S_n' = \frac{\text{Sequential time for scaled-up workload}}{\text{Parallel time for scaled-up workload}}$$

$$= \frac{\alpha W + (1-\alpha)\,nW}{W} = \alpha + (1-\alpha)\,n \tag{3.16}$$

This equation is known as *Gustafson's law*, which states that the fixed-time speedup is a linear function of n, if the workload is scaled up to maintain a fixed execution time. When the problem can scale to match available computing power, the sequential fraction is no longer a bottleneck. Note that for Gustafson's law to hold, it is important to only scale the parallelizable portion of the workload from $(1-\alpha)W$ to $(1-\alpha)nW$. The sequential portion αW should stay the same.

Scaled Speedup Now let us incorporate into Gustafson's law all the overheads. The scaled speedup becomes

$$S_n' = \frac{\alpha W + (1-\alpha)\,nW}{W + T_o} = \frac{\alpha + (1-\alpha)\,n}{1 + T_o/W} \tag{3.17}$$

Note that the parallel program with the scaled workload $W' = \alpha W + (1-\alpha)nW$ has the same fixed *computation time* as the sequential time for the original workload W. The total execution time for the parallel program also includes the overhead T_o. This overhead T_o is a function of n.

From Example 3.14, we know that T_o could have components that are increasing, decreasing, or constant with respect to n. The generalized Gustafson's law can achieve a linear speedup by controlling the overhead as a decreasing function with respect to n. But this is often difficult to achieve.

Example 3.19 Scaled speedup without consideration of overhead in APT

Consider the parallel APT program in Fig. 3.8, where the problem parameter $N=256$. The workload can be approximated as $W = 0.011N^2 + 2.8N + 2.88$ Mflop or $W = 0.00016N^2 + 0.015N + 0.04$ s, for arbitrary N. For $N = 256$, $W = 1447$ Mflop or 14.37 seconds. Estimate the fixed-time speedup when 128 nodes are used, assuming all overheads are ignored. Also indicate how many Mflop need to be executed in the scaled workload.

When all overhead is ignored, the sequential execution time is $W = 14.37$ seconds. The essentially sequential HT step accounts for $\alpha = 0.04/14.37 = 0.278\%$ of the total workload. From Eq.(3.17), the scaled speedup is

$$S_n' = \alpha + (1-\alpha)\,n = 0.00278 + 0.99722n = 127.65$$

The sequential time for the scaled workload is $\alpha W + (1-\alpha)nW = 127.65 \times 14.37 = 1834$ seconds. In other words, the time workload has been scaled up $1834/14.37 = 127.65$ times. We can substitute this value in the time workload expression to solve for N:

$$0.00016N^2 + 0.015N + 0.04 = 1834$$

which yields an N=3339. Thus N is scaled up 3339/256=13 times. Now substituting this value into the flop workload expression, we have

$$W = 0.011N^2 + 2.8N + 2.88 = 131996$$

That is, 131,996 Mflops need to be performed in the scaled workload. The flop workload has been scaled up 131996/1447=91 times.

■

Remark on Problem Size The definition of problem size must be clarified. The above example shows three different definitions: the problem parameter N, the workload in flop, or the time in seconds. These definitions lead to noticeable differences in scaling the problems. The time workload is scaled 127 times, while the flop workload is scaled only 91 times. The problem parameter N is scaled only 13 times.

Example 3.20 Scale speedup including all overheads in parallel APT

Repeat Example 3.19, but estimate the scaled speedup on 128 nodes when considering all overheads. Assume that the message length in the total exchange step is $m = 256N^2/n$ bytes. The communication overheads for the broadcast and the reduction steps do not change with N. From Eq.(3.17), the scaled speedup is ontained as follows:

$$S_n' = \frac{\alpha + (1-\alpha)n}{1 + A_o} = \frac{0.00278 + 0.99722n}{1 + T_o/W} = \frac{127.65}{1 + T_o/14.37}$$

We need to find T_o. In the previous example, we have already found $N = 3339$ for the scaled-up problem. The message length in the total exchange step is $m = 256N^2/n$ bytes = 2854/n MB. From Eq.(3.10), the overhead for total exchange is

$$T_{index} = 80\log n + 0.03n^{1.29}m \ \mu s = 0.00008\log n + 85.62n^{-0.71} \ \text{seconds}$$

The total overhead is calculated by

$$T_o = T_{index} + T_{bcast} + T_{reduce} = 85.62n^{-0.71} + 0.00445\log n + 0.0023$$

For $n = 128$, T_o is evaluated as 2.75 seconds. The scaled average overhead is $T_o/W = 2.75/14.37 = 0.1914$. From Eq.(3.17), the scaled speedup is $127.65/1.1914 = 107$, which is less than the 127.65 speedup obtained when overheads are ignored.

■

3.6.3 Sun and Ni's Law: Memory Bounding

Xian-He Sun and Lionel Ni (1993) have developed a memory-bounded speedup model that generalizes Amdahl's law and Gustafson's law to maximize the use of both CPU and memory capacities. The idea is to solve the largest possible problem, limited by memory space. This also demands a scaled workload, providing higher speedup, greater accuracy, and better resource utilization

Memory-Bound Problems   Large-scale scientific or engineering computations often require larger memory space. In fact, many applications of parallel computers are memory-bound rather than CPU-bound or I/O-bound. This is especially true in a multicomputer system using distributed memory. The local memory attached to each node is relatively small. Therefore, each node can handle only a small subproblem.

When a large number of nodes are used collectively to solve a single large problem, the total memory capacity increases proportionally. This enables the system to solve a scaled problem through domain decomposition of the data set.

Instead of keeping the execution time fixed, one may want to use up all the increased memory by scaling the problem size further. A memory-bounded model was developed under this philosophy. The idea is to solve the largest possible problem, limited only by the available memory capacity.

Example 3.21 Memory requirements of ATP benchmark programs

Table 3.24 shows the memory requirement of three STAP benchmark programs. The entries show only the memory requirement for computation. On MPPs, twice as much memory may be required, due to the need for communication buffers.

When the maximal data sets are used, a 10-Gflop/s MPP can process the APT in 1210 seconds, the HO-PD in 3326 seconds, and the GEN in 460 seconds. The corresponding memory requirement is as much as 13 GB.

Many modern processors can each sustain about 100 Mflop/s. To satisfy the memory requirement of STAP, a 10-Gflop/s, or 100-node, MPP would need 130 MB of memory per node. Such memory capacity is present in only a few supercomputers. Most MPPs installed today have less than 128 MB per processor.

■

Table 3.24 Memory Requirements of STAP Benchmark Programs

Program	Memory Requirement MB)		Workload (Mflop)
	Single Precision	**Double Precision**	
APT (min)	0.23	0.46	5
APT (norm)	18.35	36.71	1446
APT (max)	3276.80	6553.60	12,100,000
HO (min)	0.15	0.30	21
HO (norm)	50.33	100.66	12,852
HO (max)	3276.80	6553.60	33,263,288
GEN (min)	0.26	0.52	6
GEN (norm)	100	200	5326
GEN (max)	33,957	67,914	4,604,011

Memory-Bound Speedup In a multicomputer, the total memory capacity increases linearly with the number of nodes available. Let M be the memory capacity of a single node. On an n-node MPP, the total memory capacity is nM.

Given a memory-bounded problem, assume it uses all the memory capacity M on one node and executes in W seconds (e.g., the sequential computational workload is W). As usual, the workload has an essentially sequential portion and a parallelizable portion: $W = \alpha W + (1 - \alpha) W$.

When n nodes are used, a larger problem can be solved due to the increased memory capacity nM. Let us assume that the parallel portion of the workload can be scaled up $G(n)$ times. That is, the scaled workload is $W = \alpha W + (1-\alpha)G(n)W$, The factor $G(n)$ reflects the increase in workload as the memory capacity increases n times. The *memory-bound speedup* is defined as

$$S_n^* = \frac{\text{Sequential time for scaled workload}}{\text{Parallel time for scaled workload}}$$

$$= \frac{\alpha W + (1 - \alpha) G(n) W}{\alpha W + (1 - \alpha) G(n) W/n} = \frac{\alpha + (1 - \alpha) G(n)}{\alpha + (1 - \alpha) G(n) /n} \qquad (3.18)$$

The sequential time is obtained assuming an imaginary node with nM memory capacity. When all the overhead is considered, the speedup becomes:

$$S_n^* = \frac{\alpha W + (1-\alpha) G(n) W}{\alpha W + (1-\alpha) G(n) W/n + T_o} = \frac{\alpha + (1-\alpha) G(n)}{\alpha + (1-\alpha) G(n) /n + T_o/W} \qquad (3.19)$$

There are three special cases of Eq.(3.19):

(1) $G(n) = 1$. This corresponds to the case where the problem size is fixed. Thus, the memory-bound speedup becomes equivalent to Amdahl's law; i.e., Eq.(3.19) and Eq.(3.15) are equivalent when a fixed workload is given.

(2) $G(n) = n$. This applies to the case where the workload increases n times when the memory is increased n times. Thus, Eq.(3.19) is identical to Gustafson's law, Eq.(3.17), with a fixed execution time.

(3) $G(n) > n$. This corresponds to the situation where the computational workload increases faster than the memory requirement. Thus, the memory-bound model, Eq.(3.19), gives a higher speedup than both the fixed-load speedup, Eq.(3.15), and the fixed-time speedup, Eq.(3.17).

∎

Example 3.22 Comparison of three speedup factors of the parallel APT

Assume the parallel APT program (Fig. 3.8) is executed on an n-node multicomputer, where each node has 64 MB of memory. Chart the three types of speedup over n, taking into account all communication overheads. The APT program has the following additional characteristics:

- The workload varies with the problem parameter N according to the formulas

$$W = 0.011N^2 + 2.8N + 2.88 \text{ Mflop}$$

$$W = 0.00016N^2 + 0.015N + 0.04 \text{ seconds}$$

   Nominally, $N=256$ and $W=1447$ Mflops or 14.37 seconds.
- The memory requirement is $512N^2$ bytes.
- The broadcast overhead T_{bcast} and the reduction overhead T_{reduce} do not change as the workload scales up. They only depend on the machine size n.
- The total exchange overhead T_{index} will increase with the following growth rate, as the workload scales up:

$$T_{index} = 80\log n + 0.03n^{1.29} m, \quad \text{where } m = 256 (N/n)^2 \text{ bytes} \qquad (3.20)$$

(1) *Fixed-workload speedup*:

From Example 3.18, the fixed workload in terms of time is $W = 14.37$ seconds. The sequential fraction $\alpha = 0.00278$. From Eq.(3.12), the total overhead is

$$T_o = 0.5n^{-0.71} + 0.00445\log n + 0.0023$$

From Eq.(3.15), the fixed-load speedup is

$$S_n = \cfrac{1}{\alpha + \cfrac{1-\alpha}{n} + \cfrac{T_o}{W}}$$

$$= \cfrac{1}{0.00278 + \cfrac{0.99722}{n} + \cfrac{0.5n^{-0.71} + 0.00445\log n + 0.0023}{14.37}} \qquad (3.21)$$

(2) *Fixed-time speedup*.

The original workload is still $W = 14.37$ seconds. The scaled workload is $\alpha W + (1-\alpha)nW$. The corresponding N for the scaled workload is obtained by solving the following equation:

$$0.00016N^2 + 0.015N + 0.04 = \alpha + (1-a)nW$$
$$0.00016N^2 + 0.015N + 0.04 = (0.0028 + 0.997n) \times 14.37$$

which yields the value: $N = -46.875 + \sqrt{2197 + 89562n}$
From Eq.(3.12), Eq.(3.20), and this value of N, the total overhead is

$$T_o = (0.03 \cdot 10^{-6} \cdot 256 \cdot (-47 + \sqrt{2197 + 89562n})^2) n^{-0.71}$$
$$+ 0.004\log n + 0.0023$$

From Eq.(3.17), the fixed time speedup is:

$$S_n' = \frac{\alpha + (1-\alpha)n}{1 + T_o/W} = \frac{0.00278 + 0.99722n}{1 + T_o/14.37} \qquad (3.22)$$

(3) *Fixed-memory speedup.*

There are $64n$ MB memory altogether in the n-node system. The biggest problem size N that can be accommodated by this much memory is derived from

$$512N^2 = 64n \times 10^6 \Rightarrow N = \sqrt{\frac{64n \times 10^6}{512}} = 353.55\sqrt{n} \qquad (3.23)$$

The scaled workload is

$$\begin{aligned}
W^* &= 0.00016N^2 + 0.015N + 0.04 \\
&= 20n + 5.3\sqrt{n} + 0.04 = \alpha W + (1-\alpha)G(n)W
\end{aligned}$$

which yields $G(n) = 1.4n + 0.37\sqrt{n}$, from $\alpha=0.00278$ and $W=14.37$. From Eq.(3.12), Eq.(3.20), and Eq.(3.23), the total overhead is

$$\begin{aligned}
T_o &= 0.03 \times 10^{-6} \times 256 \times \left(353.55\sqrt{n}\right)^2 n^{-0.71} + 0.00445\log n + 0.0023 \\
&\quad + 0.96n^{0.29} + 0.00445\log n + 0.0023
\end{aligned}$$

From Eq.(3.19), the fixed memory speedup is:

$$\begin{aligned}
S_n^* &= \frac{\alpha + (1-\alpha)G(n)}{\alpha + (1-\alpha)G(n)/n + T_o/W} \\
&= \frac{0.00278 + 1.4n + 0.37\sqrt{n}}{1.4 + 0.37n^{-0.5} + 0.0668n^{0.29} + 0.0003\log n}
\end{aligned} \qquad (3.24)$$

The three speedup models (Eq.(3.21), Eq.(3.22), and Eq.(3.24)) are plotted in Fig. 3.9. Since $G(n) = 1.4n + 0.37\sqrt{n} > n$, the fixed-memory speedup is better than the fixed-time and the fixed-workload speedups.   ■

Summary of Speedup Laws   If the purpose is to reduce the execution time of a fixed-workload problem, the scalability of the system can be defined as the fixed-load speedup, governed by the generalized Amdahl's law [Eq.(3.15)].

However, parallel systems are often more efficiently used if the problem size (the workload) scales up with the machine size. To achieve the best scaled speedup, memory-bound scaling should be used, which is governed by Sun and Ni's law [Eq.(3.19)].

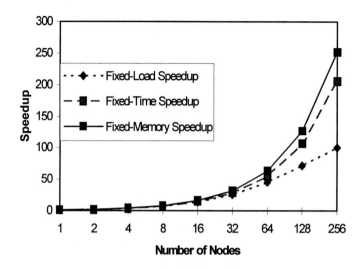

Figure 3.9   Comparison of three speedup models for the parallel APT program.

Scaling up a problem may require more execution time, even when a larger machine size is used. A compromise between fixed-load and fixed-memory scaling is Gustafson's fixed-time scaling [Eq.(3.17)], which guarantees the parallel execution time of the scaled workload will not exceed the sequential execution time of the original workload, barring any overhead.

In all cases, the achievable speedup is limited by the sequential bottleneck α and the average overhead T_o/W, where T_o is the total overhead of the parallel program for the scaled workload and W is the original workload (i.e., not scaled). The sequential bottleneck α is especially significant for fixed-load applications.

3.6.4 Isoperformance Models

Several additional metrics have been defined to characterize scalability. Three of them are discussed below. They attempt to characterize how well a parallel system scales while maintaining a fixed efficiency, per-processor speed, and utilization, respectively.

In this section, the term *system* refers to a pair comprised of an application (workload) and a parallel computer platform. It is often costly to use a large system. The three metrics allow one to first derive the scalability function on a small system and then use it to predict the performance on a large system.

Isoefficiency Grama, et al. [278] proposed an *isoefficiency* metric to characterize system scalability. The efficiency of any system can be expressed as a function of the workload W and the machine size n, i.e., $E = f(W, n)$. If we fix the efficiency to some constant (e.g., 50%) and solve the efficiency equation for workload W, the resulting function is called the *isoefficiency function* of the system.

The smaller the isoefficiency, the less the workload needs to increase relative to the increase of machine size while keeping the same efficiency, thus the better the scalability of the system.

Example 3.23 Two matrix multiplication schemes

Given two matrix multiplication schemes A and B, both have the same number of processors. The parallel execution times and the efficiencies of these systems are shown in Table 3.25. Note that the sequential time to multiply two $N \times N$ matrices is about $T_1 = cN^3$ and the efficiency is defined as $E = T_1 / (nT_n)$.

Table 3.25 Parallel Times and Efficiencies of Two Matrix Multiply Schemes

Schemes	Parallel Execution Time T_n	Efficiency E
A	$cN^3/n + bN^2/\sqrt{n}$	$E(A) = \dfrac{1}{1 + (b\sqrt{n}) / (cN)}$
B	$2cN^3/n + bN^2/(2\sqrt{n})$	$E(B) = \dfrac{1}{2 + (b\sqrt{n}) / (2cN)}$

In each of the two parallel time expressions, the first term denotes the computation time, and the second term the communication overhead. The difference between the two systems is that system B has only half the overhead of system A, but at the expense of doubling the computation time. Now which system is more scalable, assuming we want to maintain (1) $E = 1/3$ and (2) $E = 1/4$?

(1) The workload of the $N \times N$ matrix multiplication problem is approximately $W = N^3$. For system A, $E=1/3$ implies that $(b\sqrt{n}) / (cN) = 2$, thus the isoefficiency function is $W(A) = (b/(2c))^3 n^{1.5}$. Similarly, for system B, the isoefficiency function is computed to be $W(A) = (b/(2c))^3 n^{1.5}$. Thus by isoefficiency, both systems are equally scalable.

(2) For system A, $E = 1/4$ implies that $(b\sqrt{n})/(cN) = 3$, thus the isoefficiency function is $W(A) = (b/(3c))^3 n^{1.5}$. Similarly, for system B, the isoefficiency function is $W(A) = (b/(4c))^3 n^{1.5}$. Thus system B has a smaller isoefficiency and is more scalable than system A.

∎

The isoefficiency metric provides a useful tool to predict the required workload growth rate with respect to the machine size increase. For instance, the isoefficiency functions in Example 3.23 tell us that to maintain a constant efficiency, doubling the machine size requires increasing the workload by a factor of 2.82.

However, Example 3.14 indicates that we must use the isoefficiency metric with care. A more scalable system by isoefficiency does not necessarily run faster.

Isospeed Sun and Rover [594] proposed an *isospeed* concept to characterize system scalability. The basic concept is similar to isoefficiency. Instead of maintaining a constant efficiency, one can preserve a constant speed while scaling up both machine size and problem size at the same time. Interested readers may refer to their paper for details.

Isoutilization A ideal scalability metric should have the following two properties: (1) Like isoefficiency or isospeed, it predicts the required workload growth rate with respect to the increase of machine size; and (2) it is consistent with execution time, i.e., a more scalable system always has a shorter execution time. Is there a single scalable metric with these properties? The answer is yes, as shown below:

Definition 3.1 Recall from Table 3.20 that the utilization of a parallel system is defined to be $U = P_n/(n \cdot P_{peak}) = W/(n \cdot T_n \cdot P_{peak})$. If we fix the utilization to some constant (e.g., 25%) and solves the utilization equation for workload W, the resulting function is called the *isoutilization function* (or simply *isoutilization*) of the system.

∎

Like isoefficiency, the isoutilization metric can predict the workload growth rate required relative to the machine size increase, while maintaining the same utilization. Systems with a small isoutilization are more scalable than those with a large one.

Furthermore, it can be shown that under very reasonable conditions, a system with a smaller isoutilization (thus more scalable) always has a shorter execution time. The detailed derivation of this result is left as an exercise (see Problem 3.13).

Example 3.24 Isoutilization of parallel matrix multiplication schemes

Consider Example 3.23 again. The utilization of system A is

$$U = \frac{W}{nT_n(A)P_{peak}} = \frac{N^3}{n(cN^3/n + bN^2/\sqrt{n})P_{peak}}$$

$$= \frac{1/(cP_{peak})}{1 + (b\sqrt{n})/(cN)}$$

For a fixed utilization of $U = 1/(4cP_{peak})$, we have

$$b\sqrt{n}/(cN) = 3 \Rightarrow W(A) = (b/(3c))^3 \cdot n^{1.5}$$

Similarly, the utilization of system B is

$$U = \frac{W}{nT_n(B)P_{peak}} = \frac{N^3}{n(2cN^3/n + bN^2/(2\sqrt{n}))P_{peak}}$$

$$= \frac{1/(2cP_{peak})}{1 + (b\sqrt{n})/(4cN)}$$

For the same fixed utilization of $U = 1/(4cP_{peak})$, we have

$$b\sqrt{n}/(4cN) = 1 \Rightarrow W(B) = [b/(4c)]^3 \cdot n^{1.5}$$

Since $W(B) < W(A)$, we know that system B is more scalable. ■

Example 3.25 Scalability versus isoutilization

From Example 3.14, we know that system A runs faster than system B when $b = 4c$ and $N > 2\sqrt{n}$. How do we reconcile this with the conclusion of Example 3.24 that system B is more scalable? The answer is that when $b = 4c$ and $N > 2\sqrt{n}$, the utilization of system A is

$$U = \frac{1/(cP_{peak})}{1 + (b\sqrt{n})/(cN)} > \frac{1}{3cP_{peak}}$$

The same conclusion holds for system B as well. This contradicts with the fixed utilization of $U = 1/(4cP_{peak})$ in Example 3.24. Now suppose we change the assumption to a fixed utilization of $U = 1/(2.4cP_{peak})$.

For system A, we have:

$$b\sqrt{n}/(cN) = 1.4 \Rightarrow W(A) = (b/(1.4c))^3 \cdot n^{1.5}$$

For system B, we have:

$$b\sqrt{n}/(4cN) = 0.2 \Rightarrow W(B) = (b/(0.8c))^3 \cdot n^{1.5}$$

Now system A has a smaller isoutilization and is thus more scalable.

∎

This example demonstrates that a system is not necessarily always more scalable than another. It depends on the utilization level chosen. We cannot arbitrarily choose any utilization value either. The utilization of a system is always less than 100%, and often far less than that. For the two matrix multiplication systems in Example 3.23, the utilization of system A can not exceed $1/(cP_{peak})$, and that of system B can not exceed $1/(2cP_{peak})$.

3.7 Bibliographic Notes and Problems

The tutorial edited by Bhuyan and Zhang [81] covers the measurement and evaluation of multiprocessor performance. Bell [67] has characterized scalable performance from both user's and designer's perspectives. The works by Hockney and associates [313-318] contain in-depth discussions of parallel performance metrics. Karp and associates [362-364] compared various parallel Fortran dialects and indicated how to evaluate parallel application performance. The relationships between various performance metrics were discussed in by Sahni and Thanvantri [524].

Amdahl's law was first presented in 1967 [29]. Gustafson [286] presented the idea of achieving fixed execution time, by scaling up the problem size, when a larger machine is used. Gustafson and Snell recently proposed a simple benchmark called HINT that is both problem size and time scalable [287]. Sun and Ni [596] developed the memory-bound speedup model. The isoefficiency concept was developed by Kumar and associates [278, 383, 384]. The concept of isospeed was treated by Sun and associates [594-595].

Available parallelism in real programs is analyzed by Blume et al [92], Wall [632], Worley [650], and Hwang and Xu [333]. Scalable performance laws were also treated in Hwang [327], Almasi and Gottlieb [26], and Lenoski and Weber [406].

The NAS benchmark results are periodically reported by Bailey and associates [54, 525], so are the LINPACK benchmark results by Dongarra [209]. Parallel system performance is also reported by Lenoski, et al. [405, 406], Agarwal, et al. [12], Stenstrom et al. [585], Wang and Hwang [635], Athas and Seitz [45].

The STAP examples presented are based on the work of Xu and Hwang reported in [655, 656, 657]. We have developed quantitative methods to estimate the overheads in communication and parallelism management. These results can be applied in early prediction of parallel systems performance. Various performance analysis and evaluation tools are also available in [189, 301, 460, 514, 517, 564].

Homework Problems

Problem 3.1 Use the LINPACK benchmark to test the performance of a computer that is available to you. Whether the machine is parallel or not is not important. Of course, measuring the parallel performance will be more fun.

(a)   Find its R_{max}, N_{max}, $N_{1/2}$, R_{peak} values.

(b)   What is the system utilization?

(c)   What is the peak performance/cost ratio?

(d)   What is the sustained performance/cost ratio?

Problem 3.2 Use the STREAM benchmark to test your favorite computer, and find its TRIAD bandwidth and machine balance. Compare the TRIAD bandwidth obtained to the computer's peak hardware memory bandwidth.

Problem 3.3 There is currently no micro benchmark suite that is comprehensive enough for measuring how well a computer supports computing (MIPS and Mflop/s), data movement (memory and disk, MB/s), system calls, process management, communication and synchronization, etc.

(a)   A very useful and challenging class project is to develop such a suite of micro benchmarks, integrate them into a portable package, and put it in public domain on the WWW.

(b)   Compare the class project benchmark with the public-domain benchmarks: LINPACK, STREAM, LMBENCH, and PARKBENCH.

Problem 3.4 Following the style of Example 3.11, write an SPMD program to realize the ping-pong scheme for measuring the point-to-point communication latency.

Problem 3.5 Propose a method to measure collective communication (e.g., a total exchange) time, which is different from those methods discussed in Section 3.4.3. Outline the new total exchange algorithm and identify the improvement made, from the viewpoints of time complexity and implementation efficiency.

Problem 3.6 Derive the execution time T_n of the parallel APT algorithm in Fig. 3.8 on an n-node SP2, where $n \leq 256$, which should be

$$T_n = T_{comp} + T_{par} + T_{interact}$$
$$= \frac{14.33}{n} + 0.5n^{-0.71} + 0.00445\log n + 0.0423$$

The total workload is $W = 1447$ Mflop or 14.37 s on an SP2 node. In Fig. 3.8, the number of bytes communicated in the reduction step of the parallel APT program is about 0.1 MB.

Problem 3.7 Use Eq.(1.4) to derive the 50% rules for balanced design in Section 1.5.2.

(a)   For each of the load imbalance, start-up time, communication bandwidth, and parallelism overhead factors, derive an inequality.

(b)   Derive Table 1.9 and Table 1.10.

Problem 3.8 Use Example 3.16 to answer the following questions:

(a)   Symbolic programs often have smaller parallelism than numeric ones. Give two reasons why the gcc program has a smaller average parallelism than that of the sgefa program.

(b)   Suggest ways to improve the performance of parallel computers to run symbolic programs more efficiently. You can elaborate on all fronts, such as architectectural support, program restructuring, compiler optimization, etc.

Problem 3.9 The sequential time to multiply two $N \times N$ matrices is about $T_1 = cN^3$ seconds, where c is a constant. A parallel matrix multiplication scheme has a parallel execution time of $T_n = cN^3/n + bN^2/\sqrt{n}$ seconds on an n-node computer, where b is another constant. The first term denotes the computation time, and the second term the communication overhead.

(a)   Find the fixed-workload speedup and comment on the results.

(b)   Find the fixed-time speedup and comment on the results.

(c)   Find the memory-bound speedup and comment on the results.

Problem 3.10   Compare the execution times of the following application on two computers, named A and B. The application is Doppler processing consisting of 2,048 independent 8192-point FFTs (*Fast Fourier Transforms*). Suppose the user measures 50-s execution time on machine A and 30 s on machine B. Ignore all overhead times.

(a)   What is the speed of machine B in terms the number of FFTs performed per minute? How much faster is B over A?

(b)   Now suppose the user also needs to perform image rendering. The user runs the rendering code on both machines and sees that machine A needs 6 minutes to render an image frame, while machine B needs 10 minutes. The typical workload mix is 40 Doppler processing applications for each image rendering. What are the executuin times on the two machines? Which machine is faster and how much faster?

(c)   Repeat part (b), if the workload mix is changed to 10 Doppler processing applications for each image rendering.

Problem 3.11   Consider a program with a unit workload $W = 1$ and a zero sequential bottleneck ($\alpha = 0$). Estimate the fixed-time speedup factor, S_n' in Eq.(3.17), for each of the following overhead assumptions:

(a)   What is the big-O expression of the scaled speedup for overhead $T_o = O(n^{-0.5})$? Can you get better than linear speedup?

(b)   What are the possible reasons of having a superlinear speedup in real-life, parallel computing application?

(c)   Repeat Part (a) for overhead $T_o = O(1)$ and comment on the result.

(d)   Repeat Part(a) for overhead $T_o = O(\log n)$ and interpret your result. Is it possible to achieve a linear speedup, if the overhead is more than a constant.

Problem 3.12   Use Example 3.22 to answer the following questions:

(a)   Assuming a fixed efficiency of 1/5, derive the isoefficiency function of the parallel APT benchmark program running on SP2.

(b)   Assuming a fixed utilization of 1/10, derive the isoutilization function of the parallel APT benchmark program running on SP2.

Problem 3.13   Given two systems A and B, assume their isoutilization functions are $W(A)$ and $W(B)$, respectively. Derive the necessary and sufficient conditions such that $W(A) < W(B) \Rightarrow T_n(A) < T_n(B)$. That is, a more scalable system is faster.

Problem 3.14   This problem is based on the work of Sahni and Thanvantri [524]. Consider the two matrix multiplication schemes in Example 3.23. Assuming a fixed efficiency of $E = 0.25$, system B is more scalable than system A by the isoefficiency metric, for all values of $b, c > 0$. Assuming $b = 4c$, we have $W(A) = 1.33^3 \, n^{1.5}$ and $W(B) = n^{1.5}$.

(a)   How much more scalable is B over A under the above conditions?

(b)   Look the parallel execution times in Table 3.25. Derive an expression for the ratio $T_n(B) / T_n(A)$, when $N > 2\sqrt{n}$.

(c)   Can system B be slower than system A? Explain why or why not.

Part II

Enabling
Technologies

II.1 Objectives

This part consists of four chapters. Chapters 4, 5, and 6 are devoted to processor, memory, and interconnect technologies. Chapter 7 is dedicated to software support for parallelism, interactions and communication.

Chapter 4 This chapter introduces the architecture, technology, and applications of commodity microprocessors. We start with a review of hardware and software development trends. Then we assess the CISC, RISC, post RISC and possible future architectures.

We study superscalar, superpipelining, decoupled CISC/RISC, VLIW, and multimedia extensions of existing microprocessors and embedded processors. We will also check the multiprocessing support in using these microprocessors as building blocks in large systems.

Chapter 5 This chapter presents mainly distributed-memory architectures. We start with a review of hierarchical memory technology. Then we study cache coherency protocols, including the MESI snoopy and directory-based protocols. Shared-memory consistency models are introduced.

We cover the UMA, NORMA, CC-NUMA, COMA, and DSM models. Latency-tolerating techniques are presented, along with benchmark evaluations of distributed caches, data prefetching, relaxed memory consistency, and the use of multithreaded processors to hide long memory latency. Both hardware support and software implementations of DSM are presented, leaving details in case studies in Chapters 8 and 10.

Chapter 6 Gigabit networks and switched interconnects are covered in this chapter. We start with basics of interconnection networks, including network types, topologies, functional properties and performance metrics. Besides point-to-point interconnects, we also study multiprocessor busses, and crossbar and multistage switches.

High-speed networks covered include the Gigabit Ethernet, Fibre Channel, FDDI rings, Myrinet, HiPPI, SCI, and ATM technologies. Examples of using some Gigabit networks are given. In particular, we examine the details of Gigabit Ethernet, IEEE SCI Standard, and ATM switches and internetworking. Quantitative comparisons of various interconnects, networks, and switches are given.

Chapter 7 This chapter gives the readers a necessary understanding of the concepts and software support of threads, synchronization, and efficient communication on a multiprocessor or on a multicomputer.

We explain the threads in Solaris, synchronization mechanisms, PVM, MPI, BCL, MPL in SP2, TCP/IP, and UDP protocols, the pipes, and socket interface for fast communications mainly in Unix environments.

II.2 Notes to Readers

Chapters 4, 5, and 6 deal with mainly hardware technologies, except for the software implementations of DSM in Section 5.4. These three chapters must be read by computer engineers interested in architectural design and hardware organizations. Software-oriented readers may skip these chapters at the first reading.

Computer science students may want to read Chapters 5 to 6 selectively, if they lack up-to-date knowledge of hardware technology. Chapter 7 is a must for software engineers or computer science students. It should be read before Chapters 9, 12, 13, and 14. All chapters in Part II are needed to study the architecture chapters in Part III.

Chapter 4

Microprocessors as Building Blocks

Commodity microprocessors are being used to build scalable and cluster computer systems. This chapter reviews microprocessor architectures and system development trends. We start with recent advances in hardware, software, and applications of parallel processors and computer clusters. Then we proceed with the underlying VLSI technology, microarchitecture design, and graphics and multimedia extensions.

We are particularly interested in performance features, such as new mechanisms and architectural support for exploiting instruction-level parallelism, out-of-order dynamic and speculative executions, memory latency reduction, multiprocessing support, post-RISC features, and architectural trends in future microprocessors.

4.1 System Development Trends

High-performance computers evolve toward general-purpose, scalable systems with a common architecture. In this section, we examine the development trends of hardware, software, and applications in these systems. The main usage of a computer system includes the processing, storage, and transmission of information. Identified below are several shortcomings in modern computer technology.

- *The data transfer speed lags behind the processing capability.* That is, the computer is very fast in processor speed, but it is slow in moving information between processor and memory, disk, and network.

- *The software lags behind the hardware.* Software cannot allow users to fully utilize the hardware capability. Majority of system inefficiency is due to software inefficiency. Furthermore, both system and application software are always one generation behind hardware.

4.1.1 Advances in Hardware

Processors Of the components in a parallel computer system, the processor has enjoyed the most rapid technological advances. This is illustrated in Fig. 4.1 for the Intel 80x86 microprocessor family, which shows exponential performance improvement over time. In just over 17 years, the *density* (i.e., the number of transistors per microprocessor) has increased almost 200 times, the clock rate 31 times, and the peak speed 900 times!

These numbers reflect an average annual improvement rate of 36% for density, 22% for clock rate, and 49% for the Mi/s speed. The bulk of speed improvement comes from advances in processor architecture, instead of that of clock rate. We will discuss architectural advances of modern microprocessors in Section 4.3.

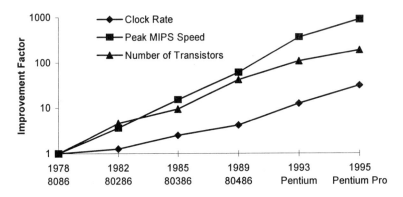

Figure 4.1   Improvement trend of Intel microprocessors.

Moore's Law The phenomenon of exponential improvement was observed as early as 1979 by Intel cofounder Gordon Moore. Moore's law is interpreted in three ways:

- The number of transistors on a microchip doubles about every 18-24 months, assuming the price of the chip stays the same.
- The speed of a microprocessor doubles about every 18-24 months, assuming the price of the processor stays the same.
- The price of a microchip drops about 48% every 18-24 months, assuming the the same processor speed or on-chip memory capacity).

Advances in Memory Memory is as important a component as the processor. Unfortunately, advances in memory technology have not kept pace with those of the processor technology, as illustrated in Fig. 4.2 for the IBM PC family. Compared to the PC/XT, the Pentium Pro PC has improved the processor speed by more than 900 times, the main memory capacity by 64 times, the hard disk capacity by 85 times, but the memory access time has improved only by less than 10 times.

Furthermore, the curves show that while the other three parameters still improve at an exponential rate, the main memory access time improves at a linear rate ever since the 386-based PC.

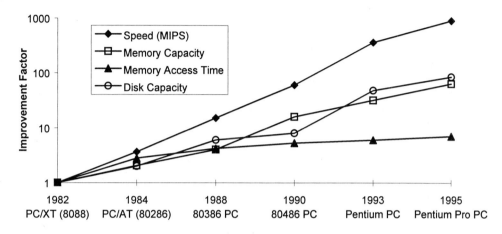

Figure 4.2   Hardware advances in IBM PC generations.

The RAM access time improves much more slowly than the processor speed. But some systems cannot even attain the already slow memory access time. For example, on the DEC 8000 server, the memory chips have an access time of 60 ns. But the *memory latency*, which is the time for the processor to read a word from the main memory, is a much higher 250 ns. In other systems, the situation is even worse. The memory latency is as high as 600 ns, using 60-80 ns as labeled on RAM chips.

Disks and Tape Units The main memory made of semiconductor chips is the *primary* storage device in a computer system. The *secondary* and the *tertiary* storage devices are made of mostly magnetic disks and tapes, respectively. They are also know as *mass storage* devices.

Because mass storage devices involve mechanical movement, their speed is much slower than that of memory. Currently, the access time and the bandwidth of memory are 2 to 3 orders of magnitude better than those for disks, and 3 to 7 orders of magnitude better than those of tape units. Furthermore, the speed of RAMs improves at faster rates than those of disk drives and tape units. The disk/tape units are often the bottleneck for I/O-centric applications.

Communications Subsystem The communication performance of Intel MPPs is compared with the processor speed in Fig. 4.3. Here the processor speed is the peak Mflop/s speed of a single processor used in Intel MPPs. The bandwidth and the start-up latency are the best numbers reported for sending a point-to-point message.

Three observations can be made from Fig. 4.3. First, all three performance metrics improve at exponential rates. Second, of the two communication performance metrics, the bandwidth improves faster than the start-up latency. Third, the processor speed (i.e., computation capability) improves at a much higher rate than the communication capability. In a decade to 1996, the processor speed of Intel MPPs has increased 5000 times, and the bandwidth 760 times, but the start-up latency has improved only 86 times.

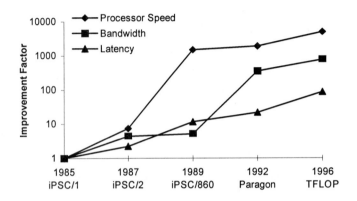

Figure 4.3   Comparison of computing and communication capabilities of Intel MPP family.

Example 4.1 Evolution of Cray and Intel Supercomputers

The evolutional trends are illustrated in Fig. 4.4 of two supercomputer families: one uses proprietary processors and the other uses Intel microprocessors. .

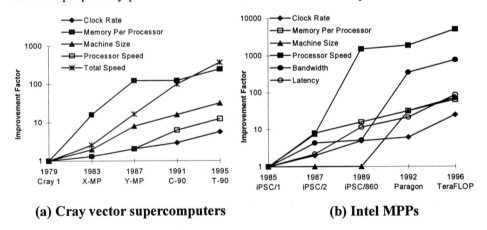

(a) Cray vector supercomputers   **(b) Intel MPPs**

Figure 4.4   Improvement trends in Cray supercomputers and Intel massively parallel processors.

Table 4.1 summarizes all the data points . The peak speed of Cray processors has improved 12.5 times in 16 years, mainly from faster clock rate. In 10 years, the Intel microprocessors has increased in speed 5000 times, of which only 25 times come from faster clock rate; the remaining 200 come from advances in the microarchitecture.

During the same period, the one-way, point-to-point communication bandwidth for Intel MPPs has increased 760 times, and the latency has improved by 86 times. Cray supercomputers use fast SRAMs as the main memory, and a custom-designed crossbar provides high bandwidth and low communication latency. As a consequence, applications running on Cray supercomputers often have higher utilizations (15% to 45%) than those (1% to 30%) on other MPPs.

Table 4.1 Evolution of Cray Supercomputer and Intel MPP Families

Vendor	Computer	Year	Clock (MHz)	Memory (MB)	Speed (Mflop/s)	Bandwidth (MB/s)	Start-up (μs)
Cray	Cray 1	1979	80	1	160	Unknown	Unknown
	X-MP	1983	105	2	210	Unknown	Unknown
	Y-MP	1987	166	256	333	Unknown	Unknown
	C90	1991	238	256	1000	9444	0.1
	T90	1995	454	512	2000	Unknown	Unknown
Intel	iPSC/1	1985	8	0.5-4.5	0.04	0.5	862
	iPSC/2	1987	16	4-16	0.3	2.2	390
	iPSC/860	1989	40	8	60	2.6	75
	Paragon	1992	50	16-128	75	175	40
	TFLOPS	1996	200	16-128	200	380	10

4.1.2 Advances in Software

Compared to the state of sequential software, the development of parallel software is in a sorry state. Parallel machines are difficult to program, since the users are faced with many new issues, discussed in Chapter 2. Furthermore, there is a lack of good software tools to aid parallel programming. Many tools become obsolete soon after their implementation, due to rapid changes in parallel system architecture. As a consequence, most parallel applications are developed using sequential Fortran or C, plus some library functions or compiler directives for process management and interaction.

In the general case, the term *software tool* (or simply *tool*) refers to any software utility that a system provides to the end user, such as

- The operating system utilities
- The application subsystems and *middleware* (e.g., database and transaction processing monitor)
- Programming languages and compilers
- The libraries (e.g., for basic computing, communication, and multithreading)
- The profiling, debugging, and visualization utilities

The parallel software crisis is becoming more acute, with high-performance computers move toward DSMs, MPPs, and clusters. However, several advances are visible, such as the following:

- We now know much more what is needed of parallel software. Key problems and issues have been identified, and effective solutions are emerging.
- A consensus is being reached that we should develop architecture-independent tools that are portable among different platforms and scalable over several machine generations. This independence principle should be maintained, even at the expense of some performance degradation.
- We should develop open, standard tools, preferably in public domain. Examples of such tools include Pthread for multithreading, HPF for data-parallel programming, MPI and PVM for message passing.
- System and application software vendors are developing parallel versions of their products. For instance, all major database vendors (IBM, Oracle, Sybase, and Informix) are now providing parallel databases.

4.1.3 Advances in Applications

In the application area, parallel software again lags far behind. There are few parallel programs developed, especially for commercial applications. Developing a parallel application is a time-consuming task. Therefore, performance, portability, and scalability must be considered during program development. We mention several notable advances and lessons learned:

- *Parallel Paradigms*: A few recurring paradigms have been identified for structuring parallel algorithms, such as *compute-interact, synchronous iteration, asynchronous iteration, pipelining, divide-and-conquer, process farm* (master/slave), and *process pool*. These paradigms were discussed in Section 2.1.2.
- *Parallel Algorithms*: The past decade has seen an explosive accumulation of parallel algorithms. Many new parallel algorithms are discovered every day.

Unfortunately, the majority of such parallel algorithms are based on the unrealistic PRAM model or its variants, which omit important "detail." To be useful for real parallel applications development, these algorithms must migrate to the more practical BSP model or the phase model in Section 2.2.2.

- *Parallel Standards*: The applications should be coded using high-level, standard languages (e.g., Fortran or C), plus standard library functions (e.g., MPI or PVM) or compiler directives (e.g., HPF).

- *Independency*: An application, once developed, should be able to execute efficiently on different machine sizes over different platforms, with little modification. The code should be independent of any specific architectural features, such as the topology of the interconnect network.

- *Coarse Granularity*: Parallel applications should exploit coarse-grain parallelism. As illustrated in Fig. 4.3, the communication overhead of parallel computers has been improving at a much slower rate than the processing speed. This trend is likely to continue. A coarse-grain parallel program has better scalability over the current generation of parallel computers.

The motivation for building scalable computers is the existence of applications of scientific, societal, and economic importance that need them. Accompanying the rapid advance of computer systems is the even faster expansion of available applications. Thanks to personal computers and the Internet, the computer resources are now accessible to ordinary people, not just experts. We discuss below several important classes of applications that need scalable computers.

Technical Applications Many scientific and engineering fields can benefit from scalable computers, especially high-performance computers. Before computers, science was studied by the theoretical and the experimental methods. Computers provide a third method, which has generated new scientific fields, such as computational biology, computational physics, and computational chemistry. The importance of computer is summarized best by the Nobel Laureate Kenneth Wilson, who stated that the supercomputer is the most important tool invented since Galileo's telescope.

We list below representative scientific and engineering applications. A common feature of such applications is that they have an insatiable appetite for computing power, especially floating-point capability. Scalable computers are always needed for solving existing problems in shorter time and for solving larger problems.

- *Biology*: drug design, protein folding, genome informatics, human body simulation, virtual human, etc.
- *Chemistry*: study of catalysts and enzymes, chemical plants simulation, etc.
- *Physics*: study of new materials, semiconductor simulation, fusion system design, study of fundamental particles, etc.

- *Astronomy*: cosmology, galaxies' formation, dynamics of sun, black holes collision, astronomy data analysis, etc.
- *Climate and Weather*: weather forecasting, climate prediction, etc.
- *Environment*: pollutant flow modeling, air and water quality modeling, ecological systems, earth observation system, etc.
- *Geophysics*: seismic data analysis, reservoir modeling, earthquake prediction, etc.
- *Engineering*: design of vehicles, buildings, and bridges; chip simulation; engine design; simulation of high-frequency circuits; etc.
- *Image and Signal Processing*: space-time adaptive processing, synthetic aperture radar, image rendering and analysis, virtual reality, etc.

Business Applications As extensive as they seem to be, scientific/engineering computing applications are only a small part of the computer applications market. A far larger portion consists of commercial applications. Such applications may not need computing power as high as that for scientific/engineering applications, but they often require large memory and disk space and a high I/O rate. They also require the computer system to have high *RAS* (reliability, availability, and serviceability). Typical commercial applications include the following:

- Database management and query
- On-line transaction processing
- Data warehousing
- Decision support system
- Data mining in manufacturing and retailing

Network Applications A recent application trend is *network-centric computing*, where an application runs on many computers connected through a network. The network could be a small LAN (local area network) or the giant Internet. Main requirements include efficient communication, coordination, interoperability, and security. Representative applications in this area include

- *World Wide Web* (WWW) services
- Multimedia processing
- Video on demand
- Electronic commerce
- Digital libraries for specific domains
- Distance learning and collaborations
- Telemedicine and health care

Throughput Processing A misconception is that only specially developed parallel applications run on parallel computers. The reality is that many users treat a parallel computer as a huge workstation with a large memory and a powerful processor. The parallel computer simultaneously executes multiple sequential programs from different users. This is known as *throughput processing*. Computer systems with a single address space (e.g., PVPs, SMPs, and DSMs) have an edge over multiple-address-space machines (e.g., clusters and message-passing MPPs).

Application Characteristics Apart from the existence of thousands of sequential application software packages, mature, portable, and efficient parallel applications appeared in only a few suites. This is due to the fact that programming parallel machines is difficult. Another excuse is the shortage of parallel software for specific applications. This is explained by the four-dimensional application space depicted in Fig. 4.5.

	Regular	Irregular	Regular	
High interaction overhead	1	2	3	4	Large grain
	5	6	7	8	Small grain
Low iteration overhead	9	10	11	12	
	13	14	15	16	Large grain
	Low parallelism	High parallelism		

Figure 4.5   Application domains divided by program characteristics.
(Courtesy of J. Worlton [651] in *Supercomputers: Directions in Technology and Its Applications*, National Academy Press, 1989.)

The application space is subdivided according to *regularity in algorithm, degree of parallelism, computational granularity,* and *interaction overhead*. These four characteristics are not totally independent. Each square represents one particular application domain.

Most of today's parallel computers are in favor of supporting applications falling in square 16, where the application has a high degree of parallelism, coarse granularity, low interaction overhead, and a regular algorithmic structure. Various computers may perform differently in different application domains.

For example, many users wish to explore square 11 or 12, where fine-grain parallelism can take advantage of massive parallelism embedded in some applications. Generally speaking, applications that have high and regularly structured parallelism with low communication overhead will be more desirable, regardless of the granularity. To explore a particular application domain is a very challenging task. It may require one to improve both computer architecture and its programming environment.

4.2   Principles of Processor Design

This section presents design principles of high-performance processors. First, we introduce a few basic processor terms. Then we review processor architectures, namely, the *CISC, RISC, superscalar, superpipelined, decoupled CISC/RISC, VLIW,* and *multimedia extensions.* Commodity microprocessor families are introduced in Section 4.3 with comparison of their performance features. Pentium Pro and Alpha 21164 are presented as case studies in Section 4.4.

4.2.1 Basics of Instruction Pipeline

The execution of a typical instruction consists of essentially four phases: *fetch, decode, execute,* and *write-back.* These execution phases are often carried out by a pipelined processor with multiple stages, called the *instruction pipeline.* The pipeline executes sequential instructions much as an industrial assembly line. Successive instructions are issued to the pipeline in program order. Out-of-sequence executions will take place when branches or interrupts are encountered. The basic idea of pipelining is to overlap the execution of successive instructions to save time and increase the throughput.

Summarized below are basic definitions associated with instruction pipeline design, operations, and evaluation:

(1) *Pipeline cycle* **or** *processor cycle*: The clock period driving the pipeline operations, often equal to the maximum time delay to complete a single-stage operation in the pipeline.

(2) *Instruction issue latency*: The number of processor cycles between the issue of two adjacent instructions in a program sequence.

(3) *CPI* **(cycles per instruction)**: The number of cycles needed to execute a given instruction through the pipeline. This quantity is instruction-dependent, thus the *average CPI* is often used for a program mix.

(4) *Instruction issue rate*: The number of instructions issued per cycle, also called the *multiplicity* of a superscalar processor.

(5) *Simple operations*: Simple operations require only one cycle to execute. They Majority RISC instructions are simple, such as *integer add, move, branch,* etc.

(6) *Complex operations*: A complex instruction requires multiple cycles to execute. Examples include *divide, logarithm, memory references,* etc.

(7) *Resource conflicts*: This refers to the situation where two or more instructions demand the use of the same pipeline stage at the same time.

Basic Pipelining Concept   A single-issue processor is defined with a baseline pipeline with four stages in Fig. 4.6, with one instruction issued per cycle and one cycle between adjacent issues. The figure illustrates the ideal case of in-sequence execution of successive instructions. *In-sequence* refers to the omission of the effects of branches and interrupts in the pipeline. Once the pipeline is filled up, each instruction requires only one cycle to emerge from the pipeline, effectively.

When branches or interrupts do occur in the program sequence, the pipeline must be drained in order to fetch the next instruction from the target location. This *out-of-sequence* execution may lead to heavy penalties paid with each branch or interrupt. *Branch prediction* is often practiced to reduce or to eliminate the out-of-sequence penalties. We shall further study these features in the next section.

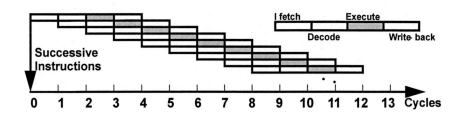

Figure 4.6   In-sequence pipelined instruction execution.

What Is a Cycle?   Sometimes the workload is expressed in terms of the number of processor clock cycles. The executable part of a high-level language program is comprised of a sequence of statements. Each statement is translated into a sequence of instructions in a binary code. An instruction is executed by a processor as a sequence of basic steps, called *machine cycles* or *pipeline cycles*.

This sequence of machine cycles executed for an instruction is called an *instruction cycle*. Each machine cycle consists of one or more processor *clock cycles* (also known as the processor *clock periods*, or simply *clocks*), which are the reciprocal of the processor clock rate. For a microprocessor, a machine cycle that involves only activities inside the processor is called an *internal cycle*, while one that needs to go outside the processor is called a *bus cycle*, since the processor bus needs to be activated.

A processor can easily have hundreds (or more) of different instruction cycles. This is because different instructions have different instruction cycles, and many processors have more than 100 instructions. Furthermore, the same instruction, when executed with different CPU states, results in different instruction cycles.

It is hard to efficiently implement hundreds of different instruction cycles directly. A fundamental technique in processor design is to define a small set (a dozen or so) of machine cycles, which can be used to realize all instruction cycles. Each machine cycle normally consists of a fixed number of clock cycles and a fixed set of control signals to be asserted at each clock cycle.

Processor Cycle versus Bus Cycle   In a computer system, different subsystems are often driven at different clock rates. For instance, an IBM PC could use a processor with a 250-MHz clock rate and only 66 MHz for the system bus. The processor clock cycle is 1000/250 = 4 ns, while the bus clock cycle is 15 ns.

Example 4.2 Instruction pipeline and cycle times of Pentium

In the Intel Pentium processor, integer instructions are executed by a five-stage instruction pipeline as shown in Fig. 4.7. Each stage executes a machine cycle. Under ideal circumstances, each pipeline stage incurs only one clock delay. Thus an integer *instruction cycle* can be as short as five pipeline clocks.

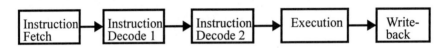

Figure 4.7   Instruction pipeline in Intel Pentium processor.

If the next instruction is in the cache, the Instruction Fetch stage executes an internal machine cycle which fetches the instruction from the cache. Such a machine cycle needs only one clock. However, if the instruction is not in the cache, the Instruction Fetch stage will execute a bus cycle to load a cache line from the memory.

This Cache-Fill bus cycle could take 5 or more bus clocks, depending on how fast the memory is. The number 5 comes from the following facts: In Pentium, there are 32 bytes in a cache line, and the data bus is 64-bit (or 8-byte) wide. In the ideal case, the Cache-Fill bus cycle spends one bus clock to send the address and other signals to the memory, and four additional bus clocks to transfer the $4 \times 8 = 32$ bytes in a cache line.

■

Example 4.3 Instruction execution on a Pentium processor

Suppose the following instruction is executed on a Pentium processor to load a data word at virtual memory location M to a working register EAX: MOV EAX, M

Assume that the instruction itself is in the I-cache, but the data word to be loaded is not in the D-cache. Then the instruction cycle consists of the following sequence of five cycles through the pipeline in Fig. 4.7.

(1) Fetch instruction from the I-cache.
(2) Perform stage 1 instruction decode.
(3) Perform stage 2 instruction decode and address translation.

(4) Fetch one cache line from the D-cache.

(5) Load the data word into register EAX.

■

Instruction Execution Ordering When the instructions are always executed by program order, it is called *in-order execution*. Early processors were all designed to execute instructions in program order. This sequential ordering often causes the pipeline to stall when an instruction cannot proceed with data or control dependencies with other instructions. In-order execution often leads to a rather poor processor efficiency if the stalling problem prevails.

Compiler techniques or special reordering mechanisms can be used to reorder the program sequence to eliminate unnecessary pipeline stalling. Such an program execution style is called *out-of-order execution*. This enables *dynamic program execution*, along with *branch prediction* and *speculative execution*. Program reordering can also eliminate unnecessary memory waits. These features will be examined later with case studies of leading microprocessors.

Pipeline Design Parameters We often refer to a processor as 16-b, 32-b, or 64-b processors. The meaning is illustrated in Fig. 4.8, where A is the width of the *address bus*, B is the width of *data bus,* C is the width of the *data path* between memory and cache, D and E are the widths of the *data path* between cache and processor (registers), D and E are also the *register widths* of integer and floating-point registers, respectively, and F and G are the *execution widths* of integer and floating-point ALUs, respectively.

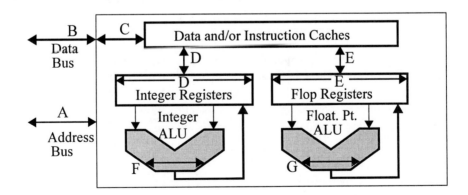

Figure 4.8   Multitude of paths and word lengths in processor design.

The magnitudes of A and B are dictated by the pin count on the processor chip as well as the width of the system bus attached to the processor. The parameters C, D, and E decide the width of the pipelined datapath. The parameter F and G decide the width of the execution units. Often the magnitudes of all of these parameters are decided jointly to maximize the clock rate and Mi/s rate, to be defined shortly.

Processor and Coprocessor   The central processor of a computer is called the *central processing unit* (CPU). We are interested only in CPUs on a single VLSI chip. A CPU is essentially an instruction processor, consisting of multiple functional units such as registers files, caches, *arithmetic and logic unit* (ALU), floating-point unit, processor control unit, system bus interfaces, etc.

One or more *coprocessors* can be attached to the CPU as shown in Fig. 4.9. The coprocessor executes special instructions dispatched from the CPU. A coprocessor may be a *floating-point accelerator*, a *vector processor* executing vector operands, a *digital signal processor* (DSP), or a *media coprocessor* executing multimedia information. Coprocessors cannot be used alone. They cannot handle regular instructions or perform I/O operations, which are performed by the CPU (processor).

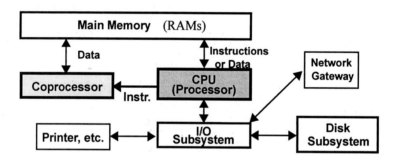

Figure 4.9   Processor and coprocessor relationship.

The CPI and MIPS Rate   The CPI indicates the number of cycles needed to execute a single instruction. Therefore, the CPI varies with respect to different instruction types. A simple operation requires mostly one cycle to execute, thus has a CPI equal to 1. A complex instruction often requires more cycles to execute, with a CPI that can easily exceed 20 processor cycles.

The *speed* or *instruction execution rate* of a processor is often measured by the MIPS (*million instructions per second*) rate, which grows linearly with the clock rate. The MIPS rate decreases directly with the *average CPI* needed to execute a program on a given processor. Consider a processor with a clock rate f MHz and an average CPI. The instruction execution rate is calculated by

$$Instruction\ execution\ rate = f\ /\ \text{CPI} \qquad (\text{in MIPS or Mi/s}) \qquad (4.1)$$

Suppose that f = 120 MHz and average CPI = 0.6. The processor in question achieved a speed of 120/0.6 = 200 MIPS. Note that the CPI value must be averaged over all instruction types and weighted with the frequencies of different instruction types in a given program mix. In Europe, the MIPS is also expressed as Mi/s in notation. In this book, we use these two notations interchangeably.

4.2.2 From CISC to RISC and Beyond

In the past 20 years, the transition from CISC (*complex instruction set computing*) to RISC (*reduced instruction set computing*) marked a radical change in ISA. We characterize below the distinctions and similarities of these two architectures. The ISA of a computer specifies the primitive commands directly executed by hardware. In other words, the ISA specifies the executable machine instructions by a given processor type.

The complexity of an ISA is attributed to instruction formats, data formats, addressing modes, general-purpose registers, opcode specifications, and flow control mechanisms used. Based on different choices in these features, CISC and RISC represent two schools of thought on ISA. At present, there are about 50 different *architectures* originated by major companies and hundreds of *implementations* manufactured by licensees of those architectures.

Architectural Distinctions Basic hardware architectures of RISC and CISC processors are compared in Fig. 4.10 This diagram shows the architectural distinctions between a pure RISC and classic CISC processors.

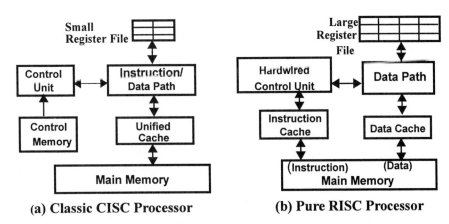

(a) Classic CISC Processor   (b) Pure RISC Processor

Figure 4.10   Architectural distinctions in RISC and CISC processors.

Early CISC architecture used a *unified cache* to hold both instructions and data. They must share the same data/instruction path and use a small register file. Recent CISC processors like the M68030 and 68040 use split caches for data and instructions. Microprogrammed control was used to implement the large instruction set in a CISC processor. But recent processors all switched to hardwired control.

In a RISC processor, split *instruction cache* and *data cache* use different access paths. A much larger register file is often used. RISC must use hardwired control for on-chip implementation. This reduces the CPI effectively to one cycle if the pipeline runs smoothly without excessive branches or interrupts.

The CISC Roadmap   Early computers started with a rather simple and small instruction set. But the complexity of the instruction set increased sharply over the years. This resulted from the sharp reduction of hardware cost and the increase of software cost in the past three decades. The net effect was that more and more functions were built by hardware support, making the instruction set larger and larger with an increasing number of complex instructions.

This CISC concept is best represented by the x86 series, Motorola M6800, Digital VAX series, and some IBM mainframes. The growth of instruction sets was also encouraged by the popularity of microprogrammed control in the 1960s and 1970s. A typical CISC instruction set can easily exceed 300 instructions. This is attributed to the use of variable instruction/data formats with 8 to 16, 32, and 64 b of word length.

Only a small set of 8 to 24 *general-purpose registers* (GPRs) is employed. CISC allows many memory-reference operations, based on more than a dozen addressing modes, including indexing and indirect addressing. Almost all computer manufacturers have heavily invested in CISC architecture since the 1960s, until challenged by the RISC chips since the 1980s.

The RISC Challenge   After three decades of growth in CISC architecture, computer users began to evaluate the performance relationship between ISA and available hardware/ software technologies. Through years of program tracing, computer scientists finally realized that only 25% of a complex instruction set is frequently used about 95% of the time. This implies that about 75% of hardware-supported instructions are rarely used at all. The basic idea of RISC is to *make the common case fast,* which led to a dramatic increase in performance over previous CISC designs,

A natural question then popped up: *Why should we waste valuable chip estate on rarely used instructions*? The answer was that complex instructions should be pushed to software rather than done by hardware, after weighting their low usage frequencies in a program mix. Pushing rarely used instructions into software vacates the chip area for possibly building the entire processor on a single VLSI chip. Even on-chip caches and floating-point units are built there in RISC processor chips.

A RISC instruction set typically contains less than 100 instructions with a fixed 32-b or 64-b instruction format, but not both. Only three to five simple addressing modes are used. All instructions are register-based, meaning the fetch of operands from registers and resulting data must be written back to registers. Only the *load/store* instructions access data from the cache or the memory hierarchy.

Operands must be loaded into working registers before they can be used. For this reason, RISC has been also called the *load/store architecture*. A RISC processor often uses two separate register files: such as 32 integer registers and 32 floating-point registers. Some RISC processors use over 100 registers. Besides using a large register file, the use of on-chip split caches also shortens the access time significantly. The basic idea in RISC is to execute the majority of instructions in one cycle by fetching operands directly from registers, reorder buffers, or data cache.

In Table 4.2, we summarize key features in pure RISC and classic CISC processors. Some architectural distinctions will disappear in the future, as both architectures change with the same underlying technology. For example, split caches and hardwired control are also in use in MC 68060 and Intel Pentium.

Table 4.2 Characteristics of Classic CISC and Pure RISC Architectures

Characteristics	Classic CISC Architecture	Pure RISC Architecture
Instruction format	Variable formats: 16 to 32 and 64 b	Fixed 32-b instructions
Clock rate*	100-266 MHz	180-500 MHz
Register files	8-24 general-purpose registers (GPRs)	32-192 GPRs, separate integer and floating-point files
Instruction set size and types	Around 300, with over four dozen instruction types	Around 100, most are register-based except *load/store*
Addressing modes	Around one dozen, including indirect/indexed addressing	Limited to 3 to 5, only *read/ store* addressing the memory
Cache design	Earlier model used unified cache, some use split caches	Most use split data cache and instruction cache
CPI, and average CPI	1 to 20 cycles, average 4 cycles	1 cycle for simple operations, average around 1.5 cycles
CPU control	Most microcoded, some use hardwired control	Most hardwired control without control memory
Representative commodity processors	Intel x86, VAX 8600, IBM 390, MC 68040, Intel Pentium, AMD 486, and Cyrix 686	Sun UltraSparc, MIPS R10000, PowerPC 604, HP PA-8000, Digital 21164

* Based on maximum clock rate in 1997.

Two-Decade Debate on ISA Using a large register file, data buffer, and separate I- and D-caches benefits internal data forwarding and eliminates unnecessary storage of intermediate results. With significantly reduced complexity of the instruction set, the entire RISC processor can be easily built on a single VLSI chip. The resulting benefits include higher clock rate, lower average CPI, lower cache miss penalty, and better opportunity for compiler optimization. However, the transition from CISC to RISC was indeed a radical change in architecture. The major sacrifice is the loss of binary compatibility to traditional CISC application codes.

Based on program trace, converting a CISC program to an equivalent RISC program increases the code length (instruction count) by only about 40%. The increase in RISC code length is much smaller than the increase in clock rate and the reduction in average CPI in a RISC processor.

The debate between RISC and CISC designers is now over. Both sides have learned a great deal from each other. The boundary between RISC and CISC architectures is rather blurred now. Quite a few processors are now built with a hybrid of having both RISC and CISC features built with the same circuit technology.

4.2.3 Architectural Enhancement Approaches

Microprocessors started with the introduction of the Intel 4004, a 4-b CPU in 1971. This eventually led to the x86 microprocessor series. Various processor families are mapped into a design space of clock rate versus CPI in Fig. 4.11. As processor technology evolves rapidly, the clock rate has increased from 10 MHz in 1980 to 300 MHz in 1997.

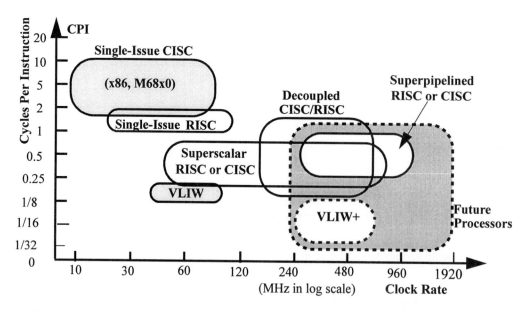

Figure 4.11   Design space of commodity microprocessors from 1980 to 2000. The shaded area at the lower right corner is most likely for designing future processors.

Single-issue CISC or RISC processors sit in upper left of the design space. It is desired to decrease the *average CPI* value from, say 10 cycles in a CISC processor to less than 1 cycle in a RISC machine. The average CPI can be lowered by improved processor architecture and compiler optimization. The whole design purpose is to lower the CPI to a small fraction of a cycle and increase the clock rate at the same time.

Superscalar versus Superpipelined Designs Today, all single-issue processors are upgraded to assume a multiple-issue *superscalar* architecture. Example CPU chips include the M68060 and Pentium assuming two-issue superscalar CISC architecture, while the PowerPC 601 and Pentium Pro are three-issue RISC processors. Current superscalar processors issue mostly 4 instructions per cycle including Alpha 21164, PowerPC 604e, HP PA-8000, MIPS R10000, and UltraSparc-2 processors.

Superpipelined processors are those that use a deeper pipeline driven by a much higher clock rate. Superpipelining is enabled by using a multiphase clocking mechanism developed with Cray supercomputers. The deeper pipeline has a lot more stages, subdivided from a few stages in earlier processor pipeline designs. Some processors applied both superscalar and superpipelining design concepts, whether they assume a CISC, or a RISC, or a hybrid architecture. This phenomenon is shown by the largely overlapped area in the middle of Fig. 4.11.

Decoupling of CISC and RISC The main advantage of this approach comes from better software compatibility for x86 users. The decoupled architecture combines the advantages of CISC, RISC, superpipelining, and superscalar designs in one microchip design, as seen in the overlapped area at the right center of Fig. 4.11. This architecture assumes neither a pure RISC nor a classic CISC, but a hybrid CISC/RISC architecture.

The best example is the Pentium Pro core studied in Section 4.4.2. The basic idea is to have a frontend section in the CPU chip to translate the x86 code into RISC-like instructions for a backend RISC core to perform superscalar and/or superpipelined execution. This has been called a *decoupled* or a *hybrid CISC/RISC architecture*.

The VLIW/Compiler Approach This processor architecture is generalized from two well-established concepts: *horizontal microprogramming* and *superscalar execution*. A typical VLIW (*very long instruction word*) instruction has hundreds of b (such as 256 or 1024 b in the Multiflow computer design [241]).

Each VLIW consists of a large number of independent operations, which can be executed by multiple functional units in parallel. The VLIW circle in lower left corner of Fig. 4.11 corresponds to the Multiflow design with an eight-way parallelism. A VLIW processor can exploit much higher *instruction-level parallelism* (ILP) in user code written in short (32-b) words. All operations in each VLIW instruction represent 8-way, 16-way, or higher degree of parallelism.

The success of VLIW must rely on an intelligent compiler to regroup and pack many independent or unrelated short instructions into a VLIW instruction. This VLIW approach may result in a much lower CPI than that of a superscalar processor. The CPI can be lowered to 1/32 for a 32-way VLIW architecture as projected by the VLIW+ circle in the lower center of Fig. 4.11 for possible future processor design. We shall study VLIW architectures in Section 4.5.3.

4.3 Microprocessor Architecture Families

With widespread use of microchips in PCs and desktops, the chip companies can afford to invest huge resources into research and development on microprocessor-based hardware, software, and applications. Consequently, commodity microprocessors are approaching the performance of custom-designed processors used in IBM mainframes or in Cray supercomputers.

4.3.1 Major Architecture Families

In Fig. 4.12, we summarize major microprocessor architectures for general-purpose and embedded applications. Today's general-purpose microprocessor assumes word length of 32 b or 64 b. Based on 1997 price quotes, they were sold between U.S. $75 and $500 in large quantity. The CISC camp is represented by the x86 and M680x0 architectures. The x86 has gone through six generations of development since 1978. In 1995, worldwide shipment of x86 chips has reached 70 million.

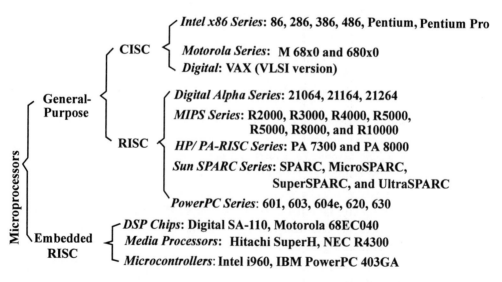

Figure 4.12   Microprocessor families and representative CPU chips.

Listed in the family tree are five major RISC architectures, namely, the Alpha originated by Digital Semiconductor, MIPS by MIPS Technologies, PA-RISC by Hewlett Packard, PowerPC by Apple, IBM, and Motorola jointly, and SPARC by Sun Microelectronics. Each architecture has been licensed to numerous manufacturers to produce different implementations for desktop, portable, or deskside systems. The classical CISC and early RISC processors were all single-issue processors using a single pipeline data path. These processors cannot exploit the ILP.

Single-Issue CISC   Classic processors like the Intel 86, 286, 386 and 486; Motorola M680x0 series; Digital's VAX/8600; and IBM 370/390 all followed the single-issue CISC architecture. These processors are all designed to issue a single instruction per 1 or 2 cycles. They were fabricated with less than 10-MHz clock in the 1970s and gradually increased to over 100 MHz in the 1980s. The 486 and M 68040 represent the last wave of CPU chips in this category.

The CPI of single-issue CISC processors varies from 1 to as high as 20 cycles, depending on the instruction type and the microarchitecture adapted. The average value lies between 1.5 and 10 cycles per instruction, depending on the ISA and processor technology. These processors are shown at the upper left corner of Fig. 4.11.

Single-Issue RISC   We show four variations of RISC architecture in Fig. 4.13, namely, the *single-issue RISC, superscalar RISC, pipelined RISC,* and *decoupled CISC/RISC.* RISC processors started with the single-issue prototype designs in Berkeley RISC and Stanford MIPS projects in 1980. Later on, the idea was picked up in Sun SPARC, MicroSPARC, MIPS R2000, and R3000, etc. These processors were fabricated with a clock of 25 MHz in 1980 and gradually upgraded to 120 MHz in 1992.

With hardwired control, the CPI of most instructions has been reduced to just 1 cycle, with an average CPI below 1.5 cycles in most applications. However, the CPI of single-issue RISC cannot be further lowered, regardless of how fast the clock is. The improvement over the CISC counterparts was also limited, until the appearance of superscalar RISC designs in Intel i860 and Digital 21064.

4.3.2 Superscalar versus Superpipelined Processors

Two basic techniques to improve processor performance are to exploit higher ILP, or to increase the clock rate by subdividing the instruction pipeline into simpler stages. These two techniques, illustrated in Fig. 4.13, are often mutually supportive and yet sometimes complementary.

Increasing the Issue Rate   In a superscalar processor, multiple instructions are issued per cycle, and multiple results are generated by multiple pipelines per cycle. Superscalar processors are designed to exploit ILP in user programs. Only independent instructions can be executed in parallel without causing a wait state.

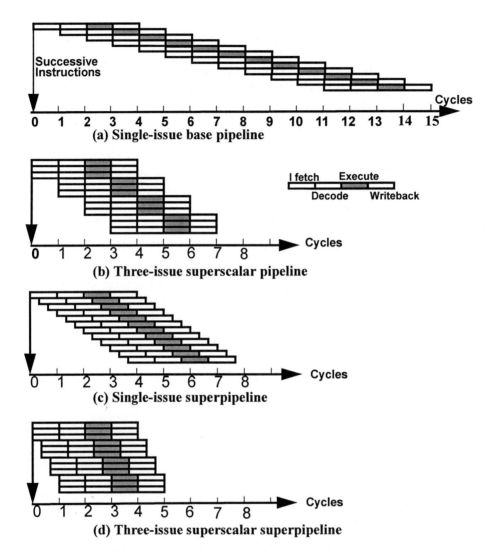

Figure 4.13 Pipelining operations in a base pipeline, superscalar, and superpipelined processors.

The amount of ILP varies from program to program. The average value of ILP was measured to be between 2 and 5 in ordinary program traces. Some special codes may contain higher ILP, such as 500 or higher in some scientific Fortran codes. For this reason, the instruction issue rate has been limited to 2 to 6 in current superscalar designs.

Single-issue RISC processors were gradually replaced by the introduction of the two-issue processors in Digital's 21064, M88110, MicroSPARC 2, Intel i860, Pentium, PA

7200, MIPS R5000, Cyrix 6x86 (M1 and M2), and M68060, etc. These microchips include both superscalar RISC and x86 superscalar CISC processors.

Pipeline Design Options   In Fig. 4.13, we show four ways of designing instruction pipelines. All four pipeline designs assume four stages and are driven by clock rates which are multiples of the base pipeline clock cycle. The superscalar designs in Cases (b) and (d) demonstrate the use of three functional pipelines in parallel. Cases (c) and (d) correspond to two superpipelined designs, where the clock cycle is only one third of that in Case (a) and Case (b).

In these pipeline operations, we demonstrate four ways to execute 12 instructions. The execution times reduce from 15 cycles to 7 cycles, 7.66 cycles, and 5 cycles from case (a) to case(d), respectively. With sufficiently large workload, Case (b) is potentially three time faster than Case (a) and so is Case (c).

Case (d) is potentially 9 times faster than Case (a) or three times faster than either Case (b) or Case (c). The throughput performances of the four processor designs can be similarly rated, given the same time limit.

In general, an m-issue superscalar processor is at most m times faster than a single-issue processor, assuming the same clock rate and same pipeline depth. A superpipelined processor with an n-times-faster clock can be at most n times faster than that of a base pipeline. In an m-issue processor driven by a n-times-faster clock [Case (d)], the combined speedup is potentially mn over thebase pipeline design in Case (a).

Superscalar x86 Architecture   Today's x86 microprocessors, including the various versions of Pentium and Pentium Pro, are all superscalar processors, as summarized in Table 4.3. Three x86 families from Intel, AMD, and Cyrix are compared in this table

The pinout correspond the packaging used. Almost all use the split I/D caches, except the Cyrix M2 using a unified 64-KB cache. The decode rate equals the issue rate in tradi tional superscalar design in Pentium and Cyrix 6x86 designs. In the designs of Pentium Pro, Pentium-2, and AMD K5/6, the decode rate is lower than the issue rate, because *micro instructions* (uops) are issued to the functional units.

Three Pentium models (P54C, P54CS, and P55C) are all two-issue superscalar designs allowing only in-order execution. The U and V pipelines in Pentium depend on each other. When one pipeline is stalled, the other must wait. This poses the serious drawback of losing cycles to potential pipeline hazards in the early Pentium models.

Based on recent benchmarks, the Pentium has improved the speed of the i486 by 2 to 3 times. Two versions of Pentium Pro (96 and 96S) further doubled the speed of Pentium. The Pentium Pro is improved from the i486 by 4 to 6 times in speed performance.

The Pentium-2 (also called Klamath) is upgraded from the Pentium Pro by adding the MMX (*Multimedia Extensions*) features. The AMD K5 and Cyrix 6x86 M1 are comparable to Pentium Pro in performance. The AMD K6 and Cyrix M2 are the corresponding MMX upgrades from the K5 and M1, respectively

Table 4.3 Key Features of Superscalar x86 Microprocessors
(Source: M. Slater, *IEEE Micro*, Dec. 1996. [566])

Vendor	Intel						AMD		Cyrix	
Model/ Feature	Pentium			Pentium Pro		Pentium-2	K5	K6	6x86	
	P54C	P54CS	P55C	96	96S				M1	M2
Max. clock rate (MHz)	120	200	200	150	200	233-266+	100	>180	150	225
Pin-out	P54C	P54C	P54C	PPro	PPro	Pentium-2	P54C	P55C	P54C	P55C
I/D caches, (KB)	8/8	8/8	16/16	8/8	8/8	N.A.	8/16	32/32	16/16	64 Ufd
MMX	No	No	Yes	No	No	Yes	No	Yes	No	Yes
Decode rate	2	2	2	3	3	3	1-4	2	2	2
Issue rate*	2 instr.	2 instr.	2 instr.	5 uops	5 uops	5 uops	4 uops	4 uops	2 instr.	2 instr.
Out-of-order execution	No	No	No	Yes	Yes	Yes	Yes	Yes	Ltd	Ltd
Die size (mm^2)	148	90	140	308	196	N.A.	181	~180	167	<200
Transistors (millions)	3.3	3.3	4.5	5.5	5.5	N.A.	4.0	8.8	3.3	6.0
Technology	Bi-CMOS		CMOS	Bi-CMOS		CMOS				
Process (um/layer)	0.5 / 4	0.35 / 4	0.28 / 4	0.5 / 4	0.35 / 4	0.28 / 4	0.35 / 3	0.35 / 5	0.44 / 5	0.35 / 5
* Indicates x86 instructions,Ufd: Unified, N.A.(not available), ++ Includes 256-KB level-two cache, Ltd: Limited, uops: micro-operation										

Superscalar RISC Processors Most superscalar designs choose the RISC approach instead of CISC. The issue rate increased to 3 in PowerPC 601, i960, Pentium Pro, Pentium-2, etc. Today's superscalar RISC designs are mostly upgraded to four-issue designs, as seen in Table 4.4. This table summarizes the superscalar designs in commodity microprocessors up to late 1997.

The increased issue rate in superscalar processors implies that the average CPI rate can be lowered to approach the minimum values of 0.5, 0.33, and 0.25 for the issue rates of 2, 3, and 4, respectively. We shall check two case studies in Section 4.4.

Table 4.4 Key Features of High-End Superscalar Microprocessors
(Source: M. Slater, *IEEE Micro*, Dec. 1996. [566])

Feature	Digital 21164	Power-PC620	Power-PC 604e	Ultra-Sparc-2	Micro Sparc-2	HP PA-8000	MIPS R10000	MIPS R5000	Pentium Pro
Maximum Clock (MHz)	500	200	225	250	110	180	200	180	200
Cache size$^+$ (KB)	8/8/96	32/32	32/32	16/16	16/8	None	32/32	32/32	8/8
Issue rate	4	4	4	4	2	4	4	2	3
Pipeline stages	7-9-12	5	6	6-9	5	7-9	5-7	5	12-14
Out-of-order execution	6 loads	16 inst.	16 Instr.	None	None	56 instr.	32 instr.	None	40 ROBs
Rename registers	None, 8 Fp	8 Int/ 8 Fp	12 Int	None	None	56 total	32 Int 32 Fp	None	40 total
Memory bandwidth (MB/s)	~400	1,200	~180	1300	~100	768	539	~160	528
Package, Pin count	CPGA-499	CBGA-625	CBGA-255	PBGA-521	CPGA-321	LGA-1085	CPGA5 27	SBGA2 72	MCM-387
Process/ (um / layers)	0.35 / 4	0.35 / 4	0.35 / 4	0.29 / 5	0.4 / 3	0.5 / 4	0.35 / 4	0.35 / 3	0.35 / 4
Die size (mm^2)	209	240*	148	149	233	345	298	84	196
Transistor count	9.3 M	6.9 M	5.1 M	3.8 M	2.3 M	3.9 M	5.9 M	3.6 M	5.5 M
Power (W)	25	30	20*	30	9	>40	30	10	35**
SPEC95^{++} integer/FP	12.6/18.3	9.0/9.0*	8.5/7.0	8.5/15	1.4/1.9	10.8/ 18.3	8.9/17.2	4.0/3.7	8.7/6.0

$^+$ Instruction cache / Data Cache / Level-2 Cache, $^{++}$ SPEC95integer / SPEC95floating- point,
* Micro Design Resources estimate, ** Includes 512-KB level-2 cache, ROB: Reorder Buffer,
Int: Integer, Fp: Floating point

Decoupled CISC/RISC   CMOS has been proven a scalable VLSI technology in fabricating microchips. The decoupled architecture incorporates positive features from both CISC and RISC architectures on the same CPU chip. The decoupled architecture became popular with the development of Intel P6, now called Pentium Pro.

We shall study details of the P6 architecture in the next section. Similar decoupled processor designs also appeared in AMD K6, Cyrix M1, and MIPS R10000. Their characteristics are given in Table 4.3 and Table 4.4.

Microprocessor Performance   In Fig. 4.14, we compare the SPEC95 performance of six superscalar, RISC or decoupled CISC/RISC microprocessors.

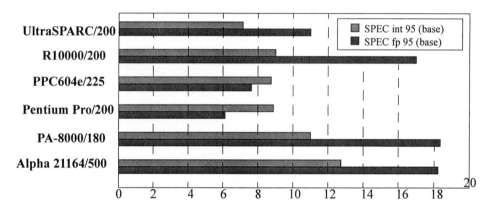

Figure 4.14   SPEC int 95 and fp 95 (base) performance of commodity microprocessors in log scale. (Source: M. Slater, *IEEE Micro*, Dec. 1996. [566])

The clock rate of each processor is shown after the slash. The Alpha 21164, PA-8000, and R10000 have comparable floating-point performance. The difference in integer performance is even smaller. The designs of PowerPC and Pentium Pro are intended mainly for business applications with better software compatibility.

Among the six, the Alpha chip certainly ranks the highest in performance and clock speed. However, Alpha ranks the lowest in performance/cost ratio. The Alpha 21164, PA-800. andR10000 have comparable SPEC fp 95 performance. The performance difference in SPEC int 95 is rather small among the R10000, PowerPC604e, and Pentium processors.

4.3.3 Embedded Microprocessors

Embedded processors have enabled digital consumer electronics. Manufacturers produce them in enormous volume. Selectively listed in Table 4.5 are some representative embedded processor models and their characteristics..

Table 4.5 Key Features of Selected Embedded Microprocessors
(Source: M. Slater, *IEEE Micro*, Dec. 1996. [566])

Manu-facturer	Digital	VLSI	NEC	Hitachi	IBM	Motorola			Intel	
Model	SA-110	ARM 710	R4300	SH 7604	PPC 403GA	860 DC	68EC 040	CF 5102	960 JA	960 HT
Archi-tecture	ARM	Strong-ARM	MIPS	SuperH	Power-PC	Power-PC	68000	Cold-Fire	i960	i960
Clock (MHz)	200	40	133	20	33	40	40	25	33	60
I/D Caches (Kbytes)	16 / 16	8 / 8	16 / 8	4 / 4	2 / 1	4 / 4	4 / 4	2 / 1	2 / 1	16 / 8
FPU	No	No	Yes	No	No	No	Yes	No	No	No
MMU	Yes	Yes	Yes	No	No	Yes	Yes	No	No	No
Bus freq. (MHz)	66	40	66	20	33	40	40	25	33	20
MIPS+	230	36	160	20	41	52	44	27	28	1008
Voltage (core/ bus)	2.0 / 3.3	5	3.3	3.3	3.3	3.3	5	3.3	3.3	3.3
Power (mW)	900	424	2200	200	265	900	4500	900	500	4500
MIPS/W	239	85	73	100	155	58	10	30	56	22
Transis-tors (million)	2.1	0.6	1.7	0.45	0.58	1.8	1.2	N.A.	0.75	2.3
Process (um/ layers)	0.35 / 3	0.6 / 2	0.35 / 3	0.8 / 2	0.5 / 3	0.5 / 3	0.65 / 3	0.6 / 3	0.8 / 3	0.6 / 4
Die size (mm^2)	50	34	45	82	39	25	163	N.A.	64	100
+ MIPS (million instructions per second) rating is based on Dhrystoen 2.1, FPU: floating point unit, MMU: memory management unit.										

Embedded RISC processors are mainly for DSP (*digital signal processing*), media processing, and microcontroller applications. These processors assume a wide range of word lengths from 4 b, to 8, 16, and 32 b. The unit price of embedded products ranges from $5 to $40 in large quantity. The 4-b and 8-b embedded chips are largely used in washing machines, cellular phones, remote control, electronic toys, and many small devices for embedded control purpose. For example, the worldwide sale of 4-b processor chips exceeded 16 billion by late 1996.

The 16-b and 32-b embedded chips are for DSP, printers, multimedia, and control applications. In 1995, the 32-b Hitachi SuperH shipment exceeded 14 million alone, because of the huge demand from the video game market.

4.4   Case Studies of Microprocessors

We study below two microprocessors which represent two extremes in evolutionary changes in RISC/CISC architecture in the past decade.

At the one extreme is Digital's Alpha 21164 processor, which adapted the traditional superscalar RISC architecture. It executes RISC instructions in order. Alpha designers have pushed the clock speed to the fastest that the CMOS technology can provide.

On the other extreme is the Pentium Pro processor adopting the decoupled CISC/RISC approach to executing reordered micro operations as described in Section 4.4.2. The Pentium approach emphasizes more on the software compatibility with existing x86 ISA applications in the PC and notebook industry.

4.4.1 Digital's Alpha 21164 Microprocessor

This four-issue Alpha CPU chip is based on Digital's 64-b AXP architecture, a performance leader among superscalar RISC microprocessors. The architecture of 21164 is shown in Fig. 4.15. There are four execution units, multiplr register files, split I/D caches, and a level-2 cache on a 0.35-um CMOS chip.

This CMOS chip has a built-in *cache control and bus interface* (called the CBOX) for accessing external level-3 cache and multiprocessor snoopy bus coherence control. The processor executes in-order programs except up to 6 reordered *load* operations. But the 21164 cannot execute *nonread* instructions out of order. Most other superscalar processor designs can execute any instruction mix out of order. In this sense, the AXP architecture is rather conservative in exploiting ILP.

Faster Clock Speed   Digital chose the maximum clock speed approach in the Alpha series, starting with the 200-Mhz 21064 in 1992, the 275-Mhz 21064A in 1994, the 300-MHz 21164 in 1995, the 500-MHz 21164A in 1996.

The company shipped the 500-MHz 21164, while most other CPU chips were still limited to 200 to 250 MHz as listed in Table 4.4. This CPU chip delivers the highest integer and floating-point SPEC95 performance, as shown in Fig. 4.14, but the performance is not proportional to the big gap in clock speed with its competitors.

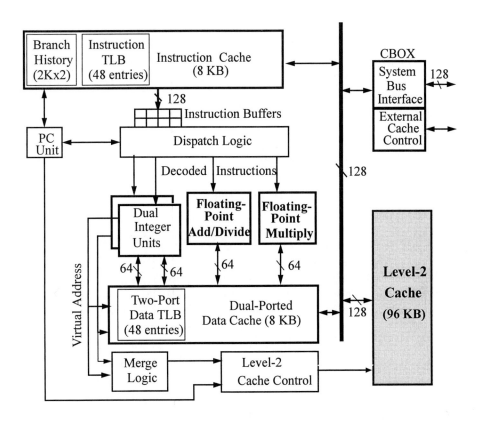

Figure 4.15   Digital's Alpha 21164 processor architecture.
(Courtesy of Digital Equipment Co. 1995 [196])

High Microchip Density   This RISC chip has the highest transistor count of close to 10 million in 0.35-um CMOS technology. But less than 2 million are allocated to the RISC logic portion, and the rest 8 million are largely for on-chip memory, including I/D caches and level-2 cache. The logic allocation is lower than that of most high-end processors.

As a contrast, the Pentium Pro has allocated 4.4 million to the CISC logic out of a total of 5.5 million transistors. Based on the 1997 standard, a CPU core typically allocates 2 to 4 million transistors to logic, with 1 to 6 million allocated to the on-chip memory.

Cache Hierarchy   Cache strategy is an area where Digital stands out among the pack as the only microprocessor with a two-level cache hierarchy on chip. Separate 8-Kbyte instruction cache and data cache can be accessed in 1 cycle even at a 500-MHz clock rate

A slower 96-KB level-2 cache provides faster access than could an external cache. With the built-in external cache control, a typical 21164 system supports an external level-

3 cache from 1 Mb to 64 MB in size. The reduced cache access time certainly helps enhance the chip performance.

Data Path Widths The integer and floating-point execution units are each 64-b wide. The instruction pipeline is 128-b wide at each stage. The bus that carries data to or from the level-2 cache is 128-b wide, including the busses to fill the level-1 caches. Finally, accessing the off-chip level-3 cache or memory is done through 128 data pins in the Alpha processor package.

Functional Pipelines: The instruction pipeline is formed from seven functional units: the *instruction F/D unit*, two *integer pipelines*, two *floating-point pipelines*, a *memory address translation unit* (MATU), and the CBOX shown in Fig. 4.15. These pipeline segments perform the following specific functions.

- The four-stage *instruction fetch/decode* (F/D) *unit* forms the frontend pipe for fetching, decoding, issuing, and retiring instructions. It also handles interrupts, exceptions, and branch prediction tasks by working with the I-cache containingthe BHT (*branch history table*) and instruction TLB (*translation lookaside buffer*). The I-cache is accessed at stage S0 and the access of *integer register file* (IRF) or *floating-point register file* (FPRF) is done at the S3 stage.

- Two *integer execution pipelines,* each having three pipeline stages, execute all integer RISC instructions and handle *load/store, branch,* and *control* instructions, except for a *conditional branch* decided by a floating-point result. The two integer pipes use the same IRF. An *integer multiplier* is attached to one of the integer pipelines. In total, integer RISC instructions execute in seven pipeline stages.

- There are two 4-stage *floating-point pipelines*: one is a *floating-point add pipe* attached with a *floating-point divider,* and another is a *floating-point multiply pipe.* In total, each floating-point instruction takes 9 cycles to flow through the instruction pipeline.

- The 21164 has an eight-stage *memory access pipeline*, which performs memory address translation and executes *load/store* and *memory barrier* instructions. This pipeline works with the data cache, level-2 cache, MATU, and CBOX to handle memory access conflicts and to implement cache coherence protocol.

Nonblocking Cache Access It takes 2 cycles to load data from the D-cache and 4 more cycles to access data in level-2 cache. It takes 2 additional cycles to fill the data/instruction cache with a cache line of 32 B from the level-2 cache at stages S10 and S11. In total, a *read/store* instruction may take 6 to 12 cycles to flow through the entire instruction pipeline. The pipeline does not stall at cache misses. It stalls only when data dependency exists between instructions. Each pipeline cycle is 2 ns for the 500-MHz clock.

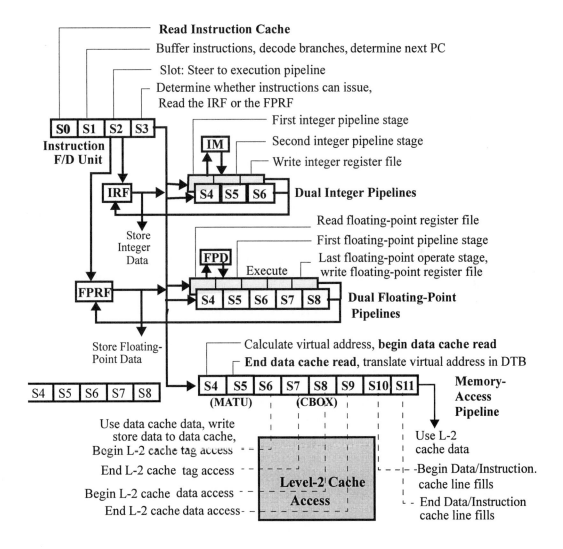

Figure 4.16   The 21164 instruction pipeline with dual integer pipelines, dual floating-point pipelines, and 6 to 12 pipeline stages for memory-access operations (IRF: Integer register file, FPRF: Floating-point register file, IM: Integer multiplier, FPD: Floating-point divider, DTB: Data translation buffer, MATU: Memory address translation unit, CBOX: Cache control/bus interface. Courtesy of Digital Equipment Co. 1995 [196])

A dependent instruction can begin execution immediately after stage S11. The MATU and the CBOX handle cache misses in a nonblocking fashion at all cache hierarchy levels. The advantages of this nonblocking design are (1) it exploits available *read* parallelism by initiating more than one access at a time and (2) it allows up to 6 *reads* to be reordered. This read reordering enables the possibility of software pipelining and data prefetching.

Other Performance Features   The 21164 uses a *branch history table* (BHT) to implement *dynamic branch prediction*. The BHT has 2048 2-b entries to record the outcome of the last two execution of each branch instruction. There are no rename registers in the IRF. However, 8 floating-point registers can be renamed to support dynamic execution.

In terms of floating-point performance, 21164 was reported with a rating of 18.3 SPEC95/FP, tied with HP PA-8000 as the leader in 1997. Its integer performance also led in the same performance ranking shown in Fig. 4.14.

With a 499-pin package, the Alpha chip has 128-b datapath for off-chip data transfer. The memory bandwidth of 400 MB/s was estimated. The power dissipation has been reduced from more than 50 W in the 21064 to now 25 W in the 21164, mainly due to the use of a 2.9 V power supply in the CMOS chip. The 48 entries in the instruction TLB and in data TLB help shorten the address translation time. Alpha chips have been used mainly in workstations , not on PCs for their high prices.

4.4.2 Intel Pentium Pro Processor

The Intel Pentium Pro directly executes x86 CISC instructions, but internally the chip implements a decoupled CISC/RISC architecture shown in Fig. 4.17.

At the front end, three x86 instructions can be decoded in parallel by an *in-order translation engine*. These decoders translate the x86 instructions into five RISC-like *micro-operations*, denoted as *uops*. Each of the two simple decoders produces one uop, while a general decoder coverts a complex instruction into 1 to 4 uops.

At the back end, a *superscalar execution engine* is capable of executing five uops out of order on five execution units in the RISC core.The P6 has five execution units: two *integer ALUs*, two *load/store units,* and one *floating-point unit*. These uops are first passed to a 40-entry *recorder buffer* (ROB), where they are stored until the required operands become available.

From the ROB, the uops are issued to a 20-entry *reservation station* (RS), which queues them until the needed execution unit is free. This design allows uops to execute out of order, making it easier to keep parallel execution resources busy. At the same time, the fixed-length uops are easier to handle in speculative, out-of-order core than complex, variable-length x86 instructions.

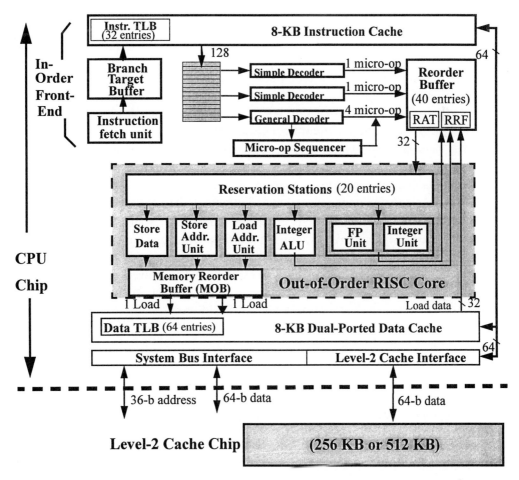

Figure 4.17   Decoupled CISC/RISC architecture of Pentium Pro®
processor.(Reprint by permission of Intel Corp., Copyright Intel
Corp. 1995, sketch from Gwennap [290])

Memory Hierarchy   Intel designed a special level-2 cache chip that is mounted in the
same package with the CPU chip. Direct connections exist between the CPU and the cache
chips. The level-2 cache chip can have either 256 or 512 KB. This level-2cache delivers
64 bits per cycle, even with a 200-MHz clock. The on-chip caches are totaled 16 KB, split
into 8 KB for instruction and 8 KB for data.

The cache hierarchy is also designed to support nonblocking cache access. The cache
traffic between the two levels occurs on a *back-side bus*, allowing the *front-side bus*
bandwidth to be entirely used for external memory and input/output accesses.

In-Order Front End Functional blocks in the P6 processor are shown in Fig. 4.18. Abbreviations of all blocks are listed below. The in-order front end is shown within the shaded area. Functions of these blocks are summarized below:

- The ***instruction fetch unit*** (IFU) containing the I-cache holds the incoming x86 instructions. It supplies a cache line of instructions to the *instruction decoder (*ID) at a time. The *micro instruction sequencer* (MIS) coordinates the generation of a long sequence of uops per each complex instruction.
- The ***branch target buffer*** (BTB) fetches the next instruction by predicting the branch target address based on branch history. With 512 entries, a two-level adaptive branch prediction scheme was implemented using associate memory. Branch misprediction is reduced to a minimum with speculative executions.
- The ***Register Alias Table*** (RAT) renames programmer visible register references to internal physical registers, done at run time. The uops are sent to the back-end RISC core under the mapping of RAT.

Out-of-Order Core This RISC core is enclosed within the center shaded box in Fig. 4.18. The RAT sends the uops to different places, seen at the bottom of the block diagram. One place is the RS, and the other is the ROB. The functions of these two mechanisms are specified below:

- The ROB is the place where reordering is done and the original program order is remembered. Because we want to retire the completed instructions in program order, we have to save their original ordering in the ROB. When a completed uop returns from an execution unit, it will check with the ROB where it belongs before retirement at the right place.
- The RS has 20 entries, each can hold one uop waiting for execution. The RS is not a *first-in, first-out* (FIFO) queue. It can handle integer, float-point, and flags; anything that allows renaming. The RS has five ports, as seen in the blowup drawing on the right.
- Port 0 is attached to the integer execution unit, floating-point adder, integer divider, floating-point divider, floating-point multiplier, and a shifter. Ports 2, 3, and 4 are dedicated to memory accesses. Port 2 generates the *load* addresses through the *address generation unit* (AGU). Ports 3 and 4 generate the *store* address. Each of these ports has its own writeback path back to the RS.

Segmented Superpipeline Design Pentium Pro is deeply pipelined to have a maximum of 14 processing stages illustrated in Fig. 4.19. To match the elasticity of the RS operations, the pipeline does not have a fixed number of stages, but the minimum number of clock cycles for an instruction to complete is 12.

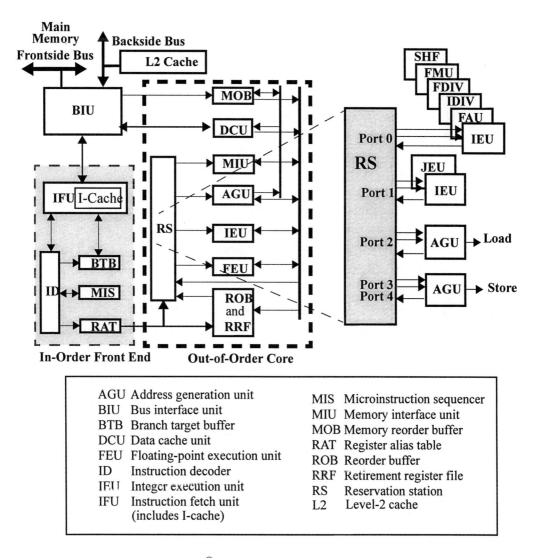

Figure 4.18 Pentium Pro® processor microarchitecture for dynamic execution. (Reprint by permission of Intel Corporation, Copyright Intel Corp. 1995, cited from Colwell and Steck [162])

The pipeline stages are divided into three segments, corresponding to processor resources at the *in-order front end* (eight I stages), *out-of-order RISC core* (three O stages), and *in-order retirement* (three R stages). The flow of uops through the pipeline is from left to right in the drawing. The slashes between segments show the queuing effect in scheduling RS or ROB resources. Stage operations are specified below from left to right.

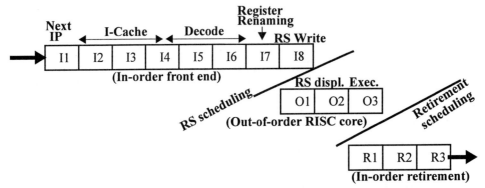

**Figure 4.19   Segmentation of 14 pipeline stages in Pentium Pro®
processor** (Reprint by permission of Intel Corporation,
Copyright Intel Corp. 1995, cited from Colwell and Steck [162])

Pipeline stage I8 may overlap with stage O1 for some micro-operations. Similarly, overlapped operation may happen between stage O3 and stage R1. That is why the minimum number of stages that an instruction must go through is 12 cycles. This segmentation of 14 stages into three shorter pipelines in a cascade provides some buffering between segments. It can also avoid the heavy penalty in using a long pipeline.

- *In-Order Decode/Translation:* The next instruction pointer (IP) is identified at stage I1, where the BTB predicts the branch direction based on history. Cache access and instruction decoding are each split across 2.5 clock cycles, including the three decoders and instruction alignment. The register rename stage is used to map virtual registers to physical registers. The last stage (I8) is for RS write, which may overlap with the O1 stage.

- *Out-of-Order Five-Way Execution:* The first two stages (O1 and O2) are for RS displacement. Depending on the waiting period, some uop may reduce to only 1 cycle for this purpose, if the operand is readily available. The third stage is actually for 1-cycle execution of most microoperations. Memory or floating-point executions may have to stretch out a few more cycles. The memory accesses are entirely different pipelined operations, which are not shown in Fig. 4.19.

- *In-Order Retirement:* The retirement process has to make sure that all uops associated with the same x86 instruction are retired as a group. Furthermore, the original program order must be restored. The result must flow back to the RS to enable some uops waiting in the RS. The results must return to the ROB to enable retirement. That is why it takes 3 cycles to do all of these correctly.

Dynamic Execution   The most impressive aspect of the Pentium Pro architecture is its binary compatibility with previous x86 generations. The chip achieves *dynamic execution* in order to exploit a higher degree of parallelism in user code. The approach is to achieve optimal adjustment of the ordering in instruction executions by predicting program flow, and speculatively executing instructions in the preferred order.

Speculative execution is done by starting the execution earlier, ahead of the normal execution sequence. Results of this speculation are stored temporarily in the ROB, since they may be discarded if the program did not flow in the predicted direction. During the retirement, these speculative results are committed from the temporary storage to permanent architectural state.

Multiprocessing Support   Up to four P6 CPUs and high-speed I/O channels can be connected to the same frontend P6 bus with no loss of performance. In fact, the four-CPU *quad SHV boards* are now available as a commodity product. This P6 bus supports cache coherence protocol and fast synchronization among the processors attached.

Furthermore, the bus also provides an extra OEM slot for users to attach any I/O or network interface devices. Intel has outlined a road map to scale the use of P6 from volume use in desktops, to workstations, servers, and even enterprise servers. Some examples will be given in Chapters 8 to 10.

Between the above two extreme processor designs, the UltraSparc-2, PowerPC 604, MIPS R10000, and HP PA-8000 in this order have built in an increasing number of *post-RISC* features over the traditional superscalar RISC design in the Alpha chip. We will compare all six architectures in Section 4.5.1, when we discuss the post-RISC concepts affecting the present and future microprocessor designs.

4.5   Post-RISC, Multimedia, and VLIW

In this section we discuss more recent development of microprocessors. In particular, we examine the continued evolution of silicon technology and instruction set architecture. We examine those new non-RISC architectural features built into recent microchips.

Multimedia extensions into existing microprocessors are reviewed to support WWW and Internet applications. Finally we assess the potential of the VLIW architecture, possibly for building future microprocessors that can exploit much higher ILP.

4.5.1 Post-RISC Processor Features

There is a reverse trend of having more and more performance features added to commodity microprocessors. Some of the added features are still RISC-like, but many others are decidedly non-RISC or even more CISC-like. A research group at Michigan State University [108] calls these added non-RISC features *post-RISC.*

Recent Performance Features As the technology advanced to increase die sizes and transistor density, RISC processor designers began to consider ways to use the available chip space. Listed below are approaches adapted by major microchip designers or noted by researchers including the Michigan State group.

- Add more registers and modify the CPU microarchitecture for multimedia.
- Enlarge on-chip caches that are clocked as fast as the processor.
- Use additional functional units for superscalar or VLIW executions.
- Add more "non-RISC" (but fast) instructions.
- Use on-chip support to accelerate floating-point operations.
- Increased pipeline depth or buffering between segmented pipelines.
- Use adaptive branch prediction and recovery schemes.
- Dynamic execution out-of program order is based on data driven.
- Add hardware code translation support at the front end.
- Put speculative execution ahead of branch events.

Many more functional features can be added to this list. As each microchip family introduces its newest processor, an interesting trend in the relationship between Specint-92 and the clock rate was observed by the Michigan State group. The performance measured by Specint-92 increases much faster than the clock rate, as shown in Table 4.6. The Specint-92 was chosen because the group must compare current processors with their predecessors using historical data.

Post-RISC Characteristics The Michigan State group has identified the following post-RISC characteristics, which are illustrated by a pipeline design in Fig. 4.20.

(1) The ISA was further augmented. Some decidedly non-RISC instructions have been introduced to enhance performance. The most significant changes are in the use of more intelligent compiler to order instructions and in using hardware to check the legality of multiple issues.

(2) Post-RISC processors are much more aggressive in the area of issuing more instructions per cycle. This is primarily done by using hardware to dynamically perform the instruction reordering. Higher ILP could be exploited by executing instructions out of program order.

(3) Out-of-order execution is not a new concept in computing -- it existed 20 years ago on IBM and CDC computers -- but it is innovative for single-chip implementations. The result is a RISC ISA with an execution core that is similar to a data flow implementation.

Table 4.6 Trends in Spec Int-92 Performance Results
(Source: Courtesy of Brehob et al. [108], 1996)

Company	Old Processor	Spec Int-92	New Processor	Spec Int-92	Percetage Increase in Clock Rate	Percetage-Increase in Spec Int-92
DEC	21064A/ 300 MHz	220	21164/ 333 MHz	400	11%	82%
HP	PA-7150/ 125 MHz	136	PA-8000/ 133 MHz	360	6%	164%
IBM	PPC-601/ 80 MHz	91	PPC-604/ 133 MHz	176	66%	93%
Intel	Pentium 166 MHz	198	Pentium Pro/ 200 MHz	320	20%	62%
Sun	Hyper-Sparc/ 125 MHz	131	UltraSPARC/ 140 MHz	200	12%	52%
MIPS	R-4400/ 200 MHz	141	R-10000/ 200 MHz	300	0%	113%

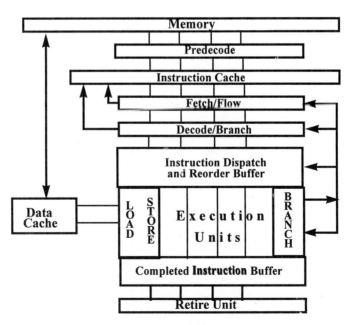

Figure 4.20   Conceptual pipeline architecture of the Michigan State post-RISC processor (Courtesy of Brehob et al. [108], 1996).

However, these processors still adhere to most of the RISC concepts. For example, the execution units of these processors are optimized to complete most instructions in a single cycle.

(4) The new components in a post-RISC implementation include a predecode unit whose results are stored with the instruction in cache, the use of rename registers to remove output and antidependencies, an instruction reorder buffer, and a retire unit.

The Michigan State Post-RISC Machine   Merging some of the nice features in the above CPU designs, the Michigan State group has proposed a conceptual post-RISC generic processor model. In their own words, this is an amalgam of features discussed above. The pipeline model is shown in Fig. 4.20. Detailed description can be found in Brehob et al. [108].

The use of predecode, reorder and completion buffers before and after the execution units is in many ways inspired by the data flow ideas. But the Michigan State group claim that this design can avoid the pitfalls of data-driven machines.

Post-RISC Features in Commodity Processors   In increasing order of having more post-RISC features, Digital Alpha 21164 chose the least aggressive RISC design, executing only in-order instructions as studied in the last section. The P6 made the most aggressive changes in having almost all of the new features identified above. The remaining four processors are between the two extremes.

The UltraSPARC Series   Sun UltraSPARC is second in lack of aggressiveness to change from the RISC influence. It does not allow out-of-order execution in general. But certain multi-cycle instructions are allowed to retire out of order. Instruction buffering is provided between two virtual pipelines in a cascade.

If either of the two pipelines stalls, the other can explore the buffer, similar to buffering roles of RS and ROB in the P6. UltraSPARC processors use register windows without support for register renaming. This approach also prevents the chip from out-of-order execution. The strength of UltraSPARC lies in very fast context switching among the register windows, variable-size pages, fast trap and interrupt handling, fast block memory move instructions, and a full set of graphics manipulation instructions, the VIS instruction set for supporting multimedia applications.

All these good features in the Ultra SPARC series were meant to improve high-end workstation and server applications, rather than for low-cost desktop PC applications.

The IBM PowerPC 604   This is a four-way superscalar RISC design. The Michigan State group ranked this CPU chip third in keeping the original RISC features. It does support instruction reordering up to 16 instructions in a reorder buffer. The 604 uses 28 rename buffers: 8 support condition registers in the branch unit, 12 support the integer register file, and 8 support the floating-point register file.

Also, many reservation stations are available at the input ports of functional units, including three integer units, the load/store unit, and floating-point unit. In addition, dynamic branch prediction and speculative execution are supported. A unique feature in 604 is the *branch-on-counter* instruction, which can implement counter-based loops.

The MIPS R10000 This CPU is more aggressive than the above three microchips in several areas. It can execute up to 32 instructions out of order. There are 32 rename registers for integer operations and another 32 for floating-point operations.

Special features include the use of a *shadow map* and a b*ranch resume cache* to minimize misprediction penalty and facilitate speculative executions. The large size of register files of 64 each for integer and floating-point instructions is another plus in enhancing the performance. R10000 is strong in supporting graphics and supercomputing applications as demanded in SGI/MIPS workstations and SMP servers.

The HP PA-8000 In terms of out-of-order execution, this is the most aggressive design in using a 56-entry *instruction reorder buffer* (IRB), 28 for arithmetic and 28 for memory operations. The PA-8000 performance almost matches that of the Alpha 21164. As a matter of fact, PA-8000 allocates 3.9 million transistors for logic, second only to the Pentium Pro CPU chip, which allocates 4.5 million for logic out of a total of 5.5 million transistors on the chip.

As a comparison, in third place is R10000 using 2.4 million for logic. The PowerPC 604 and UltraSPARC each used 2 million transistors for logic and the Alpha used only 1.8 million for logic, but more than 8 million for on-chip memory. This logic chip density is a good indicator of the amount of new or post-RISC features added, which also matches Michigan State's ranking of the six processors.

4.5.2 Multimedia Extensions

Internet, multimedia, and WWW applications have triggered the extensions of existing microprocessors. The idea is to have modest extensions to the instruction set significantly improve multimedia performance.

The approach is to have a small modification of the existing CPU logic with a small increase in die area to add some time-demanding functions, such as the MPEG encoding and decoding, audio synthesis, image processing, and modems, etc. The hope is to deliver a significant boost in performance on today's desktop applications.

Media Processing Requirements Multimedia functionality increases over time. Today's desktops are required to support multimedia information access as well as real-time communication, besides a computation tool. Multimedia represents the integration of visual, audio, textual, and sensory information, all represented digitally.

As observed by Ruby Lee of Hewlett-Packard Laboratories, media processing requires a quantum leap in performance and capacity in future computer systems. There is ample opportunity for exploiting parallelism in media processing. Listed below are some characteristics of audio, video, image, pixel processing, graphics rendering, and transformation programs:

- Audio/video processing deals with time-dependent data with precision of 16 b or less. Multiply-accumulate is the basic operation involved.

- Lots of data parallelism exists in pixel, video, image, and graphics rendering programs. Data are small integers (8 to 12 b for video/image), 16-b for intermediate results are sufficient.

- Common integer operations include add, subtract, multiplicative scaling, and simple divisions. The memory-access patterns are often predictable in video, image, and graphics operations.

- Pixel or signal processing offers great opportunities for loop vectorization, SIMD parallelism, and packing of small integers into words, all leading to subword parallelism.

- Communications require even more computations than computation itself. This opens up a new front for applying parallel processing techniques in low-cost microprocessors.

Subword SIMD/MIMD Parallelism   At the heart of most vendors' multimedia extensions is the *single-instruction, multiple-data* (SIMD) subword parallelism. By taking a 64-b ALU and allowing the carry chain to be broken at various points, essentially the same amount of logic can perform two 32-b operations, four 16-b operations, or eight 8-b operations, all in parallel.

One complication is that multiple carry bits are not available. Fortunately, however, most signal-processing operations benefit from *saturation arithmetic.* Instead of rolling over and setting the carry bit, saturation arithmetic sets the result at the minimum or the maximum value. Most multimedia extensions add saturation arithmetic as an option. Other common additions are instructions for *multiply-add* and data element *packing* and *unpacking,* etc.

In Fig. 4.21, we show the data types used in Intel's MMX (*Multimedia Extension*) technology for extending the x86 architecture. Three packed or compressed data types are defined over the conventional 64-b *quadword* shown in Fig. 4.21(d).

Part (a) shows 8 *packed bytes,* 8 b each, into one 64-b quad-word. Part (b) shows the division of 64 bits into 4 *packed words,* 16 b each.

Similarly, Part (c) shows two *packed double-word* in 64 b. Each packet element can hold a 8-b, or a 16-b, or a 32-b integer. These are the data types often used in multimedia computations.

Industrial Extensions Hewlett-Packard was the first to add the MAX extensions to its precision architecture, but HP's instructions are quite simple. Sun offers the most comprehensive set of extensions in its VIS (*Visual Instruction Set*), implemented in UltraSPARC. Sun's extensions include some relatively complex instructions, such as pixel distance, in addition to the simpler SIMD operations.

The most widely discussed set of extensions is Intel's MMX, which appeared in the Pentium P55C and Pentium-2 (Klamath) processors. Overall, 57 MMX instructions were added to the x86 *instruction architecture* (IA). Both AMD and Cyrix will offer MMX-compatible extensions as well. Intel estimates that the performance of MMX-enhanced code will be four times better for MPEG video decoding.

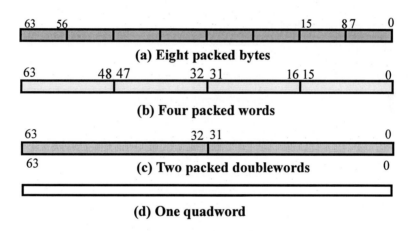

Figure 4.21   MMX data types for Intel x86 microprocessors.

At the time this book was written, no compilers can optimize on MMX codes, programmers must hand-craft the code to realize the benefits. Silicon graphics has planned the MDMX extension for its MIPS-V processor. Digital planned the MVI extensions to its Alpha chips. Details are not available for either yet.

Partitioning of Superscalar ALUs In the past, superscalar processors have been mainly used to exploit MIMD parallelism over the multiplicity of ALUs in the processor. With only minor hardware modification, the same ALU resources can be used to exploit SIMD parallelism as well.

This is demonstrated by the partitioning of two 64-b ALUs in a given superscalar processor (Fig. 4.22). Each 64-b ALU is partitioned into four 16-b ALUs, which can operate in parallel. In total, 8-way 16-b arithmetic operations can be performed per cycle in SIMD lock step. The operands come from four 64-b *multimedia registers*; each holds four 16-b words of data element.

In total, 16 data elements are processed by eight 16-b ALUs in parallel. For example, the following eight 16-b integer additions are done in 1 cycle, assuming that all operands are already loaded in the registers. Eight random-access MMX registers are built in a Pentium processor for exclusive use by MMX instructions.

$$xi + yi = xi, \quad \text{for } i = 1, 2, 3, \text{ and } 4 \tag{4.2}$$

$$yj + xj = yj, \quad \text{for } j = 5, 6, 7, \text{ and } 8 \tag{4.3}$$

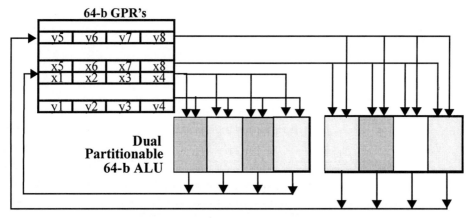

**Figure 4.22   Using two partitioned 64-b ALUs in a superscalar
processor to perform eight 16-b integer operations
per cycle.** (Courtesy R. Lee, *IEEE Micro* [400], 1996)

Packed Multiply-Add Operations   In Fig. 4.23, we show how to perform one of the frequently used MMX operations.

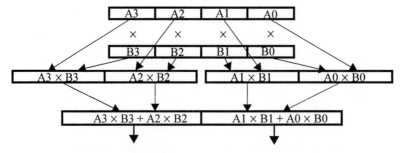

Figure 4.23   Packed multiply-add operations (Courtesy R. Lee, *IEEE
Micro* [400], 1996)

The idea is to pack multiply-add a 16-b word to 32-b double-words. Four 16-b of multiplications are performed first to produce four 32-b intermediate results in packed double-word. Two 32-b additions are then performed to produce the 32-b end results in two doublewords packed in one 64-b MMX register.

This whole sequence of six arithmetic operations is accelerated by built-in fast multiply-accumulate hardware, which is modified from the original Pentium hardware with only minor changes in the design.

Future Opportunities Media processing is becoming a new design target for computer engineers. On the PA-8000 with MAX-2 extensions, execution time has achieved 2.63 to 4.68 times speedup in various multimedia operations. Subword SIMD parallelism has opened up new opportunities for both hardware and software designers. Besides ISA extensions, one can consider designing dedicated media processors or coprocessors.

4.5.3 The VLIW Architecture

The Intel and HP joint project in developing the next generation of processors has renewed the interest to exploit more parallelism using the VLIW approach. We assess below the past failure and the future prospect of the VLIW architecture. As illustrated in Fig. 4.24(a), multiple functional units are used concurrently in a VLIW processor. All functional units share the use of a common large register file.

The operations to be simultaneously executed by the functional units are synchronized by hardware. The VLIW concept is extended from horizontal microcoding. Different fields of a long instruction carry different opcodes to be dispatched to different functional units.

A *very long instruction word* can be as long as 256 b or 1024 b as designed in the Multiflow computer [241] in early 1980s. The Multiflow VLIW processor was implemented with a microprogrammed control. The 256-b Multiflow model allowed up to seven operations to be executed concurrently. The clock rate could not be raised very high due to frequent access of control memory with microprogrammed control then.

One can view a VLIW processor as an extreme of a superscalar processor in which all independent or unrelated operations have already packed together at compile time. Programs written in conventional RISC instruction words (32 b each) must be compacted to form a VLIW instruction. This code compaction must be done by a capable compiler which can trace program flows and predict branch outcomes using trace information.

The failure of the Multiflow machine was mainly attributed to the unavailability of a truly powerful compiler to compact the short user code words into VLIW instructions efficiently. Should the P7 processor adopt the VLIW architecture, the compiler support is a necessity.

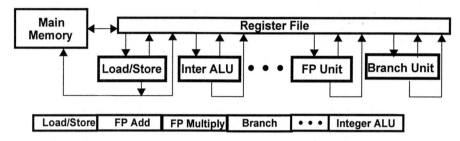

(a) A VLIW processor architecture and instruction format

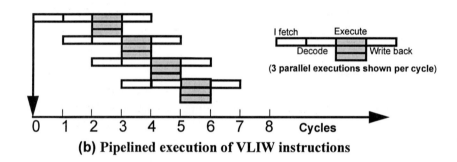

(b) Pipelined execution of VLIW instructions

Figure 4.24   The VLIW architecture and parallel execution of multiple operations at the execution stage of the processor pipeline.

Advantages of VLIW Processors   The execution of instructions by an ideal VLIW processor is shown in Fig. 4.24(b). Each instruction specifies multiple operations. The effective CPI reduces to $1/n$, for having n-way parallelism in each long word. A VLIW machine is improved over a superscalar machine in the following areas:

(1) The synchronization of parallel operations in a VLIW instruction is entirely done at compile time, making it possible to achieve higher processor efficiency than that using a superscalar processor.

(2) The code length in a VLIW program is much shorter when the available ILP is high in the short-format user codes. This implies much shorter execution time of the compiled VLIW program.

(3) Runtime resource scheduling is much simplified, because instruction parallelism and data movement in a VLIW architecture are completely specified at compile time.

VLIW Pitfalls In a VLIW architecture, random parallelism among scalar operations is exploited instead of regular parallelism exploited in a vector supercomputer or lockstep parallelism in an SIMD computer. VLIW processor cannot be used without effective code translation by an intelligent compiler.

The VLIW architecture is incompatible with today's x86 or RISC programs. For this reason, the joint HP/Intel project must deal with the issue of backward software compatibility. The success of a VLIW-based architecture depends heavily on an efficient compiler to take over the scheduling functions from hardware.

Another problem is that software compiled for one VLIW machine may have to be recompiled for another VLIW machines. In other words, new compilers must be developed when the ILP is increased between generations of VLIW machines. All of these difficulties prevented the VLIW from becoming a commercial success in the past.

4.6 The Future of Microprocessors

Future microprocessor chips will be designed with much denser and bigger microchips, higher clock rate, higher ILP exploited, lower CPI, increased power consumption, and more sophisticate software support. In this section, we assess the trends in hardware, workloads, architecture, and software support for future microprocessors. Possible processor architectural trends are introduced.

The bulk of the material is based on two special issues on modern microprocessors, appeared in *IEEE Micro,* December 1996 and in *IEEE Computer Magazine*, September 1997. We give below a global picture of the future of microprocessors. The coverage is well extended into the 21st century.

More details of these microprocessor trends can be found in the articles cited. Whether an architectural concept can become a successful commercial reality depends on many technical and business factors. Some of the factors are elaborated below.

4.6.1 Hardware Trends and Physical Limits

In 1994, the Semiconductor Industry Association (SIA) has predicted that by 2010, 800 million-transistor CPU chips will be produced with thousands of pins, a 1000-b bus, a clock rate over 2 GHz, and a power dissipation as high as 180 W.

In other words, billion-transistor microprocessors will appear in the next decade. Specifically, we study the assessment of the future from Intel and Digital Equipment, based on their prior experiences in producing high-performance microprocessors.

The Intel Projection Albert Yu, head of Intel microprocessor products, has predicted the future of microprocessors in late 1996, as summarized in Table 4.7. The prediction suggested in 10 years, the transistor count could jump to 350 million in 2006.

The square die size will increase from the 0.7 in to 1.4 in on each side. This prediction is rather conservative, because 0.1 um technology may be able to accommodate 1 billion transistors on a single chip.

Table 4.7 Prediction of Future Microprocessor Characteristics
(Source: A. Yu, "The Future of Microprocessors," *IEEE Micro*, [663],1996)

Characteristic	1996 (Actual)	Prediction	
		2000	2006
Clock speed (MHz)	200	900	4000
Transistors (millions)	6	40	350
Die size* (in)	0.7	1.1	1.4
Line width (um)	0.35	0.2	0.1
Performance: MIPS / SPEC int 95	400 / 10	2400 / 60	20,000 / 500

* Length of single side of a square die.

The increase of clock rate to 900 MHz is quite conservative. But the sharp increase to 4 GHz in 10 years is a surprise. This implies that opportunities exist for lots more functionality in future microchips. The increase of microprocessor performance will be 50 times in 10 years, according to this projection.

The DEC Projection   Digital Equipment engineers have estimated the road map of the Alpha CPU chips up to 2003, as summarized in Table 4.8. Digital's projection differs little from the Intel's. In 5 years, the clock rate could be doubled. DEC projects instruction issue rate to increase to 16 instructions per cycle by 2000 and of 32-way parallelism by 2003.

Table 4.8 Road Map for Digital's Alpha CPU Chips
(Courtesy of Digital Equipment Co. [191], 1997)

Model / Year	21264 (EV6) 1997	EV7 2000	EV8 2003
CMOS Process (um)	0.35	0.25	0.18
Die Density (million)	15	100	250
Clock rate(MHz)	500	750	1000
Instruction issue rate	8-way	16-way	32-way
SPEC int92 (projection)	800-1000	3000	Unknown

Such a high degree of ILP is rather difficult to exploit without a major breakthrough in architecture and compiler areas. One can expect much improved yield in future large-size wafers and thinner line width down to 0.1 um. However, the increase in die size, thinner line width, and much higher clock rate may create new physical limits as expressed below by leading experts in the field.

Physical Limits Martzke [435] has indicated that the most important physical limit is the fact that on-chip wires are becoming much slower than logic gates as the line width and on-chip devices shrink dramatically in the future. A single global clock may not be able to drive a billion transistors over the entire microchip. Although reduced feature size is good news, wire scaling, and clock skewing may severely sabotage the performance gains.

Only 16% of the die will be reachable within a single clock cycle in a billion-transistor processor. Sending signals across the superchip may require 20 cycles of a 1.2-GHz clock. Another limit may be imposed by the excessive heat released from a giant CPU chip. Cooling and packaging will be real problems by then. All of these physical barriers are yet to be crossed in the years ahead.

4.6.2 Future Workloads and Challenges

Processor architecture is driven by the anticipated application workload. Both general-purpose and application-specific processors will expect major changes in workload in the next two decades. For example, Internet and WWW applications have triggered the multimedia and communication extensions in microprocessors.

User interfaces will consume more power in multimedia microprocessors. Multimedia workloads will continue to grow in real-time and embedded applications. Ruby Lee of HP has indicated that subword parallelism offers opportunities for innovations in media processor architecture, cache and memory systems, subword-parallelism compilers, pictorial languages, and application algorithms.

In addition, the design, verification, and testing of microprocessors may account more than 40% of the total development cost. Fabrication plants may easily exceed $2 billion for producing high-performance microprocessors in large quantity. Manufacturers can sustain such costs only if large market demand continues flourishing.

The joint Intel/HP P7 effort in developing the next generation of processors is rumored to merge the positive features from the decoupled architecture with those of the VLIW and multimedia architectures. It will be interesting to watch the actual architectures of the EV7, EV8, and P7, all targeted for use by the turn of the century.

Besides hardware advances, other challenges in future microprocessors include the exploitation of ILP in new processors dedicated for specific workloads. Compiler and runtime system software are both needed to automatically extract parallelism in future architecture. The issue of backward compatibility with legacy software also hinders the architectural innovation.

4.6.3 Future Microprocessor Architectures

Possibly for future billion-transistor microprocessors, Burger and Goodman [115] have surveyed seven candidate architectures in Fig. 4.25. The current microprocessors sit on top the evolution path. The farther down the architectures are, the more they depart from current programming models.

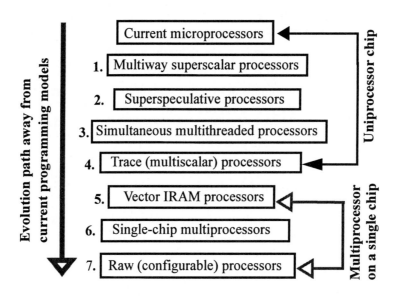

Figure 4.25   Projected evolution of future processor architectures
(Courtesy Berger and Goodman, *IEEE Computer Magazine*
[115], September 1997.)

These architectural trends are loosely ordered along the vertical axis. Those factors discussed in previous sections determine the actual growth path of the future architecture. Among the eight architectures in Fig. 4.25, the current microchips and the top 4 architectures are uniprocessor on a single chip. The remaining three are multiprocessor chips.

The trend to depart from existing ISA is counterbalanced by the need of retaining the current programming models. If the market continues to insist on compatibility with legacy code, the market will prevent architectures from evolving toward those at the bottom of the road map.

These architectures are mostly still at the research stage. We briefly introduce them below. Major research groups in the U.S., that are developing these architectures, are identified for readers to follow up with their future development.

(1) ***Multiway superscalar processors***: Today's superscalar processors are mostly 3-issue or 4-issue designs. The research group at the University of Michigan led by Yale Patt [491] identified that the instruction supply, data memory supply, and an implementable execution core are the key problems preventing current superscalars from scaling up to 16-way or 32-way designs.

They purpose to use out-of-order fetching, multi-hybrid branch predictors, and trace caches to improve the instruction supply and to use huge on-chip caches and data value speculation to enhance the data supply. They advocate a large, out-of-order-issue instruction window (2000 instructions), clustered banks of functional units, and hierarchical scheduling of ready instructions. They want to retain software compatibility with current uniprocessor chips.

(2) ***Superspeculative processors***: The group led by John Shen at the Carnegie-Mellon University [418] focuses on using massive speculation at all levels to improve performance. They proposed a Superflow microarchitecture with fetch width of 32, a reorder buffer of 128, and a 128-entry store queue for various memory configurations.

The use a weak-dependence model to achieve a superspeculative performance of up-to-19 *instructions per cycle* (IPC) for some benchmarks and a harmonic mean of 9 IPC for the SPEC95 integer suite, without requiring recompilation or change to the ISA. This effort complements, in many ways, the multiway superscalar architecture at Michigan.

(3) ***Simultaneous multithreaded processors***: *Simultaneous multithreading* (SMT) represents multiple-context uniprocessor at the University of Washington led by Susan Eggers [227]. The SMP approach departs from those single-threaded uniprocessor architectures.

The SMT processors share an aggressive pipeline among multiple threads out of multiprogrammed workloads. The success of this approach depends heavily on the availability of high ILP at the thread level. So far, only simulation results have demonstrated some performance gains of this approach.

(4) ***Trace (multiscalar) processors***: This concept was proposed by Smith and Vajapeyam [572] at the University of Wisconsin. The idea is to use a trace processor consisting of multiple on-chip processor cores, each of which simultaneously executes a difference trace of code. All but one core executes the traces speculatively, using branch prediction to select traces.

The Wisconsin group argues that future *multiscalar* processors will rely heavily on replication, hierarchy, and prediction to dramatically increase the execution speed of ordinary sequential programs. For an example, four processing elements can execute 16 or more instructions per cycle, if each element issues four instructions per cycle.

(5) ***Vector IRAM processors***: IRAM stands for *intelligent random-access memory*. This is a research project at the University of California at Berkeley led by David Patterson [377]. They argue that memory system will be the major performance bottleneck in the future. They explore DRAM technology to build scalable multiprocessor embedded within a large-scale memory array on chip.

The resulting on-chip memory capacity and high bandwidths should allow cost-effective vector processors to reach performance levels much higher than traditional architectures. The approach requires explicit compilation as maturely done in current supercomputers. They believe that future workloads will contain more vectorizable components.

(6) ***Single-chip multiprocessor***: This is Stanford University project [296] developing a *chip multiprocessor* (CMP), which implements 4 to 16 fast processors on one chip. Each processor is tightly coupled to a small level-1 cache and all processors share a large level-2 cache. The processors may collaborate on a parallel job or run independent tasks.

To make CMP a success, either programmers or compilers will have to make code explicitly parallel. The old ISA will not be compatible with this CMP architecture, although they can run slowly or inefficiently on one of the small processors. The Stanford researchers expect thread and process parallelism to become widespread in the future.

(7) ***Raw (configurable) processors***: This project at MIT Laboratory for Computer Science [631] offers the most radical architecture, away from the traditional or future architectures we have studied so far. The idea is to implement highly parallel architecture with hundreds of very simple processors, each with some reconfigurable logic on a single chip.

Parallel execution and communication are entirely resorting to software controlling and coordination. This approach eliminates the traditional instruction set interfaces. It exposes the replicated architecture directly to the compiler. This configurable architecture allows the compiler to customize the hardware to each application.

4.7 Bibliographic Notes and Problems

Advanced microprocessors are surveyed in the book by Tabak [603]. Crawford describes the i486 architecture in [173]. The M68040 processor is described in Edenfield, et al. [224]. A special issue marking the 25th anniversary of microprocessors is the December 1996 issue of *IEEE Micro*. A tutorial on RISC computers is given in Stallings [581]. A survey of RISC architecture is given in Hennessy and Patterson [305]. Hwang's book [327] covers processors used in all computer classes.

The history of Intel 4004, the very first microprocessor, is given in Faggin et al. [234]. The Berkeley RISC was reported by Patterson and Sequin in [493]. The MIPS R2000 was discussed in [359], MIPS R4000 in Mirapuri et al [450], and MIPS R10000 in Gwennap [291] and Yeager [660]. Superscalar and superpipelined machines are characterized by Jouppi and Wall [356]. Johnson [352] covers most of the design issues of superscalar microprocessors.

SPARC architecture is characterized in Weaver and Germond [637]. Optimizing compilers for SPARC appears in Munchnick [456]. UltraSPARC is treated in Greenley et al. [281] and Tremblay and O'Connor [617]. Multiprocessing support is given in [23] for the PowerPC 601. The PowerPC 604 is described in Denman et al. [202] and in Song et al. [578]. The PowerPC 620 is described in Levitan et al. [408].

The HP precision architecture is assessed in Lee [398]. The PA-8000 is described in Hunt [322]. Using microprocessors in MPPs is assessed by Smith [571]. The Alpha AXP architecture is described in Digital manuals [191], [196], Edmondson et al. [225], and Sites. RISC processor development at IBM was reported in [160], [335].

The Pentium architecture is described in Anderson and Shanley [31]. A good source of the i860 architecture can be found in Margulis [428]. Pentium Pro (P6) architecture is described in Colwell and Steck [162] and in Gwennap [290]. Multimedia extensions of microprocessors can be found in Lee [400], [401], Peleg [495], and Weiser [640]. Diefendorff and Dubey [205] have assessed how the multimedia wordloads may change the microprocessor design.

The VLIW architecture was first proposed by Fisher [241]. The post-RISC material is based on the report by Brehob et al. [108]. The Hot-Chips Symposium Series [321] often presents the latest development on processor chips. The impact of microprocessors on society is discussed in Markoff [429].

An assessment of future microprocessors appears in Yu [663]. Billion-transistor architectures are surveyed in Burger and Goodman [115]. Matzke [435] asseses the physical limits on processor scalability. The seven future microprocessor architectures are reported in Patt et al. [491], Lipasti and Shen [418]. Smith and Vajapeyam [572], Eggers et al.[227], Kozyrakis et al. [377], Hammond et al. [296], and Waingold et al.[631].

Homework Problems

Problem 4.1 Define the following basic terms related to processor technology:

(a)   Instruction issue rate

(b)   Addressing modes

(c)   Unified cache versus split caches

(d)   Hardwired control versus microprogrammed control

Problem 4.2 Answer the following questions on designing scalar RISC or superscalar RISC processors:

(a)   Why do most RISC integer units use 32 general-purpose registers? Explain the concept of register windows implemented in the SPARC architecture.

(b)   What are the design tradeoffs between a large register file and a large D-cache? Why are reservation stations or reorder buffers needed?

(c)   What is the relationship between the integer unit and the floating-point unit in most RISC processors with scalar or superscalar organization?

Problem 4.3 Based on the discussion of advanced processors in Section 4.2, answer the following on RISC, CISC, superscalar, and VLIW architectures.

(a)   Compare the instruction set architecture in RISC and CISC processors in terms of instruction formats, addressing modes, and cycles per instruction.

(b)   Discuss the advantages and disadvantages of using a common cache or separate caches for instructions and data. Explain the support from data paths, MMU and TLB, and memory bandwidth in the two cache architectures.

(c)   Explain the difference between superscalar and VLIW architectures in terms of hardware requirements and software support.

Problem 4.4 Characterize the following five microprocessor architectures by describing one example commodity or research microarchitecture in each case.

(a)   Superscalar RISC with out-of-order execution

(b)   Post-RISC with multimedia extensions

(c)   Superspeculative microarchitecture

(d)   IRAM or intelligent RAM architecture

(e)   Single-chip multiprocessor (CMP).

Problem 4.5 Answer the following two questions on multimedia extensions of commodity microprocessors:

(a)   What are the major multimedia data types that can exploit SIMD parallelism on today's 64-b superscalar RISC processor?

(b)   What is the maximum speedup in using a 64-b, four-issue, superscalar processor with four integer ALUs for packed-byte SIMD/MIMD multimedia processing? Assume that the processor is fully equipped with multimedia extensions. All comparisons are against an identical processor without multimedia extensions.

Problem 4.6 Consider the execution of an object code with 200,000 instructions on a 40-MHz processor. The program consists of four major types of instructions. The instruction mix and the CPI needed for each instruction type are given below, based on the result of a program trace experiment.

(a) Calculate the average CPI when the program is executed on a uniprocessor with the above trace results.

(b) Calculate the corresponding MIPS rate based on the CPI obtained in part (a).

Instruction Type	CPI	Instruction Mix
Arithmetic and logic	1	60%
Load/store with cache hit	2	18%
Branch	4	12%
Memory reference with cache miss	8	10%

Problem 4.7 A 40-MHz processor was used to execute a benchmark program with the following instruction mix and clock cycle counts. Determine the effective CPI, MIPS rate, and execution time for this program:

Instruction Type	Instruction Count	Clock Cycle Count
Integer arithmetic	45,000	1
Data transfer	32,000	2
Floating point	15,000	2
Control transfer	8,000	2

Problem 4.8 Consider the execution of a program of 15,000 instructions by a linear pipeline processor with a clock rate of 25 MHz. Assume that the instruction pipeline has five stages and that one instruction is issued per clock cycle. The penalty due to branching is ignored in (a), but not in (b).

(a) Calculate the speedup factor in using this pipeline to execute the program compared to the use of an equivalent nonpipelined processor with an equal amount of flow-through delay.

(b) What are the efficiency and throughput of this pipelined processor, if 25% additional cycles are needed to execute the same code, if the branching effects are included ?

Problem 4.9 Consider a superpipelined superscalar processor, that can issues m instrucyions per cycle and is driven by a clock n times faster than that of a base processor that issues one instruction per cycle.

(a)   Derive a speedup expression $S(m, n)$ of the superpipelined superscalar processor over the base processor. Note that the speedup is a function of m and n.

(b)   What are the practical limitations preventing the growth of the m-issue superscalar processor?

(c)   What are the practical limit on the growth of the superpipeline with an n-times faster clock?

Problem 4.10   A nonpipelined processor X has a clock rate of 25 MHz and an average CPI of 4. Processor Y, an improved successor of X, is designed with a five-stage pipeline. However, due to latch delay and clock skew effects, the clock rate of Y is only 20 MHz.

(a)   If a program containing 100 instructions is executed on both processors, what is the speedup of processor Y compared with that of processor X?

(b)   Calculate the MIPS rate of each processor during the execution of this program.

Problem 4.11   Consider the post-RISC features in some of today's microprocessors.

(a)   Explain at least four post-RISC features that differ from the pure superscalar RISC features. Characterize each feature and explain why the added feature can enhance the performance of microprocessor.

(b)   For each post-RISC feature identified in part (a), identify one processor example having the feature and justify your claims.

(c)   Rank the following 10 processors in order of having more built-in post-RISC features identified in Parts (a) and (b):

PowerPC 620: _____   AMD K6: _____

PA-8000: _____   Alpla 21164: _____

Pentium-2: _____   Pentium P55C: _____

UltraSparc-2: _____   MIPS R10000: _____

PowerPC 604: _____   Alpha 21064: _____

Chapter 5

Distributed Memory and Latency Tolerance

In a modern computer platform, the cost of storage devices (caches, RAM chips, hard disks, tape units, and other peripheral devices) has far exceeded that of the processor core. In this chapter, we assess recent advances in memory technology. In particular, we concentrate on distributed-memory architectures and the underlying latency tolerance techniques. Both hardware and software approaches are studied to reduce, avoid, or hide latencies incurred with remote memory accesses.

We are interested in both hardware and software mechanisms for supporting memory sharing, cache coherence, memory consistency, and latency hiding in scalable parallel computers. These topics on memory architecture affect the performance, scalability, and programmability of multiprocessors and multicomputers choosing various distributed-memory architectures such as CC-NUMA, COMA, software-implemented DSM, etc. Our study will cover memory sharing at all granularity levels, ranging from memory word to cache line and virtual pages.

5.1 Hierarchical Memory Technology

Memory characteristics and operational properties are summarized below for a typical memory hierarchy. Then we present a linear programming model for capacity planning of such a memory hierarchy.

5.1.1 Characteristics of Storage Devices

Storage devices such as registers, external caches, main memory, disk drives, and tape units are often organized as a hierarchy depicted in Fig. 5.1. We use M_i to refer the memory at the ith level. In Fig. 5.1, we show a hierarchy with four levels, where M_0 is the registers or internal cache in the CPU, M_1 the external cache, M_2 the main memory, M_3 the disk storage, and M_4 the tape unit. The memory technology and organization at each level are characterized by five parameters defined below.

- The *access time* t_i in μs to access the memory M_i
- The *memory capacity* s_i of M_i in MB (megabytes)
- The *cost per byte* c_i of M_i in cents per byte
- The *data transfer rate* or *bandwidth* b_i in MB/s between M_i and M_{i+1}
- The *unit of transfer* x_i in bytes, or words, cache lines, pages, or segments between adjacent levels M_i and M_{i+1}

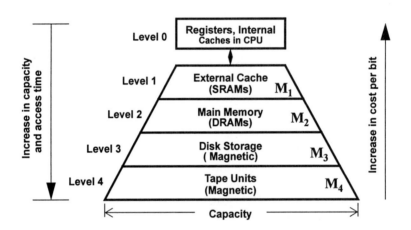

Figure 5.1   A four-level memory hierarchy for a large computer system.

A memory device at a lower level is faster to access, smaller, more expensive per byte, higher in transfer bandwidth, and it uses a smaller unit of transfer compared with those parameters at a higher memory level. In other words, we have the following inequalities for i = 1, 2, ..., n in an n-level memory hierarchy:

$$t_{i-1} < t_i, \quad s_{i-1} < s_i, \quad c_{i-1} > c_i, \quad b_{i-1} > b_i, \quad x_{i-1} < x_i \qquad (5.1)$$

A modern processor has typically 32 integer registers and 32 floating-point registers. These are called *general-purpose registers* (GPRs) or *register files.* At compile time, virtual register assignment is made. They are mapped into physical registers at run time.

The registers and on-chip caches are part of the CPU chip. In this memory model, the off-chip external cache is considered the M_1 at level 1. The main memory is at level 2, hard disk at level 3, and tape unit at level 4 in Fig. 5.1.

Multilevel Caches The cache itself could have a subhierarchy of several levels. On-chip cache, often accessed in 1 cycle by the processor, is considered level 1. Off-chip cache is called the *level-2 cache.* For simplicity in our analysis, we consider only the effective access time of the entire cache subhierarchy as one level. Off-chip caches are often built with *static random-access memory* (SRAM) integrated-circuit chips.

Most on-chip I-cache and D-cache have 32 to 128 KB in combined capacity. The off-chip cache may vary from 128 KB to 64 MB, depending on the external cache controller designed with the processor type. The *memory management unit* (MMU) is now built as part of the CPU chip. The MMU is programmer-transparent. Most caches are transparent to users with a few exceptions. In other words, most caches are not user-programmable.

Main Memory The main memory is often built with multiple *dynamic random-access memory* (DRAM) chips packaged on small memory boards, called SIMMs (*small in-line memory module*). DRAMs are available in 4, 8, 16, 32, and 64 Mbs per chip in 1997. The 128- to 1-Gb DRAMs are soon to appear. Typical main memory capacity rages from 8 to 64 MB in today's PCs, 32 to 128 MB in workstations, and 128 MB to 4 GB in SMP servers. These capacities will continue to increase in the future.

Main memory is often organized as *interleaved memory modules* for fast pipelined access or for parallel access of a large number of memory words at the same time. The *memory bandwidth* is defined as the number of words or millions of bytes that can be transferred over a memory bus or a memory switch. Interleaved memory modules can be also grouped together as *memory banks* to provide not only more memory bandwidth, but also higher tolerance of memory failures isolated to each memory bank.

Main memory is managed by the MMU in cooperation with the operating system. Efficient memory management between physical memory and the huge disk space is one of the major OS functions. Options are often provided to extend the main memory by adding memory boards, or CPU/memory boards if the physical SIMMs are distributed to different CPU boards. The main memory can be also organized by multiple levels using different speed memory technologies at different levels. The purpose is to reduce the total memory cost.

Disk Drives Disk drives and tape units are handled by the OS with limited user intervention. The disk storage is considered the highest level of *on-line memory.* It holds the system programs such as the OS and compilers, as well as user programs and their data sets. Magnetic disks appear in two types: *floppy disks* and *hard disks,* both using mechanical rotation of 2 to 20 platters coated with magnetic recording material. Floppy disks are made from flexible mylar substance, often 1.44 MB per floppy diskette, and are removable from the floppy drive.

Hard disks are made from metal and cannot be removed from the disk drive, once installed. Hard disks have been reduced in diameter from 14 to 1.8 in in products over the past 15 years. The formatted data capacity of these disks varies from 20 MB for a 1.8-in Integral 1820 disk drive to 22.7 GB in a 10.88-in disk drive in the IBM 3090 mainframe computer.

The disk access time varies from 10 to 25 ms, which is quite slow due to mechanical rotations. The data transfer rate varies from 1.9 to 4.2 MB/s. Magnetic disks are nonvolatile and worry-free from power interruption. But they are prone to crushes or mechanical failures. That is why information stored on disks should be backed up on the next level of memory, the tape units.

Tape Units   The magnetic tape units are used off-line as a backup storage. The tape unit access time could be as slow as seconds. But the capacity can easily reach 100 GB or even in TB. The tape units hold copies of present and past user programs, executed results, and processed files. Information files stored on tapes are considered achieved and not on-line. To retrieve them, computer operators must intervene.

Table 5.1 compares the physical characteristics of main memory, disk drive, and tape unit. The unit capacity shows the information storable in a unit area. The system capacity shows the capacity of a high-end workstation computer. The current values are the advertised numbers by workstation/server vendors in 1997..

Table 5.1   Memory, Disk Drive and Tape Unit for A High-end Workstation

Parameter	Value	Memory	Disk Drive	Tape Unit
Capacity	Unit value	64 Mb/die	0.2 Mb/mm^2	$0.1\text{-}0.4 \text{ Mb/mm}^2$
	System	32-128 MB	2-4 GB	8-16 GB
	Improve rate	2X per 2 yr	25-60% per yr	10X per 10-21 yr
Access time	Current value	60 ns	10-25 ms	seconds
	Improve rate	3X per 12 yr	2X per 10 yr	N/A
Bandwidth	Current value	500-1000 MB/s	1-5 MB/s	0.5-4 MB/s
	Improve rate	N/A	2X per 5 yr	2X to 6X per 10 yr

5.1.2 Memory Hierarchy Properties

Described below are three fundamental properties, *inclusion, coherence,* and *locality* for designing or using an efficient memory hierarchy. The design rules of a memory hierarchy are essentially derived from these properties. These properties set the relationship between adjacent memory levels. They also govern the operation of the memory hierarchy.

Inclusion Property The *inclusion property* is stated by the following set inclusion relations among n memory levels:

$$M_1 \subset M_2 \subset M_3 \subset \dots \subset M_n \qquad\qquad (5.2)$$

The set inclusion relation between two adjacent levels implies that all information objects (words, cache lines, or pages) stored in level M_i form a subset of those stored in level M_{i+1} for all $i = 1, 2, \dots, n$. Initially, the outermost level M_n contains everything needed to execute the programs. During the execution period, subsets of M_n are copied into the next lower level M_{n-1}. Similarly, subsets of M_{n-1} are copied into M_{n-2}, and so on until the required instructions and data move to the innermost memory M_1, ready for execution by the processor.

If a piece of information is found in M_i, then copies of the same information can be also found in higher levels M_{i+1}, M_{i+2}, ..., M_n. However, a word stored in M_{i+1} may not be found in M_i. A word miss in M_i implies also a miss at all lower levels M_{i-1}, M_{i-2}, ..., M_1. The highest level is the backup storage, where everything can be found. As far as information copies are concerned, M_i is a proper subset of M_{i+1}.

Data Transfer Units This refers to the unit of information transfer between adjacent memory levels. Memory *words*, 4 b each for 32-b machines and 8 B for 64-b machines, are the transfer unit between CPU registers and internal on-chip caches. *Cache lines*, typically 32 B per cache line, are the transfer unit between internal cache and external cache at M_1. Some recent machines have extended the cache line size to 64 or 128 B.

The main memory at M_2 is often divided into fixed-size *pages,* say 4 KB per page. Different machines may choose different page sizes. A page is the unit of data transfer between the disk and main memory. Scattered pages are often logically grouped into a *segment* of files in the disk. Therefore, segment is the transfer unit between the tape unit and the disk. It should be clear that a word is part of a cache line, which is a part of a page. Similarly, a page is part of a segment in a segmented paging memory environment.

Coherence Property The coherence property requires that copies of the same information items be consistent at different memory levels. If a word or a cache line is modified in the lower level, copies of that word or cache line must be updated immediately or eventually at all higher memory levels. Currently used instruction/data items often accumulate quickly into the cache from higher-level memories.

Thus the coherence problem extends all the way from the cache at M_1 to the outermost memory M_4. Maintaining memory coherence among successive levels is a nontrivial task, often demanding extra bus cycles or prolonging the memory latency.

In general, there are two strategies to maintain the coherence in a memory hierarchy. The first method is called *write-through* (WT), which requires immediate update in M_{i+1} if a word is modified in M_i, for $i = 1, 2, \dots, n - 1$. The second method is *write-back* (WB), which delays the update in M_{i+1}, until the memory unit in M_i containing the modified information item is replaced by another new information item. Memory design is thus often divided into WT and WB.

In practice, the WT policy is often used for updating internal caches (I-cache or D-cache), because on chip update will cause extra-cycle delays. The WB policy is often used in updating external level-2 cache, labeled M_1 memory in our model. In other words, the WB policy will not update the higher levels immediately, not until the block or page replacement time.

Well-known memory replacement policies include the LRU (*least recently used*), LFU (*least frequently used*), FIFO (*first-in-first-out*), and random algorithm, etc. Details of these replacement policies can be found in most computer architecture books cited in Section 1.6.

Locality of Reference   The memory hierarchy is developed based on program behavior known as *locality of reference*. Memory references are generated by the CPU for either instruction or data access. These accesses tend to be clustered in certain regions in time, or space, or ordering as defined below:

(1) *Temporal locality.* Recently referenced items (instructions or data) are likely to be referenced again in the near future. This is often caused by special program constructs such as iterative loops, process stacks, temporary variables, or subroutines. Once a loop is entered or a subroutine is called, a small code segment will be referenced repeatedly many times. Thus temporal locality tends to cluster around in the most recently used code or data segments.

(2) *Spatial locality.* This refers to the tendency for a software process to access information items whose addresses are near one another. For example, operations on tables or arrays involve accesses of a certain clustered area in the address space. Program segments, such as routines and macros, tend to be stored in the same neighborhood of the memory space.

(3) *Sequential locality.* In a typical program, the execution of instructions follows a sequential order, called the *program order*, unless branch instructions create out-of-sequence executions. The ratio of in-sequence execution to out-of-sequence execution is roughly 5 to 1 in an ordinary program. Besides, the access of a large data array also follows a sequential order, such as row-major access of matrix elements.

Memory Design Implications   The sequential locality in program behavior contributes to all types of locality, because sequentially coded instructions and array elements are often stored in adjacent memory locations and referenced at about the same time. Each type of locality affects the design of the memory hierarchy.

• The temporal locality acts in favor of the use of the LRU replacement algorithm for memory design. It also leads to *working set* concept to be introduced later.

- The spatial locality assists us in determining the size of unit data transfers between adjacent memory levels. The temporal locality also helps determine the size of memory at successive levels.
- The sequential locality affects the determination of grain size for optimal scheduling (grain packing). Prefetch techniques are affected by these locality properties.

5.1.3 Memory Capacity Planning

The performance of a memory hierarchy is determined by the *effective access time* T_{eff}, to any level in the hierarchy. It depends on the *hit ratios* and *access frequencies* at successive levels.

We formally define these terms below. Then we present a model to optimize the capacity of a memory hierarchy subject to a given cost constraint.

Hit Ratios *Hit ratio* is defined between any two adjacent levels of a memory hierarchy. When an information item is found in M_i, we call it a *hit*; otherwise, a *miss*. Consider an *n*-level memory hierarchy. The *hit ratio* h_i is the probability that an information item is found at M_i. It is a function of the characteristics of levels M_{i-1} and M_i. The *miss ratio* is defined as $1 - h_i$.

The hit ratios at successive levels are a function of the memory capacities, management policies, and program behavior. Successive hit ratios in the hierarchy are independent random variables with values between 0 and 1. To simplify the derivation, we assume $h_0 = 0$ and $h_n = 1$, meaning the CPU initiates an access to the M_1 after a miss within the CPU chip; and the access to the outermost memory M_n always results in a hit.

The *access frequency* to M_i is defined as:

$$f_i = (1 - h_1)(1 - h_2) \ldots (1 - h_{i-1}) \, h_i \qquad (5.3)$$

This is indeed the probability of having a hit at level M_i when there are i - 1 misses at all the lower levels. Due to the locality property, the access frequencies decrease very rapidly from low to high levels; that is,

$$f_1 \gg f_2 \gg f_3 \gg \ldots \gg f_n \qquad (5.4)$$

This implies that the inner levels of memory are accessed much more often than the outer levels. Typically, the first level f_1 is higher than 95%, and the remaining levels add up to the remaining 5% in decreasing orders of magnitude.

Effective Access Time   In practice, we wish to achieve as high a hit ratio as possible at M_1. Every time a miss occurs, a penalty must be paid to access the next higher level of memory. The misses have been called *cache line misses* in the cache or *page faults* in the main memory, because cache lines and pages are the units of transfer between these levels.

The time penalty for a page fault is much longer than that for a cache line miss due to the fact that $t_1 < t_2 < t_3$. Stone (1990) has pointed out that a cache miss is 2 to 4 times as costly as a cache hit, but a page fault is 1000 to 10,000 times as costly as a page hit. Using the access frequencies f_i for $i = 1, 2, ..., n$, we formally define the *effective access time* of a memory hierarchy as follows:

$$
\begin{aligned}
T_{eff} &= \sum_{i=1}^{n} f_i \cdot t_i \\
&= h_1 t_1 + (1 - h_1)\, h_2 t_2 + (1 - h_1)\,(1 - h_2)\, h_3 t_3 + ... \\
&\quad + (1 - h_1)\,(1 - h_2)\, ...\, (1 - h_{n-1})\, t_n
\end{aligned}
\tag{5.5}
$$

The first several terms in Eq.(5.5) ominate the total value, because f_1 and f_2 are much greater than the others. The effective access time depends on the program behavior and memory design choices. Only after extensive program traces can one estimate the hit ratios and the value of T_{eff} accurately.

Hierarchy Optimization   The total cost of a memory hierarchy is estimated as follows:

$$
C_{total} = \sum_{i=1}^{n} c_i \cdot s_i
\tag{5.6}
$$

This implies that the cost is distributed over n levels. Since $c_1 > c_2 > c_3 > ... > c_n$, we have to choose $s_1 < s_2 < s_3 < ... s_n$. The optimal design of a memory hierarchy should result in a T_{eff} close to t_1 of M_1 and a total cost close to c_n of memory level M_n. In reality, this is difficulty to achieve due to the tradeoffs among n levels.

The optimization process can be formulated as a linear programming problem, given a ceiling C_o on the total cost. In other words, we want to minimize T_{eff}, subject to the following constraints:

$$
s_i > 0, \quad t_i > 0 \qquad \text{for } i = 1, 2, ..., n
$$

$$
C_{total} = \sum_{i=1}^{n} c_i \cdot s_i < C_o
\tag{5.7}
$$

As shown in Table 5.2, the unit cost c_i and capacity s_i at each level M_i depend on the speed requirement or the access time t_i required. Therefore, the above optimization involves tradeoffs among t_i, c_i, s_i, and f_i or h_i at all levels i = 1, 2, ..., n. The following example shows how to design a memory hierarchy by solving the linear programming problem.

Example 5.1 The design of a memory hierarchy

Consider the design of a three-level memory hierarchy with the following specifications for memory characteristic. The design goal is to achieve an effective memory access time T_{eff} = 10.04 us with a cache hit ratio h_1 = 0.98 and a main memory hit ratio h_2 = 0.9. Also, the total cost of the memory hierarchy has as an upper bound $15,000. The memory hierarchy cost is calculated as $c = c_1 s_1 + c_2 s_2 + c_3 s_3 < 15,000$.

Table 5.2   Parameters in the Example Memory Subsystem

Memory Level	Access Time	Capacity	Cost/KB
Cache	$t_1 = 25$ ns	$s_1 = 512$ KB	$c_1 = \$1.25$
Main memory	$t_2 = $ unknown	$s_2 = 32$ MB	$c_2 = \$0.2$
Disk array	$t_3 = 4$ ms	$s_3 = $ unknown	$c_3 = \$0.0002$

The maximum capacity of the disk is thus obtained as s_3 = 39.8 GB without exceeding the budget. Next, we want to choose the access time t_2 of the RAMs to build the main memory. The effective memory access time is calculated as

$$T_{\text{eff}} = h_1 t_1 + (1 - h_1)h_2 t_2 + (1 - h_1)(1 - h_2)h_3 t_3 < 10.04 \text{ us} \tag{5.8}$$

Substituting all known values, we have $10.04 \times 10^{-6} = 0.98 \times 25 \times 10^{-9} + 0.02 \times 0.9 \times t_2 + 0.02 \times 0.1 \times 1 \times 4 \times 10^{-3}$. Thus the access time of the disk is chosen as t_2 = 903 ns.

Suppose we want to double the main memory to 64 MB, at the expense of reducing the disk capacity, under the same budget limit. This change will not affect the cache hit ratio. But it may increase the hit ratio in the main memory if a proper page replacement algorithm is used. The total effective access time of the entire hierarchy will be reduced by doing so.

∎

5.2   Cache Coherence Protocols

Cache coherence protocols are studied in this section. The choice of a coherence protocol will affect the shared-memory performance as well as the correctness in program execution. The *coherency problem* refers to inconsistency of distributed cached copies of the same cache line addressed from the shared memory.

5.2.1 Cache Coherency Problem

A memory subsystem is considered *coherent* at the cache line level if the access (for read or write) to any cached copy of a cache line X in memory always returns with the following caching results (see Patterson and Hennessy [492], p. 656):

- A *read* following a *write* of X by processor P, with no *writes* of X by other processors, will return the value written by P.
- A *read* by processor P following a *write* to X by another processor Q will return the value written by Q, if the *read* and the *write* are sufficiently separated and there are no *writes* of X by other processors in between.
- *Writes* to the same cache line X by different processors are always serialized to present the same order of memory events, seen by all processors.

Sources of Incoherence   Cache incoherence has three possible causeses, as listed below:

(1) The *write* by different processors into their cached copies of the same cache line in memory, asynchronously
(2) *Process migration* among multiple processors without alerting each other
(3) *I/O operations* bypassing the owners of cached copies

These sources are illustrated below with simple examples on a shared-memory dual-processor system without any cache coherence control. Each processor has a private cache. We consider the effects on *write-through* (WT) and *write-back* (WB) caches separately. In the case of a WT cache, the memory cache line is always consistent with the latest modification in the cache. The WT cache consumes more bus cycles to update the memory after each write.

For a WB cache, the memory copy will not be updated until replacement. Therefore, the cache line in memory may differ from the cached copy, immediately after a write hit in the cache. The memory write-back is really a write-through delayed until it is absolutely necessary to do so. Therefore, the WB caches are more economical to implement on the memory bus, because this will free up some useful bus cycles for data transfers.

Example 5.2 Incoherent caches caused by write of shared data

The changes in cache state before and after a cache *write* are shown in Fig. 5.2. Before the update, assume that the two cached copies (labeled **x**) in the two processors are consistent with the cache line **x** in the shared memory.

In using WT caches, after processor P_1 changes its cached copy to **x'**, the memory copy is also changed to **x'** ; making it different from the other cached copy. In the case of WB caches, the change in cache 1 will not cause the change in cache 2 or in the memory copy, making them both different from the latest write in cache 1. This example clearly demonstrates the need to either invalidate the caches or to update them immediately after each cache write operation, because writable data are often shared in a multiprocessor system.

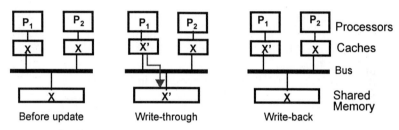

Figure 5.2   Cache incoherence by sharing of writable data

■

Example 5.3 Incoherent caches caused by process migration

Now, assume P_1 has a cached copy of line **x** in memory. P2 writes into the same line with a new content **y**, after a process migration from P_1 to P_2. This will create an inconsistency between cache 1 and memory in using WT caches as shown in the middle of Fig. 5.3.

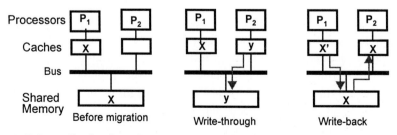

Figure 5.3   Cache incoherence caused by process migration.

On the other hand, the old value **x** may be migrated to cache 2 with a read operation, if the migration is done ahead of the write-back operation, when WB caches are used as shown on the right. In both cases, process migration causes cache incoherence.

■

Example 5.4 Incoherence caused by I/O operations bypassing caches

Input/output operations may bypass the caches to create incoherence as illustrated in Fig. 5.4, where the I/O box is an input device in using the WT caches and an output device for the WB caches. Initially, two cached copies are consistent with the cache line in memory.

The input device may load the memory line with a new content **x'**, making it different from neither of the two cached copies. In case of output with WB caches, cache 1 could have been modified to **x'**, but the old value **x** gets read by the output device. The message being conveyed is that I/O operations should alert the cache controller to the changes.

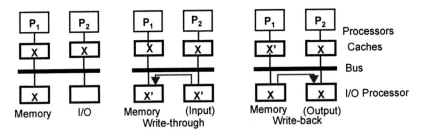

Figure 5.4   Incoherence due to I/O operations bypassing the cache

■

5.2.2 Snoopy Coherency Protocols

The above examples imply the need to apply some coherency control in writing of shared data, process migration, and I/O operations in a multiprocessor environment. In using private caches accessed by different processors tied to a common bus, two classes of coherency protocol are the *write-invalidate* and *write-update* protocols.

Essentially, the write-invalidate protocol invalidates all other cached copies when a local cached copy is updated. The write-update protocol broadcasts the newly cached copy to update all other cached copies with the same line address. These cache coherency protocols have been implemented with the use of *snoopy buses*.

Snoopy protocols require a broadcast mechanism such as that provided by a bus or ring. The idea is to design the bus to constantly monitor the caching events across the bus between processor and memory modules. For this reason, the name *snoopy coherency protocol* was created.

Two snoopy protocols are specified below and illustrated in Fig. 5.5 for the case of WB caches. When WT caches are used, the difference lies only in the timing to update the memory line after each *write* operation.

Write-Invalidate Protocol   In Fig. 5.5(a), we show the existence of three cached copies of a cache line **x.** When a processor wants to write to **x,** it must first obtain exclusive right to access **x.** It then writes the cache line with a new content **x'** and invalidates all cached copies of **x** in other processors. With WB caches, the memory line is also invalidated as shown in Fig. 5.5(b). If WT caches are used, the memory line will be filled with the changed value **x'** immediately..

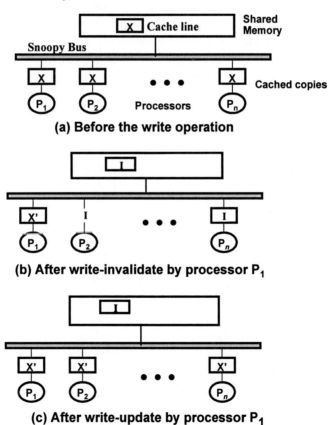

(a) Before the write operation

(b) After write-invalidate by processor P$_1$

(c) After write-update by processor P$_1$

Figure 5.5   **Write-invalidate and write-update snoopy protocols in using write-back caches** (**x**: original cached copy, **x'**: modified cache line, and **I**: cache line invalidated).

The invalidated cache copies are marked by I in Fig. 5.5(b). When any other processor tries to access the modified cache line **x'**, it will find a *cache miss* and will get the new value **x'** from the cached copy in processor P$_1$, if WT caches are used. In case WB caches are used, the new value will be retrieved from the modified cache, and the memory line must be filled (updated) at the same time.

Write Update Protocol   In this case, a processor writes to cache line **x**, it must update (i.e., broadcast the new value **x'** to) all cached copies of **x** in all processors. When any other processor tries to access the newly cached copy **x'**, it will encounter a *cache hit* and find the new value in its local cache as shown in Fig. 5.5(c). The memory line will be updated as shown in Fig. 5.5(c), with the use of WB caches. But the memory can be also updated immediately depending on the implementation choice.

Obviously, the write-update protocol will enforce a higher degree of coherency at all times. But write update is a very expensive operation. It may consume many bus cycles to update all caches as well as the cache line in memory. Therefore, most multiprocessor designers choose to implement the write-invalidate cache coherency protocol using WB caches. We will show the MESI invalidate protocol in the next subsection.

Besides snoopy protocols on a memory bus, cache coherency can be implemented with *directory-based protocols*, mainly on network-connected multiprocessors with NUMA or DSM models to be specified in Section 5.4.2.

5.2.3 The MESI Snoopy Protocol

The MESI is a write-invalidate snoopy protocol. It keeps track of the state of a cache line, considering all *read* or *write*, *cache hit* or *cache miss*, and *snoopy events* detected on the bus. We choose to illustrate the MESI protocol implemented in a Pentium-based multiprocessor shown in Fig. 5.6.

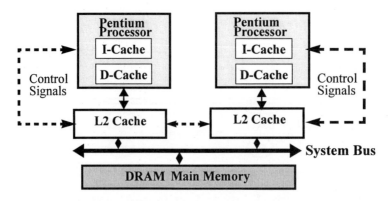

Figure 5.6   A dual-Pentium multiprocessor with level 2 caches

The Pentium MESI supports both WB and WT caching events controlled by an external signal. The protocol chooses a *non-write-allocate policy*, meaning there is no cache line fill from the memory on a *write miss*.

The MESI protocol is shown in Fig. 5.7 with a four-state transition diagram. All memory events and snoopy signals correspond to edge labels between states. Three possible transition edges are shown in the diagram. The thick dash lines apply only to WT protocol. The dot lines are for WB caching events only. The solid lines are for either WB or WT protocols.

Every line in the data cache is always in one of the following four possible states:

- *Modified* (**M** state): The cache line has been updated with a *write hit* by the local processor in the cache.
- *Exclusive* (**E** state): The cache is valid, and it is not valid in any other cache. Furthermore, the memory line has not been updated yet.
- *Shared* (**S** state): The line is valid, but it may also be valid in one or more remote caches or in the cache line in memory.
- *Invalid* (**I** state): This is the initial cache state after reset, or the cache line has been invalidated by a *write hit* by another cache with the same address.

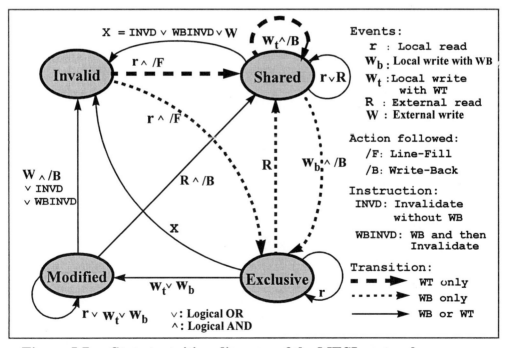

**Figure 5.7   State transition diagram of the MESI protocol
for data cache in a Pentium-based multiprocessor.**

The name *MESI* is taken from the first letters in the above four states. Transition from one state to another is caused either by a local processor *read* or *write* or by an external snoopy operation initiated by another bus master. The application of the *invalidate instructions* (INVD or WBINVD) is initiated by programmers. This offers the option for system programmers to flush the cache at reboot or context switching. User programmers usually are not allowed to use these instructions.

To clarify the variations of different snoopy protocols, we summarize in Table 5.3 the the caching and snoopy events, corresponding to all MESI state transitions in Fig. 5.7.

Table 5.3 Caching and Snoopy Bus Events

Event Symbol	Interpretation of Event, Action, and Consequence
M	Modified: dirty copy; read or write hit; write-back needed if being invalidated
E	Exclusive: sole clean copy; read hit; change to M state after a local write
S	Shared: clean copy; read hit; write to invalidate other caches; two or more copies exist in the system for a WB caching event
I	Invalid: not cached; read or write miss
r	A local read operation
w_t	A local write to a WT cache line
w_b	A local write to a WB cache line
R	An external snoopy hit on read
W	An external snoopy hit on write
/F	Fill this cache line from main memory
/B	Write back: invalidate all other caches and update the main memory
INVD	System instruction to flush its own cache
WBINVD	System instruction to flush its own cache after write-back
X	The joint event of INVD or WBINVD or W

Because of the support of both WT and WB caching events, Fig. 5.7 is really the combination of two transition diagrams: one for WT caching and another for WB caching events. They do share some of the states and transition edges. For simplicity and clarity, we describe these two cases separately.

Write-Through Protocol For a WT cache line, the initial state is **I,** and the only other possible state is **S.** When there is a *miss*, the clean cache line comes from either the main memory or another cache, and the cache line ends at the shared state. All transitions are enclosed within these two states. A two-state subgraph, formed by **S** and **I** and their surrounding thick-dash and solid transition edges at the top half of Fig. 5.7, is called the *SI protocol* for WT caching events exclusively.

A cache line will be filled from cache after a *read miss*. Also, other processors may read and cache this line (multiple copies exist) until some processor writes this line again. Upon this *write* operation, the writing processor first invalidates all other cache copies and then writes-through the new datum into its own cache and the memory. The writer's status remains at the **S** state, while those of others go to the invalid state **I.**

Write-Back Caches In a WB cache, a line is brought into the cache in the **E** state, which implies that this is the sole cache copy in the system. After another one or more processors read and cache this block, all their states change to **S,** meaning two or more cache copies exist. Similar to WT, when a *local write* occurs in **S** state, the writer invalidates all other states, and changes its own state to **M.**

The **M** state indicates the cache line that needs to be written back when another processor requests to read or write it. However, before that happens, the local processor can arbitrarily read or write the line without informing other caches.

The situation for external snoopy events is quite simple. All *remote writes* change other caches to the **I** state (write-back first if necessary). All *remote reads* are hit at the state **S.** Note that there is no out-going edge of *write* from the invalid **I,** because of Pentium's non-write-allocate policy.

The MSI Protocol for WB Caches In order to understand the MESI protocol, we examine Fig. 5.7 more closely. If we merge the **E** state into the **S** state to have a single **S** node, what we yield a 3-state *MSI protocol* for use in some WB caches only.

Subtracting the MSI subdiagram from the MESI state diagram yields the following three transitions:

$$I \xrightarrow{r} E \ , \tag{5.9}$$

corresponding to the sole-copy status of the cache line;

$$E \xrightarrow{R} S, \tag{5.10}$$

indicating that two or more cached copies exist;

$$S \xrightarrow{w_b} M, \tag{5.11}$$

corresponding to the WB operation ocurred only at the first *write* but not in subsequent *writes*.

Intuitively, Eq. 5.11 implies *write once* when there are multiple *writes* to the same line by one or more processors (the potential consumers) before the next *read*. We leave the reader to examine how the MESI protocol outperforms the MSI protocol in a home work problem based on the analysis of the producer-consumer sharing pattern.

The SI Protocol for WT Caches Furthermore, one can simplify the three-state MSI protocol to a two-state SI protocol for WT caches. Even a one-state protocol is possible, corresponding to no caching al all. On the snoopy protocol spectrum, the number of states of a coherency protocol represents the capability to classify the cache lines by their degree of sharing.

Summary of Snoopy Protocols We have seen at least three snoopy write-invalidate protocols for different types of caches and different line-fill and write-allocate policies. In general, the more states a coherency protocol employs, the more categories it can separate the cache events with more possible transitions.

Of course, higher cost is expected to result in more complex protocols in terms of implementation efficiency. The MESI protocol specified in Fig. 5.7 can be applied to the level-2 cache with minor modifications, under the assumptions that the write-through policy is always used at level-1 data cache, and that the write-back is always used from the level-2 cache to the memory.

5.3 Shared-Memory Consistency

The *reads* and *writes* of the shared memory face a consistency problem if these memory events are not properly ordered. To cope with this problem, an efficient memory model must be developed for shared-memory multiprocessors. Such a model affects the performance, correctness, programmability, as well as portability of parallel applications.

5.3.1 Memory Event Ordering

The *consistency* problem is caused by different ordering of *read/write* events initiated by the same or different processors. We outline the basic conditions and use examples to illustrate the need to achieve controlled consistency in memory events. Shared-memory behavior is determined by both program order and memory access order.

The Basic Concept On a multiprocessor, concurrent instruction streams (threads or processes) may execute on different processors simultaneously. Memory events performed by one process may create data to be used by another process. Memory events correspond to *read* or *write* of the shared memory. A *memory consistency model* specifies how the memory events initiated by one process should be observed by other processes in the machine.

Event ordering can be used to declare which memory access is allowed at a given time and which should wait for a later access, when several processes compete for the access of the same set of memory locations. A *memory order* generated by a given consistency model directs multiple processors to access the memory in a specific order for correct and speedy execution of the concurrent processes.

Memory Events in a Multiprocessor As illustrated in Fig. 5.8(a), a uniprocessor system follows the sequential *program order* to determine the *memory order*. There a *read* to the memory always returns the last *write* to the same location. But this is not the case of an MIMD multiprocessor. In using a multiprocessor system to execute concurrent programs that are related, local dependence checking is necessary but may not be sufficient to preserve the intended outcome of the concurrent execution.

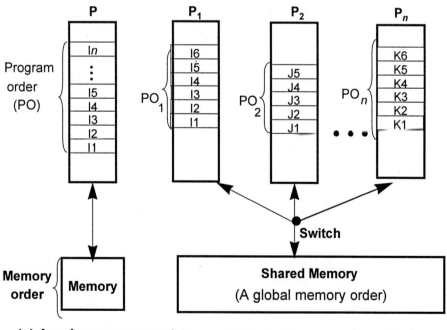

(a) A uniprocessor system   **(b) A multiprocessor system**

Figure 5.8   **Program order versus memory order in a uniprocessor system and in a multiprocessor system**

This is illustrated in Fig. 5.8(b), where three instruction streams (labeled I, J, and K strings) compete to issue *read/write* requests to access the shared memory through a single-port switch control. Maintaining correctness and predictability of the execution results is rather complex in an MIMD multiprocessor system for the following two reasons: The two examples to follow will illustrate these points.

- The program orders of individual instruction streams may have to be modified because of interaction among them. If no synchronization among the instruction streams exists, then a large amount of instruction interleaving is possible. Finding the optimum *global memory order* is an NP hard problem.
- If accesses are not atomic with multiple copies of the same data existing in a cache-based system, then different processors can individually observe different ordering. In this case, the total number of possible instantiations of multiple programs becomes even larger.

Example 5.5 Memory-event ordering in a three-processor system

To illustrate possible ways of interleaving concurrent instruction streams from different processors updating the same memory, we examine the case of three streams in Fig. 5.9.

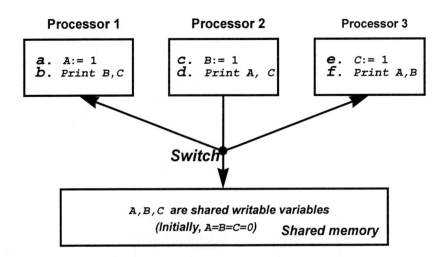

Figure 5.9   **The execution of three programs with *write* and *print* operations initiated by three processors asynchronously.**

The shared variables are initially set to have a value of zero. Assume that the *print* statement in each stream reads both variables indivisibly during the same cycle to avoid confusion. To check the global memory ordering, we concatenate in the program orders in processors P_1, P_2, and P_3 to form a 6-tuple of binary strings.

There are $2^6 = 64$ possible output combinations. If all processors execute instructions in their own program orders, then the order (*a, b, c, d, e, f*) yields the output string 001011. Another interleaving, (*a, c, e, b, d, f*) also preserves the individual program orders, yielding a different output string 111111.

If processors were allowed to execute instructions out of program order, assuming that no data dependencies exist among reordered instructions, then the interleaving (*b, d, f, e, a, c*) yields the output string 000000. Out of 6! = 720 possible permutations, only 90 preserve the individual program orders.

From these 90 interleavings, not all 6-tuple combinations are considered correctly executed. For example, the outcome 000000 is not possible if processors execute instructions in program order only. Another example shows that the outcome 011001 is allowed if different processors can observe memory events in different orders. This example clearly shows that memory consistency can be relaxed to give fewer constraints in accessing the shared memory.

∎

5.3.2 Memory Consistency Models

The traditional uniprocessor or multiprocessor follows a *sequential consistency* model as characterized below. One can relax the sequential constraints to yield the *weak consistency, processor consistency,* and *release consistency* shared-memory models, as proposed in recent years. From the above examples, the order of *reads* and *writes* from two different processors certainly makes a big difference. Therefore, the relative event ordering must be made below to define a memory consistency model.

Sequential Consistency This memory consistency model requires that the *reads*, *writes*, and *swaps* by all processors appear to execute serially in a single global memory order that conforms to the program orders in individual processors. This means the global order must preserve all individual program orders, regardless of how instruction streams are interleaved.

Definition 5.1 Lamport [391] has defined a multiprocessor system as *sequentially consistent* (SC) if the result of any execution is the same as if the operations of all the processors were executed in some sequential order, and the operations of each individual processor appear in this sequence in the order specified by its own program.

∎

The SC memory model does not appeal to program optimization in exploiting parallelism in a multiprocessor. For example, the *write* and *read* in the following program are not constrained by program orders residing on two processors.

Weak Consistency Dubois, Scheurich, and Briggs [218] have derived a *weak consistency* (WC) memory model by relating memory order to synchronization points in the program. This model is specified by three memory access conditions:

(1) All previous synchronization accesses must be performed before a *read* or a *write* access is allowed with respect to any other processor.

(2) All previous *read* and *write* accesses must be performed before a synchronization access is performed with respect to any other processor.

(3) Synchronization accesses are sequentially consistent with respect to one another.

These conditions provide a weak ordering of memory access events in a multiprocessor. The sequential ordering is limited to only hardware-recognized synchronizing variables. Between two synchronization points, not all *read/write* operations have to follow the program order. This opens up opportunities for buffering of reordered *write* operations except for those *writes* that are part of synchronization operations. Buffering memory accesses in multiprocessors can enable pipelined memory access, as mentioned earlier.

Processor Consistency Goodman [273] introduced the *processor consistency* (PC) model in which *writes* issued by each individual processor are always in program order. However, the *writes* from different processors can be out of program order. In other words, sequential consistency in *writes* is observed in each processor, but the order of *reads* from each processor is not restricted as long as it does not involve other processors.

The PC model relaxes from the SC model by removing some restrictions on *writes* from different processors. This opens up more opportunities for write buffering and pipelining.

The following two conditions related to other processors are required to ensure processor consistency:

- Before a *read* is allowed to perform with respect to any other processor, all previous *read* accesses must be performed.
- Before a *write* is allowed to perform with respect to any other processor, all previous *read* or *write* accesses must be performed.

These conditions allow *reads* following a *write* to bypass the *write*. To avoid deadlock, the implementation should guarantee that a *write* appearing previously in program order will eventually be performed.

Release Consistency The *release consistency* (RC) model was introduced by Gharac-horloo et al. [264]. Release consistency requires that synchronization accesses in the program be identified as either *acquires* (locks) or *releases* (unlocks). An *acquire* is a *read* operation (which can be part of a *read-modify-write*) that gets permission to access a set of data, while a *release* is a *write* operation that gives away such permission. This information is used to provide the flexibility in buffering and pipelining of memory accesses between synchronization points.

The main advantage of a relaxed memory model is the potential of better memory performance by hiding as much *write latency* as possible. The main disadvantage is the increased hardware complexity and a more complex programming model to ordinary users. The following three conditions ensure the RC shared-memory model:

- Before an ordinary *read* or *write* access is allowed to perform with respect to any other processor, all previous *acquire* accesses must be performed.
- Before a *release access* is allowed to perform with respect to any other processor, all previous ordinary *read* and *store* accesses must be performed.
- *Special accesses* are processor-consistent with one another. The ordering restrictions imposed by weak consistency are not present in release consistency. Instead, the RC model requires processor consistency and not sequential consistency.

Nitzberg and Lo [468] have characterized the relationships among the above four memory consistency models, as summarized in Fig. 5.10.

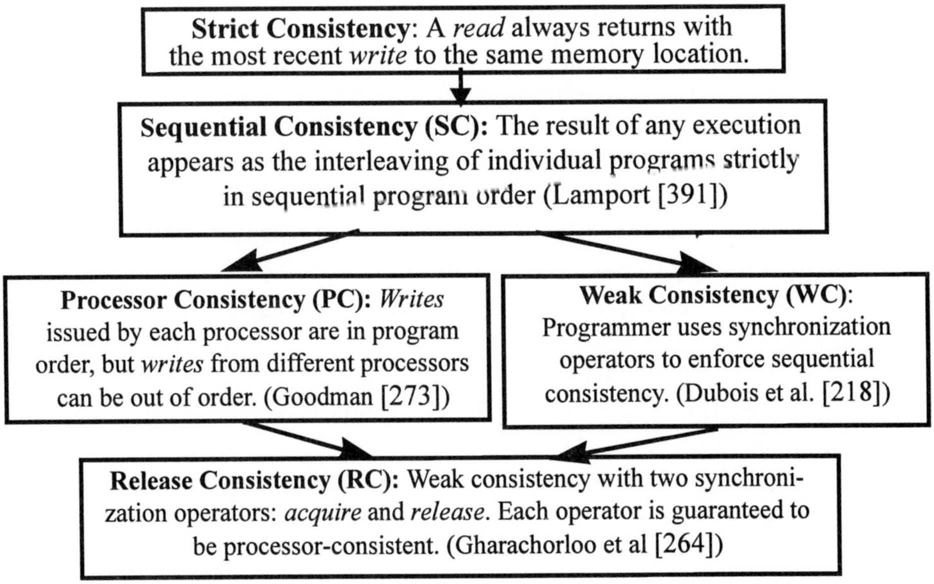

Figure 5.10   Intuitive definitions of four memory consistency models.

The RC model has combined the advantages of both the PC and WC models as seen in Fig. 5.10. The RC memory access can be satisfied by (1) stalling the processor on an *acquire* access until it completes and (2) delaying the completion of *release* access until all previous memory accesses are completed.

The cost of implementing the RC model over that of the SC arises from the extra hardware/software needed to provide a lockup-free cache and to keep track of multiple outstanding requests. Although this cost is not negligible, the same features are also required to support prefetching and multiple contexts.

5.3.3 Relaxed Memory Models

The main goals of relaxed memory models are to alleviate the overhead of cache coherence control. The decrease in coherence overhead should preserve the correctness in program execution. Relaxed memory models are meant to achieve both. Two relaxed memory characteristics are suggested below by Adve and Gharachorloo [8]:

- Memory models differ on the basis of how they relax the order from a *write* to a *read* following the *write*, between two *writes*, and from a *read* to a *read* or *write* following the *read*. These relaxations apply only to operation pairs with different addresses and are similar to the optimizations for sequential consistency for an uncached architecture.

- Some memory models allow a *read* to return the value of another processor's *write* before the *write* is made visible to all other processors. This relaxation applies only to a cache-based system.

Commercial Memory Models   Relaxed memory consistency models relax program order between all operations to different locations, allowing a *read* or *write* to be reordered with respect to a following *read* or *write*. Thus, they may violate sequential consistency.

The memory operations following a *read* operation may be overlapped or reordered with respect to the *read*. This flexibility allows hardware to hide the latency of *reads* with either statically (in-order) or dynamically (out-of-order) scheduled processors.

Six relaxed memory models are assessed by Adve and Gharachorloo in Table 5.4. The *weak ordering* (WO) model, two flavors of the *release consistency* model (RCsc and RCpc), and three models are adopted in Digital Alpha, Sparc *relaxed memory order* (RMO), and IBM PowerPC. Except for Alpha, these models also allow the reordering of two *reads* to the same location.

An X in the table indicates that the relaxation is allowed in an implementation of a given memory model. It also indicates that the relaxation can be detected by the programmer (by affecting the results of the program). However, the *read-own-write-early* relaxation is not detectable with the SC, WO, Alpha, or PowerPC models.

Table 5.4   Various Relaxed Memory Consistency Models

(Courtesy Adve and Gharachorloo, *IEEE Computer* [8], 1996.)

Relaxed Memory Model	W --> R Order	W --> W Order	R --> RW Order	Read-Others-Write-Early	Read-Own-Write-Early	Safety Nets Ref.
SC					X	No [391]
IBM370	X					Serialization [341]
TSO	X				X	Read-modify-write [539]
PC	X			X	X	Read-modify-write [264]
PSO	X	X			X	Read-modify-write [579]
WO	X	X	X		X	Synchronization [218]
RCsc	X	X	X		X	Release,
RCpc	X	X	X	X	X	acquire, nsync, read-modify-write [264]
Alpha	X	X	X		X	MB, WMB, [563]
RMO	X	X	X		X	Variou MEMBARs, [579]
PowerPC	X	X	X	X	X	Sync [434]
Representative computer systems	AlphaServer 8400, Cray T3E, NUMA-Q,Sparc-Center 2000, Convex SPP, Ultra Servers	AlphaServer 8400, Cray T3D/T3E, Convex SPP in WO mode	Cray T3D	AlphaServer 8400, Cray T3D/T3E, SparcCenter 2000, Ultra Ent. Servers		

The *read-others-write-early* relaxation is possible and detectable with certain complex implementations of the RCsc model. Blanks correspond to those cases where the access constraints will not be applied to a given memory model.

The relaxed memory consistency models are designed to provide better performance than the sequential consistency models. According to Adve and Gharachorloo [8], the increase in processor speed relative to memory and communication speeds will only increase the potential benefit from these models. In addition to gains in hardware performance, relaxed memory consistency models play a key role in enabling compiler optimizations.

For these reasons, many commercial architectures, such as the Digital Alpha, Sun Sparc, and IBM PowerPC, support relaxed consistency as identified in Table 5.4. Detailed interpretation of each of the six relaxed memory consistency models can be found in the references in the rightmost column. The safety nets refer to mechanisms for overriding the default relaxations. For example, explicit fence instructions such as *read-modify-write* and various *synchronization* operators may be used to override program order relaxations.

Relationship with Cache Coherence Protocol   Any cache coherence protocol must observe the memory access constraints imposed by a given memory consistency model. For example, consider a *read* hit in a processor's cache. Reading the cached value without waiting for the completion of previous *write* operations may violate sequential consistency. The cache coherence protocol provides a mechanism to propagate a newly written value. The memory consistency model places an additional constraint on when the value can be propagated to a given processor.

Detecting *write* completion plays a crucial role in the successful implementation of a combined cache protocol/memory consistency model. A *write* must be eventually visible to all processors. All *writes* to the same location must be seen in the same order by all processors in order to enforce sequential consistency. Multicache updating process is inherently a nonatomic memory operation. Maintaining *write* atomicity is another important requirement. These issues are yet to be perfected in real system designs.

Compiler Support   This offers another level of optimization. The idea is to reorder the memory operations to satisfy the access constraints imposed by a given caching protocol/memory consistency scheme.

- First, the compiler must preserve program order among shared-memory operations.
- Second, no compiler-generated reordering can violate the access constraints imposed by the consistency model. In fact these requirements apply to hardware-generated reordering as well.

To support a given memory consistency model, a compiler for explicit parallelism is much easier to build than one for implicit parallelism. At present, none of the commercial compilers can support relaxed memort consistency with implicit parallelism.

5.4 Distributed Cache/Memory Architecture

Physically distributed memories could be logically shared or unshared, or both. The term *shared-memory* used in academia and industry may have different meanings. To avoid confusion, the shared-memory concept must be made clear. First, we must differentiate shared-memory *architecture* from shared-memory *programming environment*.

A shared-memory architecture (multiprocessor) can support both shared-memory and message-passing programming models. A shared-memory programming model can be implemented on both shared-memory or non-shared-memory architectures (e.g., multicomputers). It is generally agreed that SMPs have shared-memory architecture, while traditional MPPs (e.g., Intel Paragon) do not. But there are many distributed-memory architectures in between. These distributed shared-memory systems could have quite different architectural and programming characteristics.

5.4.1 NORMA, NUMA, COMA, and DSM Models

Various multiprocessors and multicomputers are classified in Fig. 5.11, where each class is exemplified by one or more commercial systems. The tree classification is based on all memory models used in our study. We first clarify the terminology of shared-memory architecture by examining several common cases from the user's viewpoint.

A system has a *shared-memory architecture* (multiprocessor) if a process running on any processor can directly access any local or remote memory in the entire system. Otherwise, we have a non-shared-memory architecture (multicomputer).

The key word is *directly*, meaning an instruction such as *load* or a *store* can access any memory location in the entire multiprocessor system. Memory sharing is possible in a non-shared-memory architecture. However, remote memory access is not direct, but through a software layer such as runtime or user-callable library routines. This translates to differences in access latency. Current multiprocessors have a remote access latency of hundreds of nanoseconds and multicomputers tens of microseconds.

Central Memory versus Distributed Memory A parallel computer has either a central-memory or distributed-memory architecture. Distributed-memory systems include *non-uniform memory access* (NUMA) and *no-remote memory access* (NORMA) architectures. Some people use a broader definition of NUMA to include NORMA, as memory accesses in these systems are also nonuniformly done.

Central memory systems are also known as *uniform memory access* (UMA) systems. In a UMA architecture, all memory locations are at an equal distance away from any processor, and all memory accesses roughly take the same amount of time. There are two types of UMA systems: the *parallel vector processor* (PVP) and the *symmetric multiprocessor* (SMP). They are briefly discussed in Section 1.3.4. PVPs are also called *vector supercomputers*. Details of SMP system characteristics will be given in Chapter 8.

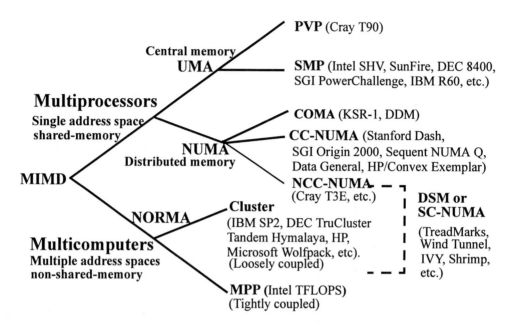

Figure 5.11   Different types of memory architecture for building parallel, distributed, and cluster computers.

Distributed-Memory Architectures   A distributed-memory computer contains multiple nodes, each having one or more processors and a *local memory*. Memories in other nodes are called *remote memories*. Several types of distributed-memory architectures are used in current commercial and research parallel systems. They are NORMA, *non-cache coherent NUMA (*NCC-NUMA), *cache-coherent NUMA* (CC-NUMA), and *cache-only memory architecture* (COMA). Their main differences are illustrated in Fig. 5.12.

The NORMA model can be further divided into loosely-coupled clusters and tightly-coupled MPPs, as already discussed in Section 1.3.4. Clusters and MPPS will be discussed in detail in Chapters 10 and 11, respectively.

In a NORMA machine, the node memories have separate address spaces. A node cannot directly access remote memory. The only way to access remote data is by passing messages. In Fig. 5.12(a), node P wants the data A in node Q. This is accomplished by node Q executing a send routine and node P executing a corresponding receive routine. In the end, the value of A is copied into variable B in node P's local memory.

The other three architectures all have special hardware that glues all the local memories into a single address space, enabling any processor to access any memory location. For instance, they all can use a load instruction to get the remote data A as shown in Fig. 5.12. However, their acessing mechanims differ.

A typical NCC-NUMA machine is Cray T3E (to be detailed in Section 11.2). Besides the local memory, each node has a set of node-level registers called *E-registers* [Fig. 5.12(b)]. An instruction loads the value of *A* into an E-register *E1*. The value can then be transferred to a processor register or the local memory. Other NCC-NUMA systems may allow loading a remote value directly into a processor register (e.g., by an instruction "load R1, A"). Note that the cache block of *A* is not automatically copied into the local processor cache or the local memory.

In a CC-NUMA system, an instruction loads the value of *A* into a local processor register *R1* [Fig. 5.12(c)]. At the same time, the cache block of *A* is automatically copied into a node-level cache called *remote cache* (RC). However, this cache block is not copied into the local memory. Some CC-NUMA systems do not have remote caches (e.g., SGI Origin 2000). Then the cache block of *A* is copied into the node's level-2 cache.

In a COMA machine, all local memories are structured as caches (we call them *COMA caches*). Such a cache has much larger capacity than the level-2 cache or the remote cache of a node. An instruction loads the value of *A* into a local processor register *R1* [Fig. 5.12(d)]. At the same time, the cache block of *A* is automatically copied into the local memory (now called COMA cache). COMA is the only architecture that provides hardware support for replicating the same cache block in multiple local memories.

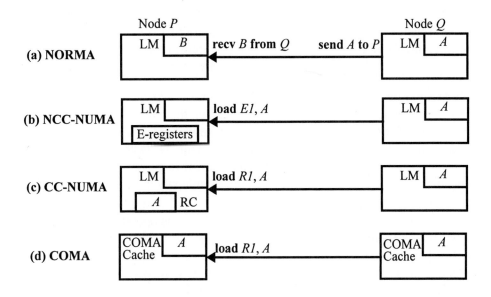

Figure 5.12   Comparison of four distributed-memory architectures.
(LM: local memory, *A* and *B*: memory addresses, RC: remote cache, E1: an E-register, R1: a general-purpose register)

NCC NUMA versus CC-NUMA/COMA An NCC-NUMA system does not have hardware support for cache coherency. Both the CC-NUMA and the COMA architectures provide cache coherency support by hardware. For this reason, it is easier to build a scalable NCC-NUMA system than either a CC-NUMA or a COMA system.

The best known example is the Cray T3D/T3E system, which totally relies on software to keep the address space coherence. This is both an opportunity and a challenge. Since the architecture details are exposed, the system software and even the application has the flexibility to manage the address space to maximize performance. However, most users do not want to see such details. It is not easy for the system software to efficiently maintain a coherent address space.

The Cray T3E chooses a compromise. For performance-oriented users, the T3E provides a low-level shared-memory libary for the user to optimally manage data sharing. For users that do not want to be bothered, the T3E provides a high-level programming tool called Craft. Then the compiler/runtime system maintain memory coherency.

We shall study in Section 5.4.3 the Stanford Dash, one of the pioneering CC-NUMA prototype systems. Industrial systems include the SGI Origin 2000, HP/Convex Examplar X-Class, Sequent NUMA-Q2000, and Data General Multi-server NUMA system. Some of these commercial CC-NUMA systems will be studied in Chapter 8.

The COMA model has been implemented in the Kendell Square KSR-1/2 systems [116] and in the *data diffusion machine* (DDM) developed at the Swedish Institute of Computer Science [294]. The KSR systems are multiprocessors built with a hierarchy of one to three levels of slotted rings. The memory architecture in KSR machine has been called ALLCACHE, meaning all local memories are constructed as caches.

CC-NUMA versus COMA In a CC-NUMA multiprocessor, the main memory consists of all the local memories. In a COMA multiprocessor, the main memory consists of all the COMA caches. The COMA architecture demands more hardware support to maintain tags and state information for the COMA caches. The same hardware must integrate COMA cache management with virtual memory management (e.g., how does one bring a page from the disk into the COMA caches?). All this complexity makes a COMA system more expensive to implement than a NUMA machine.

When a program is loaded to run in a CC-NUMA, the operating system assigns each cache line a *home node*, meaning its local memory is where the cache line is allocated and permenantly resides. When a node P accesses the cache line, P temporarily stores the cache line in the local cache of P [see Fig. 5.12(c)]. At write-back time, the cache line has to be stored back to the local memory of the home node.

To alleviate such write-back traffic, a CC-NUMA operating system can replicate or migrate the page to the local memory of the node where the page is referenced most frequently. This is detailed in Example 8.2.

A main difference of COMA from CC-NUMA is that COMA handles replication and migration more efficiently. First, COMA uses hardware to realize replication and migration. To use a COMA machine, data can be allocated anywhere. At run time, the cache line can migrate to wherever it is needed, a concept called *attraction memory* by Hagersten et al. [293]. Second, the granularity of replication or migration is a cache line in COMA, compared to page granularity in CC-NUMA. This implies *false sharing* is less severe in COMA than in CC-NUMA. False sharing refers to the situation where two nodes access two different data variables which happen to reside in the same cache line or the same page.

Overall, the performance of CC-NUMA and COMA architectures depends on two major factors: the *size of the working set* and the *communication-to-computation ratio* (CCR). With low CCR and small working set, both architectures perform well, because most memory references can be satisfied by local processor caches. With a large working set and high CCR, both architectures will perform poorly. This is attributed to the fact that the large working set causes more misses in both architectures and the high CCR implies greater communication demand in the given application.

With high CCR and small working set, the CC-NUMA machine tends to perform better than the COMA machine, since COMA has higher miss penalty. With low CCR and large working set, the COMA machine is expected to perform better than the CC-NUMA machine, since the COMA caches have much larger capacity than remote caches in CC-NUMA. These results are based on an analysis made by Stenstrom et al. [585].

Software-Implemented DSM To enable shared-memory computing on NORMA and NCC-NUMA, researchers have proposed the *software-coherent NUMA* (SC-NUMA) memory model, also known as the *distributed shared-memory* (DSM) model. Examples include the TreadMarks project at Rice, the Wind Tunnel project at Wisconsin, and the SHRIMP project at Princeton. Some of these DSM machines will be treated in Chapter 10.

A software-implemented DSM relies mainly on software extensions to achieve single address space, data sharing, and coherence control. One approach is called *shared virtual memory* (SVM). The virtual memory management mechanism in a traditional node operating system is modified to provide data sharing at the page level. The IVY and SHRIMP projects take this approach. Modification of applications is not needed.

The other approach does not modify the operating system. It uses compilers and library functions to convert single-address-space codes to run on multiple address spaces. The application codes must be modified to include data sharing, synchronization and coherence primitives The TreadMarks and the Wisconsin Wind Tunnel take this approach.

Example 5.6 Tradeoffs in shared-memory planning

When *shared-memory* is mentioned, computer vendors and users often mean different things. Suppose a company is marketing a machine with eight nodes and 4-GB total memory. The sales representative may emphasize that the system provides shared-memory mechanisms whereby all the memory can be utilized by the operating system and applications.

The user may buy the machine on the spot, because he needs a 3-GB memory space to run the Mathematica software in very large numerical computations. But he may be soon disappointed to find out that the *shared-memory* mechanisms require him to execute one of the following pretasks, whether he likes it or not:

(1) Link his application to a vendor-provided run-time library.

(2) Recompile his application using a vendor-supplied compiler.

(3) Insert library function calls in his application, and recompile and relink.

(4) Use a new programming language to rewrite his application.

The user cannot execute task 1 because he does not have the object code of the Mathematica. He cannot execute tasks 2 to 4 because he does not have the Mathematica source code. Even if he has the source, he may not want to execute tasks 3 or 4 because it is too time-consuming. He just wants to get on with his own work, not program an unfamiliar machine using never-heard-of language constructs or library functions.

■

The above problems will disappear if the Mathematica is ported to the machine by the software vendor. This example shows that to gain wide user acceptance, a system vendor must enable applications including third-party applications. A practical criterion to test if a system is truly shared-memory is to see if it enables *sequential, binary* codes to use all its memory. In today's commercial systems, only UMA and CC-NUMA systems can achieve this level of sharing. How to design an effective SC-NUMA scheme to meet this criterion is still an open research topic.

Distributed Memory Summary We summarize the five distributed memory architectures in Table 5.5. Coherency is enforced by hardware in CC-NUMA and COMA, by OS or runtime library in SC-NUMA, and by user-level software in NORMA or NCC-NUMA.

Table 5.5 Characteristics of Five Distributed-Memory Architectures

Attribute	NORMA	NCC-NUMA	SC-NUMA	CC-NUMA	COMA
Granularity	Object	Object	Page	Cache-line	Cache-line
False sharing	No	No	Yes	Yes	Yes
Replication and migration	By user at object level	By user, object level	By OS or runtime at page level	By OS at page level	By hardware at cache-line level
Coherency	By user application	By user-level software	By OS or runtime	By hardware at cache-line level	By hardware at cache-line level

In general, NORMA and NCC-NUMA systems share data at object (variable) level. A node gets whatever variables it needs. There is no false sharing and spatial locality is not exploited. COMA and CC-NUMA machines normally share data at cache-line level. SC-NUMA systems usually share data at page level. The last three types of systems can exploit spatial locality, at the expense of some degree of false sharing.

Data allocation, replication, and migration are done by user software in NORMA and NCC-NUMA at data object granularity. Replication and migration is done at page level by the operating system in CC-NUMA, at cache-line level by hardware in COMA, and at object or page level by operating system or runtime library in SC-NUMA.

5.4.2 Directory-Based Coherency Protocol

The snoopy protocol is based on the broadcast capability on the memory bus. Other cache coherency protocols may not use broadcast. The directory-based protocols are not designed to use broadcast on a snoopy bus. We introduce the basic concepts of directory-based protocol in this section. A concrete example of directory-based protocol will be given in the next section.

Cache Directory The directory approach is to use a directory to record the locations and states of all cached lines of shared data. This list of cache locations is called a *cache directory*.

A directory entry for each cache line of data contains a number of pointers to specify the locations of all remote copies of the same line. Each directory entry also contains a dirty bit to specify whether a unique cache has permission to write the associated line of data.

Tang [606] proposed the first directory scheme for cache coherency control. The idea is to use a central directory to record all cache conditions, including the presence information and all cache line states. This central directory is only suitable for cache coherency control in a small-scale SMP with centralized shared memory.

The main difficulty is that the central directory itself requires a huge amount of memory to implement and it must be associatively searched to reduce the update time. This scheme is not suitable for use in a large-cache SMP, NUMA, or any distributed-memory platform.

Distributed Directories The distributed directory scheme was proposed by Censier and Feautrier [135] Each memory module maintains a separate directory which records the states and presence information of all cached lines.

The directory entries contain the states of cache lines plus the presence information pointing to the locations of remote caches that have also copies of the same cache line. The following example will clarify the ideas behind distributed cache directories.

Example 5.7 Distributed directories for cache coherency control

In Fig. 5.13, the shared memory is formed with multiple memory modules. Each memory module M_i for i = 1, 2, ..., m, maintains a cache directory D_i. The caches are denoted by C_i for processor P_i. The example demonstrates the sharing of the same cache line between caches C_1 and C_2.

The interconnection network is not restricted to a bus anymore. A multistage or crossbar switch, or any other point-to-point *system-area network* (SAN) or *local-area network* (LAN), will apply. These switches and networks will be covered in Chapter 6. The solid arrows indicate the signal paths for cache coherency control. The hollow arrows are data paths.

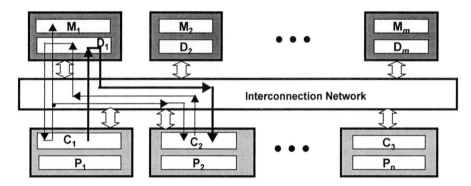

Figure 5.13 Directory-based cache coherence scheme (Courtesy of Censier and Feautrier, *IEEE Trans. Computers* [135],1978).

The memory/cache update works as follows: A *read-miss* (thin lines in Fig. 5.13) in cache C_2 results in a request sent to directory D_1 that indicates the presence of a clean copy in C_1. The memory controller retransmits the request to cache C_1. This cache returns with its clean copy to M_1 and C_2. In case of a *write-hit* at C_1 (thick lines), a command is sent to the memory controller, which sends an invalidation to all caches (C_2 in this case) marked in the presence bits in directory D_1.

■

Cache Directory Structures Different types of directory protocols fall under three categories: *full-map directory, limited directory,* and *chained directories.*

- A *full-map cache directory* contains information of all cache lines shared on a global basis. Each directory entry contains N pointers, where N is the number of processors. The pointers are identified by bit vectors. A full directory repeats in every node, occupying a lot of memory space itself. Only a small-scale multiprocessor or multicomputer can afford the use of the full-map directory approach.

- A *limited cache directory* uses a much-reduced, fixed number of pointers per directory entry, regardless of the system size. This will reduce the memory demand from the cache directories, thus is more economical to implement if the memory sharing is low. But this may slow down the cache/memory update process if the degree of memory sharing is excessive.

- *Chained cache directories* emulate the full-map scheme by distributing the directory information to small local directories. To get a global picture of memory sharing, one must search through the linked lists of cache directories. The IEEE SCI Standard has specified the structure of chained directories. We will discuss SCI cache coherence protocol in Section 6.6.3.

5.4.3 The Stanford Dash Multiprocessor

This is an experimental CC-NUMA multiprocessor system built at Stanford University under the leadership of John Hennessy. The name Dash is abbreviated from the full name of *directory architecture for shared memory*. The major contribution of the Dash project lies in that it is possible to build a scalable parallel computer with a single address space, using distributed coherent caches and a distributed-memory hierarchy. Dash pioneers this approach to building a CC-NUMA architecture, while maintaining the scalability of a message-passing multicomputer.

The Dash Prototype   The hardware organization of the Dash architecture is illustrated in Fig. 5.14(a). It incorporates up to 64 microprocessors in 16 SGI SMP nodes, each containing 4 MIPS R3000/R3010 processors. The node architecture is slightly modified from Silicon Graphics, 4D/340 Powerstation.

Two special daughterboards were built to plug into each Powerstation node. These boards house the network interface circuitry and the cache directory shared by all four processors attached to the same snoopy bus. The interconnection network among the 16 SGI nodes is a pair of wormhole-routed, two-dimensional 4 × 4 mesh networks. The meshes are essentially built with flat wire strips and wormhole routers. The channel width in the meshes is 16 b with a 50-ns fall-through time and a 35-ns cycle time.

One *request mesh* network is used to request remote memory, and the other is a *reply mesh*. The small squares at mesh intersections are the wormhole routers developed by the Caltech research group led by Charles Seitz.

Special Hardware   The Dash designers claimed scalability for the Dash approach. although the prototype is limited to 16 node clusters (a 4 × 4 mesh), due to the limit on memory address space (256 MB) of each 4D/340 node. In theory, the system should be scalable to support hundreds of processors. To use the 4D/340 in the Dash, the Stanford team made minor modifications to the existing system boards and designed a pair of new boards to support the directory memory and internode message passing.

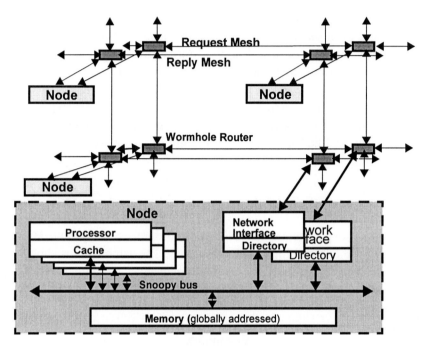

(a) The Dash architecture

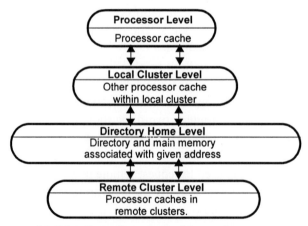

(b) Distributed-memory hierarchy

Figure 5.14   Stanford Dash: a CC-NUMA multiprocessor project.
(Courtesy of D. Lenoski et al., *Proc. 19th Int. Symp. Computer Architecture* [405], May 1992.)

The main modification to the existing CPU boards was to add a bus retry signal that is used when a request requires service from a remote node. The central bus arbiter has also been modified to accept a mask from the directory. The mask holds off a processor's retry until the remote request has been serviced. This effectively creates a split-transaction bus protocol for requests of remote services.

The new directory controller boards contain the directory memory, the intercluster coherence state machines and buffers, and a local section of the global interconnection network. The directory logic is split between the two logic boards along the lines of the logic used for outbound and inbound portions of intercluster transactions.

The mesh networks support a scalable local and global memory bandwidth, which rewards locality. The single address space with coherent caches makes it easier for compilers and programmers, permits incremental porting or performance tuning of applications, and exploits temporal and spatial locality. Other factors contributing to improved performance include relaxed memory consistence mode and data prefetch mechanisms for hiding remote memory access latency.

Dash Memory Hierarchy   Memory sharing in Dash is done at the cache-line level. Dash implemented a write-invalidation coherency protocol using distributed cache directories. A cache line in memory or a cached copy in local caches may be in one of three states:

- **Uncached** -- not cached in any node cluster
- **Shared** -- an unmodified state in the caches of one or more node clusters
- **Dirty** -- modified in a single cache of some node cluster

The directory keeps the summary information for each cache line, specifying its state and the node clusters that are caching it. The Dash memory system is logically divided into a four-level hierarchy, as illustrated in Fig. 5.14(b).

The first level is each individual processor's cache that is designed to match the processor speed and support snooping on the local bus. It takes only one clock to access the processor's cache. A request that cannot be serviced by the processor's cache is sent to the caches within a local node cluster.

The prototype Dash allows 30 processor clocks to access the caches at a local node. This level includes all other processors' caches within the requesting processor's node domain. Otherwise, the request is sent to the home cluster level.

The home level consists of the node cluster that contains the directory and physical memory for a given memory address. It takes 100 processor clocks to access the directory at the home level. For many memory accesses (for instance, most private data references), the local and home clusters are the same, and the hierarchy collapses to three levels. In general, a request travels through the mesh network to the home cluster.

The home node cluster can usually satisfy the request immediately, but if the directory entry is in a dirty state, or in a shared state when the requesting processor requests exclusive access, the fourth level must be accessed. The remote cluster level for a cache line consists of the node clusters marked by the directory as holding a copy of the cache line. It takes 135 processor clocks to access processor caches in a remote node cluster in the prototype mesh-connected Dash design.

5.4.4 Directory-Based Protocol in Dash

The directory memory relieves the processor caches from snooping on memory requests. In the home node, there is a directory entry for each cache line frame. Each entry contains one presence bit per processor cache. In addition, a state bit indicates whether the line is uncached, shared in multiple caches, or held exclusively by one cache (i.e., whether the line is dirty).

Using the state and presence bits, the memory can tell which caches need to be invalidated when a line is written. Likewise, the directory indicates whether memory's copy of the cache line is up-to-date or which cache holds the most recent version. If the memory and directory are partitioned into independent units and connected to the processors by a scalable interconnect, the memory system can provide a scalable memory bandwidth.

By using the directory memory, a node writing a location can send point-to-point invalidation or update messages to those processors that have cached that line. This is in contrast to the invalidating broadcast required in a snoopy protocol. The scalability of the Dash depends on this ability to avoid broadcasts.

Another important attribute of the directory-based protocol is that it does not depend on any specific interconnection network topology. As a result, one can readily use any of the low-latency scalable networks, such as mesh or hypercube, that have been developed for message-passing multicomputers.

Example 5.8 Cache directory protocol in the Dash multiprocessor

In Fig. 5.15(a), we illustrate the flow of a *read* request to remote memory with the directory entry in a dirty state. The *read* request is forwarded to the owning dirty node. The owning node sends out two messages in response to the *read*. A message containing the data is sent directly to the requesting cluster, and a sharing write-back request is sent to the home cluster. The sharing write-back request *writes* the cache line back to the memory and updates the directory.

This protocol reduces latency by permitting the dirty node to respond directly to the requesting node cluster. In addition, this forwarding strategy allows the directory controller to simultaneously process many requests (i.e., to be multithreaded) without the added complexity of maintaining the states of all outstanding requests.

Serialization is reduced to the time of a single intercluster bus transaction. The only resource held while intercluster messages are being sent is a single entry in the originating node's remote-access cache.

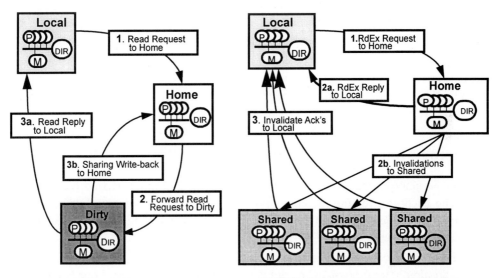

(a) Read of dirty remote cache line   **(b) Write to shared remote cache line**

Figure 5.15   The directory-based cache coherency protocol in Dash.

In Fig. 5.15(b), we show the corresponding sequence for a *write* operation that requires remote service. The write-invalidate protocol requires the processor (actually the write buffer) to acquire exclusive ownership of the cache line before completing the *write*. Thus, if a *write* is made to an uncached line or has been cached only in a shared state, the processor issues a *read-exclusive* request on the local bus.

In this case, no other cache holds the line entry dirty in the local node cluster. So a RdEx Request (message 1) is sent to the home cluster. As before, a remote-access cache entry is allocated in the local cluster.

At the home cluster, the pseudo-CPU issues the read-exclusive request to the bus. The directory indicates that the line is in the shared state. This causes the directory controller to send a RdEx Reply (message 2a) to the local cluster and invalidation requests (Inv-Req, message 2b) to the sharing cluster.

The home node cluster owns the cache line, so it can immediately update the directory to the dirty state, indicating that the local node now holds an exclusive copy of the cache line. The RdEx Reply message is received in the local node by the reply controller, which can then satisfy the *read-exclusive* request.

To ensure consistency at release points, however, the remote-access cache entry is deallocated only when it receives the number of invalidate acknowledgments (Inv-Ack, message 3) equal to an invalidation count sent in the original reply message.

■

An important feature of the directory-based protocol is its forwarding strategy. If a node cannot reply to a directory for a given request, it forwards the responsibility of the request to a node cluster that should be able to respond.

This technique also minimizes the serialization of requests since no requests are blocked while internode messages are in progress. Forwarding allows the directory controller to work on multiple requests concurrently (i.e., making it multithreaded) without having to retain any additional state about forwarded requests.

5.5 Latency Tolerance Techniques

Future scalable systems are most likely to use a distributed shared-memory architecture. The access to remote memory may experience a long latency. Furthermore, the processor speed is increasing at a much faster rate than the speed increase of memory and interconnection network.

Scalable multiprocessor or large-scale multicomputer clusters must rely on the use of latency reduction, latency avoidance, and latency hiding mechanisms. Four latency-hiding mechanisms are studied and assessed below for enhancing the scalability and programmability of future systems.

5.5.1 Latency Avoidance, Reduction, and Hiding

Three approaches to solving the memory latency problem have been suggested in the past. We introduce these approaches with simple examples.

Latency Avoidance   This technique tries to organize user applications at architectural, compiler, or application levels to achieve data/program locality. The purpose is to avoid the long latency in remote data or program access. This is possible only when applications exhibit either temporal locality or spatial locality. Many techniques have been developed to enhance the temporal and spatial localities.

These techniques fall into the following three categories:

- *Architectural Support*. Architectural support includes various types of cache coherency protocols, memory consistency models, fast message passing, and synchronization hardware that can be built in a scalable system.
- *User Support*. The programmer (user) can write the application specifically to enhance locality. Some languages, such as *High Performance Fortran* (HPF), have facility to help users explore locality. The appreoach is to instruct the compiler how to best allocate the data set to provide local accesses. The idea is illustrated in Example 5.9, and will be discussed in detail in Section 14.3.2.

- *Software Support.* The system software, such as the compiler, can perform certain transformations to enhance locality. This is illustrated in Example 5.10, and will be discussed in detail in Section 12.3.5.

Example 5.9 Locality of distributed data structures

Arrays of complex numbers are frequently used in scientific computation and signal processing applications. Two popular methods to represent a complex array follow:

```
/****** Method (1) Array of Structure *********/
typedef struct { double real,image;} COMPLEX;
COMPLEX data[N1][N2][N3];

/****** Method (2) Separate Arrays *********/
double data_real[N1][N2][N3], data_image[N1][N2][N3];
```

These two methods show quite different performance, even when the same algorithm is used to solve a problem. Method 1 has been used on most supercomputing applications, because it has better locality properties, thus results in a smaller cache miss ratio.

■

Example 5.10 Improvement in data locality by loop transformation

A traditional algorithm to perform matrix multiplication $C = A * B$ is shown below as a nested loop:

```
for (i = 0; i < N; i++)
  for (j = 0; j < N; j++)
    for (k = 0; k < N; k++)
      c[i][j] += a[i][k] * b[k][j];
```

A transformation, called *loop interchange*, can be performed by an optimizing compiler to exploit greater data locality. As shown below, the k loop and the j loop are interchanged. This transformation does not change the result computed, but could improve the performance by more than an order of magnitude.

```
for (i = 0; i < N; i++)
  for (k = 0; k < N; k++)
    for (j = 0; j < N; j++)
      c[i][j] += a[i][k] * b[k][j];
```

■

Latency Reduction   As Burton Smith [569] commented, data locality may be limited, difficult to discover, or changing dynamically. Most sorting algorithms have all three characteristics. When locality cannot be explored to completely avoid latency, we need techniques to reduce long latencies caused by remote references. A main design goal of a scalable computer is to make the communication subsystem efficient. Toward this end, we must have

- *Efficient interconnect hardware.* This will be treated in Chapter 6.
- *Efficient network interfaces.* This will be discussed in Chapter 6.
- *Fast communication software.* This will be treated in Chapters 7 and 14.

Latency Hiding   This technique refers to hiding communication latency within computation, i.e., through some overlapping techniques. Latency hiding can be accomplished with four complementary approaches:

- *Prefetching techniques* bring instructions or data close to the processor before they are actually referenced. In other words, data prefetching hides the *read* latency.
- *Distributed coherent caches* supported by hardware reduce cache misses or shorten the time to retrieve the clean copy.
- *Relaxed memory consistency* models use buffering and pipelining of memory reference operations. Using relaxed memory consistency is meant to hide the *write* latency.
- *Multiple-context processors* allow a processor to switch from one context to another when a long-latency operation is encountered. We will study hardware-supported multithreading in Section 5.6.

Even the most advanced avoidance and reduction techniques cannot eliminate the latency in a large parallel system for all applications. The speed of light also imposes a hard lower bound on latency.

Latency hiding techniques aim at achieving high performance in the presence of long latencies. A basic idea for tolerating latency is to keep the processor busy doing useful computation while data access is in progress, i.e., overlapping computation and communication. The first three latency hiding techniques are presented in this section. The multithreading approach will be treated in Section 5.6.

5.5.2 Distributed Coherent Caches

Using distributed caches that are coherent not only will reduce local cache misses but also will offer an effective way to achieve DSM at the cache-line level. As shown in previous sections, both hardware and software implementations are possible to support distributed coherent caches. We show below some benchmark results to see the effectiveness in performance enhancement of having distributed coherent caches.

Coherent Cache Projects and Products   The MIT Alewife, KSR-1, and Stanford Dash have all implemented directory-based coherence protocols. It should be noted that distributed caching offers a solution to the *remote-load* problem, but not the *synchronizing-load* problem. Multithreading offers a solution to remote loads and possibly to synchronizing loads as well. However, the two approaches can be combined to solve both types of remote-access problems.

While the cache coherence problem is easily solved for small bus-based multiprocessors through the use of snoopy coherence protocols, the problem is much more complicated for large-scale multiprocessors that use point-to-point or switched interconnects. As a result, some existing large-scale multiprocessors do not provide hardware support for caches (e.g., BBN Butterfly or Cray T3D/T3E). Others provide caches that must be kept coherent by software (e.g., IBM RP3).

Four recent NUMA multiprocessor systems, SGI Origin 2000, HP/Convex Exemplar X Class, Sequent NUMA-Q2000, and Data General NUMA server, are all learned from the experiences on CC-NUMA projects at universities. They are all implemented with directory-based cache coherence protocol.

Benchmark Conditions   We evaluate the benefits when both private and shared data are cacheable, as allowed in the Dash hardware coherent caches, compared to the case where only private data are cacheable. In Fig. 5.16, we show a breakdown of the normalized execution times with and without caching of shared data for each of the three applications. Private data are cached in both types of caches. The left bar for each benchmark corresponds to no caching. The right bars show the caching effects of shared read/write data.

The three benchmark programs are MP3D, a particle-based, three-dimensional simulator used in aeronautics research; an LU, a decomposition program in solving large-scale linear system of equations; and PTHOR, a digital logic simulation program. These benchmark programs were tested on the Stanford Dash multiprocessor, and the performance results are analyzed here with respect to each type of latency hiding mechanism.

The execution time of each application is normalized with respect to the case when shared data are not cached. The bottom section of each bar represents the busy time or useful cycles executed by the processor. The section above it represents the time that the processor is stall waiting for *reads* to complete. The section above that is the amount of time the processor is stall waiting for *writes* to complete. The top section, labeled *synchronization,* accounts for the time the processor is stalled due to the use of *locks* and *barriers*.

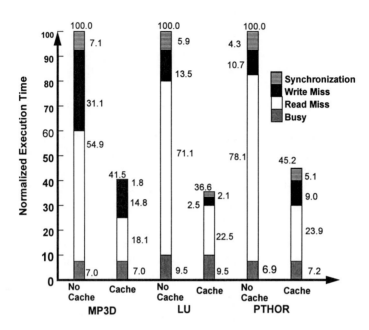

Figure 5.16   Effect of caching shared data in simulated Dash benchmark experiments. (Courtesy Gupta et al. *Proc. Int'l Symp. on Computer Architecture,* [285], Toronto, May 1991.)

Benefits of Caching   As expected, the caching of shared read-write data provides substantial gains in performance, with benefits ranging from 2.2- to 2.7-fold improvement for the three Stanford benchmark programs. The largest benefit comes from a reduction in the number of cycles wasted with read misses. The cycles wasted with write misses are also reduced, although the magnitude of the benefits varies across the three programs due to different write-hit ratios.

The cache-hit ratios achieved by MP3D, LU, and PTHOR are 80, 66, and 77%, respectively, for shared-read references, and 75, 97, and 47% for shared-write references. It is interesting to note that these hit ratios are substantially lower than the usual uniprocessor hit ratios. The low hit ratios arise from several factors:

The data set size for engineering applications is large, parallelism decreases spatial locality in these applications, and communication among processors results in invalidation misses. Still, hardware cache coherence is an effective technique for substantially enhancing the performance. This improvement is made with no assistance from the computer operator or from the programmer.

5.5.3 Data Prefetching Strategies

Date prefetching uses knowledge about the expected misses in a program to move the relevant data closer to the processor before they are actually referenced. Perfetching can be controlled by hardware or software or by both.

Prefetch Types Prefetching can be classified as binding or nonbinding. With *binding prefetch*, the value of a later reference (e.g., a register load) is bound or loaded directly into the working registers at the time when the prefetch completes. This places restrictions on when a binding prefetch can be issued, since the value will become stale if another processor modifies the same location during the interval between prefetch and actual reference. Binding prefetching may result in a significant loss in performance.

In contrast, *nonbinding prefetch* brings the data to the cache only, remaining visible to the cache coherence protocol, and is thus kept consistent until the processor actually reads the value. *Hardware-controlled prefetch* includes schemes such as long cache lines and instruction lookahead. The effectiveness of long cache lines is limited by the reduced spatial locality in multiprocessor applications, while instruction lookahead is limited by branches and the finite lookahead buffer size.

With *software-controlled prefetch*, explicit *prefetch* instructions are issued in user code. Software control allows the prefetching to be done selectively (thus reducing bandwidth requirements) and extends the possible interval between prefetch issue and actual reference, which is very important when latencies are large. The disadvantages of software control include the extra instruction overhead required to generate the *prefetches*, as well as the need for sophisticated software intervention. In our study, we concentrate on *nonbinding software-controlled prefetching*.

Benefits of Prefetching The benefits of prefetching come from several sources. The most obvious benefit occurs when a prefetch is issued early enough in the code that the line is already in the cache by the time it is referenced. However, prefetching can improve performance even when this is not possible (e.g., when the address of a data structure cannot be determined until immediately before it is referenced). If multiple prefetches are issued back to back to fetch the data structure, the latency of all but the first prefetched references can be hidden due to the pipelining of the memory accesses.

Prefetching offers another benefit in multiprocessors that use an ownership-based cache coherence protocol. If a cache line is to be modified, prefetching it directly with ownership can significantly reduce the write latencies and the ensuing network traffic for obtaining ownership. Network traffic is reduced in *read-modify-write* instructions, since prefetch with ownership avoids first fetching a read-shared copy.

Stanford Benchmark Results Stanford researchers (Gupta, Hennessy, Gharachorloo, Mowry, and Weber, 1991) have reported some benchmark results for evaluating various latency hiding mechanisms.

The effect of prefetching is illustrated in Fig. 5.17 based on benchmarking the MP3D code on simulation runs on the Stanford Dash multiprocessor.

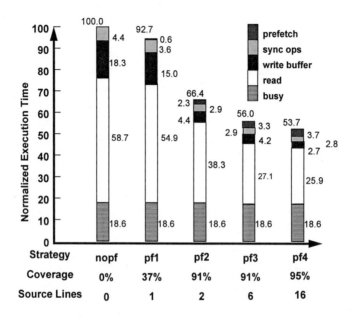

Figure 5.17   Effect of various prefetching strategies in benchmarking MP3D on the Dash multiprocessor. (Gupta et al. *Proc. Int'l Symp. on Computer Architecture* [285], May 1991.)

The simulation runs involved 10,000 particles in a $64 \times 8 \times 8$ space array with five steps. Five prefetching strategies were tested (*nopf, pf1, pf2, pf3,* and *pf4* in Fig. 5.17. Theses strategies range from no prefetching (*nopf*) to prefetching of the particle record in the same iteration or pipelined across increasing numbers of iterations (*pf1* through *pf4*).

The bars show the execution times normalized with respect to the *nopf* strategy. Each bar shows a breakdown of the time required for prefetch, synchronization operations, using write buffers, reads, and busy computing.

The end result is that prefetches are issued for up to 95% of the misses that occur in the case without prefetching (referred to as the *coverage factor* in Fig. 5.17). The prefetching causes significant time reduction in synchronization operations, using write buffers, and performing read operations.

The best speedup achieved is 1.86, when the *pf4* prefetching strategy is compared with the *nopf* strategy. Still the prefetching benefits are application-dependent. To introduce the prefetches in the MP3D code, only 16 lines of extra code were added to the source code consisting of thousands of instruction lines.

5.5.4 Effects of Relaxed Memory Consistency

In Fig. 5.18, we show the breakdown of execution times under SC (*Sequential Consistency*) and RC (*Release Consistency*) for the three benchmark applications. The execution times are normalized with respect to those shown in Fig. 5.18 with shared data cached. As seen from the results, The RC memory model removes all idle time due to write-miss latency.

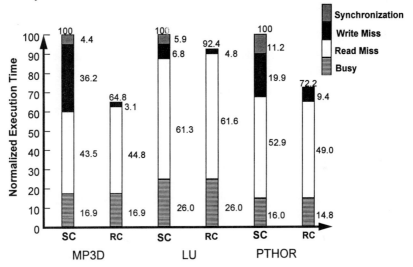

Figure 5.18   Effect of relaxing the shared-memory model from sequential consistency (SC) to release consistency (RC). (Gupta et al. *Proc. Int'l Symp. on Computer Architecture* [285], May 1991.)

The performance gains are large in MP3D and PTHOR since the write-miss time constitutes a large percentage in using the SC memory model (35% and 20%, respectively), while the gains are small in LU due to the relatively small write-miss time under the SC memory model (7%).

5.6 Multithreaded Latency Hiding

Multithreading demands that the processor be designed to handle multiple contexts simultaneously on a context-switching basis. In this case, a processor is switched to execute another thread of program, while the current thread is waiting for a data/ instruction access to complete. We outline the architecture of multiple-context processors. The effects of multithreading and improvement of processor efficiency are studied below.

5.6.1 Multithreaded Processor Model

A scalable parallel computer is modeled by a network of processors (P) and memory (M) nodes as depicted in Fig. 5.19(a). The distributed memories form a global address space. Four machine parameters are defined to analyze the performance of this system:

- *Latency L*: This is the communication latency experienced with a remote memory access. The value of L includes the network delay, cache-miss penalty, and delays caused by contentions in split transactions.

- *Number of threads N*: This is the number of threads that can be interleaved in a processor. A *thread* is represented by a *context* consisting of a program counter, a register set, and the required context status words.

- *Context-switching overhead C*: This refers to the cycles lost in performing context switching in a processor. This time depends on the switch mechanism and the number of processor states needed to maintain active threads.

- *Interval between context switches R*: This refers to the run length in terms of cycles between context switches triggered by remote references. The inverse, $p = 1/R$, is called the *rate of requests* for remote accesses. This reflects a combination of program behavior and memory system design.

In order to increase processor efficiency, one approach is to reduce the rate of requests by using distributed coherent caches. Another is to eliminate processor waiting through multithreading. The basic concept of multithreading is described below.

Multithreaded Computations   Bell [67] has described the structure of the multi-threaded parallel computations model shown in Fig. 5.19(b). The computation starts with a sequential thread (1), followed by supervisory scheduling (2) where the processors begin multiple threads of computation (3), by intercomputer messages that update variables among the nodes when the computer has a distributed memory (4), and finally by synchronization prior to beginning the next unit of parallel work (5).

The communication overhead period (3) inherent in distributed-memory structures is usually distributed throughout the computation and is possibly overlapped. Message-passing overhead (*send* and *receive* calls) in multicomputers can be reduced by specialized hardware operating in parallel with computation.

Communication bandwidth limits granularity, since a certain amount of data has to be transferred with other nodes in order to complete a computational grain. Message-passing calls (4) and synchronization (5) are nonproductive. Fast mechanisms to reduce or to hide these delays are needed. Multithreading is not capable of speedup in the execution of single thread, while weak ordering or relaxed consistency models are capable of doing so.

Multithreaded systems are constructed with *multiple-context* (or *multithreaded*) processors. In this section, we study an abstract model based on the work of Saavedra et al. [521]. We discuss the processor efficiency issue as a function of *memory latency (L)*, the *number of contexts (N)*, and *context-switching overhead (C)*.

Processor Efficiency A conventional single-thread processor will wait during a remote reference having a latency of L cycles. A multithreaded processor, as modeled in Fig. 5.20, will suspend the current context and switch to another. Only experiencing a few cycles of context-switching overhead, the processor will again be busy doing useful work, even though the remote reference is outstanding. Only if all the available contexts are suspended (blocked) will the processor be idle.

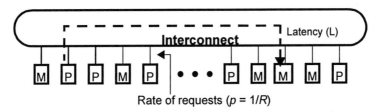

(a) An abstract model of scalable parallel system

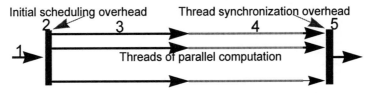

(b) The concept of multithreading in an MPP system

Figure 5.19 Multithreaded computation model for a scalable multiprocessor system.

Clearly, our objective is to maximize the fraction of time that the processor is busy. We define the *efficiency* of a multithreaded processor as follows:

$$\text{Efficiency} \;=\; \text{busy} \,/\, (\,\text{busy} + \text{switching} + \text{idle}) \qquad (5.12)$$

where *busy, switching,* and *idle* represent the amounts of time that the processor is in the corresponding states. The basic idea behind a multithreaded processor is to interleave the execution of several contexts in order to reduce the value of *idle* time without increasing the magnitude of the *switching* time.

The state of a processor depends on the disposition of various contexts in the processor. During its lifetime, a context cycle goes through the following states: *ready, running, leaving,* and *blocked.* There can be at most one context running or leaving. A processor is *busy* if there is a context in the running state; it is *switching* while making the transition from one context to another, i.e., when a context is leaving.

Otherwise, all contexts are blocked, and we say the processor in the *idle* state. A running context keeps the processor busy until it issues an operation that requires a context switch. The context then spends C cycles in the *leaving* state, then goes into the *blocked* state for L cycles, and finally reenters the *ready* state. Eventually the processor will enable a context, and the cycle repeats.

Abstract Processor Model   A multithreaded processor is modeled in Fig. 5.20. For simplicity, we assume one thread per context, and each context is represented by its own hardware resources: the *program counter* (PC), *register file*, and *process status word* (PSW). Assume that each context is fully executable with these dedicated resources. With multiple register files, one to each context, the context-switching overhead could be as short as one or two processor cycles.

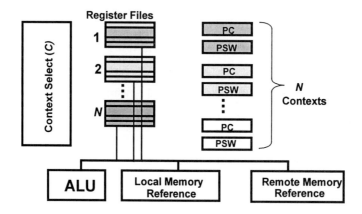

Figure 5.20   Multiple-context processor model with one thread per context. (PC: program counter, PSW: process status word.)

5.6.2 Context-Switching Policies

Different multithreaded architectures are distinguished by the context-switching policies adopted. Specified below are four switching policies:

    (1) ***Switch on cache miss.*** A context is switched when it encounters a cache miss. In this case, R is taken to be the average interval between cache misses (in cycles), and L is the time required to recover from the cache miss.

    (2) ***Switch on every load.*** This policy allows switching on every *load* operation, independent of whether it will cause a miss or not. In this case, R represents the

average interval between *loads*. A general model assumes that a context is blocked for *L* cycles after a switch; but in the case of a *switch-on-load* processor, this happens only if the *load* causes a *cache miss*.

The general model can be employed if it is postulated that there are two sources of latency (L_1 and L_2), each having a particular probability (p_1 and p_2) of occurring on every switch. If L_1 represents the latency on a cache miss, then p_1 corresponds to what is normally referred to as the *miss ratio*. Latency L_2 is a zero-cycle memory latency with probability p_2.

(3) **Switch on every instruction.** This policy allows switching on every instruction, independent of whether or not it is a *load*. In other words, it interleaves the instructions from different threads on a cycle-by-cycle basis. Successive instructions will be independently executed, which will benefit pipelined execution.

However, the cache miss may increase due to breaking of locality. It has been verified by some trace-driven experiments that cycle-by-cycle interleaving of contexts provides a performance advantage over switching at a *cache miss* in that the context interleaving could hide pipeline dependences and reduce the context-switch cost.

(4) **Switch on block of instruction.** Blocks of instructions from different threads are interleaved. This will improve the cache-hit ratio due to preservation of some locality. It will also benefit single-context performance.

Single-Threaded Processor Efficiency   A single-threaded processor executes a context for *R* cycles until a remote reference is issued. It becomes idle until the remote reference completes in *L* cycles. There is no context switch and obviously no switch overhead. We can model this behavior as a recycling process having a cycle of *R + L*, where *R* and *L* correspond to the processor's being *busy* and *idle*, respectively. Thus the efficiency of a single-threaded processor is defined by

$$E_1 = \frac{R}{R+L} = \frac{1}{1+L/R} \tag{5.13}$$

This shows clearly the performance degradation of a processor in a parallel system with a very large memory latency *L*. In that case, E_1 is always low because the run length $R \ll L$ in general.

Multithreaded Processor Efficiency   With multiple contexts, memory latency can be hidden by switching to a new context with *C* cycles of overhead. Assuming the run length between switches is constant with a sufficient number of contexts, there is always a context ready to execute when a switch occurs, so the processor is never idle. The processor efficiency is analyzed below under two different conditions as illustrated in Fig. 5.21.

Saturated Region   In the saturated region illustrated above, the processor operates with a maximum utilization. The cycle of the repeated runs in this case is $R + C$, and the efficiency is simply defined by

$$E_{sat} = \frac{R}{R+C} = \frac{1}{1+C/R} \qquad (5.14)$$

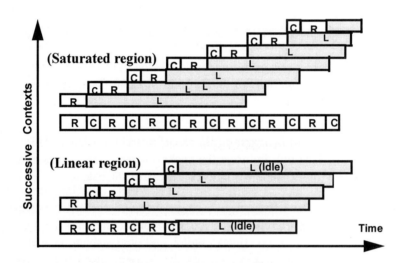

Figure 5.21   Snapshots of context switching in a multithreaded processor under saturation and linear regions.

The efficiency in saturation is independent of the latency L and also does not change with a further increase in the number of contexts. Saturation is achieved when a processor spends more time to service other threads than the time required to process a single thread, i.e., when $(N-1)(R+C) > L$. This gives the saturation point under a constant run length:

$$N_d = \frac{L}{R+C} + 1 \qquad (5.15)$$

Linear Region   When the number of contexts is below the saturation point, there may be no ready contexts after a context switch, so the processor will experience some idle cycles. The times required to switch to a ready context, to execute it until a remote reference is issued, and to process the reference add to $R + C + L$. During this time all other contexts have a turn in the processor. Thus, the efficiency is given by

$$E_{lin} = \frac{NR}{R + C + L} \qquad (5.16)$$

In Fig. 5.22, the two efficiency expressions in Eq.(5.14) and Eq.(5.16) meet at the saturation point N_d in Eq.(5.15). Note that the processor efficiency increases linearly with the number of contexts until the saturation point. Beyond that point, the curve for E_{sat} gives the utmost limit on the processor efficiency. It should be noted that multithreading increases both processor efficiency and network traffic. Trade-offs do exist between these two opposing goals as discussed in Agarwal [9].

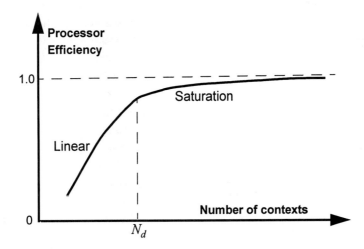

Figure 5.22   Processor efficiency plotted against the context number.
(Courtesy of a presentation by R. Saavedra in 1992)

Example 5.11 Hiding remote memory latency in a multithreaded processor

Consider a parallel computer with a remote memory latency L ranging from 50 to 200 processor cycles. Assume three cases of context-switching overhead with $C = 1, 4,$ and 16 cycles. Saavedra et al. [522] have obtained some numerical results on the processor efficiency under the assumption that the typical run length of a thread is $R = 16$ cycles. These results are shown in Table 5.6.

Table 5.6 Processor Efficiency for Two Multithreaded Architectures
(Courtesy R Saavedra et al. [521], *Proc. ACM Symposium*
Parallel Algorithms and Architecture, July 1990)

Context- Switching Overhead (C)	N = 2 Contexts per Processor				N = 6 Contexts per Processor			
	L = 50	L =100	L = 150	L = 200	L = 50	L = 100	L = 150	L = 200
1	0.48	0.29	0.22	0.15	0.94	0.76	0.68	0.45
4	0.45	0.27	0.21	0.15	0.80	0.68	0.56	0.44
16	0.36	0.25	0.16	0.14	0.50	0.49	0.45	0.41

The results shown here are obtained from substituting the numerical values of the key parameters into Eq.(5.14) to Eq. (5.16). Two architectural cases are being evaluated: one using two-context processors and the other using six-context processors. The table entries reveal a number of interesting observations:

- Using two-context processors, the processor efficiency drops sharply from 0.48 to 0.14 with respect to increasing memory latency and higher context switching overhead. In other words, switching between two contexts is not enough to hide the memory latency beyond 50 cycles.

- The drop in processor efficiency is much slower for six-context processors than the case two-context processors. More than twice the efficiency can be achieved by using the six-context processors over the two-context processors.

- Under extremely low overhead of context switching ($C = 1$ processor cycle), using six-context processors can hide 50 to 150 cycles of memory latency with an efficiency between 0.94 and 0.68.

- The case of $C = 16$ cycles leads to less than 0.50 processor efficiencym, which is considered unacceptable in both types of processor architecture.

- With a context-switching overhead $C = 4$ cycles, the processor efficiency converges to that of the ideal case of $C = 1$ cycle as the memory latency increases.

Based on the above numerical results, one can extrapolate a processor design with 85% efficiency. The architecture is suggested to have no more than 2 cycles of context-switching overhead. The number of contexts in each processor should be 4 in order to hide a memory latency of up to 75 processor cycles.

■

Agarwal [9] indicated that only switching among a small number of contexts benefits the system performance. More threads may compete for limited resources to cause excessive context switching, which may severely degrade the performance.

A good example of a multithreaded machine is the Tera multiprocessor system [27]. Interested readers may want to follow Tera's development for actual design and benchmark experiences. No multithreaded machines have proved to be a commercially viable in computer industry.

5.6.3 Combining Latency Hiding Mechanisms

The effect of combining various latency hiding mechanisms is illustrated by Fig. 5.23 based on the MP3D benchmark results released from Stanford University. The effect of using *multiple-context* (MC) processors is presented along with other latency hiding mechanisms such as PF (*prefetching*), CC (*coherent cache*), RC (*release consistency*).

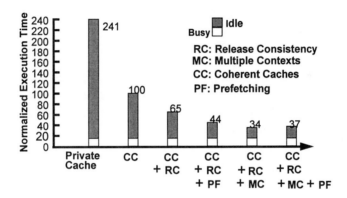

Figure 5.23 **Effect of combining various latency hiding techniques based on MP3D benchmark results on the Stanford Dash multiprocessor.** (Gupta et al. *Proc. Int'l Symp. on Computer Architecture* [285], May 1991.)

The busy parts of the execution time are equal in all combinations of mechanisms. This corresponds to the CPU busy time in executing the MP3D program. The idle parts in the bars correspond to memory latency including all cache-miss penalties. All the times are normalized with respect to the execution time (100 units) required in a cache-coherent system.

The leftmost bar (with 241 units) corresponds to the worst case of using a private cache exclusively without shared *reads* or *writes*. Long overhead is experienced in this case due to excessive cache misses. The use of a cache-coherent system shows a 2.41-fold improvement over the private case. All the remaining cases are assumed to use hardware coherent caches.

The use of *release consistency* shows a 35% further improvement over the coherent system. The adding of prefetching reduces the time further to 44 units. The best case is the combination of using coherent caches, RC, and MC. The rightmost time bar is obtained from applying all four mechanisms. The combined results show an overall speedup of 4 to 7 over the case of using private caches.

The above benchmark results reported at Stanford suggest that a coherent cache and relaxed consistency uniformly improve the performance. The improvement from prefetching and multiple contexts is also appreciable, but much more application-dependent.

Combinations of various latency hiding mechanisms generally attain a better performance than applying each mechanism alone. In other word, the four mechanisms are mutually supportive, rather than cancel each other's effects.

5.7 Bibliographic Notes and Problems

The concept of a working set is introduced in Denning [203]. The linear programming optimization strategy is described in Chow [157] for memory capacity planning. Cragon [171] has discussed memory systems for pipeline processor design. The book by Przybylski [510] deals with cache memory architecture. The books by Flynn [245], Hwang [327], and Patterson and Hennessy [305] treat the memory hierarchy design for different requirements in sequential and parallel architectures.

The MESI snoopy protocol is based on the one used in Pentium processor [31]. Censier and Feautrier [135] proposed the first cache directory structure. Dubois et al. [219] characterize synchronization, coherence, and event ordering in multiprocessors. Smith [568] assessed cache memories developed up to the early 1980s. Chaiken et al. [137] and Agarwal et al. [11] have evaluated a directory-based protocol in the MIT Alewife multiprocessor. Other developments in cache architecture can be found in Dubois and Thakkar [220] and Stone [590].

Adve and Gharachorloo have provided a survey of shared-memory consistency models [8]. The concept of sequential consistency was developed by Lamport [391]. The weak consistency memory model was due to Dubois et al. [218]. The formal specifications of the TSO and PSO weak models are given in Sindhu et al. [558]. Adve and Hill [7] have studied the weak ordering issue.

Processor consistency model is due to Goodman [273]. The Stanford release consistency model can be found in Gharachorloo et al. [264]. Commercial relaxed memory models are specified for IBM System/370 [341], Sun SPARC [579], Cray GigaRing [539], Digital Alpha [563], and IBM PowerPC [434]. Four shared-memory models are evaluated by Chong and Hwang [155].

The Stanford Dash multiprocessor is reported in Lenoski et al. [405]. The KSR-1 is described in Burkhardt et al.[116]. The Swedish DFM is reported in Hagersten et al. [293]. Bisiani and Ravishankar [83] developed the PLUS DSM system. Nitzberg and Lo [468] have surveyed the earlier DSM systems. The IVY is described in Li [413], the TreadMarks at Rice [30], the Wind Tunnel at Wisconsin [515], and Shrimp at Princeton [237].

Stenstrom et al. compare the performance of CC-NUMA and COMA architectures [585]. Latency tolerance through software-controlled prefetch is studied in Mowry and Gupta [454]. Gupta et al. have evaluated the effects of distributed coherent caches [285]. Saavedra et al. [522] have evaluated various data prefetching strategies.

A good survey of multithreaded architecture is given by Nikhil [467]. Saavedra et al. [521] have modeled multithreaded architectures for parallel computing. Smith [569] discussed data locality issues in relation to multithreaded computing. Various latency hiding techniques are also treated in Hwang [327] and in Culler et al. [181].

Homework Problems

Problem 5.1 Define the following technical terms associated with a memory hierarchy design and management:

(a)   Virtual address space

(b)   Physical address space

(c)   Address translation

(d)   Cache line size

(e)   Cache hit ratio

(f)   Page fault

(g)   Memory replacement policies

Problem 5.2 Consider a two-level memory hierarchy, M_1 and M_2. Denote the hit ratio of M_1 as h. Let c_1 and c_2 be the costs per KB, s_1 and s_2 be the memory capacities, and t_1 and t_2 be the access times, respectively.

(a)   Under what conditions will the average cost of the entire memory system approach the value of c_2?

(b)   What is the effective memory-access time, t_a, of this hierarchy?

(c)   Let $r = t_2 / t_1$ be the speed ratio of the two memories. Let $E = t_1 / t_a$ be the access efficiency of the memory system. Express E in terms of r and h.

(d)   Plot E against h for $r = 5$, 20 and 100 on a graph paper.

(e)   What is the required hit ratio h to yield an efficiency $E > 0.95$ if $r = 100$?

Problem 5.3 You are asked to perform capacity planning for a two-level memory system. The first level, M_1, is a cache with three capacity choices of 64 KB, 128 KB, and

256 KB. The second level, M_2, is a main memory with a 4 MB capacity. Let c_1 and c_2 be the costs per byte and t_1 and t_2 the access times for M_1 and M_2, respectively. Assume $c_1 = 20\ c_2$ and $t_2 = 10\ t_1$. The cache hit ratios for the three capacities are assumed to be 0.7, 0.9, and 0.98, respectively.

- (a) What is the average access time t_a in terms of $t_1 = 20$ ns in the three cache designs? (Note that t_1 is the time from CPU to M_1 and t_2 is that from CPU to M_2, not from M_1 to M_2).

- (b) Express the average cost of the entire memory hierarchy given the per-KB cost $c_2 = \$\ 0.2/\text{KB}$.

- (c) Compare the three memory designs and indicate the order of average costs and the order of average access times, respectively. Choose the optimal design based on the product of the average cost and the average access time.

Problem 5.4 Explain the *inclusion property* and *memory coherence* requirements in a multilevel memory hierarchy. Distinguish between *write-through* and *write-back* policies in maintaining the coherence in adjacent levels. Also explain the basic concepts of *paging* and *segmentation* in managing the physical and virtual memories of a hierarchy.

Problem 5.5 Reconsider Prob. 5.3 with the following parameter values. Suppose that $h_1 = 0.95$, $t_1 = 20$ ns, $s_1 = 512$ KB, $c_1 = \$10/\text{KB}$, $c_2 = \$\ 5/\text{KB}$, t_2 and s_2 are unknown. The total cost of the memory hierarchy has $15,000 as an upper bound.

- (a) Derive a formula showing the effective access time, t_{eff}, of this memory system.

- (b) Derive a formula showing the total cost of this memory system.

- (c) How large a capacity of M2 ($s_2 = ?$) can you acquire without exceeding the budget limit?

- (d) How fast a main memory ($t_2 = ?$) do you need to achieve an effective access time of $t_{\text{eff}} = 40$ ns?

Problem 5.6 Answer the following questions on locality properties:

- (a) Explain the *temporal locality*, *spatial locality*, and *sequential locality* associated with program/data access in a memory hierarchy.

- (b) What is the *working set*? Comment on the sensitivity of the observation window size to the size of the working set. How will this affect the main memory hit ratio?

- (c) What is the *90-10 rule* and its relationship to the *locality of references*?

Problem 5.7 Consider a RISC-based shared-memory multiprocessor with p processors, each having an instruction cache and a data cache. The peak execution rate of each processor (assuming a 100% hit ratio in both caches) is x MIPS. You are required to derive a performance formula, taking into account the cache misses, shared-memory accesses, and synchronization overhead.

Assume that on the average, @ percent of the instructions executed are for synchronization purposes, and the penalty for each synchronization operation is an additional t_s us. The average number of memory references per instruction is assumed to be m. Among all memory references by the CPU, let f_i be the percentage of references to instructions. Assume that the instruction cache and data cache have hit ratios h_i and h_d, respectively, after a long period of program tracing on the machine. On cache misses, instructions and data are accessed from the shared memory with an average access time of t_m us.

(a) Derive an expression for approximating the effective MIPS rate of this multi-processor in terms of p, x, m, f_i, h_i, h_d, t_m, @, and t_s. Ignore the cache access time and other system overheads in your derivation.

(b) Given the values $m = 0.4$, $f_i = 0.5$, $h_i = 0.95$, $h_d = 0.7$, @ $= 0.05$, x $= 5$, $t_m = 0.5$ us, and $t_s = 5$ us. Determine the minimum number of processors needed in the above multiprocessor system to achieve an effective total execution rate of 25 MIPS. Ignore all other memory interferences caused by cache coherence, access conflicts, or page faults.

Problem 5.8 Reconsider Prob.5.7, given the following cost information. Suppose the total cost of all caches and shared memory has $25,000 as an upper bound. The cache memory costs $4.70/KB, and the shared memory costs $0.4/KB. With $p = 16$ processors, each having an instruction cache of $s_i = 32$ KB and a data cache of $s_d = 64$ KB, what is the maximum shared-memory capacity c_m (in MB) that can be acquired within the given budget limit?

Problem 5.9 Specify the following three snoopy protocols with state transition diagrams. These protocols apply to three cache architectures with increasing numbers of cache states. All cache states, and caching events, and snoopy bus events must be specified.

(a) A two-state SI protocol for WT (*write-through*) caches, where S stands for *shared state* and I for *invalid state* of a cache line.

(b) A three-state MSI protocol for WB (*write-back*) caches, where S and I are defined as above and M is the *modified state* of a cache line.

(c) A four-state MESI protocol for a combined WT/WB cache architecture, where the added E stands for the *exclusive* state in which the cache line has been written only once.

Problem 5.10 Consider the concurrent execution of two programs (P_0 and P_1) on two processors with a shared memory. Assume that variables *A, B, C, D* are initialized to have the value 0 and that a *Print* statement prints two variables indivisibly at the same cycle. The output forms a 4-tuple as either *ADBC* or *BCAD*.

P_0 :	P_1 :
a. *A* = 1	d. *C* = 1
b. *B* = 1	e. *D* = 1
c. *Print A, D*	f. *Print B, C*

   (a)   List all execution interleaving orders of the six statements that will preserve the individual program order.

   (b)   Assume program orders are preserved and all memory accesses are atomic, i.e., a *store* by one processor is immediately seen by the other processor. List all the possible 4-tuple output combinations.

   (c)   Assume program orders are preserved but memory accesses are nonatomic; i.e., a *store* by one processor may be buffered so that some other processors may not immediately observe the updated result. List all possible 4-tuple output combinations under this condition.

Problem 5.11 After studying the Stanford Dash memory hierarchy and the directory protocol, answer the following questions with an analysis of potential performance:

   (a)   Define the cache states used in the Dash multiprocessor architecture.

   (b)   How are the cache directories implemented in the memory hierarchy?

   (c)   Explain the Dash directory-based coherence protocol when reading a remote cache block that is dirty in a remote cluster.

   (d)   Repeat Part (c) for the case of writing to a shared remote cache line.

Problem 5.12 Characterize the following NUMA and NORMA memory architectures for building scalable multiprocessors, multicomputers, or cluster of computers. In each case, choose an example computer system as a case study to explain the special architectural features built. Compare their relative strengths and weaknesses.

   (a)   The CC-NUMA architecture

   (b)   The COMA architecture

   (c)   The NCC-NUMA architecture

   (d)   Software-implemented DSM architecture

   (e)   Loosely coupled cluster NORMA architecture

   (f)   Tightly coupled NORMA architecture for MPP

Problem 5.13 Compare the performance of NUMA and COMA architectures under small or large working set and low or high *communication-computation ratio* (CCR):

(a) Explain why both architectures can perform well with a low CCR and a small working set in a given application.

(b) With a large working set and a low communication-to-computation ratio, explain why the COMA architecture performs better than the NUMA architecture.

(c) With a small working set and a high communication-to-computation ratio, explain why the COMA machine performs worse than the NUMA architecture.

Problem 5.14 Briefly explain the basic ideas in each of the following latency tolerance techniques. Both hardware and software approaches should considered.

(a) Latency reduction techniques

(b) Latency avoidance techniques

(c) Latency hiding techniques

(d) Distributed coherent caches

(e) Data prefetching to hide *read* latency

(f) Multithreading to hide *write* latency

Problem 5.15 Compare the memory access constraints on the following four consistency models for the shared memory of a multiprocessor system. You should cover both *read* and *write* accesses, as well as *synchronization* accesses per thread.

(a) The *sequential consistency* (SC) memory model specified by Lamport

(b) The *weak consistency* (WC) memory model specified by Dubois et al.

(c) The *processor-consistent* (PC) memory model specified by Goodman

(d) The *release consistency* (RC) memory model by Gharachorloo et al.

Problem 5.16 Consider a scalable multiprocessor with p processing nodes with a distributed shared memory. Let $1/R$ be the rate of each processing node generating a request to access the remote memory through the interconnection network. Let L be the average latency for a remote memory access. Derive expressions for the *processor efficiency E* under each of the following conditions:

(a) Each processor is single-threaded, uses only a private cache without coherence support or other hardware support for latency hiding. Express E as a function of R and L.

(b)   Suppose a coherent cache is supported by hardware with proper data sharing and h is the probability that a remote request can be satisfied by a local cache. Express E as a function of R, L, and h.

(c)   Now assume each processor is multithreaded to handle N contexts simultaneously. Assume a context-switching overhead of C. Express E as a function of N, R, L, h, and C.

(d)   Now consider the use of a two-dimensional $r \times r$ torus with $r = p$ and bidirectional links. Let t_d be the time delay between adjacent nodes and t_m be the local memory access time. Assume that the network is fast enough to respond to each request without buffering. Express the latency L as a function of p, t_d, and t_m. Then express the efficiency E as a function of N, R, h, C, p, t_d, and t_m.

Problem 5.17   Compare the four context-switching policies: *switch on cache miss, switch on every load, switch at each instruction cycle,* and *switch on a block of instructions.*

(a)   What special hardware and software support is needed to implement each context-switching policy ?

(b)   What are the advantages and disadvantages of each context-switching policy?

(c)   Rank the four policies according to processor complexity and efficiency.

(d)   Under what program behavior can each of the policies perform satisfactorily?

Chapter 6

System Interconnects and Gigabit Networks

In this chapter, we assess available network technologies and system interconnects. The focus is placed on standard interconnects, commodity networks, and their design and application principles. We study the topologies, protocols, and connectivity of buses, crossbar switches, multistage switches, shared-media networks, cell switched and packet switched networks. These topics pave the way to studying parallel or cluster computer architectures in subsequent chapters.

6.1 Basics of Interconnection Network

Switching fabrics are the core of a modern computer network. We introduce below basic network terms, environments, classifications, standard protocols, and performance metrics of today's high-bandwidth networks. In this book, *networks* refer to the connections among multiple computers. The *system interconnect* glues together subsystems of an integrated computer system. A multicomputer cluster needs to use both types of interconnection structures.

6.1.1 Interconnection Environments

In a scalable multiprocessor, multicomputer cluster, or a distributed system, the building blocks (desktop hosts, servers, mainframes, switches, networks, adapter cards, and peripheral devices etc.) are interconnected at five networking environments: *System or I/O buses or crossbar/multistage switches* are system interconnects often found on the backplane of a single rack of computer platform, or confined within a few interconnected racks in the same room. A dozen network technologies are mapped into a two-dimemsion space in Fig. 6.1.

A *system-area network* (SAN) connects processors, memory, and interface boards with microstrip cables within a short distance. A *local-area network* (LAN) is confined within a building, a campus, or an enterprise. A *metropolitan-area network* (MAN) covers a whole district or within a city limit. A *wide-area network* (WAN) often appears as an internetwork of a number of smaller networks, thus it can be extended to long distance.

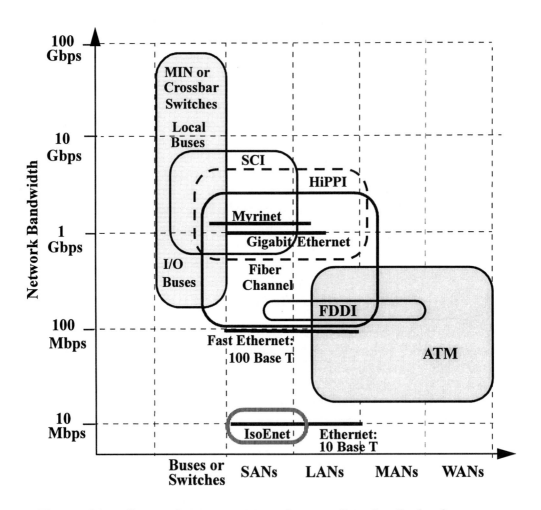

Figure 6.1   System interconnects and network technologies for building scalable parallel and cluster platforms.

These network technologies will be described in subsequent sections. The mapping is based on 1997 technology. The maximum network distance increases from left to right along the horizontal axis. The interconnection environments do overlap in operating distance. We did not mark the distance scale in Fig. 6.1 due to the overlapping situation. We use Fig. 6.2 to define a set of terms for buses and networks that are widely used in industry and academic. The pins of a processor chip form the *processor bus*. The bus that connects the processor and the memory modules is called the *local bus*, or *memory bus*, the length of which is within tens of centimeters.

A *system bus* (also called *I/O bus*) provides *slots* to connect I/O devices, such as disk drives, tape drives, network interface cards, etc. It is connected to the local bus through an I/O bridge circuit. A system bus is often built on a motherboard, a backplane or center-plane of a system rack, thus limited to 2 m.

Examples of system bus include the AT bus (ISA), the EISA bus, the PCI bus, and the Micro Channel used in IBM PCs, the Sbus in Sun workstations. Note that the term *I/O bus* is often used to refer to the SCSI (*small computer system interface*) bus. Depending on the cabling technology, an internal I/O bus rarely exceeds 3 m and an external SCSI cable can be extended to almost 20 m.

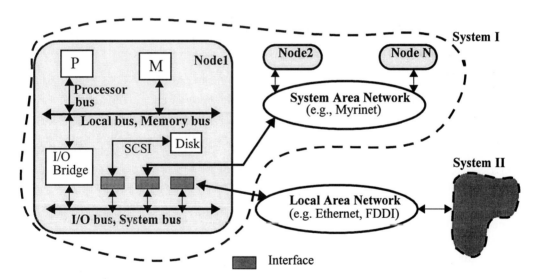

Figure 6.2   Illustration of local bus (memory bus), system bus (I/O bus), system-area network, and local-area network.

Many multiprocessor or multicomputer systems divide their hardware into a number of *nodes*. The network connecting these nodes together to form a single system is called a SAN. In contrast, a LAN is used to connect multiple systems. Of course, a technology (e.g., Ethernet) can be used in SANs and LANs.

What constitutes a system will be made clear in Chapter 9, when we discuss clusters and single system image in detail. A SAN spans 3 to 25 m. A LAN can extend from 25 m to 500 m or 2 km. A MAN normally extends to 25 km or longer.

A WAN is essentially unlimited in distance. It can extend beyond international boundaries. The vertical axis corresponds to increasing *network bandwidth*, defined as the maximum amount of information transferable through the interconnect or network per unit time. Some professionals call it simply the *speed* of the network, often measured as Mbps (*Million bits per second*) or Gbps (*Giga bits per second*). The interconnection technologies shown in Fig. 6.1 yield a bandwidth from 10 Mbps to 100 Gbps.

We shall examine the speed of buses, crossbar switch, and various network technologies in subsequent sections. Specific technology may be upgraded further with higher bandwidth or wider cabling distance. Therefore, some of the circled ranges may move toward the upper right corner of Fig. 6.1. A Tbps (*Tera bits per second*) network speed is entirely possible beyond the year 2000.

6.1.2 Network Components

All switched networks are basically built with three components: *links* (*channels* or *cables*), *switches* (*routers*), and *network interface cards.* Shared-media networks will not use switches. We briefly introduce these network components below before getting into other characteristics of networks.

Links, Channels, or Cables   All three terms refer to the physical connections between two hardware units in a computer system. A *link* can be implemented with copper wires or fiber-optic cables. The simplest link is an *unshielded twisted pair* (UTP) of copper wires. Using copper links is cheaper but limited by a short cabling distance due to a signaling problem.The cabling distance can be extended to some extent, if *shielded twisted pair*s (STPs) are used. Fiber-optic cables are more expensive, but do offer a much higher bandwidth and a longer cabling distance.

Some authors call a link a *channel* or a *cable.* A link may connect two switches or connect a switch and a network interface attached to a host node. A *short link* may contain only one logic signal at a time, while a *long link* behaves as a transmission line, allowing a string of logical signals to propagate along the line at the same time. A *serial link* or a *thin link* has one bit line to be shared by data, and control signals in a multiplexed manner.

A *wide link* or a *parallel link* has multiple bit lines, allowing parallel transmission of data and control information. The links are driven by one of two clocking mechanisms: synchronous or asynchronous. *Synchronous clocking* means the source and destination ends operate with the same global clock. *Asynchronous links* use some embedded clock encoding mechanism to allow both ends to handshake with different clock rates.

Switches   *Switches* are needed in building switched networks. In general, a switch has multiple input and output ports. Each input port has a receiver and input *buffer* to handle the arriving packets or cells. Each output port has a transmitter to pass the outgoing data signal to a communication link connected to another switch or a network interface. A four-input and four-output switch is shown in Fig. 6.3.

An internal *crossbar* is used to establish *n* connections between *n* inputs and *n* outputs simultaneously. Each *crosspoint* can be switched on (as marked by dots in Fig. 6.3) or off under program control. This number *n* is often called the *degree* of the switch. Multiple switches and links are often used to build a large switched network following a chosen topology.

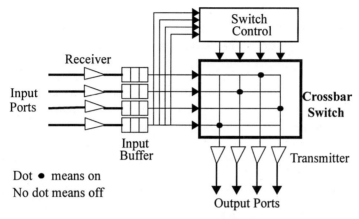

**Figure 6.3   A four-port crossbar switch providing any permutation
connection between 4 inputs and 4 outputs.**

Network Interface Circuitry   A *network interface circuitry* (NIC) is often used to connect a host computer to a LAN or to another network. Thus some professionals call it a *host interface.* The NIC must be able to handle two-way traffic between the host and the network. Therefore, the architecture of a NIC depends on the network and the host. Different hosts may require different interface cards, even connected to the same network.

A typical NIC contains an embedded processor, some input and output buffering, and some control memory and logic. Its functions include the formatting of packets or cells, routing path selection, coherence checking, flow and error control, etc. Therefore, the cost of a NIC is determined by the port size, storage capacity, processing power, and control circuitry; often resulting in a higher complexity than that of a routing switch.

6.1.3 Network Characteristics

We distinguish below between basic network terms; often these terms appear as pairs with two opposite characteristics. Or one can consider each characteristic a dimensionality of the entire network design space.

Static versus Dynamic Networks   Based on connectivity and control strategy, we classify interconnection networks into several categories. *Static networks* are built with point-to-point links that will not change during program execution. In other words, static networks have fixed connections among processing nodes. Static networks are also known as *direct networks*, because only one host computer is connected to each node switch. On the other hand, a *dynamic network* is implemented with switched channels, which are dynamically configured to meet the communication requirements in user programs.

Dynamic networks include system buses, crossbar switches, multistage networks, and various SANs, LANs, MANs, and WANs. Dynamic networks are also called *indirect networks*, where multiple hosts are connected to specific node switches and routing is done by a sequence of switching decisions. Some hybrid networks having both types of nodes support either static or dynamic connections upon program control.

Shared-Media Networks In a *shared-media network,* physical links (copper or fiber) are accessible by all computers connected to the network. For this reason, shared-media networks are also known as *multiaccess networks*. However, only one requester is granted access at a time. In other words, the bandwidth is shared by multiple requesters. The most challenging problem in designing a shared-media network is to allocate equal bandwidth to all competing requesters.

Shared-media networks can be implemented with FDDI, Fast Ethernet (100Base-T), Fiber Channel technologies. Since no switches are used, shared-media networks often cost less than the corresponding switched networks. The allocation of available bandwidth to all requesters is known as the *bandwidth management* problem. The selection of a network access protocol will greatly affect the efficiency of a bandwidth management system.

Switched Networks In contrast, a *switched network* uses switching fabric to allocate or deallocate the media resources to one requester at a time. In other words, the access to a media link is not shared by more than one user at a time. Many shared-media networks can be converted to a switched design at an increased cost. For example, most Ethernet and FDDI rings share the media, but switched Ethernet or switched FDDI rings are also available. On the other hand, all ATM networks are inherently switched networks.

Switched networks are often designed with three options: *circuit switching, packet switching,* and *cell switching,* as characterized below.

Circuit Switched Networks Switching is needed to determine the routing path between a pair of nodes for point-to-point communication or among multiple nodes for collective communications. In *circuit switching*, the entire path (of links and buffers) from the source node to the receiving node is reserved for the entire period of transmission.

This is similar to the railroad transportation by a long train on a reserved track, where no other train is allowed to share the same track at the same time. If the train is very long, the reserved track may be held for a long period before the next train can pass through.

Packet Switched Networks For a *packet network*, a long message is broken into a sequence of small packets. Each packet contains the routing information and a segment of data payload. Packets from the same message can be routed separately on different paths. This is similar to use smaller trucks (or cars) in dividing a whole trainload of goods.

Packet switching requires disassembly of the message into packets before the transmission and message reassembly upon arrival of all packets. Packet switching is supposed to yield better utilization of network resources, because many packets from different messages can share the same network links and buffers more efficiently.

Cell Switched Networks In a packet network, packets from different messages may differ in length. In other words, long and short packets may be serialized in a long string of packets. A small packet following a long packet may experience excessive delays, until the long packet is removed from the shared path. The *cell switching* is improved over *packet switching* by partitioning a long packet into fixed-size small *cells.*

The purpose is to alleviate the blocking imposed by long packets as depicted in Fig. 6.4. In practice, a long packet may be as long as 4096 B and a short packet may have only a few bytes. The waiting time is large when the small packet of an isochronous traffic such as voice traffic is waiting behind a large packet. In a *cell switched network*, a typical cell is 56 B. Constant transmission delay is possible in a cell network.

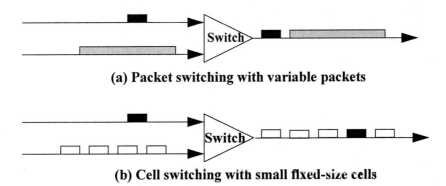

(a) Packet switching with variable packets

(b) Cell switching with small fixed-size cells

Figure 6.4 **Serialization of packets and cells in two types of switched networks.** (Courtesy of C. Patridge, *Gigabit Networking*, (Figs. 3.2 and 3.3 on p. 45) , ©1994 Addison-Wesley Pub. Co. [489], Peprinted by permission of Addison Wesley Longman).

Another advantage of cell networks over packet networks lies in the simplified hardware design of the cell switches, because of handling small, fixed-size cells instead of variable-length packets. The cell switching became popular with the introduction of ATM networks. Fragmenting data messages into many cells may lead to very low network performance, since the cells could be lost in transit and retransmission is often not allowed. We shall deal with these problems when ATM cell switches and ATM networks are discussed in Section 6.5.

6.1.4 Network Performance Metrics

Latency and bandwidth are two fundamental metrics used to assess the performance of a network or of a system interconnect. We describe below various aspects of these two performance metrics.

Communication Latency   We consider the total time needed to transmit a message from a source node to a destination node in a multicomputer system. This latency consists of four time components:

(1) The *software overhead* associated with sending and receiving the message at both ends of the network

(2) The *channel delay* caused by the channel occupancy (or the total message length divided by the channel bandwidth)

(3) The *routing delay* caused by the time spent in the successive switches in making a sequence of routing decisions along the routing path

(4) The *contention delay* caused by traffic contentions in the network

The software overhead is largely contributed by the host kernels in handling the message at both ends. The channel delay is often determined by the bottleneck link or channel. The routing delay is proportional to the *routing distance* (or the *path length* or the number of *hops* between the ends). The contention delay is most difficult to predict, because it depends on the traffic conditions in the network.

Network Latency   Both the software overhead and the contention delay depend on the program behavior. For this reason, hardware designers consider only the sum of the middle two terms, channel delay and switch delay, as the *network latency.* This quantity is totally determined by the network hardware characteristics, independent of the program behavior and traffic conditions.

In a light-traffic (contention-free) message-passing networked system, the network latency (usually around 1 us) is much smaller than the software overheads or contention delay (in tens or hundreds of microseconds). Overall, the communication latency is a function of the *kernel overheads* at both ends, the *message length* (including the header), the *channel bandwidth*, the *switch delay* (or the routing algorithm used), the *path length*, and the *network traffic* (or program behavior*)*.

In order to reduce the latency, a major effort should be conducted at reducing or hiding the software overhead time. Various latency reduction or hiding techniques will be discussed at the end of this chapter.

Per-Port Bandwidth   We have already defined the *per-port bandwidth* earlier as the maximum number of bits (or bytes) that can be transmitted per second from any port to any other port in a network. For example, the IBM SP2 *high-performance switch* (HPS) has a per-port bandwidth of 40 MB/s.

In a *symmetric network*, all the ports are topologically equivalent. In other words, the per-port bandwidth is independent of the port location in a symmetric network. Otherwise, we have an *asymmetric network*. The per-port bandwidth for an asymmetric network is defined as the minimum of all per-port bandwidths.

Aggregate Bandwidth   The *aggregate bandwidth* of a given network is defined as the maximum number of bits (or bytes) that can be transmitted from one half of the nodes to another half of the nodes per second. For example, the HPS is a symmetric network consisting of *n* nodes (ports), where *n* has an upper bound of 512.

With a per-port bandwidth of 40 MB/s. The aggregated bandwidth of a 512-node HPS is calculated as: $(40 \times 512) / 2 = 10.24$ GB/s. The division by 2 is attributed to the fact that bidirectional traffic should be counted only once.

Bisection Bandwidth   The aggregate bandwidth is also related to a topological term, called *bisection bandwidth*. This is defined as the maximum number of bits (or bytes) per second crossing all the wires on the bisection plane dividing the network into two equal halves.

Let *b* be the number of links crossing the bisection plane and *w* be the number of wires per link (called the *link width* or *channel width*). The product *bw* is called *bisection width*, representing the total number of wires crossing the bisection plane. If each wire transmits *r* b/s, the bisection bandwidth is defined as $B = bwr$ b/s. The time to move M B across a network has as a lower bound the ratio M/B.

In general, the network bandwidth is sensitive to the *network topology*, the *channel width*, the *network size* (or the port number), the *channel number*, the *switch degree*, and network *clock rate*. Only hardware architecture affects the network bandwidth, which is independent of program behavior and traffic patterns. This is also true for network latency. The communication latency is affected by both machine and program behaviors.

6.2   Network Topologies and Properties

Basic network properties are revealed along with their architectural implications. Before examining network topology, we define parameters often used to estimate the complexity, communication efficiency, and cost of a network.

6.2.1   Topological and Functional Properties

Basically, an interconnection network is represented by a *directed* or an *undirected graph* of a finite number of *nodes* linked by *edges*. A node could be a switch, or a host, or a device. An edge may refer to a link or channel. The number of nodes in the graph corresponds to the *network size*.

Node Degree   The number of edges (links or channels) incident on a node is called the *node degree d*. In case of unidirectional channels, the number of channels into a node is the *in degree*, and that out of a node is the *out degree*. The node degree is the sum of the two. It reflects the number of I/O ports required per node, and thus the cost of a node. The node degree should be kept as small as possible, to reduce the implied cost.

Network Diameter   The *diameter D* of a network is the maximum path between any two nodes. The path length is measured by the number of links traversed. The network diameter indicates the maximum number of distinct hops between any two nodes, thus providing a simple figure to estimate the communication delay on a data network.

Blocking versus Nonblocking Networks   Message passing in a network can be designed to be blocking or nonblocking.

A *blocking network* allows a message to be buffered for later transmission in case of channel conflicts among multiple messages. Therefore, blocking networks and *buffered networks* are used interchangeably.

In a *nonblocking network*, no buffers are used. Alternate paths are used to resolve conflicts among messages. Most nonblocking networks use switches to route messages.

Synchronous versus Asynchronous Networks   In a *synchronous network*, both the sender and receiver must be synchronized in time and space. This is similar to a telephone network requiring synchronization between caller and callee. Most synchronous networks do use buffers and thus are blocking networks.

Asynchronous networks do not use buffers and are nonblocking in nature. No synchronization of sender and receiver is needed in an asynchronous network. This is similar to the postal delivery network. Some communication networks are built to support both synchronous and asynchronous communications.

Functionality and Complexity - *Functionality* refers to special features or mechanisms built inside a network to support packet routing, interrupt handling, fast synchronization, message combining, data coherence, switch conflicts, channel conflicts, flow control, fault tolerance, lost packet handling, data prefetching, latency tolerance, bandwidth management, etc.

The complexity affects the network construction cost. Complexity is often determined by ports, cables, switches, buffers, connectors, arbitrators, and interface logic required in a network construction. The control circuitry, buffer memory, and power supply are also part of the network complexity.

Symmetry and Scalability   We call a network *symmetric* if the topology looks the same from any node in the network. Otherwise, the network is called *asymmetric*. Symmetric networks are easier to lay out on a VLSI circuit or on a PC board. It is also easier to route messages in a symmetric communication network.

Network *scalability* refers to the ability of a network to be modularly expandable. The expansion must be accompanied by proportional increase in expected performance. Network scalability is limited not only by topological properties such as node degree, diameters, symmetry, bisection bandwidth, and complexity, but also by physical technology and management requirements such as packaging, power dissipation, cooling, programmed control, traffic management, etc.

6.2.2 Routing Schemes and Functions

Two different ways of routing packets in a switched network are introduced below: The *store-and-forward* scheme was used in the first generation of message-passing multi-computers; the *cut-through* or *wormhole routing* scheme is implemented in more of today's multicomputer or cluster systems.

Store-and-Forward Routing The entire packet must be stored in a *packet buffer* in a node before it is forwarded to an outgoing link. Thus, successive packets are transmitted sequentially without overlapping in time. In Fig. 6.5(a), we show the transmission of a packet consisting of four cells from node 1 to node 4 via two intermediate nodes (node 2 and node 3). Each packet consists of four cells. The header is labeled as **h**, followed by three data cells labeled as **a, b,** and **c.**

The header is used to determine the routing path. Each node consumes 4 time units (equal to the packet length) to pass the packet to the next node. Therefore, it takes 16 time units (cycles) to pass the packet from N1 to N4 if there are no resource conflicts or deadlock situation along the routing path. When the network is jammed, the routing time could be even longer.

Cut-Through Routing Each node uses a *flit buffer* to hold one *flit,* one cell of the packet, at a time. The flit is automatically forwarded to an outgoing link, once the header is decoded. All data flits in the same packet follow the same path that the header traverses. Therefore, each node forwards the packet one flit down the path per unit time. Successive packets are forwarded in a pipelined fashion.

In Fig. 6.5(b), it takes only 7 time units for node 4 to receive the entire packet of 4 flits, compared with 16 cycles needed in the store-and-forward routing. Let L be the packet length and N be the number of nodes from source to destination. In general, the store-and-forward scheme takes NL time units, while the wormhole routing takes only $L + N - 1$ time units to pass through a path of $N-1$ hops. A speedup of $NL / L+N-1$ is expected. For a sufficient long packet, the limiting speedup approaches N times faster for wormhole routing over the traditional routing scheme.

A wormhole router overlaps the transmission of successive packets by using the network resources (links, switches, and network interfaces) along the selected path as a transmission pipeline (Fig. 6.5). The packet bypasses an intermediate node without copy and storage delays.

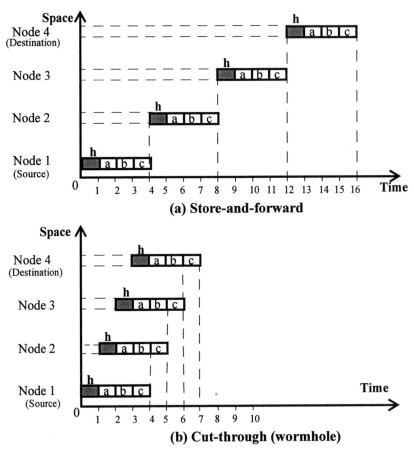

Figure 6.5   Store-and-forward versus cut-through routing in a packet switched network (h: header; a, b, and c are data elements).

The wormhole routing ʾuces the connection latency from 500 μs in store-and-forward scheme to 20 μs ir   ᵐercial switch designs. Interested readers may refer to Dally and Seitz [185] for details of wormhole routing.

Data Routing Functions   A *data network* is used for internode data communications. We specify below primitive data routing functions implemented on a routing network. Commonly seen data-routing functions include *shifting, rotation, permutation* (one-to-one), *broadcast* (one-to-all), *multicast* (many-to-many), *personalized communication* (one-to-many), *shuffle,* ʾ *ʾange*, etc.

Permutations   For *n* oᵦjects, there are *n*! permutations by which *n* objects can be reordered. The set of all permutations forms a *permutation group* with respect to a composition operation. One can use the cycle notation in group theory to specify a permutation function.

For example, the permutation $\pi = (a, b, c)$ (d, e) stands for a bijection mapping: $a \rightarrow b$, $b \rightarrow c$, $c \rightarrow a$, $d \rightarrow e$, and $e \rightarrow d$. The cycle (a, b, c) has a period of 3, and the cycle (d, e) a period of 2. Combining the two periods, the permutation has a total period of $2 \times 3 = 6$. If one applies the permutation 6 items, the identity mapping $I = (a), (b), (c), (d), (e)$ is obtained.

One can use a crossbar switch to implement any of the permutation patterns. Multi-stage networks can implement some of the permutations in one or multiple passes through the network. Permutations can also be implemented with shifting or broadcast operations. When n is large, the permutation speed may decide the basic performance of a data routing network.

Perfect Shuffle   *Perfect shuffle* is a special permutation function suggested by Harold Stone [588] for parallel processing applications. The mapping from a perfect shuffle is shown in Fig. 6.6a. The *inverse perfect shuffle* is defined in Fig. 6.6.

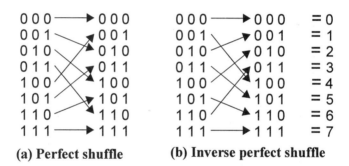

(a) Perfect shuffle   **(b) Inverse perfect shuffle**

Figure 6.6   **Perfect shuffle and its inverse mapping over eight objects.**
(Courtesy of H. Stone; reprinted with permission from *IEEE Trans. Computers* [588], 1971)

Shuffling n cards (objects) evenly is performed by a game dealer. One can represent each card in a deck of $n = 2^k$ cards by a k-b binary number $x = (x_{k-1}, \ldots, x_1, x_0)$. The perfect shuffle maps card x to card y, where the representation $y = (x_{k-2}, \ldots, x_1, x_0, x_{k-1})$ is obtained by shifting x one bit to the left and wrapping around the most to the least significant position. This gives a binary method to generate the perfect shuffle pattern.

Hypercube Routing Functions   A three-dimensional binary cube is shown in Fig. 6.7(a). Three routing functions are defined by three bits in the node address. For example, one can exchange the data between adjacent nodes differing in the least significant bit C_0, as shown in Fig. 6.7(b). Similarly, two other routing patterns can be obtained by checking the middle bit C_1 [Fig. 6.7(c)] and the most significant bit C_2 [Fig. 6.7(d)], respectively. In general, an n-dimensional hypercube has n routing functions, defined by each bit of the n-bit address.

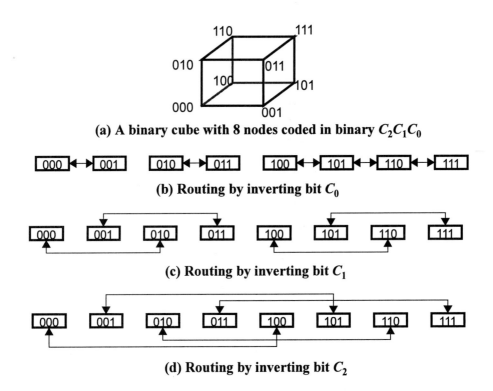

(a) A binary cube with 8 nodes coded in binary $C_2 C_1 C_0$

(b) Routing by inverting bit C_0

(c) Routing by inverting bit C_1

(d) Routing by inverting bit C_2

Figure 6.7   Three routing functions on a binary cube.

Broadcast and Multicast *Broadcast* is a one-to-all mapping. This can be easily achieved in an SIMD computer using a broadcast bus. A message-passing multicomputer uses a *multicast* mechanism to broadcast messages. Multicast corresponds to mapping from one subset to another subset of nodes (many-to-many communication). *Personalized broadcast* sends personalized messages to only selected receivers. Broadcast is often treated as a global operation in a multicomputer.

6.2.3 Networking Topologies

Static networks use point-to-point links that are fixed once connected. This type of network often shares the media and is more suitable for predictable traffic patterns. These topologies can be also applied to build switched networks.

Linear Array   This is the simplest network in which N nodes are connected by $N-$ links forming a single cascade [Fig. 6.8(a)]. Internal nodes have degree 2. The diameter is $N-1$, which is rather long for a large N. The bisection width $b = 1$.The structure is not symmetric and poses a communication inefficiency when N becomes very large.

For the degenerate case of $N = 2$, it is rather economical to implement a linear array. As N increases, another topology may be better. It should be noted that a *linear array* is very different from a *bus,* which is timeshared through switching among many nodes attached to it. A linear array allows concurrent use of different sections (links) of the cascade by different users.

Ring   A *ring* is obtained by connecting the two terminal nodes of a linear array with one more link to form a closed loop [Fig. 6.8(b)]. A ring can be unidirectional or bidirectional. It is symmetric with a constant node degree of 2. The diameter equals $N/2$ for a bidirectional ring, and $N - 1$ for a unidirectional ring.The IBM *Token Ring* has this topology, in which messages circulate along the ring until they reach the destination with a matching token.

Chordal Ring   By increasing the node degree from 2 to 3 or to 4, we obtain two *chordal rings* shown in Fig. 6.8(c) and Fig. 6.8(d), respectively. One and two extra links are added to produce these two chordal rings, respectively. In general, the more links added, the higher the node degree and the shorter the network diameter. Chordal rings with higher node degree require a lot of more links to implement.

Comparing the 16-node ring with the two chordal rings, we see the network diameter drops from 8 to 5 and to 3, respectively. In the extreme, a *completely connected network* in Fig. 6.8(f) has the highest node degree of $N - 1$ with the shortest possible diameter of 1. The tradeoffs between node degree and network diameter affect the cost/performance ratio, as well as the scalability and fault tolerance of a given interconnection network.

Barrel Shifter   As shown in Fig. 6.8(e), the *barrel shifter* is modified from the ring by adding extra links from each node to distance nodes with integer power of 2 hops away. This implies that node i is connected to node j if $|j - i| = 2^r$ for some $r = 0, 1, 2, ..., n-1$ and the network size is $N = 2^n$. Such a barrel shifter has a node degree of $d = 2n - 1$ and a diameter $D = n/2$.

Obviously, the connectivity in the barrel shifter is increased over that of any chordal ring of lower node degree. For $N = 16$, the node degree is 7 and a diameter of 2. All the topologies presented have lower complexity than that of a completely connected network, which is used only if the network size is rather small.

Tree   A *binary tree* of 7 nodes in three levels is shown in Fig. 6.9(a). The node degree of a binary tree is 3. A k-level binary tree has $N = 2^k - 1$ nodes. A binary tree can be extended to a *multiway tree* with more than two offsprings per node. In general, an m-way tree has $N = m^k - 1$ nodes. Each nonleaf (parent) node may have up to m child nodes.

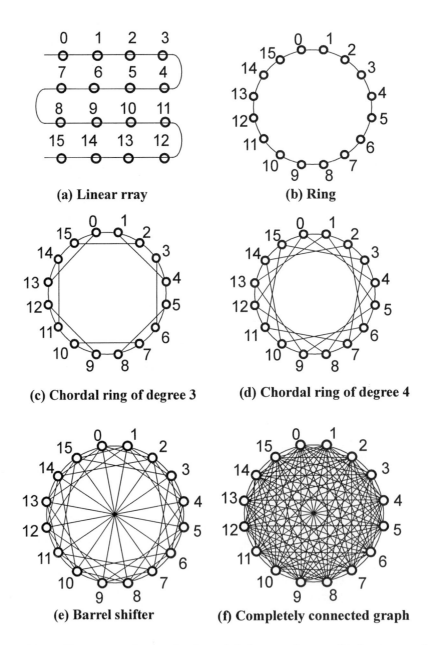

Figure 6.8   Six network topologies with increasing node degree and
connectivity from a linear array to a ring, two chordal
rings, a barrel shifter, and a completely connected graph.

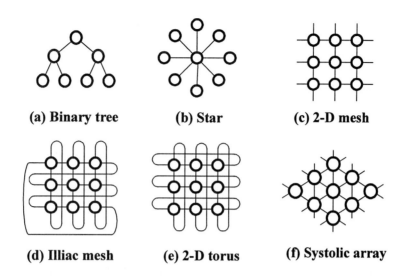

(a) Binary tree   (b) Star   (c) 2-D mesh

(d) Illiac mesh   (e) 2-D torus   (f) Systolic array

Figure 6.9   Six additional topologies for network construction.

Star   The *star* [Fig. 6.9(b)] is a two-level, multiway tree with the highest node degree equal to $d = N - 1$ and the smallest diameter of 2. One problem in using a tree or a star network is the bottleneck effect towards the root of a tree or towards the center node of a star network. Traffic becomes much heavier at the root or the center node, when leaf nodes communicate with each other.

Mesh   A two-dimensional, 3×3 *mesh* network is shown in Fig. 6.9(c). This is a popular architecture which has been implemented in a number of systems in the past, such as the Goodyear MPP and Intel Paragon.

In general, a k-dimensional mesh with n nodes on each dimension has an interior node degree of $2k$, and the network diameter is $k(n - 1)$. Note that a mesh is not a symmetric network, because the boundary nodes are different from the interior nodes.

Illiac Mesh   In Fig. 6.9(d) we show a variation of the mesh by allowing wraparound connections vertically and a snakelike loop horizontally. This drawing is a reduced version of the *Illiac network* [105]. The original Illiac network assumed an 8×8 mesh with a node degree of 4 and a diameter of 7.

In general, an $n \times n$ Illiac mesh should have a diameter of $d = n - 1$, which is only half of the diameter of a 2-D mesh. This Illiac mesh shown in Fig. 6.9(d) is topologically equivalent to a chordal ring of degree 4 shown in Fig. 6.8(d).

Torus   A 2-dimensional *torus* is shown in Fig. 6.9(e). This architecture extends from the mesh by having wraparound connections. A torus would have about half of the diameter of a mesh. Or one can say that this topology combines the advantages of ring and mesh. Both mesh and torus can be extended to higher dimensions.

For example, Cray Research Inc. chooses to implement a 3D torus network in their T3D/T3E series. The torus is a symmetric topology, whereas a mesh is not. All added wraparound connections help reduce the torus diameter and restore the symmetry.

Systolic Array   This is a special class of application-driven array architectures designed for VLSI implementation of certain data flow algorithms. What is shown in Fig. 6.9(f) is a systolic array specially designed for matrix-to-matrix multiplication.

Static systolic arrays are pipelined with multidirectional flow of data streams. With a fixed structure, a systolic array is supposed to match the communication structure of a given algorithm. The structure cannot implement other algorithms efficiently once it is optimized for a given application.

Fat Tree   To alleviate the bottleneck towards the root of a tree structure, Leiserson [403] introduced the multiway *fat tree* in 1985. The number of links between parent and child nodes in a fat tree increases as we traverse from a leaf node toward the root.

The fat tree is more like a natural tree in that the trunks get thicker toward the root. Fat tree architecture has been implemented in Connection Machine CM5 system.A four-way fat tree was adapted in building the data network in CM5. The fat tree built in CM5 allows each node to have 4 child nodes and 2 or 4 parent nodes as illustrated by the internode connections in Fig. 6.10.

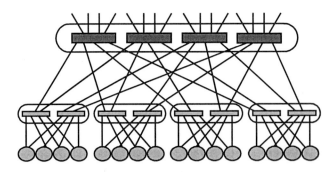

Figure 6.10   Four-way fat tree implemented in the CM5 (Courtesy
C. Lesiserson et al.[403], *IEEE Trans. Computers*, Oct. 1985),

Hypercube   Formally, a hypercube should be called a *binary n-cube*. Binary refers to the fact that each dimension of the cube has two nodes, where n is called the *dimensionality* of the hypercube. In general, an *n-cube* consists of $N = 2^n$ nodes spanning along n dimensions. A *3-cube* with 8 nodes is shown in Fig. 6.11(a). A *4-cube* is formed by interconnecting the corresponding nodes of two 3-cubes as illustrated in Fig. 6.11(b). The node degree of an *n*-cube equals n, and so does the network diameter. In fact, the node degree increases linearly with the dimensionality, making it difficult to scale up the size of a hypercube architecture.

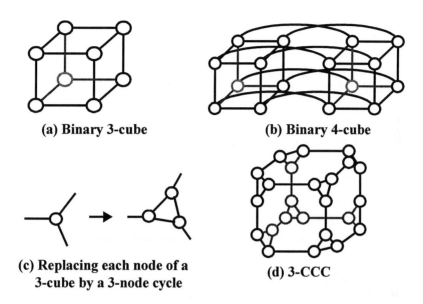

(a) Binary 3-cube   (b) Binary 4-cube

(c) Replacing each node of a   (d) 3-CCC
3-cube by a 3-node cycle

Figure 6.11   Binary hypercube and cube-connected cycles (CCCs).

Binary hypercube has been a very popular architecture for research and development in the 1980s. Intel iPSC/1, iPSC/2, nCUBE, and Connection Machine's CM2 machines were built with the hypercube architecture. The architecture has dense connections. Binary trees, meshes, and many other lower-dimension topologies can be embedded in a hypercube. With poor scalability and difficulty in packaging high-dimension cubes, the hypercube architecture is gradually replaced by lower-dimension architectures.

For example, Connection Machine CM5 chooses the fat tree over the hypercube implemented in CM2. Intel Paragon chooses a two-dimensional mesh over its hypercube predecessors. Topological equivalence has been established among a number of network architectures.

The bottom line for an architecture to survive lies primarily in cost/performance trade-offs in packaging, efficiency, scalability, and performance. With the introduction of wormhole routing and pipelined network operations, the choice of a regular topology becomes less important than before. In other words, architectural topology becomes less sensitive or less critical to determine the performance of an interconnection network. It is the media and switching technology that dominate.

Cube-Connected Cycles   This architecture is modified from the hypercube. As illustrated in Fig. 6.11(c), a 3-cube is modified to form *3-cube-connected cycles* (3-CCCs) shown in Fig. 6.11(d). The idea is to cut off each corner node of the 3-cube and replace it by a *cycle* of 3 nodes.

In general, one can construct a *k-cube-connected cycle* from a *k*-cube with $n = 2^k$ cycle nodes. The idea is to replace each vertex of the *k*-dimensional hypercube by a ring of *k* nodes. A *k*-cube can be thus transformed to a *k*-CCC with $k \times 2^k$ nodes.

The 3-CCC shown in Fig. 6.11(d) has a diameter of 6, twice that of the original 3-cube. In general, the network diameter of a *k*-CCC equals 2*k*. The major improvement of a CCC over a hypercube lies in a fixed node degree of 3, which is independent of the dimension of the underlying hypercube.

For example, a 64-node CCC can be formed by replacing the corner nodes of a 4-cube with cycles of 4 nodes, corresponding to the case of *n* = 6 and *k* = 4. The CCC has a diameter of 2*k* = 8. But the CCC has a node degree of 3, smaller than the 6 of a 6-cube. In this sense, the CCC is a better architecture for building scalable systems if latency can be tolerated in some way.

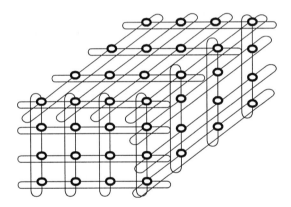

Figure 6.12   A 4-ary 3-cube network, hidden parts not shown.

k-ary n-Cube Networks   Rings, meshes, tori, binary *n*-cubes (hypercube), and Omega networks are all topologically isomorphic to a family of *k-ary n-cube* networks. In Fig. 6.12 we show a 4-ary 3-cube network with *k* = 4 nodes at each dimension. For clarity, the hidden nodes and hidden links are not shown.

There are $N = k^n$, ($k = \sqrt[n]{N}$, $n = \log_k N$) nodes in this network. A node in a *k*-ary *n*-cube is identifiable by an *n*-digit radix-*k* number $A = a_0\, a_1\, a_2\, ...\, a_n$, where a_i represents the node's position in the i-th dimension. Traditionally, low-dimensional *k*-ary *n*-cubes are called *tori*, such as for *n* = 2 or 3. Higher-dimension binary *n*-cubes are called *hypercubes*. Both fat trees and *k*-ary *n*-cube networks are universal in simulating other networks.

Dally [183] has revealed some interesting properties of *k*-ary *n*-cube networks. The cost of most networks is dominated by the amount of wires, rather than by the number of switches required. If limited by a constant wire bisection, low-dimension networks with wider channels provide lower latency, less contention, and higher throughput than high-dimension networks with narrower channels.

Summary of Network Topologies In Table 6.1, we summarize important characteristics of the above network topologies. Those networks with node degree of 4 or less are more desirable. For example, the INMOS Transputer chip is a microprocessor with four ports linking to four channels in a 2D mesh architecture.The completely connected and star networks are both very bad for having a high node degree. The hypercube node degree increases with $\log_2 N$, which is also bad when N becomes very large.

Table 6.1   Topological Properties of Static-Connection Networks

Network Type	Node Degree d	Network Diameter D	No. of Links l	Bisection Width B	Symmetry	Network Size and Remarks
Linear array	2	$N-1$	$N-1$	1	No	N nodes
Ring	2	$N/2$	N	2	Yes	N nodes
Completely connected	$N-1$	1	$N(N-1)/2$	$(N/2)^2$	Yes	N nodes
Binary tree	3	$2(h-1)$	$N-1$	1	No	Tree height $h = [\log_2 N]$
Star	$N-1$	2	$N-1$	$N/2$	No	N nodes
2D Mesh	4	$2(r-1)$	$2N-2r$	r	No	$r \times r$ mesh for $r = \sqrt{N}$
Illiac Mesh	4	$r-1$	$2N$	$2r$	No	Chordal ring with $r = \sqrt{N}$
2D Torus	4	$2[r/2]$	$2N$	$2r$	Yes	$r \times r$ torus for $r = \sqrt{N}$
Hyper-cube	n	n	$nN/2$	$N/2$	Yes	$N = 2^n$ nodes,
CCC	3	$2k-1+[k/2]$	$3N/2$	$N/(2k)$	Yes	$N = k\,2k$ nodes with cycle $k \geq 3$
k-ary n-cube	$2n$	$n[k/2]$	nN	$2k^{n-1}$	Yes	$N = k^n$ nodes

With pipelined wormhole routing, the network diameter becomes less critical than before, because the hardware delay between any two nodes is almost independent of the path length. The number of links used affects the network cost most. The bisection width and channel width dominate the network bandwidth. Network symmetry is often desired to enhance scalability and routing efficiency.

6.3   Buses, Crossbar, and Multistage Switches

Dynamic connections are implemented by placing electronic switches, routers, concentrators, distributors, arbitrators at the crosspoint of multiple communication paths. Three families of dynamic connections are *buses, crossbar switches*, and *multistage switches.* As seen from Fig. 6.1, these interconnects currently yield a wide range of data transfer rates (bandwidth) between 200 Mbps and 100 Gbps.

6.3.1 Multiprocessor Buses

A *bus* is essentially a collection of wires and connectors for data transactions among processors, memory modules, and peripheral devices. A *system bus* is used for data transfer between a *master* device such as a processor and a *slave device* such as a memory board. The bus arbitration logic grants the bus access one request at a time, thus the name *contention bus.*

Many bus standards have been established, such as the PCI, VME, Multibus, SBus, MicroChannel, and IEEE Futurebus. Most standard buses offer the lowest cost in building uniprocessor systems. We will focus on *multiprocessor buses* and *hierarchical buses* for building large SMP, NUMA, or DSM machines. These scalable buses are often equipped with hardware support for cache coherence, fast microprocessor synchronization, and interrupt handling in split transactions, etc.

In Fig. 6.13, we show the structure of a typical multiprocessor bus system. The system bus is often implemented on the backplane or centerplane. Each processor (P) or each *I/O processor* (IOP) can be a master generating a request to access a particular slave device (memory or disk drive, etc.). The system bus consists of data path, address lines and control lines, providing a common medium for communication among all plug-in functional boards. Specialized *interface logic* (IF) and special functional *controllers* (C) are used in different plug-in boards.

Local buses are found on CPU, I/O, or network interface boards. The local bus on a memory board is particularly called a *memory bus*. A typical *I/O bus* can be the SCSI or other I/O channels connecting local disks, printers, and other peripheral devices attached to host machines.

Important issues of designing a multiprocessor bus system include bus arbitration, interrupts handling, protocol conversion, fast synchronization, cache coherence protocol, split transactions, bus bridging, hierarchical bus extensions, etc. Often, snoopy protocols are built in a multiprocessor bus as discussed in Section 5.2.2.

Hardwired barrier synchronization lines can be added to a multiprocessor cluster bus as proposed in [552]. These hardwired lines can reduce the software-based synchronization time from several thousands of microseconds to a few hundreds of nanoseconds.

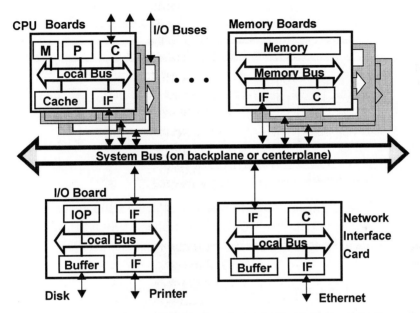

Figure 6.13 A multiprocessor bus consisting of a system bus, memory buses, and local buses on various functional boards.

We consider below two case studies of system buses developed in commercial SMP servers: the Sun Microsystems *Gigaplane bus* and the SGI *POWERpath-2 bus*.

Example 6.1 Sun Microsystems 2.5 Gbps Gigaplane bus interconnect

As illustrated in Fig. 6.14, Sun's Ultra Enterprise server (SunFire) uses a new system bus, called Gigaplane, to interconnect CPU/memory boards and SBus I/O boards, yielding a maximum bandwidth of 2.5 GB/s, 6 times faster than Sun's XDBus used in many servers. The Gigaplane bus data-path has a 256-b packet switching design with 42-b address lines to address 2 TB of shared memory.

The bus clock rate is 83.4 MHz. With a cache line of 64 B transferred per 2 bus cycles, the total burst data transfer rate is 2.6 GB/s. Instead of connecting processors, memory, and I/O buses directly to the Gigaplane, each board uses a UPA (*ultra port architecture*) interconnect switch to fan out the Gigaplane to individual units.

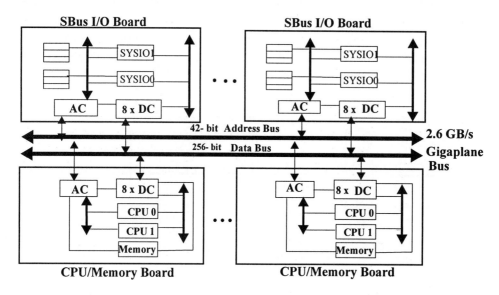

Figure 6.14   The Gigaplane bus in Sun Enterprise x000 servers. (AD: Address Controller, DC: Data Controller, SYSIO: SBus-to-UPA interconnect, Courtesy Sun Microsystems [600], 1997)

Example 6.2 Silicon Graphics POWERpath-2™ Bus

This is a 256-b synchronous bus with a maximum memory bandwidth of 1.2 GB/s. Up to 36 MIPS® R10000® processors, 16 GB of memory, graphics, and six I/O buses can be connected to the POWERpath-2 bus shown in Fig. 6.15.

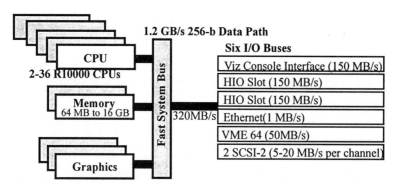

Figure 6.15   The POWERpath-2™ bus architecture in Silicon Graphics servers. (Courtesy Silicon Graphics Computer Systems, 1996)

The system bus uses a split transaction protocol. Four R10000s are mounted on each CPU board. With 320 MB/s per I/O board, up to 500-MB/s sustained data transfer rate can be yielded by the I/O subsystem. High-performance graphics support visualization in supercomputing applications.

■

Many proprietary multiprocessor buses were also built in IBM mainframes and Digital and Hewlett-Packard multiprocessor servers. IEEE has also developed VME bus (IEEE Standard 1014-1987), Multibus II (IEEE Standard 1296-1987), and Futurebus (IEEE Standard 896.1-1991) for multiprocessors. Readers may refer to published bus standards for specifications of electrical, mechanical, and functional protocols.

Hierarchical Buses The single SMP bus has limited scalability to accommodate large-scale systems. A hierarchical bus structure can alleviate this problem to some extent. We show in Fig. 6.16 a conceptual bus hierarchy designed for interconnecting multiple multi-processor clusters to build a CC-NUMA machine.

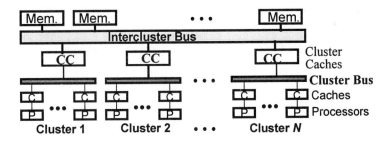

Figure 6.16   A hierarchy of buses for connecting multiprocessor clusters to build a CC-NUMA machine (Courtesy A.W. Wilson, *Proc. Int'l Symp. Computer Architecture*, [644] 1987)

All processors belonging to the same cluster (such as a SMP server) are tied to a common *cluster bus*. A *cluster cache* (CC) can be used as a second-level cache to be shared by all processors in the same cluster. Multiprocessor clusters communicate with each other through an *intercluster bus* connecting all globally shared-memory modules.

Multiple levels of buses must be equipped with a bridging mechanism to maintain cache coherence among all private and shared caches and to interface among the clusters. The IEEE Futurebus [344] has developed special bridges, cache and memory agents, message interfaces, and cable segments for building hierarchical bus systems.

Shortcomings in Bus Interconnects Bus interconnects are for time-shared use by many processors. Even when the bus bandwidth is high, *per-processor bandwidth* is only a fraction of the total bus bandwidth. Furthermore, bus is prone to failure for lack of redundancy. Bus also has limited scalability. These shortcomings are primarily constrained by packaging technology and the cost involved.

Buses are often confined within a small rack. Clock skewing and global clocking are difficult to overcome in a hierarchical bus extension beyond several racks. Upgrading bus architecture with crossbar switches or multistage networks will overcome these shortcomings to some extent, as studied in the next two subsections.

6.3.2 Crossbar Switches

Much higher bandwidth can be provided by a crossbar switched network for the same datapath width per port and equal number of connecting ports. A crossbar is a single-stage switched network. Like a telephone switchboard, the crosspoint switches provide dynamic connections between all (source, destination) pairs. The cross-point switch can be set on or off dynamically upon program control.

Two ways of using crossbar switches are illustrated below: one for interprocessor communication in symmetric multiprocessors or in multicomputer clusters; and another for *inter-processor-memory* accesses in SMP servers or in vector supercomputers.

Example 6.3 Digital's GIGAswitch/FDDI Crossbar Switch

In Fig. 6.17, we show the crossbar design in Digital's GIGAswitch/FDDI. This crossbar was designed as a cluster interconnect among multiple FDDI rings of Alpha workstations and servers. Using a *FDDI full-duplex technology* (FFDT), it can connect up to 22 FDDI ports in 100 Mbps each way.

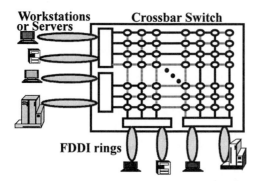

Figure 6.17   GIGAswitch/FDDI: A crossbar switch built for Alpha workstation/server farms (Courtesy Digital Equipment Co., [194], Sept. 1994).

With cut-through routing, latency is reduced to less than 20 us. Two or four port line cards are used to yield the 22-port crossbar switch. As a whole, the GIGAswitch/ FDDI delivers up to 3.6 Gbps of bandwidth, 360 times that of an Ethernet. The GIGAswitch/FDDI has been used in Digital's TruCluster.

The GIGAswitch/FDDI is general networking product that can be used in any FDDI ring-connected system. This is a good example of an internode communication system based on a crossbar design.

∎

In general, the crossbar complexity increases directly with N^2, where N is the number of ports. Limited by cost, one cannot expect to build very large-scale crossbar switches in the future. The interprocessor crossbar provides permutation connections among N processors. Recently, Sun Microsystems upgraded the Gigaplane bus to a *Gigaplane-XB interconnect* in the Ultra Enterprise 10000 (StarFire) SMP server.

This is a crossbar interconnect using a packet switched scheme with separate address and data paths. Up to 64 processors are interconnected by four snoopy address buses and one 16 × 16 data crossbar switch. The crossbar is dedicated for point-to-point data communication among processors, while address distribution is handled by broadcast routers on the snoopy buses. We shall further study the SunFire system in Chapter 8.

Crossbar Chip Design Crossbar networks are built with crossbar switch chips. These crossbar chips can be also used to build multistage networks. We study below a crossbar chip design in IBM SP2 multicomputer system.

Example 6.4 IBM Vulcan crossbar chip design

The IBM Vulcan crossbar switch chip design is shown in Fig. 6.18, This switch has been used as the building block of the multistage network in the IBMSP2. Each chip has eight input ports and eight output ports.

The buffered wormhole routing is used. An 8 × 8 crossbar enables eight packet cells (called *flits*) to pass through the switch in every 40-MHz cycle if there is no conflict. When there is a *hot-spot* of memory access conflict, only one crosspoint switch will be enabled at a time. The blocked flits are buffered in a central queue. This buffering frees up the input ports to receive subsequent flits from the previous switch stage.

The central queue is implemented by a dual-port RAM that can perform one read and one write in a clock cycle. To match the maximum bandwidth, each input port first deserializes eight flits from its FIFO into a chunk and writes the entire 64-b chunk to the central queue in one cycle.

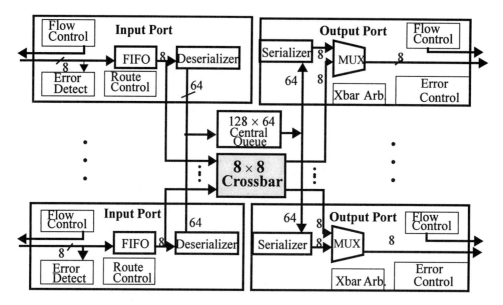

Figure 6.18   An 8 × 8 crossbar (Vulcan) chip design for SP2 HPS
(Copyright 1995, International Business Machines Corp. Reprinted
with permission from C.B. Stunkel et al. *IBM Systems Journal,*
Vol. 34, No.2, [592], 1995)

■

Processor-Memory Crossbar Switch A bus-connected multiprocessor is limited by its bus bandwidth. Very often, the bus becomes the bottleneck in accessing the shared memory by a large number of processors. A better approach is to replace the interprocessor memory bus by a crossbar switch as illustrated in Fig. 6.19. This opens up the possibility of parallel memory accesses of multiple memory banks.

This is essentially a memory access network. The major benefit comes from a sharp increase of memory bandwidth or data transfer rate between processors and shared-memory modules. Each memory module can be accessed by only one processor at a time. When multiple requests collide at a module, the crossbar must resolve the conflicts.

The behavior of each crosspoint switch is very similar to that of a contention bus. However, the processor can generate a sequence of addresses to access different memory modules in a pipelined fashion or simultaneously. In traditional SMP servers, almost all models use the bus interconnect between processors and memory banks. In recent years, IBM and Sun Microsystems among other SMP suppliers have pushed for crossbar switched interconnects in their scalable SMP servers.

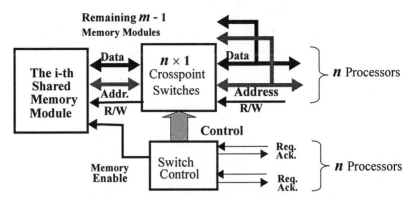

Figure 6.19 **Partial circuitry of a $n \times m$ crossbar switch between n processors and m memory modules.**

6.3.3 Multistage Interconnection Networks

To build a larger switched network, the single-stage crossbar must be extended to multiple stages, often called a *multistage interconnection network* (MIN). MINs have been built in both MIMD and SIMD computers. Multiple switch modules are used in each stage. Fixed interstage connections are used between switches in adjacent stages. The switch connections can be dynamically set on or off to establish the desired connection patterns between inputs and outputs.

Switch Modules An $n \times n$ switch module has n inputs and n outputs. A binary switch corresponds to a 2×2 switch module as shown in Fig. 6.20(a). Each input can be connected to any of the output ports. But only one-to-one or one-to-many mappings are allowed; no many-to-one mappings are allowed due to out put conflicts.

Omega Network In general, an $n \times n$ crossbar switch can achieve $n!$ permutation connections. Different classes of MINs differ in the switch modules and *interstage connection* (ISC) patterns used. The ISC patterns often used include *perfect shuffle, butterfly, multiway shuffle, crossbar, cube connection*, etc.

In Fig. 6.20(a), we show four possible connections of a 2×2 switch used in constructing the Omega network. A 8×8 Omega network is shown in Fig. 6.20(b) using three stages of 2×2 switches.

An n-input Omega network requires $\log_2 n$ stages of 2×2 switches. Omega network has been implemented in the Cedar multiprocessors system at the University of Illinois [382]. Next, we show two MIN design examples in Cray Y/MP, a vector multiprocessor, and in IBM SP2 multicomputer.

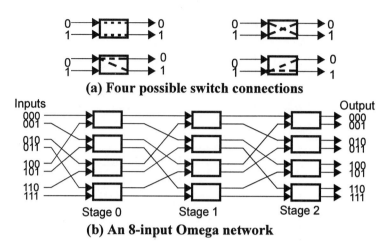

(a) Four possible switch connections

(b) An 8-input Omega network

Figure 6.20 **Use of 2 × 2 switches and perfect shuffle to construct a 8 × 8 Omega network** (Courtesy D.H. Lawrie, *IEEE Trans. Computers,* [395] 1975).

Example 6.5 IBM high performance switch (HPS)

All the switch hardware in the node and switch frames forms the *high-performance switch* (HPS). Each *frame* consists of 16 processing nodes (N0 to N15) connected by a 16-way *switchboard.* Eight frames are interconnected by an extra stage of switchboards. Each thin line represents an 8-b bidirectional link. Each thick line represents four 8-b links. There are two stages of switch chips on each switch board. In total, this MIN has four switch stages.

The HPS is a packet switching, multistage Omega network with buffered wormhole routing, driven by a 40-MHz clock. An example HPS connecting 128 nodes is shown in Fig. 6.21. Each switchboard has two stages of switch chips. The left links of a switchboard are used to connect nodes within a frame. The right links provide interframe connections. Each link is 8-b, bidirectional.

Messages are divided into packets, which are in turn divided into 8-b *flow control digits* (flits). When there is no contention, it takes only 5 clocks, or 125 ns for a flit to travel through one stage (i.e., one switch chip) of the HPS. Thus the contention-free hardware latency of the HPS is small, only 875 ns or less for up to 512 nodes.

The actual latency seen by application processes is much higher: It takes at least 40 µs for a process to send an empty message to another process. A large portion of this message-passing latency is due to software overhead. Between a pair of nodes, the HPS provides a per-port bandwidth of 40 MB/s.

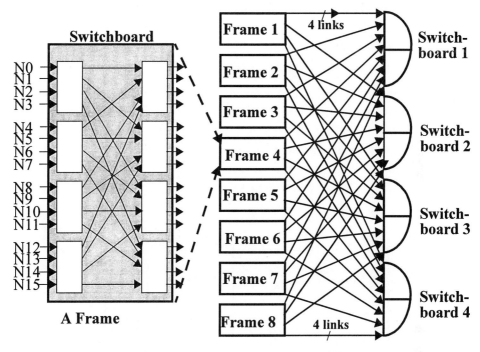

Figure 6.21 A 128-way high-performance switch built with four stages of 16-way switches in the IBM SP2 (Copyright 1995, International Business Machines Corp. Reprinted with permission from C. B. Stunkel et al. *IBM Systems Journal,* Vol. 34, No.2, [592], 1995)

■

Two important problems need to be solved in designing any high-performance cluster systems. It is desirable for the clocks in all nodes to be synchronized. For instance, when two nodes call the Unix function *gettimeofday()* at the same time, the two calls should return the same value. Of course, perfect synchronization is impossible. How to align the clocks as closely as possible (e.g., within a small drift range of a few microseconds) is the *clock synchronization* problem.

The second problem is the *long-wire* problem. The nodes of a cluster cannot be packed as closely as those of a traditional MPP, because a cluster node is more complex than an MPP node. It is not uncommon that internode communication wires are more than one m long. Long wire length can adversely affect both latency and bandwidth.

The SP2 solves the clock synchronization problem by using a synchronous communication network. The entire HPS is synchronized by a clock from a common 40-MHz oscillator. The synchronous network enables each node to maintain a local time-of-day clock that is in synchrony with the clocks in other nodes. IBM developed a software called *Worm* to tune the node clocks to be synchronized within a few microseconds.

Example 6.6 Cray Y/MP Multistage Network

This network is built with 4 × 4 and 8 × 8 crossbar switches and 1 × 8 demultiplexors in three stages shown in Fig. 6.22.

The network is designed to support data streaming between 8 vector processors and 256 memory banks. The network can avoid memory-access conflicts by the eight processors.

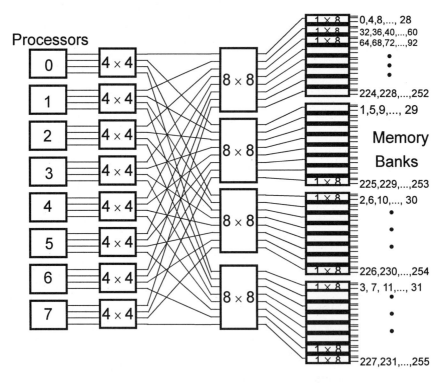

Figure 6.22   The multistage crossbar network for inter-processor-memory connections in Cray Y-MP/816 (Courtesy Cray Research, Inc. [174], 1992)

6.3.4 Comparison of Switched Interconnects

We compare below hardware requirements and performance potentials of system buses, multistage networks, and crossbar switches for building dynamic system interconnects in scalable platforms or cluster systems. Table 6.2 summarizes the wiring/switching complexity, per-processor bandwidth, and aggregate bandwidth of switched interconnects.

Hardware Complexity   A bus interconnect costs the lowest among the three with limited wiring and switching complexities. The wiring complexity is primarily determined by the data path width and by the address width of the bus design. 256 data lines and 42 address lines represent state of the art in today's buses. The address lines can be hidden as part of the data path, such as 64 shared address/data lines in the 256-b Futurebus.

The bus switching complexity is determined by the number n of connection taps on the bus, which is also limited by a small number of processors, memory and I/O boards attached to a system bus. Let w be the data path width on the bus. The hardware complexity of a bus interconnect increases linearly with respect to both n and w, as shown by $O(n + w)$ in Table 6.2.

Table 6.2 Complexity and Bandwidth of Switched Interconnects

Interconnect Characteristics	System Bus	Multistage Network	Crossbar Switch
Hardware complexity	$O(n + w)$	$O(nw \log_k n)$	$O(n^2 w)$
Per-processor bandwidth	$O(wf/n)$ to $O(wf)$	$O(wf)$	$O(wf)$
Reported aggregate bandwidth	2.67 GB/s for Gigaplane bus in SunFire server	10.24 GB/s for 512-node HPS in IBM SP2	3.4 Gb/s for Digital's GIGAswitch

The crossbar switch is the most expensive one to build, because its hardware complexity increases with respect to the product $n^2 w$, where n^2 corresponds to the number of cross-points in the crossbar switch and w is the channel width in the crosspoint switch design. For the same data path width, an $n \times n$ crossbar switch is almost n^2 more expensive than that of a bus interconnect.

An n-input multistage network has a hardware complexity of $O[(n \log_k n)w]$, where $n \log_k n$ crosspoints to the number of switches used, assuming $k \times k$ switches are used as building blocks. Here, w is the link width in the MIN design. This complexity lies between the two extremes of buses and crossbar switches. One can roughly estimate that MIN is $n/\log n$ less expensive than that of a crossbar switch for the same channel width.

Per-Processor Bandwidth   A bus is time-shared by n processors in an SMP server. Therefore, n processors compete for the bus bandwidth. Assuming the same clock rate f, each unit data transfer takes only 1 cycle on all three interconnects. The *per-processor bandwidth* of the bus varies within the range $O(wf/n)$ and $O(wf)$. Both MIN and crossbar have a wider per-processor bandwidth varying linearly with respect to the function $O(wf)$. This clearly indicates the bottleneck effect in the bus structure.

Even with the same clock rate f, the bus takes less time (usually 1 or 2 cycles) to transfer a unit piece of data, compared with many cycles needed to transfer a piece of data through the multiple switch stages in a MIN. Therefore, the actual per bus bandwidth may not be much lower than that of a MIN. In all cases, the crossbar has the highest per-processor bandwidth, because of a short latency (1 or 2 cycles as well) and conflict-free connections between the input and output ports.

Aggregate Bandwidth   In the last row of Table 6.2, we show the aggregate bandwidths of three representative bus, MIN, and crossbar interconnects available from the commodity market. In this case, the Gigaplane bus has an aggregate bandwidth of 2.67 GB/s = 21.36 Gbps. Assuming $n = 24$ processors sharing the bandwidth, the per-processor bandwidth can be as low as 21.36/24 = 0.89 Gbps.

On the other hand, the GIGAswitch crossbar has an aggregate bandwidth of 3.4 Gbps. This means that the per-processor bandwidth of GIGAswitch is potentially 3.8 times higher than that of the Gigaplane bus. Among the three, the IBM HPS has a much larger configuration of $n = 512$ ports, resulting in an aggregate bandwidth of 10.24 Gbps = 8192 Gbps. With this reasoning, we conclude that the MIN is more scalable than either the bus or the crossbar interconnect.

Switch Selection   Among commercial systems, there is no doubt that the bus is still the most cost-effective system interconnect for an SMP system with less than 40 processors. The crossbar interconnect should be chosen if performance outweighs price tag, or when the system is rather small, say, having less than 16 processors. This is evidenced by the fact that Sun's StarFire SMP server is built with an interconnect having four address buses and a 16×16 crossbar switch among 64 processors.

The major advantage of MINs lies in their scalability with modular construction. However, the latency increases with log n, the number of stages in the network. Also, costs due to increased wiring and switching complexity are another constraint in building large MINs in commodity networks. For this reason, the commercial SP2 system goes only to a maximum size of 128 nodes.

To build future MPP or clustered systems, point-to-point topologies are more flexible and even more scalable than regularly structured networks. With advances in optical interconnects and digital signal processing microelectronics, large-scale MINs or crossbar networks may become economically feasible in larger multiprocessors or multicomputers. Based on 1997 technology, 64×64 crossbar and 512-input MIN have been built in commercial systems.

6.4 Gigabit Network Technologies

Today's network technologies are assessed in the next three sections. These technologies have been used to implement shared-media, point-to-point, and switched networks. We characterize below the functional capabilities, architectural connections, and applications of these Gigabit network technologies.

6.4.1 Fiber Channel and FDDI Rings

We first distinguish between channels and networks. Then we study Fiber Channel and FDDI rings; both explore the fiber-optic technology, besides using copper wires or coaxial cables in establishing network links.

Channels and Networks These are two basic types of data communication among processors or between processors and peripherals. A *channel* provides a direct or switched point-to-point connection between the communicating devices. A channel is typically hardware-intensive and transports data at high speed with low overhead. Channels operate among only a few devices with predefined addresses. HiPPI, IBM, and SCSI all have well-defined data channel standards.

In contrast, a *network* is an aggregation of distributed nodes (like workstations, file servers or peripherals) with its own protocol that supports interaction among these nodes. A network has relatively high overhead since it is software-intensive, and consequently, slower than a channel.

Networks can handle a more extensive range of tasks than channels since networks operate in an environment of unanticipated connections. Well-established network standards include the IEEE 802, TCP/IP, and ATM protocols.

Fiber Channel ANSI X3T11 has specified *Fiber Channel* (FC) as an integrated set of channel and network standards for networking, storage, and data transfer among workstations, mainframes, supercomputers, storage devices, and displays. The FC standard addresses the need for very fast transfers of large volumes of information. It is intended to relieve system manufacturers of the burden of supporting the variety of channels and networks currently in place, as it provides one standard for networking, storage and data transfer.

Fiber Channel attempts to combine the best of channel and network methods of communication into a new I/O interface that meets the needs of channel users and network users. In order to complement the standard for assurance of interoperability among different vendors' products, a committee, called the *Fiber Channel Systems Initiative* (FCSI), was formed by Hewlett-Packard, IBM, and Sun Microsystems.

Fiber Channel Technology Fiber Channel can be a shared medium. It can be also built as a switched technology. Currently, Fiber Channel operates with a speed range of 100 to

133, 200, 400, and 800 Mbps. The FCSI vendors are also pushing for higher speed of 1, 2 and 4 Gbps in the future. Fiber Channel has been used to support point-to-point, loop, and switched star connections, besides providing client-server solution or hub solution in LAN applications.

Fiber Channel operates up to 50 m using STP copper at 100 Mbps and up to 10 km using single-mode fiber. With multimode fiber, a Fiber Channel LAN can span to 2 km at 200 Mbps. Today's high software-driver overhead limits the ultimate performance of Fiber Channel to less than 255 Mbps in most implementations using optical links. The usefulness of Fiber Channel is evidenced by the fact that some Gigabit LANs (such as Gigabit Ethernet to be studied in Section 6.4.2) are based on Fiber Channel technology.

Five Layers of FC Standards   Fiber Channel architecture consists of five layers of standards as summarized in Table 6.3. These layers define the physical media and transmission rates (FC-0), the data encoding and decoding scheme (FC-1), framing protocol and flow control (FC-2), common services and features selection (FC-3), and upper-layer protocol and application interfaces (FC-4) to various data channels and network standards.

Table 6.3 Five Layers of Fiber Channel Standards (Courtesy S. Saunders (ed.), *The McGraw-Hill High-Speed LANs Handbook* [530], 1996)

Standard	Data Channels			Network Protocols			OSI Layer
FC-4	HiPPI	IBM	SCSI	IEEE 802	TCP/IP	ATM	Data link layer
FC-3	Common services						
FC-2	Framing / Flow control / Service classes						
FC-1	Encode / Decode 8 B/10 B						Physical layer
FC-0	100 Mbps	200 Mbps	400 Mbps	800 Mbps	Higher rates in the future		

The IBM channel refers to the Escon (*enterprise systems connection*) interfaces. The FC-3 layer characterizes four distinct service classes of FC connections, ranging from hardwired (circuit switched) to frame switched, constant-bit-rate isochronous, and intermix transmissions. The lower two layers combined correspond to OSI physical layer, while the upper three layers resemble the OSI data link layer.

The lower three layers are collectively called *Fiber Channel Physical Standard* (FCPH). The advantage of Fiber Channel lies in its flexibility to transport both channel and network protocols simultaneously over a single link. It offers a universal interface for both channel and network data communications. It can work with FDDI, serial HiPPI, SCSI, IPI (*intelligent peripheral interface*), IP (*internet protocol*), IEEE 802.2, etc.

Fiber Channel Topologies   Flexibility in networking topology is a major asset of Fiber Channel. It supports point-to-point, arbitrated loop, and switched fabric connections:

- The *point-to-point connection* [Fig. 6.23(a)] can connect computer to computer or computer to disk with the highest possible bandwidth of the three topologies.

- The *arbitrated loop* [Fig. 6.23(b)] connects up to 126 devices in a token-driven ring. This is good for the interconnection of a large number of storage devices. The available bandwidth is shared among all devices. The advantage of this ring is its low cost, because no switch is required.

- The switched fabric topology [Fig. 6.23(c)] gives the greatest throughput. Many different-speed devices can be connected to the central fabric switch.

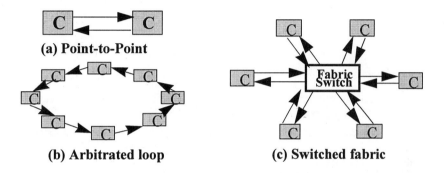

(a) Point-to-Point

(b) Arbitrated loop

(c) Switched fabric

Figure 6.23   Three Fiber Channel interconnection topologies (Courtesy E.M. Frymoyer, *Data Communications* [257], 1995).

The buffered wormhole routing is used. An 8×8 crossbar enables eight packet cells (called *flits*) to pass through the switch in every 40-MHz cycle, if there is no conflict. When there is a hot-spot conflict, only one crosspoint switch will be enabled at a time. The blocked flits are buffered in a central queue. This buffering frees up the input ports to receive subsequent flits from the previous switch stage.

The central queue is implemented by a dual-port RAM that can perform one read and one write in a clock cycle. To match the maximum bandwidth, each input port first deserializes eight flits from its FIFO into a chunk and writes the entire 64-b chunk to the central queue in 1 cycle.

FDDI   Digital Equipment developed the shared-media FDDI (*fiber distributed data interface*) technology. FDDI uses dual, optic-fiber token rings to provide 100-200 Mbps transmission between workstations. The use of two rings rotating in opposite directions provides redundant paths for reliability purposes. It has the capacity to interconnect a large number of devices up to 100 m over copper, 2 km over multimode fiber, and 60 km over single mode fiber.

Dual-attached multimode fiber FDDI rings can be extended to 200 km without using repeaters or bridges. This makes it possible to use FDDI rings in both LAN and MAN applications. FDDI rings are also superior in fault-tolerant operations. FDDI concentrators make the network more reliable by isolating failures. Mission-critical servers can be also connected to two concentrators to provide even higher fault tolerance.

A 100 Mbps backbone FDDI network is shown in Fig. 6.24. Special routers are used to link the FDDI rings to an Ethernet hub connecting to a large number of desktops. The FDDI rings are often used in environments demanding frequent additions, deletions, or moves of attached host or devices without causing any network disruption. While the token-ring FDDI is popular in shared-media applications, FDDI awaits a switched upgrade to provide higher flexibility to support workstation clusters.

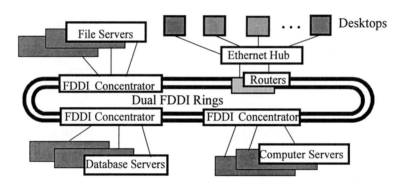

Figure 6.24   Dual FDDI rings as a backbone network (Courtesy D.A. Stamper, *FDDI Handbook,* Addison Wesley Pub. Co.[582] 1994).

A drawback of FDDI is its inability to support multimedia traffic, because traditional FDDI operates only asynchronously. This may weaken FDDI in the competition with ATM technology. However, synchronous FDDI products have appeared to cope with time-sensitive traffic. This will enable FDDI to retain some of its user groups in future applications. Digital's FFDT technology makes it possible to use FDDI in full duplex mode, which also enhances the competitiveness of FDDI.

6.4.2 Fast Ethernet and Gigabit Ethernet

In this section, we examine three Ethernet generations. We introducethe Myrinet as a result of academic research projects at university environments.

Ethernet Generations   The 10-Mbps Ethernet, first shipped in 1982, is no longer adequate to satisfy today's bandwidth demand in multicomputer clusters or in Internet applications. In 1994, two 100-Mbps versions of Fast Ethernet (the 100BaseT and 100VG-AnyLAN) were developed. Recently, the IEEE 802.3 working group announced the availability of 1-Gigabit Ethernet in 1997. The evolution of Ethernet technology is summarized in Table 6.4 in speed (bandwidth), cabling distance, and representative applications..

Table 6.4 Three Generations of Ethernet Development.
(Source: Gigabit Ethernet Alliance [267], Feb. 1997)

Generation	Ethernet 10 BaseT	Fast Ethernet 100 BaseT	Gigabit Ethernet
Year of introduction	1982	1994	1997
Bandwidth (speed)	10 Mbps	100 Mbps	1 Gbps
UTP twisted pair	100 m	100 m	25-100 m
STP/coaxial cable	500 m	100 m	25-100 m
Multimode fiber	2 km	412 m in half duplex, 2 km in full duplex	500 m
Single-mode fiber	25 km	20 km	2 km
Major applications	File sharing, printer sharing	Workgroup computing, client-server architecture, large database access	Large image files, multimedia, Intranet, Internet, data warehousing

Cabling Distance   The maximum distance of an Ethernet spans up to 25 km depending on cabling technology used. The network distance is reduced to the range of 25 m to 2 km for Gigabit Ethernet. Table 6.4 shows a sharp distance increase from twisted pairs (Category 5 UTP), to STP coaxial cables, and to single-mode fibers. The Gigabit Ethernet is targeted for campuses or buildings requiring greater bandwidth.

The *100Base-T* operates 100 Mbps over 100 m over copper, 20 km over single-mode fiber, and 2 km over full-duplex multimode fiber. Another Fast Ethernet technology is the *100VG-AnyLAN,* which offers 100 Mbps speed over 100-150 m of twisted-pair and 4 km of fiber-optic cables. The Ethernet mostly assumes the bus or star topologies, while Fast Ethernet supports mostly the star topology. For investment protection, Gigabit Ethernet does not require to change the network infrastructure, management, and applications from the earlier Ethernet generations.

Gigabit Ethernet Migration   Interoperability and backward compatibility are the two major leverages of Gigabit Ethernet. It retains the CSMA/CD access protocol used in Ethernet for many years. With fiber-optic cabling, the initial per-port price is rather high. Advances in silicon technology and digital signal processing may enable cost-effective support of Gigabit Ethernet over category 5 UTP wiring in the future.

Example 6.7 A Gigabit Ethernet LAN backbone

In Fig. 6.25, we show how to upgrade a switched Fast Ethernet backbone to a Gigabit backbone LAN. High-performance server farms can be directly connected to this Gigabit backbone with Gigabit Internet NICs. This upgrade will increase the throughput of a multiserver cluster for Internet users.

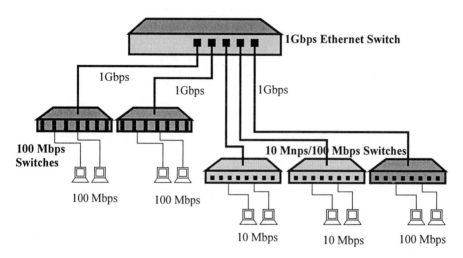

Figure 6.25   The construction of a Gigabit Ethernet LAN backbone
(Source: Gigabit Ethernet Alliance [267], Feb. 1997)

■

Four Ethernet upgrade scenarios are identified below:

- Upgrading switches to server links to achieve high-speed access to applications and file servers.
- Upgrading switch-to-switch links to obtain 1-Gbps pipes between 100-Mbps and 1-Gbps switches.
- Upgrading a switched Fast Ethernet backbone to aggregate Fast Ethernet switches with a Gigabit Ethernet switch or repeater.
- Upgrading a shared FDDI backbone to install FDDI switch or Ethernet-to-FDDI switches/routers with Gigabit Ethernet switches or repeaters.

The Gigabit Ethernet Standard, IEEE 802.3z, will be complete in 1998. All Ethernet generations use the MAC (*media access control*) protocol CSMA/CD. At the physical layer, both Fiber Channel and twisted-pair technologies are employed in cabling and connection interfaces.

Several vendors have already produced Gigabit Ethernet switches. For example, Plaintree Systems has produced the WaveSwitch 9200, a nonblocking switch capable of supporting 16 Gigabit Ethernet ports, or 64 FDDI SAS ports, or 128 100Base-TX/FX ports.

6.4.3 Myrinet for SAN/LAN Construction

Myrinet is a Gbps packet switched network by Myricom, Inc. The goal of Myricom is to make system interconnect a commodity product for building clusters of computers. We summarize below Myrinet technology, switches, links, interfaces, software support, and suggested applications.

Myrinet is based on multicomputer and VLSI technology developed at California Institute of Technology and ATOMIC/LAN technology developed at the University of Southern California. It is designed to construct either in-cabinet SAN clusters or LAN-based clusters of desktop hosts and server farms.The Myrinet can assume any topology, it is not necessarily restricted to a mesh of switches or any regular topology.

Myrinet is defined at the data link level by its variable-length packet format, flow control and error control on every link, the use of cut-through crossbar switches to route packets, and custom-programmable host interfaces. Myrinet SAN has a lower cost than a Myrinet LAN for its reduced physical size and parts. At the physical level, Myrinet uses full-duplex SAN links up to 3 m with a peak speed of 1.28 + 1.28 Gbps. To use as LAN links, 0.64 + 0.64 Gbps can be achieved up to 25 m on electrical cables and to 500 m on ribbon fiber as planned.

Myrinet Switches Blocking cut-through (wormhole) packet routing is used in Myrinet switches, similar to those used in Cray T3D and Intel Paragon. Multiple-port switches are connected by links to other switches or to a single-port host interfaces in any network topology. Inside each switch is a pipelined crossbar with flow control and input buffers similar to that shown in Fig. 6.3. A packet is advanced to a selected outgoing channel as soon as the header is received and decoded.

Following a deadlock-free routing scheme, multiple packets can flow through a Myrinet simultaneously. At present, an 8-port Myrinet switch achieves a bisection bandwidth of 10.24 Gbps, with at most a 300 ns latency in path formation and 6 to 11 w in power consumption. Sixteen-port and 32-port Myrinet switches are in the planning stage. The payload of a Myrinet packet is of arbitrary length. It can carry any type of packet (such as IP) without an adaptation layer.

Myrinet Host Interfaces The host interface is built around a 32-b, custom-VLSI processor, called LANai chip, with Myrinet interface, packet interface, DMA engine, and fast static RAM. The SRAM is used to store the *Myrinet control program* (MCP) as well as for packet buffering. This microarchitecture provides a flexible and high-speed interface between a generic bus and a Myrinet link.

At present, Myricom is shipping Myrinet/SBus interfaces for Sun SPARC workstations, and Myrinet/PCI interfaces for PCI-based PCs. Optic-fiber interfaces are also underway. The MCP software is executed on the interface processor, avoiding the OS overhead. However, the device driver and OS are still executed on the host. Myricom delivers both a standard TCP/IP and UDP/IP interfaces, and a streamlined Myrinet API.

Example 6.8 A Myrinet-connected LAN/cluster configuration

In Fig. 6.26, we show the use of four Myrinet switches to build a Myrinet LAN for connecting desktop workstations, PCs, an in-cabinet multicomputer cluster, and a single-board multiprocessor cluster. The two switches in the multicomputer cabinet form a SAN. Network RAM and disk arrays are attached to a Myrinet..

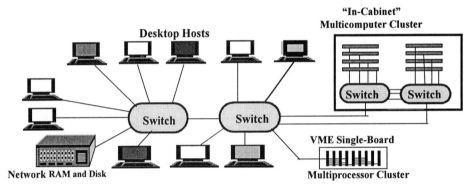

Figure 6.26   A Myrinet cluster built with four 8-port switches
(Courtesy N. J. Boden, et al. *IEEE Micro* [97], 1995).

■

In summary, Myrinet has a great potential to support cluster-computing applications. However, the bus-dependent host interfaces still pose a limitation in connecting a large variety of hosts to a Myrinet. The cabling distance of less than 25 m over copper poses the restricted use of Myrinet as essentially SAN, until fiber-optic links become available.

6.4.4 HiPPI and SuperHiPPI

HiPPI technology has been widely used in networking of heterogeneous computers and their peripherals. HiPPI is not just for supercomputers anymore. We assess the present status and future perspectives of HiPPI below:

HiPPI Technology *High-Performance Parallel Interface* (HiPPI) is a standard propounded by Los Alamos National Laboratory in 1987 with the intention of unifying the interface of all mainframes and supercomputers from different vendors. HiPPI was adopted in the mainframe and supercomputer industry as a high-speed I/O channel for short-distance system-to-system and system-to-peripheral connections. In 1993, the ANSI X3T9.3 committee ratified the HiPPI standard, which covers the physical and data link layers. Anything on top of it is up to the users.

HiPPI is a simplex point-to-point interface for transferring data at a speed of 800 Mbps to 1.6 Gbps. When HiPPI networking products were first introduced, the price was too high for popularization – costing up to $30,000 per node. As technology advanced and the price has lowered – as low as $4000 per node, it started to draw attention from outside the supercomputing community. Moreover, the HiPPI interoperability with ATM, Fiber Channel, and Sonet has been worked out. In the light of this, the role of HiPPI in high-speed networking is quite viable.

Interface and Cabling Requirements The basic interface was 50-b wide, out of which, 32 b is data and 18 b is control signals. A 32-b word emitted every 40-ns makes up a total of 800-Mbps in speed. The physical specification makes use of a 50-pair, shielded twisted pair for distance up to 25 m.

Such a small distance suffices possibly for small networks, but hardly for mainstream LANs. Multimode fiber cable works for 300 m in HiPPI standard without using fiber extenders. For greater distances, fiber extenders can support up to 10 km for multimode or 20 km for single-mode.

In simplex HiPPI channels, double cables are needed for two-way communications. Four cables have to work together, to achieve full-duplex 1.6 Gbps. HiPPI makes use of 50-pair UTP, which cannot be found in common enterprise network infrastructures. This poses a distinct disadvantage compared with other high-speed networks like ATM and Fiber Channel.

Example 6.9 A typical HiPPI internetworking configuration

Bidirectional HiPPI crossbar switches have been developed to establish dedicated connections among various mainframes, servers, and supercomputers as illustrated in Fig. 6.27.

The HiPPI-Serial allows connections over fiber to 10 km. Two HiPPI switches are linked to form the HiPPI backbone. If the switches are not equipped with serial interfaces, HiPPI fiber extenders are required for long links.

Internetworking with TCP/IP networks is addressed in RFC1347. Most HiPPI switch vendors also offer TCP/IP drivers. However, HiPPI's connection-oriented nature necessitates a compromise of the implementation of broadcast transmission. Whether to use an address server like ATM LANE or to handle the broadcast request intrinsically in the switches remains to be resolved.

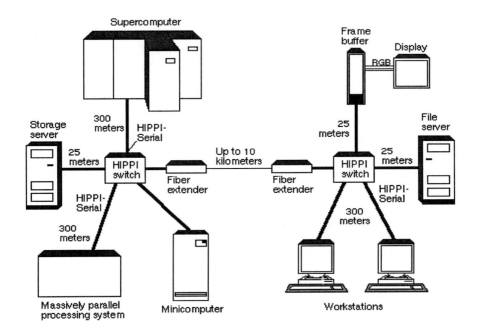

**Figure 6.27   Use of HiPPI channels and switches as the basic backbone
of LAN interconnect** (Courtesy.Tolmie and Flanagan, *Data
Communications*, HiPPI Networking Forum [613], 1997.)

HiPPI Channels and Switches   The HiPPI channels were introduced as high-speed I/O
or peripheral channels. But they do not support multicasting. In commercial machines,
HiPPI channels and HiPPI switches are used in the SGI PowerChallenge server cluster,
IBM 390 mainframes, and Cray Y/MP, C-90, and T3D/T3E systems. Like Fiber Channel,
HiPPI is not suitable for low-latency, dynamic, interactive usage.

Summarized below are some operational features, some existing and some desired, of
HiPPI channels or of HiPPI switches:

- *Very high-speed data transfers.* At present, HiPPI can be configured at either
  800-Mbps or 1.6-Gbps speed, both simplex and full duplex.

- *Very simple signaling sequences.* Basically, HiPPI connections can be set up with
  three messages: *request* (a source asks for a connection), *connect* (the destination
  indicates a connection established), and *ready* (the destination indicates it's
  readiness to accept a stream of packets).

- *Protocol independence.* HiPPI channels can handle a so-called raw HiPPI (data formatted with the framing protocol, without using any upper-layer protocols), TCP-IP datagrams, and IPI-3 (*intelligent peripheral interface*) framed data. IPI-3 is the protocol used to connect peripherals like RAID (*redundant array of inexpensive disks*) devices to computers. Thus, HIPPI is equally adaptive at inter-networking with Ethernet and FDDI and high-speed data storage and retrieval.

- *Physical-layer flow control.* HiPPI offers a credit-based system for reliable and efficient communications between devices operating at different speeds. In effect, the source keeps track of the *ready* signals and only sends data when the destination can handle them.

- *Connection-oriented circuit switching.* Nonblocking circuit switches allow multiple conversations to take place concurrently. Thus, the aggregate bandwidth of HiPPI switches can be multiples of 800-Mbps or of 1.6-Gbps, equal to the per-port bandwidth times the number of ports.

- *Compatibility with copper and fiber.* HiPPI uses 50-pair STPs for short distances. It works with single mode and multimode across the campus or metropolitan area. Sonet is used for long-distance communication.

HiPPI Protocols The HiPPI protocol is defined at several layers by a collection of standards. The HiPPI-PH standard defines the mechanical, electrical, and signaling, of a single point-to-point connection at the physical layer. Since HiPPI is a simplex protocol, a full-duplex link would be achieved by another HiPPI connection in the opposite direction.The HiPPI-FP (*Framing Protocol*) describes the format and content (including header) of each packet of user information.

HIPPI-SC was developed as one workable solution to allow many computers to share data. It allows for a switching mechanism to be built which could allow multiple point-to-point connections to occur simultaneously. HiPPI-SC does not specify any switching hardware, only provides functional mechanisms for the hardware to be used.

To map HiPPI into other protocols, three standards have been developed for link encapsulation, as listed below and summarized in Table 6.5.

- HiPPI-LE (*link encapsulation*) provides mapping of IEEE 802.2 LLC headers to the D1_Area and the beginning of the D2_Area.
- HiPPI-FC (*FiberChannel*) maps FiberChannel products to the HiPPI-FP standard.
- HiPPI-IPI (*disk & tape commands*) maps IPI-x standard command sets into HiPPI-FP headers. This standard may be incorporated into the IPI standards.:

SuperHiPPI Recently, a SuperHiPPI technology was developed to offer a potential speed of 6.4 Gbps, 8 times faster than HiPPI with a much lower latency. This SuperHiPPI technology has renewed some hope of traditional HiPPI users. It offers an entirely different implementation and thus lacks backward compatibility with traditional HiPPI channels. This implies a fresh start in upgrading the HiPPI channels or switches.

Table 6.5 The HiPPI Protocol Suite (Courtesy.Tolmie and Flanagan,
Data Communications, HiPPI Networking Forum [613], 1997.)

HiPPI-LE Link Encapsulation (IEEE 802.2)	HiPPI-FC (Fiber Channel Protocols)	HiPPI-IPI (IPI-3 Command Sets)
HiPPI-FP Framing Protocol (Packet format, packet headers)		
HiPPI-PH, Mechanical, Electrical, Signaling (Physical Layer)		HiPPI-SC, Switch Control (for Physical Layer)
HiPPI-Serial Extension, Fiber-based HiPPI-PH 10 km Extender (Not an ANSI Standard)		

Both SGI and Los Alamos National Laboratory have explored HiPPI technology to build as high as 25.6-Gb/s HiPPI switches. In *ACM Supercomputing, 94* held in Washington D.C., it was demonstrated that an all-fiber HiPPI backbone, consisting of 18 miles of multimode cable and 16 exhibitors, can deliver an aggregate bandwidth of over 90 Gbps collectively.

6.5 ATM Switches and Networks

Asynchronous transfer mode (ATM) is defined by ATM Forum (founded in 1991) and ITU standards. The ATM technology, cell format, and ATM layers are described below, along with an internetworking example using ATM switches and fiber-optic cable. ATM networks started with Telecom Companies and appeared as MAN or WAN in the Internet backbone. Now, major computer companies have their own ATM networking support for Intranet or LAN support.

6.5.1 ATM Technology

ATM is a medium-independent information transmission protocol that segments traffic into short fixed-length 53-B cells. Its underlying technology is based on cell switching discussed in Section 6.1.3. The goal of ATM is to incorporate both real-time (i.e., delay-sensitive) and burst data (i.e., non-delay-sensitive) transmission capabilities into one single network technology.

ATM provides dynamic bandwidth allocation and high-speed switching required for greater economy and usability in broadband networking applications. ATM adopts the switching paradigm in that end stations do not share a common medium and thus it supports both LAN and WAN requirements. The fact that ATM uses short cells makes it possible to build the switches on a silicon fabric, from which the high speed comes.

ATM Cell Format As depicted in Fig. 6.28, a long packet muse be broken into cells before they are transmitted through an ATM network. At the receiving end, the cells are reassembled to restore the original packet format. The first 5 B is the cell header for routing purposes. The payload contains at most 48 B of data information. The use of small cells, virtual paths, and virtual channels make the ATM cell switch design atrtractive.

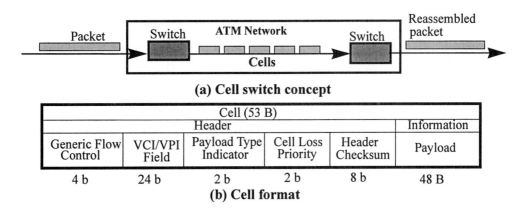

(a) Cell switch concept

Cell (53 B)					
Header					Information
Generic Flow Control	VCI/VPI Field	Payload Type Indicator	Cell Loss Priority	Header Checksum	Payload
4 b	24 b	2 b	2 b	8 b	48 B

(b) Cell format

Figure 6.28   The concept of ATM cell switching and cell format.
(Courtesy E.W. Zegura, *IEEE Communications* [664]. 1993).

The ATM cells can be routed using either a *source routing* scheme or a *hop-by-hop* scheme. For source routing, the entire routing path information must be included in the cell header. Therefore, the route must be limited in length in source routing. For hop routing, only the *hop ID* is needed in the header, resulting in a more flexible selection of the routing paths. Most ATM switches choose to implement the hop-by-hop routing scheme using the cell format shown.

Cell Switch Design Apart from traditional use of packet switched networks, ATM offers a new standard for building cell switched networks. It is designed for high-speed transmission of all media types from voice to image, video and data over one network. ATM should be incorporable with existing LANs, MANs, and WANs using standard communication protocols. ATM cell networks have been used to transmit broadband and multimedia information. In particular, ATM must be able to perform multicast operations efficiently.

In Fig. 6.29, we show the cell switch design concept based on the hop routing scheme. On each incoming link of an ATM switch, an arriving cell's *virtual path identifier* (VPI) and *virtual channel identifier* (VCI) values uniquely determine the path and channel a cell will traverse. A new virtual identifier will be placed in the cell header on the outgoing link. The routing tables are initialized by the virtual-path establishment procedure.

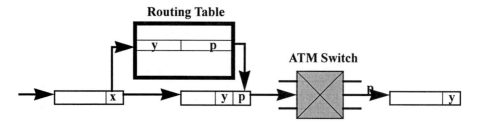

Figure 6.29   ATM cell switching using hop-by-hop routing (Courtesy E.W. Zegura, *IEEE Communications* [664]. 1993).

Upon entering an ATM switch, a cell's VPI field (**x**) is used to select an entry from the routing table. This entry carries the output port number (**p**) the cell should be forwarded to. A new VPI value (**y**) will be placed in the cell to be forwarded to the next switch. Over the same virtual path, two hosts may multiplex many application streams, using the VCI fields in cells to distinguish among a bundle of virtual channel streams.

ATM Speed   Presently, ATM networks support a variety of speeds ranging from 25 to 51, 155, and 622 Mbps. The lower the speed, the lower will be the cost of ATM switches and links used. The ATM speed may increase further in the future. With 622 Mbps per port, the FORE System ASX 1000 ATM switch can yield an aggregate bandwidth of 10 Gbps.

6.5.2 ATM Network Interfaces

ATM cells are switched in a two-level hierarchy: *virtual path* and *virtual circuit*. Several virtual paths can reside in a physical link. Within each virtual path, there can exist more than one virtual circuit. Virtual circuits are recognized by those ATM switches connecting to hosts, while between switches, the virtual path is used instead. As such, two network interfaces are defined: the UNI (*user-network interface*) and the NNI (*network-network interface*) as depicted in Fig. 6.30.

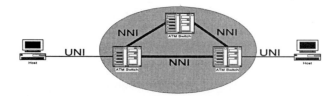

Figure 6.30   User-network-iInterface (UNI) and network-network-interface (NNI) in an ATM network (Courtesy of ATM Forum [46],*Version UNI 3.1 Specification*, 1994).

Such a two-level switching hierarchy has the advantage that when a virtual path is rerouted by some switches due to a hardware failure, all its associated virtual circuits are rerouted automatically, without reestablishing the routes individually. The 5-B ATM layer header has different formats for the UNI and NNI as shown in Fig. 6.31.

ATM layer header at the UNI:

GFC (4-bit)	VPI (8-bit)	VCI (16-bit)	PTI (3-bit)	CLP (1-bit)	HEC (8-bit)

ATM layer header at the NNI:

VPI (12-bit)	VCI (16-bit)	PTI (3-bit)	CLP (1-bit)	HEC (8-bit)

GFC: Generic Flow Control   PTI: Payload Type Identifier
VPI: Virtual Path Identifier   CLP: Cell Loss Priority
VCI: Virtual Channel Identifier   HEC: Header Error Check

Figure 6.31   Header formats for UNI and NNI in ATM cells. (Courtesy of ATM Forum [46],*Version UNI 3.1 Specification*, 1994).

The major difference lies in the presence of a GFC (*Generic Flow Control*) field in UNI and a bigger VPI (*Virtual Path Identifier*) field in NNI. The GFC field identifies the end station, and the switch prioritizes one cell over another when they are contending for a virtual circuit. The default setting is all zeros. It will be overwritten by the first switch the cell reaches, incapacitating its end-to-end significance.

The larger VPI field for NNI cells enables switches to make use of a greater number amount of available paths for routing. The ATM Forum has also developed a *Private Network-Network Interface* (PNNI) standard to enable switches from different suppliers.

Some ATM switch and network products are listed in Table 6.6 in terms of speed, port numbers, and UNI or NNI support. We will show later how to use the FORE Systems ATM switches and links in building a LAN backbone.

6.5.3 Four Layers of ATM Architecture

We describe next the layered ATM network, its correspondence with the OSI (*open systems interconnection*) standards, and the ATM cell formation process.

The Layered ATM Model Conceptually, the ATM network is described by a four-layered, three-dimensional model as shown in Fig. 6.32. The physical layer is subdivided into two ATM sub-layers (TC and PMD). The ATM layer corresponds to a network layer

Table 6.6 Example Gigabit ATM Switches and Networks

Product	Aggregate Bandwidth	No. of Ports	UNI and NNI Support
Bay Networks Centillion 100 ATMSpeed/155	1.2 Gbps		
Cisco LightStream 1010	5 Gbps, shared memory nonblocking	64	v3.0 and v3.1
DEC GIGAswitch/ATM	3.6 Gbps	22	
FORE ASX-1000	10 Gbps, nonblocking	96	Full v3.0
2230 Nways 600 and 650	5 Gbps, nonblocking	8	v3.0 and v3.1

that has the characteristics to support end-to-end virtual circuits switching. It is connection-oriented. It does not provide any mechanism of acknowledgments.

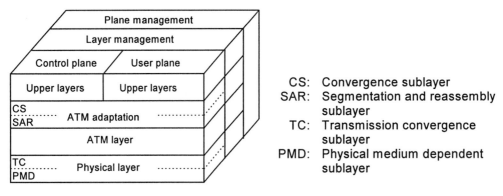

Figure 6.32 The ATM network architecture (Courtesy S. Saunders, ed., *The McGraw-Hill High-Speed LANs Handbook* [530], 1996)

The ATM adaptation layer is divided into CS and SAR sublayer. The upper layers are toward the user plane and control plane. The third dimension shows the layer and plane management structures. The rationale behind no ATM cell retransmission was based on using fiber optic networks that are highly reliable. It was considered sufficient to leave the duty of error control to upper layers. Moreover, for real-time applications such as audio and video, it is better to tolerate missing cells than to retransmit the entire image.

Corresponding OSI Layers Table 6.7 defines the functionality of the ATM sublayers and their correspondence with the seven layers of OSI network standards by ISO (*International Standards Organization*). Interested readers may refer to ATM and ISO standards for details.

Table 6.7 Functionality of ATM Layers with Corresponding OSI Layers (Courtesy of ATM Forum [46],*Version UNI 3.1 Specification*, 1994)

OSI layer	ATM layer	ATM sublayer	Functionality
3/4	AA1 (ATM adaptation Layer)	CS	Providing the standard interface (convergence)
		SAR	Segmentation and reassembly
2/3	ATM		Flow control, cell header generation/extration, virtual circuit/path management, cell multiplixing/demultiplexing
2	Physical	TC	Cell rate decoupling, header checksum generation and verification, cell generation, packing/unpacking cells from the enclosing envelope
1		PMD	Bit timing, physical network access

Cell Formation In an outbound cell, the ATM layer accepts the 48-B payload from the higher level, AAL, adds a 5-B header, and passes it to the physical layer. For inbound traffic, it does exactly the reverse. This cell formation is depicted in Fig. 6.33.

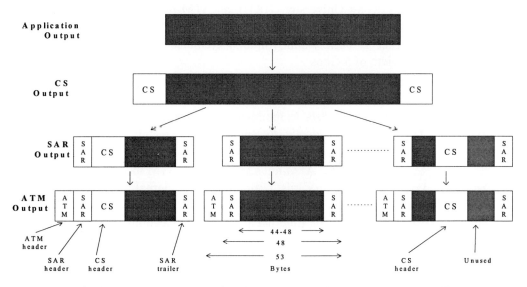

Figure 6.33 Passage of application payload in an outbound cell to successive ATM layers. (Courtesy of ATM Forum [46], *Version UNI 3.1 Specification*, 1994)

The ATM layer guarantees the exact order of arrived cells along the same virtual circuit. Though the network is entitled to discard cells in case of congestion, it is not allowed to reorder cells sent on a single virtual circuit. Such a guarantee is not extended to cells sent on different virtual circuits.

When an intermediate switch drops a cell due to transient overload, the entire packet is damaged and a long latency may arise. Presently, reliable cell delivery must be done on top of the ATM layer. Furthermore, the optimal cell size is still debated among different user groups. ATM was mainly used as a telephone network technology in the past. But the trend is to use ATM technology in smaller and localized networks.

6.5.4 ATM Internetwork Connectivity

Various ATM switches, links, and adapters can be used for different networking requirements. For example, one can build dedicated ATM switches for connecting PCs, workstations and servers within a workroom. An ATM switch can be used to build a LAN backbone, or to access legacy LANs (such as Ethernet, Token rings, or FDDI rings), or to distribute video to a cluster of workstations or to an Intranet with Internet access. We show below the design of an ATM-based Intranet for distributed multimedia processing in providing dedicated Internet service applications.

Example 6.10 PearlCluster: An ATM-based multicomputer cluster at HKU

At the Hong Kong University (HKU), an ATM-based cluster of workstations and servers has been established. In Fig. 6.34, we show the ATM network configuration. The cluster is built around three interconnected ATM switches with an aggregate bandwidth of 5 Gbps.

Optical cables using 155-Mbps multimode fibers are used to interconnect six SMP servers and over 40 Sun and SGI workstations and PCs, all of which are located in the same building. All hosts in the cluster can access outside resources via the HARNET (*Hong Kong Academic Research Network*) and Internet backbone to access the WWW (*World-Wide-Web*) resources.

Both ATM links and non-ATM links are used. All ATM incompatible hosts must be attached with ATM adapter cards. Different-speed external links are connected to or from the Pearl cluster. But internally within the cluster, only 155-Mbps fiber- optic cables are used besides existing Ethernet and FDDI connections.

Most hosts are connected to ATM backbone switches (ASX-1000, ASX-200BX). Some Ethernet-attached hosts are indirectly connected through an Ethernet hub (PowerHub 7000). The LAX-20 LAN access switch connects Ethernet and FDDI ring. ATM switches, 155-Mbps fiber links, and adapter cards are basic components used to build the HKU cluster.

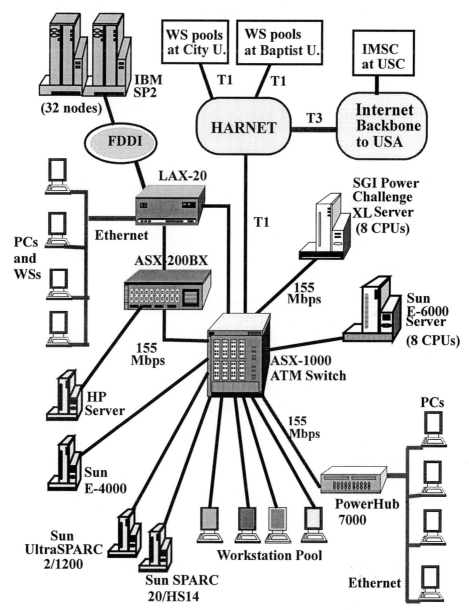

Figure 6.34 PearlCluster: an ATM cluster of servers/workstations at the University of Hong Kong, dedicated for distributed multimedia and metacomputing applications. (HARNET: Hong Kong Academic Research Net, IMSC: Integrated Multimedia System Center, WS: workstation, PC: personal computer)

Applications of the PearlCluster: Research projects based on the PearlCluster facilities cover special cluster architectural development, distributed multimedia processing, metacomputing, financial digital library, etc.

(1) On the Pearl cluster, some load sharing facility (LSF) is applied to achieve SSI (*single system image*) operations in *financial digital library* (FDL) applications.

(2) Development of an *MPI-Java Interface* (MJI) for high-speed message passing using active messages and JVM (*Java Virtual Machine*) shells in addition to the MPI (*Message Passing Interface*) and PVM (*Parallel Virtual Machine*) standards to be discussed in Chapter 13.

(3) The cluster is designed with a bilingual WWW index and search engine for high-speed information retrieval of both English and Chinese documents or multimedia files from the Web resources.

(4) Extending TPC (*Transaction Processing Council*) benchmarks and generating new DMS (*distributed multimedia System*) benchmarks for cluster performance testing. These TPC and DMS benchmarks are designed to evelaute the potential of using a clustered superserver in commercial applications.

∎

Niches and Shortcomings ATM networks have the potential to see widespread use in digital telephony, medical imaging, transaction processing, video on demand, and distributed multimedia processing. The telecommunications industry has already chosen ATM to realize many broadband ISDN (*Integrated Services Digital Network*) services. The future of ATM rests on its ability to integrate with LAN and WAN technologies.

On the negative side, ATM networks lack a commonly accepted cell management and network protocols. Telephone companies and the computer industry apply different standards at this time. ATM Forum is the international organization whose purpose is supposedly to unify the standards. For an example, an ATM network cannot guarantee the reliable delivery of a cell or a packet, because it offers a data-link layer protocol.

Unless the computer, multimedia, and telecommunications industries agree to the same set of ATM standards, these shortcomings may outweigh the benefits of ATM technology. Even ATM has demonstrated its usefulness in broadband or multimedia-oriented MANs, WANs, and internetworking areas. The potential widespread use of ATM technology in LANs is limited by its high cost, compared with Gigabit Ethernet and other low-cost LAN technologies.

6.6 Scalable Coherence Interface

Those Gigabit and ATM networks we have discussed above have one thing in common: they are all connected to the I/O bus of a node (see Fig. 6.2), and communication is done by passing messages among the computer nodes. Gustavson and Li [289] pointed out that this type of *I/O communication* is inferior to *shared-memory communication* in a bus-based SMP.

Shared-memory communication has low latency because a communication is no more than one processor executing a *store* instruction and another processor executing a *load* instruction. In an I/O communication, hundreds to thousands of instructions need to be executed. Furthermore, it is difficult for I/O communication to take advantage of cache coherency information, which is generated by hardware in a bus-based SMP.

For these reasons, a high-performance network should be interfaced to the memory bus (local bus). Traditional networks have two main advantages over a bus.

- They are standards and relatively stable, while a memory bus is not. Ethernet has been standardized for more than 10 years, and is widely used today. But a memory bus may change every 18 months when a new processor comes out.

- They are space-scalable to tens to hundreds of meters, while a memory bus can rarely exceed 1 m. The short length limits the number of processors an SMP system can support. Is it possible to design a *standard* interconnect structure that maintains the advantages of a bus and also has the space scalability of a traditional network?

The answer is the IEEE/ANSI Standard 1596-1992: *Scalable Coherence Interface* (SCI), which extends from conventional backplane bus to a fully duplex, point-to-point interconnect structure, providing a coherent cache image of distributed shared memory. SCI is a rich protocol (the base specification [343] is 248 pages long). We discuss below how SCI is evolved from the bus, the basic data transfer protocol, and how SCI enforces cache coherence. Example system implementations of SCI will be discussed in Chapter 8.

6.6.1 SCI Interconnects

The SCI is designed to provide a low-latency (less than 1 μs) and high-bandwidth (up to 8 GB/s) point-to-point interconnect. Once fully developed, the SCI can connect up to 64K nodes. The most recent SCI standard (IEEE 1596-1996) specifies bandwidths from 250 MB/s to 8 GB/s. SCI links can be implemented with copper as well as fiber-optic cables. The interface chips are implemented by CMOS, BiCMOS, and GaAs technologies.

The cache coherence feature of SCI is implemented with a linked list of SCI modules. Each processor node is attached to one SCI module. When a processor updates its cache, the cache state is propagated along all SCI modules sharing the same cache line. This linked list for distributed coherent caches is scalable, because a larger system can be created by inserting more SCI modules along the linked list.

Interconnect Topologies   The SCI defines the interface between nodes and the external interconnect. The initial goal was to use 16-b links with a bandwidth of 1 GB/s per link.

As a result, backplane buses have been replaced by unidirectional point-to-point links. The SCI interconnects are topology-independent in general.

Each SCI node can be a processor with attached memory and I/O devices. The SCI interconnect can assume any topology. Each node has an input link and an output link that are connected from or to the SCI ring or crossbar. The bandwidth of SCI links depends on the physical standard chosen to implement the links and interfaces.

In an SCI environment, the concept of broadcast, bus-based transactions has been abandoned. Coherence protocols are based on point-to-point transactions initiated by a requester and completed by a responder. A ring interconnect provides the simplest feedback connections among the nodes.

From Bus to SCI Ring   In Fig. 6.35, we show a typical SCI ring interconnect extending from the bus intereconnect.

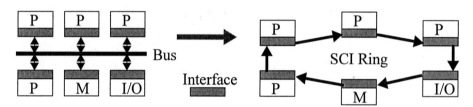

Figure 6.35   Evolution of an SCI ring from a digital bus (Courtesy D. V. James, et al. *IEEE Computer* [348], 1990)

To understand the motivation for SCI, we need to look at three problems of the bus (see Fig. 6.35) when building a scalable system.

- *Signaling Problem.* Bus transmission lines are not perfect because they have taps to connect various processor, memory, and I/O devices. This causes reflections and introduces noise, especially since bus drivers require lots of current.

- *Bottleneck Problem.* Bus is a shared medium which can be used by only one transmitter at a time. Split-transaction protocols help a little. But bus arbitration and addressing always need to be done for each transaction.

- *Size Problem.* Due to signaling difficulties, a fast bus must be short. The short bus limits the number of devices that can be connected to it.

SCI adopts the following techniques to overcome the three problems associated with the bus, as shown in Fig. 6.35:

- **Point-to-Point Link**. SCI views various processor, memory, and I/O devices as nodes, and uses a point-to-point link from one sender node to a receiver node, with differential signaling. There are no connection taps any more, and the reflection/noise problem is significantly alleviated, allowing a large increase in signaling speed. SCI does not rule out more complex nodes that contain processors, memory, and I/O devices.

- *Unidirectional Ring.* The links are run continuously and in one direction. This makes the driver current remain constant, further reducing noise. Since every node must have at least one input link and one output link, a unidirectional ring is the simplest topology.

- *Parallelism.* Unlike the bus where only one transaction can use it at a time, multiple nodes can inject and extract packets to the SCI ring simultaneously. Communication Protocols

From Rings to a Mesh Other topologies such as the 2D mesh that can be constructed using multiple rings, as illustrated in Fig. 6.36. A mesh of rings is bridged with the interface modules. The bandwidth, arbitration, and addressing mechanisms of an SCI interconnect are expected to significantly outperform backplane buses. The main advantage of SCI lies in its scalability.

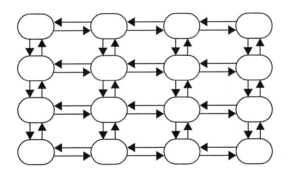

Figure 6.36   A 2D mesh built from multiple SCI rings (Circles are computer nodes, Courtesy Bryhni and Wu [114], 1994).

6.6.2 Implementation Issues

The SCI standard has gained support from many computer manufacturers; notable examples include HP/Cray, IBM, Data General, Sequent, and SGI/Convex. Most of the SCI architectural applications will be studied in Chapter 8.

In this section, we explain some SCI implementation considerations, such as link format, packet format, node interfaces, and SCI clocking. Toward the end, we show an exmaple SCI implementation.

Link and Packet Formats   SCI uses split-transaction protocols based on packets. The formats of SCI links and of packets used in SCI interconnects are given in Fig. 6.37.

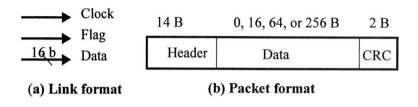

(a) Link format   **(b) Packet format**

Figure 6.37   SCI link and packet formats (Courtesy of
IEEE SCI Standard 1596-1992. [343])

SCI Node Interfaces   The node interface is detailed in Fig. 6.38. An SCI link is 18-b wide with 1 clock bit, 1 flag bit, and 16 data bits. Such an 18-b entity is called a *symbol*. A packet contains a 14-byte header, a 2-B CRC for error handling, and 0, 16, 64, or 256 B of data.

At each SCI clock cycle, one symbol (2 B) of a packet can be transmitted. The clock and the flag bits are generated by hardware. Therefore, a packet can have 8, 16, 40, or 136 symbols.

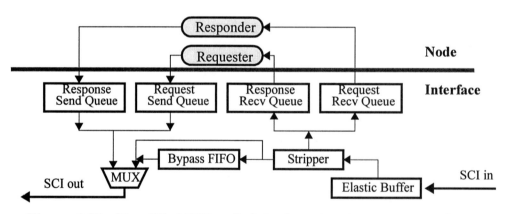

Figure 6.38   Simplified SCI node interface structure (Courtesy of
IEEE SCI Standard 1596-1992. [343])

Suppose a source node wants to read data from a destination node. The source (called the *requester*) injects a *request* packet into the SCI ring through the *request send queue*. The header of the request contains the addresses, the command, and other information. This packet passes along the ring from node to node until it reaches the destination node. How can one tell if a node is the destination?

The *stripper* of a node takes out those incoming packets whose address matches the node ID and puts them in one of the receive queues. The destination node (called the *responder*), after interpreting the read request, fetches the data, forms a response packet and injects it into the SCI ring. The response packet then travels along the ring to the source node, just as the request packet did.

When a packet arrives at a node (say, node *A*) that is not the intended destination, it goes directly to the SCI output link to go to the next node. However, the link may be occupied by node *A* itself sending data out. Then the packet is put into the bypass FIFO, waiting for the output link to become available.

SCI Clocking   SCI is *synchronous* in that logically speaking, all links and interface circuits are driven by the same clock (usually 500 MHz). To deal with clock variations in different nodes, the *elastic buffer* at a receiver node is used to insert and delete *idle symbols* in incoming packets.

When a sender node has no packets to transmit, it sends idle symbols to keep the receiver synchronized. Later on, when the sender needs to send packets, it can do so without having to first send a synchronization preamble as in Ethernet. The SCI communication protocol paid attention to ensure the fair use of the network and to avoid deadlock and livelock.

An example is the use of separate send/receive queues for request and response. Without separate queues, it would be possible for excessive requests to prevent the sending of a response, resulting in deadlock. We show below an example SCI-connected CC-NUMA multiprocessor sysytem

Example 6.11 .Data General SCI-connected multiserver cluster

A Data General SCI multiprocessor architecture is shown in Fig. 6.39. Two SCI rings are used to interconnect four SMP server nodes.

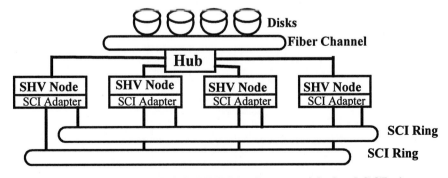

Figure 6.39   Data General CC-NUMA cluster with dual SCI rings
(Courtesy of Data General Corporation, 1995)

Each server node has four Intel commodity Pentium Pro SHV (*standard high volume*) processors with 4 GB of local memory. The dual SCI rings provide a total of 1-GB/s bandwidth supporting up to eight SHV nodes. This cluster is designed to build large-scale enterprise servers for scalable database applications.

All distributed memories can form a *global coherent memory* (GCM), following a cache-coherent NUMA model. The Data General SCI design can support up to eight SHV server nodes with a total DSM of 32 GB. A Fiber Channel ring is used to connect the shared disk arrays to all server nodes through the Hub switch.

This system promotes distributed coherent caches and develops both DG/UX Unix operating system attributes to support the CC-NUMA architecture and an SCI interconnect adapter.

The system has a reported speed of 14,000 transactions per second by running an OLTP-like workload over four SHV nodes of 16 Pentium Pro processors. This is a typical example of building an SMP server with commodity processors, shared memory, and open SCI architecture.

■

6.6.3 SCI Coherence Protocol

The cache coherence protocols used in the SCI are directory-based. A sharing list is used to chain the distributed directories together for reference purposes.

Sharing-List Structures   Sharing lists are used in the SCI to build chained directories for cache coherence use. The length of the sharing lists is effectively unbounded. Sharing lists are dynamically created, pruned, and destroyed. Each coherently cached block is entered onto a list of processors sharing the block. Communications among sharing processors are supported by shared-memory controllers, as shown in Fig. 6.40.

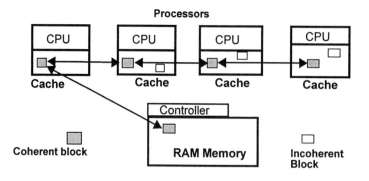

Figure 6.40   SCI cache coherence protocol with distributed directories.
(Courtesy D. V. James, et al. *IEEE Computer* [348], 1990)

Processors have the option of bypassing the coherence protocols for locally cached data. By distributing the directories, the SCI avoids scaling limitations imposed by using a central directory.

Other blocks may be locally cached and are not visible to the coherence protocols. For every block address, the memory and cache entries have additional tag bits which are used to identify the first processor (head) in the sharing list and to link the previous and following nodes.

Doubly linked lists are maintained between processors in the sharing lists, with forward and backward pointers as shown by the double arrows in each link. Incoherent copies may also be made coherent by page-level control. However, such higher-level software coherence protocols are beyond the scope of the published SCI standard. The backward pointers support independent deletions of entries in the middle of the linked list.

Sharing-List Creation The states of the sharing list are defined by the state of the memory and the states of list entries. Normally, the shared memory is either in a home (uncached) or a cached (sharing-list) state. The sharing-list entries specify the location of the entry in a multiple-entry sharing list, identify the only entry in the list, or specify the entry's cache properties, such as clean, dirty, valid, or stale.

The head processor is always responsible for list management. The stable and legal combinations of the memory and entry states can specify uncached data, clean or dirty data at various locations, and cache writable or stale data. The memory is initially in the home state (uncached), and all cache copies are invalid. Sharing-list creation begins at the cache where an entry is changed from an invalid to a pending state.

When a read-cache transaction is directed from a processor to the memory controller, the memory state is changed from uncached to cached and the requested data are returned. The requester's cache entry state is then changed from a pending state to an only-clean state. Multiple requests can be simultaneously generated, but they are processed sequentially by the memory controller.

Sharing-List Updates For subsequent memory access, the memory state is cached, and the cache head of the sharing list has possibly dirty data. A new requester (cache A) first directs its read-cache transaction to memory but receives a pointer to cache B instead of the requested data.

A second cache-to-cache transaction is directed from cache A to cache B. Cache B then sets its backward pointer to cache A and returns the requested data. Any sharing-list entry may delete itself from the list. However, the addition of new sharing-list entries must be performed in a first-in-first-out order to avoid potential deadlock.

The head of the sharing list has the authority to purge other entries from the list to obtain an exclusive entry. Others may reenter as a new list head. Purges are performed sequentially. The chained-directory coherence protocols are fault-tolerant in that dirty data are never lost when transactions are discarded.

6.7   Comparison of Network Technologies

In this final section, we compare the various network technologies studied in previous sections. First, we compare network architectures, cabling media, and protocol standards. Then we compare their relative performances in speed, cabling distance, and representative applications.

6.7.1 Standard Networks and Perspectives

Network architectures, standards, and future perspectives are elaborated below:

Network Standards   In Table 6.8, we classify networks by shared-media versus switched architectures or both. Cabling media include copper wires and fiber optic-cables. Protocols and standards are identified with ANSI, IEEE, Alliance, or Forum bodies.

Table 6.8   Network Architectures, Cabling Media and Standards

Technology	Gigabit Ethernet	HiPPI	Fiber Channel	FDDI	SCI	ATM	Myrinet
Architecture	Shared media, switched	Switched	Shared media, switched	Shared media	Shared media	Switched	Switched
Media type	UTP, Coaxial, fiber	Fiber, 50-pair STP	STP, Coaxial, fiber	Fiber, copper	STP, fiber	UTP, STP, fiber	Electric links, fiber planned
Standard body	IEEE 802.3z, Gigabit Ethernet Alliance	ANSI X3.183, X3.210, X3.218, X3.222	ANSI X3T11	ANSI X3T9.5	IEEE 1596-1992	ATM Forum, IETF, ITU-TSS	Myricom

UTP: unshielded twisted pair, STP: shielded twisted pair.

Future Perspectives   Fast Ethernet and Gigabit Ethernet have the highest potential to dominate the LAN and Intranet market for their backward compatibility with the well-established Ethernet Standards. The ATM technology was and is still heavily supported by telecommunications industry. ATM is especially suitable for internetworking, Internet, and multimedia applications.

It could penetrate into the MAN and LAN domains if the production cost of ATM product could be lowered in the future. For MAN and WAN applications, ATM and FDDI are the primary candidates. The 10 BaseT Ethernet, Fiber Channel, and HiPPI rank the next in cabling distance.

The migration of ATM technology to LAN applications is limited by ATM's high cost, compared with Ethernet products. Most of today's ATM networks are set at low speeds such as 25 Mbps or 155 Mbps.

The reason is to lower the cost of ATM links, switches, and adapters at a lower level. For example, the equipment price of a 622-Mbps ATM network with multimode fiber is $6600 per port, 8.4 times higher than $785 per port of a switched Fast Ethernet in 1996. The higher price tags of ATM and of Gigabit Ethernet may limit their future growth.

The SCI, HiPPI, Myrinet, Gigabit Ethernet, and Fiber Channel are all good candidates for Gbps SAN, LAN, and clustering applications.The choice of a particular system interconnect is largely determined by cost/performance tradeoffs. The general trend is to upgrade networks from using shared media to a switched architecture. Fiber-optic cables are more desirable for longer distance as well as higher speed. Again, the final choice is often decided by budget rather than performance.

6.7.2 Network Performance and Applications

We compare below the *bandwidth* (speed), *latency*, and *cabling distance* of major network technologies. These technologies have been used to build clusters or scalable systems. All numerical values given correspond to the fastest speed in Mbps or Gbps.

The longest distance is expressed in m or km using copper wires (UTP or STP) or using fiber optical cables in single-mode or multimode fibers. The table entries may change in value in the future, since networking technology changes rapidly. Table 6.9 summarizes the network speed and distance performance reported up to 1997.

Performance Spectrum Gigabit speed will become the bottom-line requirement in future networks. This is in response to the increasing demand for much higher bandwidths in Internet, Intranet, multimedia, and commercial applications. The usage of commodity networks in multicomputer clusters or scalable multiprocessors has increased rapidly. Terabit/s networks are needed to support Tflop/s computer systems in the future.

Gigabit bus/switch interconnects have appeared in many commercial servers, mainframes, clusters, and MPPs. Gigabit speed is achieved in networks built with SCI, HiPPI, Fiber Channel, Myrinet, and Gigabit Ethernet technologies.

Commercial ATM, FDDI, Fast Ethernet, IsoEnet, and Ethernet network products are still confined within the low speed range from 10 Mbps to 100 or 622 Mbps for economic reasons. For SAN and LAN applications, all the above bus, switch, and network technologies are suitable.

The choice of a specific network depends on the cost/performance ratio set for the given application. The latency is largely attributed to software overhead in setting up the connection path, not the hardware delay in routing messages through the switches or networks. The numeric entries in Table 6.10 may vary in the future, as network technology improves rapidly.

Table 6.9 Network Performance and Cabling Distance

Technology	Bandwidth	Distance over Copper	Distance for Single-Mode Fiber	Distance over Multimode Fiber
Gigabit Ethernet	1 Gbps	25-100 m	2 km	500 m
HiPPI	800 Mbps, 1.6, 3.2 Gbps, 6.4 Gbps for SuperHiPPI	25 m	20 km with fiber extender	300 m for direct links, 10 km with extenders
Fiber Channel	100, 200, 400, and 800 Mbps, 1, 2, and 4 Gbps	50 m at 100 Mbps	10 km	2 km at 200 Mbps
FDDI	100 Mbps, 200 Mbps	100 m	60 km	2 to 200 km
100Base-T	100 Mbps	100 m	20 km	412 m in half duplex, 2 km in full duplex
100VG-AnyLAN	100 Mbps	150 m	2 km	4 km in diameter
ATM	25 Mbps to 51, 155, 622 Mbps	Variable distance in using repeaters or fiber extenders for internetworking connections		
SCI	800 Mbps to 8+ Gbps	10 m	> 100 m under experiments	
Myrinet	1.28 Gbps	25 m	Unknown	500 m

Application Potentials   Based on the above studies, we highlight below the application potentials of major network technologies:

- Fiber Channel transports data in five layers of FC protocols that work with various channel and network standards for the storage, networking, and data transfer among heterogeneous hosts and peripheral devices.

- Primarily a shared-media standard with low speed, but higher reliability and flexibility. Needs a major speed upgrade as well as moving into switched FDDI compatible with other standard.

- Gigabit Ethernet is targeted in Intranet, multimedia, large image files, LAN, workgroup computing, client-sever architecture, database and warehousing applications with a smooth upgrade path from the well-established Ethernet and Fast Ethernet user groups.

- Myrinet is attractive mainly in SAN and clustering applications at this time. But its potential to cut into the Gigabit LAN arena is increasing, if the interface and protocol barriers could be alleviated with common standards.

- HiPPI and SuperHiPPI will continue to serve as a high-speed medium for large data movement among mainframes, SMP servers, and supercomputers. With protocol independence, HiPPI is adaptive to internetworking applications as well.

- ATM will continue to dominate WAN internetworking, multimedia, Internet and broadband ISDN applications. Its slow acceptance by LAN users will be driven by cost effectiveness in future development. Most likely it will lose the battle, but win the war with a significant cut in the production cost.

- SCI is gaining popularity among scalable interconnect systems, especially in SMP and NUMA multiprocessor or MPP clusters recently announced by Sun/Cray, SGI/Cray, HP/Convex, Data General, etc.

6.8 Bibliographic Notes and Problems

Feng has given a survey of earlier interconnection architectures [238] up to 1981. Siegel's book covers multistage interconnection networks [555]. Networks for multiprocessors and multicomputers are discussed in Varma and Raghavendra [624]. Kung et al. characterized network-based multicomputers [386]. Recent interconnection networks are covered in the book by Duato, Yalamanchili, and Ni [214]. Hwang assessed Gigabit networks for building scalable systems and clusters [328].

Three special issues on hot interconnects appeared in *IEEE Micro Magazine* in February 1995, February 1996, and January 1997. Myrinet was described in Boden et al. [97]. Sun Gigaplane bus is described in Singhal et al. [561]. SGI POWERpath-2 is described in [549]. IBM Vulcan crossbar and SP2 switch design are discussed in [592]. Digital GIGAswitch is described along with Advantage clusters in [194].

The memory channel technology is described in Gillett and Kaufmann [268]. Applications of the perfect shuffle are given in Stone [588]. The Omega network was proposed by Lawrie [395] and built in the Cedar project [382]. Preparata and Vuillemin developed the CCC architecture [508]. The concept of systolic arrays was developed by Kung and Leiserson [385]. The chordal ring was treated in Arden and Lee [43]. The k-ary n-cube networks are studied in Dally [183]. Fat trees are introduced by Leiserson [403]. Hwang and Ghosh proposed the hypernet [331] for massively parallel computing.

Stallings surveyed recent advances in LAN, MAN, and ISDN in [579] and [580]. Kung assessed Gigabit LANs in [387]. ATM and cell switching technology are treated in the *Gigabit Networking* by Partridge [489]. Some of the network data are taken from the McGraw-Hill handbook on high-speed LANs by Saunders [530]. Johnsson revealed collective communication properties in network architectures [353]. Ni and McKinley have surveyed the wormhole routing techniques [465].

Information on Gigabit Ethernet can be found in [267] and in Roberts [518]. Fiber Channel is assessed in Frymoyer [257], FDDI ring in [582], HiPPI in Tolmie and Flanagan [613], and SCI in Bryhni and Wu [114]. The SCI protocol is specified in IEEE Standard 1596-1992 [343] and introduced by James et al. [348]. Myrinet technology and products are described in Boden et al [97].

Parallel fiber-optic SCI links are introduced by Engerbretsenh et al. [231]. The ATM Forum UNI 3.1 is specified in [46]. Zegura [664] covers the basic issues in designing ATM switched system. Various network standards and protocols can be found in publications from ANSI, IEEE, ATM Forum, and Gigabit Ethernet Alliance.

Homework Problems

Problem 6.1 Define the following basic terms for interconnection networks:

(a) Node degree
(b) Network diameter
(c) Bisection bandwidth
(d) Static connection networks
(e) Dynamic connection networks
(f) Nonblocking networks
(g) Multicast and broadcast
(h) Mesh versus torus
(i) Symmetry in networks
(j) Multistage networks
(k) Crossbar switches
(l) Cell switched networks

Problem 6.2 You are asked to design a direct network for a multicomputer with 64 nodes using a three-dimensional *torus*, a six-dimensional binary *hypercube*, and *cube-connected-cycles* (CCCs) with a minimum diameter.

The following address the relative merits of these network topologies. Your answers must be based on analytical reasoning.

(a)   Let d be the node degree, D the network diameter, and L the total number of links in a network. The quality of a network is measured by $(d \times D \times L)$ - 1. Rank the three architectures according to this quality measure.

(b)   A mean internode distance is the average number of hops (links) along the shortest path from one node to another. The average is calculated for all (source, destination) pairs. Order the three architectures based on their mean internode distances, assuming that the probability that a node will send a message to all other nodes with distance i is $(D - i + 1)$, where D is the network diameter.

Problem 6.3 Consider an Illiac mesh (8 × 8), a binary hypercube, and a barrel shifter, all with 64 nodes labeled N0, N1, . . . , N63. All network links are bidirectional.

(a)   List all the nodes reachable form node N0 in exactly three steps for each of the three networks.

(b)   Indicate in each case the tightest upper bound on the minimum number of routing steps needed to send data from any node Ni to another node Nj.

(c)   Repeat part (b) for a larger network with 1024 nodes.

Problem 6.4 Compare buses, crossbar switches, and multistage networks for building a multiprocessor system with n processors and m shared-memory modules. Assume a word length of w b and 2×2 switches are used in building a multiprocessor system with n processors and m shared-memory modules.

The comparison study is carried out separately in each of the following four categories:

(a)   Hardware complexities such as switching, arbitration, wires, connector, or cable requirements.

(b)   Minimum latency in unit data transfer between processor and main memory.

(c)   Bandwidth range available to each processor.

(d)   Communication capabilities such as permutations, data broadcast, blocking handling, etc.

Problem 6.5 Answer the following questions related to multistage networks:

(a)   How many legitimate states are there in a 4×4 switch module, including both broadcast and permutations? Justify your answer with reasoning.

(b)   Construct a 64-input Omega network using 4×4 switch modules in multiple stages. How many permutations can be implemented directly in a single pass through the network without blocking?

(c)   What are the number of one-pass permutations compared and the total number of permutations achievable in one or more passes through the network?

Problem 6.6 Answer the following questions for k-ary n-cube network:

(a)   How many nodes are there?

(b)   What is the network diameter?

(c)   What is the bisection bandwidth?

(d)   What is the node degree?

(e)   Explain the graph-theoretic relationship among k-ary n-cube networks and rings, meshes, tori, binary n-cubes, and Omega networks.

(f)   Explain the difference between a conventional torus and folded torus.

(g)   Under the assumption of constant wire bisection, why do low-dimension networks (tori) have lower latency and higher hot-spot throughput than high-dimension networks (hypercubes)?

Problem 6.7   This is an example of a design specification of a backplane bus for a shared-memory multiprocessor with 4 processor boards and 16 memory boards under the following assumptions:

- Bus clock rate = 20 MHz.
- Memory word length = 64 b; processors request data in block of four words.
- Memory access time = 100 ns.
- Shared address space = 240 words.
- Maximum number of signal lines available on the backplane is 96.
- Synchronous timing protocol.
- Neglect buffer and propagation delays.

(a)   Maximum bus bandwidth

(b)   Effective bus bandwidth (worst case)

(c)   Arbitration scheme

(d)   Name and functionality of each of the signal lines

(e)   Number of slots required on the backplane

Problem 6.8   Consider the use of 8 × 8 crossbar switches to build a 512-input multistage Omega network. How many stages are required in this network? How many crossbar switches are needed to construct the network? If the network is expanded to have 4096 nodes, how many additional 8 × 8 crossbar switches are needed ?

Problem 6.9   Answer the following two questions on shared-media, packet switched, ATM switched networks, giving qualitative and quantitative justifications.

(a)   What are the differences between shared-media networks and switched networks? Suggest two example network technologies in each network class and explain their advantages and disadvantages.

(b)   What are the major differences between an ATM switched network and the conventional packet switched network? Comment on the advantages and disadvantages of ATM switched networks.

Problem 6.10   Fill in the following table entries with architectural descriptions, speed ranges, and cabling distances for the listed network technologies.

Note that your entry could be a range, depending on the time frame within which your assessment is made..

Network Technology	Network Architecture	Bandwidth (Speed Range)	Distance over Copper	Distance over Optic Fibers
Gigabit Ethernet				
Myrinet				
HiPPI				
SCI				
Fibre Channel				
ATM				
FDDI				

Problem 6.11 The SCI technology is becoming a popular choice to build scalable, shared-memory multiprocessor. Answer the following specific questions on SCI:

(a) Explain the cache coherence protocol used in IEEE SCI standard to achieve *distributed shared memory* (DSM) among processing nodes interconnected by SCI links.

(b) Why are the SCI connections more scalable than most shared-media networks employed to build DSM multiprocessors?

(c) Explain how to use SCI links in the construction of a CC-NUMA multiprocessor systems.

(d) What are today's limits and shortcomings in using the SCI interconnect to build scalable parallel computers?

Problem 6.12 The Cedar multiprocessor at Illinois is built with a clustered Omega network as shown in Fig. 6.41.

Four 8×4 crossbar switches are used in stage 1 and four 4×8 crossbar switches are used in stage 2. There are 32 processors and 32 memory modules, divided into four clusters with eight processors per cluster

(a) Figure out a fixed priority scheme to avoid conflicts in using the crossbar switches for nonblocking connections. For simplicity, consider only the forward connections from the processors to the memory modules.

(b) Suppose both stages use 8×8 crossbar switches. Design a two-stage Cedar network to provide switched connections between 64 processors and 64 memory modules

(c) Further expand the Cedar network to three stages using 8×8 crossbar switches as building blocks to connect 512 processors and 512 memory modules. Show all connections between adjacent stages from the input end to output end.

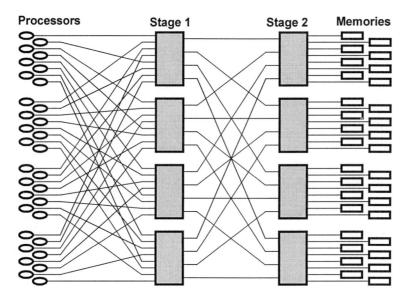

Figure 6.41   The Omega network in the Cedar multiprocessor.

Chapter 7

Threading, Synchronization, and Communication

This chapter discusses various hardware and software mechanisms to support process/thread management, synchronization, and communication. These mechanisms are used in various multiprocessor and multicomputer platforms to be studied in Part III. With the advent of VLSI technology, some of these mechanisms are even incorporated in single-processor systems.

We first discuss in Section 7.1 the *thread* concept, as opposed to the process concept discussed in Chapter 2. Section 7.2 discusses various synchronization mechanisms. Section 7.3 and Section 7.4 deal with communication. We first discuss the popular TCP/IP standard, then present a number of more efficient communication techniques.

7.1 Software Multithreading

It should be made clear: *hardware multithreading* means the use of multiple-context processors to hide memory latency as discussed in Section 5.6. *Software multithreading* is a very different concept. Two important advances should be observed concerning modern processor architecture and operating system.

- First, current RISC processors are designed to have high integer and floating-point computing performance, but have not paid equal attention to operating system supports.
- Second, more architectural supports are required for the added functionality and modularity of the operating systems in scalable computing platforms.

These advances even apply to the Intel 80x86 and Pentium processors, which dominate today's PC market. These processors have a CISC origin, and their architectures have incorporated many RISC ideas. The gap between processor architecture and operating system requirements was discussed by Anderson et al. [36]. Some results of their study are tabulated below.

Table 7.1 Operating System Support of Processor Architectures (Courtesy: Anderson et al. *Proc. 4th Int'l Conf. Archiitectural Support for Programming Languages and Operating Systems*, 1991 [36])

Attribute		Absolute Value			Relative to CVAX	
		DEC CVAX	MIPS R2000	Sun SPARC	MIPS R2000	Sun SPARC
OS function performance (μs)	Null sys call	15.8	9	15.2	1.8	1
	Trap	23.1	15.4	17.1	1.5	1.4
	Page table entry change	8.8	3.1	2.7	2.8	3.3
	Context switch	28.3	14.8	53.9	1.9	0.5
OS function instruction count	Null sys call	12	84	128	7	10.6
	Trap	14	103	145	7.4	10.3
	Page table entry change	11	36	15	3.3	1.4
	Context switch	9	135	326	15	33.2
Number of 32-bit words in processor state	Registers	16	32	136	2	8.5
	F.P. state	0	32	32	-	-
	Misc. state	1	5	6	5	6

Although R2000 and SPARC improved the computing speed (measured by the SPECmark benchmark) of VAX more than 4 times, the speed for operating systems functions has improved far less. This gap between processor architecture and operating system will keep widening, unless attention is paid by both processor and operating system designers to address it. A key innovation in recent operating system research is *threads*. In this section, we discuss how threads can be implemented efficiently. How to *use* threads to write multithreaded programs will be discussed in Chapter 12.

7.1.1 The Thread Concept

We use a simplified model based on *Solaris threads* [583], a widely used commercial product that has a lot in common with the IEEE POSIX Threads (Pthreads) [345]. A *thread* (or *thread of control*) is a sequence of instructions being executed. The thread concept is illustrated in Fig. 7.1.

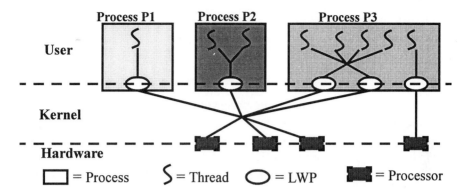

Figure 7.1   Multithreading in the Sun Solaris operating system.

The main feature of the Solaris threads model is a *two-level* architecture. A thread that works in the user mode (user space) is called a *user-level thread* (or *user thread*, or simply *thread*). A thread that works in the kernel mode (kernel space) is called a *kernel thread*, or *lightweight process* (LWP).[1] A user-level, dynamically linked *thread library* implements the threads application programming interface and manages user threads, while the kernel manages LWPs.

Each process has one or more threads and is allocated one or more LWPs. One can think of an LWP as a virtual CPU. The number of LWPs allocated to a process is called the *concurrency level* of the process. The thread library schedules the threads of a process to run on its pool of LWPs, just as the kernel schedules the LWPs from all processes to run on the pool of physical processors. Threads are not visible outside their process. For instance, the number and the identities of threads in a process are unknown to the kernel or another process. The kernel cannot manage user-level threads, only LWPs.

A thread that is scheduled to run on the pool of LWPs within a process is called an *unbound thread*. A thread that is permanently bound to an LWP is called a *bound thread*. From a programmer's viewpoint, a bound thread is just like an unbound thread. But from the operating system's viewpoint, a bound thread behaves as an LWP. For instance, while unbound threads are scheduled within a process, a bound thread is scheduled with respect to all active threads, a useful feature for real-time applications.

A thread can execute any application code and system calls just as a process can. The threads of a process share its address space: its code, most of its data, and most of its

---

1. The kernel threads and LWPs are two different concepts. For simplicity, we regard LWPs as kernel threads that are used to execute user processes. Kernel threads are used for other purposes, such as handling interrupts. Unless otherwise stated, we will ignore the differences between kernel threads and LWPs in this book.

process descriptor information. Threads execute independently and asynchronously with one another. When one thread is blocked (e.g., waiting for an I/O operation to finish), other threads can still proceed. Multiple threads of a process can execute on multiple processors at the same time.

Example 7.1 Concepts of process, thread, LWP, and processor

The system in Fig. 7.1 has three processes running on four processors. Process P1 is a traditional Unix process with a single thread and a single LWP. Process P2 has two threads running on a single LWP. Process P3 has five threads running on three LWPs. The concurrency levels of P1, P2, and P3 are 1, 1, and 3, respectively. The thread in P1 and the rightmost thread in P3 are bound threads. All others are unbound.

■

Thread operations are less expensive than process operations because they are implemented by the thread library in the user space, and threads in a process do not have separate address spaces (lightweight). Therefore, thread operations such as creation, termination, and synchronization are all done in the same user space, without entering the kernel. The overheads incurred for process operations due to copying, context switching, and crossing protection boundaries, are significantly reduced.

7.1.2 Threads Management

Thread Creation Creating a thread is much simpler than creating a process, as there is no need to create a new address space. Only the following state information needs to be maintained for a specific thread:

- The thread ID
- The processor state (the registers, including the program counter and the stack pointers)
- A thread specific stack
- The priority of the thread, which is used in thread scheduling
- A thread local storage area for holding the thread-specific data
- A thread specific signal mask

This information is created and maintained by the thread library in a data structure called the *thread structure*. The memory area for the stack is either automatically allocated by the thread library (in the heap area of the process's address space) or passed in by the application code at the creation of the thread.

Variables in a process are shared by all threads of the process. Thus if one thread modifies a variable, the effect is visible to all threads of the process. Each thread also has a *thread local storage* that is used to hold variables private to this particular thread, which cannot be directly accessed by other threads. The thread structure itself is also allocated in the thread local storage.

Thread Scheduling The thread library provides a thread scheduler that multiplexes threads to execute on the pool of LWPs. Each thread is in a certain state at any time, as shown in Fig. 7.2. A newly created thread starts at the RUNNABLE state, waiting for an idle LWP to become available. Then the library dispatches it to run on the LWP, and it enters the ACTIVE state.

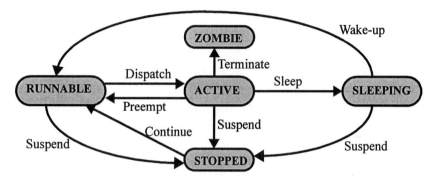

Figure 7.2 **Thread state transition diagram controlled by the thread library.**

When an ACTIVE thread is blocked on a process-local synchronization object (i.e., one that is not shared among processes), the LWP puts it into the SLEEPING state and the LWP switches to execute the next RUNNABLE thread with the highest priority. However, if the thread is a bound thread, its blocking will block the underlying LWP as well. A sleeping thread is woken up when the synchronization object becomes available, i.e., when the blocking condition is removed.

Solaris threads support *preemption*. A higher-priority thread can preempt an ACTIVE thread and take its LWP. The programmer can suspend a thread and transfer it to the STOPPED state using library functions such as thr_suspend(). Such a thread is alive but cannot become RUNNABLE until a thr_continue() is called. Finally, when a thread terminates (when the thread code ends or when a thr_exit() is executed), it enters the ZOMBIE state. The thread library has a reaper thread that "buries" these ZOMBIE threads by freeing up their memory space (stacks, etc.).

An ACTIVE thread is one that is already dispatched to run on an LWP. We say the thread is *attached* to the underlying LWP. Each LWP has at most one attached thread at any given time. The underlying LWP of an ACTIVE thread can be in one of four states at any time, as shown in Fig. 7.3, which can be considered a detailed description of the thread ACTIVE state. Note that the LWP is always in the RUNNING state when the thread transfers from/to another thread state.

LWPs are scheduled by the kernel. A RUNNING LWP transfers to the RUNNABLE state when being preempted or timed out by the kernel. It can block by executing a blocking system call, which forces the LWP to wait until a certain condition becomes true. An LWP can be suspended by an LWP library function call to enter the STOPPED state, and can return to the previous state by calling a continue library function. When a thread terminates and there is no more RUNNABLE thread, the underlying LWP enters an *idle* state by waiting for a global LWP condition variable. When a thread becomes RUNNABLE, the global variable is signaled to wake up an idling LWP.

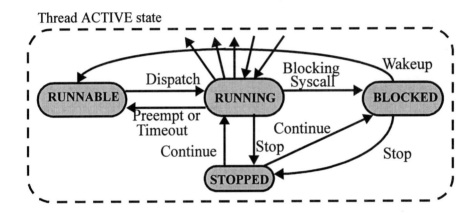

Figure 7.3   LWP state transition diagram controlled by the kernel.

7.1.3 Thread Synchronization

The Solaris threads library implements two types of synchronization objects: *process-local* and *process-shared*. A *process-local* synchronization object is accessible to the threads within the same process, and it is used to synchronize threads of the same process. The thread library maintains a sleep queue for each synchronization variable. When a thread executes a synchronization operation on a synchronization variable and blocks, it is put to sleep in the queue of that variable.

If the thread is unbound, the underlying LWP is switched to execute the next RUNNABLE thread. If the thread is a bound thread, the underlying LWP also blocks by waiting on an LWP semaphore associated with the bound thread. We say the LWP is *parked* on the thread. When the blocking variable becomes available (the blocking condition disappears), the blocked thread is woken up and put into the RUNNABLE state, to be scheduled by the library. If the woken-up thread is a bound thread, its LWP will be unparked and scheduled by the kernel.

A *process-shared* synchronization object is accessible by multiple processes, and it can be used to synchronize threads in different processes. Process-shared synchronization primitives rely on LWP synchronization primitives to put blocking threads to sleep in the kernel while still attached to their LWPs. Such objects must be initialized so that synchronization primitives can recognize them as process-shared.

7.2 Synchronization Mechanisms

Synchronization is a rich concept and an active research field that spans many computer areas. We have found students puzzled by why they have to study the seemingly similar material of synchronization in multiple courses, such as computer architecture, database, operating system, parallel processing, and programming languages. Four aspects must be understood when one is studying synchronization:

(1) The synchronization problems faced by the user, such as *mutual exclusion, atomicity, producer-consumer, dining philosophers, barrier synchronization,* etc. Andrews [37] is a good reference for various synchronization problems. The mutual exclusion and the atomicity problems are especially important, and will be studied in Section 7.2.1.

(2) The language constructs employed by the user to solve the synchronization problems. The constructs can be in the form of intrinsic constructs, library routines, or compiler directives. Popular constructs in current shared-memory programming environments include locks, semaphores, events, critical regions, barriers, etc. These constructs are called *high-level constructs*, and will be discussed in detail in Section 7.2.2 and in Chapter 12.

(3) The synchronization primitives available in the multiprocessor architecture, such as test and set, fetch and add, compare and swap, etc. These are often directly provided by the system hardware, thus called *low-level constructs*. They will be studied in Section 7.2.3.

(4) The algorithms used to implement the high-level constructs with the low-level ones. In many current systems, locks are still the fundamental construct, upon which other constructs are built. Therefore, we will discuss in subdequent sections how to build efficient locks using low-level primitives.

7.2.1 Atomicity versus Mutual Exclusion

As already mentioned in Section 2.4.1, there are three types of synchronization operations: *atomicity*, *data synchronization*, and *control synchronization*. A more detailed hierarchy is shown in Fig. 7.4.

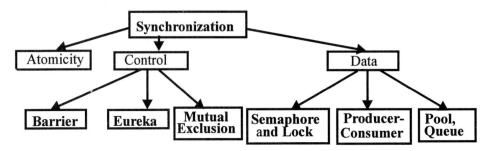

Figure 7.4   Various types of synchronization.

The native programming models of current multiprocessors support barrier, mutual exclusion, lock/semaphore, and producer-consumer synchronizations. They do not adequately support atomicity and eureka synchronizations.

Informally, barrier synchronization ensures a group of processes all reach a certain point in their execution. Eureka synchronization is the opposite of barrier: When one process reaches a certain point, it immediately informs all other processes asynchronously (Eureka! I got it!). This is useful in applications such as parallel search.

Synchronization of a pool requires that an item not be fetched from an empty pool or put into a full pool. Synchronization for a queue is similar, except that there is some order (e.g., FIFO queue, priority queue). All synchronizations are implemented usually by some locking primitives on shared-memory machines (PVPs, SMPs. and DSMs) and by message-passing primitives on MPPs and clusters.

Mutual exclusion and atomicity are two related, but different concepts. Mutual exclusion simply says that two things cannot be in the same place at the same time. Atomicity says that a sequence of operations should be executed as a single, finite, indivisible step.

The Mutual Exclusion Problem   The mutual exclusion problem can be more formally specified using the following code:

```
parfor (i=0; i<n; i++)              // there are n processes
{                                   // each executing a parfor iteration
    while ( Condition ) {
        independent computation, noncritical section
        critical {                  // entry point of mutually exclusive code
            critical section
        }                           // exit point of mutually exclusive code
        independent computation, noncritical section
    }
}
```

A *critical section* is a piece of code that *should* be executed in mutually exclusive fashion. *Condition* is a logic expression that may be modified by the critical section. The keyword **critical** specifies a *critical region*, a structured language construct that *guarantees* the code enclosed in the parentheses { and } (the critical section) is executed mutually exclusively. A solution to the mutual exclusion problem must satisfy the following three properties:

- *Mutual Exclusion*: Only one process is allowed to be executing the critical section at a time.

- *Guaranteed Progress*: If multiple processes are trying to enter the critical region to execute their critical section, at least one will eventually succeed. In other words, the program must not deadlock or livelock.

- *Starvation Freedom*: A process attempting to execute its critical section should eventually succeed. It cannot be starved by the conspiracy of other processes.

A good solution should also be *efficient*. For instance, the performance of the mutual exclusion algorithm should not depend on n, the total number of processes, but only on m, the number of processes that are competing to execute the critical section. The value $n - m$ is the number of *noncompeting* processes, which are executing their noncritical code or have terminated.

As another example, if no process is executing the critical section, a process attempting to enter the critical region should be allowed to do so immediately. It should not be delayed by the other noncompeting processes.

Lamport's Pets Problem   A classic presentation of the mutual exclusion problem was suggested by Lamport [390]:

> *Alice and Bob are neighbors sharing a yard. Alice has a pet cat and Bob has a pet dog. They want their pets to use the yard. But because the two animals don't like each other, only one pet at a time can be in the yard. Therefore, Alice and Bob must find a way to allow their pets to use the yard while avoiding fighting.*

This is a two-process problem, which can be easily generated to n processes. Mutual exclusion says that only one pet can be in the yard at a time. Guaranteed progress says that *some* pet will be able to use the yard. Starvation freedom says that no pet will wait forever to use the yard.

Atomicity   The atomic operation concept can be defined formally using a state machine model. In imperative programming, a program can be considered a state machine (or an *automaton*). A program starts with an initial state. It then goes through a sequence of indivisible steps, each transforming a current state into a next state. Such an indivisible state transformation step is called an *atomic operation*. More formally, we give the following definition:

Definition 7.1 An *atomic operation* in a program is a sequence of instructions that has the following three properties:

- *Finite*: The time is finite from the start of an atomic operation to its finish.
- *Indivisible*: From the programmer's viewpoint, an atomic operation is executed as a single, indivisible step. Only the final result of the atomic operation may be visible to the rest of the program. Any intermediate results generated by the atomic operation are not visible. If for some reason the atomic operation aborts in the middle, it must undo all partial effects and *roll back* to the starting state.
- *Serializable*: A corollary of indivisibility is *serializability*, meaning that when several atomic operations are executed in parallel, the end result should be as if these operations were executed one after another in some arbitrary order.

■

In this definition, atomic operations are defined from the programmer's viewpoint, and the unit of activity is the instruction. A computer architect may want to use atomicity at more detailed levels, such at bus cycles or clock cycles. Usually, the architecture and the programming environment of a computer system will provide some basic atomic operations, such as an arithmetic/logic operation, memory read, memory write, test-and-set, compare-and-swap, fetch-and-add, and so on.

The *atomicity problem* is how to make a single atomic operation out of a sequence of such basic operations. Such a larger-grain atomic sequence is also called a *transaction*. A solution to the atomicity problem must ensure atomicity, guaranteed progress, and starvation freedom. The atomicity concept is further elaborated using the following banking problem.

The Banking Problem A big company has three accounts X, Y, and Z with a bank. These accounts can be accessed by any number of company employees concurrently through three atomic operations, as exemplified in Fig. 7.5.

During any period of time many employees may be accessing these accounts simultaneously. We want these transactions to be safe and efficient. For simplicity, we will ignore error-handling codes needed in withdrawing and transfers when account X does not have sufficient funds. The codes in Fig. 7.5 use a C-like notation, with a new construct added to specify atomicity.

This *atomic region* construct "**atomic** { region body }" starts with a reserved word **atomic** and states that all operations in the region body should be executed as a single atomic operation. Atomic region is a special form of a more powerful structured construct, called *coherent region*, which is discussed in Section 12.3.5.

Atomicity and Mutual Exclusion Atomic operations and mutually exclusive operations have the following differences:

```
                Withdraw $100 from account X:
                    atomic {
                        if ( balance[X] > 100 ) balance[X] = balance[X] – 100 ;
                    }


                Deposit $100 to account Y:
                    atomic { balance[Y] = balance[Y] + 100 ; }


                Transfer $100 from account X to account Z:
                    atomic {
                        if ( balance[X] > 100 ) {
                            balance[X] = balance[X] – 100 ;
                            balance[Z] = balance[Z] + 100 ;
                        }
                    }
```

Figure 7.5   Three transactions in the banking problem.

- An atomic operation is finite, but finiteness is not an inherent property of a mutually exclusive operation, such as a critical section. This issue is ignored in existing parallel languages. It is possible for a critical section to contain an infinite loop not detectable by the compiler or run-time support system.
- Mutual exclusion can guarantee indivisibility. However, indivisibility can be realized by other, non-mutually exclusive means.
- Mutual exclusion forces sequential execution, but atomicity permits parallel execution. In the mutual exclusion problem, only one of the n processes can execute the critical section at a time. In the banking problem, multiple transactions can be executed simultaneously, as long as atomicity is enforced.

Atomicity is traditionally realized by mutual exclusion. Since the introduction of the semaphore construct by Dijkstra, locking has become the predominant technique to support mutual exclusion, thus atomic operations. The mutually exclusive locking solution to atomicity is very natural in the setting of operating systems, where multiple, concurrently executing processes *compete* for shared resources.

A main concern there is to ensure safe resource sharing. For instance, suppose two processes each want to print a file on the same printer. These two printing operations must be synchronized to avoid a printout with mixed lines from both files. It is natural for a process to first lock the printer, allowing itself mutually exclusive access, and unlock the printer only after it has finished the printing operation.

Things are different in parallel computing, where multiple processes *cooperate* in computing a function. Resource sharing is not a main concern for the programmer, and mutual exclusion is not a natural way to realize atomicity anymore. In fact, mutual exclusion is counter to the objective of parallel computing, because it forces sequential execution. However, current multiprocessors still use mutual exclusion constructs such as locks and critical regions as the basic support for atomic operations.

Mutual exclusion is not the only or the best way to achieve atomicity in parallel computing applications. New parallel language constructs and architecture supports are being investigated to realize atomicity more efficiently and safely. However, we must first break the mind-set that atomicity can only be modeled as a critical section problem, which requires mutual exclusion, thus requires locking.

Example 7.2 Lamport's pets problem revisited

To illustrate how easy it is to be trapped in a mind-set, let us revisit Lamport's pets problem. When a computer scientist is asked to solve this problem, it is very likely that a locking solution such as using a pair of **lock** and **unlock** will be offered.

- *Alice and Bob can work out a schedule as to which pet can use the yard at what time.* Indeed, in synchronous parallel computers such as SIMD and PRAM, the mutual exclusion problem disappears.
- *Alice or Bob can move to another neighborhood.* This amounts to modification of the algorithm of the parallel program, so that a shared variable is not shared any more.
- *Alice and Bob can build a fence in the yard so that each pet can have its own half of the yard.* This solution is widely used in parallel programming. The idea is to partition a shared data structure into smaller pieces, so that each process can work on its own piece of data without contention.
- *Alice and Bob can raise a balloon to tell the neighbor that her or his pet wants to use the yard. The pet can go to the yard if the neighbor's balloon is not already raised.* This is similar to Lamport's solution using flags [390].
- *A pet can go to the yard anytime. However, if it finds the yard occupied, it will go home and try later.* This idea appears in *lock-free* synchronization techniques [306]. This solution has a variant: Let the two animals fight briefly and the loser will be chased away. In grown-ups' terminology, this amounts to allowing priority and preempting.

Figure 7.6   Mona's solutions to Lamport's pets problem.

Now let us see how a child would solve the same problem. Xu's 10-year-old daughter Mona was asked to solve the above problem. Mona proposed 12 solutions. Some solutions are not really valid, such as letting the pets wear bells while in the yard, using walkie-talkies, etc. Mona did give five valid solutions, which are summarized in Fig. 7.6.

It is interesting that although the problem was presented to Mona deliberately as a mutual exclusion problem (not an atomicity problem), Mona did not suggest a mutually exclusive solution using lock or critical region. Her fourth solution, though, is in that direction. We also note that it is easy to explain the concept of mutual exclusion to a child: Two things cannot be in the same place at the same time. However, it is difficult to explain the atomicity concept, or the concept of atomic operations.

■

7.2.2 High-Level Synchronization Constructs

Parallel programming environments of current shared-memory multiprocessors provide four types of synchronization primitives: event, barrier, lock/semaphore, and critical region. Event operations are used to realize producer-consumer synchronization. Barrier is used in barrier synchronization. Locks and critical regions are mainly used to realize atomicity in a mutually exclusive fashion. These last two constructs will be evaluated in this section.

Semaphores and Locks A *semaphore* S is a nonnegative integer variable that can be manipulated by only two atomic operations: **P**(S) and **V**(S).

- The **P**(S) operation is used to delay a process until S becomes greater than 0. Then it decrements S by 1.
- The **V**(S) operation simply increments S by 1.

A *binary semaphore* S (i.e., S is either True or False) is also called a *lock*. The **P**(S) and the **V**(S) operations for a lock S are often written as **lock**(S) and **unlock**(S), respectively. A common use of lock is to turn a critical section into an atomic operation through mutual exclusion, as demonstrated by the following example.

Example 7.3 The use of lock and unlock operators

We use a single lock S in all transactions. A process must first *acquire* the lock S by executing **lock**(S), before it can do any withdraw, deposit, or account transfer, and it must *release* the lock by executing **unlock**(S) after finishing the transaction. A process is *holding* a lock if it has acquired the lock but has not released it yet. An account transfer transaction can be realized by the code below:

```
// Account Transfer -- Version 1
    lock(S) ;                          // Entry code
    if ( balance[X] > 100 ) {          // critical section
        balance[X] = balance[X] – 100 ;
        balance[Z] = balance[Z] + 100 ;
    }
    unlock(S) ;                        // Exit code
```

■

Problems of Locking   The main advantage of locking is that it is already supported by most multiprocessors. Much research has been done on efficient locking, as discussed in Section 7.2.4. Locking is a very flexible mechanism that can be used to realize almost any synchronization. However, mutually exclusive locking techniques have some serious drawbacks when used to realize atomicity, which may create the following problems.

(1) *Unstructuredness*: Locks are not structured constructs, and using them is error-prone. The compiler cannot check if a lock/unlock statement is missing or redundant, since the erroneous code is still legal. It is easy to write code with *nonserializable* and *deadlock* problems, as discussed below.

(2) *Overspecification*: Locking is not really what the user wanted. The user cares only that a transaction is done atomically. Locking is only *one* way to achieve atomicity. This over-specification excludes possibly more efficient *lock-free* techniques discussed in Section 7.2.3. Overspecification with one particular implementation also hurts portability, and could make the code difficult to understand.

(3) *State Dependence*: A transaction is just an unconditional atomic operation. But the locking solution needs to introduce a semaphore S and use a conditional atomic operation **lock**(S). Whether a process can pass **lock**(S) or not depends on the value of the semaphore variable S. In general, such data state-dependent synchronizations are more difficult to understand than state-independent ones. Although this is not really a problem in the simple account transfer code, there are a lot of more complex cases where locking will cause a problem.

(4) *Sequential Execution*: Suppose n processes each want to execute a transaction. By the mutually exclusive nature of the locking solution, these n transactions must be executed one at a time, even if they are accessing different accounts. Again, this sequential execution is not what the user wants. An often suggested idea to increase parallelism is to use a separate semaphore for each account. However, as we will see shortly, this idea does not always work.

(5) *Locking overhead*: There is also the additional overhead of sequentially executing $O(n)$ **lock**(S) and **unlock**(S) operations. Furthermore, when n processes each

execute a **lock**(S) operation, at most one of them can succeed (finding S=True). The others have to retry by repeatedly accessing S. This could generate a lot of memory traffic. Competing for shared variable S may cause the *hot-spot* problem [499] to occur, which further degrades performance.

(6) *Priority Inversion*: This problem occurs when a low-priority process is preempted while holding a lock needed by a higher-priority process. The latter cannot proceed since it is blocked by the lock [307].

(7) *Convoying Blocking*: When a process holding a lock is interrupted by page fault or timing out, other processes waiting for the lock cannot proceed [306].

(8) *Nonserializability*: In Example 7.3, no other process can access any account while an account transfer transaction is in progress. An intuitively better solution would be to use a separate lock for each account. The idea is that when one account is locked by a process, other accounts can still be accessed by other processes.

The following code is based on this idea:

```
Account Transfer -- Version 2
    lock(S[X]) ;                    // Lock (the semaphore for) account X
    if ( balance[X] > 100 ) {
        balance[X] = balance[X] - 100 ;
        unlock(S[X]) ;              // Finished using account X
        lock(S[Y]) ;               // Lock (the semaphore for) account Y
        balance[Y] = balance[Y] + 100 ;
        unlock(S[Y]) ;             // Finished using account Y
    }
    else
        unlock(S[X]) ;             // Make sure account X is unlocked
```

It is easy to forget the **else** branch, which will cause an error that cannot be detected by the compiler. Then when there is not sufficient fund in account X, the true branch of the **if** statement, including **unlock**(S[X]), will not be executed. Therefore, account X will be locked forever, preventing any process from accessing it in the future.

This code, even after we ensured that account X is always unlocked, still does not work. This time, the bug is not so obvious. The problem is that the above code is not always *serializable*, meaning if several transactions are executed concurrently, the result will not necessarily be as if these transactions were executed one after another in some arbitrary order. Nonserializability violates atomicity.

(9) *Deadlock*: There are locking schemes that ensure serializability, one of which is known as the *two-phase locking* protocol, originally developed for concurrency control in database systems. With this scheme, all **lock** operations must be executed before the first **unlock** in a transaction. Shown below is a code using two-phase locking:

Account Transfer -- Version 3

```
lock(S[X]) ;                    // Lock (the semaphore for) account X
if ( balance[X] > 100 ) {
    balance[X] = balance[X] – 100 ;
    lock(S[Y]) ;                // Lock (the semaphore for) account Y
    balance[Y] = balance[Y] + 100 ;
    unlock(S[Y]) ;              // finished using account Y
}
unlock(S[X]) ;                  // finished using account X
```

This code is serializable and slightly simpler than Version 2. However, it limits concurrency: When one process has finished processing account X and is processing account Y, account X is still locked. Therefore, although account X is not needed anymore by this process, no other processes can use it. Of course, Version 3 still admits more concurrency than Version 1, which does not allow any concurrency at all.

A far more serious problem of the Version 3 code is the possibility of *deadlock*. Suppose while process *P* is transferring $100 from account X to account Y, another process *Q* is transferring $50 from account Y to account X. Since *P* and *Q* are executing in parallel, we can get into a no-win situation: While *P* holds the lock for X and is trying to acquire the lock for Y, *Q* holds the lock for Y and is trying to acquire the lock for X. According to the protocol of two-phase locking, neither process will release a lock before it gets all the locks. Therefore none of them can get the desired lock. They are deadlocked with *P* waiting at **lock**(S[Y]) and *Q* at **lock**(S[X]).

There are several ways to deal with such deadlocks, but none of them is attractive for parallel programming. For instance, with the *deadlock detection* approach, the run-time support system can monitor the execution of lock/unlock operations and construct a *wait-for graph*. A node is created for each transaction that is executing. When transaction *P* executes **lock**(S) to wait for a lock S that is currently held by transaction *Q*, an arc is drawn from *P* to *Q*. A deadlock occurs if the wait-for graph has a cycle.

Once a deadlock is detected, some transaction in the cycle can be aborted, and the program state is *rolled back* to undo the effect of that transaction. This approach might be efficient for database concurrency control when the transaction load is light. But for parallel computing programs, the run-time overhead of creating the wait-for graph and of aborting and rollback is likely to be too great.

Another approach is *deadlock prevention*. That is, the program is written in such a manner that processes will never enter a state of circular wait. One way to achieve this is to order all the locks (e.g., using lexical order of semaphore names), and to ensure that any transaction needing several locks will always acquire them in that order. However, this scheme further limits concurrency and puts additional burden on the programmer. The resultant code may be unduly complicated.

A number of constructs have been derived to overcome these problems, as discussed below. We compare these constructs in Table 7.2 from a user's viewpoint. As far as the user is concerned, the best atomicity construct is atomic region, and the best mutual exclusion construct is critical region.

Table 7.2 Comparison of Constructs for Multivariable Transactions

Construct Problem	Semaphore or Lock	Critical Region	Test & Set	Compare and Swap	Transaction Memory	Fetch &Add
Unstructured	XXX		XXX	XXX	XXX	XXX
Overspecified	XXX	XX	XXX	XXX	XXX	XXX
State dependent	XXX		XXX	XXX	XXX	XXX
Sequential execution	XXX	XXX	XXX	X	XXX	
Overhead	XXX	XXX	XX	X	X	
Priority inversion	XXX	XXX	XXX	X	X	
Convoying blocking	XXX	XXX	XXX	X	X	
Nonserializable	XX		XX	XX	X	XXX
Deadlock	XX		XX	XX	X	XXX

XXX: Severe, or extremely difficult to overcome; XX: Less severe, or can be alleviated if user carefully writes the code; X: Slight, or can be eliminated if user follows certain well-defined rules; (blank): Not a problem, taken care of by the system.

Critical Region A structured construct to ensure the mutual exclusion of a critical section is *critical region*, which has a syntax similar to the following:

```
critical_region    resource
{                               // entry point
    S1; S2; ... Sn;                // critical section
}                               // exit point
```

where *resource* denotes a set of shared variables. The idea is that all critical regions sharing the same resource must be executed mutually exclusively. This requirement will be enforced by the system (compiler and run-time support). Again, this construct was originally proposed for operating system applications. Two modifications are made when it is used for parallel programming. First, the resource part is not really useful and thus is dropped. The critical region construct used in real multiprocessors has the following syntax:

```
critical_region                     Equivalent locking code
{                                   lock(S)
    S1; S2; ... Sn;                 S1; S2; ... Sn;
}                                   unlock(S)
```

Second, critical region in those multiprocessors is meant to be a structured way of locking, as shown above. The system would automatically declare and initialize an implicit semaphore *S* and generate the correct lock/unlock statements.

Critical region has many advantages over locking. It is structured and state-independent, therefore easier to use. Transactions using critical regions are serializable and deadlock-free. It is less overspecifying and less architecture-dependent than locking. A critical region just says that a piece of code will be executed mutually exclusively. It does not say that locking must be used. Although locking is current the most popular technique used to realize mutual exclusion, there are other ways that could be better.

7.2.3 Low-Level Synchronization Primitives

The hardware of many multiprocessors guarantee the atomicity of individual read or write operations on an elementary variable. In addition, most multiprocessor hardware provides some atomicity instructions, each of which implements a single read-modify-write operation on an elementary variable. Such instructions can be used together to realize larger-grain atomic operations. We discuss below four low-level constructs and show how they can be used to realize locks, in the order of increasing complexity.

Test-and-Set   A test-and-set instruction, denoted by TAS(S, temp), is an atomic operation that reads the value of a shared variable S into a local variable temp, and then sets S to True. A major use of test-and-set is to implement locks, as shown below:

```
while ( S ) ;                    // These three lines
TAS(S, temp) ;                   // implement a
while (temp) TAS( S, temp) ;     // lock(S) operation
if ( balance[X] > 100 ) {        // critical section
    balance[X] = balance[X] – 100 ;
            balance[Z] = balance[Z] + 100 ;
}
S = False ;                      // unlock(S)
```

Each execution of a TAS(S, temp) needs to write to the shared variable S. This could lead to heavy memory access traffic. The above implementation of **lock**(S) uses the operation *test-test-and-set*. The first **while** loop checks a copy of S in a local cache repeatedly. A process will leave the first **while** loop only when it detects that lock S has been released (unlocked) by another process.

Compare-and-Swap A compare-and-swap instruction, CAS(S, Old, New, Flag), is an atomic operation which compares the value of a shared variable S with the value of the local variable Old. If these two variables agree in value, the value of a local variable New is stored in S, and a local Flag is set to True indicating that S is modified. If the two values disagree, the variable Old gets the value of S, and the Flag is set to False. The intended use of compare-and-swap is demonstrated by the following account deposit code:

```
Old = balance[X] ;                          // read shared variable
do {
    New = Old - 100 ;                       // modify
    CAS( balance[X], Old, New, Flag ) ;     // write
} while ( Flag == False ) ;
```

As a comparison, the same account deposit operation can be realized by locking:

```
lock(S) ;
balance[X] = balance[X] – 100 ;             // read-modify-write
unlock(S) ;
```

The lock implementation makes the entire sequence (of read, modify, and write) mutually exclusive. However, the compare-and-swap implementation allows multiple processes to concurrently execute reading and modifying. Only the writing is done as a mutually exclusive operation (the CAS instruction). Thus an advantage of using compare-and-swap is that the length of a critical section can be reduced to just one instruction.

Compare-and-swap has several disadvantages as well. First, it is a low-level construct and difficult to use. Second, it is hard, though possible, to use compare-and-swap to implement a transaction that accesses more than one shared variable. That is why we used the account deposit example in this section. The compare-and-swap code to implement an account transfer transaction would be much more complex. The third disadvantage is the following *ABA problem*.

An implicit assumption regarding the CAS(S, Old, New, Flag) instruction is that when Old=S, S has not been modified. But this assumption does not always hold. For instance, suppose process P reads a value of A dollars in account[X]. Then process Q deposits \$200 to X, changing the balance to $B = A + 200$. Then process R withdraws \$200 from X, changing the balance back to A. So afterward when process P executes a compare-and-swap, it will find that account X still has A dollars and will conclude that the account has not been changed, while in reality it has been modified. This ABA problem does not affect the correctness of the simple account deposit code. However, in other cases, especially when the shared variable is a pointer, it may lead to incorrect results.

A solution to the ABA problem was proposed in [349], which was later implemented in many modern RISC processors. This technique provides two atomic instructions. A process P can execute a *read-and-reserve* instruction RAR(Local,S) to read the value of a shared variable S into a local register Local and the address of S into a local *reservation register*. This instruction also sets to 0 a 1-bit tag associated with the reservation register, indicating that S has not been modified by other processes.

Process *P* can then compute a new value New for S locally. Afterward, *P* updates S by executing a *write-if-reserved* instruction WIR(S, New, Flag). This instruction first checks if the tag is 0. If so, S is successfully updated with the New value. Otherwise the update fails. The Flag will be set to 1 if the update succeeds, and to 0 otherwise. If S is modified by another process before *P* executes the write-if-reserved instruction, the tag of the reservation register will be automatically set to 1 by hardware.

Transactional Memory and Oklahoma Update The reservation mechanism just discussed is suitable for the atomic updating of a single shared variable. A recent result is a multi-reservation scheme, which allows lock-free, atomic updating of multiple shared variables. This scheme was independently developed by Herlihy and Moss of DEC [307] and by Stone et al. of IBM [589].

The DEC version is called *Transactional Memory*, while the IBM version is called *Oklahoma Update*. We will discuss the basic idea from a programmer's viewpoint, without getting into implementation details. Both groups have shown that the multi-reservation scheme can be implemented efficiently on multiprocessors, taking advantage of existing cache coherency hardware.

We will use the term *transaction memory* when referring to the multi-reservation scheme, because the DEC paper is more programmer-oriented. Transaction memory is intended to provide efficient lock-free implementation of transactions with the following atomic instructions:

- A *read-and-reserve* instruction RAR(Local, S), as discussed above.
- A *tentative-write* instruction TW(S, Local), which tentatively writes a Local value to a shared variable S. However, this new value is not visible to other processes until the entire transaction successfully *commits*.
- A *commit* instruction COMMIT, which succeeds and returns a True if other processes have not updated the shared variables accessed in this transaction. Otherwise it fails and returns a False. If a COMMIT succeeds, it has in effect instantaneously made all tentative writes visible to all other processes. If it fails, all tentative writes are discarded.

Using these instructions, an account transfer transaction can be implemented as shown in Fig. 7.7. The program first performs reads, modifies, and tentative writes of shared variables. It then tries to commit the transaction and exits the while loop if the transaction commits successfully. Otherwise, it loops back to the beginning of the transaction and tries again.

The commented lines implement a well-known technique for reducing memory contention, called *exponential backoff*. The idea is that once a transaction aborts (i.e., fails to commit), it will be delayed for a certain amount of time before its next attempt. This WaitTime will be doubled for each successive attempt.

```
while ( True ) {                              // This loop represents a transaction
        RAR(oldX, balance[X]) ;               // read shared variable into local oldX
        if ( oldX <= 100 ) break ;            // insufficient fund, transaction terminates
            newX = oldX - 100 ;               // compute the local newX value
            TW(balance[X], newX) ;            // tentative write to account X
            RAR(oldY, balance[Y]) ;           // read shared variable into local oldY
            newY = oldY + 100 ;               // compute the local newY value
            TW(balance[Y], newY);             // tentatively write to account Y
            if ( COMMIT() ) break ;           // terminate if successfully commits
//          WaitTime=random()%(01<<backoff)   // otherwise do an
//          while ( WaitTime -- ) ;           // exponential backoff
//          if (backoff<BACKOFFMAX) { backoff++; }
}                                             // and loop back to try again
```

Figure 7.7   Account transfer transaction using transaction memory.

Fetch-and-Add The techniques discussed above (lock, critical region, test-and-set, compare-and-swap, and transaction memory) implement atomicity *sequentially.* When n transactions are executed, they must be executed one at a time. This is true even with lock-free schemes like compare-and-swap and transaction memory, because although all n processes can execute all the n transactions in parallel, only one transaction can success-fully commit, and the other $n-1$ processes must try again. Thus, executing n transactions needs $O(n)$ time.

A fetch-and-add instruction Result = F&A(S,V) is an atomic operation that returns the value of a shared variable S to a local variable Result, and then adds a local value V to S. The fetch-and-add instruction is meant to allow real parallel executions of multiple transactions. The code for account deposit using fetch-and-add is just one instruction:

F&A(balance[X], 100) ;

This code is not only simpler, but also faster than previous codes. With a technique called *combining*, it is possible for n processors to simultaneously execute n fetch-and-add instructions (which access the same shared variable) in $O(\log n)$ time [380]. The combining technique has been implemented by hardware in NYU Ultracomputer [276] and IBM RP3 [500]. It can also be implemented through software, which is then known as *software combining* [272].

Fetch-and-Φ   Kruskal et al. [380] have shown that combining is a general technique, not restricted to just *fetch-and-add*. They presented a general formulation of *read-modify-write* operations and of combinations of these operations. In fact, a lot of *fetch-and-Φ* operations can be combined, where Φ can be one of a large set of simple operations, such as any Boolean or arithmetic operations. It was shown that if *n fetch-and-Φ* operations accessing the same shared variable are executed simultaneously, they can be combined and take only $O(\log_2 n)$ time to finish.

There are three main criticisms of using *fetch-and-Φ* operations to support atomicity. These problems are stated below:

- The first is that *fetch-and-Φ* operations are not as widely supported as, for instance, locks in existing parallel computers, due to the need for an expensive combine network. This criticism is countered by recent advances, which show that the combine network can be implemented efficiently and inexpensively [204], and that combining can be implemented in software [272].
- The second criticism is that a *fetch-and-Φ* can only atomically update a single shared variable. It would be much more difficult, for instance, to realize the account transfer transaction using *fetch-and-Φ*.
- The third criticism is more serious: *Fetch-and-Φ* is a low-level and subtle construct, more complex than *test-and-set, compare-and-swap*, or transaction memory, and therefore more difficult for an ordinary programmer to use correctly.

7.2.4 Fast Locking Mechanisms

Two types of synchronization operations are predominant in current shared memory multiprocessors: locks and barriers. This book concentrate on locks, since barriers are a control synchronization construct, thus have much simpler semantics. Many efficient or hardwired synchronization implementations exist. Examples can be found in Johnson et al. [351], Scott [540], Shang and Hwang [552]. Also, other synchronization constructs in current parallel programming languages are often implemented by locks.

Recently, there has been much progress in efficient locking, aimed at lowering the locking overhead by reducing memory or network traffic and alleviating the hot-spot problem. These locking mechanisms are often used to solve the mutual exclusion problem. Thus they are also called *fast mutual exclusion algorithms*. A common idea is to let processes wait on a lock in cache or local memory most of the time.

There are two types of locking algorithms: Algorithms that are based on some form of atomic read-modify-write primitives (e.g., test-and-set, fetch-and-add) are called *hardware algorithms*, while those that do not use atomic read-modify-write primitives are called *software algorithms*. Current commercial systems mostly use hardware algorithms. We discuss two hardware algorithms to demonstrate the basic ideas. A good recent paper on mutual exclusion is Zhang et al. [665], where the authors compare the performance of several representative hardware and software mutual exclusion algorithms. Kontothanassis et al. [374] evaluated various hardware synchronization algorithms in the presence of multiprogramming, e.g., multiple processes are allowed to execute on a processor.

The Ticket Algorithm   The basic idea is that each process first takes a ticket and then waits for its turn to execute the critical section, as shown in the following code adapted from Zhang et al. [665]:

```
int Ticket=0, Turn=0 ;              // shared variables, initialized to 0
parfor (i=0; i<n; i++)              // there are n processes
{  int Myticket ;                   // local variable
   while ( Condition ) {
       independent computation, noncritical section
       Myticket = FAA(Ticket, 1) ;{     // these three lines
       while (Myticket != Turn)          // are equivalent to
           Delay(Myticket-Turn) ;        // a lock operation
       critical section
       Turn = Turn+1 ;                   // unlock
       }                             // exit point of mutually exclusive code
       independent computation, noncritical section
   }
}
```

Each process *P* uses three variables, as shown in Fig. 7.8. The local variable Myticket is stored in the cache or local memory and cannot be accessed by other processes. The two global variables, Ticket and Turn, are in the shared memory accessible by all processes.

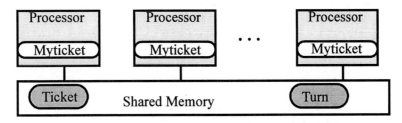

Figure 7.8   **Variable distribution in the ticket algorithm.**

To enter a critical section, a process *P* first executes an atomic *fetch-and-add* instruction to get the Ticket number into local variable Myticket and to increment Ticket by 1. The process then spin-waits until its turn arrives. To reduce the number of times to read the shared variable Turn, the process *P* delays for a time period before reading Turn again. This delay time is proportional to the difference between Myticket and Turn, which is the number of processes waiting in front of process P to execute the critical section. After executing the critical section, process P unlocks by incrementing Turn. Note that although Turn=Turn+1 is a read-modify-write operation, it does not have to be atomic.

The ticket algorithm is a simple *spin-locking scheme* suitable for multiprocessors that provide an atomic *fetch-and-add* (or *fetch-and-increment*) primitive. It has the *FIFO property*: Processes obtaining a smaller ticket number will execute the critical section first. One must ensure that the two integer variables, Ticket and Turn, do not overflow.

Array-Link-Based Algorithm   With the ticket algorithm, the n processors need to access two shared variables Ticket and Turn. To reduce contention, the algorithm uses a fetch-and-add to access and increment Ticket. All the n processors can get a ticket in $2t_M \cdot \log n$ cycles using the combining technique, where t_M cycles are needed for a shared-memory access.

Each processor can get a copy of Turn into a local register or cache, and spin on the local copy, until Turn is modified by another processor that releases the lock. The proportional delay feature is also attractive.

To further reduce contention, the lock can be distributed, as shown in the following array-link-based algorithm by Zhang et al. [665], which is adapted from an algorithm developed by Mellor-Crummey and Scott [441]. The array-link-based algorithm uses the following data structures. The shared variable Lock is initialized to the value -1.

```
typedef struct MLINK {
    struct MLINK * Successor; short Get_Lock;
} Mlink;
short Lock, NodeID ;
Mlink * Lock_q[Num_Proc – 1 ] ;
int NodeCount ;
```

Processor NodeID using the two procedures **Acquire_Lock**(NodeID, Lock) and **Release_Lock**(NodeID, Lock) to acquire and to release a Lock. The details of these procedures are shown in Fig. 7.9. How to use these procedures is left in Problem 7.5.

7.3   The TCP/IP Communication Protocol Suite

Communication over a network is usually realized as a stack of *protocol* layers. A protocol is a set of rules governing the data format and how the data are transmitted over the network. A protocol stack is also known as a *protocol suite*. In this section, we discuss a standard communication protocol suite called the *TCP/IP suite*. Nonstandard, but more efficient protocols are discussed in Section 7.4.

```
Acquire_Lock(short NodeID, short Lock)
{       int Pred ;
P:      Lock_q[NodeID]->Successor = Nil ;
Q:      Pred = Fetch_and_Store(Lock, NodeID) ;
R:      if (Pred != −1) {
            Lock_q[NodeID]->Get_Lock = 0 ;
            Lockq[Pred]->Successor = Lock_q[NodeID] ;
            while (Lock_q[NodeID]->Get_Lock != 1) ;
        }
}
Release_Lock(short NodeID, short Lock)
{       int Pred ;
X:      if (Lock_q[NodeID]->Successor != Nil)
            Lock_q[NodeID]->Get_Lock = 1 ;
        else {
Y:          Pred = Fetch_and_Store(Lock, −1) ;
Z:          if (NodeID != Pred ) {
                Pred = Fetch_and_Store(Lock, Pred) ;
                while (Lock_q[NodeID]->Successor == 0) ;
                Lock_q[NodeID]->Successor->Get_Lock = 0 ;
            }
        }
}
```

Figure 7.9   Array-link based locking primitives. (Courtesy: Zhang et al. *IEEE Parallel and Distributed Technology*, 1994 [665])

We first summarize the main features of the TCP/IP suite in Section 7.3.1. We then discuss in Section 7.3.2 three most frequently used protocols (UDP, TCP, and IP) and how they are implemented on Ethernet. These protocols are just rules. To realize them in any program or application, the programmer needs an application programming interface (API). In Section 7.3.3, we present a de facto standard API called *sockets*, which can be used to utilize either UDP/IP or TCP/IP.

7.3.1 Features of The TCP/IP Suite

Two widely known protocol stacks are shown in Fig. 7.10. The seven-layer, Open-System Interconnect (OSI) stack was developed by the International Standard Organization (ISO). This standard, although not widely implemented, is influential in clarifying the concepts and terminology of communication architectures. The Internet communication stack, commonly known as the *TCP/IP suite* and developed by the Internet Activities Board (IAB), is widely used and has become a de facto standard.

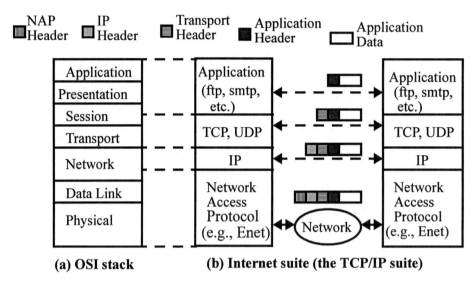

(a) OSI stack   **(b) Internet suite (the TCP/IP suite)**

Figure 7.10   The OSI and Internet communication protocol stacks.

Transmitting a message between two computers (often called *hosts*) in a network is a complex activity. The TCP/IP suite uses a layered structure to cope with this complexity. It has several features common to all communication protocol suites:

- *Layers*: A stack of protocol layers is used in transmitting a message. Each protocol layer is dedicated to performing a specific set of functions, and interacts with the layers immediately below and above.

  The TCP/IP suite itself has three layers: the *application layer* (containing protocols such as FTP, TELNET, SMTP, and SNMP), the *transport layer* (containing the protocols TCP and UDP), and the *network layer* (containing the IP). The suite operates on a network access layer, which depends on the communication hardware.

- *Data Encapsulation*: When an application process wants to transmit a piece of data over a network, some extra information must be attached to the data to form a *message*. This extra information is usually called a *header*, although part of it can be attached after the data. In a protocol stack such as the TCP/IP suite, a lower layer treats the entire message (data and header) from the higher layer as its data, and attaches a header to form its own message.

  The lower layer does not care about or understand the message format of the higher layer. This is called *data encapsulation*, as illustrated in Fig. 7.11. The

TCP layer attaches a 9-field TCP header to the application message to form a transport-layer message called a *TCP segment* (or several segments, if the message is large).

The IP layer attaches an 11-field IP header to the TCP segment to form a network-layer message called an *IP datagram*. The NAP layer attaches a 6-field Ethernet header to the IP datagram to form a NAP-layer message called an *Ethernet frame*. Note that the CRC field comes after the data in an Ethernet frame.

- *Peer-to-Peer*: Each layer in one host talks to the same layer in the other host, without caring about the other layers. For instance, one host may be connected to an Ethernet while the other is connected to an FDDI, but the IP layers sees that IP datagrams are transmitted according to the same protocol.

- *Open Interfaces*: Each layer provides an open interface to the layer above. Most importantly, application layer can use either the *sockets* or the *TLI* (*Transport Level Interface*) to access TCP/IP services. Since TCP, UDP, and IP are usually implemented in the kernel, these interfaces manifest as a set of system calls.

Application Protocols An application-layer protocol provides the user interface of the functions implemented by the protocol suite. It also deals with application-specific syntactic and semantic issues, assuming messages sent from a source process will reach the correct destination process.

These issues include ensuring a little-endian host and a big-endian host see the same logic data from the transmitted message, enforcing the ACID properties in a transaction processing system, etc. There exist many application protocols that are supported by the TCP/IP suite. The following is a list of the most widely used ones:

- FTP (*File Transfer Protocol*), for transferring files between a local process and a file system in a remote host
- TELNET, for logging into a remote host
- SMTP (*Simple Mail Transfer Protocol*), for email services
- SNMP (*Simple Network Management Protocol*), for controlling and monitoring a computer network (e.g., a LAN)
- HTTP (*Hyper-Text Transfer Protocol*), for WWW applications

Network Addressing The application-layer protocol must also tell the underlying protocols the address of the destination (and source) process, which may be in a remote host. This is given as an (*IP address*, *port*) pair shown below:

159.226.43.150, 23   host IP address, port number

The IP address uniquely identifies any destination host, while the port number indicates the destination process or protocol. Existing port numbers on a local Unix host can be found by typing "more /etc/services".

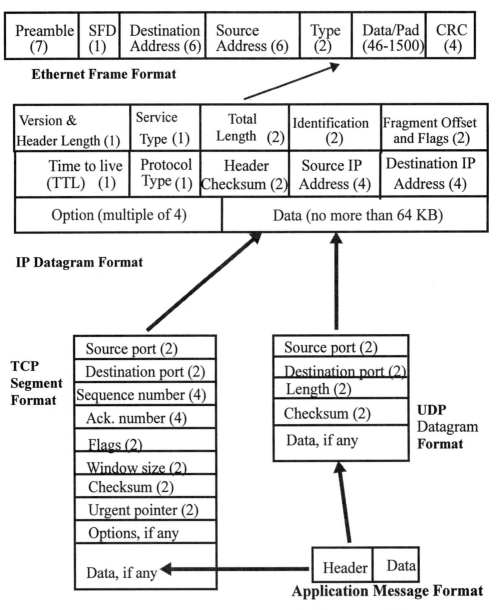

Figure 7.11   Various data formats in the TCP/IP suite on Ethernet.
(Numbers in parentheses indicate the field's length in bytes)

A port is a transport-layer communication endpoint. Transport-layer communication is always from one port to another. The port number is a 16-bit integer, implying that there can be up to 64 K ports per host. Port numbers from 0 to 1023 are *reserved* for the root user. Most reserved port numbers are *well-known*, in that everybody agrees that such a port number is used for a specific protocol or service, and for nothing else. For instance, port number 23 is a well-known port number. It is used for the TELNET protocol. Any process wanting to use the system supplied TELNET protocol must use port number 23.

7.3.2 UDP, TCP, and IP

The TCP/IP suite contains two transport-layer protocols: TCP and UDP. They both operate on top of a single network-layer protocol called IP.

The User Datagram Protocol (UDP) The UDP is a very simple transport-layer protocol. Its simplicity offers efficiency but little functionality. The simplicity is reflected in its simple datagram format, as shown in Fig. 7.11. The UDP header contains only four 16-bit fields: the source port address, the destination port address, the length of the entire UDP datagram in bytes, and the checksum of the UDP datagram.

The destination port and the datagram length are always required, while the other two are optional. The source port is needed only when a reply is expected. When not used, the source port and the checksum fields are filled with zeros. When a source host wants to send a piece of data to a destination host using UDP, it first forms a UDP datagram as shown in Fig. 7.11.

The UDP software on both hosts then cooperates to transmit the datagram to the destination host, which then can extract the data. Note that a UDP datagram has limited length. If the data to be transmitted have a large size, the sender needs to break them into smaller pieces and transmit them as separate datagrams. The destination host receives these datagrams also as separate messages and needs to reassemble them into the original data. The datagrams may arrive at the destination out of order, or never arrive at all.

These properties make UDP a *connectionless* and *best-effort* protocol. It is like sending information by post mail. The UDP is connectionless in that the sender sends out data without establishing a connection. Rather, large amounts of data must be broken up and wrapped as individual datagrams (letters), which are then transmitted as separate entities. In the TCP to be discussed shortly, a connection is established first, and then the senders sends out the data as a continuous stream, not as individual datagrams. We say that the UDP *maintains message* datagram *boundary*, while the TCP does not.

The UDP is a best-effort protocol in that like a post office which tries its best to deliver the mail, there is no guarantee. A UDP datagram may be corrupted or simply lost during transmission. The receiving side UDP will detect the error (via checksum) and drop the corrupted datagram. UDP is appropriate when guaranteed delivery is not an important issue, such as transmitting video/audio messages, when occasional dropping of packets does not affect the performance much. If reliable messaging is required (e.g., transmitting program and data files where not even a single bit can be wrong), one has to realize the reliable messaging functions at the application layer, or switch to TCP.

The Transmission Control Protocol (TCP)   The TCP is a more powerful and more complex transport-layer protocol than UDP. TCP is a *connection-oriented* protocol that guarantees reliable message delivery. It is like sending a message by telephone. The two parties first establish connection, and then the message is transmitted. Through error detection, acknowledgment, and retransmission mechanisms, the message is guaranteed to be reliably transmitted and correctly ordered.

TCP is a two-way protocol: Data-carrying segments are sent from the source to the destination, and acknowledgment segments with a null data field are transmitted in the reverse direction. The TCP header (Fig. 7.11) contains nine fields. The source port, the destination port, and the checksum fields are similar to those in the UDP header.

We will ignore the flags and the options fields. The urgent pointer field indicates the number of data bytes in the segment that should be immediately delivered to the destination process once the data are received by the destination TCP.

There is no *segment length* field in the TCP header, because segment length is calculated by TCP. The TCP transmits messages as a *stream*: Once a TCP connection is established, one or more messages can be sent over the connection as a stream of bytes. TCP is responsible for dividing this stream into TCP segments and to reassembling the segments into messages at the destination. TCP is also responsible for dealing with out-of-order segments, duplicated segments, reliable transmission, and flow control.

The remaining three fields are used to achieve these goals. The *sequence number* is filled in by the source TCP and indicates the byte offset of the first byte of the segment from the base address of the stream. It is used by the destination TCP to take care of out-of-order or duplicated segments. The *acknowledgment number* is filled in by the destination TCP to indicate which segment is received.

The *window size* is also filled in by the destination TCP, indicating the amount of data beyond the currently acknowledged segment that the destination TCP is prepared to accept. This value is used for flow control and depends on the available space in the destination TCP buffer for this connection. TCP uses this information to choose the length of the next segment to be sent. The length selection algorithm attempts to maintain high bandwidth while controlling destination side congestion.

TCP uses an *acknowledgment with retransmission* scheme for reliable messaging. A timer at the source side is activated when a segment is sent. The destination TCP sends an acknowledgment back when it receives the segment. The timer is canceled when the source receives the acknowledgment before the timer expires. If no acknowledgment has been received when the timer expires, the source resends the same segment.

This scheme has several problems. First, bandwidth will suffer if the source waits for the arrival of an acknowledgment before sending out the next segment. One may think that the source could continuously send out segments without waiting for acknowledgment. But this can quickly saturate the destination buffer and cause severe network congestion, which increases latency and reduces bandwidth. Moreover, retransmission causes multiple copies of a segment to be sent to the destination, further worsening the congestion.

TCP uses a *sliding-window* flow control mechanism to solve these problems. The source is allowed to continuously send segments out, which have the window size as an are upper bound, therefore, will not cause severe destination side congestion. The destination does not acknowledge every segment sent. Let us look at Fig. 7.12, assuming a constant segment size of 10 KB and an initial window size of 100 KB, or 10 segments.

The initial window covers segments 1 to 10. After segments 1 to 10 are sent out, the source receives acknowledgment indicating segments 1 to 5 have been received. The destination also sends with the acknowledgment that it wants the new window size to be 8 segments. Now the window is slid right to cover segments 6 to 13. The source then proceeds to send out segments 11 to 13. Note that segments 6 to 10 are in network, meaning they have been sent but not yet acknowledged. They can be en route to the destination, they may have arrived but the destination has not sent back acknowledgment, or the acknowledgment was sent but has not arrived at the source yet.

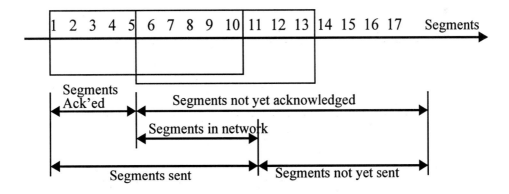

Figure 7.12   The sliding window mechanism in TCP flow control

The Internet Protocol (IP)   The main function of the IP is to route messages from one host to another in an internetwork (or Internet) that may consist of more than one LAN. That is, the message may pass through one or more intermediate nodes called *routers* (also known as *gateways*).

Any host on the Internet has a unique IP address. A router is different from an ordinary host in that it is connected to at least two LANs and has more than one IP address. A router's main function is to route IP datagrams in an internet, while a host is usually used for other purposes (e.g., a desktop PC, a file server, a compute server, etc.).

The IP is similar to the UDP in that it is a connectionless, best-effort protocol. IP datagrams may arrive out of order or may be dropped, and IP lets the layers above solve these problems. We illustrate the mechanism of the IP by describing how host A sends a datagram to host Z in Fig. 7.13.

The IP software module in host A, once it receives the request, looks up a table locally to find that host Z is not on the same LAN and that the datagram must be sent to Router I. This router determines further the routing path and forwards the datagram to either Router II or Router III, which then sends the datagram to the final destination host Z. The routing function uses the source and the destination IP addresses in the datagram (Fig. 7.11). Besides routing, the IP layer provides the following two functions:

- *Fragmentation and Reassembly.* The LANs in an internetwork may be of different types and impose different limits on maximum packet sizes. The IP therefore needs to fragment a large datagram into several smaller packets and reassemble them. This function utilizes the following fields of an IP datagram: total length, identification, fragment offset and flags, and source/destination IP addresses.
- Error Notification. In case of errors, the hosts involved must be informed.

The *service type* field specifies the demands on precedence (priority), reliability, delay, and throughput. The *time-to-live* field gives the datagram a lifetime, which is useful for destroying expired datagrams. Otherwise, a datagram may circulate endlessly and clog the networks, due to improper routing decisions. The *protocol type* field specifies which transport protocol (TCP, UDP, etc.) is using the IP. The *identification* field, together with the *protocol type* and the *source/destination IP addresses*, uniquely identifies a datagram.

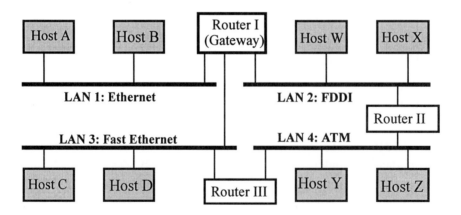

Figure 7.13   An Internet with four LANs connected by three routers

7.3.3 The Sockets Interface

Sockets is an API (i.e., a set of data types and functions) for using the TCP/IP suite. Sockets was originally implemented in the Berkeley Unix (BSD 4.2). Now it has become a de facto standard, implemented on almost all Unix systems, as well as Microsoft Windows platforms. A *socket* is a communication endpoint.

When two processes want to communicate using the sockets interface, they each first create a socket and specify which transport protocol to use (i.e., UDP or TCP). Then they communicate by reading/writing to their respective socket. The sockets software is responsible for performing the actual communication.

We illustrate the key ideas of the sockets interface with a much simplified domain name server. A server program and a client program are constructed. The client sends the name of an arbitrary host to the server, which looks up the IP address of the host and sends it back to the client.

Connectionless Implementation (UDP) The program skeleton for the UDP implementation is shown in Fig. 7.14. Both the server and the client first call for the service of the *socket* function.

```
main ()
{
    Mysocket = socket(AF_INET, SOCK_DGRAM, ...) ;
    bind( Mysocket, ... ) ;
    recvfrom( Mysocket, Hostname, ...) ;
    host_IP = NameToIP(Hostname) ;
    sendto( Mysocket, host_IP, ...) ;
}
```

(a) The server code skeleton

```
main (int argc, char *argv[])
{
    Mysocket = socket(AF_INET, SOCK_DGRAM, ...) ;
    sendto(Mysocket, argv[1], ...) ;
    recvfrom(Mysocket, host_IP, ...) ;
}
```

(b) The client code skeleton

Figure 7.14   Code skeleton of a simplified domain name server using UDP sockets.

The purpose is to create a socket, pointed to by the variable *Mysocket*. The three dots indicate arguments we are not interested in here. The constant AF_INET indicates that internetwork domain address (the IP address) will be used as the socket addressing format, and SOCK_DGRAM indicates that the UDP will be used. The server then binds this socket to a port number by a *bind* function call. The client does not have to explicitly bind, but can leave it to the system to automatically choose a port number to bind.

After the sockets are established, the client sends the Hostname to the server by a *sendto* function. The server receives the message by a *recvfrom* function, translates it into a corresponding IP address by executing a local function NameToIP, and sends the IP address back to the client by a *sendto*. The client receives the IP address by a *recvfrom*.

Connection-Oriented Implementation (TCP) The code skeleton for the TCP implementation is shown in Fig. 7.15. The socket creation and binding are similar to the UDP implementation, except that the constant SOCK_STREAM indicates that the TCP will be used. Unlike the UDP, the TCP is connection-oriented. The server tells the kernel that it is ready to accept connection by executing a *listen* function. Then the server waits to accept a connection request from the client by executing an *accept* function. This function will not return until a client connects by executing a *connect* function.

The two-way connection established behaves just as an ordinary file, and is pointed to by a file pointer *fp*. Now the server can communicate to the client by reading/writing to the connection, just like accessing a file. After all communication is down, the server closes the connection by executing close(fp).

7.4 Fast and Efficient Communication

The raw network bandwidth has been improving at an impressive exponential rate, almost increasing 10 times every 5 years. Gigabit networks are quite common in current *high-performance computing* (HPC) systems. Terabit networks have been demonstrated in laboratories (e.g., Fujitsu achieved 1.1-Tb/s transmission in February 1996; see http://endeavor.fujitsu.co.jp/hypertext/news/1996/Feb/29-3e.html).

A large gap may exist between the raw bandwidth and what the user can realistically achieve. Table 7.3 compares the peak and effective bandwidths of several HPC systems. The asymptotic bandwidth and the latency are measured with MPI point-to-point communication. The effective bandwidth is estimated with a 128-B message by the formula:

$$\text{Effective bandwidth} = \frac{\text{Message length}}{\text{Communication time}} = \frac{128}{t_o + 128/r_\infty} \qquad (7.1)$$

```
            main ()
            {
                Mysocket = socket(AF_INET, SOCK_STREAM, ...) ;
                bind( Mysocket, ... ) ;
                listen( Mysocket, ...) ;
                fp = accept( Mysocket, ...) ;
                read (fp, Hostname, ...) ;
                host_IP = NameToIP(Hostname) ;
                write(fp, host_IP, ...) ;
                close(fp) ;
            }


            (a) The server code skeleton


            main (int argc, char *argv[])
            {
                Mysocket = socket(AF_INET, SOCK_STREAM, ...) ;
                connect( Mysocket, ... ) ;
                write (Mysocket, argv[1], ...) ;
                read (Mysocket, host_IP, ...) ;
            }


            (b) The client code skeleton
```

**Figure 7.15   Code skeleton of a simplified domain name server
using TCP sockets.**

Data in Table 7.3 are obtained with communication hardware and software that are supposed to be efficient. The gap between the hardware peak and the effective bandwidth will be much wider if one uses commodity networks (e.g., Ethernet) and standard protocols (e.g., sockets). To bridge this gap, we need to understand the factors that critically impact the communication performance.

7.4.1 Key Problems in Communication

Many factors affect the performance of the communication subsystem. The important ones are listed below:

- The communication hardware, including the node memory and I/O architecture, the network interface, and the network

Table 7.3 Peak and Effective Bandwidths of Four HPC Systems

System	Hardware Peak (MB/s)	Asymptotic r_∞ (MB/s)	Latency t_0 (μs)	Effective (MB/s)
Intel Paragon	200	52	40	3.01
IBM SP2	40	35	46	2.58
Cray T3D	300	70	10	10.82
DEC TruCluster with Memory Channel	100	61	6.9	14.22

- The communication software, including the protocol structure and the algorithms
- The user environment, including whether to support multiuser, multiprocess, multiprogramming, whether to use the SPMD paradigm, etc.
- The communication services provided, including message delivery, flow control, fault handling, protection, etc.

Communication Hardware   A typical communication hardware architecture is shown in Fig. 7.16. In a *loosely coupled* architecture, the NIC is attached to the I/O bus (e.g., PCI) of a node. To go from one node to another, a message travels from the sender node memory, through the memory bus, the I/O bus, the NIC to the network. At the receive side, it needs to travel the same path in reverse order. The communication bandwidth achievable is limited by the slowest component along the communication path.

Example 7.4 Bottlenecks in a cluster of computers

A cluster uses nodes with an 800-MB/s memory bus, a 133-MB/s I/O bus, and a 200-MB/s NIC DMA. The following two questions are being asked:

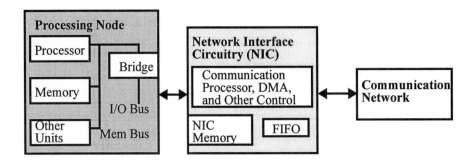

Figure 7.16   A typical communication hardware architecture.

(1) What is the hardware limit on communication bandwidth if the communication network has a peak speed of 1 Gb/s?

(2) What if the network is upgraded to a 400-MB/s one?

Answers :

(1) The Gigabit network is the hardware bottleneck. So the limit is 128 MB/s.

(2) The I/O bus is the hardware bottleneck. So the limit is 133 MB/s. To get around this bottleneck, *tightly-coupled* MPPs (e.g., Intel Paragon) directly attach the NIC to the memory bus. Then the NIC becomes the bottleneck, with a bandwidth limit of 200 MB/s.

■

The limits given in Example 7.4 are the hardware limits that can never be exceeded. The real achieved communication bandwidth is often much lower. Stricker and Gross [591] noted that the communication performance is often limited by memory bandwidth rather than network bandwidth. For instance, many communication schemes require the DMA engine to access a restricted memory region (called the *DMA buffer*). The user data must be first copied to the DMA buffer, before they are sent out. The memory copying bandwidth is usually much smaller than the memory bus peak bandwidth. Table 7.9 shows the bandwidths of various components of three MPPs. The I/O bus items for Paragon and T3D are missing because the NIC is directly attached to the memory bus in these two systems. The sustained bandwidth is what an application user can achieve.

Table 7.4 Bandwidth (MB/s) Measured in Three MPPs (Courtesy: Stricker and Gross, *Proc. of 22nd Annual Symp. Computer Architecture,* [591], 1995)

MPP System	Memory Bus	Copying	I/O Bus	Network	Sustained
Intel Paragon	400	68	-	200	52
IBM SP2	2100	> 120	80	40	35
Cray T3D	320	93	-	300	70

Different architectures have been proposed for the network interface circuitry (NIC). However, several common features have emerged:

- The NIC should have DMA functionality. Programmed I/O not only incurs significant processor overhead, but also leads to long latency and low bandwidth. One to three DMA engines have been used in real NICs.

- The NIC should have its own processor, called the *NIC processor, coprocessor, microcontroller,* or *adapter processor.* The main processor in the node is called the *node processor* or *host processor*. The NIC processor is needed for DMA initialization, packet packing/unpacking, checking for protection, etc.

- In many systems, all DMAs are initiated by the adapter processor, not the node processor. This has the advantage that data can be moved without involving the node processor; i.e., the node processor is *decoupled* from the communication network.

- The NIC should have a memory to store NIC code and to temporarily buffer messages, among other things. This memory can be accessed by the node processor through programmed I/O. The size of this NIC memory ranges from 100 KB to a few megabytes in current systems. The NIC should also contain FIFOs to buffer packets.

Communication Software   The software overhead often dominates communication time in current clusters and MPPs. It follows that communication time cannot be significantly reduced, even with very efficient network and NIC, if we do not have efficient communication software. The software overhead comes mainly from three sources:

- The software needs to traverse several protocol layers. A common technique to reduce this type of overhead is to simplify the protocol structure.

- Message communication may involve a number of memory copying instances, which call for a zero-copy protocol.

- Communication software may cross protection boundaries several times in transmitting a message. To deal with this problem, *user-level* techniques are emerging that perform all communication completely in user space.

Common protocol structures are illustrated in Fig. 7.17. The familiar *socket* interface has been used in many distributed systems applications, such as networked file systems, remote procedure calls, ftp, telnet, http, etc.

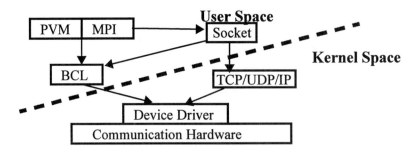

Figure 7.17   Various structures of communication protocols

Communication libraries such as PVM and MPI may be realized on top of sockets. At the sender, a message travels down the socket layer, the TCP/IP (or UDP/IP) layers, to the driver and network hardware layer. At the receiving end, the same process will be repeated but in reverse order. Sockets can be implemented directly on a low-level *base communication layer* (BCL), bypassing TCP/IP. An example is the Berkeley Fast Socket, which uses Active Messages as the BCL.

One can also implement the PVM/MPI on top of BCL, bypassing the socket/TCP/IP layers. The main objective of the BCL is to expose as much raw hardware performance as possible to the application user. To evaluate a communication subsystem, the performance of MPI, PVM, and sockets are more important than that of that of BCLs.

A large portion of the total communication time is spent on copying data from one memory region to another. It is important to reduce the number of memory-copying events and the time for each copying. Ideally, a *zero-copy protocol* should be used, whereby the message is moved directly from the send buffer in the source node to the receive buffer in the destination node, without being buffered in another memory region.

Example 7.5 The IBM SP communication software

The memory copying involved in sending a message in the IBM SP multicomputer is illustrated in Fig. 7.18. The message to be transmitted is originally stored in a send buffer, which is usually denoted by a variable in the user application. For instance, suppose the user application wants to send an array *A* in one node to array *B* in another node. Then *A* denotes the send buffer and *B* the receive buffer. The pipe buffer and the I/O queues are used by the communication software internally .

The SP communication protocol requires the sender node processor to copy data from the send buffer to the pipe buffer, and then from the pipe buffer to the output queue. The receive node processor performs the same actions in reverse order. Altogether, four memory copy operations are performed. To reduce copying, SP allows a long message to bypass the pipe buffer, shown by dashed lines in Fig. 7.18.

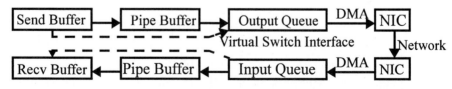

Figure 7.18 Data movement in IBM SP communication software.
(Copyright 1995, International Business Machines Corp. Reprinted with permission from M. Snir, et al. *IBM Systems Journal,* Vol. 34, No.2, [574], 1995)

In a traditional communication software, transmitting a message involves crossing the protection boundaries at least four times, to enter and to leave the kernel at the send and the receive sides. On many modern platforms (e.g., RISC processor and Unix), entering the kernel takes at least 10 μs. Thus, performing messaging completely at the user space could greatly reduce the latency.

Communication Environment Another factor that affects the design of the communication subsystem is the environment the subsystem is to work in. The issues that should be considered are listed in Table 7.5.

Table 7.5 Communication Environment Issues

Issue	Less General	More General
Application	Technical computing	Commercial/distributed
Tasking	Single-tasking	Multitasking
Homogeneity	SPMD	MPMD
Threading	Single-threading	Multithreading

For commercial and networked/distributed applications, the communication subsystem should efficiently support standard protocols such as TCP/IP in a multiprogrammed environment. Technical computing applications on dedicated MPPs are simpler, as the communication subsystem only needs to support PVM and MPI in many cases.

While SPMD is used in many technical computing applications, commercial and distributed applications are often MPMD. Node processes on early MPPs are single-threaded, but multithreaded processes are used in modern MPPs and clusters, especially when SMP nodes are used.

A crucial issue is, Is there one or multiple processes on a node? Single-tasking (or single-programming) was used in early MPPs, where there is only one process per node that runs until completion. In Cray T3D/T3E, the processes of a job can be rolled out in the middle of computation, and the processes from another job rolled in to the just vacated nodes. The processes of a job are gang-scheduled, so that when one process of a job is in a node, all other processes of the job are in other nodes.

With multitasking, multiple processes, probably from different jobs, may reside on a node at the same time. Multitasking complicates the design of the communication subsystem considerably. For instance, the following issues must be addressed:

- While the NIC is getting a message via DMA from process *P*'s DMA buffer, the buffer is swapped out by another process *Q* that requires more memory space. Then NIC will transmit wrong data, because the original DMA buffer now holds data for process *Q*. To solve this problem, the DMA buffer must be *pinned* down, i.e., never swapped out.

- Since the node is time-shared by multiple processes, any process may be timed out at any time nondeterministically, and it is not predictable how long the process will stay descheduled. A message sent to a descheduled process may block the receive side, and the back pressure caused by the receive end congestion may eventually block the entire network.
- Multitasking also makes protection more challenging, since multiple processes may share the NIC and associated communication ports, queues, buffers, etc. This difficulty compounds with supports for user-level communication.

Example 7.6 The communication subsystem in the IBM SP

The communication subsystem of the IBM SP offers a compromise in resolving the communication environment issues. It supports both computing and commercial/ distributed applications. A parallel application could be SPMD or MPMD. Each node is just like a workstation, allowing multitasking.

Two protocols are available for communication over the high-speed HPS network, the standard IP and a proprietary *user space* (US) protocol, which corresponds to the BCL in Fig. 7.17. The NIC to the HPS can be shared by multiple processes in a node. However, only one process per node can use the US protocol.

∎

Communication Services What services are provided can greatly impact the performance of a communication subsystem. For instance, a BCL provides very limited services in a restricted environment but could show excellent performance. However, MPI/PVM built on top of this BCL may perform poorly, since the implementation must provide additional services. The following services are expected of a communication subsystem:

- *Guaranteed delivery*: Once a send function call returns, the communication subsystem guarantees to deliver the complete message to the intended destination node in normal situations. The message will not go to a wrong node, and no packets of the message will be lost. A less reliable alternative is the *best-effort* delivery, where the subsystem does its best but offers no guarantee.
- *Flow control*: The communication subsystem should perform flow control to avoid buffer overflow, deadlock, or severe congestion. In particular, it should address the *receive-side congestion* problem.

  Suppose multiple nodes concurrently send messages to a receive node, while the receive process is busy computing and will not execute any matching receive for some time. The incoming messages will soon fill the receive side buffer, and the saturation may soon back-propagate to saturate the entire network.
- *Fault handling and reliable delivery*: When a fault occurs in the communication subsystem, especially the network, there are three ways to deal with the fault. It

can be ignored, reported, or corrected. In a distributed system using TCP over an Ethernet, the last method is used because the medium is not reliable. A received packet is checked for error, and is retransmitted after an error is detected.

In MPPs and clusters using a special network, the first two methods can be used, as the network error rate is often very low. For instance, the DEC Memory Channel network has a bit-error rate of only 10^{-16}, i.e., one error bit for every 10^{16} bits transmitted. A protocol offers *reliable delivery* if it guarantees that correct messages will be delivered to the correct destination.

- *Ordered delivery*: When a source node sends two messages to the same destination, executing a receive operation at the destination will receive the first message.

7.4.2 The LogP Communication Model

Culler et al. [178] proposed a *LogP* model to characterize a parallel computer with four parameters: *L, o, g,* and *P.* A short message is a word, a double-word, etc. The LogP model is useful to develop communication algorithms as illustrated in Fig. 7.19.

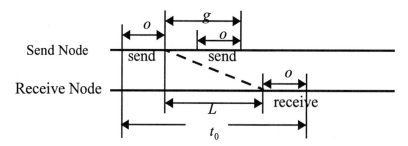

Figure 7.19   Message passing under the LogP model.
(Courtesy of Culler et al. *ACM Symp. on Principles & Practice of Parallel Programming*, 1993[178])

The time to communicate a word from one node to another (i.e., the start-up time t_0 in Section 1.3.2) is subdivided into three terms. That is, $t_0 = o + L + o$. The first o is called the *send overhead*, which is the time for the sending node to execute a message send operation to inject a word into the network. The second o is called the *receive overhead*, which is the time for the receiving node to execute a message receive operation to get a word from the network.

For simplicity, the two overheads are assumed equal and called the *overhead o*. In other words, the overhead o is the length of a time period that a node is engaged in sending or receiving a word. During this time, the node cannot perform other operations (e.g., overlapping computations). The *latency* L is the hardware network latency, i.e., the time for one word to propagate through the network from the sending node to the receiving node. The *gap* parameter g is defined to be the minimum time interval required between two consecutive message sends or receives at a node.

The LogP model is later extended to LogGP [22], where the added parameter G is called the *bulk gap*, equivalent to the reciprocal of the bandwidth. That is, $G = 1/r_\infty$.

Example 7.7 LogP model parameters of two parallel computers

On most existing parallel computers, the overhead o is much larger than the latency L. However, there are machines that show just the opposite. This is shown in Table 7.6 which lists the approximate values for the LogGP parameters of two real parallel computers.

Table 7.6 LogP Parameters of Two Parallel Computers

(Courtesy of Martin et al. *Int'l Symp. Computer Architecture*, 1997 [431])

Machine	L	o	g	P	t_0	r_∞
IBM SP2	1.25 µs	22.38 µs	24 µs	400	46 µs	35 MB/s
Meiko CS-2	10 µs	3.8 µs	13.8 µs	64	17.6 µs	43 MB/s

∎

The LogP model can be used to identify communication bottlenecks. Recently, Martin et al. [431] proposed an enhanced LogGP model to address the following two questions: Of the four parameters (network latency L, processor overhead o, gap g, and bandwidth $1/G$),

(1) Which ones affect application performance most significantly?

(2) How much do they impact the performance?

These are important questions, because a designer of the communication subsystem wants to devote the limited resource to improving the most critical parameter. Martin et al. study the impact by measuring the performance of a suite of 10 parallel computing benchmark programs on a 32-node workstation cluster. These programs are implemented in Split-C, a language that provides a shared-memory environment on top of a message-passing parallel computer. The experiment platform allows latency, overhead, and gap to be adjusted independently. The default configuration of the cluster platform has the following parameter values: $L = 5$ µs, $o = 2.9$ µs, $g = 5.8$ µs, and $1/G = 38$ MB/s. Their study revealed several interesting points:

- Applications are most sensitive to overhead. Applications can be slowed down by up to 50 times when the overhead increases from 2.9 μs to 103 μs. Even lightly communicating applications can slow down by a factor of 3 to 5. The execution time of an application is observed to be linearly proportional to the added overhead. Of course, different applications have different coefficient values.

- Applications are also sensitive to gap, but not as much as the overhead. When the gap changes from 5.8 μs to 105 μs, some applications are not affected, and the worst application is slowed down by a factor of 16. The execution time of an application is observed to be linearly proportional to the gap.

- Most applications are fairly insensitive to network latency. In the worst case, the slowdown is no more than a factor of 4 when the latency is increased to over 100 μs. Note that latency in the LogP model is the network hardware latency, not the start-up time t_0. In commercial as well as research parallel computers, the network latency is no more than 10 μs. A 100-μs latency is very long.

- Most applications are fairly insensitive to bulk bandwidth of the communication system. In the worst case, the slowdown is no more than a factor of 3 when the bulk bandwidth is reduced to only 1 MB/s. The performance of all the benchmarks measured stays about the same until the bandwidth is reduced to 15 MB/s.

The results indicate that the overhead is the most critical, followed by the gap, while latency and bandwidth are the least critical. The importance of processor overhead has been realized by many people, and all the techniques to be discussed in Section 7.4.3 attempt to reduce overhead. However, as is discussed in Section 7.4.3, the SP2 communication subsystem design follows a different guideline, putting bandwidth as the first priority.

A somewhat surprising result of this study is that a bulk bandwidth of 15 MB/s should be enough. This study provides just one data point for parallel computing applications. It remains to be seen whether the same conclusions hold for other parallel computing benchmarks, throughput processing, or commercial applications. The study does not say how the results will change when the processor speed increases. This point will be addressed in Section 11.5.

7.4.3 Low-Level Communications Support

From Fig. 7.17, we see that BCL plays a critical role in improving the communication performance. In this section, we discuss three representative BCLs: the SHRIMP VMMC (zero-copy), the Fast Messages (1-copy), and the SP2 user-space (US) protocol (2-copy). Other BCLs include Active Messages, U-Net, Meiko CS-2, Cray T3D/T3E, DEC TruCluster/Memory Channel, Intel Paragon/TFLOP, etc.

When evaluating the performance of a communication subsystem, we will list the point-to-point communication performance at three levels:

- The hardware-level performance, which is the best performance that the communication hardware can deliver to the BCL. It is not the same as the "hardware peak", which is the rated peak performance. The latter will be used if the former data are not available.

- The BCL best performance, which is not necessarily achievable by applications.

- The application-level best performance, which is represented by the best performance using PVM, MPI, or RCP.

The SP2 Communication Software   The structure of the IBM SP2 system communication software is shown in Fig. 7.20. The SP2 supports two protocols: a kernel space, IP-based protocol and a user-space protocol called US. For the kernel space protocol, an AIX *network interface driver* (NID) connects the IP to the communication adapter.

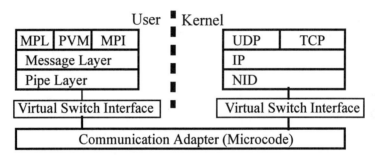

Figure 7.20   SP communication software (Copyright 1995, International Business Machines Corp. Reprinted with permission from M. Snir, et al. *IBM Systems Journal,* Vol. 34, No.2, [574], 1995)

The NID uses a *virtual switch interface*, which consists of two queues in the node system memory and a few control registers in the adapter memory. The two queues, one for input and the other for output, provide the system DMA buffer from which the adapter can DMA data in and out. There are two separate virtual switch interfaces, one for the IP and the other for the US protocol.

While the IP-based protocol is standard, the US protocol is proprietary, designed for high performance. As can be seen in Table 7.10, the US protocol significantly improves the performance over the IP. It is impressive that the US protocol can deliver over 80% of the hardware peak bandwidth to the application user.

The US protocol is designed based on the following facts and decisions:

- The HPS must support not only message passing communication used for technical computing, but also commercial applications, system services, and I/O (e.g., file) operations which often use the TCP/IP. Thus the adapter must be shared

Table 7.7 Performance of Point-to-Point Communication in SP2

(MPL is an application level message passing library similar to MPI)

	Latency t_0 μs	Bandwidth r_∞ MB/s	Half-Peak Length $m_{1/2}$ B
Hardware	<1	40	<40
MPL (US)	39	35.5	1385
MPL (UDP/IP)	277	10.8	2991

between a single process that uses the US protocol and any number of processes using the IP.

- Of the three communication performance parameters, bandwidth is considered the most important, latency the second, and node processor overhead the least.
- The main node processor (a POWER2 processor) is more powerful than the communication processor in the adapter (an i860). Therefore, it pays to let the node processor perform a large portion of message processing.
- It is slow for the node processor to access the adapter memory through programmed I/O, thus it is important to restrict such access.

The message layer and the pipe layer in Fig. 7.20 form the base communication library (BCL) of Fig. 7.17. The message layer consists of a few simple, non-blocking, point-to-point communication library functions.

All higher level message-passing functions in MPL, MPI, and PVM are implemented using these primitives. The pipe layer maintains a pair of pipes to provide a reliable, flow-controlled, and ordered byte stream between any pair of send/receive processes. The basic send-receive algorithm follows, referring to Fig. 7.18, Fig. 7.20, and Fig. 7.21:

- The source node processor executes a message-layer send operation to copy data from the send buffer to the pipe buffer. (If no room is available in the pipe buffer, the send blocks.) The message layer then invokes the pipe-layer code to copy data from the pipe buffer to the output queue.
- The source adapter moves data from the output queue to the adapter by DMA, then transfers the data via the network to the destination adapter.
- The destination adapter transfers the incoming data to the destination node's input queue (in the virtual switch interface) by DMA.
- The destination node processor executes the pipe layer to copy data from the input queue to the pipe buffer. The pipe layer then invokes the message layer to copy data from the pipe buffer to the final receive buffer, if a matching receive is already posted.

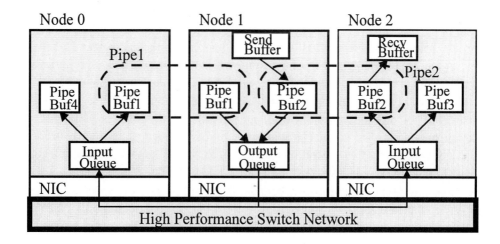

Figure 7.21   Buffer management in the SP2 communication subsystem

The last step needs elaboration. How is the destination processor made to execute the pipe layer to drain the network? This is a very important question, since receiver-side congestion can severely degrade communication performance. In the IBM SP2, only the pipe layer can drain data from the input queue and only the node processor can execute the pipe layer. The IBM SP2 uses a combination of two approaches to drain the network: *polling* and *interrupts*, as discussed below:

With the polling approach, the pipe layer is invoked by the message layer. The message layer is invoked by the destination process executing a receive operation. The message layer invokes the pipe layer to poll each pipe, and the pipe layer moves data from the input queue to a pipe buffer. If a matching message has arrived at a pipe buffer, the message layer copies data to the final receive buffer. If the message has not arrived, the receive operation is posted for later handling.

When a message arrives at the input queue of a destination process, the process may be busy doing something else without having executed a matching receive. Then the message will stay in the input queue. To guarantee progress in the pipe layer, the SP2 US protocol provides two *interrupt* mechanisms to drain the network: The pipe layer is periodically invoked by timer interrupts, and the pipe layer is triggered by an adapter interrupt when adapter buffers have filled.

The communication subsystem establishes a *pipe* for every pair of send/receive processes, which consists of a pair of pipe buffers, such as the two Pipe Buf2 in Fig. 7.21. There is one input queue and one output queue for each process. These queues are maintained in nonswappable memory area, shared by all pipes of the process.

If a parallel program consists of n processes, one per node, there will be at most one output queue and n outgoing pipe buffers at a sender node, and one input queue and n incoming pipe buffers at a receive node.

The node processor executes the computing code, the message layer, and the pipe layer in a coroutine fashion, with all three codes residing in the same user process space. So there are no system calls or kernel crossing. All DMAs are initiated and performed by the adapter. All data transfers are performed by DMA. Only control and status information is passed between the node processor and the adapter by programmed I/O.

The SP2 US protocol provides guaranteed, ordered, reliable, and flow-controlled delivery of messages. The adapter provides the function to translate logical process IDs used in US communication operations to physical addresses, while guaranteeing that communication will be contained within the allocated partition.

A token protocol is used for flow control: Data will be sent out from the source side of a pipe (i.e., data are copied from the pipe buffer to the output queue) only when there is available space at the destination side. When the receive node executes the pipe layer to drain data from the input queue to the pipe buffer, a token is released back to the source side. Acknowledgment packets with a time-out and retransmission scheme are used for reliable delivery.

Fast Message   Fast Messages (FM) is a public-domain BCL developed at the University of Illinois [480] for low-latency, high-bandwidth communication on clusters and MPPs. It has been implemented on Cray T3D and a network of Sun Sparc workstations connected by a Myrinet. A main objective of FM is to provide enough functionality in the messaging layer so that higher level package (e.g., sockets, PVM, and MPI) can realize communication performance close to the hardware limit.

On the T3D, FM achieved a latency of 6.1 μs and a bandwidth of 112 MB/s, out of a hardware limit of 130 MB/s. Table 7.10 shows the performance of FM on a cluster of SPARCstation 20 workstations connected by a Myrinet. It is impressive that an MPI implementation based on FM (called MPI-FM) achieved a latency of 19 μs and a bandwidth of 17.3 MB/s, out of a 23-MB/s hardware peak. A user-level socket interface achieves a 35-μs latency and a 11.3-MB/s bandwidth.

Table 7.8 Performance of Point-to-Point Communication with Fast Messages

Scheme	Latency t_0 μs	Bandwidth r_∞ MB/s	Half-Peak Length $m_{1/2}$ B
Hardware	<1	23	12
FM	14	17.6	246
MPI-FM	19	17.3	296
Socket-FM	35	11.3	386

The FM API is very similar to that of Active Messages (see Section 10.5.1), but is simpler, containing just a few functions. The important primitives are shown in Table 7.9. FM assumes that at most one process runs on a node that uses FM to communicate.

Table 7.9 Fast Messages Application Programming Interface

FM Primitive	Meaning
FM_initialize()	Initialize Fast Messages
FM_send(*dest, h, buf, size*)	Send a long message from memory
FM_send_4(*dest, h, i0, i1, i2, i3*)	Send a 4-word message from registers
FM_extract()	Extract received messages and invoke handlers

The function FM_initialize is the first that must be called by all node processes before FM can be used for communication, i.e., before any FM constants and global variables are accessed and before any communication function is invoked. This function initializes global variables used by the FM library, sets up the message queues and the DMA buffer, and assigns a node ID. It also performs a barrier synchronization.

The FM_send_4(*dest, h, i0, i1, i2, i3*) sends a four-word message to the destination node *dest*, calling the handler *h* and passing it up to four 32-bit integer arguments *i0, i1, i2, i3* and the sender's node ID.

The FM_send(*dest, h, buf, size*) function is similar to FM_send_4, but sends a message of *size* bytes at memory location *buf* to the destination node *dest*. The handler *h* will be called with *buf, size*, and the sender's node ID as arguments. A call to M_extract() processes all pending messages and invokes their handlers.

Note that an FM_send or FM_send_4 only specifies that a handler should be called; the actual invocation of the handler is done by an FM_extract. If no messages are available, FM_extract will return immediately, like a null function call.

A handler can contain arbitrary computation. In particular, a handler can call FM functions. Deadlock can happen because no resources are freed until the handler exits. FM users are responsible for preventing deadlock. One simple way is to force a handler to perform only local computation.

Example 7.8 A message-passing program using FM primitives

The following SPMD code shows how to send a message of *Size* bytes at location *Buf* in node 0 to location *A* in node 1:

```
#include "fm.h"
define handler H, which moves message into location A
main ()
{   ...
    FM_initialize() ; ...
    if (FM_NODEID == 0)
        FM_send(1, H, Buf, Size) ;
    else
        FM_extract() ; ...
}
```

Here FM_NODEID denotes the node ID assigned by FM_initialize. Different nodes will have different IDs.

■

The FM communication protocol is illustrated in Fig. 7.22. Four FIFO queues are used. The Myrinet adapter memory contains a send queue and a receive queue. The node kernel memory contains a larger receive queue in the nonswappable DMA region. A reject queue resides in the user space memory for flow control.

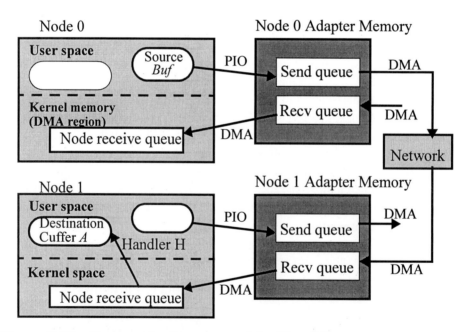

Figure 7.22   Communication protocol for Fast Messages
(DMA region in kernel space)

FM has been implemented on a cluster of SPARCstation 20 workstations connected by an 80-Mb/s Myrinet. The Myrinet NIC adapter contains a 128-KB SRAM and is loosely coupled to the I/O bus (SBus) of the workstation node. It contains three DMA engines, one to transfer from the adapter to the host, and two to transfer between the adapter and the network.

FM is a user-level communication layer. After each node executes FM_initialize, the adapter memory and the node receive queue are exposed to the user level. Subsequent FM_send (FM_send_4) and FM_extract do not have to cross protection boundaries by entering the kernel. Let us return to Example 7.8. When Node 0 executes FM_send(1, *H*, *Buf*, *Size*) to send a message to Node 1, the following sequence of steps is executed. First, the processor of Node 0 fetches data from *Buf*, packs the data into packets, and stores them directly into the send queue of the adapter by programmed I/O.

The remaining data movements are all via DMA. The packets are injected into the network by Node 0's adapter. When a packet arrives at Node 1, it is put into the adapter receive queue and later into the node receive queue. None of these DMAs involves the node processor. In particular, the last DMA from the adapter to the node receive queue can happen while the destination process is descheduled. This decoupling helps prevent receive side congestion.

FM_send and FM_send_4 are *blocking* sends: When a send function call returns, FM guarantees that *Buf* can be reused. The receive node (Node 1) should execute FM_extract sufficiently often to ensure that the node receive queue will not become full. Each FM_extract processes all pending messages from the node receiving queue, and invokes the associated handlers. In Example 7.8, the FM_extract executed by Node 1 processes the message sent from Node 0 and invokes the handler *H*, which will move the message data into the user memory space.

FM on Myrinet FM on Myrinet offers ordered, guaranteed delivery with flow control, but ignores fault handling, as the Myrinet is highly reliable, with a bit error rate less than 10^{-15}. Message order is guaranteed by the facts that all queues and Myrinet network buffers are FIFO, that there is a unique path from any source node to any destination node, and that the rest of protocol for flow control and guaranteed delivery does not cause messages to be out of order.

The Myrinet network uses wormhole routing. When an output port is busy, packets are blocked in the network buffer. Packets blocked for more than 50 ms are dropped. Because FM decouples the node and the network, the receive node adapter can drain the network even when the receive process is busy doing computation or is swapped out, as long as the destination node's receive queue is not full.

Therefore, packets will not be dropped and guaranteed delivery will be achieved if the destination receive queue never fills. This is achieved in FM by an end-to-end flow control scheme. Each sender process is allocated with a portion of the node receive queue, called *credit*, and the sum of all portions cannot exceed the total receive queue size.

A sender process can send a message to a receive node only when the sender still has credits. When a receive process removes a packet from the node receive queue, it sends a credit back to the sender process. If for some reason a receive process neglects to remove packets from the receive queue, eventually all sends will block. But a packet already sent is ensured that there is always space available in the destination node's receive queue, and the packet will not be dropped.

SHRIMP VMMC The Princeton SHRIMP project proposed a communication mechanism called *virtual memory-mapped communication* (VMMC), which is one of the few BCLs that offers a true zero-copy communication protocol. Other BCLs that use a similar approach include Meiko CS-2 and DEC TruCluster with Memory Channel.

All these systems need to modify the OS source code to accommodate new memory management requirements. The VMMC mechanism has been implemented on a PC cluster interconnected by a Myrinet, and on a PC cluster interconnected by custom communication hardware. The BCL level performance is very impressive, as shown in Table 7.10.

Table 7.10 Performance of Point-to-Point Communication with VMMC

Scheme	Latency t_0 μs	Bandwidth r_∞ MB/s	Half-Peak Length $m_{1/2}$ B
Hardware	5	110	550
VMMC	9.8	108	1058
vRPC	33	33	1089

In the SHRIMP system, a node is connected to the Myrinet through a PCI bus. The Myrinet and the PCI bus have a peak bandwidth of 160 MB/s and 133 MB/s, respectively. However, 110 MB/s is the highest bandwidth that the communication hardware can deliver to VMMC. The VMMC achieves over 98% of the hardware bandwidth. The vRPC row shows the performance of a VMMC-based RPC. No performance data are available for MPI/PVM on VMMC. However, communication schemes like VMMC can lead to good MPI/PVM performance. To demonstrate this, we show the performance of DEC TruCluster in Table 7.10, which uses an approach similar to VMMC. The BCL of the TruCluster system is called UMP, for *Universal Message Passing*.

The good performance of VMMC comes mainly from two techniques: *protected user level communication* and true *zero-copy protocol*. The VMMC mechanism is illustrated in Fig. 7.23. Three pieces of trusted software are used on each node: a VMMC daemon, a VMMC device driver, and an adapter control program which implements VMMC and maps the Myrinet network topology.

Let us return to Example 7.8. When Node 0 executes FM_send(1, *H*, *Buf*, *Size*) to send a message to Node 1, the following sequence of steps is executed. First, the processor

**Table 7.11 Performance of Point-to-Point Communication
with DEC TruCluster Memory Channel**

Scheme	Latency t_0 μs	Bandwidth r_∞ MB/s	Half-Peak Length $m_{1/2}$ B
Hardware	2.9	64	186
UMP	5.8	61	354
MPI	6.9	61	421
PVM	8	43	344

of Node 0 fetches data from *Buf*, packs the data into packets, and stores them directly into the send queue of the adapter by programmed I/O.

The VMMC daemons talk to each other over the Ethernet to establish an *import-export* relationship between the sender and the receiver nodes. Then a process in node 0 can send a message in its (user-space, virtual-address) source buffer directly to the destination buffer of another process in node 1. There is no need to go through one or more kernel space DMA buffers as in Fast Messages and the IBM SP2.

The adapter in Fast Messages and the IBM SP2 sees physical addresses of the message buffers, while with VMMC, the adapter sees virtual addresses. Therefore, the adapter must maintain page tables to perform address translation. The page tables and the protection mechanism are set up during the import-export phase.

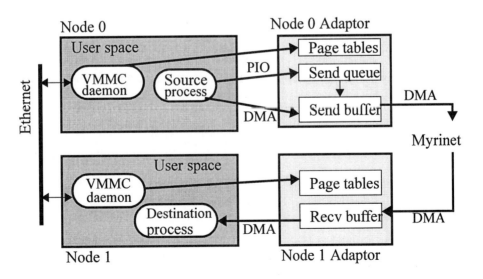

Figure 7.23 Communication protocol of SHRIMP VMMC.

7.4.4 Communication Algorithms

We have discussed the techniques for efficient support of point-to-point communications in Section 7.3.3. This section discusses how to support collective communications, which are further grouped into regular and irregular ones. For reasons of generation scalability and heterogeneity scalability, the parallel community have been moving toward using a common machine model to develop communication algorithms.

This model consists of n nodes, and any node can directly send a point-to-point message of m-byte to any other node in $t_0 + m \cdot t_c$ μs, where t_0 and t_c are the start-up and the per-byte times, respectively. In other words, the interconnect topology is ignored. Another less-used model assumes that a node can broadcast a message to all nodes, which is an abstraction of Ethernet-connected systems.

Regular Communications There are many types of regular communications, as discussed in Example 2.7. Such communications can be further classified into two classes: A *rooted communication* is an all-to-one or one-to-all communication where one node serves as the *root* node. Examples include broadcast, gather, scatter, and reduction. A *homogeneous* communication is a collective communication which does not have the notion of a distinct *root* node. Instead, the operation is specified the same for all nodes. Examples include total exchange, barrier, and combine (where every node receives the reduction sum). We discuss below how total exchange is realized.

There are three main algorithms for *total exchange,* as briefly introduced below.

(1) The *direct-exchange* algorithm, first proposed by Take [604], is very simple and theoretically optimal for large messages.

(2) The *standard-exchange* algorithm, proposed by Johnsson and Ho [354], works well for short messages.

(3) The *multiphase-complete-exchange* algorithm, proposed by Bokhari [99,100], combines the advantages of both the direct exchange and the standard exchange algorithms.

These algorithms were all originally proposed for hypercube multicomputers. But Bokhari [100] showed that the multiphase algorithm performs well on non-hypercube machines such as IBM SP2, Intel Paragon, and Meiko CS-2. We will illustrate the direct and the standard algorithms.

Example 7.9 The direct-exchange approach to implementing total exchange

Suppose a multicomputer consists of $n = 2^d$ nodes interconnected by a hypercube network. Each node has an ID that can be obtained by calling function myid(). Initially, node i ($0 \leq i < n$) contains n data blocks, denoted by $i{:}0, i{:}1, ..., i{:}n-1$. At the end of the total exchange, data block $i{:}j$ should be in node j. The direct-exchange algorithm follows, where each node executes the same code:

```
i = myid();
for (k=1; k<n; k++) {
    destination = i | k ;              /* i exclusive_or k */
    send block (i:destination) to destination node
}
```

■

This algorithm has three advantages: It is very simple; it transmits the minimum amount of data; and the exclusive-or operation guarantees that there is no link contention in a hypercube if synchronous dimension-order routing is used.

Although the direct-exchange algorithm transmits the minimum amount of data, it needs to execute $(n-1)(n-1)$ send/receives in total, and each message contains one block. For short messages, the cumulative start-up overhead could be high.

Example 7.10 The standard-exchange algorithm for total exchange

The standard-exchange algorithm combines several short messages into larger chunks, and each node only needs to send $\log n$ messages. This reduces the cumulative startup overhead. Each message contains $n/2$ blocks. So the total transmitted data amount is $(n^2 \log n)/2$ blocks. The algorithm follows:

```
i = myid();
for (k=d; k>=0; k--) {
    if (bit k of i is 1) superblock = blocks (i:0) to (i:n/2-1) ;
    else superblock = blocks (i:n/2) to (i:n-1) ;
    destination = i | 2^k;            /* i exclusive_or 2^k */
    send superblock to destination node
    shuffle blocks locally
}
```

■

Irregular Communications An abstraction of irregular communications is the h-relation described in Definition 1.6. If we know that each node sends at most h words out and receives at most h words, and at least one node sends or receives h words, the communication is an h-relation but not an $(h-1)$-relation. We discuss a simplified h-relation problem and a solution by Bader et al. [51]. Their solution is applicable to the general h-relation problem.

Suppose a set of N *elements* is initially distributed among n nodes evenly, such that no node holds more than N/n elements. Each element is a pair *<data, dest>*, where *data* are to be routed to the *dest* node. Any node is the destination of at most h elements; N divides n and h divides n.

The algorithm by Bader et al. is based on a scheme originally proposed by Leslie Valiant [623]. The idea is that an irregular communication should be realized as a two-phase procedure to reduce network contention. In the first phase, each node sends its message to an arbitrary node. In the second phase, the messages are sent to their final destination. The algorithm has the following steps:

(1) Each node P_i keeps n bins. Its elements are assigned to one of the n bins as follows: The elements are looked at one by one. An element is placed into bin $(i+j) \bmod n$ if it is the first element with destination j. Otherwise it is placed into bin $(b+1) \bmod n$, where b is the bin into which the last element with destination j was placed.

(2) The nodes perform a total exchange communication, where node P_i sends the contents of its bin j to node P_j. Each bin contains no more than $N/n^2 + n/2$ elements.

(3) Each node rearranges the elements into n bins according to their final destinations.

(4) The nodes perform a total exchange communication, such that node P_i sends the contents of its bin j to node P_j. Each bin contains no more than $h/n + n/2$ elements.

Bader et al. have shown that this algorithm has a parallel time complexity of $O(h + t_0 + (h + n^2 + N/n) t_c)$. They have also given experimental data on several MPPs to show that their algorithm is faster than other existing algorithms. In particular, their two-phase algorithm is faster than single-phase implementations, where a node sends an element directly to its final destination. Their h-relation algorithm is used to develop an efficient parallel sorting algorithm for message-passing machines.

7.5 Bibliographic Notes and Problems

The concept and implementation of Solaris threads were described in Sun's publications [132, 583, 601]. Similar threading techniques have been used in commercial operating systems [193, 340]. Multithreading is a feature of Java programming language.

The Lazy Threads approach [270] tries to make thread creation as efficient as making a sequential procedure call. The work-stealing scheduler in the Cilk thread library [94] schedules threads to limit space, time, and communication requirements. The Nexus run-time system [250] shows an efficient way to integrate communication and threading.

Lamport's 1983 paper [390] presents mutual exclusion and Lamport's pet problem. The ideas of *fetch-and-add* and combining are proposed by Gottlieb et al. [276], while Kruskal et al. [380] gives a formal treatment. A software combining technique is presented by Goodman et al. [272]. The shortcomings of traditional locking synchronization are discussed by Herlihy et al. [306, 307] and by Stone et al. [589], where techniques such as wait-free synchronization, transactional memory, and Oklahoma Update are suggested.

Another hardware synchronization construct is *Queue-On-Lock-Bit*, or *QOLB*, proposed by Goodman et al. [272]. QOLB has been implemented in the IEEE SCI standard. Kagi et al. [358] recently show that QOLB outperforms other locking schemes, and can be realized by software in current CC-NUMA machines.

Xu and Hwang [654, 658] discuss the disadvantages of locking from a programmer's viewpoint, and proposed a new language construct called *coherent region* to deal with the problems. Coherent region will be discussed further in Section 12.3.5. For recent advances in synchronization, see Mellor-Crummey and Scott [441], Anderson and Moir [32], and Kontothanassis et al. [374].

The TCP/IP suite is descibed in Comer's book [163] in details. The communication overhead of current MPPs is analyzed in Karamcheti and Chien [360]. The IBM SP2 communication subsystem is described in Snir et al. [574]. The Illinois Fast Messages (FM) is described in Lauria and Chien [394], and Pakin, et al. [480].

The communication speedup efforts in Princeton SHRIMP are described in Blumrich et al. [95, 96, 215, 216, 237, 342]. Other treatments of fast communication include Active Message [179, 426, 535, 626, 628], Myrinet [97], and UNet [627].

The LogP model was presented by Culler et al. [178], and later extended to the LogGP model [22]. Efficient algorithms for regular communications can be found in Bala et al. [59], Bokhari [100], Bruck et al. [112], and Johnsson and Ho [354, 355]. Techniques for irregular communications are discussed in Leighton [402] and Bader et al. [51].

Homework Problems

Problem 7.1 Answer the following questions related to Solaris threads:

(a)   What are two advantages of using the two-level structure (i.e., threads and LWPs instead of just threads)? Briefly explain.

(b)   What is the advantage of having ZOMBIE threads? When should the reaper thread be executed?

Problem 7.2 Compare parallel processing at the threading and process levels.

(a)   Use concrete examples to explain four reasons why multithreading can be beneficial compared to exploiting parallelism at the process level.

(b)   Also give one counterexample to explain the situation where multithreading will not benefit, as far as parallel performance is concerned.

Problem 7.3 Consider the following parallel code using lock and unlock:

> **parfor** (i=0; i<n; i++)
> {
>   noncritical section
>   **lock**(S) ;
>   *critical section*
>   **unlock**(S) ;
> }

Assume the noncritical section takes T_{ncs}, the critical section T_{cs}, and a **lock**(S) t_{lock} seconds. The **unlock**(S) overhead is negligible. The corresponding sequential code needs $n(T_{ncs} + T_{cs})$ seconds to complete. You can ignore the parallelism overhead in your solutions:

(a)   What is the total parallel execution time ?

(b)   What will be the speedup in using *n* processors?

Problem 7.4 To reduce contentions in accessing two shared variables Ticket and Turn, the ticket algorithm in Section 7.2.4 has suggested several techniques.

(a)   Briefly describe three such techniques.

(b)   Assume there are *n*=4 processors, each read or write of the shared memory takes 100 cycles, and executing the critical section once needs 1000 cycles. The delay function let the processor wait for $k \times$ (Myticket−Turn) cycles.

All other operations take zero time. How many cycles are needed to execute the following program if *k*=100?

```
int Ticket=0, Turn=0 ;          // shared variables, initialized to 0
parfor (i=0; i<n; i++)          // there are n processes
{  int Myticket ;               // local variable
   Myticket = FAA(Ticket, 1) ;{    // these three lines
   while (Myticket != Turn)        // are equivalent to
       Delay(Myticket-Turn) ;      // a lock operation
   critical section
   Turn = Turn+1 ;              // unlock
}
```

(c)   What is the optimal value of k such that the execution time in (b) is minimized? You have to justify the choice with reasoning.

Problem 7.5 Solve the following problems regarding the use of fast locking techniques described in Section 7.2.4.

(a)   Write a test-test-and-set locking algorithm with exponential backoff.

(b)   Write a mutual-exclusion program similar to that in Problem 7.4(b) but using the array-link-based algorithm.

Problem 7.6 Explain the following concepts regarding the TCP/IP suite:

(a)   A connection-oriented protocol versus a connectionless protocol

(a)   The TCP versus the UDP

(b)   The IP. Focus on its functionality differences from TCP and UDP.

(c)   The sockets interface

Problem 7.7 Explain the following concepts for efficient communication:

(a)   Node processor versus NIC processor. What are they used for?

(b)   Basic communication library (BCL) versus MPI.

(c)   Protected user-level communication. What are its advantages?

(d)   One-copy protocol versus zero-copy protocol. For each protocol, explain how many memory-copying operations are performed from the start of a send to the end of a corresponding receive.

(e)   Guaranteed delivery versus ordered delivery

Problem 7.8 Protocol processing, memory copying, and protection boundary crossing are three sources of communication software overhead. Use a concise table to contrast how the SHRIMP, the Fast Messages, and the SP2 US protocols tried to reduce the three types of overheads. In more detailed explanatory text, compare their advantages and disadvantages.

Problem 7.9 A binary hypercube multicomputer of 8 nodes has a point-to-point start-up time of $t_0 = 50$ μs and an asymptotic bandwidth of $r_\infty = 100$ MB/s. The binary hypercube interconnect uses dimension-order routing. Ignore all other time.

(a)   What is the time to perform a total exchange of a 100-MB array using the direct-exchange algorithm?

(b) What is the time to perform a total exchange of a 100-MB array using the standard-exchange algorithm?

(c) What is the time to perform 100 total exchanges of a 1-MB array using the direct-exchange algorithm?

(d) What is the time to perform 100 total exchanges of a 1-MB array using the standard-exchange algorithm?

Problem 7.10 Use the algorithm by Bader et al. in Section 7.4.4 to realize the following h-relation on a 4-node system. Initially, each node has 3 elements:

Node 0	Node 1	Node 2	Node 3
<D0,0>	<D4,1>	<D7,2>	<D10,3>
<D1,2>	<D5,2>	<D8,2>	<D11,2>
<D2,3>	<D6,3>	<D9,3>	<D12,3>

The element <D0, 1> indicates that data D0 should be sent to the destination node 1. Draw a sequence of figures to show the contents of each node in the four steps of the algorithm. What is the value of h?

Part III

Systems Architecture

In recent years, we see a surge of interest in the research and development of scalable parallel systems. A good indicator is the proliferation of more and more new products and professional conferences in this area. Some research prototype machines built at universities are being transferred for industrial production. Business applications of scalable and cluster platforms increase rapidly.

III.1 Motivations

Market demand and application trend are two major factors that have motiviated the system architecture development over the past.

Market Demand Almost all major computer manufacturers, from Compaq to IBM, are competing to push out parallel or cluster systems products. Software companies from Microsoft to Oracle are rushing to announce "parallel-ready" systems and applications software products. Market researchers are predicting a healthy growth of low- to mid-range parallel servers.

Eighty percent of *information technology* (IT) managers are now considering parallel systems when planning future deployment, while a few years ago, only 15 percent of them did. There is a growing consensus that there will be only two types of computers by year 2000:

- Various inexpensive *client machines*; such as desktops, laptops, set-to-boxes, and *network computing* (NC) machine

- Various *scalable servers,* including everything we will cover in this part.

Application-Driven Trends The explosive growth in Internet, Intranet, and World Wide Web further accelerates the research and development of scalable systems. When millions of users throughout the world are connected, they will demand more than just a simple E-mail service.

Internet telephony, video-conferencing, media servers, information servers, electronic commerce, and Internet-based supercomputing (*metacomputing*), all demand parallel servers that are scalable and highly available, the *superservers*. Several application trends of superservers are identified below:

- While technical computing will still be highly visible, applications on parallel systems will be dominated by commercial database operations (e.g., OLTP, OLAP, and data warehousing) and network-based applications (e.g., Web servers, contents servers, metacomputing).

- A recent concept of using parallel machines is *server consolidation* or *LAN consolidation*, where a large parallel server replaces multiple discrete small servers, to offer better manageability and scalability.

- Applications and systems software development is becoming more important than hardware installations.

- Scalability, availability, and manageability become increasingly more important than just the speedup in solving a single large problem.

- While mainframes and vector supercomputers are struggling to survive, SMPs, CC-NUMA machines, and clusters are taking a major share of the server market for data warehousing and supercomputing applications.

- More and more commodity components will be used, in both hardware and software. However, innovative components are still needed for supporting new applications and performance-critical subsystems.

- Communication and I/O capabilities, not computing capability, are becoming the system bottleneck to require more R/D effort.

III.2 Part Objectives

Four classes of scalable systems are presented in this part. While some research projects are included, we emphasize commercial systems that are representative of the industry and have innovative features. For each system, we describe the hardware architecture, the system software, and special features that make the system unique. We pay special attention to the underlying technology, which can be used even when a particular machine is discontinued.

Chapter 8 This chapter discusses SMP and CC-NUMA systems. These systems have a hardware-supported single memory space with a large number of existing applications. SMPs make up the server market today. CC-NUMA machines extend the SMP architecture to hundreds of processors.

Chapter 9 This chapter covers the principles of clustering. We start with cluster architectures and design issues. The we discuss hardware and software support for availability and single system image. Finally, we study job management issues in clusters.

Chapter 10   This chapter describes various cluster systems starting with a grand tour of research and commercial clusters. Then we study the details of the Berkeley NOW, IBM SP2, and Digital TruCluster.

Chapter 11 This chapter presents several MPP architectures. MPP machines are expensive but have the best size scalability. Grand challenge applications still require Tflop/s speed over a system of thousands of processors.

III.3 Notes to Readers

Chapter 8 is of crucial importance to server designers, because they still dominate the parallel market. The two cluster chapters should be read in order, Chapter 9 ahead of Chapter 10. SMP, NUMA, and clusters are the future of parallel and distributed computing. MPPs in Chapter 11 suggest a rebirth of the supercomputer industry, since vector multiprocessors have limited scalability.

The four chapters in Part III are on the must-read list by computer architects, systems designers, and application programmers. We consider them the prerequisite for those who want to read the programming chapters in Part IV. However, those hardware engineers who have no time to program the machines they build can rest with Part III, perhaps go fishing instead.

Chapter 8

Symmetric and CC-NUMA Multiprocessors

In this chapter, we study the architecture of scalable, shared-memory multiprocessors. Design experiences of recent *symmetric multiprocessors* (SMPs) and *coherent-cache nonuniform memory-access* (CC-NUMA) machines are reviewed. In particular, we study the details and unique features in two SMP systems: the Intel SHV system board and the Sun Ultra E-10000 server; and three CC-NUMA systems: the HP/Convex Exemplar X-class, SGI/Cray Origin 2000, and Sequent NUMA-Q 2000. A technical comparison of four CC-NUMA multiprocessors is given at the end.

8.1 SMP and CC-NUMA Technology

SMP architecture has been adopted in almost all parallel servers used today. NUMA machines are natural extensions of SMP systems. The architectural characteristics, a survey of SMP systems, and two case studies of SMP systems are given below. NUMA machines are discussed in subsequent sections.

8.1.1 Multiprocessor Architecture

A common architecture of shared-memory systems is shown in Fig. 8.1(a). Shared-memory system (SMP or CC-NUMA) has a number of advantages as listed below:

- *Symmetry:* Barring normal operating system protection, any processor can access any memory location and any I/O device.
- *Single Address Space*: This feature brings several benefits. *Single system image* is naturally supported, since there is only one copy of operating system, database, application, etc., residing in the shared memory. The single OS schedules

processes to run on multiple processors according to their workload, easily achieving *dynamic load balancing*. All data reside in the same shared-memory space, thus users do not have to worry about data distribution and redistribution.

- *Caching:* Data locality is supported by a hierarchy of caches.
- *Coherency*: Cache coherency is enforced by hardware. However, different SMPs or CC-NUMA machines may support different shared-memory models, some of which may demand programmer's intervention to guarantee consistency.
- *Memory Communication:* Shared-memory communication has low latency because it is done with simple load/store instructions and the hardware generates coherency information. This is in contrast to *I/O communication* in multicomputers (see Section 6.6), which needs to execute many instructions and does not utilize the cache coherency information.

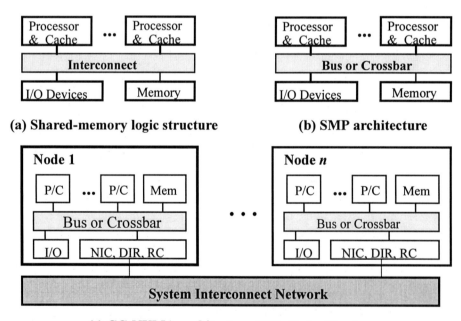

(a) **Shared-memory logic structure**   (b) **SMP architecture**

(c) **CC-NUMA architecture** (DIR: Cache Directory,
NIC: Network Interface Circuitry, RC: Remote Cache)

Figure 8.1   Common architectures for SMP and CC-NUMA systems

Symmetric Multiprocessors   Most commercial SMP systems assume the bus or crossbar architecture shown in Fig. 8.1(b). Most SMPs use a bus interconnect, and cache coherency is enforced through some MESI-like snoopy protocol. Bus-connected SMPs have become commercially successful, taking a great portion of the parallel computer market today. Listed below are basic issues that SMPs must address:

- *Availability:* This is probably the biggest problem of any SMP. Failure of the bus, the memory, or the operating system will break the entire system.

- *Bottleneck:* All processors and I/O controllers compete for the memory bus and the shared memory, which become the bottleneck. To alleviate this problem, most SMPs are using techniques such as split transactions (packet switched bus) and multiple nonblocking outstanding requests.

- *Latency*: Compared to multicomputers, SMPs have small latency. However, the latency is still large compared to the processor speed. Contentions could further lengthen the latency. Even we ignore contention, the *minimum memory latency* of a current SMP is often hundreds of processor cycles, long enough for executing thousands of instructions. The reduction of latency has not kept pace with the increase in processor speed.

- *Memory Bandwidth*: The memory bandwidth has not kept pace with the increase in processor speed or that of memory capacity. John Mashey of SGI has estimated that memory and disk capacities increase 4 times every three years and the SMP memory bus bandwidth increases only two times every three years [432].

- *I/O Bandwidth*: The increase rate of individual I/O bus bandwidth is even lower. One way to increase the system bus bandwidth is to replace it with a crossbar switch network, connecting all processors to all memory and I/O devices.

- *Scalability*: A bus is not scalable (see Section 6.6.1), limiting the maximum number of processors to tens. For many applications environments, no more than four processors can be effectively used, due to the contention on the shared bus/memory. Three approaches have been used to scale to larger systems: (1) using a bus/crossbar interconnect (e.g., Sun Ultra E-10000 discussed in Section 8.1.2), (2) using a CC-NUMA architecture (as discussed later in this chapter), and (3) using clustering (to be discussed in Chapters 9 and 10).

Memory Latencies According to Larry McVoy of SGI, there are several types of memory read latency. The first measures the time when the read request leaves the processor chip to the time the data return from the memory to the processor chip. This is essentially the time of a memory read bus cycle. Normally, this ranges from a few to tens of bus clocks.

In an Intel SHV, for example, a memory read bus cycle takes 13 bus clocks. With the bus running at 66 MHz (15 ns), that is 195 ns. The second definition of latency measures the time when the processor starts to execute the memory access operation to the time data return to the appropriate register, assuming there is no paging overhead, contention over the bus, or interference from previous outstanding requests.

This is called the *minimum memory latency*. This latency could be several times of a memory-read bus cycle. The third latency definition is similar to the second one, except contention and interference overheads are included.

Example 8.1 The effective memory bandwidth in an SMP system

In 1997, the bus used in a typical SMP system runs at a clock around 66 MHz. Only a few exceed 100 MHz. Suppose that we use an enhanced bus of k-B width running at $f = 100$ MHz. A bus transaction needs 1 clock for addressing, 1 clock for arbitration, 1 clock for cache coherency, and $\lceil m/k \rceil$ clocks to transmit a block of m-B data. When $k = 64$ b $= 8$ B and $m = 16$ B, the manufacturer-published data sheet claims a bus bandwidth of $k \times f = 8 \times 100 = 80$ MB/s. However, the effective bandwidth is $(m/(3 + \lceil m/k \rceil)) \times f = (16/(3 + 16/8)) \times 100 = 320$ MB/s at its maximum value.

Now the manufacturer increases the bus width to 256 b, or 32 B, and claims a $32 \times 100 = 3200$ MB/s bandwidth. Will the effective bandwidth also increase by four times? The answer is no. The maximum effective bandwidth is increased to only $(16/(3 + \lceil 16/32 \rceil)) \times 100 = 400$ MB/s, or by 25%. Of course, the 3-clock overhead can be amortized by using a larger block size m.

In practice, m often has an upper bound of the cache line size, which is around 32 to 256 B in many high-end systems. When the line size m is enlarged to 64 B, the effective bandwidth becomes $(64/(3 + \lceil 64/32 \rceil)) \times 100 = 1280$ MB/s, still far less than $32 \times 100 = 3.2$-GB/s hardware peak. The effective bandwidth so calculated is the *maximum* achievable. The sustained memory bandwidth could be even less.

■

CC-NUMA Systems   As shown in Fig. 8.1(c), a CC-NUMA machine extends SMPs by connecting several SMP nodes into a larger system. Most commercial CC-NUMA multiprocessor systems employ a directory-based, cache-coherent protocol. While maintaining the advantages of the SMP architecture, the CC-NUMA machine alleviates the scalability problem of conventional SMPs.

The distributed shared-memory architecture enhances scalability. Not only processors, but also memory capacity and I/O capabilities can be increased by adding more nodes. The contention and bandwidth problems are alleviated because an application can access multiple local memories simultaneously most of the time, taking advantage of data locality.

Some CC-NUMA machines even attempt to solve the availability problem of SMP. For instance, the SGI Cellular IRIX operating system implements the *cell* concept, whereby multiple copies of a portion of the operating system can run on multiple nodes, and the failure of one will not disrupt the entire system.

A notable advantage of CC-NUMA machines over multicomputers or NCC-NUMA machines is that the programmer does not have to explicitly distribute data structures over the nodes. The system hardware and software will automatically distribute data among the nodes initially. During runtime of the application, the cache coherency hardware will automatically move the data to the nodes where they are referenced. However, this advantage is not always realized effectively, as revealed in the following example:

Example 8.2 The remote caching problem in a CC-NUMA system

Suppose a program has two processes P and Q executing the following code that accesses data arrays A and B:

	P:	Q:
Phase 1:	use(A)	use(B)
Phase 2:	use(B)	use(A)

The system initially allocates data as shown in Fig. 8.2(a), which is ideal for phase 1. There is no need for communication, and the remote caches are empty. Note that the *remote* caches are so named because they are used to cache data whose *home* is in a remote memory.

In phase 2, the hardware will automatically move B to node 1 and A to node 2 as shown in Fig. 8.2(b). The programmer does not have to do a data redistribution, which is required in a multicomputer.

Now comes the problem. The remote cache is much smaller than the local memory in a node. For instance, in Sequent NUMA-Q, the remote cache is 32 MB, while the local memory is 4 GB. What if the work set of the program in phase 2 is much larger than remote cache?

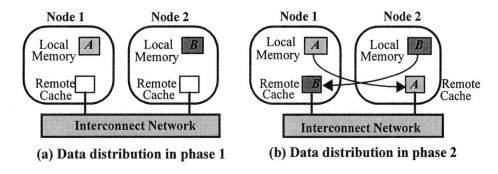

(a) Data distribution in phase 1   (b) Data distribution in phase 2

Figure 8.2   Data relocation using remote caching in a CC-NUMA multiprocessor system.

The consequence is that a capacity miss will occur, and the contents of remote caches must be written back to the home memory in a remote node. Satisfying such capacity misses requires communicating over the network.

To solve this *remote-caching problem*, operating systems in CC-NUMA machines incorporate a number of techniques. For instance, the *affinity scheduling* technique will reschedule the two processes in phase 2 so that process *P* is moved to run in node 2 to be close to its data, while process *Q* is moved to run in node 1. The *page migration* technique will migrate the pages of data *B* to the local memory of node 1, and those of *A* to node 2.

■

The innovative techniques in CC-NUMA system hardware and software can be very effective in exploiting data locality and enhancing scalability. There is some evidence showing that for commercial applications, the majority of data accesses are satisfied within local nodes, and most communications over the network are not for transmitting data, but for cache invalidation. Remote caching will be treated further in Section 8.4.2.

8.1.2 Commercial SMP Servers

SMP servers have become the most successful parallel computers in use today. We summarize five systems in Table 8.1. We then discuss two representative SMP systems in detail. The Intel SHV four-processor server represents the low end, and it has been widely used as a building block in many NUMA systems. The Sun Ultra Enterprise 10000 represents the high end, expandable to 64 processors and is treated in Section 8.2

The parameters in Table 8.1 are all maximum or best values. Some features are common to all SMPs and are explained below:

- *Performance*: All SMPs strive to provide scalable performance, including not only more and faster processors, but also bigger cache/memory/disk capacity, more I/O slots, higher memory and I/O bandwidth, and lower memory access latency.

- *Resilience:* The system must be resilient to component failure and maximize application uptime. Known techniques include using reliable, redundant, and hot-swappable components (power supplies being one); using a service processor to perform predictive diagnostics, system monitoring and testing; making the system *cluster-ready*, meaning several SMPs can be easily clustered to provide higher availability; and doing extensive testing and integration with databases and applications.

- *Integration:* A system should provide interoperability glue to facilitate integration between applications, middleware, operating systems, and networks within

Table 8.1 Comparison of Five Commercial SMP Systems

System Characteristics	DEC Alphaserver 8400 5/440	HP 9000/ T600	IBM RS6000/R40	Sun Ultra Enterprise 6000	SGI Power Challenge XL
No. processors	12	12	8	30	36
Processor type	437 MHz Alpha 21164	180 MHz PA 8000	112 MHz PowerPC 604	167 MHz UltraSPARC I	195 MHz MIPS R10000
Off-chip cache per processor	4 MB	8 MB	1 MB	512 KB	4 MB
Max memory	28 GB	16 GB	2 GB	30 GB	16 GB
Interconnect bandwidth	Bus 2.1 GB/s	Bus 960 MB/s	Bus + Xbar 1.8 GB/s	Bus + Xbar 2.6 GB/s	Bus 1.2 GB/s
Internal disk	192 GB	168 GB	38 GB	63 GB	144 GB
I/O channels	12 PCI buses, each 133 MB/s	N/A	2 MCA, each 160 MB/s	30 Sbus, each 200 MB/s	6 Power Channel-2 HIO, each 320 MB/s
I/O slots	144 PCI slots	112 HP-PB slots	15 MCA	45 Sbus slots	12 HIO slots
I/O bandwidth	1.2 GB/s	1 GB/s	320 MB/s	2.6 GB/s	320 MB per HIO slot

mixed-vendor environments. An ideal server should be able to seamlessly integrate with mainframe, Unix, Windows NT, and other platforms.

- *Security*: Many SMP servers will be working in a networked environment, instead of in an isolated mainframe plus terminals configuration. Security is especially important with the popularization of Internet, Intranet, and data centers.

- *Management:* The system should provide capability of managing configuration, faults, resource, performance, and accounting.

8.1.3 The Intel SHV Server Board

A typical configuration of the Intel *standard high-volume* (SHV) server board is shown in Fig. 8.3. It consists of four Pentium Pro (P6) packages, two PCI bridges, one or more memory controllers, and an OEM bridge. These components are called bus agents and are all attached to a 66-MHz, 64-b memory bus. A P6 package integrates a P6 processor, a 256- or 512-KB level-2 cache, an *advanced programmable interrupt controller* (APIC), and an L2/bus controller into a single package at the board level.

The PCI bridges interface two PCI buses to the memory bus. These two PCI buses provide connections to disks, CD-ROMs, and other I/O devices. The memory controllers manage DRAMs up to 4 GB. The OEM bridge is open to be built by third-party developers, who can put whatever they want there (e.g., a level-3 cache). A popular use is glue logic that connects multiple SHV boards into a large cluster or CC-NUMA machine.

The Memory Bus The Intel SHV memory bus incorporates several features that are common to many SMP buses. It supports a hardware cache-coherence protocol, which in the SHV is the MESI protocol. It uses separate signal lines to perform different tasks, such as bus arbitration, addressing and request, and response. Memory accesses can be pipelined. Split transactions and outstanding requests are supported to provide high sustained bandwidth. A bus with this last feature is also called a *packet-switching bus*.

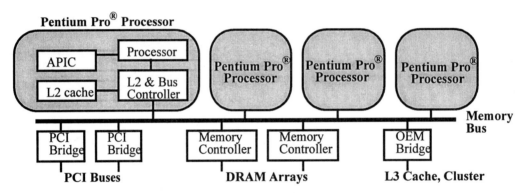

Figure 8.3   **The Intel SHV SMP board architecture.** (Reprint by permission of Intel Corp. Copyright Intel Corp. 1997, from [346])

The SHV memory bus has a bus clock of 15 ns, or 66.7 MHz. Normally, the SHV memory bus performs *pipelined, in-order* transactions, as shown in Fig. 8.4. *In-order* means that the bus requests are served in first-come, first-served basis. The bus signal lines can be divided into six individual groups (subbuses). A read memory bus cycle starts by asserting the arbitration subbus to acquire the request subbus. This takes two clocks. Then a request is issued on the request subbus across two adjacent clocks: The first contains address, and memory type, and the second contains a unique transaction ID, request length, etc. Three clocks after the request is issued, the error subbus is checked. Any error will cause the request to be reissued.

Four clocks after the request is issued, the request arrives at the designated bus agent and causes it to indicate if it wants to respond. If not, wait states are inserted in a multiple of 2 clocks. If the agent decides to complete, it asserts the completion subbus, and two cycles later, asserts the response subbus and puts data on the data subbus. Suppose there is a cache line length of 32 B. Then 4 clocks is needed to transmit the data. Altogether, a read bus cycle takes 13 clocks. We can view the six subbases as pipeline stages. A subsequent bus transaction can begin every 4 clocks, instead of every 13 clocks.

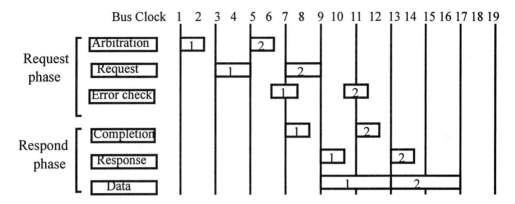

Figure 8.4   Pipelined Intel SHV bus read cycles. (Reprint by permission of Intel Corp. Copyright Intel Corp. 1997, from [346])

The bus supports *split transaction* to allow multiple agents to share the bus. A transaction is split into a request phase (arbitration, request, and error check) and a respond phase (completion, response, data). The bus can execute the request phase of a transaction, then execute other transactions before executing the respond phase of the first transaction. The bus supports up to four outstanding requests, while executing other transactions.

Example 8.3 Split transactions in using Intel SHV SMP boards

Suppose a CC-NUMA machine uses a number of SHV boards as building blocks. The designer of the CC-NUMA machine uses the OEM bridges to connect the SHV boards through an SCI ring. Suppose two read bus cycles are executed, and the first read bus cycle does not find the data in local memory. It has to go to a remote memory, which takes 130 bus clocks. The second read bus cycle always finds data in the local memory and should take only 13 clocks. However, without split transaction, the second bus cycle must wait till the first one finishes, meaning its latency is increased to 130 clocks as well.

With split transaction, the first transaction tells the bus that it could not complete immediately and yields the bus to allow the second transaction to finish in 13 clocks. Meanwhile, the outstanding request of the first transaction is kept in the OEM bridge, waiting for the remote access to complete. When the data arrive, the OEM bridge puts the data on the bus and finishes the transaction. There is no confusion because every transaction has a unique ID.

■

8.2   Sun Ultra Enterprise 10000 System

The Ultra Enterprise 10000 is Sun's high-end SMP server, also called StarFire. This server incorporates several novel techniques to enhance performance and availability. It can scale up to 64 processors, 64 GB of memory, and over 20 TB of on-line disk. The system runs all Solaris applications and provides connectivity and interoperability supports to integrate with mainframe environments.

8.2.1   The Ultra E-10000 Architecture

The Enterprise 10000 has made significant architectural improvement in scalability, reliability, availability, and serviceability over the Enterprise 6000 system. The block diagram of the Enterprise 10000 is shown in Fig. 8.5. The system is built around a high-bandwidth centerplane as described below.

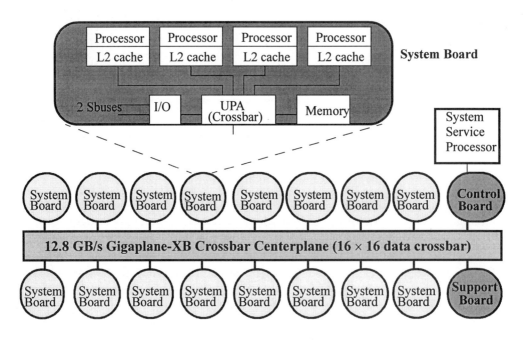

Figure 8.5   The Sun Ultra Enterprise 10000 architecture.
(Courtesy of Sun Microsystems, [600], 1997)

Gigaplane-XB Interconnect   The Gigaplane-XB interconnect is improved by 10 times in bandwidth from the Gigaplane system bus in Enterprise x000 servers (Example 6.1). The Gigaplane-XB interconnect is based on two levels of crossbar. This interconnect glues all memories and I/O devices in the system boards into a single address space, such that any processor can access any memory or I/O device in the entire system.

The interconnect adheres to the *Ultra Port Architecture* (UPA) standard. The UPA bus, the primary bus for the Ultra-1 desktop workstations and Enterprise servers, is used as an intermediate bus to connect CPU/memory boards and I/O boards to Gigaplane bus. The UPA bus runs at 83.3 MHz, with a peak bandwidth of 1.3 GB/s.

The UPA in a system board is a crossbar that allows the four processors and the I/O module to talk to the memory simultaneously. The UPA has a data width of 128 b at each port. The UPA is also responsible for routing data packets to/from the global Gigaplane crossbar.

Physically, the Gigaplane-XB interconnect is implemented as a circuit board center-plane with two symmetric sides, each side mounting up to eight system boards, a center-plane support board, and a control board. The Gigaplane-XB consists of a 16-B wide 16×16 *data crossbar*, four *address buses*, and two *power distribution buses*.

Enhanced Clock Rates   The UltraSPARC-I processor requires the system clock rate and the processor clock rate to be either a 1:2 or 1:3 ratio. The initial E-10000 system uses a 83.3-MHz system clock and 250-MMz processors. However, the design is capable of using a 100-MHz system clock to match the use of future faster processors.

Four Address Buses   The UPA defines a separate address and data interconnect. Usually on a bus-based system, only about 70 percent of the wire bandwidth is available for data, with the rest used for address and control. Separating the two functions results in a better utilization of the available wire bandwidths. Snoop addresses need to be broadcast simultaneously to all the boards, while data packets can be sent point-to-point through the crossbar switch.

The four global address buses broadcast addressing requests to all system boards. One address bus is used for each of the four memory banks on a system board. Each bus is independent, meaning that there can be four distinct address transfers simultaneously. An address transfer takes 2 clock cycles.

A Global 16×16 Data Crossbar   There are 16 data paths that can be established in the crossbar, and each datapath is 16-B or 256-b wide, allowing a separate connection to each system board. The Enterprise 10000 system uses four snoop paths to supply enough address bandwidth to match the data bandwidth. The data crossbar routes data packets between the 16 system boards. Data are organized in 64-B packets, and 4 system clocks are needed to transmit a packet.

With the 16×16 crossbar, the Gigaplane provides an aggregated bandwidth as high as 12.8 GB/s. The Enterprise 10000 system has a two-stage routing topology based upon the physical board partitioning. Local many-to-one routes gather on-board requests and connect them to one off-board port. A global-data crossbar connects one port from each board together.

8.2.2 System Board Architecture

The system contains a *control board*, a *support board*, and up to 16 *system boards*, plugged into the Gigaplane-XB centerplane. The control board contains system-level logic used by all system boards, such as the system clock generator and temperature/airflow monitoring. The control board also interfaces through an Ethernet to the *system service processor* (SSP, the system console), which handles booting, shutdown, monitoring, diagnostics, and other system administration tasks.

System Boards   Each system board contains four *processor modules*, a *memory module*, and an *I/O module*, interconnected by the UPA crossbar. The system board architecture is shown in Fig. 8.6. Each processor module contains a 250-MHz UltraSPARC microprocessor and a 1-MB level-2 cache. The memory module contains four interleaved banks of 64-Mb ECC DRAM chips with a capacity of up to 4 GB per module.

The I/O module bridges two standard Sbuses (IEEE 1496-1993), and each Sbus has two slots for networking and I/O. The Enterprise 10000 uses virtual addresses in its I/O scheme, and each Sbus interface includes its own memory management unit for virtual-to-physical address translation.

While the internal clock of the processors is 250 MHz, the rest of the system board works at an 83.3 MHz (12-ns) system clock. The memory module can perform a 64-B cache-line read or a write every 4 system clocks (48 ns), providing a bandwidth of 1.3 GB/ s. The 64-b Sbus operates at 25 MHz and sustains a 100-MB/s bandwidth.

Data Routing   Data routing in the Enterprise 10000 system is conducted at two levels: global and local. The global data router is an 18-B-wide, 16×16 crossbar that steers data packets between the 16 system boards. With the 16×16 crossbar, any port can be connected to any other throughout the centerplane. Of the 18 B, 16 for data and the remaining 2 B is for error correction.

The StarFire's address routing is implemented over a separate set of four global address buses. Although called "address buses" to convey that addresses are broadcast, the implementation is really a point-to-point router. The significance of this is that routers have greater inherent reliability than a bus. The buses are 48 b wide including error-correcting code bits.

Each bus is independent, meaning that there can be four distinct address transfers simultaneously. An address transfer takes 2 clock cycles, equivalent to a snoopy rate of 167 million snoops per second on address buses. Should a failure occur on an address bus, degraded operation is possible using the remaining buses.

8.2.3 Scalability and Availability Support

Special hardware and software support is identified below for enhancing the scalability and availability in the StarFire system.

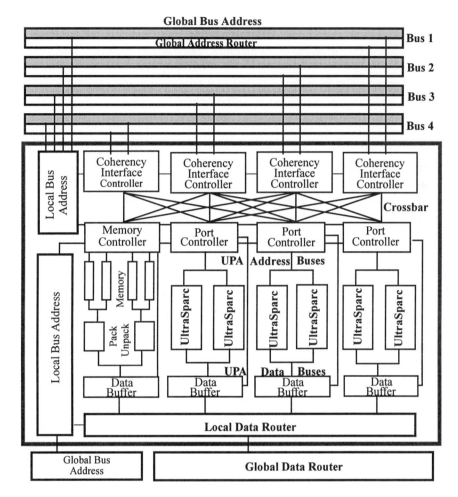

Figure 8.6   System board architecture of Sun StarFire server.
(Courtesy of Sun Microsystems [600],1997)

Scalability Support   Larger SMP servers are in demand for the widespread use of powerful workstations and networked personal computers. Sun has claimed the following advantages toward scalable SMP development:

- The Gigaplane-XB interconnect and cache coherency mechanism played a crucial role in enhancing the scalability of Sun's SMP servers. Another important role is played by Sun's Solaris operating system.

- The StarFire can be configured from 16 to 64 CPUs, 2 to 64 GB of memory, and to over 20 TB of on-line disk storage. There are no slot tradeoffs between processors, memory, and I/O.

- All the 250-MHz processor modules, SIMMs, and SBus boards used in the Enterprise 3000, 4000, 5000, and 6000 servers are common to the Enterprise 10000 server. Therefore, when upgrading to a larger configuration, customers can move these components from the existing chassis to the new chassis.

Availability Support The Enterprise 10000 provides several enhanced RAS (*reliability, availability,* and *serviceability*) features as listed below:

- *Error-correction hardware:* Data and address buses in the memory modules, the I/O modules, and the Gigaplane are protected by a combination of error correcting codes and parity.
- *Redundant, hot-swappable components:* Power supplies and most board-level system components, for example.
- *Service console*: The system service processor connects to the Enterprise 10000 server via a conventional Ethernet, enabling system diagnostics, testing, monitoring, and administration from a single control point and a remote location.

8.2.4 Dynamic Domains and Performance

Dynamic System Domains A unique feature of the Enterprise 10,000 is the concept of *dynamic system domains*, whereby sets of system boards can be partitioned into multiple *domains*, and each domain runs an independent copy of Solaris. Each domain is completely isolated from hardware or software errors that might occur in another domain. The domains can be formed and reconfigured dynamically at run time. Sun believes dynamic system domains have several advantages over a set of smaller individual servers:

- *Server consolidation*: A single Enterprise 10000 server can carry out the duties of several smaller servers. It preserves the independence and isolation of the discrete servers but is easier to administer and offers the flexibility to freely shift resources from one "server" to another.
- *Simultaneous development, production, and test*: These three functions can safely coexist within a single Enterprise 10000 system. The isolation of the domains enables development and test work to continue with production.
- *Software migration*: One can update systems or applications software one domain after another in a rotating fashion, without having to take the entire system off line, as is needed for earlier SMP systems.

Performance Comparison The performance of the SunFire Enterprise 6000 and the StarFire Enterprise 10000 is compared in Table 8.2.

Both systems use the 250-MHz UltraSPARC II processors, with 1 MB of external cache per processor; and both run with the Solaris 2.5.1 operating system. The improvement in performance is attributed to the following upgrades in architectural

Table 8.2 Performance Comparison of Two Sun SMP Servers

Performance Type	Enterprise 6000 (The SunFire System)	Enterprise 10000 (The StarFire system)
System bus speed	2.5 GB/s	10.4 GB/s for 83.3 MHz
Peak system bandwidth	2.6 GB/s	12.2 GB/s for 100 MHz
SPECrate_int95 SPECrate_fp95	2317 for 30 CPUs 1782 for 30 CPUs	N/A for 16 to 64 CPUs
System bus throughput	2.5 GB/s	10.4 GB/s
Memory latency	300 ns	500 ns
Network bandwidth	Up to 622 MB/s	Up to 622 MB/s
I/O performance	1-30 SBus, 200 MB/s	2-32 SBus, 200 MB/s
	Up to 10 independent SBus	Up to 32 independent SBus
SCSI performance	20 MB/s	20 MB/s
UltraSCSI performance	40 MB/s	40 MB/s
Fiber Channel	25 MB/s	25 MB/s

features: from 30 to 64 CPUs, from 30 to 64 GB of maximum memory, from 45 to 64 SBus slots, from 162 to 190 GB internal disk, from >10 TB to >20 TB of total disk.

Obviously the improvement in system interconnects has enhanced the scalability in StarFire over SunFire. The significantly added RAS features have enhanced the availability of the StarFire over the SunFire. The increase in memory latency is due to the much enlarged configuration and memory capacity in SunFire.

8.3   HP/Convex Exemplar X-Class

Exemplar is a CC-NUMA machine, introduced by Hewlett-Packard/Convex in 1996 as the *Scalable Parallel Processor* (SPP). This machine provides a single shared-memory space using hardware support for cache/memory coherency over physically distributed memory modules and caches. The Exemplar is a successor of the Convex Exemplar SPP 1600 system. It is designed as a general-purpose platform for technical computing, mass data storage management, and network servers, etc. The architecture can scale to 512 processors and 512 GB of physical memory.

8.3.1   The Exemplar X System Architecture

The architecture of an Exemplar X is shown in Fig. 8.7. An Exemplar X system consists of a number of *hypernodes*, connected through a *Coherent Toroidal Interconnect*

(CTI). Each hypernode is an SMP system (called the S-Class by HP/Convex) having up to sixteen 64-b PA-8000 processors clocked at 180 MHz, and each processor has a 1-MB instruction cache and a 1-MB data cache.

The processor/caches are shown as P/C in Fig. 8.7. The system hardware and software provide a globally shared memory, allowing any processor and any I/O device controller (e.g., a DMA engine) to access any memory in the entire system, including those in a different hypernode. Similarly, any processor can access any device attached to any hypernode. The hardware maintains systemwide cache coherency.

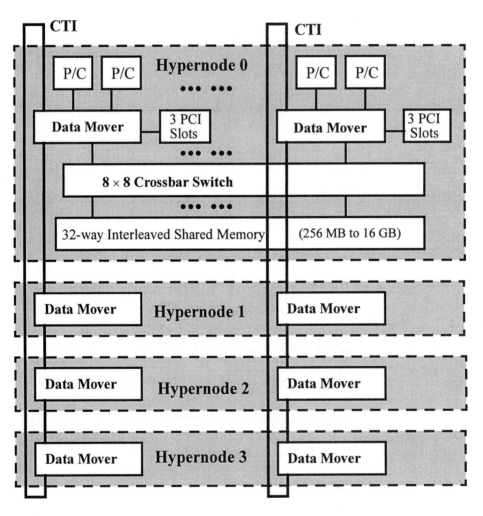

Figure 8.7   Architecture of the HP/Convex Exemplar X-Class SPP
(P/C: Processor/ Cache, CTI: Coherent toroidal interconnect, PCI: Peripheral component interconnect. Courtesy of HP/Convex [165])

Physically, there is a four-level memory hierarchy in the same address space: Most of the time, a processor needs only to access its own caches. In case of a cache miss, the processor accesses the local shared memory in the same hypernode through the crossbar switch. If the data are not found in the local memory, the processor gets the data from the memory of another hypernode through the CTI.

To reduce remote memory access, each hypernode contains a *CTI cache* (also known as cluster cache) that caches data previously accessed from other hypernodes. The CTIcache is physically indexed and tagged with the global physical address. The main parameters of the Exemplar X-Class system are summarized in Table 8.4..

Table 8.3 Resource and Performance Attributes of Exemplar X-Class

Attribute	A Hypernode	An X-Class System
Number of processors	4-16	16-64
Peak speed (Gflop/s)	11.5	46
Cache memory capacity (MB)	8-32	32-128
Physical memory capacity (GB)	0.256-16	1-64
Peak memory bandwidth (GB/s)	15.38	61
Number of PCI controllers	1-24	1-96
Internal disk capacity (GB)	148	592
Peak I/O bandwidth (GB/s)	1.92	7.68
Peak CTI link bandwidth (MB/s)	-	450

The CTI interconnect is derived from the SCI standard. It provides systemwide coherent access to all the shared memories in the hypernodes. CTI is implemented as a series of point-to-point unidirectional links. Like SCI, CTI uses a split-transaction protocol to permit multiple outstanding requests. Suppose a source hypernode requests a cache line from a target hypernode. The source hypernode hardware sends a request packet out over a CTI ring. The request packet travels over the ring to reach the target hypernode, which removes the packet from the interconnect.

The memory subsystem on the target hypernode then retrieves the cache line from its local memory. At the same time, the target hypernode sends an acknowledgment packet to the source hypernode. When the target hypernode has retrieved the cache line, it sends a response packet containing the data to the source hypernode. Finally, the source hypernode sends an acknowledgment packet to the target hypernode to indicate that data are received.

A key feature of the Exemplar X system is the use of specially designed crossbar and data mover hardware to enable high-bandwidth data transfers among processors, memories, and I/O devices. The crossbar provides nonblocking access from CPUs and I/O devices to the memory subsystem. Each of the eight crossbar ports on the processor side connects to a single data mover, which supports a pair of PA-8000 processors, a 240-MB/s I/O channel that provides three PCI slots, and a 450-MB/s interface to the CTI.

On the memory side of the crossbar, each port connects to a four-way interleaved memory board. The crossbar operates at 120 MHz. The data path width of a crossbar port is 64-b. Thus the peak bandwidth of a crossbar port is 960 MB/s in each direction, and the aggregate bandwidth of the entire crossbar is 15 GB/s if counting both directions. Various I/O devices, such as disks, tapes, RAID disks, and networks, can be attached to any hypernode through the PCI slots. Supported devices include Ethernet, FDDI, ATM, HiPPI, Ultra SCSI, etc.

8.3.2 Exemplar Software Environment

The software environment of the Exemplar X-Class is illustrated in Fig. 8.8. The X-Class system can be viewed as a huge workstation capable of simultaneously executing many sequential programs, providing large *throughput*. This feature is adopted in more and more modern parallel systems.

The SPP-UX Operating System   The operating system is called SPP-UX, which can be viewed as an extension of HP-UX, HP's version of Unix. A notable feature is that the Exemplar parallel platform supports *sequential* applications. In fact, SPP-UX provides an environment (shown in the shaded area in Fig. 8.8) that is identical to an HP-UX as found in an HP workstation.

Thus many sequential applications, binary codes existing on HP workstation and SMP server platforms, can be executed without any modification or recompiling. SPP-UX runs a microkernel on each hypernode, which provides the most fundamental kernel functionality, such as virtual memory management and scheduling of processors.

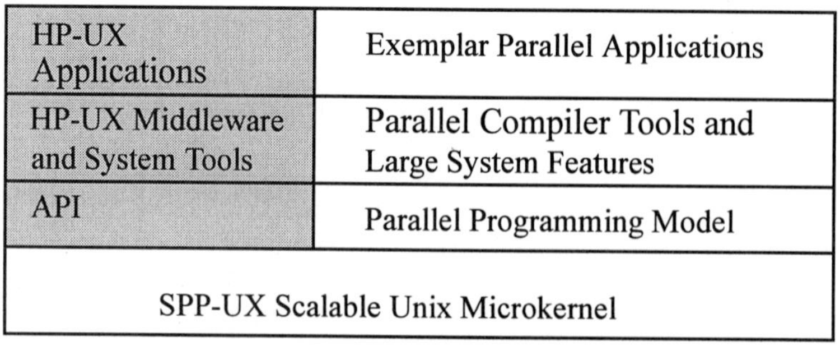

HP-UX Applications	Exemplar Parallel Applications
HP-UX Middleware and System Tools	Parallel Compiler Tools and Large System Features
API	Parallel Programming Model
SPP-UX Scalable Unix Microkernel	

Figure 8.8   The SPP-UX operating system and software environment.
(Courtesy Brewer and Astfalk, *Proc. of COMPCON* [110], 1997)

The majority of the operating system is implemented in server tasks and in protected modules that run in user space, to provide standard functionality such as file system management, device management and network services. SPP-UX provides a number of *large system features* that are needed for data centers and enterprise systems, such as 64-b features to support Terabyte files and file system, 64-b data types and pointers, batch job scheduling, checkpointing, and flexible resource allocation functionality.

Parallel Programming Model The SPP-UX provides a single parallel programming model that supports C, C++, Fortran 77, and Fortran 90, plus compiler directives and libraries for both shared-memory and message-passing programming. The supported libraries include PVM, MPI, and Pthreads. A parallel program can automatically adapt itself to the number of processors available to the user at run time.

When a program begins execution, there is one thread of execution per available processor, and all threads are idle except for thread zero (called the *root thread* or *base thread*), which executes alone until it encounters a parallel construct. It then activates all the idle threads of the program. At the end of the parallel construct the threads synchronize at a barrier, and all threads except thread zero become idle, waiting for work.

Compiler Support The Exemplar compilers provide a hierarchy of automatic optimizations. The lowest level is the *code generation* level, where scalar performance is enhanced by advanced RISC optimization techniques, such as instruction scheduling, software pipelining, tiled global register allocation, loop blocking, etc. The next level is *automatic parallelization* on data-independent loops. The next level utilizes compiler directives to explicitly control parallelization and data distribution.

Graphics and Application Tools The SPP-UX environment includes a graphic tool called *Subcomplex Manager*, which allows the dynamic partitioning of the processors into several groups, called *subcomplexes*. (Subcomplexes are called *pools* or *partitions* in other systems.) For instance, we may partition the 64 processors into four subcomplexes, one for interactive jobs, one for batch jobs, one for file servers, and one for visualization.

Within a subcomplex, message passing is implemented through shared memory. Across subcomplexes, message passing is implemented through conventional Unix sockets. SPP-UX also provides a set of environment tools, called CXtools, to help users in debugging, testing, and optimizing parallel applications. This tool set includes a parallel debugger; a parallel performance profiler/analyzer; and a parallel event analyzer.

8.4 The Sequent NUMA-Q 2000

Sequent started the NUMA-Q design in 1992, and the first NUMA-Q 2000 system was shipped in the Fall of 1996. This system is designed primarily for commercial applications. It adopts the CC-NUMA machine architecture.

The uniqueness lies in the efficient use of commodity and standard components to build a large system. This allows Sequent to devote resources to developing those key glue (or hook) components that greatly impact functionality and performance and that are not commodities yet. An example is the SCI interface called *IQ-Link*.

8.4.1 The NUMA-Q 2000 Architecture

The NUMA-Q 2000 architecture is illustrated in Fig. 8.9, and the important architectural parameters are listed in Table 8.4.

The basic building block of NUMA-Q is a four-processor SMP board called a *Quad*. The system can have up to 63 Quads, or 252 processors. It is a CC-NUMA machine. All the local memories in the Quads are glued together by SCI-based interconnect (IQ-links and an IQ-Plus) to form a global memory, which is accessible from any processor. Cache coherence is enforced by a hardware directory-based protocol.

The Use of Intel SHV Quad Sequent uses commodity and standard components extensively in building NUMA-Q. The Quad is a modification of the off-the-shelf Intel SHV server board. A Quad in NUMA-Q consists of four Pentium Pro processors and a local memory, connected by a 64-b, 66 MHz local bus.

Also connected to the local bus are two Intel *Orion PCI bridges* (OPB) and an SCI interface card called IQ-Link. Each of these 32-b, 33-MHz PCI buses has four slots, for a total of eight slots. One slot is connected through a PCI-EISA bridge to a *management and diagnostic processor* (MDP), leaving a total of seven PCI slots available.

Table 8.4 Resource Attributes of Sequent NUMA-Q 2000

Attribute	Quad	System
Maximum number of processors	4	252
Cache memory capacity (MB). L2 cache is on processor package, and L3 the remote data cache	L2: 0.512 L3: 32	L2:31.5 L3: 2016
Physical memory capacity (GB)	0.5-4	31-252
Aggregated peak memory bandwidth (GB/s)	0.533	33.5
Number of I/O ports	7 PCI	441
Aggregated peak I/O bandwidth (GB/s)	0.266	16.7

The Intel four-processor board provides a number of features that facilitate its use as a building block in a CC-NUMA system. For instance, the board is an open system with a published specification. The board allows a third-party designated bus agent to control the local bus. In NUMA-Q, this agent is the IQ-Link card, which is the glue to integrate multiple Quads into a large system. The board has limited capability of supporting multiple outstanding memory requests, so that processors will not be blocked by a single memory request that is being served.

To improve performance and merge with SCI, Sequent has made a number of modifications and enhancements to the Intel Quad board. Redundant power supplies and greater fault isolation capability are added. Unused PC logic is removed from the board.

The System Interconnect As shown in Fig. 8.9, five types of standard system interconnect are used, of which four are commodity ones and interface a Quad through the two PCI buses. The IQ-Link interconnect is Sequent proprietary, but follows the SCI standard. It is connected directly to the local bus of a Quad, and supports inter-Quad communication needed for distributed shared memory. The optional IQ-Plus is an SCI ring integrated in a central box. Alternatively, individual links can be used to connect the IQ-Links into a ring. The links connecting the IQ-Links and the IQ-Plus are copper cables with up to 15 m.

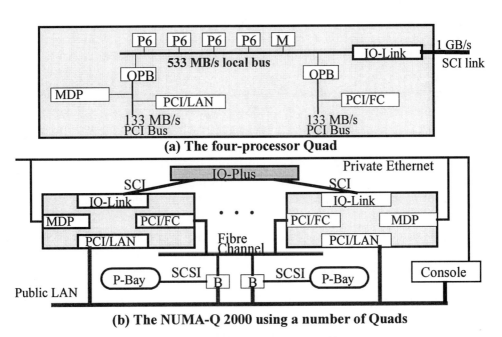

(a) The four-processor Quad

(b) The NUMA-Q 2000 using a number of Quads

Figure 8.9   The NUMA-Q 2000 architecture diagram. (OPB: Orion PCI Bridge, MDP: Management and Diagnosis Processor; PCI/LAN card, PCI/FC: PCI Fiber Channel Interface; B: Fiber Channel Bridge, P-Bay: Peripheral Bay. Courtesy: Lovett and Clapp, *Proc. of the 23rd Annual Int'l Symp. Computer Architecture* 1996 [420])

A private Ethernet connects the system console to the Sequent-designed MDP (*management and diagnostic processor*) on every Quad. This facilitates the controlling, testing, and monitoring of Quads. A public LAN (Ethernet, Fast Ethernet, FDDI, etc.) connects the Quads and the console to the outside world. Client PCs, workstations, and terminals can access the NUMA-Q system through this LAN.

Connection to mass storage devices (disks, RAID, tapes, and CD-ROMs) is based on one or more 100-MB/s Fiber Channels, each being an optical cable up to 500 m. Sequent-developed Fiber Channel bridges connects the Fiber Channel to fast/wide SCSI buses, where the storage devices are attached.

The IQ-Link   A main technical innovation in NUMA-Q is the SCI interface card (the IQ-Link shown in Fig. 8.10), which is crucial for efficiently realizing the CC-NUMA architecture. Even here, Sequent uses off-the-shelf components whenever possible. Physically, the IQ-Link consists of four components: an *Orion bus interface controller* (OBIC) chip from LSI Logic, an SCI *cache link interface controller* (SCLIC) chip, a DataPump chip from Vitesse Semiconductor, and synchronous DRAM arrays.

Only the SCLIC chip is custom-made by Sequent. The DataPump is a GaAs chip operating at 500 MHz. The other components use CMOS technology and operate at 66 MHz, the local bus clock rate. The three chips are nontrivial, each containing more than 140,000 gates and as many as 550 pins.

Logically, the IQ-Link consists of three subsystems. The DataPump directly drives the SCI network, and is similar to the interface suggested in the SCI standard (see Fig. 6.37). The bus-side interface manages the remote cache data arrays and bus snooping logic.

The OBIC chip supports up to four outstanding remote accesses and two incoming remote requests. The directory controller maintains the SCI network-side Local Memory directory and the network-side remote tags and uses them in the directory-based cache coherency protocol. The *programmable protocol engine* (SCLIC) performs this task.

A remote memory request can be issued by any bus agent, such as a processor or one of the I/O controllers. Suppose a processor issues a memory request which needs to fetch data from a remote node. Then it goes through the following steps:

(1) The processor makes an intra-package memory request which misses both the L1 and L2 caches.

(2) The processor places a memory read request on the memory bus to retrieve the missing cache line.

(3) The OBIC executes a snoop cycle to check if the address is within the range for the local memory. As the address is outside the local memory space, the remote cache tags are checked.

(4) The request misses the remote cache, causing the OBIC to indicate on the memory bus that the reference will not be completed immediately. This allows

subsequent independent transactions to proceed without waiting. The OBIC then allocates a line in the remote cache and passes the request to the SCLIC.

(5) The protocol engine spawns a task to construct a packet containing a CACHE_FRESH command, the address of the cache line, and the ID of the line's home node.

(6) The task passes the packet to the DataPump. The task then allocates a line in an internal tag cache and goes to sleep.

(7) The DataPump transmits the request packet over the SCI ring to the home node, which retrieves the data and sends back a response.

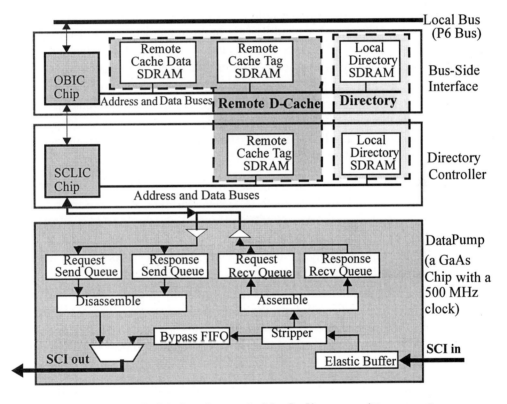

Figure 8.10   The IQ-Link schematic block diagram. (Courtesy: Lovett and Clapp, *Proc. of the 23rd Annual Int'l Symp. Computer Architecture* [420], 1996)

(8) When the response containing the data arrives from the home node, the DataPump captures the response and passes it to the directory controller.

(9) The response uses its transaction ID to wake up the task gone to sleep in step 6, which passes the data to the OBIC, updates the cache state, and terminates.

(10) The OBIC arbitrates for the bus and then places the response on the memory bus. At the same time, the data are written into the remote cache and the cache state is updated.

(11) The memory read transaction which starts in step 2 finishes, and the cache line is loaded into the L1/L2 caches and an appropriate register.

8.4.2 Software Environment of NUMA-Q

The NUMA-Q machine runs DYNIX/ptx, which is Sequent's version of the Unix operating system enhanced for supporting the Symmetry and NUMA-Q 2000 series of multiprocessor systems. The programming environment includes standard sequential languages (C, C++, Cobol, Fortran, and Pascal), mainstream database software (e.g., from Informix, CA/Ingres, Oracle, Progress, Sybase, and Unify), a parallel debugger called ptx/Debug, and a comprehensive set of library routines for parallel programming.

The DYNIX system was first developed in 1984 and has been used on the Symmetry multiprocessor series for more than a decade. It can support large relational database applications with thousands of users and many TB of data. Designed for commercial enterprise applications, the DYNIX/ptx system provides a number of features for scalability, availability, and manageability.

Scalability Support   To support large numbers of users, large applications, and large data sets, DYNIX/ptx includes enhancements to system utilities, such as larger tables and hashing. DYNIX/ptx supports up to 16 GB of physical memory and tens of terabytes of disk storage, and thousands of users simultaneously. To minimize degradation in response time, DYNIX/ptx automatically distributes and schedules tasks to the less-loaded processors, realizing dynamic load balancing.

The CC-NUMA architecture and the SCI ring help reduce latency and increase bandwidth. Sequent scheduler allows a form of *gang scheduling* (processor-group) and attempts to schedule processes to be close to their data (*processor affinity*). Finally, when more scalability is required than a single NUMA-Q machine can provide, several systems can be clustered together, using the ptx/CLUSTERS software.

Availability Support   The Sequent Volume Manager enables on-line replacement of disks and tapes in the NUMA-Q 2000 peripheral bays. It also provides disk mirroring capabilities, redundant boot disks, and the ability to mirror the root and swap devices. The Enhanced File System reduces file system recovery time. File systems can be recovered in as little as 5 s. Further availability is provided by the ptx/CLUSTERS software, which includes a distributed lock manager and a cluster integrity manager to provide coordinated error recovery.

Manageability   Sequent provides a number of tools to strive for mainframe-class manageability. These tools include the following:

- A menu-based system administration software called ptx/ADMIN, which handles daily management tasks such as system setup, disk backup, software installation, user account management, configuration management, etc. The ptx/ADMIN interface is extensible, allowing user-defined functions like data base start-up and system shutdown.
- A remote system monitoring software called SequentLINK, which connects a NUMA-Q machine at the customer site to a Sequent service center. This allows Sequent engineers to remotely monitor the system's daily operation, detect problems, and address service and configuration needs.
- An integrated software set called CommandPoint, which manages a networked system based on the SNMP protocol.
- An advanced disk backup software called ptx/ESBM (Enterprise Scalable Backup Manager), which supports a number of high-performance tape drives and manages full backups, incremental backup, backup aging, file catalogs, etc.

The Sequent NUMA-Q is an open system, supporting many third-party software packages, such as software for job management, database management, network management, etc.

8.4.3 Performance of the NUMA-Q

Sequent has predicted the performance of the NUMA-Q 2000 by running the TPC-B and the TPC-D benchmarks on Symmetry 2000 and 5000, and by using hardware counters, logic analyzers, and operating system event counters to collect performance data. A simulation model is then constructed based on these performance data to generate various simulation results.

Sequent believes the workload characteristics are representative of OLTP and *decision support system* (DSS) applications, and the simulation results are predictive of the NUMA-Q behavior under such workloads. The predicted NUMA-Q system uses thirty-two 133-MHz Pentium Pro Processors (8 Quads), where each processor has a 512-KB level-2 cache.

An observation made in the NUMA-Q development is that for commercial applications, the majority of data accesses can be satisfied within a Quad, which has a large (4 GB) local memory and a large (32-MB) remote cache. Consequently, most traffic on the inter-Quad network (the SCI ring) is for cache-coherency protocol (invalidation), not for transferring data.

Therefore, the *data transfer* bandwidth the system provides is close to the *sum* of all local Quad data bandwidths. This observation is worthy of further study. If it is generally true, communication bandwidth will not be a significant limiting factor, and latency hiding will become more important.

Workload Characteristics The cache access behavior is shown in Table 8.5. The 32-processor NUMA-Q has 64-B cache lines, 32-MB remote caches, and 512-KB level-2 caches. For TPC-B, executing 10,000 instructions will generate 228 level-2 cache misses on average, of which 20 misses will result in an invalidation being sent to all Quads that hold the missed cache line. For TPC-D, there are only 18 misses and 1.6 cache lines to be invalidated.

Table 8.5   Cache Access Profile of TPC-B and TPC-D Benchmarks on a 32-Processor NUMA-Q.
(Courtesy: Lovett and Clapp, *Proc. of the 23rd Annual Int'l Symp. Computer Architecture* [420], 1996)

Cache Event	Rate	TPC-B	TPC-D
L2 Cache miss	Per instruction	0.0223	0.0018
Hit to local memory	Per L2 miss	50.4%	51.6%
Hit to Remote Cache	Per L2 miss	27.2%	27.6%
Local cache-to-cache transfer	Per L2 miss	1.3%	1.6%
2-hop remote	Per L2 miss	3.1%	3.6%
4-hop remote	Per L2 miss	9.1%	10.9%
Local hit, remote invalidate	Per L2 miss	8.9%	4.5%
Sharing list length	Per write to shared line	1.2	1.6
Invalidate line	Per instruction	0.002	0.00016
Remote invalidate	Per invalidate	96%	109%

For TPC-B, 50.4% of L2 misses will find the data in the local Quad memory, 27.2% in the remote cache in the same Quad, and 1.3% in another L2 cache of the same Quad. All these cases do not need inter-Quad data transfer, although inter-Quad communication (8.9%) may still be needed for invalidation.

The TPC-D column shows numbers for the TPC-D Query-6. These two benchmarks have quite different characteristics. The TPC-D is a data warehousing, decision support system benchmark, where infrequent read accesses of large regular data chunks are normal.

The TPC-B benchmark is similar to the TPC-C benchmark described in [501]. It makes many *read/write* accesses to small data blocks (database records). Consequently, TPC-B has higher L2 cache miss rate than TPC-D, and TPC-B has to perform many more invalidation operations. The other characteristics of the two benchmarks are similar.

Simulation Results The simulation results show that NUMA-Q is balanced in that data transfer bandwidth is not a problem. The utilization of the local bus in a Quad never exceeds 40%, and the utilization of the SCI ring is less than 33%. The average latencies on L2 cache misses (i.e., the L2 cache miss penalty) are shown in Fig. 8.11.

The latencies in most cases are lower than 2 μs, which are much better than latencies experienced in clusters and MPPs. The figure also shows that processing of the cache coherency protocol in custom-designed hardware still contributes a significant latency.

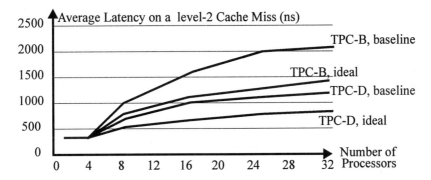

Figure 8.11   Latencies of the TPC benchmarks running on NUMA-Q. (The baseline cases use the NUMA-Q protocol engine. The ideal cases assume protocol processing. Courtesy: Lovett and Clapp, *Proc. of the 23rd Annual Int'l Symp. Computer Architecture* [420],1996).

The speedup behavior of NUMA-Q is shown in Fig. 8.12. Almost linear speedup is observed in all four cases (baseline and ideal TPC-C and TPC-D).

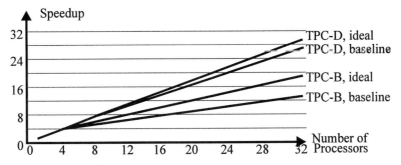

Figure 8.12   NUMA-Q performance relative to one processor. The baseline cases use the NUMA-Q protocol engine. The ideal cases assume protocol processing. (Courtesy: Lovett and Clapp, *Proc. of the 23rd Annual Int'l Symp. Computer Architecture* 1996 [420])

However, due to larger latencies, the TCP-B benchmark shows performance much worse than the TCP-D benchmark. Because TPC-B has ten times larger L2 cache miss rate than TPC-D (0.0223 versus 0.00018 per instruction, see Table 8.5), handling cache misses efficiently is more important.

In particular, Fig. 8.12 suggests that improving the processing of cache coherency protocol could boost the performance of TPC-B by as much as 43%. Sequent is investigating using two protocol engines in the SCLIC, which simulations show could improve TPC-B performance by 20%.

8.5   The SGI/Cray Origin 2000 Superserver

SGI/Cray announced Origin 2000 systems in October 1996. This series of systems adopts the CC-NUMA architecture. The architecture has been known as the *Scalable Shared-memory MultiProcessing* (S^2MP, or S2MP) system internally at SGI/Cray. We study first the design goals. Then we proceed with the architecture and software environments developed for scalable computing.

8.5.1 Design Goals of Origin 2000 Series

The Origin 2000 system is designed to be a highly scalable superserver with the following specific marketing goals:

(1) The system is size scalable from 1 to 1024 processors, while maintaining proportional data storage, memory access, communication, synchronization, and I/O capabilities. However, only systems up to 128 processors are being shipped in 1997.

(2) The system retains the cache-coherent, globally addressable memory model of the PowerChallenge server series to support existing applications.

(3) The entry level and incremental cost should be lower than those of a high-end SMP, approaching those of a cluster of workstations.

To achieve these goals, the SGI Origin design team has made the following decisions, under heavy influence from the experiences learned from the Stanford Dash project:

•   The system should employ a CC-NUMA architecture, with distributed shared memory and hardware enforced cache coherency. The cache coherency protocol is an improved version of the directory-based protocol used in the Stanford Dash machine.

•   Every processor should be able to directly access all memory in the system, as well as all I/O ports. Furthermore, any I/O device can DMA to and from all memories in the system, not just the local memory.

- The system should use a modular physical architecture that allows incremental growth in machine size.
- Efficient hardware support must be provided for internode communication, synchronization, and I/O. In particular, accesses to remote memory or I/O devices should have low latency, as low as twice those of the local accesses. Internode bandwidth for memory and I/O operations should be high, and the aggregate bandwidth should grow linearly with the machine size.
- The system should be backward compatible with SGI PowerChallenge series.
- The operating system should efficiently support the CC-NUMA architecture, providing locality and availability support previously lacking in an SMP machine.

The resulting Origin 2000 architecture supports up to 1024 processors and 1-TB main memory. Systems with 1 to 64 processors are called SGI Origin 2000, while systems with 128 or more processors are called Cray Origin 2000.

8.5.2 The Origin 2000 Architecture

The Origin 2000 architecture is illustrated in Fig. 8.13, and currently available architectural configurations are summarized in Table 8.6. The Origin 2000 has a modular physical structure. The minimum deskside system contains 1 to 4 nodes (1 to 8 processors), 1 to 32 MB of secondary cache, 64 MB to 16 GB of main memory, and up to 14 I/O controllers. Nodes, main memory, I/O controllers, and interconnect can be incrementally added to a system, eventually scaling to a maximal rack system with 128 processors, 512-MB level-2 cache, 256-GB main memory, and 208 I/O controllers.

The Node Card   Each node consists of 1- or 2- MIPS R10000 processors clocked at 195 MHz, each with a secondary cache of up to 4 MB. Each R10000 processor is capable of 2 flop in a cycle, leading to a peak speed of 390 Mflop/s. The Origin 2000 main memory is a DSM, physically partitioned among the nodes but accessible by all nodes. The hardware enforces cache coherency through a directory-based scheme.

The Origin 2000 uses a crossbar ASIC, called HUB, to connect the processors, the memory, the interconnection fabric, and the I/O subsystem, as shown in Fig. 8.14. The HUB ASIC has four bidirectional ports. With a 195-MHz clock, a port provides a one-way peak bandwidth of 780 MB/s, and a full-duplex peak bandwidth of 1.56 GB/s.

The HUB is used to route both intranode and internode communications. It is responsible for converting internal messages format to/from the external format used by the XIO or the CrayLink interconnect. This is done by the four individual control circuits called *interfaces*. Each interface has two FIFOs, one for *incoming messages* (ififo) and one for *outgoing messages* (ofifo). The Origin 2000 provides a global real-time clock. One HUB is designated as the clock master, which sends a global clock out through the interconnect fabric to other HUBs.

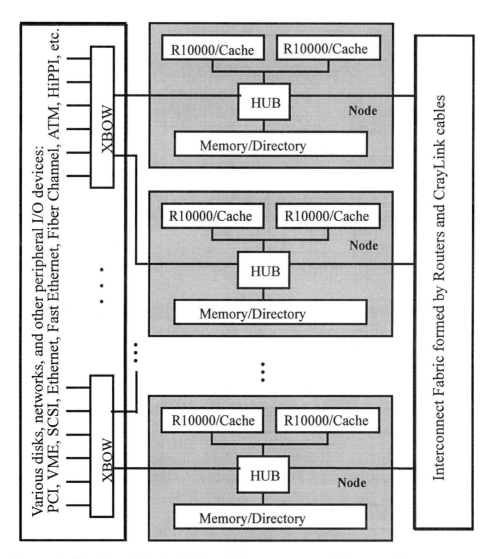

Figure 8.13   The Origin 2000 multiprocessor architecture diagram.
(Courtesy of Laudon and Lenoski, *Proc. of the 24th Annual Int'l Symp. Computer Architecture* [393], 1997)

The I/O Subsystem   The Origin 2000 I/O subsystem is designed to be scalable up to 208 I/O ports, accessible from any processor. The I/O data are organized as messages to be passed to various devices. The message format, routing, and control are governed by an I/O protocol called XIO. Each node card has an XIO port, which can be extended to six ports via a crossbar ASCI called XBOW (for crossbow).

Table 8.6 Architecture Configurations of SGI Origin 2000

Attribute	Desksides	Rack
Number of processors	1-8	2-128
Peak speed (Gflop/s)	3.12	49.92
Cache memory capacity (MB)	1-32	2-512
Physical memory capacity (GB)	0.064-16	0.064-256
Aggregated peak memory bandwidth (GB/s)	3.12	49.92
Number of I/O ports	14	208
Aggregated peak I/O bandwidth (GB/s)	6.2	102

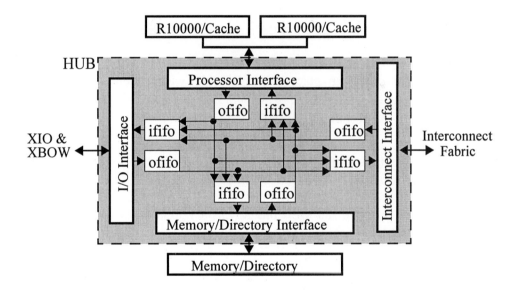

Figure 8.14   The crossbar HUB chip in an Origin 2000 node. (Courtesy of Laudon and Lenoski, *Proc. of the 24th Annual Int'l Symp. Computer Architecture* [393], 1997)

The I/O subsystem is illustrated in Fig. 8.15. The XBOW is an eight-port crossbar switch ASIC. All ports use the XIO protocol. Two ports are used to connect to nodes, while the remaining six connect directly to XIO devices or to ASICs called *widgets*, which are an interface that converts the XIO protocol to the protocol used by a graphics device or a PCI bus.

Through the 32-b or 64-b PCI buses, different types of I/O devices can be connected to the Origin 2000, such as ATM, HiPPI, Ethernet, FDDI, Fiber Channel, SCSI, VME bus extension, etc. The XBOW determines the control and destination information

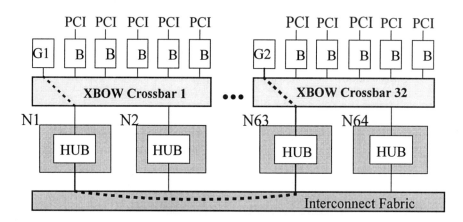

Figure 8.15   **Diagram of the Origin 2000 I/O subsystem.** (N1-N128: nodes, G: graphics widget, B: bridge widget. Courtesy of Laudon and Lenoski, *Proc. of the 24th Annual Int'l Symp. Computer Architecture* [393], 1997)

Each of the eight ports of a XBOW is connected to a node or widget by a bidirectional XIO link, 16-b wide in each direction. Physically, the XIO link uses the same technology as the CrayLink, but the XIO protocol is optimized for I/O traffic. The peak bandwidth of an XIO link is 780 MB/s one-way, and 1.56 GB/s full duplex. The Origin 2000 provides single I/O space: Any processor can access any I/O device.

For instance, a processor in node N1 can access the graphics device G1 through its local HUB and the first XBOW. If the same processor wants to access a remote device (e.g., G2) that is not directly attached to the local XBOW, it needs to communicate with the device via the local HUB, the interconnect fabric, the remote HUB in node N63, and the XBOW 32, as shown in Fig. 8.15.

The Interconnect Fabric The interconnect fabric is made of CrayLink cables connecting a number of nodes and routers (Fig. 8.16), according to a hypercubelike topology (Fig. 8.17). Each CrayLink cable is a double-shielded bundled twin-axial one containing 50 differential signal pairs, for a total of 100 conductors (wires). The maximum cable length is 5 m, with a minimal bend radius of 1.25 in.

The SGI SPIDER router chip provides reliable, buffered wormhole routing of messages. It is a six-port crossbar switch built with the state-of-the-art technology. The SPIDER chip contains more than 1 million gates and has more than 600 pins. Each port consists of a pair of unidirectional, 16-b links clocked at 390 MHz. Such a high frequency is made possible with advanced technology: the differential *positive-shifted ECL* (PECL) and the *SGI transistor logic* (STL), which is a low-swing CMOS signaling technology capable of a speed as high as 400 MHz (2.5-ns cycle time).

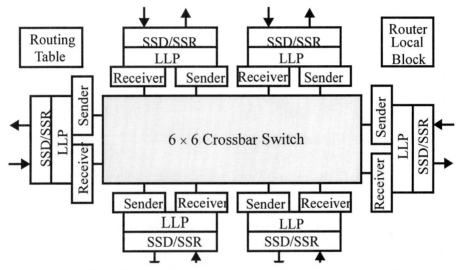

Figure 8.16   The SPIDER router chip used in the Origin 2000 interconnect fabric. (Courtesy of Laudon and Lenoski, *Proc. of the 24th Annual Int'l Symp. Computer Architecture,* [393], 1997)

These ports are connected to nodes or other routers via CrayLink cables. The router allows all six ports to operate simultaneously at full duplex. The end result is impressive communication performance: Each port or CrayLink can provide a one-way peak bandwidth of 780 MB/s, and 1.56 GB/s at full duplex. The router provides an aggregated peak bandwidth of 9.36 GB/s.

Inside the router chip, the six ports are identical. Although the external frequency is 390 MHz, the internal frequency of the router chip is only 97.5 MHz. The *source synchronous driver* (SSD) and the *source synchronous receiver* (SSR) are responsible for converting external 390-MHz, 16-b data from/to internal 97.5-MHz, 64-b data.

The *link-level protocol* (LLP) circuitry takes care of reliable message delivery. Errors are detected by using CRC code, and corrected by retransmission via a sliding window protocol. Besides sending data to the LLP, the sender manages flow control via a credit scheme. The receiver forwards data from the LLP to the routing table and the sender. It also manages virtual channels and message buffers.

The SPIDER router realizes wormhole routing with a small hardware latency. The pin-to-pin latency is only 41 ns. It also uses a number of techniques to alleviate contention and congestion. Four virtual channels are used per physical channel. Congestion control logic allows messages to switch adaptively between two virtual channels; 256 levels of message priority are supported. A packet's priority increases as it ages (blocked longer), so it will have a better chance being transmitted.

The Fat Hypercube Topology   Another interesting feature of the Origin 2000 router is that is reconfigurable (programmable). The default topology follows a hypercubelike pattern (called *fat hypercube*), as shown in Fig. 8.17.

However, other topology can be constructed by modifying the routing tables of the routers in the interconnect fabric. In particular, the number of processors does not have to be a power of 2. Configuring a router is done through the status and control registers in the router.

The fat hypercube is modified from the traditional binary hypercube. The interesting property of this topology is that it maintains the advantage of linear bisection bandwidth of hypercube while avoiding an increasing node degree. Each router chip has six ports, of which two are used to connect nodes, leaving four for network connection.

A node degree of 4 is only enough to construct a 16-vertex hypercube, or a 32-node Origin 2000 if the hypercube topology is used. However, the fat hypercube topology allows the Origin 2000 to scale to 512 nodes. How to do this is partially explained below and partially left as an exercise for the reader.

As shown Fig. 8.17(a), the smallest configuration has one node, which contains two processors and is connected to an XBOW crossbar switch attaching I/O devices. A *SPIDER router* (R) connects two nodes [Fig. 8.17(b)], and two routers connect 4 nodes [Fig. 8.17(c)]. This pattern keeps going up to 32 nodes [Fig. 8.17(f)], similar to the growth of the binary hypercube, if we view the routers as the vertices in a hypercube.

The differences from hypercube are:

- The Origin 2000 topology is "bristled," in that two nodes are connected to a router, and

- The unused router ports can be connected by the *express links* to reduce latency and increase bisection bandwidth. This is illustrated by dotted lines in Fig. 8.17(d).

Beyond 32 nodes, the Origin 2000 employs a *fat hypercube* topology using extra routers (called *metarouters*). There are eight metarouters physically implemented in a box called the *Cray Router*. These metarouters are used solely for intermediate routing, not connected to any nodes.

The 64-node (32 vertices, or routers) fat hypercube topology is shown in Fig. 8.17(g), where each vertex in four eight-vertex subcubes is connected to a metarouter. The four vertices at the same corner of four subcubes connect to the same metarouter. This fat hypercube scales up to 1024 processors, by replacing each of the metarouters in Fig. 8.17 (g) with a 5-dimemnsional hypercube.

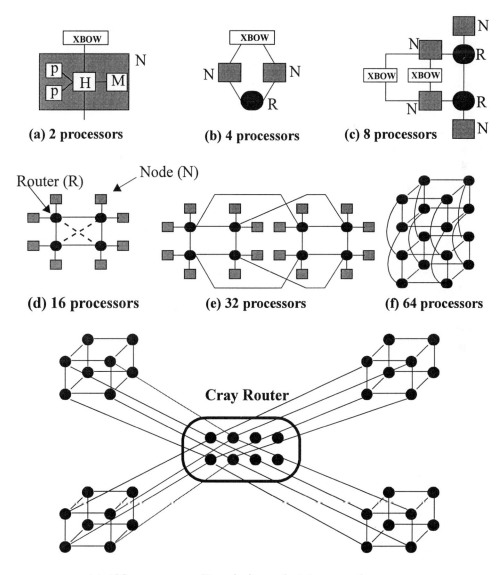

(a) 2 processors   **(b) 4 processors**   **(c) 8 processors**

(d) 16 processors   **(e) 32 processors**   **(f) 64 processors**

(g) 128 processors (For clarity, only 16 connections
to the left four metarouters in the Cray router are shown)

Figure 8.17   The fat hypercube topology adapted in the Origin 2000.
(N: Node in square, M: Memory, P: Processor, H: HUB, XBOW:
Crossbar I/O Switch, R: Router in shaded circle. Courtesy of
Laudon and Lenoski, *Proc. of the 24th Annual Int'l Symp.
Computer Architecture* [393],1997)

Distributed Shared Memory   The Origin 2000 system has a distributed shared-memory subsystem. The memory modules on the same node as the processors are called *near* memory, while those on other nodes are called *remote memory*. The main memory and the directory memory in a node are implemented using synchronous DRAM parts, and organized as up to eight banks.

Each memory bank provides 4-way interleaving, for a maximum of 32-way interleaving when all 8 banks are populated. The memory subsystem is fast enough to achieve full memory bandwidth (780 MB/s) with 4-way interleaving.

The hardware enforces a directory-based coherency protocol maintained by write-back and invalidation. The memory model is sequential consistency. At any time, a memory block can be in one of the following seven *coherency states*, of which four are stable states and three are transient states. These states are kept in the directories of the processing nodes:

- *Unowned*: The memory block is not cached in any cache memory.
- *Exclusive*: The memory block is cached in exactly one cache for read and write by one processor.
- *Shared*: The memory block may be cached in multiple caches in the system for read only.
- *Poisoned*: The data pages have been migrated to another node, and the TLB needs to be updated.
- *Busy Shared, Busy Exclusive, Wait*: These are transient states to deal with when multiple requests are pending for a memory location.

Initially, a memory block is unowned. When a processor P1 accesses this block, it is brought to P1's cache and the state becomes exclusive. The processor P1 can read and write the cached version of this block without concerning the rest of the system, as long as the block stays in the exclusive state. When another processor P2 wants to access the same block, the state becomes shared, and the node numbers of P1 and P2 are saved in the directories of P1 and P2.

If P1 has modified the cache block, it is marked "dirty" and written back to the main memory. So P2 will load the most up-to-date copy of the data from the main memory, instead of getting a stale copy. The two processors can read the shared data block from their respective caches many times.

However, when P2 wants to write to the block, it has to send an invalidation message to all other nodes that share this block. This information is maintained in P2's directory. Once invalidation is done, the block in P2's cache becomes exclusive again, and P2 can go ahead and write to it.

8.5.3 The Cellular IRIX Environment

The Origin 2000 runs the *Cellular IRIX* operating system, which provides a number of interesting features to offer new functionalities and to efficiently exploit the S2MP architecture. It attempts to extend the advantages of SMP (e.g., single system image, easy to manage) with the benefits of MPPs and clusters (e.g., scalability and availability).

The 1996 version, Cellular IRIX 6.4, is a 64-b, multithreading, multiprogrammed operating system that runs on all Origin and Onyx2 Impact systems. The design goals of Cellular IRIX are as follows:

- *Compatibility*: Cellular IRIX is compiled with an open-system standard such as X/Open and POSIX and is compatible with SGI's previous IRIX 5 and IRIX 6 operating systems. Existing applications on SGI systems such as PowerChallenge can be executed on an Origin 2000 without change.

- *Scalability*: Although the operating environment is compatible with old systems, the architecture of Cellular IRIX has been redesigned to enable the Origin 2000 system to scale to 128 or more processors. In contrast, SMP systems usually cannot efficiently scale to more than 10 processors.

    Cellular IRIX is a 64-b operating system which provides a 1-TB virtual memory address space and supports files as large as 9 million TB. This enables Origin 2000 to solve large problems.

- *Availability*: A number of techniques are used to enhance the availability, which has always been a weak point of SMP systems.

- *Throughput*: Origin 2000 is meant to be used not only as a supercomputer to solve large individual computing problems, but also as a high-performance network server to satisfy a large number of simultaneous user requests. Techniques are used in Cellular IRIX to enhance the system throughput, so one slow job will not slow down the entire system.

The Cell Concept for Scalability   Two approaches are used in Cellular IRIX to enhance scalability: the *cell concept* which can modularize the operating system kernel and the *memory management software* that efficiently exploits the S2MP hardware.

The Cellular IRIX architecture distributes kernel text and data into SMP-sized chunks called *cells*, where each cell contains a local copy of the kernel text and data structure, and is responsible for controlling a machine module consisting of a set of processors, memory and I/O devices. A key feature of the cell concept is that operating system services are distributed and localized.

Applications running on an Origin 2000 do not all go to a central, monolithic kernel for services, which would become a bottleneck as the machine size increased. Instead, a cell provides all services that access the machine module it controls. Lock contentions for resources are usually localized to one cell.

Multiple cells can reside on an Origin 2000 system, allowing applications to access multiple machine modules simultaneously and independently. When intercell coordination is needed, the Cellular IRIX architecture ensures that no additional data copying or context switching overhead needs to incur.

Memory Management The Cellular IRIX includes new dynamic memory *replication*, *migration*, and *placement* mechanisms to take advantage of the S2MP hardware architecture. The basic unit of memory management is *page*.

Cellular IRIX supports multiple page sizes (16 KB, 64 KB, 256 KB, 1 MB, 4MB, 16MB), dynamically selectable by user processes. Using a large page size can reduce *table lookaside buffer* (TLB) miss overhead, especially for large database applications.

Memory Replication This is the mechanism to replicate a copy of a read-only page in the memory location closest to each of the processors that share the page. When a page is read-only, the hardware will detect the sharing and invoke the memory management subsystem to perform memory duplication.

Memory Migration This is a mechanism to move a page close to the processor that accesses (reads or writes) it most frequently. When a memory page is being accessed, hardware counters record the off-node references. When the number or proportion of the references passes some predefined relative or absolute threshold, the memory management subsystem will migrate the page.

Since migration has significant overhead, Cellular IRIX incorporates techniques to prevent excessive migration. These include *bounce detection*, which prevents a page from being migrated back and forth (the ping-pong effect); freezing certain pages (i.e., disabling migration) and subsequently melting them when needed.

Memory Placement This refers to the initial allocation of data and threads to exploit locality. This can be done using system-provided defaults, compiler directives, or application-driven *memory policies*. The Cellular IRIX environment allows the user to tell the system to place pages within a specific distance of other pages, to map threads to pages, and to locate threads a specific distance from hardware resources such as graphics and I/O.

Memory Locality Domains (MLDs). An MLD is a set of virtual pages that should be placed nearby. How close is measured by the radius of the MLD, which is 1 plus the number of routers that may need to cross. The concept of using MLD is illustrated in the Example 8.4.

Multiple MLDs can be grouped together in an MLDset, which is mapped to the machine topology by special placement commands. A *policy module* in Cellular IRIX enables the user to tell the operating system what policies (paging, placement, replication, migration, etc.) should be used for an address range.

Example 8.4 Mapping MLD and MLDset to an Origin 2000

A program runs on a 8-node Origin 2000, where each node has 64-MB memory. The MLDset mapping is shown in Fig. 8.18. The program has six threads which use four large data structures A, B, C, and D of 48, 32, 24, and 32 MB, respectively.

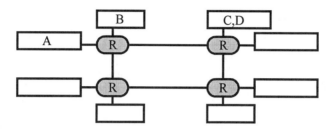

Figure 8.18   Mapping MLD and MLDset to an Origin 2000.

Of access to A and B, 95% are from the first four threads, while 90% of accesses to C and D are from the other two threads. From these data, we can define an MLDset consisting of two MLDs. The first is an MLD of radius 2 consisting of A and B, while the second an MLD of radius 1 consisting of C and D.

∎

Availability Support   The Origin 2000 hardware is designed with availability supporting features, such as ECC memory, redundant power and cooling, hot-pluggable disks, RAID, etc. The interconnect fabric can isolate failed modules, so that a portion of the system can be powered down for maintenance without affecting the rest of the system. The interconnect network supports multiple paths between nodes, and a hardware link-level protocol is provided to detect and retransmit faulty packets.

The IRIX system software includes features to further enhance availability, such as file system journaling for fast rebooting; a software called *availmon* for monitoring and communication of downtime events, duration, and symptoms; disk mirroring and RAID; software called *IRIS FailSafe* for application failover.

The Origin 2000 provides access protection rights so that memory and I/O devices cannot be accessed by unauthorized nodes. The availability of operating system services are aided by the cell concept, which contains random hardware and software failures within a single cell. Cellular IRIX is one of a few operating systems that provide a checkpoint-restart mechanism that is kernel-based and application-transparent.

Users can checkpoint existing binaries without modification while supporting sessions, processes, process groups, open files, pipes, and accounting. Additional support is planned for highly available network, database, and other applications, through support of queues, sockets, raw disks, X-windows, graphics, and special devices.

Throughput Support — Optimized Scheduling Cellular IRIX provides a number of mechanisms to efficiently utilize systems and enhance the system throughput, including a scheduler that support large numbers of processors and user processes and a 64-b journaled file system called XFS which supports high throughput and guaranteed I/O rate.

The kernel pauses every few milliseconds to make scheduling decisions. Processes are each given a guaranteed time slice to execute without being preempted. At the end of a time slice (time-out), the kernel chooses which process to run next, based on process priority.

A running process may also yield to another process voluntarily by making a system call that causes it to sleep. Cellular IRIX 6.4 supports three types of jobs: *timesharing (interactive), real-time,* and *background (batch).* A real-time process have a priority in the range from 0 (low) to 255 (high).

Background processes always have priority 0 and run only when free CPU cycles are available. Interactive processes have a priority from 0 to 40. The kernel schedules interactive processes based on a *currency-based* mechanism. The chosen process runs on the processor with a nondegradable default priority of 20 until it sleeps on a system call or time-out.

The currency-based scheduling mechanism is motivated by a deficiency in traditional Unix scheduling that assigns a priority to each user. Newly spawned processes inherit the same priority, which degrades as the process runs to allow lower-priority processes an opportunity to run. With this scheme, a single user with a relatively high priority can spawn multiple threads with the same priority, monopolizing the system unfairly. A user with 10 processes would receive 10 times the CPU time of a user with one equivalent process and an equal priority.

With currency-based scheduling, the system administrator assigns each user a proportion of total CPU time (*currency*). If there are more processes than processors, the process with the highest currency is scheduled to run and is charged for the CPU usage (i.e., its currency decreases). Processes with lower currency are not running, but maintain their currency.

Eventually the currency of a waiting process will exceed that of a previously running process, allowing it to execute. The main advantage of currency-based scheduling is long-term assurance of fairness while allowing the scheduler to make short-term optimizations (e.g., taking advantage of cache or memory affinity).

In summary, Cellular IRIX provides the following scheduling mechanisms:

- *Frame Scheduling*: The REACT/Pro frame scheduler can completely take over scheduling and dispatch processes on one or more CPUs to guarantee precise rates of forward progress.
- *Gang Scheduling*: Cellular IRIX can schedule a group of threads that communicate with each other using locks or semaphores to start in parallel. This helps

ensure that a thread holding a lock is scheduled in the same time interval as another thread that is waiting on the lock.

- *Processor Affinity*: Cellular IRIX automatically notes the CPU on which a process last ran and attempts to run a process on that same CPU, based on the assumption that some data and code remain in cache or local memory.

Throughput Support—The XFS File System XFS is a 64-b journaled file system supporting high throughput and guaranteed I/O rate. Aggregate read/write performance of more than 500 MB/s has been demonstrated. High throughput is achieved with the *extent* concept and the *journaling* technology. Guaranteed I/O rate is achieved by allowing applications to reserve file system bandwidth. The XFS supports logical block sizes ranging from 512 B to 64 KB. It incorporates the *extent* concept, allowing data contained in an extent to be placed on disk in contiguous blocks.

An XFS extent can have over 1 million (2^{20}) blocks. Use of extents helps reduce disk seeks and rotational delays, thus increases I/O throughput. *Journaling* (also known as *logging*) is a technique to record all file system changes in an append-only log file. This file is later written to disk in large chunks.

The journaling technology also helps reduce system recovery time. The system references a small log of recently updated file system transactions after a system crash, without the need to performing time-consuming file system check as in Unix file systems.

8.5.4 Performance of the Origin 2000

The Origin 2000 architectural design pays great attention to having a balanced system, so that memory, I/O, and interconnect capabilities all proportionally scale up as the machine size increases. This is largely achieved at the hardware level, as shown in Table 8.7 and Table 8.8.

The Origin 2000 clock frequency is 195 MHz, or about 5-ns cycle time. From these tables, we see that the aggregate bandwidth (memory, I/O, and interconnect) increases almost linearly with more processors, while the latency only grows logarithmically. The low latency and high bandwidth make the Origin a more tightly coupled system than other DSM machines. Initial performance evaluation shows that the Origin 2000 achieves good speedup on parallel scientific computation benchmarks such as NPB and SPLASH.

8.6 Comparison of CC-NUMA Architectures

We compare four CC-NUMA architectures in Table 8.9, according to their node architecture, internode connection, cache coherency protocols, and other performance enhancement features. Each system represents a different approach to implementing CC-NUMA. The Dash is a research prototype, while the others are commercial products.

Table 8.7 Read Latency in Origin 2000 Memory Hierarchy
(Courtesy of Laudon and Lenoski, *Proc. of the 23rd Annual Int'l Symp. Computer Architecture* [393],1997)

Data Location	Read Latency (Clocks)
Registers	0
Level-1 cache (on-chip)	1-3
Level-2 cache (off-chip)	10
Local memory	61
Remote memory (1 router away)	117
Remote memory (2 routers away)	137
Remote memory (3 routers away)	157

Table 8.8 Origin 2000 Bandwidth (in GB/s) and Latency (in ns)
Courtesy of Laudon and Lenoski, *Proc. of the 23rd Annual Int'l Symp. Computer Architecture* [393], 1997)

System Configuration I/O:Nodes:CPUs	Aggregate Peak Memory Bandwidth	Aggregate Peak I/O Bandwidth	Interconnect Bisection Bandwidth	Memory Read Latency
1:1:2	0.78	1.56	-	313
2:2:4	1.56	3.12	1.56	497
2:4:8	3.12	6.24	3.12	601
4:8:16	6.24	12.48	6.24	703
8:16:32	12.48	24.96	12.48	805
16:32:64	24.96	51.2	12.48	908
32:64:128	49.92	99.84	24.96	1112

Node Architecture Most CC-NUMA systems employ SMP nodes, where a node contains multiple processors connected to the local memory and the I/O subsystem of the node through a bus. Intranode cache coherency is enforced by a snoopy protocol. Other systems that adopted this include Stanford Dash, Sequent NUMA-Q, Data General NUMALiiNE, and HAL S1, etc.

This node architecture is popular because it is simple to implement, especially with the availability of commodity SMP boards such as the Intel four-processor SHV board (see Section 8.1.3). However, there are several disadvantages of using snoopy bus-based SMP nodes in a CC-NUMA system [393]:

Table 8.9 Comparison of Four CC-NUMA Architectures

Architecture Features	Stanford Dash	Sequent NUMA-Q	HP/Convex Exemplar	SGI/Cray Origin 2000
Node architecture	4-CPU SMP node with snoopy bus	4-CPU SMP node with snoopy bus	8-CPU SMP node with crossbar	2-CPU non-SMP node with HUB
Internode connection	2D mesh	SCI ring	Multiple 2D SCI rings	Fat hypercube
Cache coherency protocol	Snoopy within each node and Directory globally	SCI linked list coherence protocol	Modified from SCI protocol	Modified from Dash protocol
Other performance features	Intranode cache-to-cache sharing	Node cache, gang scheduling, processor affinity	Node cache	Gang scheduling, page migration, placement, and replication

- *Long remote memory access latency:* Because the snoopy bus needs to support multiple processors, it is difficult to achieve high frequency. Furthermore, a remote read must go through the local snoopy bus, the interconnect, and the remote snoopy bus, and finally get the data in the remote memory.

- *Limited remote memory bandwidth:* For the above reasons, the remote memory access achieves a bandwidth only half or one-third of local memory bandwidth, as shown by the Dash experience. The Exemplar uses an intranode crossbar to increase both local and remote memory to alleviate this problem. However, this does not help reduce the remote access latency much.

- *Large entry system cost*: Even when users want an entry-level system with one or two processors, they have to pay for all the cost of the SMP node. We should caution the readers, though, that cost is not mainly determined by technology. Market volumes is often the most important factor. An expensive node architecture, if produced in high volume (such as Intel's SHV SMP boards), could end up being cheaper than a less expensive node technology.

To overcome these problems, the SGI 2000 adopts a non-SMP node architecture. Although each node has two processors, they do not form an SMP. The two processors share a bus to the HUB (a crossbar) in a multiplexed way, much like the early Intel 8088 processor that uses the same physical bus to multiplex address and data buses. The purpose is to save the HUB pin counts.

The shared bus to the HUB is not a snoopy bus. Intranode cache coherency is done by the directory-based coherency hardware. In fact, a two-processor Origin 2000 node functions like two one-processor nodes. The current implementation packages two 1-processor nodes into a single node to reduce cost.

Using these techniques, the Origin 2000 achieves better remote access latency and bandwidth than the other systems. The bandwidth for accessing a remote memory is about the same as the local memory bandwidth. The best-case ratio of local to remote read latency is about 2:1 (see Table 8.7). That is, it takes twice as long to access a remote memory one hop away as to access the local memory. The latency grows logarithmically as the number of hops. For other systems, local-to-remote latency ratio is from 5:1 to 10:1.

System Interconnect The Dash uses a two-dimensional mesh topology. The NUMA-Q uses a single SCI ring, while the Exemplar uses multiple SCI rings. The Origin 2000 uses a fat hypercube topology. For large configurations, the fat hypercube topology provides better bisection bandwidth, thus is more scalable than the other systems.

But most real parallel computers in use today have small configurations (e.g., no more than 20 processors). Only through actual design, production, and market test, can one tell whether a less scalable topology is sufficient.

A common trend in the interconnect design of the CC-NUMA machines is that nodes are not required to pack very closely together to achieve low latency and high bandwidth, which is a main feature in previous-generation MPPs and PVPs.

In these tightly packaged systems, the communication links or cables must have short and regular lengths. In contrast, the SCI ring links in the NUMA-Q can be as long as 15 m, and the Craylink in the Origin 2000 can be 5 m long.

Coherency Protocol All current CC-NUMA systems use some type of directory-based cache coherency protocol. However, the NUMA-Q and the Exemplar use the one defined in the IEEE SCI standard, while the Dash and the Origin 2000 use proprietary protocols. The Origin protocol is further modified from the Dash protocol. The Exemplar protocol is only slightly modified from the IEEE SCI standard.

Other Features An SMP node has multiple processors, each having an off-chip cache (usually called the level-2 cache). Using SMP nodes allows intranode data sharing among the caches, which is realized in the Dash architecture. The NUMA-Q and the Exemplar use a third-level cache per node, called the *remote cache* in the NUMA-Q and the *cluster cache* in the Exemplar.

This node cache is used to cache remote data that have been brought into the node from previous misses, so that a node does not have to frequently go out to get remote data. The Dash and the Origin 2000 do not use such a node cache, believing it will prolong remote latency and reduce bandwidth. Instead, they use techniques such as *page migration* and *page replication* to reduce misses.

8.7   Bibliographic Notes and Problems

The issues of shared-memory multiprocessing are discussed in Alexander et al. [21], Amarasingbe et al. [28], Baron et al. [62], Catenzaro [132], Culler et al. [181], Dubois and Scheurich [217], Gajski and Peir [258], Hwang [327], Lenoski and Weber [406], Shang and Hwang [552], Syanarayanan [534], Thakkar et al. [610], and Yew and Wah [662].

Holt et al. [319] analyze applications and the architecture bottlenecks in DSM machines. Hwang et al. [325] present an orthogonal multiprocessor architecture for scientific parallel processing. Performance models of multiprocessors are discussed in Marson, et al [430], and Bhuyan and Zhang [81]. Using multiprocessors to solve partial differential equations is discussed in Wang and Hwang [635].

The Intel SHV boards are discussed in the Intel Web sites [346]. The SHV boards have been used in a number of commercial shared-memory systems, as described in Clark and Alnes [159], Lovett and Clapp [420], and Weber et al. [639]. The Sun SMP servers are described in Catenzaro [132], Cekleov et al. [134], and Fenwick et al. [239].

The Gigaplane used in the Sun Ultra Enterprise 10000 is discussed in Singhal et al. [561]. The HP/Convex Exemplar is described in Convex [165,166], Brewer and Astfalk [110], and Sterling et al. [586]. The Sequent NUMA-Q 2000 is discussed in Lovett and Clapp [420] and Lovett et al. [421]. The SGI Origin 2000 system is discussed in Laudon and Lenoski [393], SGI [550], and Whitney et al. [643].

Homework Problems

Problem 8.1   Define the following terminology:

(a)   Split transaction bus

(b)   Outstanding memory request

(c)   Remote cache (also known as node cache, cluster cache)

(d)   Home node of a cache line

Problem 8.2   Consider the Intel SHV SMP board in Section 8.1.3.

(a)   What is the OEM port? List three things that can utilize the OEM port.

(b)   Assume a cache line size of 32 B and a main memory cycle time of 60 ns. Find the minimum memory access latency when the board is populated with only one processor.

(c)   Assume a cache line size of 32 B and a main memory cycle time of 60 ns. Find the maximum memory bandwidth achievable when the board is populated with four processors.

Problem 8.3 Consider the Sequent NUMA-Q 2000 system in Section 8.4.

(a)   A main strategy in the NUMA-Q design is to use commodity and standard components. List five commodity components used in NUMA-Q, and list two proprietary ones.

(b)   Why are there two remote cache tag SDRAMs in a node?

Problem 8.4 Refer to Fig. 8.17 and answer the following questions:

(a)   Redraw Fig. 8.17(e) by adding all express links.

(b)   Redraw Fig. 8.17(g) by adding all missing links.

(c)   Show the fat hypercube topology of a 128-node (256-processor) Origin 2000.

(d)   Show the fat hypercube topology of a 512-node (1024-processor) Origin 2000.

Problem 8.5 Name five architectural differences between Sequent NUMA-Q 2000 and SGI Origin 2000. These should not be differences in capacities, such as machine size, memory capacity, etc. For each architectural difference, explain which system is superior.

Problem 8.6 List and briefly explain four techniques that have been used to exploit data locality by at least one of the systems described in this chapter.

Problem 8.7 List and briefly explain four techniques that have been used to hide latency by at least one of the systems described in this chapter.

Problem 8.8 List and briefly explain five techniques that have been used to enhance resource scalability by at least one of the systems described in this chapter.

Problem 8.9 List and briefly explain six techniques that have been used to enhance availability by at least one of the systems described in this chapter.

Problem 8.10 List and briefly explain four techniques that have been used to enhance manageability by at least one of the systems described in this chapter.

Chapter 9

Support of Clustering and Availability

Clusters are known to computer professionals by many names, such as *cluster of computers*, *cluster of workstations* (COW), *network of workstations* (NOW), etc. Clustering is becoming a trend in developing cost-effective, scalable parallel computers [498]. Basic concepts of cluster computing were introduced in Section 1.4.

In this chapter, we study cluster design principles and review hardware and software support needed to enable the clustering of multiple computers. We address the three major issues of system availability, single system image, and cluster job management. In the next chapter, we will show how these clustering techniques are integrated to build commercial or research clusters of computers.

9.1   Challenges in Clustering

We first clarify the cluster concept by giving a taxonomy of clusters. Then we elaborate the issues that must be addressed by the designer and the user of any cluster system. Techniques for solving these problems will be discussed in subsequent sections.

9.1.1 Classification of Clusters

Clusters have been classified in various ways in the literature, often with confusing synonyms. We classify clusters using four orthogonal attributes: *packaging*, *control*, *homogeneity*, and *security*. In keeping with the mainstream and for simplicity, we allow only two values for each attribute, as shown in Table 9.1. Currently, only two combinations are used in real clusters, termed *dedicated clusters* and *enterprise clusters*. But conceptually, all 16 combinations are possible.

Table 9.1 Attributes Used in Cluster Classification

Attributes	Attribute Value	
Packaging	Compact	Slack
Control	Centralized	Decentralized
Homogeneity	Homogeneous	Heterogeneous
Security	Enclosed	Exposed
Example	Dedicated cluster	Enterprise cluster

Packaging The cluster nodes can be compactly or slackly packaged. In a *compact* cluster, the nodes are closely packaged in one or more racks sitting in a room, and the nodes are not attached to peripherals (monitors, keyboards, mice, etc.). Gorden Bell calls such nodes *headless* workstations [67]. In a *slack* cluster, the nodes are attached to their usual peripherals (i.e., they are complete SMPs, workstations, and PCs), and they may be located in different rooms, different buildings, even geographically remote regions.

Packaging directly affects the communication wire length, and thus the selection of the interconnection technology. While a compact cluster can utilize a high-bandwidth, low-latency communication network that is often proprietary, nodes of a slack cluster are normally connected through standard LANs or WANs.

Control A cluster can be either controlled or managed in a *centralized* or *decentralized* fashion. A compact cluster normally has centralized control, while a slack cluster can be controlled either way. In a centralized cluster, all the nodes are owned, controlled, managed, and administered by a central operator. In a decentralized cluster, the nodes have individual owners.

For instance, consider a cluster comprised of an interconnected set of desktop workstations in a department, where each workstation is individually owned by an employee. The owner can reconfigure, upgrade, or even shut down the workstation at any time. This lack of single point of control makes the system administration of such a cluster very difficult. It also calls for special techniques for process scheduling, workload migration, checkpointing, accounting, etc.

Homogeneity A *homogeneous* cluster means that the nodes adopt the same platform. That is, they use the same processor architecture and the same operating system. Often, the nodes are from the same vendors. A *heterogeneous* cluster uses nodes of different platforms. Interoperability is an important issue in heterogeneous clusters.

For instance, process migration is often needed for load balance or availability. In a homogeneous cluster, a binary process image can migrate to another node and continue execution. This is not feasible in a heterogeneous cluster, as the binary code will not be executable when the process migrates to a node of a different platform.

Security Intracluster communication can be either *exposed* or *enclosed*. In an exposed cluster, the communication paths among the nodes are exposed to the outside world. An outside machine can access the communication paths, and thus individual nodes, using standard protocols (i.e., TCP/IP). Such exposed clusters are easy to implement, but have several disadvantages:

- Being exposed, intracluster communication is not secure, unless the communication subsystem performs additional work to ensure the privacy and security.
- Outside communications may disrupt intracluster communications in an unpredictable fashion. For instance, heavy BBS traffic may disrupt production jobs.
- Standard communication protocols tend to have high overhead.

In an enclosed cluster, intracluster communication is shielded from the outside world, which alleviates the above problems. A disadvantage is that there is currently no standard for efficient, enclosed intracluster communication. Consequently, most commercial or academic clusters realize fast communications through one-of-a-kind protocols.

Dedicated Versus Enterprise Clusters Louis Turcotte [620] divided clusters into two classes: *dedicated clusters* and *enterprise clusters*. A dedicated cluster has the following characteristics:

- It is typically installed in a deskside rack in a central computer room.
- It is typically homogeneously configured with the same type of nodes.
- It is managed by a single administrator group like a mainframe.
- It is typically accessed via a front-end system.

Dedicated clusters are used as substitutes for traditional mainframes or supercomputers. A dedicated cluster is installed, used and administered as a single machine. Many users can log into the cluster to execute both interactive and batch jobs. The cluster offers much enhanced throughput, as well as reduced response time.

An *enterprise cluster* is mainly used to utilize idle resources in the nodes. It has the following characteristics:

- Each node is usually a full-fledged SMP, workstation, or PC, with all necessary peripherals attached.
- The nodes are typically geographically distributed, not necessarily in the same room or even in the same building.
- The nodes are individually owned by multiple owners. The cluster administrator has only limited control over nodes, as a node can be turned off at any time by its owner. The owner's "local" jobs have higher priority than enterprise jobs.
- The cluster is often configured with heterogeneous computer nodes. The nodes are often connected through a low-cost Ethernet.

9.1.2 Cluster Architectures

The nodes of a cluster can be connected in one of three ways, as shown in Fig. 1.2, and redrawn in Fig. 9.1. The shared-nothing architecture is used in most clusters, where the nodes are connected through the I/O bus. The shared-disk architecture is in favor of small-scale *availability clusters* in business applications. When one node fails, the other node takes over.

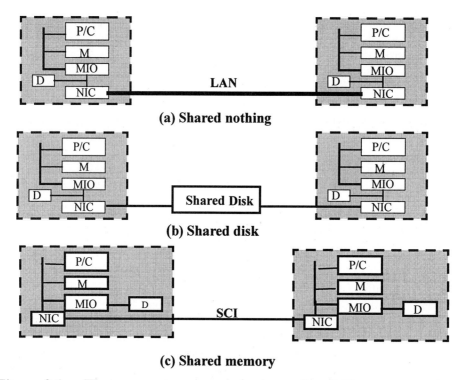

(a) Shared nothing

(b) Shared disk

(c) Shared memory

Figure 9.1   **Three ways to connect cluster nodes.** (P/C: processor and cache; M: memory; D: disk; NIC: network interface circuitry; MIO: memory-I/O bridge.)

Shared-memory clusters are just emerging. An example is the Sequent NUMA-Q computer [345], which uses four Intel Pentium Pro processors in a node, and the nodes are connected through an SCI ring. The SCI is connected to the memory bus of the node through an NIC module.

In the shared-memory architecture, the interconnect is attached to the memory bus of each node. In the other two architectures, the interconnect is attached to the I/O bus. The shared-memory architecture is more difficult to implement than the others, for the following reasons:

- The memory bus usually operates at a higher frequency than the I/O bus.
- There is no widely accepted standard for memory bus. But there are such standards for I/O buses. One recent, popular standard is the PCI I/O bus standard. So if one implements an NIC card to attach a faster Ethernet to the PCI bus of a cluster node, one can be assured that this card can be used in other systems that use PCI as the I/O bus.
- The I/O bus evolves at much slower rate than the memory bus. Consider a cluster that uses connection through PCI bus. When the processors are upgraded, or when the node architecture is changed to a different platform, the interconnect and the NIC do not have to change, as long as the new system still uses PCI. In a shared-memory cluster, changing the processor implies the redesign of the node board and the NIC card.

9.1.3 Cluster Design Issues

Several issues must be considered in developing and using a cluster. Although much work has been done, they are still active research and development areas. Four important issues: *availability support, single system image, job management,* and *efficient communication,* are introduced below.

Availability Support Clusters can provide cost-effective high availability with lots of redundancy in processors, memories, disks, I/O devices, networks, operating system images, etc. However, to realize this potential, availability techniques are needed. These techniques will be illustrated later when we discuss how the DEC clusters (Section 10.4) and the IBM SP2 (Section 10.3) attempt to achieve high availability.

Single System Image A set of workstations connected by an Ethernet is not necessarily a cluster. A cluster is a single system. Take a simple example. Suppose a workstation has a 300-Mflop/s processor, 512-MB memory, 4-GB disk and can support 50 active users and 1,000 processes.

By clustering 100 such workstations, can we get a single system that is equivalent to one huge workstation, or a *megastation,* that has a 30-Gflop/s processor, 50-GB memory, and 400-GB disk and can support 5000 active users and 100,000 processes? This is an appealing goal, but very difficult to achieve. *Single system image* (SSI) techniques are aimed at achieving this goal.

Job Management Clusters try to achieve high system utilization, out of traditional workstations or PC nodes that are normally not highly utilized. Job management software is needed to provide batching, load balancing, parallel processing, and other functionality. This issue will be discussed in Section 9.5.

Efficient Communication   It is more challenging to develop an efficient communication subsystem for a cluster than for an MPP. There are several reasons:

- Because of higher node complexity, cluster nodes cannot be packaged as compactly as MPP nodes.

- The internode physical wire lengths are longer in a cluster than in an MPP. This is true even for centralized clusters. Long wire implies larger interconnect network latency. But more importantly, longer wires have more reliability, clock skew, and cross-talking problems. These problems call for reliable and secure communication protocols, which increase overhead.

- Clusters often use commodity networks (e.g., Ethernet, ATM) with standard communication protocols such as TCP/IP. Commodity components enjoy Moore's law, but TCP/IP protocols have high overhead. Low-level communication protocols, such as those discussed in Section 7.4.3, are much more efficient. But currently there is no accepted standard for low-level communication protocol.

Communication techniques discussed in Section 7.4.3 can be used in clusters. We will show how efficient communication is achieved in the Berkeley NOW project (Section 10.5), in the DEC clusters (Section 10.4), and in the IBM SP2 (Section 10.3).

Idealized Cluster Architecture   We envision an ideal cluster as having an architecture as shown in Fig. 9.2. The processing nodes are workstations, PCs, SMP servers, and even supercomputers. The node operating systems are multiuser, multitasking, and multi-threaded systems. The nodes can be either homogeneous or heterogeneous.

The nodes are interconnected by one or more fast commodity networks. These networks use standard communication protocols and operate at a speed that is two orders of magnitude higher than that for current TCP/IP over an Ethernet.

Each node's network interface circuit is connected to the standard I/O bus (e.g., PCI) of the node. When the processor or the operating system is changed, only the driver software needs to change, but not the network or the network interface.

A set of platform-independent software subsystems sits on top of the node's platforms. There is an availability subsystem offering high-availability services. A single system image layer provides a single entry point, a single file hierarchy, a single point of control, and a single job management system. Single memory may be realized with the help of compiler or runtime library technology. Single process space is not necessarily supported.

The usability of such a cluster is supported by three types of subsystems:

- First, conventional databases and OLTP monitors offer users an environment exactly like that in a workstation.

- Second, in addition to the conventional sequential programming environment, the cluster supports parallel programming based on standard languages and communication libraries. The environment also includes tools for debugging, profiling, monitoring, etc.

- Third, a user-interface subsystem combines the advantages of the Web interface and the windows GUI. Fox et al. [252] call such an environment a *WebWindows*. It should also provides user-friendly links to various programming environments, job management tools, hypertext and search support so that users can easily get help, tutorials, examples, demonstrations, and answers to frequently asked questions..

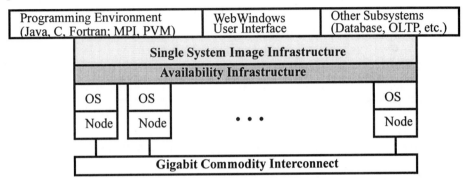

Figure 9.2   **The architecture of a working cluster with full support for availability and single system image.**

9.2 Availability Support for Clustering

In designing robust, highly available systems, three terms often go together: *reliability, availability,* and *serviceability* (abbreviated as *RAS*). Availability is the most interesting measure since it combines the concepts of reliability and serviceability as defined below:

- *Reliability* measures how long a system can operate without a breakdown.
- *Availability* indicates the percentage of time that a system is available to the user, that is, the percentage of system uptime.
- *Serviceability* refers to how easy it is to service the system, including hardware and software maintenance, repair, upgrade, etc.

Example 9.1 Availability of an average computer system

The demand for RAS is driven by practical market needs. A recent Find/SVP survey found the following figures among Fortune 1000 companies: An average computer is down 9 times a year with an average downtime of 4 hours.

The average loss of revenue per hour downtime is $82,500. With such hefty penalty for downtime, many companies are striving for systems that offer *24 × 365 availability*, meaning that the system is available 24 hours a day, 365 days a year.

■

9.2.1 The Availability Concept

As shown in Fig. 9.3, a computer system operates normally for a period of time before it fails. The failed system is then repaired, and the system returns to normal operation. This operate-repair cycle then repeats.

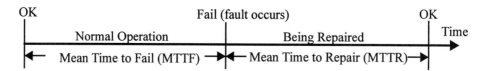

Figure 9.3   The operate-repair cycle of a computer system.

There are sometimes confused definitions and metrics for *reliability, availability,* and *serviceability* (RAS). We distinguish these terms as follows:

Definition 9.1 A system's reliability is measured by the *mean time to failure* (MTTF), which is the average time of normal operation before the system (or a component of the system) fails. The metric for serviceability is the *mean time to repair* (MTTR), which is the average time it takes to repair the system and restore it to work condition after it fails. The availability of a system is then defined by:

$$Availability = \frac{MTTF}{MTTF + MTTR} \tag{9.1}$$

■

Planned versus Unplanned Failure When studying RAS, we call any event that prevents the system from normal operation a *failure*. This includes:

- *Unplanned Failure*: The system becomes broken, caused by operating system crash, hardware failure, network disconnection, human operation errors, power outage, etc. All these are simply called failures. The system must be repaired to correct the failure.

- *Planned Shutdown*: The system is not broken, but is periodically taken off normal operation for upgrade, reconfiguration, and maintenance. A system may also be shut down for weekends or holidays. The MTTR in Fig. 9.3 for this type of "failure" is the planned downtime.

Table 9.2 Availability of Computer System Types
(Courtesy: Digital Equipment Co. [197], 1996)

System Type	Availability (%)	Down Time in a Year
Conventional workstation	99	3.6 days
High-availability system	99.9	8.5 hours
Fault resilient system	99.99	1 hour
Fault-tolerant system	99.999	5 minutes

Table 9.2 shows the availability values of several representative systems. For instance, a conventional workstation has an availability of 99%, which means that it is up and running 99% of the time in a year. The downtime is 1% of a year, which is approximately 3.5 days.

An optimistic definition of availability does not consider the planned downtime, which may be significant. For instance, many supercomputer installations have a planned downtime of several hours per week, while a telephone switch system cannot tolerate a downtime (planned or not) of more than a few minutes per year.

Transient versus Permanent Failures   A lot of failures are *transient* in that they occur temporarily and then disappear. They can be dealt with without replacing any components. A standard approach is to roll back the system to a known state and start over.

For instance, we all have rebooted our PC to take care of transient failures such as frozen keyboard and window. *Permanent failures* cannot be corrected by rebooting. Some hardware or software component must be repaired or replaced. For instance, rebooting will not work if the system hard disk is broken.

Partial versus Total Failures   A failure that renders the entire system unusable is called a *total failure*. A failure that only affects part of the system is called a *partial failure* if the system is still usable, even at a reduced capacity. A key approach to enhancing availability is to make as many failures as possible partial failures, by systematically removing *single points of failure*, which are a hardware or software components whose failure will bring down the entire system.

Example 9.2 **Single points of failure in an SMP and in clusters of computers**

In an SMP as shown in Fig. 9.4(a), the shared memory, the OS image, and the memory bus are all single points of failure. On the other hand, the processors are not. In a cluster of workstations interconnected by an Ethernet [Fig. 9.4(b)], there are multiple OS images, each residing in a workstation.

This avoids the single point of failure caused by OS as in the SMP case. However, the Ethernet now becomes a single point of failure, which is eliminated in Fig. 9.4(c), where a high-speed network is added to provide two paths for communication.

When a node fails in the clusters in Fig. 9.4(b) and Fig. 9.4(c), not only will the node applications all fail, but also all node data cannot be used until the node is repaired. The shared disk cluster in Fig. 9.4(d) provides a remedy. The system can store persistent data on the shared disk, and periodically *checkpoint* applications to save intermediate results on the disk. When one node fails, the data are not lost.

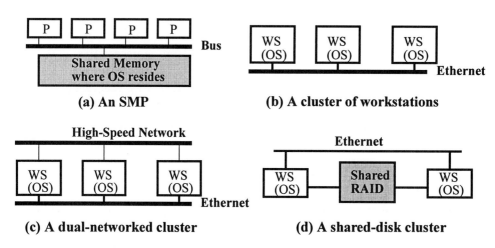

(a) An SMP   (b) A cluster of workstations

(c) A dual-networked cluster   (d) A shared-disk cluster

Figure 9.4   **Single points of failure in an SMP and in three clusters**

■

Example 9.3 **Availability of a cluster of computers**

Consider the cluster in Fig. 9.4(d). Assume only the nodes can fail. The rest of the system (e.g., interconnect and the shared RAID disk) is 100% available. Also assume that when a node fails, its workload is switched over to the other node in zero time. The following questions are posed on the cluster:

  (1) What is the availability of the cluster if planned downtime is ignored?
  (2) What is the availability if the cluster needs 1 hour per week for maintenance?

(3) What is the availability if it is taken down 1 hour per week for maintenance, but one node at a time?

Solutions:

(1) From Table 9.2, a workstation is available 99% of the time. The time both nodes are down is only $0.01 \times 0.01 = 0.0001$, or 0.01%. Thus the availability is 99.99%. It is now a fault-resilient system, with only 1 hour downtime per year.

(2) The planned downtime is 52 hours a year, that is, $52/(365 \times 24) = 0.0059$. The total downtime is now 0.59% + 0.01% = 0.6%. The availability of the cluster becomes 99.4%.

(3) Suppose we ignore the unlikely situation where the other node fails while one node is being maintained. Then the availability is 99.99%, the same as in (1).

∎

9.2.2 Availability Techniques

From Eq.(9.1), there are basically two ways to increase the availability of a system: increasing MTTF or reducing MTTR. Increasing MTTF amounts to increasing the reliability of a system. The computer industry has strived to make reliable systems, and today's workstations have MTTF in the range of hundreds to thousands of hours. However, to further improve MTTF is very difficult and costly.

Clusters offer a high-availability solution based on reducing the MTTR of the system. A multinode cluster has a lower MTTF (thus lower reliability) than a workstation, and failures occur more often. However, the failures are taken care of quickly to deliver higher availability. We now discuss several availability techniques used in clusters.

Isolated Redundancy A key technique to improve availability in any system is to use redundant components. When a component (the *primary* component) fails, the service it provided is taken over by another component (the *backup* component). Furthermore, the primary and the backup components should be *isolated* from each other, meaning that they should not be subject to the same cause of failure.

Clusters provide high availability with redundancy in power supplies, fans, processors, memories, disks, I/O devices, networks, operating system images, etc. In a carefully designed cluster, redundancy is also isolated. Isolated redundancy provides several benefits:

- First, a component designed with isolated redundancy is not a single point of failure, and the failure of that component will not cause a total system failure.
- Second, the failed component can be repaired while the rest of system is still working.
- Third, the primary and the backup components can mutually test and debug each other.

The IBM SP2 communication subsystem is a good example of isolated-redundancy design. All nodes are connected by two networks: an Ethernet and a high-performance switch (HPS). Each node uses two separate interface cards to connect to these networks. There are two communication protocols: a standard IP and a *user-space* (US) protocol; each can run on either network. If either network or protocol fails, the other network or protocol can take over.

Example 9.4 *N*-version programming to enhance software reliability

A common isolated-redundancy approach to constructing a mission-critical software system is called *N-version programming* (NVP). The software is implemented by *N* isolated teams who may not even known the others exist. Different teams are asked to implement the software using different algorithms, programming languages, environment tools, even different platforms.

In a fault-tolerant system, the *N* versions all run simultaneously and their results are constantly compared. If the results differ, the system is notified that a fault has occurred. But because of isolated redundancy, it is extremely unlikely that the fault will cause a majority of the *N* versions to fail at the same time. So the system continues working, with the correct result generated by majority voting.

In a highly available, but less mission-critical system, only one version needs to run at a time. Each version has built-in self-test capability. When one version fails, another version can take over.

■

Example 9.5 Elimination of single failures in the IBM HACMP

The design of the IBM HACMP (*high-availability cluster multiprocessing*) system carefully removes all single points of failures, as shown in Fig. 9.5.

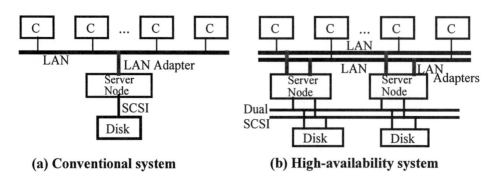

(a) Conventional system   (b) High-availability system

Figure 9.5   Duplication of cluster resources to eliminate all single points of failure in the IBM HACMP.

A conventional client-server configuration is shown in part (a), where multiple *clients* (C) access the services provided by the *server node* through a LAN such as an Ethernet. The server node is connected to an external disk through an SCSI bus.

This conventional system has five possible single points of failure:

(1) LAN network
(2) LAN adapter of the server node (shown as a thick shaded line)
(3) Server itself
(4) SCSI bus
(5) External disk

The high-availability system in part (b) eliminates all single points of failure by using an aggressively redundant design. The redundancy comes from the duplication of all of the five system resources as listed above; namely two LANs, two server nodes, two LAN adaptor cards per server node, two sets of SCSI buses, and two disk arrays used. This is a quite expensive way to achieve high availability, but the IBM HACMP adopted upon customer's demand.

∎

Redundant components can be configured in different ways, according to different cost, availability, and performance requirements. The following three configurations are frequently used:

- *Hot Standby*: A primary component provides service, while a redundant backup component stands by without doing any work, but is ready (hot) to take over as soon as the primary fails. A more economical design uses one standby component to back up multiple primary components. As the backup component is idle, it can take over a failed primary component quickly.
- *Mutual Takeover*: All components are primary in that they all actively perform useful workload. When one fails, its workload is redistributed to other components. This scheme is cost-effective because there are no "normally idle" components. However, the takeover is more time-consuming.
- *Fault-Tolerant*: This was already mentioned in the NVP of Example 9.4. This is the most expensive configuration, as N components deliver the performance of only one component, at more than N times the cost. However, the failure of $N-1$ components is masked (not visible to the user).

Failover   This is probably the most important feature demanded in current clusters for commercial applications. When a component fails, this technique allows the remaining system to take over the services originally provided by the failed component. A failover mechanism must provide several functions, such as *failure diagnosis*, *failure notification*, and *failure recovery*.

Failure diagnosis refers to the detection of a failure and the location of the failed component that causes the failure. A commonly used technique is *heartbeat*, where the cluster nodes send out a stream of heartbeat messages to one another. If the system does not receive the stream of heartbeat messages from a node, it can conclude that either the node or the network connection has failed.

Example 9.6 Failure diagnosis and recovery in a dual-network cluster

A cluster uses two networks to connect its nodes [Fig. 9.4(c)]. One node is designated as the *master node*. Each node has a *heartbeat daemon* that periodically (every 10 s) sends a heartbeat message to the master node through both networks. The master node will detect a failure if it does not receive messages for a beat (10 s) from a node and will make the following diagnoses:

- A node's connection to one of the two networks failed if the master receives a heartbeat from the node through one network but not the other.
- The node failed if the master does not receive heartbeat through either network. It is assumed that the chance of both networks failing at the same time is negligible.

■

The failure diagnosis in this example is simple, but has several pitfalls. What if the master node fails? Is the 10-s heartbeat period too long or too short? What if the heartbeat messages are dropped by the network (e.g., due to network congestion)? Can this scheme accommodate hundreds of nodes? Practical high-availability systems must address these issues.

A popular trick is to use the heartbeat messages to carry load information, so that when the master receives the heartbeat from a node, it knows not only that the node is alive, but also the resource utilization status of the node. Such load information is useful for load balancing and job management.

Once a failure is diagnosed, the system notifies the components that need to know the failure event. Failure notification is needed because the master node is not the only one that needs to have this information. For instance, in case of the failure of a node, the *domain name server* (DNS) needs to be told so that it will not connect more users to that node. The resource manager needs to reassign workload and to take over the remaining workload on that node. The system administrator needs to be alerted so that it can initiate proper actions to repair the node.

Recovery Schemes Failure recovery refers to the actions needed to take over the workload of a failed component. There are two types of recovery techniques. In *backward* *recovery,* the processes running on a cluster periodically save a consistent state (called a *checkpoint*) to a stable storage. After a failure, the system is reconfigured to isolate the failed component, restores the previous checkpoint, and resumes normal operation. This is called *rollback.* Backward recovery is relatively easy to implement in an application-independent, portable fashion, and has been widely used. However, rollback implies wasted execution.

If execution time is crucial, such as in real-time systems where the rollback time cannot be tolerated, a *forward recovery* scheme should be used. With such a scheme, the system is not rolled back to the previous checkpoint upon a failure. Instead, the system utilizes the failure diagnosis information to reconstruct a valid system state and continues execution. Forward recovery is application-dependent and may need extra hardware.

Example 9.7 MTTF, MTTR, and failure cost analysis

Consider a cluster that has little availability support. Upon a node failure, the following sequence of events takes place:

(1) The entire system is shut down and powered off.
(2) The faulty node is replaced if the failure is a hardware one.
(3) The system is powered on and rebooted.
(4) The user application is reloaded and rerun from the start.

Assume one of the cluster nodes fails every 100 hours. Other parts of the cluster never fail. Steps 1 to 3 in total take 2 hours. In average, the mean time for step 4 is 2 hours. What is the availability of the cluster? What is the yearly failure cost if each 1-hour downtime costs $82,500 (Example 9.1).

Solution:

The cluster's MTTF is 100 hours, the MTTR is 2 + 2=4 hours. From Eq.(9.1), the availability is 100/104 = 96.15%. This corresponds to 337 hours downtime in a year, and the failure cost is $82500 × 337, i.e., more than $27 million.

∎

Example 9.8 Availability and cost analysis of a cluster of computers

Repeat Example 9.7, but the cluster now has much increased availability support. Upon a node failure, its workload automatically fails over to other nodes. The failover time is only 6 minutes. Meanwhile, the cluster has *hot swap* capability: The faulty node is taken off the cluster, repaired, replugged, and rebooted, and it rejoins the cluster, all without impacting the rest of the cluster. What is the availability of this ideal cluster, and what is the yearly failure cost?

Solution:

The cluster's MTTF is still 100 hours, but the MTTR is reduced to 0.1 hours, as the cluster is available while the failed node is being repaired. From Eq.(9.1), the availability is 100/100.5 = 99.9%. This corresponds to a 8.75 hours downtime per year, and the failure cost is $82,500 × 8.75 = $722,000, a 27M/722K = 38 times reduction in failure cost from the design in Example 9.7.

∎

9.2.3 Checkpointing and Failure Recovery

These are two techniques that must be developed hand in hand to enhance the availability of a cluster system. We start with the basic concept of checkpointing.

The Basic Concept *Checkpointing* is the process of periodically saving the state of an executing program to stable storage, from which the system can recover after a failure. Each program state saved is called a *checkpoint*. The disk file that contains the saved state is called the *checkpoint file*. Although all current checkpointing software saves program states in a disk, research is underway to use node memories in place of a stable storage [33] in order to improve performance.

Checkpointing techniques are useful not only for availability, but also for program debugging, process migration, and load balancing. Many job management systems and some operating systems support checkpointing to a certain degree. The Web Resource contains pointers to numerous checkpoint-related Web sites, including some public domain software such as Condor and Libckpt.

In this section, we discuss the important issues for the designer and the user of checkpoint software. We first consider those issues that are common to both sequential and parallel programs, then we discuss the issues pertaining to parallel programs.

Kernel, Library, and Application Levels Checkpointing can be realized by the operating system at the *kernel level*, where the OS transparently checkpoints and restarts processes. This is ideal for the users. However, checkpointing is not supported in most operating systems, especially for parallel programs.

A less transparent approach links the user code with a checkpointing library in the user space. Checkpointing and restarting are done by this run-time support. This approach is used widely because it has the advantage that user applications do not have to be modified. A main problem is that most current checkpointing libraries are static, meaning the application source code (or at least object code) must be available. It does not work if the application is in the form of an executable code.

A third approach requires the user (or the compiler) to insert checkpointing functions in the application; thus the application has to be modified, and the transparency is lost. However, it has the advantage that the user can specify where to checkpoint. This is helpful to reduce checkpointing overhead. Checkpointing incurs both time and storage overheads. These overheads are defined below.

Checkpoint Overheads   During a program's execution, its states may be saved many times. Denote by t_c the time consumed to save one checkpoint. The storage overhead is the extra memory and disk spaces required for checkpointing. Both time and storage overheads depend on the size of the checkpoint file. The overheads can be substantial, especially for applications that require a large memory space. A number of techniques have been suggested to reduce these overheads.

Choosing an Optimal Checkpoint Interval   The time period between two checkpoints is called the *checkpoint interval*. Making the interval larger can reduce the checkpoint time overhead. However, this implies a longer recomputation time after a failure. Wong and Franklin [648] derived an expression for optimal checkpoint interval, based on a theoretical model:

$$\text{Optimal checkpoint interval} = \sqrt{(MTTF \cdot t_c)/h} \qquad (9.2)$$

Here *MTTF* is the mean time to failure of the system, t_c is the average time consumed to save one checkpoint, and h is the average percentage of normal computation performed in a checkpoint interval before the system fails. The parameter h is always in the range $0 \leq h \leq$. After a system is restored, it needs to spend $h \times$ (checkpoint interval) time to recompute. These time parameters are illustrated in Fig. 9.6.

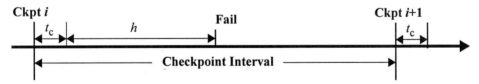

Figure 9.6   Time parameters between two checkpoints.

Incremental Checkpoint   Instead of saving the full state at each checkpoint, an *incremental checkpoint* scheme [503, 504, 505] saves only the portion of the state that is changed from the previous checkpoint. However, care must be taken of *old* checkpoint files. In full-state checkpointing, only one checkpoint file needs to be kept in the disk. Subsequent checkpoints simply overwrite this file. With incremental checkpointing, old files needed to be kept, because the complete state may span several files. Thus the total storage requirement is larger.

Forked Checkpointing   Most checkpoint schemes are blocking, in that the normal computation is blocked (stopped) while checkpointing is in progress. If there is enough memory space, checkpoint overhead can be reduced by making a copy of the program state in memory, and involving another asynchronous thread to perform checkpointing concurrently with normal computation.

A simple way to overlap checkpointing with computation is to use the Unix fork( ) system call. The forked child process will duplicate the parent process's address space and checkpoint it. Meanwhile, the parent process continues execution. Overlapping can be achieved since checkpointing is disk-I/O intensive. A further optimization is to use the copy-on-write mechanism [415].

Checkpoint Compression   Another idea is to shrink the checkpoint using standard compression algorithms. However, compression only works if the compression overhead is small and the checkpoint can be significantly reduced. There is some evidence indicating that compression is not effective for uniprocessor, and is effective for parallel systems with disk contention [505].

User-Directed Checkpointing   The checkpoint overheads can sometimes be substantially reduced if the user inserts codes (e.g., library or system calls) to tell the system when to save, what to save, and what not to save. See [503] for an overview of user-directed checkpointing.

What to Checkpoint   What should be the exact contents of a checkpoint? It should contain just enough information to allow a system to recover. As discussed in Section 2.2, the state of a process includes its data state and control state. For a Unix process, these states are stored in its address space, including the text (code), the data, the stack segments, and the process descriptor.

Saving and restoring the full state are expensive and sometimes impossible. They are also often unnecessary. For instance, the process ID and the parent process ID are not restorable, nor do they need to be saved in many applications. Most checkpointing systems save a partial state. For instance, the code segment is usually not saved, as it does not change in most applications.

Generality   What kind of applications can be checkpointed? Current checkpoint schemes require programs to be *well-behaved*, the exact meaning of which differs in different schemes. At the minimum, a well-behaved program should not need the exact contents of state information that is not restorable, such as the numeric value of a process ID.

The Condor package [608] can checkpoint only those programs that do not create children processes, and do not communicate with other processes using any of the Unix interprocess communication methods such as signals, pipes, sockets, files, etc.

The Libckpt checkpoint library [503] saves only the opened files table, but not other system state information. It is important to know what programs are well-behaved when using any checkpointing software, as ill-behaved programs cannot be correctly checkpointed.

Usability How easy is it to use checkpoint software? Ideally, the checkpointing functionality should be integrated into the operating system, so that user applications can be saved and restored automatically and transparently. But most current operating systems do not support checkpointing well.

The next best thing is a runtime support library that the user can link with the running application. But whether the library is static or dynamic can make a big difference for users. In early versions of Condor, users must link the static checkpoint library with their object code. This is impossible if the application is a third-party executable code. In version 6 of Condor, a dynamic linkable library is provided for Sun Solaris to eliminate this limitation.

The above schemes are called *transparent checkpointing*, as the user does not have to modify his code. Other schemes are not transparent, the simplest of which is illustrated in the Libckpt library, which requires a user to just change one line of the source code, namely, to change the main() function to ckpt_target().

Other schemes that is not transparent require users to insert checkpoint statements in the source code. Although this is inconvenient to the user, this inconvenience should be weighted against possible gains from user-directed checkpointing.

Checkpointing Parallel Programs We now turn to checkpointing parallel programs. The state of a parallel program is usually much larger than that of a sequential program, as it consists of the set of the states of individual processes, plus the state of the communication network. Parallelism also introduces various timing and consistency problems.

Example 9.9 Checkpointing a parallel program

Checkpointing of a three-process parallel program is illustrated in Fig. 9.7. The arrows x, y, and z represent point-to-point communications among the processes. The three thick lines a, b, and c represent three *global snapshots* (or simply *snapshots*), where a global snapshot is a set of checkpoints (represented by dots), one from every process.

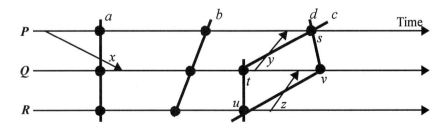

Figure 9.7   Consistent and inconsistent checkpoints in a parallel program.

In addition, some communication state may need to be saved. The intersection of a snapshot line with a process's time line indicates where the process should take a (local) checkpoint. Thus the program's snapshot c consists of three local checkpoints: s, t, u for processes $P, Q,$ and R, respectively, plus saving the communication y.

$\blacksquare$

Consistent Snapshot A global snapshot is called *consistent* if there is no message that is received by the checkpoint of one process, but not yet sent by that of another process. Graphically, this is equivalent to saying that no arrow should cross a snapshot line from right to left. By this definition, snapshot a is consistent, because arrow x is from left to right. But snapshot c is inconsistent, as y goes from right to left.

Netzer and Xu [461] proved a necessary and sufficient condition for any two checkpoints to belong to a consistent snapshot, when all checkpoints are given. The condition is that there should not be any *zigzag path* between these two checkpoints. For instance, checkpoints u and s cannot belong to a consistent global snapshot. A stronger definition of consistent snapshot requires that no arrows cross the snapshot. By this definition, only snapshot b is consistent in Fig. 9.7.

Coordinated versus Independent Checkpointing Checkpointing schemes for parallel programs can be classified into two types. In *coordinated checkpointing* (also called *consistent checkpointing*), the parallel program is frozen, and all processes are checkpointed at the same time. An example is CoCheck [608]. In '*independent checkpointing*, the processes are checkpointed independent of each other.

These two types can be combined in various ways. Coordinated checkpointing is difficult to implement and tends to incur large overhead. Independent checkpointing has small overhead and can utilize existing checkpointing schemes for sequential programs. However, it must solve the *Domino effect* problem.

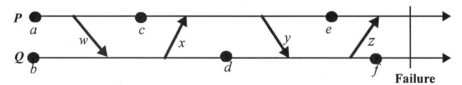

Figure 9.8   Illustration of the domino effect in a sequence of checkpoints.

The Domino Effect Consider Fig. 9.8. Suppose the system fails and process P rolls back to its local checkpoint e. It needs a message to be resent from process Q by communication z. To do this, Q has to be rolled back to checkpoint d. Now Q needs a message to be resent from process P by communication y. Therefore, P needs to be rolled back to c. In the end, both P and Q need to be rolled back to the initial states a and b, making checkpointing useless.

To avoid the domino effect, independent checkpointing is augmented with *message logging*. The idea is that during checkpointing, not only do individual processes save their local checkpoints independently, but also the messages are saved in a log. In Fig. 9.8, a global snapshot just before the failure consists of the local checkpoints e and f and the message log for communication z. At recovery, the processes roll back to their checkpoints e and f, and communication z is replayed.

9.3 Support for Single System Image

Single system image (SSI) does not mean a single copy of operating system image residing in memory, as in an SMP or a workstation. Rather, it means the *illusion* of a single system as characterized below:

- *Single System*: The entire cluster is viewed by the users as one system, which has multiple processors. The user could say: "Execute my application using five processors." This is different from a distributed system.
- *Single Control*: Logically, an end user or system user utilizes services from one place with a single interface. For instance, a user submits batch jobs to one set of queues; a system administrator configures all the hardware and software components of the cluster from one control point.
- *Symmetry*: A user can use a cluster service from any node. In other words, all cluster services and functionalities are symmetric to all nodes and all users, except those protected by the usual access rights.
- *Location Transparent*: The user is not aware of the whereabouts of the physical device that eventually provides a service. For instance, the user can use a tape drive attached to any cluster node as if it were physically attached to the local node. There may be some performance difference, though.

9.3.1 Single System Image Layers

The main motivation to have SSI is that it allows a cluster to be used, controlled, and maintained as a familiar workstation is. The word *single* in *single system image* is sometimes synonymously called *global* or *central*.

For instance, a global file system means a single file hierarchy, which a user can access from any node. A single point of control allows an operator to monitor and configure the cluster system.

Although there is an illusion of a single system, a cluster service or functionality is often realized distributively, through the cooperation of multiple components. A main requirement (and advantage) of SSI techniques is that they provide both the performance benefits of distributed implementation and the usability benefits of a single image.

Definition 9.2 From the viewpoint of a process *P*, cluster nodes can be classified into three types. The *home node* of a process *P* is the node where *P* resided when it was created. The *local node* of a process *P* is the node where *P* currently resides. All other nodes are *remote nodes* to *P*.

∎

Definition 9.3 Cluster nodes can be configured to suit different needs. A *host node* serves user logins through Telnet, rlogin (or even ftp and http). A *compute node* is one that performs computational jobs. An *I/O node* is one that serves file I/O requests. If a cluster has large shared disks and tape units, they are normally physically attached to I/O nodes.

∎

There is one home node for each process, which is fixed throughout the entire life of a process. At any time, there is only one local node, which may or may not be the host node. The local node and remote nodes of a process may change when the process migrates. A node can be configured to provide multiple functionalities. For instance, a node can be designated as a host, an I/O node, and a compute node at the same time.

The literature uses confusing names. For instance, sometimes *host* is used to refer to any type of node, e.g., host is equivalent to node. A host or I/O node is sometimes called a *server node*, to indicate that it provides certain services. Some even call a compute node a server node.

The illusion of an SSI can be obtained at several layers, four of which are discussed below. Note that these layers may overlap with one another.

- *Application Software Layer*: Two examples are Web browsers and various parallel databases. The user sees a single system image through the application and is not even aware that he is using a cluster. This approach demands the modification of workstation or SMP applications for clusters.
- *Hardware or Kernel Layer*: Ideally, SSI should be provided by the operating system or by the hardware. Unfortunately, this is not a reality yet. Furthermore, it is extremely difficult to provide a single system image over heterogeneous clusters. With most hardware architectures and operating systems being proprietary, only the manufacturer can use this approach.
- *Above-Kernel Layer*: We believe the most viable approach is to construct an SSI layer just above the kernel, as shown in Fig. 9.2. This approach is promising because it is platform-independent and does not require the modification of the applications. Many cluster job management systems have already adopted this approach, to be discussed in Section 9.5.

9.3.2 Single Entry and Single File Hierarchy

Single system image is a very rich concept, consisting of single entry point, single file hierarchy, single I/O space, single networking, single control point, single job management system, single memory space, and single process space.

Single Entry Point *Single entry point* enables users to log in (e.g., through telnet, rlogin, or http) a cluster as one virtual host, although the cluster may have multiple physical host nodes to serve the login sessions. The system transparently distributes the user's login and connection requests to different physical hosts to balance load.

One may wonder why we need a cluster serving thousands of user sessions. The answer is that clusters could substitute for mainframes and supercomputers. Also, in an Internet cluster server, thousands of http or ftp requests may come simultaneously.

Establishing a single entry point with multiple hosts is not a trivial matter. Many issues must be resolved. The following is just a partial list:

- *Home Directory*: Where to put the user's home directory?
- *Authentication*: How to authenticate user logins?
- *Multiple Connections*: What if the same user opens several sessions to the same user account (e.g., two Telnet's, one ftp)?
- *Host Failure*: How to deal with the failure of one or more hosts?

Example 9.10 Realizing single entry point in a cluster of computers

A simple way to realize single entry point is illustrated in Fig. 9.9. Four nodes of a cluster are used as host nodes to receive users' login requests. Although only one user is shown, thousands of users can connect to the cluster in the same fashion.

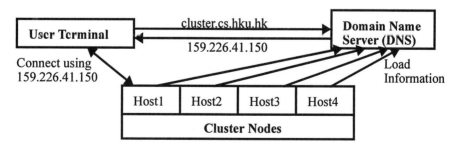

Figure 9.9   Realize single entry point using a load-balancing domain name server.

When a user wants to log into the cluster, he issues a standard Unix command such as "telnet cluster.cs.hku.hk", using the symbolic name of the cluster system. The symbolic name is translated by the DNS, which returns with the IP address 159.226.41.150 of the least-loaded node, which happens to be node Host1. The user then logs in using this IP address.

The DNS periodically receives load information from the host nodes to make load-balancing translation decisions. In the ideal case, if 200 users simultaneously log in, the login sessions are evenly distributed among the four hosts, 50 users per host. This gives the illusion of a single host that is 4 times as powerful.

■

Example 9.11 Solutions of the home directory problem

One trivial solution to the home directory problem is to maintain all users' home directories in a stable storage of the cluster (the concept of stable storage will be discussed further when we talk about single file hierarchy). The problem is that accessing the home directory becomes a remote operation, thus is not efficient.

An alternative is to maintain a copy of each user's home directory in every host. However, this solution has two major shortcomings:

- The disk requirement could be too large. Assume the cluster needs to support 1000 user accounts, and each user has a disk quota of 20 MB for his home directory. This implies that the local disk of each host must have at least 20 GB capacity, which is likely to be too great.

- To maintain consistency, the most recent home directory should be broadcast to all hosts when a user disconnects.

■

Single File Hierarchy We use the term *single file hierarchy* in this book to mean the illusion of a single, huge file system image that transparently integrates local and global disks and other file devices (e.g., tapes). In other words, all files a user needs are stored in some subdirectories of the root directory /, and they can be accessed through ordinary Unix calls such as open, read, etc. This should not be confused with the fact that multiple file systems can exist in a workstation as subdirectories of the root directory.

The functionalities of single file hierarchy have already been partially provided by existing distributed file systems such as *Network File System* (NFS) [527] and *Andrew File System* (AFS) [453]. From the viewpoint of any process, files can reside on three types of locations in a cluster, as shown in Fig. 9.10. *Local storage* is the disk on the local node of a process. The disks on remote nodes are *remote storage*. *Stable storage* is characterized by the following:

- It is *persistent*, which means that data, once written to the stable storage, will stay there at least for a period of time (e.g., a week), even after the cluster shuts down.

- It is fault tolerant to some degree, by using redundancy and periodical backup to tapes.

Stable storage is shown in Fig. 9.10 as a logically centralized (global) unit. Files in stable storage are called *global files*, those in local storage *local files*, and those in remote storage *remote files*. Stable storage could be implemented as one centralized, large RAID disk. But it could also be realized distributively using the local disks of cluster nodes. The first approach uses a large disk, which is a single point of failure and a potential performance bottleneck. The latter approach is more difficult to implement, but it is potentially more economic, more efficient, and more available.

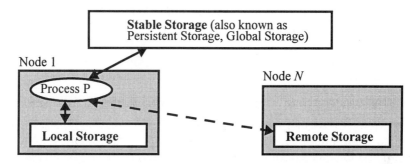

Figure 9.10   Three types of storage in a single file hierarchy. (Solid lines show what process *P* can access. Dashed line shows what *P* may be able to access.)

Example 9.12 A single file hierarchy in a Unix cluster

On many cluster and MPP systems, it is customary for the system to make visible to the user processes the following directories in a single file hierarchy:

- The usual *system directories* as in a traditional Unix workstation, such as /usr and /usr/local, etc.
- The user's *home directory* ~/. that has a small disk quota (1-20 MB). The user store his code files and other files. But large data files must be stored elsewhere.
- A *global directory* /scratch (or /globalscratch, or some other name) that are shared by all users and all processes. This directory has a large disk space of multiple gigabytes. Users can store their large data files here.
- On a cluster system, a process can access a special directory on the local disk, often denoted /localscratch. This directory has medium capacity (e.g., a few hundred megabytes) and are faster to access than the /globalscratch directory.

■

Visibility of Files   The term *visibility* here means that a process can use traditional Unix system or library calls such as fopen, fread, and fwrite to access files. Note that there are multiple local scratch directories in a cluster. The local scratch directories in remote nodes are not in the single file hierarchy, and are not directly visible to the process. A user process can still access them with commands like *rcp* or some special library functions, by specifying both the node name and the filename.

The name scratch indicates that the storage is meant to act as a scratch pad for temporary information storage. Information in a local scratch could be lost once the user logout. Files in a global scratch will normally persist even after the user logs out, but will be deleted by the system if not accessed in a predetermined time period. This is to free disk space for other users. The length of the period can be set by the system administrator, usually ranging from one day to several weeks. Some systems back up the global scratch space to tapes periodically or before deleting any files.

Other Desired SSI Properties   It is desired that a single file hierarchy have the SSI properties discussed, which are reiterated for file systems as follows:

- *Single System*: There is just one file hierarchy from the user's viewpoint.
- *Symmetry*: A user can access the global storage (e.g., /scratch) using a cluster service from any node. In other words, all file services and functionalities are symmetric to all nodes and all users, except those protected by access rights.
- *Location-Transparent*: The user is not aware of the whereabouts of the physical device that eventually provides a service. For instance, the user can use a RAID attached to any cluster node as if it were physically attached to the local node. There may be some performance difference, though.

A file operation should be a transaction satisfying the ACID properties (Definition 1.3). A cluster file system should maintain the *Unix semantics*: Every file operation (fopen, fread, fwrite, fclose, etc.) is a transaction. When an fread accesses a file after an fwrite modifies the same file, the fread should get the updated value. However, existing distributed file systems do not completely follow the Unix semantics. Some of them update a file only at close or flush.

A number of alternatives have been suggested to organize the global storage in a cluster. One extreme is to use a single file server that hosts a big RAID. This solution is simple and can be easily implemented with current software (e.g., NFS). But the file server becomes both a performance bottleneck and a single point of failure. Another extreme is to utilize the local disks in all nodes to form a global storage. This could solve the performance and the availability problems of a single file server. However, no mature software is readily available for this purpose today.

Implementing Single File Hierarchy   To improve performance, *caching* is widely used in most file systems. A popular approach is *client-side caching*: The node issuing file operation requests caches the frequently used file segments. However, caching brings about the *coherency* (or *consistency*) problem. Cache-coherency techniques developed for DSM can be used here.

There are several ways to implement a single file hierarchy image. Three approaches are suggested below:

(1) **NFS:** The *Network File System* is an open standard developed by Sun Micro-systems, and it has gained wide acceptance among Unix platforms. File systems in remote nodes are attached (*mounted*) to a single hierarchy. When a process issues a file operation request, NFS converts it into a *remote procedure call* (RPC) and forwards to the remote node through standard TCP/IP.

The file operation is executed by a remote node and the result is returned by the RPC. This is done transparently. The user can access remote files as easy as accessing a local file. The main problem of NFS is its poor scalability.

(2) **AFS:** The *Andrew File System* aims to enhance file system scalability. It divides a large distributed system into a hierarchy of segments, each with its own file servers. To reduce network traffic, AFS performs extensive caching and employs a *state protocol*, instead of the stateless protocol in NFS. It reduces the server load by allocating functions that used to be performed by the file server in NFS to the clients.

It has been demonstrated that AFS can scale to thousands of nodes over several continents. However, accessing a local file through AFS is much slower than ordinary Unix file operations. (This is analogous to a parallel system that exhibits high speedup but poor single-node performance.)

(3) **Solaris MC:** Both NFS and AFS are mature, commercial systems. If the resources are available, a third approach is to modify the workstation file system to create a single file hierarchy. Fortunately, most Unix systems provide a standard *vnode* interface that allows one to add a new file system to the kernel without the need of the kernel source code.

This approach has the advantage that one can tailor the new file system to suit the needs. In fact, AFS was originally implemented with this approach. The down side is that developing a file system demands many man-months. The Solaris MC offers an example of this approach to be discussed in Section 9.4.

9.3.3 Single I/O, Networking, and Memory Space

To achieve SSI, we desire single control point, single address space, single job management system, single user interface, and single process control. These concepts are depicted in Fig. 9.11. In this example, each node has exactly one network connection. Two of the four nodes each have two I/O devices attached.

Single Networking   A properly designed cluster should behave as one system (the shaded area). In other words, it is like a big workstation with four network connections and four I/O devices attached. Any process on any node can use any network and I/O device as if it were attached to the local node. Single networking means any node can access any network connection.

Single Input/Output   Assume the cluster in Fig. 9.11 is used as a Web server. The Web information database is distributed between the two RAIDs. An httpd daemon is started on each of the nodes to handle Web requests, which come from all four network connections. Single I/O space implies that any node can access the two RAIDs. Suppose most requests come from the ATM network. It would be beneficial if the functions of the http on node 3 could be distributed to all four nodes.

Single Point of Control   The system administrator should be able to configure, monitor, test, and control the entire cluster and each individual node from a single point. Many clusters aid this through a system console that is connected to all nodes of the cluster, as shown in Fig. 9.11. The system console is normally connected to an external LAN (not shown in Fig. 9.11) so that the administrator can login remote to the system console from anywhere in the LAN to perform administration work.

Note that single point of control does not mean that all system administration work should be carried out solely by the system console. In reality, many administrative functions are distributed across the cluster. It means that controlling a cluster should be no more difficult than administering an SMP or a mainframe.

It implies that administration- related system information (such as various configuration files) should be kept in one logical place. The administrator monitor the cluster with one graphics tool, which shows the entire picture of the cluster, and the administrator can zoom in and out at will.

Single point of control (or *single point of management*) is one of the most challenging issues in constructing a cluster system. Techniques from distributed and networked systems management can be transferred to clusters. Several de facto standards have already been developed for network management. An example is the SNMP (*Simple Network Management Protocol*).

Major computer vendors have developed sophisticated, proprietary network management software, such as the IBM NetView and the HP OpenView. HP also developed a software called ClusterView specifically for HP clusters management. Nevertheless, it is still an active research and development topic to develop an easy-to-use, powerful, efficient cluster management software that seamlessly integrates with the rest of cluster software, such as the availability support system, the file system, and the job management system.

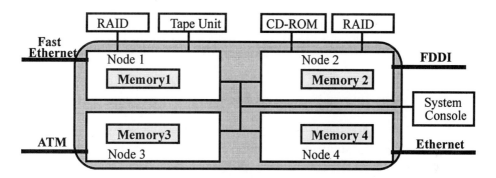

Figure 9.11   Illustration of single networking, single I/O space, single memory, and single point of control.

Single Memory Space   *Single memory space* gives users the illusion of a big, centralized main memory, which in reality may be a set of distributed local memories. PVPs, SMPs, and DSMs have an edge over MPPs and clusters in this respect, because they allow a program to utilize all global or local memories.

A good way to test if a cluster has a single memory space is to run a *sequential* program that needs a memory space larger than any single node can provide. Suppose each node in Fig. 9.11 has 256-MB memory available to users. An ideal single memory image would allow the cluster to execute a sequential program that needs 1-GB memory. This would enable a cluster to feel like an SMP.

Several approaches have been attempted to achieve single memory space on clusters. The software DSM approach has already been discussed in Chapter 5. Another approach is to let the compiler distribute the data structures of an application across multiple nodes. It is still a challenging task to develop a single memory scheme that is efficient, platform-independent, and able to support sequential binary codes.

Other Single Images   The ultimate goal of single system image is for a cluster to be as easy to use as a workstation or SMP. We list below additional types of single system image, which are present in workstations:

- *Single Job Management System:* All cluster jobs can be submitted from any node to a single job management system. There are many such systems available today, as discussed in Section 9.5.

- *Single User Interface*: The users should be able to use the cluster through a single graphical interface. Such an interface is available for workstations (e.g., the *Common Desktop Environment*, or CDE) and PCs (e.g., Microsoft Windows 95). A good direction to take in developing a cluster GUI is to utilize the Web technology.

- ***Single Process Space***: All user processes, no matter on which nodes they reside, belong to a single process space and share a uniform process identification scheme. A process on any node can create (e.g., through a Unix fork) or communicate with (e.g., through signals, pipes, etc.) processes on remote nodes.

9.4   Single System Image in Solaris MC

Sun Microsystems has developed Solaris MC (MC stands for *multicomputer*), a prototype extension of the single-node Solaris kernel [369]. This is shown in Fig. 9.12. The nodes correspond to Sun workstations running the Solaris operating system.

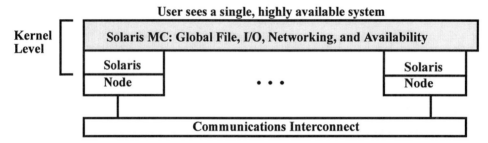

Figure 9.12   Conceptual view of Solaris MC by Sun Microsystems.
(Courtesy Khalidi et al. Sun Microsystems, also *Proc. of the 1996 USENIX Conference,* [369])

Solaris MC provides single system image and high availability at the kernel level, and is believed to be more efficient than other above-kernel approaches. Solaris MC is implemented through object-oriented techniques. It extensively uses the object-oriented programming language C++, the standard COBRA (*Common Object Request Broker Architecture*) object model and its *Interface Definition Language* (IDL).

We discuss below how Solaris MC realizes single file hierarchy, single process space, single networking, and single I/O space. The availability functionality of Solaris MC was not fully developed at this book's writing. Solaris MC has been implemented on a hardware prototype consisting of a network of 16 dual-processor SPARCstations.

9.4.1 Global File System

Solaris MC uses a global file system called *Proxy File System* (PXFS). The main features claimed include single system image, coherent semantics, and high performance. The PXFS is illustrated in Fig. 9.13.

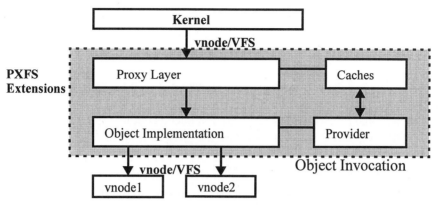

Figure 9.13   Proxy File System in Solaris MC. (Courtesy Khalidi et al. Sun Microsystems, also *Proc. of the 1996 USENIX Conference* [369])

The Proxy File System   The PXFS makes file accesses transparent to process and file locations: Any process can access any file on any node with the same path name. This makes a cluster file system exactly like a workstation file system. PXFS achieves this single system image by intercepting file access operations at the vnode/VFS interface.

Modern Unix systems (Solaris included) perform file operations through a *virtual file system* (VFS) interface [621], where a vnode is an abstraction of ordinary data and code files, as well as other special files such as directories, symbolic links, special devices, streams, swap files, etc. On a workstation, the Solaris kernel accesses files through VFS/vnode operations.

On a cluster, the process and the file may reside on different nodes. When a client node performs a VFS/vnode operation, Solaris MC proxy layer first converts the VFS/vnode operation into an object invocation, which is forwarded to the node where the file resides (the server node). The invoked object then performs a local VFS/vnode operation on the Solaris file system of the server node. This implementation approach needs no modification of the Solaris kernel or the file system.

File Caching   To improve performance, PXFS uses extensive caching on the clients to reduce remote object invocations. A *cached* object is maintained on a client to manage the cached data, and a *cached* object on the server maintains file semantic coherency. At any time, the pages of a file reside on the stable storage of the server node and on caches of one or more client nodes. PXFS also has a "bulkio" object handler to perform zero-copy, large data transfers between nodes.

PXFS uses a token-based coherency protocol to allow a page to be cached read-only by multiple nodes or read-write by a single node. If a process needs to write to a shared page, it has to first acquire the token. If this dirty page is transferred to another node, it is first written to the stable storage on the server node to avoid losing updates. The token is also used to enforce atomicity of read/write system calls.

9.4.2 Global Process Management

Solaris MC provides a single process space. A process can reside on any node, although all threads of a process must reside on the same node. An arbitrary process can be located by a global *process identifier* (pid) that encodes in its top bits the home node of the process. The home node is the node where the process is "born" (created).

A process may migrate to other nodes, but its current location is always maintained in its home node. When a user or system program wants to locate a process *P*, it first looks in the local cache with the pid of *P*. If *P* is not cached, it goes to the home node of *P*.

Distributed Node Managers Solaris MC realizes the single process space by adding a global process layer on top of Solaris kernel layer, as illustrated in Fig. 9.14. There is a node manager for each node and a *virtual process* (vproc) object for each local process. The vproc maintains information of the parent and children of each process.

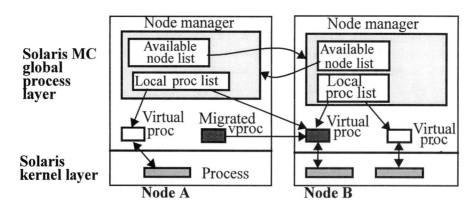

Figure 9.14   Data structures used in Solaris MC process management.
(Courtesy Khalidi et al. Sun Microsystems, also *Proc. of the 1996 USENIX Conference* [369].)

The node manager keeps two lists: the *available node list* and *local process list*, including migrated ones. When a process migrates to another node, a shadow vproc is still kept on the home node. Operations received by the shadow vproc are forwarded to the current node where the process resides. Additional work must be done to realize the global process management.

For instance, all process-related system calls are redirected to the global process layer, some hooks are added to the Solaris kernel to call the global layer (which means modification of the kernel), the global layers of different nodes need to communicate through the IDL interface, local /proc file systems are merged into a global /proc so that Unix operations such as **ps** can access the local /proc on all nodes by looking in just one global /proc.

Khalidi et al. [369] found that global process management was made difficult by the lack of an open, standard interface such as the vnode/VFS interface for file systems. Establishing such a standard would greatly facilitate the development of single process space support for a cluster system.

9.4.3 Single I/O System Image

Two techniques have been developed to support single image in I/O subsystems of a cluster; one is *uniform device naming,* and the other is *single networking.*

Uniform Device Naming Solaris MC provides a single I/O subsystem image with uniform device naming. A device number consists of the node number of the device, as well as the device type and the unit number. A process can access any device by using this uniform name as if it were attached to the local node, even if it is attached to a remote node.

The device drivers are dynamically loadable and configurable. The configuration is kept consistent by a distributed device server, which is notified whenever a new device is added to any node in the cluster. When a process on any node invokes a device driver, the driver is loaded on that node. Solaris MC is implemented so that all Solaris device drivers can run without change.

Single Networking Solaris MC ensures that existing networking applications need not be modified and see the same network connectivity, regardless of which node the applications run on. The networking structure is illustrated in Fig. 9.15.

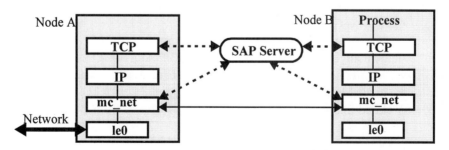

Figure 9.15 **Networking structure in Solaris MC.** (Solid lines for data communication, and dashed lines for control communication. SAP: Service access point. Courtesy Khalidi et al. Sun Microsystems, also *Proc. of the 1996 USENIX Conference* [369])

Just like a single workstation, a Solaris MC cluster may have several network devices (e.g., one each for Ethernet, Fast Ethernet, ATM, FDDI, etc.) attached to multiple nodes. However, any process can use any network device as if it were attached to the local node. All the processes have the same view of the network from any node.

The network is physically connected only to node A through a network device le0. However, a process on node B can use le0 to access the network as if it is physically attached to node B.

Suppose the process wants to send out a message; this is accomplished as follows:

(1) All protocol processing (e.g., TCP/IP) is done on the local node (i.e., node B). This prevents node A from becoming a bottleneck.

(2) The mc_net module on node B finds out the location of le0 and routes packets to the mc_net on node A.

(3) The mc_net on node A performs local network operations to send the packets out to the physical network.

Service Access Point Network services are accessed through a *service access point* (SAP) server. For TCP/IP, an SAP is just a port. The SAP name space database is maintained in an SAP server, which is provided on one or more nodes of the cluster. All processes goes to the SAP server to locate which node an SAP is on. The SAP server also ensures that the same SAP is not simultaneously allocated to different nodes.

Solaris MC allows multiple nodes to act as replicated SAP servers for network services. For instance, *rlogin*, *telnet*, and *http* servers are by default replicated on every node. The user sees a single entry point for each service. Suppose a Solaris MC cluster is used as a Web server. When a user accesses the cluster (not a node) using a single URL, the system will automatically connect the user to the http server on a least busy node. This way, all nodes can serve http requests in parallel, which increases the throughput and decreases the response time.

9.5 Job Management in Clusters

Job management is also known as *workload management, load sharing,* or *load management*. We first discuss the basic issues facing a job management system and summarize the available software packages. We then detail a widely used job management systems: LSF (Load Sharing Facility), a commercial software from Platform Computing.

9.5.1 Job Management System

Saphir et al. [528] analyzed the job management requirements for the *Numerical Aerodynamic Simulation* (NAS) parallel systems and clusters at NASA Ames Research Center. A *Job Management System* (JMS) should have three parts:

- A *user server* lets the user submit jobs to one or more queues, specify resource requirements for each job, delete a job from a queue, inquire about the status of a job or a queue.
- A *job scheduler* performs job scheduling and queuing according to job types, resource requirements, resource availability, and scheduling policies.
- A *resource manager* allocates and monitors resources, enforces scheduling policies, and collects accounting information.

Additional functionalities include job migration and checkpointing.

JMS Administration and Use   The functionality of a JMS is often realized distributively. For instance, a user server may reside in each host node, and the resource manager may span all cluster nodes. However, the administration of a JMS should be *centralized*. All configuration and log files should be maintained in one location. There should be a single user interface to use the JMS. It is undesirable to force the user to run PVM jobs through one software, MPI jobs through another, and HPF jobs through yet another. PVM, MPI, and HPF will be discussed in programming chapters in Part IV.

The JMS should be able to dynamically reconfigure the cluster with minimal impact on the running jobs. The administrator's prologue and epilogue scripts should be able to run before and after each job for security checking, accounting, and cleanup. Users should be able to cleanly kill their own jobs. The administrator or the JMS should be able to cleanly suspend or kill any job.

Clean means that when a job is suspended or killed, all its processes must be included. Otherwise some "orphan" processes are left in the system, which wastes cluster resource and may eventually render the system unusable. NASA users have found the lack of ability to detect and clean up orphaned processes and foreign jobs the single biggest problem with the current JMS.

Job Types   Several types of jobs execute on a cluster. *Serial jobs* run on a single node. *Parallel jobs* use multiple nodes. *Interactive jobs* are those that require fast turnaround time, and their input/output is directed to a terminal. These jobs do not need large resources, and the users expect them to execute immediately, not made to wait in a queue. *Batch jobs* normally need more resources, such as large memory space and long CPU time. But they do not need immediate response. They are submitted to a job queue to be scheduled to run when the resource becomes available (e.g., during off hours).

While both interactive and batch jobs are managed by the JMS, *foreign jobs* are created outside the JMS. For instance, when a network of workstations is used as a cluster, users can submit interactive or batch jobs to the JMS. Meanwhile, the owner of a workstation can start a foreign job at any time, which is not submitted through the JMS.

Such a job is also called a *local job*, as opposed to *cluster jobs* (interactive or batch, parallel or serial) that are submitted through the JMS of the cluster. The characteristic of a local job is fast response time. The owner wants all resources of the workstation devoted to execute his job, as if the cluster jobs did not exist.

Characteristics of Cluster Workload   To realistically address job management issues, we must understand the workload behavior of clusters. It may seem ideal to characterize workload based on longtime operation data on real clusters. However, this is extremely difficult. Arpaci et al. [44] used the following approach in the Berkeley NOW project. They collected 2 months real sequential workload traces from a network of 53 workstations, and 1 month's worth of parallel workload traces from a 32-node CM-5.

The parallel workload traces include both development and production jobs. These traces are then fed to a simulator to generate various statistical and performance results, based on different sequential and parallel workload combinations, resource allocations, and scheduling policies. The Arpaci study shows the following workload characteristics, which are consistent with measurements from other researchers:

- Roughly half of parallel jobs are submitted during regular working hours.
- Almost 80% of parallel jobs run for 3 minutes or less.
- Parallel jobs that run for 90 minutes or more account for over 50% of the total execution time.
- The sequential workload shows that 60% to 70% of workstations are available to execute parallel jobs at any time, even during peak daytime hours.
- On a workstation, 53% of all idle periods are 3 minutes or less, but 95% of idle time is spent in periods of time that are 10 minutes or longer.
- A 2:1 rule applies, which says that a network of 64 workstations, with a proper JMS software, can sustain a 32-node parallel workload in addition to the original sequential workload. In other words, clustering gives a supercomputer half of the cluster size for free!

Job Scheduling   Job scheduling issues are similar to process scheduling discussed in Section 2.2.6. In addition, jobs may be scheduled to run at a specific time (*calendar scheduling*) or when a particular event happens (*event scheduling*). Various schemes to resolve job scheduling issues on a cluster are listed in Table 9.3.

Jobs are scheduled according to *priorities*, which are calculated based on submission time, resource (nodes, execution time, memory, disk, etc.) requirements, job type, and user identity. With *static priority*, jobs are assigned priorities according to a predetermined, fixed scheme. A simple scheme is to schedule jobs in a first-come, first-served fashion. Another scheme is assign different priorities to different users.

With *dynamic priority*, the priority of a job may change over time. An often used scheme is to assign higher priorities to short, interactive jobs in daytime, in order to reduce turnaround time, and during evening hours, to use a technique called *tiling* (see Example 9.3 below) to improve the system utilization.

Table 9.3 Job Scheduling Issues for Cluster Nodes

Issue	Scheme	Key Problems
Job priority	Non-preemptive	Delay of high-priority jobs
	Preemptive	Overhead, implementation
Resource required	Static	Load imbalance
	Dynamic	Overhead, implementation
Resource sharing	Dedicated	Poor utilization
	Space sharing	Tiling, large job
	Time sharing	Process-based job control, Context switch overhead
Scheduling	Independent scheduling	Severe slowdown
	Gang scheduling	Implementation difficulty
Competing with foreign (local) jobs	Stay	Local job slowdown
	Migrate	Migration threshold Migration overhead

The job resource requirement can be *static* or *dynamic*. For instance, the number of nodes needed by a job can be fixed; the same nodes will be assigned to the job for its entire life, even when they become idle during the Job's execution. Static resource is widely used in current clusters. However, it underutilizes the cluster resource and forces the user to do load balancing. Furthermore, it cannot handle the situation when the needed nodes become unavailable, such as when the workstation owner shuts down the machine.

Dynamic resource allows a job to acquire or release nodes during execution. However, it is much more difficult to implement, requiring cooperation between a running job and the JMS. The job needs to make asynchronous requests to the JMS to add/delete resources, and the JMS needs to notify the job when resources become available.

The synchrony means that a job should not be delayed (blocked) by the request/ notification. Cooperation between jobs and the JMS requires modification of current programming languages/libraries. Primitive mechanism for such cooperation exists in PVM, and some work is underway for MPI.

Dedicated Mode   Three schemes are used to share cluster nodes. In the *dedicated mode*, only one job runs in the cluster at a time, and at most one process of the job is assigned to a node at a time. The single job runs until completion before it releases the cluster to run other jobs. Note that even in the dedicated mode, some nodes may be reserved for system use and not be open to the user job. Other than that, all cluster resources are devoted to run a single job. This may lead to poor system utilization.

Space Sharing In the *space-sharing* mode, multiple jobs can run on disjoint partitions (groups) of nodes simultaneously. At most one process is assigned to a node at a time. Although a partition of nodes is dedicated to a job, the interconnect and the I/O subsystem may be shared by all jobs. Space sharing must solve the tiling problem and the large-job problem, as demonstrated by the following example.

Example 9.13 Job scheduling by tiling over cluster nodes

The *tiling technique* is illustrated in Fig. 9.16. In part (a), the JMS schedules four jobs in a straightforward first-come first-served fashion on a four-node cluster. Jobs 1 and 2 are sequential, each needing only one node. They are assigned to node1 and node 2. Jobs 3 and 4 are parallel, each needing three nodes. When job 3 comes, it cannot run immediately. It must wait until job 2 finishes to free the needed nodes.

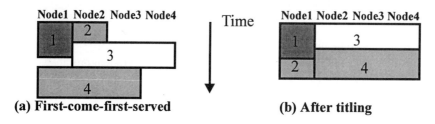

(a) First-come-first-served   (b) After titling

Figure 9.16   The tiling technique for scheduling jobs to cluster nodes.

Tiling will increase the utilization of the nodes, as shown in Fig.9.16(b). The overall execution time of the 4 jobs are significantly reduced after repacking the jobs over the available nodes.

■

The *large-job problem* occurs when a large job takes too long to finish preventing small jobs from using cluster resources. This problem cannot be solved in dedicated or space-sharing modes. However, it can be alleviated by timesharing.

Time sharing In the dedicated or space-sharing models, only one user process is allocated to a node. However, the system processes or daemons are still running on the same node. In the *time sharing* mode, multiple user processes are assigned to the same node. Time-sharing introduces the following parallel scheduling policies:

(1) ***Independent Scheduling***: The most straightforward implementation of time sharing is to use the operating system of each cluster node to schedule different processes as in a traditional workstation. This is called *local scheduling* or

independent scheduling. However, the performance of parallel jobs could be significantly degraded.

The reason is that processes of a parallel job need to interact. For instance, when one process wants to barrier-synchronize with another, the latter may be scheduled out. So the first process has to wait. When the second process is rescheduled, the first process may be swapped out.

(2) **Gang Scheduling:** The *gang scheduling* scheme schedules all processes of a parallel job together. When one process is active, all processes are active. The Berkeley study shows that gang scheduling can improve the performance by one to two orders of magnitude.

The cluster nodes are not perfectly clock-synchronized. In fact, most clusters are asynchronous systems, not driven by the same clock. When we say, "All processes are scheduled to run at the same time," they do not start at the exact same time. *Gang-scheduling skew* is the maximum difference between the time the first process starts and the time the last process starts.

The Berkeley study indicates that the execution time of a parallel job increases as the gang scheduling skew becomes larger, leading to longer execution time. A NOW should use either a larger time slice (1 s) or a small gang scheduling skew (10 ms). The Berkeley study used a homogeneous cluster, where gang scheduling is effective. However, gang scheduling is not yet realized in most clusters, because of implementation difficulties.

(3) **Competition with Foreign (Local) Jobs:** Scheduling becomes more complicated when both cluster jobs and local jobs are running. The local jobs should have priority over cluster jobs. With one keystroke, the owner wants command of all workstation resources.

There are basically two ways to deal with the situation: The cluster job can either stay in the workstation node or migrate to another idle node. A *stay scheme* has the advantage of avoiding the migration cost. Several stay schemes have been suggested. For instance, the cluster process can be run at the lowest priority. The workstation's cycles can be divided into three portions, for kernel processes, local processes, and cluster processes [213].

However, to stay slows down both the local and the cluster jobs, especially when the cluster job is a load-balanced parallel job that needs frequent synchronization and communication. This leads to the migration approach to allow the jobs to flow around the available nodes, mainly for balancing the workload.

Migration Schemes A *migration scheme* must consider the following three issues:

- *Node Availability*: Can the job find another available node to migrate to? The Berkeley study indicates the answer is likely to be affirmative: Even during peak hours, 60% of workstations in a cluster are available.

- ***Migration Overhead***: What is the effect of the migration overhead? The migration time can significantly slow down a parallel job. The Berkeley study indicates a slowdown as great as 2.4 times. Therefore, it is important to reduce the migration overhead (e.g., by improving the communication subsystem) or to make migration rare.

    The slowdown is much less if a parallel job is run on a cluster of twice the size. For instance, for a 32-node parallel job run on a 60-node cluster, the slowdown caused by migration is no more than 20%, even when the migration time is as long as 3 minutes. This is because more nodes are available, thus the demand of migration becomes less frequent.

- ***Recruitment Threshold***: What should be the recruitment threshold? The worst scenario is when a process just migrates to a node - the node is immediately claimed by its owner. Thus the process has to migrate again, and the cycle continues. The recruitment threshold is amount of time a workstation stays unused before the cluster considers it an idle node. The Berkeley study indicates that a 3-minute recruitment threshold maximizes parallel job performance.

Specifications of Scheduling Policy Because user's resource requirements are complex, a JMS should have a flexible scheduler interface to implement complex scheduling rules. A menu of options is not flexible enough. The interface should also provide a default scheduling policy that is suitable to most users.

9.5.2 Survey of Job Management Systems

More than twenty JMS packages, both commercial and public domain, are currently in use. Baker et al. [56] evaluated these packages according to a set of well-defined criteria. Table 9.4 summarizes the top five of these packages ranked by their criteria. The Web Resource contains pointers to all these packages.

- ***The DQS***: Of the public domain packages, *Distributed Queueing System* (DQS) is more mature (now in version 3) and widely used now. DQS is created and maintained by Florida State University. It has a commercial version called Codine, that is offered by GENIAS GmbH in Germany. GENIAS claims Codine has become a de facto standard in Europe.

- ***The LSF***: *Load Sharing Facility* (LSF) from Platform Computing Corporation in Canada is probably the most widely used, with more than 20,000 licenses sold worldwide. We will discuss LSF in detail in Section 9.5.3.

- ***The NQS***: There are two other packages, Condor and *Network Queueing System* (NQS), that are not listed in Table 9.4 but are very influential. Condor is a public domain software package that was started at University of Wisconsin. It is one of the first systems that utilize idle workstation cycles and support checkpointing and process migration. NQS was marketed in the early 1980s by Stirling Software, which contracted with the NASA Ames Research laboratory to produce a JMS for its NAS sand box of DEC, IBM, and Cray Research supercomputers.

- *The Connect:Queue*: NQS has become a de facto JMS standard, and it is still widely used in various supercomputers for batch job processing. Most of the current JMS packages are labeled *NQS-compatible*, meaning they support NQS commands. The commercial NQS has evolved into *Connect:Queue* from Sterling Software. There is also a public domain NQS implementation, called Generic NQS, now maintained at University of Sheffield in the United Kingdom.

Table 9.4 Overview of Job Management Systems. (Courtesy of M. Baker et al. [56], Northeast Parallel Architectures Center , Syracuse University, 1995)

JMS Name	Commercial Packages			Public Domain Packages	
	Connect: Queue	LSF	NQE	DQS	PBS
Vendor/creator	Sterling	Platform	Cray	FSU	NASA
Platform (Unix)	Most	Most	Most	Most	Several
Batch support	Yes	Yes	Yes	Yes	Yes
Interactive support	No	Yes	Yes	Yes	Yes
Parallel support	Yes	Yes	Yes	Yes	Yes
Impact on owner	Yes	Yes	Variable	Small	Variable
Load balancing	Yes	Yes	Yes	Yes	Yes
Checkpointing	No	Yes	No	Yes	No
Process migration	No	No	No	No	No
Job monitoring	Yes	Yes	Yes	Yes	Yes
Suspend/resume	No	Yes	Yes	Yes	Yes
Dynamic resource	Yes	Yes	Yes	Yes	Yes
User interface	GUI	GUI	GUI, WWW	GUI	cmmd-line
Fault tolerance	Yes	Yes	Some	Yes	Little
Security	Unix	Kerboros	US DoD	Kerboros	Kerboros

The packages in Table 9.4 have the following features, some of which are not explicitly listed. These features are representative of other packages. Some of these features will be elaborated when we discuss LSF in Section 9.5.3.

- User supports are provided, but commercial packages generally have better user supports.
- They all support heterogeneous clusters. Each node can be any of the major Unix platforms. However, none supports Windows NT, Windows 95, or Windows 3.
- All packages do not require additional hardware or software.
- All support parallel and batch jobs. However, Connect:Queue does not support interactive jobs.
- All provide multiple, configurable types of job queues.
- When such packages are used in an enterprise cluster, cluster jobs managed by the JMS will impact the owner of a workstation in running the local jobs. However, NQE and PBS allow the impact to be adjusted. In DQS, the impact can be configured to be minimal.
- All packages offer some kind of load-balancing mechanism to efficiently utilize cluster resources.
- Some packages support checkpointing.
- No package in Table 9.4 supports dynamic process migration. They support static migration: a process can be dispatched to be executed on a remote node when the process is first created. However, once it starts execution, it stays in that node. A package that does support dynamic process migration is Condor.
- All packages allow dynamic suspension and resumption of a user job by the user or by the administrator.
- All packages allow resources (e.g., nodes) to be dynamically added to or deleted from a resource pool.
- Most packages provide both a command-line interface and a graphic user interface. The NQE even provides a web-like user interface.
- Besides the usual Unix security mechanisms, most packages use the Kerboros authentication system.

9.5.3 Load-Sharing Facility (LSF)

LSF is a commercial workload management system from Platform Computing. It is evolved from the Utopia system [668] developed at the University of Toronto. LSF emphasizes job management and load sharing of both parallel and sequential jobs. In addition, it has support for checkpointing, availability, load migration, and single system image.

The LSF is highly scalable and can support a cluster of thousands of nodes. LSF has been implemented for various Unix and Windows/NT platforms from PCs, workstations, SMPs, to the IBM SP2 MPPs. Recently, LSF has been extended to support wide-area networks. Theoretically, LSF can now be used to manage jobs running on multiple Internet sites.

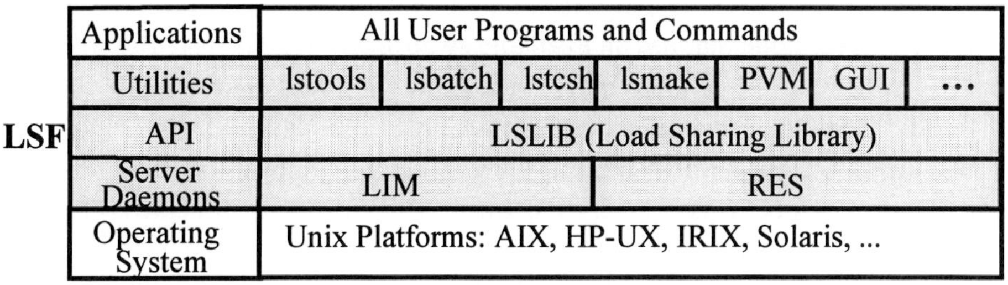

Applications	All User Programs and Commands						
Utilities	lstools	lsbatch	lstcsh	lsmake	PVM	GUI	...
API	LSLIB (Load Sharing Library)						
Server Daemons	LIM			RES			
Operating System	Unix Platforms: AIX, HP-UX, IRIX, Solaris, ...						

(The leftmost label **LSF** spans the Utilities, API, and Server Daemons rows.)

Figure 9.17   Layered architecture of LSF. (LIM: Load Information Manager; RES: Remote Execution Server; LSLIB: Load Sharing Library.Courtesy S. Zhou, Platform Computing Corp. [667], 1996)

LSF Architecture   A layered representation of the LSF architecture is shown in Fig. 9.17. LSF supports most Unix platforms and uses the standard IP for JMS communication. Because of this, it can turn a heterogeneous network of Unix computers into a cluster.

There is no need to change the underlying OS kernel. In other words, LSF is platform independent. The end user utilizes the LSF functionalities through a set of utility commands. PVM is supported, and MPI supports are planned. Both a command-line interface and a GUI are provided. LSF also offers skilled users an API that is a runtime library called LSLIB (*load sharing library*). Using LSLIB explicitly requires the user to modify the application code, whereas using the utility commands does not.

Two LSF daemons need to be started on each server node in the cluster. The *load information manager* (LIMs) periodically collect and exchange load information. The *remote execution server* (RES) provides for transparent remote execution of arbitrary tasks.

LSF Utilities   The LSF literature calls a computer a *host*. We will use the term *node* to be consistent with the rest of the book. Note that a node may have multiple processors (an SMP), but always runs only a single copy of the operating system.

- LSF supports all the four combinations of interactive, batch, sequential, and parallel jobs. A job that is not executed through LSF is called a *foreign job*. A *server node* is one which can execute LSF jobs. A *client node* is one that can initiate or submit LSF jobs but cannot execute them. Only the resources on the server nodes can be shared. Server nodes can also initiate or submit LSF jobs.

- LSF offers a set of tools (*lstools*) to get information from LSF and to run jobs remotely. For instance, *lshosts* lists the static resources (see below) of every server node in the cluster. The command *lsrun* executes a program on a remote node.

When a user types the command line:% lsrun -R 'swp>100' myjob at a client node, the application myjob will be automatically executed on the most lightly loaded server node that has an available swap space greater than 100 MB.

- The *lsbatch* utility allows users to submit, monitor, and execute batch jobs through LSF.
- The *lstcsh* utility is a load-sharing version of the popular Unix command interpreter *tcsh*. Once a user enters the *lstcsh* shell, every command issued will be automatically executed on the most suitable node, including the local node. This is done transparently: the user sees a shell exactly like a tcsh running on the local node.
- The *lsmake* utility is a parallel version of the Unix make utility, allowing a makefile to be processed in multiple nodes simultaneously.

Example 9.14 Application of the LSF on a cluster of computers

Suppose a cluster consists of 8 expensive server nodes and 100 inexpensive client nodes (workstations or PCs). The server nodes are expensive due to better hardware and software, including application software. A license is available to install a Fortran compiler and a CAD simulation software, both valid up to four users.

Using a JMS such as LSF, all the hardware and software resources of the server nodes are made available to the clients in a transparent way. Users sitting in front of a client's terminal feel as if the client node has all the software and the speed of the servers locally. By typing "lsmake my.makefile", the user can compile his source codes on up to four servers. The LSF selects the nodes with the least amount of load.

Using LSF also benefits resource utilization. For instance, a user wanting to run a CAD simulation can submit a batch job. LSF will automatically check the availability of the CAD license, and will schedule the job as soon as it becomes available. Users do not have to wait. Note that in many organizations, it is likely that a client is either idle or executing some local jobs (word processing, Web browsing, etc.).
■

Resource Information The LIM on each server node collects both static and dynamic resource information of this node. Static resource information does not change over time and is obtained by the LIM at start-up. Dynamic information is represented as a set of *load indices* in a *load vector*, which changes over time and is collected by the LIM periodically.

- ***Static Resources***: Some of the static resources reported by LSM are listed in Table 9.5. The node type is determined by the cluster configuration and set by the LSF administrator. It is a character string, such as Alpha, RS6000, MIPS, etc. The

CPU factor is the relative speed of the node compared to the slowest node in the cluster, which has a CPU factor of 1.

Table 9.5 Static Resources in LSF (Courtesy of S. Zhou, Platform Computing Corp.[667],1996)

Index	Measures	Units	Determined by
type	Node type	String	Configuration
cpuf	CPU factor	Relative	Configuration
maxmem	Maximum RAM available to users	MB	LIM
maxswp	Maximum available swap space	MB	LIM
maxtmp	Maximum available space in /tmp	MB	LIM

The next three indices are the maximum memory space of the node, the maximum swap space on the local disk, and the maximum space in the /tmp file system of the node. Other static resources include the license and availability of a software, where the node is a file server, etc.

- *Load Indices*: Some of the load indices are listed in Table 9.6. These indices are updated by the LIM every 1, 30, or 120 s. The node status can be *ok* (available to execute LSF jobs), *unavailable* (LIM not responding), *busy* (load exceeding a threshold), and some other status. The *r15s* run queue length is the average number of processes ready to use the CPU during the past 15 s. The CPU utilization is the average percentage in the past minute that the CPU is executing user or system codes..

- *Paging Rate*: The paging rate refers to the virtual memory paging rate, i.e., the number of pages read from or written to the disk per second. A node that is paging heavily (having a high paging rate) will respond slowly to interactive jobs. The *ls* index measures the number of users that have logged in to this node. The idle time measures how many minutes have elapsed since the last input or output action, such as a key stroke, a mouse click, or a printout.

The LSF has developed some load-sharing policies, availability support, and load-balancing strategies as described below:

Load-Sharing Policies Zhou et al. [668] evaluated various load-sharing and job placement policies and chose the following ones for LSF. When the cluster size is small (e.g., no more than a few tens), one of the LIMs will be elected as the *master LIM*, and the

Table 9.6 Some of the Load Indices Included in an LSF Load Vector
(Courtesy of Zhou, Platform Computing Corp. [667], 1996)

Index	Measures	Units	Average over	Update Interval
status	Node status	Character string	N/A	15 s
r15s	Run queue length	Processes	15 s	15 s
ut	CPU utilization	Percent	1 min	15 s
pg	Paging rate	# (pages) per sec.	1 min	15 s
ls	Logins	Users	N/A	30 s
lt	Idle time	Minutes	N/A	30 s
swp	Available swap space	MB	N/A	15 s
mem	Available memory	MB	N/A	15 s
tmp	Available space in /tmp	MB	N/A	120 s

rest are *slave LIMs*. The slave LIMs periodically send their load vectors to the master LIM, which combines them into a *load matrix* of the cluster. When a node submits an LSF job (e.g., through *lsrun*), where to execute the job is determined by the master LIM.

Master LIM In other words, the master LIM is a central manager, which maintains load information from all servers and dispatches all cluster jobs. For larger clusters (e.g., hundreds of server nodes), this single-master policy is inadequate (see Problem 9.12).

LSF divides the large cluster into a number of smaller subclusters. There is still a single master LIM within a subcluster, but the master LIMs exchange load information and collectively make inter-subcluster load-sharing decisions. As shown in Table 9.7, the load sharing overhead is low.

Table 9.7 Job Response Time, Master LIM Overhead, and Network Traffic by LSF. (Courtesy S. Zhou, Platform Computing Corp.[667],1996)

Cluster Configuration	Number of Nodes Requested			Master LIM CPU Time %	Network Traffic (KB/s)
	1	5	10		
One 15-server cluster	3.8 ms	4.6 ms	5.0 ms	0.08	0.04
One 33-node cluster	4.5 ms	7.7 ms	8.4 ms	0.15	0.09
One 60-node cluster	5.8 ms	14.2 ms	18.5ms	0.29	0.17
Twenty 50-server subclusters	N/A			1.1	1.2

Example 9.15 Estimation of LSF overhead

For instance, for a cluster consisting of 60 server nodes, the measured time from when a job request is issued to when the job starts is just a few milliseconds, when up to 10 nodes are requested; the master LIM consumes only 0.29% of the node CPU time; and the network bandwidth consumed is only 0.17 KB/s.

For a cluster consisting of 20 subclusters, each containing 50 nodes, the master LIMs CPU overhead is only 1.1%, and the consumed network bandwidth is 1.2 KB/s.

∎

Batch Support The LSF batch system architecture is shown in Fig. 9.18. The LSF provides extensive supports for batch jobs. It is incorporable with NQS, which is a popular batch queuing system for supercomputers.

LSF uses a number of *batch server nodes* to execute a batch job. The set of batch servers is a subset of the set of all servers in the cluster. A *slave* batch daemon *sbatchd* runs on every batch server. The cluster has a single *master* batch daemon *mbatchd* that where the master LIM is running. This *master node* maintains all the batch job queues. All batch job requests are sent to the master batch daemon, which is responsible for scheduling and dispatching the jobs to slave batch server nodes to execute.

Batch Job Submission An LSF job is submitted by executing a bsub command such as

bsub -q *night* -n 10 -i *jobin* -o *jobout* -e *joberr* -R "mem>20" *myjob*

which submits *myjob* to the queue called *night*. The input, the output, and the error files of *myjob* are *jobin*, *jobout*, and *joberr*, respectively. The job is to be executed on 10 nodes, where each node should have at least 20-MB memory available.

An LSF batch job's life cycle may go through the state transitions shown in Fig. 9.19. When a job is submitted to an LSF batch queue, it enters the pending (PEND) state. It is transferred to the RUN state when the job is dispatched to a node that satisfies all the resource and scheduling conditions.

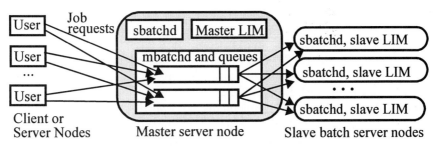

Figure 9.18 The LSF batch system architecture. (Courtesy of Zhou, Platform Computing Corp.[667],1996)

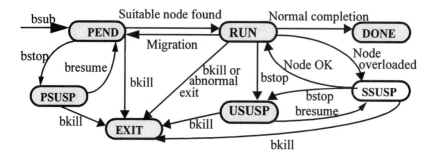

Figure 9.19   LSF batch job state transition diagram. (Courtesy of Zhou, Platform Computing Corp. [667],1996)

The job may be killed from any state by executing the LSF *bkill* command. A running job may abnormally exit when it crashes. A job may be suspended by its owner or the LSF administrator through a *bstop* command, which turns the state from PEND to PSUSP, or to USUSP if the job is already dispatched. The lsbatch system may also automatically suspend the job if the node is overloaded.

LSF supports a variety of scheduling policies, including first-come first-served, fair-share, preemptive, and exclusive. The fair-share policy schedules the job first that has consumed less than its share of resources. Preemptive scheduling allows jobs in a preemptive queue to preempt lower-priority jobs in a node, even if they have not completed. An exclusive job runs alone on a node. Once it is started, the *lsbatch* locks the node so that no other LSF jobs, either interactive or batch, can use this node until the exclusive job finishes.

Availability and Checkpointing   LSF incorporates a number of techniques to enhance availability and load balancing. As a result, an LSF cluster can still work as long as at least one server node is alive. LSF supports checkpointing of batch jobs through a static, user-level library on most systems. Kernel-level checkpointing is realized for Convex OS platforms.

A batch job is not lost even when the master node fails. It can be restarted either from the beginning or from a previous checkpoint. Jobs can migrate to another node dynamically, as long as the new node has the same architecture and operating system.

Master Selection in LSF   Most LSF clusters have a single master node, which may become a single point of failure. LSF implements a *master election* scheme (Fig. 9.20) to cope with this problem. In an LSF cluster, the LSM daemons LIM, RES, and the slave batch daemon *sbatchd* are automatically started on every server node at boot time. When a server is elected the master, its *sbatchd* starts the *mbatchd* daemon on it. Shown in Fig. 9.20 is the state transition of one node.

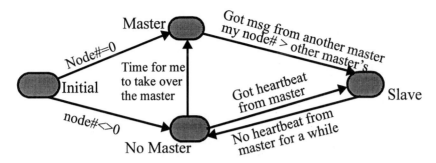

Figure 9.20   Illustration of the master-election scheme in LSF.
(Courtesy Zhou et al. *Software-Practice and Experience* [668], 1993)

Any LSF cluster maintains a configuration file, which orderly lists all the server nodes in the cluster. Each node has a node number, with the first node in the list numbered 0. Upon initialization, node 0 becomes the master if it is up. The master periodically sends heartbeat messages to the other server nodes. Any other node first enters the "no master" state where it waits until it receives a heartbeat from the master. Then it becomes a slave.

When the master node fails, it will not send out heartbeat messages. Each slave detects this and enters the "no master" state. It waits for a time period proportional to the node number and offers to become the new master. If more than two nodes compete, the node with the smallest node number wins.

This master election scheme is designed to ensure high availability as well as load balancing among all server nodes available in the cluster.

9.6   Bibliographic Notes and Problems

Gregory Pfister's book [497] provides an intuitive introduction of clustering concepts. Martin et al. [431] have discussed communication issues in clusters. Turcotte [620] surveys various software tools for cluster computing. Walker et al. [633] have developed an open single system image software package for clusters and MPPs. We have also [657] compared clusters and MPPs for scalable computing.

The IBM HACMP cluster is described in Berstin [77] and [338]. Baker et al. [56] review more than a dozen cluster job management systems. The job scheduling issues for clusters have been addressed in Arpaci et al. [40], Feitelson and Rudolph [236], Saphire et al. [529], and Du and Zhang [213]. Load balancing and fault tolerance are treated in Petri and Langendorfer [496], Rosenblum et al. [520], and Kumar et al. [384].

The availability data in typical computer systems (Table 9.2) are from Digital [197]. Availability systems are studied in Gary and Siewiorek [259], Bowen et al. [106], and Azagury et al. [47]. The Sun Solaris MC is described in Khalidi et al. [369]. The LSF is described in Zhou et al. [667, 668].

Elnozahy et al. [228] and Deconinck et al. [199] have given comprehensive surveys of checkpointing and rollback recovery techniques. These techniques are also studied in Li et al. [415], Cao et al. [125], Plank et al. [503,504,505], Netzer and Xu [461], and Wong and Franklin [648]. Tananenbaum and Litzkow [608] discussed the Condor system.

Distributed file systems have been reviewed in Levy and Silberschartz [409], Santa-yaraman [526], and Vahalia [621]. NFS Version 3 is described in Pawlowski et al. [494], and the Web version of NFS is described in Callaghan [120]. AFS has evolved from Andrew described in Morris et al. [453] to the DFS specified in the DCE of OSF [474].

Homework Problems

Problem 9.1 Differentiate and exemplify the following terms related to clusters.

(a)   Compact versus slack clusters

(b)   Centralized versus decentralized clusters

(c)   Homogeneous versus heterogeneous clusters

(d)   Enclosed versus exposed clusters

(e)   Dedicated versus enterprise clusters

Problem 9.2 List four features of an enterprise cluster that are also found in an LAN. Then list two features that make the cluster not an ordinary LAN. Briefly explain the common and the different features.

Problem 9.3 List two other situations that are not listed in Section 9.2.1 but can cause unplanned failures of a system. Suggest a technique to solve such problems.

Problem 9.4 The configuration in Fig. 9.5(b) is criticized as being expensive.

(a)   List and briefly explain two other disadvantages.

(b)   Design your own high availability system as a compromise between Fig. 9.5(a) and Fig. 9.5(b), and explain why your design is better than either of them. Assume that the LAN and SCSI are highly reliable, and so are their node controllers; a RAID disk itself is reliable, but the disk controller is not.

(c)   Assume an SCSI has eight ports. Generalize your design to a seven-node system.

(d)   Assume an SCSI has eight ports. Modify your design to suit a 16-node system.

Problem 9.5 This problem refers to Example 9.3. Assume that when a node fails, it takes 10 s to diagnose the fault and another 30 s for the workload to be switched over:

(a)   What is the availability of the cluster if planned downtime is ignored?

(b)   What is the availability of the cluster if the cluster is taken down 1 hour per week for maintenance, but one node at a time?

Problem 9.6 Compare Condor and Libckpt, and discuss their respective advantages and shortcomings with respect to generality and usability.

Problem 9.7 Assume a sequential computer has 512-MB main memory and enough disk space. The disk read/write bandwidth for large data block is 1 MB/s. The following code runs and needs to be checkpointed:

```
do 1000 iterations
    A = foo (C from last iteration)    /* this statement takes 10 minutes */
    B = goo (A)                        /* this statement takes 10 minutes */
    C = hoo (B)                        /* this statement takes 10 minutes */
end do
```

The A, B, and C are arrays of 120 MB each. All others (e.g., code, operating system, libraries) take at most 16 MB of memory. Assume the computer fails exactly once, and the time to restore the computer itself is ignored.

(a)   What is the worst-case execution time for the successful completion of the above code if checkpointing is performed?

(b)   What is the worst-case execution time for the successful completion of the above code if plain transparent checkpointing is performed? Here *plain* means no optimization (incremental, forked, and so on).

(c)   Is it beneficial to use forked checkpointing with (b)?

(d)   What is the worst-case execution time for the successful completion of the above code if user direct checkpointing is performed? Show the code where user directives are added to minimize the worst-case execution time.

(e)   What is the worst-case execution time for the successful completion of the above code if forked checkpointing is used with (d)?

Problem 9.8 Give a practical scheme to realize single entry point, and show how it solves the home directory problem and the multiple-connection problem. Your scheme should support popular protocols, such as ftp, telnet, http.

Problem 9.9 Explain and exemplify the following terms of single system image:

(a) Single file hierarchy

(b) Single control point

(c) Single memory space

(d) Single process space

(e) Single I/O and networking

Problem 9.10 Consider the Solaris MC system and answer the following questions:

(a) Which single system image features in Problem 9.9 are supported by Solaris MC, and which are not?

(b) For those features Solaris MC does support, explain how it does it.

Problem 9.11 Use examples to explain and contrast the following terms regarding cluster job management systems:

(a) Serial jobs versus parallel jobs

(b) Batch jobs versus interactive jobs

(c) Cluster jobs versus foreign (local) jobs

(d) Cluster processes, local processes, versus kernel processes

(e) Dedicated mode, space-sharing mode, and timesharing mode

(f) Independent scheduling versus gang scheduling

Problem 9.12 This exercise is for LSF.

(a) Give an example of each of the four types of LSF jobs.

(b) Give an example of a foreign job.

(c) For a 1000-server cluster, give two reasons why the LSF load- sharing policy is better than (1) the entire cluster has one master LIM or (2) all LIMs are masters.

(d) In the LSF master-election scheme, a node in the "no master" state waits for a time period proportional to the node number before offering to become the new master. Why is the wait time proportional to the node number?

Chapter 10

Clusters of Servers and Workstations

In recent years, we have seen a surge in computer cluster products and their applications in business, enterprises, and research institutions. Most clusters emphasize higher availability and scalable performance. We assess below the field of cluster computing and review the trend in supporting clusters from high-end to low-cost systems.

We study the details of several cluster systems: the Microsoft Wolfpack, the SGI POWER CHALLENGEarray, the Marathon fault-tolerant cluster, the IBM SP, the Digital TruCluster, the Berkeley NOW project, and the Rice TreadMarks DSM cluster. These case studies reveal the design and application experiences of today's cluster products and cluster research projects.

Wolfpack is an industrial project to develop cross-vendor software standard for clustering of Windows NT server clusters. The SP represents a successfully marketed commercial cluster/MPP for both scientific and business applications. The TruCluster is a cluster of SMP servers leveraging on the Memory Channel technology.

NOW is a university research project to explore some new mechanisms for clustering of Unix workstations. TreadMarks is a software-implemented DSM cluster of workstations. For comparison purpose, other cluster systems and projects are also briefly introduced in this chapter.

10.1 Cluster Products and Research Projects

As estimated by Pfister [498], over 100,000 computer clusters are in use worldwide. These include both commercial products and custom-designed clusters. Nodes in these clusters are mostly PCs, workstations, and SMP servers. The cluster sizes are mostly in the order of tens; only a few clusters exceeded 100 nodes.

Most clusters use commodity networks such as Fast or Gigabit Ethernet, FDDI rings, ATM or Myrinet switches to interconnect the nodes besides regular LAN connections among them. A few also use proprietary high-bandwidth networks such as the IBM SP2 and Compaq NT cluster.

Some public-domain and commercial software packages are available for supporting single system image, higher availability, and distributed job management. The basic clustering technology exists. The momentum is catching up rapidly to begin volume production of cluster products in business applications.

10.1.1 Supporting Trend of Cluster Products

The support for clustering of computers is moving from the high-end market to the high-volume market. As depicted in Fig. 10.1, clustering support started with the linking of large mainframe computers together, like the IBM Sysplex [339] for *system complex* and the SGI POWER CHALLENGEarray [548]. At the highend, the purpose is to ease the demand of cooperative group computing and higher availability in critical enterprise Internet and data mining applications.

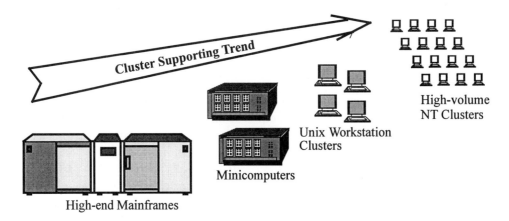

Figure 10.1   The cluster-supporting trend from high-end mainframes to Unix workstations and high-volume PC systems.

Gradually, the trend moves towards the support of multiple minicomputers as in Digital's Open VMS cluster [551], in which multiple AXPs and VAXs are attached to the same set of disk/tape controllers. Tandem's Himalaya [605] was designed as a high-end cluster for fault-tolerant, *on-line transaction processing* (OLTP). With the increasing trend of using desktop or deskside workstations, clustering of Unix workstations became very popular among major workstation suppliers.

Some commercial cluster systems are listed in Table 10.1. This table is by no means exhaustive in covering all vendors' cluster products. The table does not include many locally assembled clusters. We briefly introduce below these preassembled commercial clusters. More detailed information can be found from the WWW home pages, or references identified at the right column.

Table 10.1 A Sample of Commercial Cluster System Products

Company	System Name	Brief Description and References
DEC	VMS-Clusters	High-availability clusters for VMS [551]
	TruClusters	Unix cluster of SMP servers [127], [397]
	NT Clusters	Alpha-based clusters for Windows NT [195]
HP	Apollo 9000 Cluster	Computational cluster [497]
	MC/ServiceGuard	HP NetServer for NT cluster solution [573].
IBM	Sysplex	Shared-disk mainframe cluster for commercial batch and OLTP [339]
	HACMP	High-availability cluster multiProcessor [338]
	Scalable POWERparallel (SP)	Workstation cluster built with POWER2 nodes and Omega switch as a scalable MPP [14]
Microsoft	Wolfpack	An open standard for clustering of Windows NT servers [446]
SGI	POWER CHALLENGEarray	A scalable cluster of SMP server nodes built with HiPPI switch for distributed parallelism [548]
Sun	Solaris MC	Extension of Solaris Sun workstation cluster [369]
	SPARCcluster 1000/2000 PDB	High-availability cluster server for OLTP and database processing [599]
Tandem	Himalaya	Highly scalable and fault-tolerant cluster with duplexed nodes for OLTP and database [605]
Marathon	MIAL2	High-availability cluster with complete redundancy and failback [425]

Good example workstation clusters include Sun's SPARCclusters [599], HP's Apollo 9000 computational cluster, Sequent Symmetry 5000, and Digital's Alpha workstation farms [194], etc. More recently, the trend moves further down to the clustering of Intel-based PCs. Clustered products appear as integrated systems, software tools, availability infrastructure, and operating system extensions, etc.

The clustering trend moves from supporting large rack-size, high-end computer systems to high-volume, desktop or deskside computer systems, matching with the downsizing trend in computer industry. Supporting clusters of smaller nodes will increase the sales in volume as predicted by many cluster advocates.

This trend is triggered by the widespread use of Intel-based PCs and low-cost workstations. From IBM and Digital to Sun, SGI, HP, Compaq, and Microsoft, the whole industry is excited with the niches of clustering. The availability of Intel's SHV system, a four-way Pentium Pro-based motherboard, and of Windows NT's symmetric multiprocessing capabilities has greatly enhanced market potential of low-cost server clusters.

Clustering can improve both availability and performance. As shown by the entries in Table 10.1, most systems are built as *availability clusters*. Others are used as *compute clusters* that emphasize scalable performance. The two cluster goals are not necessarily in conflict. Some high-availability clusters aggressively use hardware redundancy for scalable performance, as exemplified by Tandem Himalaya system.

Most availability clusters use software techniques to exploit the redundancy already built into cluster hardware. High-performance clusters are often supported by specially designed communication subsystem, including high-performance network, network interface, and communication software. Two examples are DEC TruCluster that utilizes the *memory channel* technique and the IBM SP2 that employs a *high-performance switch*.

10.1.2 Cluster of SMP Servers

A visible industrial trend is to cluster a number of homogeneous SMP servers together as an integrated superserver. The Sun Ultra Enterprise 10000 superserver presented in Section 8.2 is a good example. Another superserver example is the SGI POWER CHALLENGEarray introduced below.

Example 10.1 SGI SMP cluster: the POWER CHALLENGEarray

A schematic block diagram of the SGI superserver cluster is shown in Fig. 10.2. This system links two to eight SMP nodes, called POWERnodes, to form a cluster as a superserver. Each POWERnode is a Power Challenge SMP server with up to 36 MIPS R10000 processors and a 16-GB shared memory.

In total, the superserver cluster can have up to 128 GB of main memory, more than 4 GB/s of sustained disk transfer capacity, and 28-TB disk capacity. The RAID storage capacity can reach 139.2 TB.

The nodes are interconnected by a crossbar HiPPI switch to support high bandwidth communication. The bisection bandwidth is estimated at 1.6 GB/s for the maximum configuration. The system can be accessed via an Ethernet from Indy workstations. The system administration console is a single point of control. It has a direct RS232 connection to each node (the dashed lines) via an ST-1600 box.

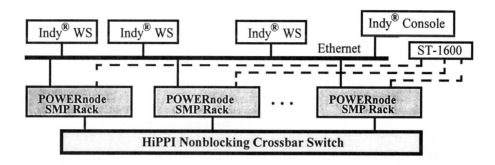

Figure 10.2   Silicon Graphics cluster of SMP servers. (Courtesy Silicon
Graphics Computer Systems, [548], 1996)

This cluster operates with shared memory within each POWERnode. Message
passing is fully supported among the POWERnodes. The Silicon Graphics cluster
consists of 2- to 288- MIPS® R10000® processors packaged into two to eight
POWERnode racks.

Notable new hardware and software features of the Silicon Graphics superserver
are summarized below:

- This cluster of SMP servers offers a modular approach to distributed parallel
  processing. The peak performance is 109 Gflop/s if fully utilized.
- It combines shared-memory and message-passing techniques. This provides
  software compatible with the traditional SMP servers.
- With high-bandwidth HiPPI crossbar connection among the SMP nodes, the
  communication-to-computation ratio is reduced for irregular parallelism.
- All POWERnodes run Silicon Graphics® IRIX™ 6.1 enhanced 64-b Unix® OS,
  offering high-performance graphics support for interactive supercomputing.
- There is less than 1 μs of memory latency in accessing the shared memory within
  each POWERnode and less than 1 μs crossbar switch connection time.
- It uses a hierarchy of programming models: shared-memory programming with
  parallelizing Fortran, HPF, and C compilers and message passing with PVM and
  MPI support.

■

10.1.3 Cluster Research Projects

Clustering is an active research field. Fast communication, job scheduling, single tative cluster research projects are listed in Table 10.2. Again, this table cannot cover all projects in progress. What are shown here are some sample projects highlighting the new progress being made.

Table 10.2   Representative Cluster Research Projects

Project/Reference	Special Features to Support Clustering
Berkeley NOW Project [33]	A serverless network of workstations featured with active messaging, cooperative filing, and global Unix extensions (Section 10.5)
Princeton SHRIMP [237]	Commodity components, efficient communication and shared virtual memory through special network interfaces
Karsruhe Parastaion [636]	Efficient communication network and software development for distributed parallel processing
Rice TreadMarks [30]	Software-implemented distributed shared-memory cluster of workstations
Wisconsin Wind Tunnel [516]	Distributed shared memory on a cluster of workstations interconnected with a commodity network
The NSCP [133] http://nscp.upenn.edu	National Scalable Cluster Project for metacomputing over three local clusters linked through Internet.
Argonne Globus [200]	Developing metacomputing platform and software through an ATM-connected WAN of 17 sites in North America
Syracuse WWVM [207]	A World-Wide Virtual Machine for high-performance computing with Internet and HPCC technologies
PearlCluster [328] http://pearl.cs.hku.hk	A research cluster for distributed multimedia and financial digital library applications (Example 6.10)
Virginia Legion [282]	Developing metacomputing software towards a national virtual computer facility

We briefly characterize these cluster projects below. Each of the listed cluster projects has developed some unique features worthy of mentioning. The features cover architectural innovations as well as software support for single system image, availability, and distributed benchmark applications with Intranet and/or Internet resources.

- The NOW project addresses almost the whole spectrum of cluster computing issues, including architecture, software support for Web servers, single system image, I/O and file system, efficient communication, and enhanced availability. We will study the details of NOW project [33] in Section 10.5.

- The Princeton SHRIMP project [237] is focused on supporting efficient communication and shared memory in a cluster made of inexpensive PC-based systems. Special network interface boards are developed to achieve shared virtual memory at page migration level.

- The Wisconsin Wind Tunnel project is investigating how to realize cache-coherent shared memory on a loosely coupled cluster, built from commodity nodes and network [516].

- The Rice University TreadMarks [30] is a good example of software-implemented shared-memory cluster of workstations. The memory sharing is implemented with a user-space runtime library. We will study details of TreadMarks in Section 10.6.

- The *National Scalable Cluster Project* (NSCP) investigates metacomputing systems [133], which are heterogeneous clusters that span several cities in distant geographic regions. The NSCP prototype system consists of three university clusters in Chicago, Maryland, and Pennsylvania, which are connected through the Internet. See Web site http://nscp.upenn.edu for details.

- Similar metacomputing projects include the *Globus* project, which was implemented on an ATM network connecting supercomputers, mass storage systems, and visualization devices at 17 sites in North America [200, 248, 249].

- The Legion project develops metacomputing software [282, 410] toward a national virtual computer facility on the U.S. mainland.

- The WWVM (*World-Wide Virtual Machine*) project aims at realizing high-performance computing using Internet and HPCC (*High Performance computing and Communication*) technologies [207, 252].

- The PearlCluster at Hong Kong University was introduced in Example 6.10. This is an ATM-connected cluster of Unix-based servers and workstations from Sun, HP, Alpha, and SGI. The unique features include middleware support for SSI, bilingual Internet search engine, and Java-interfaced MPI communications subsystem development. These features are specially tailored for financial digital library and distributed multimedia applications.

10.2 Microsoft Wolfpack for NT Clusters

At the time this book was completed, Microsoft released the initial version of a cross-vendor standard for clustering Windows NT servers. Initially, the project involved the core members of Intel, Compaq, Digital, HP, NCR, and Tandem, besides Microsoft.

The strategic partnerships are meant to fashion a new cluster standard with the code name *Wolfpack*, inspired by the name *Dogpack* introduced by Pfister.

10.2.1 Microsoft Wolfpack Configurations

The Wolfpack project started in late 1995 to develop an open standard of clustering software for Windows NT servers. It is also the name of the standard clustering software. Wolfpack has been targeted to have the following unique and promising features:

- Wolfpack is an *open standard*, providing users, system vendors, and third-party software developers open specifications of application programming interfaces. This will definitely accelerate the development of application software, system software, and cluster hardware.
- Wolfpack is geared toward commodity PC or Intel-based server platforms, SCSI storage buses, and standard networks. This is to reduce development time and provide cost-effectiveness. However, it does not exclude the use of high-performance hardware, such as a low-latency network.
- The long-term goal of Wolfpack emphasizes not only availability, but also scalability and manageability of clusters of PCs, workstations, and high-volume servers, whether they run with NT or non-NT operating systems.

Wolfpack Components Wolfpack describes a set of cluster-aware APIs, NT cluster support, and a cluster solution as briefly introduced below:

(1) *The Wolfpack API*: This *application programming interface* (API) standard and the *software development kit* (SDK) are designed to enable applications' *cluster awareness*. Accessing the API can speed up failure recovery, notify users of failure events, reacquire nonstandard resources, and monitor complex faults besides simple crash or lock-up. Potentially, the Wolfpack API can promote scalability and dynamic load balancing on an NT cluster.

(2) *NT Cluster Support:* This is the software support to make an NT server application *Wolfpack compliant.* This implies that Wolfpack will do basic failover recovery of any NT server application. A wrapper *dynamic link library* (DLL) has been developed to notify the cluster manager and to create basic heartbeat signals. It seems that Microsoft wants to add cluster support to NT servers in the same way that Windows NT has symmetric multiprocessing support today.

(3) *Clustering Solution*: Wolfpack is to be developed in several phases. Support for 16-node clusters goes into Beta test in mid-1998. Microsoft is committed to making Wolfpack available to Alpha-, PowerPC-, and MIPS-based servers besides Intel-based systems. Today's NT server architecture supports up to 32-way SMP operations. Wolfpack will attempt to scale up to 32 nodes or more in the future.

Availability Cluster Configurations The initial Wolfpack cluster solution is targeted to provide availability support for two server nodes with three ascending levels of availability: *hot standby, active takeover,* and *fault-tolerant* as introduced in Section 9.2.2. The three Wolfpack cluster configurations are specified in Table 10.3 in terms of *recovery time, failback feature,* and *node activeness.*

Table 10.3 Three Availability Cluster Configurations Supported by Wolfpack (Courtesy M. Smith, *Windows NT Magazine,*[573], 1997)

Config-uration	Recovery Time	Failback Capability	Node Activeness	Multiserver Cluster Vendors
Hot Standby	40-200 s	No	Not for standby node	Vinca, IBM, Octopus, Compaq
Active Takeover	15-90 s	Yes	Yes	Microsoft, Digital, Compaq, Tandem, NCR, HP, Amdahl, and Stratus
Fault-Tolerant	~ 1 s	Yes	Yes / No	Marathon

The level of availability increases from standby to active and fault-tolerant cluster configurations. The shorter the recovery time, the higher the cluster availability. Failback refers to the ability of a recovered node to return to normal operations after repairing or maintenance. Activeness refers to whether the node is used in active work during normal operations. These clusters will be described in subsequent sections.

10.2.2 Hot Standby Multiserver Clusters

In a *hot standby* cluster, only the primary node is actively doing all the useful work normally. The standby node is powered on (hot) and running some monitoring programs to communicate heartbeat signals to check the status of the primary node, but not actively running other useful workload. The primary node must mirror any data to a shared disk storage, which is accessible by the standby node. The standby node requires a full second copy of all the data.

In Fig. 10.3, a typical standby cluster architecture is shown. The double bus connections to both servers remove the single point of failure on a single bus, giving higher availability of the cluster.

In case of failure in the primary node, the job running there can be failed over to the standby node, which becomes a new primary node while the old one is under repair. However, the failover is not!reversible in most standby clusters.

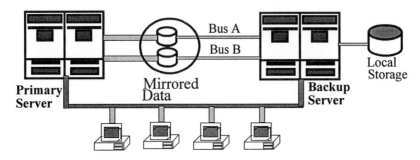

Figure 10.3   The hot standby cluster configuration. (Courtesy M. Smith, *Windows NT Magazine* [573], 1997)

Vendors supporting standby clustering are summarized in Table 10.4. Standby cluster is lower in cost with a limited failure recovery capability. Some of these standby clusters can automatically switch users of Windows NT, Windows 95, and Windows for Workgroups (WFW) 3.11 to the new active server. Mac, OS2, and Windows 3.1 users must manually switch to the standby server. .

Table 10.4 Industrial Standby Server Clusters (Source: M. Smith, *Windoes NT Magazine*, August, 1996)

Cluster Vendor	Cluster Model	Server/Disk Platform	Cluster Support Provided
Vinca	Vanica standby server	Intel-based servers and SCSI controllers	Support NetWare, OS/2, and NT
IBM	PC server high availability solution	Bundling IBM PC server hardware with Vinca's standby server platform	OS/2, NT, Notes cluster, DB2 cluster
Octopus Technologies	Automatic switch over (ASO)	Intel and Alpha servers	NT and N-way failover
Compaq	Standby recovery server and On-line recovery server	Compaq servers, external storage box, SCSI cards, and hardware interconnect card	Failover through proprietary hardware, rather than software

10.2.3 Active Availability Clusters

In Fig. 10.4, we show the architecture of an availability cluster which follows the *active takeover* configuration.

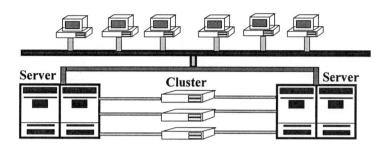

Figure 10.4   The architecture of a dual-node active cluster. (Courtesy
M. Smith, *Windows NT Magazine* [573], 1997)

In this case, the architecture is symmetric between the two server nodes. Both servers
are primary, doing useful work normally. Both failover and failback are often supported on
both server nodes. When a node fails, the user applications fail over to the available node
in the cluster. Eight vendors listed in Table 10.5 supply software and hardware support for
building active clusters.

The products from Microsoft and Digital support mainly NT clusters of Intel-based
and Alpha-based servers, respectively. The remaining vendors have expressed the will to
support Wolfpack compliance, once the NT clustering package can be successfully ported
to their own hardware/network platforms.

Depending on the time required to implement the failover, users may experience
some delays or may lose some data that were not saved in the last checkpoint. Manual load
balancing can minimize the added burden incurred with the failover conditions, or can
ease the overloading situation for more than two nodes in an active cluster. The feature of
automatic failback poses a major improvement over the basic standby clusters.

Tandem has no NT cluster solution, but the merge of Tandem into Compaq will
enhance their joint expertise in marking active NT clusters or fault-tolerant Tandem
clusters. Himalaya clusters support over 1000 business-critical applications including
OLTP, electronic commerce, Internet/WWW, data warehousing, decision support,
retailing, health care, and transport markets.

Some of the Wolfpack features are being extended to support clusters with non-NT
server nodes. The NCR LifeKeeper, HP MC/Service Guard, Amdahl's EnVista Servers,
and Stratus RADIO are such active cluster examples. Failover and failback capabilities
exist in all these server clusters. The RADIO is especially impressive for its complete
redundancy in that two compute servers, dual storage devices, and double networking are
used to eliminate all single points of hardware, software, and network failure

Table 10.5 Active Cluster Vendors and Their Offerings

Vendor	Cluster Products or Technology	Node/Network Platform	Clustering Support and Wolfpack Compliance
Microsoft	Wolfpack	Intel-based now. Alpha, MIPS and PowerPC-based servers later	Support Windows NT, API, SDK, DLL in Wolfpack
Digital	NT Clusters, NT Cluster Plus Pack	Alpha NTFS, Microsoft SQL Server 6.5, Oracle7/8 servers	Failover support for Intel- and Alpha-based Servers
Compaq	On-line recovery server	Compaq SCSI Switching and Tandem ServerNet	No automatic failback, Wolfpack compliance, ServerNet interconnect
Tandem	NonStop ServerWare servers, ISV Portfolio, ServerNet	Compaq NT servers, Himalaya clusters of SMP servers	Port Wolfpack to Compaq, SQL database, TUXEDO, CICS APIs, OLTP, and business applications
NCR	LifeKeeper with recovery kits for variety of servers	Intel-based servers, Oracle 7 parallel server, TCI/IP, NetBEUI, and SQL servers	Windows NT, Wolfpack support with automatic failback and automatic reconnection.
HP	MC/Service Guard, HP Wolfpack	HP 9000 Unix, NetServer running NT server, Oracle parallel server	Will support NT cluster solution in Wolfpack
Amdahl	Scaling multi-server cluster	Intel SHV-based EnVista servers, Fast Ethernet, 40 MB/s Fujitsu Interconnect	EnVista availability manager, failover, will support Wolfpack
Stratus	RADIO cluster	Two Pentium-based nodes, two PCI fast wide SCSI-2 Disk drives, and two100 Base T hubs	Isis availability manager supporting all active cluster features to overcome any single failure in hardware, software, or network

10.2.4 Fault-Tolerant Multiserver Cluster

This architecture represents the highest level development of availability clusters. Fault tolerance is achieved with complete redundancy and with active takeover capabilities. All single points of failure are duplicated to provide the fault tolerance capability. Critical components, such as CPUs, operating systems, memory buses, network interfaces, and disk drives, power supplies, and system-area networks are all duplicated.

Only a few computer companies such as Tandem and Marathon have pushed out fault-tolerant clusters as products. The Tandem Himalaya was marketed for high-end users. The Marathon couster is given below as an example.

Example 10.2 The Marathon fault-tolerant multiserver cluster

In Fig. 10.5, we show the architecture of the Marathon MIAL 2, a fault-tolerant cluster built with off-the-shelf components. The cluster requirement is to have every cluster part be active and redundant with another component. Failover time of the cluster should be maintained within 1 s. The goal is to achieve an availability of 99.999% with no more than 6 minutes of downtime per year. This is a very demanding request and often rather costly.

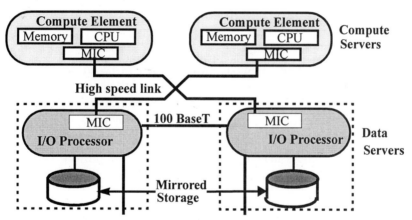

Figure 10.5 The Marathon MIAL 2 fault-tolerant cluster.
(Courtesy of Marathon Corp. [425], 1997)

The MIAL 2 has two compute servers built with the latest CPU chips. Two data servers are connected to the two compute servers with full duplexed, high-speed links and MICs (Marathon interface cards). The compute servers write information to both data servers simultaneously to provide real-time data protection. Proprietary 100 BaseT links exist between the two data servers with mirrored storages disks. If one data server fails, the user application can continue on the second data server. This is achieved by hardware-assisted data integrity checks and 32 MB/s throughput.

When the failed server is repaired, the available data server will automatically resynchronize the recovered server by replicating the entire disk storage to the recovered server. This replication takes place in the background at a speed of 10 minutes per 1 GB of the mirrored storage area. With everything duplicated and active, the MIAL can tolerate all types of single failures. Marathon claims that its cluster system requires no APIs, no scripts, and no special version of NT. But Marathon wants its cluster to participate as a node in a Wolfpack scaling cluster.

■

10.3   The IBM SP System

The IBM SP is special in that it uses a cluster approach to building MPPs. Its design experiences are well documented (e.g., *IBM System Journal*, February 1995) and provide many insights in designing large-scale, high-performance systems. After giving an overview of IBM's design goals and strategies, we discuss the architecture and software environment of the SP cluster in detail.

10.3.1 Design Goals and Strategies

IBM decided to enter the MPP business in the fall of 1991, launching the Scalable POWERparallel (SP) project [646]. A development team was assembled in February 1992 to bring out the first product in no more than 18 months. In fact, the first product, called SP1, was delivered to users in April 1993. The follow-on 120-node SP2 systems were delivered in July 1994. Now, the system is simply called SP by IBM distributers.

By 1998, more than 3000 SP systems were installed worldwide. The SP has been applied in the worldclass chess match in the IBM Deep Blue project. Listed below are factors affecting the success of SP systems that appear as either clusters or MPPs.

- *Time to market*: High-performance computers follow a trend similar to Moore's law (Section 1.4.1). To attain leadership price-performance, it is imperative that a product be developed within a short time.

- *General purpose*: SP must be a general-purpose system that supports different technical and commercial applications, prevalent programming models, and different operation modes.

- *High performance*: SP should provide high sustained performance, not just peak speed. This relies on not only many fast processors, but also on a fast memory system, high-speed communication, good compilers, libraries, etc.

- *Availability*: SP must exhibit good reliability and availability so that customers can run their commercial production codes on it.

To meet these goals, the IBM team adopted the strategies of using a flexible cluster architecture; custom interconnect; standard system environment with enhanced high-performance services and availability supports; standard programming models; and selected single system image supports.

Cluster Architecture The most difficult and important decision made by the IBM team was to select the *cluster architecture*. The main features include:

- Each node is an RS/6000 workstation, with its own local disk.
- A complete AIX (IBM's Unix) resides on each node.
- The nodes are connected through a high-speed network. The network interface is loosely coupled: a node connects to the network through its I/O bus (Micro-channel), not through the local memory bus.

IBM believes that such an architecture has two main advantages: *simplicity* and *flexibility*. It is simple to build, yet can be configured to match different application needs. It can use the RS/6000 workstation technology and the Vulcan distributed-memory techniques already developed in IBM (Vulcan is the name of an IBM internal research multicomputer prototype).

The SP series use standard workstation components as much as possible. Custom software and hardware are developed only where standard technology cannot meet the desired performance requirement of a scalable system. Choosing the cluster architecture is the key to enable SP to achieve the four design goals, especially the time-to-market and the general-purpose goals.

Flexible Architecture Having a flexible architecture means that SP must allow different configurations. More specifically:

- The system should be size-scalable, from a few nodes to hundreds of nodes.
- The nodes should be individually configurable (in both hardware and software) to meet the user's requirements of application and environment.

Custom Interconnect IBM believes that existing commodity networks such as Fiber-Channel and ATM are not adequate for providing the required communication latency and bandwidth of an MPP. Therefore, IBM decided to use the multistage network technology developed in the Vulcan project, which becomes the *High-Performance Switch* (HPS) interconnect used in constructing scalable SP systems.

Standard Environment SP uses a standard, open, distributed Unix environment. This avoids the difficult task of developing a machine-specific OS software for an MPP. Being a workstation cluster, SP can utilize the existing standard software for systems management, job management, storage management, database, and message passing that is available in a distributed Unix environment. All this software already exists in the IBM workstation AIX operating system.

High-Performance Services For those existing or new applications that cannot be executed efficiently in a conventional distributed AIX environment, SP provides a set of high-performance services. They include:

- The high-speed interconnection network (the HPS)
- An efficient user-level communication protocol (the US protocol)
- An optimized *Message-Passing Lbrary* (MPL)
- A parallel program development and execution environment
- A parallel file system
- Parallel databases (e.g., parallel DB2)
- A high-performance I/O subsystem

Standard Programming Models To achieve the general-purpose goal, IBM decided to support prevalent programming models in three areas:

- *Sequential Computing*: Although SP is a parallel computer, it must allow existing sequential programs to run without change on a single node. Because of the cluster architecture and the standard environment, this task is easily accomplished in SP2. Besides providing conventional C, C++, and Fortran, SP enables more than 10,000 applications developed for the RS/6000 workstations to execute without requiring any modification.
- *Parallel Technical Computing*: SP currently supports message-passing (MPL, MPI, PVM) and data-parallel programming (HPF) models. Support for the shared-memory model is planned.
- *Parallel Commercial Computing*: To support commercial applications, IBM is parallelizing a few key database and transaction monitor subsystems. A parallel version of the IBM DB2 database system (called DB2 Parallel Edition, or DB2 PE) has been implemented on SP2.

System Availability An SP system consists of thousands of commodity parts. These hardware and software components were originally developed for inexpensive workstations, not for large, fault-tolerant systems. They are bound to fail frequently enough to crash the entire system if availability measures are not taken. SP has adopted the following techniques to enhance system availability:

- The cluster architecture implies a separate operating system image in each node. The failure of one image does not have to disable the entire system. This is advantageous over an SMP architecture, where a single operating system image resides in the shared memory, and the failure of the OS will crash the entire system.
- The SP design systematically removes single points of failure that would disable the entire system. For instance, the nodes are connected by two networks: the HPS and an Ethernet. When the HPS fails, the nodes communicate through the Ethernet.
- A software infrastructure provides services for failure detection, failure diagnosis, system reconfiguration, and failure recovery. This infrastructure enables SP to degrade gracefully.

Selected Single-System Image In a distributed system, the user sees a number of individual, separate workstations. In a system with a true single image, the user sees on giant workstation (e.g., one operating system image). IBM has examined different user and environment requirements and found that a true single image is difficult to implement and not a crucial requirement in some commercial applications.

Consequently, IBM decided on a compromise between the two extremes. The SP system has implemented the SSI features of single entry point, single file hierarchy, single control point, and single job management system. The single addrtess space was not implemented in the SP system, among other desired SSI features. This will become clear as we proceed with the SP architecure.

10.3.2 The SP2 System Architecture

A simplified logical architecture of an SP is shown in Fig. 10.6. An SP consists of 2 to 512 nodes, each having its own local memory and local disk. Each SP also requires an RS/6000 workstation to serve as the system console.

The System Interconnects The nodes are interconnected by two networks: a conventional Ethernet and the HPS. The Ethernet, though slow, provides the following benefits:

- The Ethernet can be used for program development where communication performance is not critical. The developed code can use the HPS for production runs.
- It serves as a backup when the HPS fails. Without the Ethernet, the HPS would be a single point of failure.
- It helps reduce SP development time by taking advantage of *concurrent engineering*: while the HPS and associated software are under development and improvement, the Ethernet allows the rest of the system to be developed, debugged, tested, and used.
- The Ethernet can also be used for system monitoring, booting, loading, testing, and other system administration.

We have already studied the HPS architecture (Example 6.5) and the crossbar switch design (Example 6.4) in Chapter 6. Details will not be repeated here.

Node Architecture As shown in Fig. 10.6, each node has a private memory and a local disk. The connections to the Ethernet and the HPS are *loosely coupled*, i.e., they are through the Micro Channel I/O bus. This is in contrast to Intel Paragon or Cray T3D, where the NIC is tightly coupled to the memory bus.

SP provides three physical node types: *wide node, thin node,* and *thin node 2* to efficiently support configuration flexibility (see Example 1.1). These three node types are compared in Fig. 10.7. All nodes use a POWER2 microprocessor clocked at 66.7 MHz. Each processor has a 32-KB instruction cache, a data cache, an instruction and branch control unit, two integer units, two floating-point units where each can execute a multiply and an add in one clock cycle. This gives the POWER2 processor a peak speed of $4 \times 66.7 = 267$ Mflop/s.

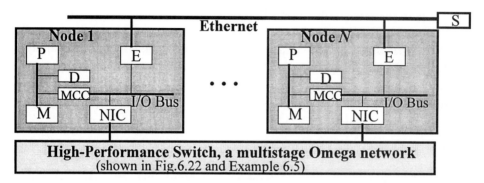

Figure 10.6   The SP system architecture. (P: processor, M: memory, D:disk, MCC: MicroChannel controller, NIC: network interface switch, E: Ethernet adapter, S: system console. Copyright 1995, International Business Machines Corp. Reprinted with permission from T. Agerwala, et al. *IBM Systems Journal,* Vol. 34, No.2, [14], 1995)

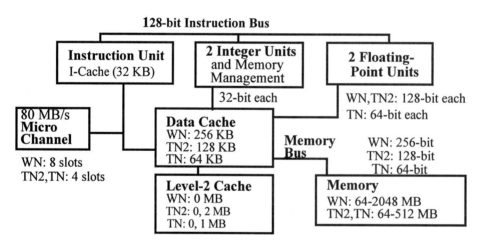

Figure 10.7   The processor node structure in wide node (WN), thin2 node (TN2), and thin node (TN) of the SP system. (Copyright 1995, International Business Machines Corp. Reprinted with permission from T. Agerwala, et al. *IBM Systems Journal,* Vol. 34, No.2, [14], 1995)

POWER2 is a superscalar processor, capable of executing 6 instructions per clock cycle. With its short instruction pipelines, sophisticated branch prediction techniques, and register renaming techniques, the POWER2 can perform in a clock cycle two loads/stores, two floating-point multiply-add, an index increment, and a conditional branch.

The three node types differ in the capacity of the memory hierarchy, the data path width, and the number of I/O bus slots. For instance, a wide node has up to 2 GB of main memory, a 256-KB data cache, and eight I/O slots on the MicroChannel. The memory bus is 256-b wide and provides a peak bandwidth of 2.1 GB/s.

The four-way set associative data cache can supply data to the two floating-point units at the peak speed of four 64-b operands per cycle. The thin nodes have smaller maximum memory/cache capacity and bandwidth, and allow up to four I/O slots. A thin node (thin node 2) can have a 1-MB (2-MB) level-2 cache.

The large memory/cache capacity and bandwidth, the superscalar design, and a good compiler enable the SP to give good per-node sustained performance. This partially explains why the utilization of SP is better than that of other MPPs in computing the NAS benchmarks (see Fig. 3.2).

10.3.3 I/O and Internetworking.

The I/O subsystem, network interfaces, and internetworking requirements of the SP system are given below:

SP I/O Subsystem The architecture of the SP I/O subsystem is shown in Fig. 10.8. The I/O subsystem is essentially built around the HPS with an LAN gateway to other machines outside of the SP system. .

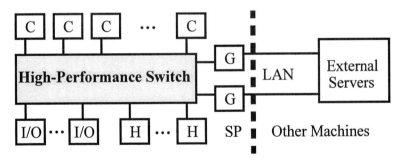

Figure 10.8 **The SP I/O Subsystem** (C: compute node, G: gateway node, H: host node, I/O: standard or parallel I/O server node, LAN:Ethernet, Token-Ring, FDDI, HiPPI, ATM, etc. Copyright 1995, Internaional Business Machines Corp. Reprinted with permission from T. Agerwala, et al. *IBM Systems Journal,* Vol. 34, No.2, [14], 1995)

The SP nodes can be configured as four classes: The *host nodes* (H) are used to handle various user login sessions as well as interactive processing. The *I/O nodes* mainly perform I/O functions, such as global file servers. The *gateway nodes* (G) serve networking functions. The *compute nodes* (C) are devoted to computing. These four classes may overlap. For instance, a host node can be a compute node, and an I/O node a gateway node. The external servers are additional machines external to the SP, such as file servers, network routers, and visualization systems.

Network Interface   Each SP node is connected to the HPS through a network interface circuitry, called the *switch adaptor*, or *communication adapter*, as shown in Fig. 10.9. The adapter includes an 8-MB DRAM and is controlled by a 40-MHz Intel i860 micropro-cessor. The adapter is attached to the MicroChannel, which is a standard I/O bus used for connecting peripherals to RS/6000 workstations and IBM PCs.

Each adapter is connected to an HPS link via a chip called the *memory and switch management unit* (MSMU). A bidirectional link physically consists of two channels, each 8 bits wide, that are connected to an incoming and an outgoing FIFOs in the MSMU. Besides the two FIFOs, it contains a number of control/status registers, and it also serves as the i860 bus controller, checking and refreshing the DRAM. A 4-KB bidirectional FIFO (2 KB each direction) called the BIDI connects the Micro Channel and the i860 bus.

The adapter uses a large size (8-MB) DRAM to accommodate different protocols that need large amounts of local message buffering. The node processor can directly access the adapter DRAM and MSMU through programmed I/O instructions. However, data transfers using DMA must go through the BIDI.

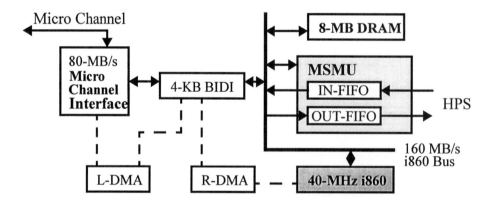

Figure 10.9   The SP communication adaptor (MSMU: memory and switch management unit, BIDI: buffer memory, DMA: direct memory-access engine. Copyright 1995, International Business Machines Corp. Reprinted with permission from T. Agerwala, et al. *IBM Systems Journal,* Vol. 34, No.2, [14], 1995)

Sending of data from a node to the HPS proceeds as follows: The node processor tells the adapter about the data to be sent. The i860 writes to the BIDI a header, which contains the necessary information for a *direct memory access (*DMA) transfer. When the header reaches the head of the BIDI, the *left-hand DMA engine* (L-DMA) takes over and transfers the data from the node (Micro Channel) into the BIDI.

At the completion, the L-DMA increments a hardware counter. The i860 then writes another header to the *right-hand DMA engine* (R-DMA), which transfers the data from the BIDI to the OUT-FIFO in the MSMU, which then transfers the data to the HPS.

Receiving data are similar. After data arrive, the MSMU informs the i860, which writes a header to initiate a right-hand DMA. The R-DMA takes over to transfer data from the IN-FIFO to the BIDI. At the completion, the i860 writes a header to the BIDI. The L-DMA extracts the header when it reaches the head of the BIDI and transfers data from the BIDI to the MicroChannel. The adaptor architecture allows concurrent operations.

Example 10.3 The 400-node SP2 Multicomputer at MHPCC

One of the world's largest SP2 configurations is shown in Fig. 10.10.

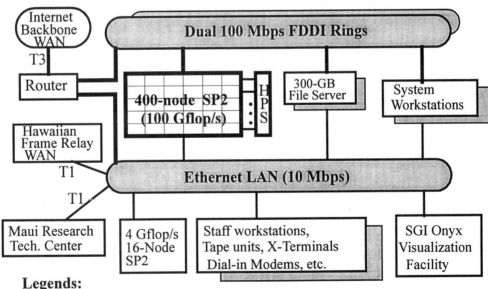

Legends:
HPS = High-performance switch, WAN = wide-area network
FDDI = Fiber Distributed Data Interface, T3 line (45 Mbps), and
T1 lines (1.54 Mbps). Thin links are 10-Mbps Ethernet connections

**Figure 10.10 The 400-node IBM SP2 system installed at the Maui High
Performance Computing Center.** (HPS : High performance
switch, Courtesy MHPCC, Maui, Hawaii [445].)

This is a 400-node cluster installed at the Maui High Performance Computer Center (MHPCC) in 1994. MHPCC is located on the island of Maui in Hawii islands. Another bigger sytem is the 512-node SP system at Conell Theory Center.

Besides the HPS, all nodes are also connected by a dual FDDI ring and by an 10 Mbps Ethernet. The FDDI (100 Mbps) rings provide hardware redundancy needed for high availability.

Message communications in a parallel application has three alternatives: HPS, FDDI, and Ethernet. All other communications, such as for system management, availability, file operations, and network accesses, can go through either the FDDI or the Ethernet networks. This SP2 system has been applied in many scientific super-computing and large-scale signal-processing applications

■

10.3.4 The SP System Software

The SP system software hierarchy is shown in Fig. 10.11. The core is the IBM AIX operating system. The SP reuses most of the RS/6000 workstation environment, such as:

- More than 10,000 sequential applications;
- Database management systems (e.g., DB2);
- On-line transaction processing monitors (e.g., CICS/6000);
- System management and job management;
- Fortran, C, C++ compilers;
- Mathematical and engineering libraries (e.g., ESSL);
- The standard AIX operating system

Applications			
Application Subsystems (Databases, Transaction Monitors, etc.)			
System Management	Job Management	PE	Compilers, etc.
Global Services (providing single-system image)			
Availability Services			
High-Performance Services	Standard Operating System (AIX)		
Standard RS/6000 Hardware (processors, memory, I/O devices, adapters)			

Figure 10.11 SP system software hierarchy. (PE: processing environment.Copyright 1995, International Business Machines Corp. Reprinted with permission from M. Snir et al. *IBM Systems Journal,* Vol. 34, No.2, [574], 1995)

The SP system only adds some new software and improves some existing software that is required for a scalable parallel cluster system.

The Parallel Environment (PE) The AIX PE provides a platform for users to develop and execute their parallel programs, as shown in Fig. 10.12. It has four components: a *Parallel Operating Environment* (POE), a *Message-Passing Library* (MPL), a *Visualization Tool* (VT), and a *parallel debugger* (pdbx).

The Parallel Operating Environment The POE is used to control the execution of parallel programs. Its structure is shown in Fig. 10.12. The execution is controlled by a *Partition Manager* process running on a *home node*, which is an RS/6000 workstation connected to the SP or an SP node. The home node is where the user invokes the parallel program, which runs as one or more tasks on *compute nodes* of the SP.

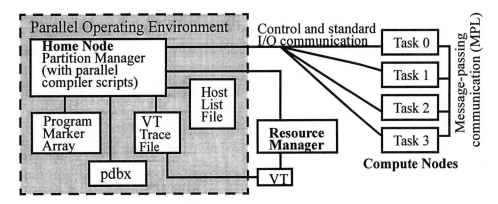

Figure 10.12 The PE and POE built in the IBM SP cluster. Copyright 1995, International Business Machines Corp. Reprinted with permission from M. Snir et al. *IBM Systems Journal,* Vol. 34, No.2, [574], 1995)

The home node provides standard Unix I/O devices (e.g., stdin, stdout, and stderr). It performs standard I/O communications with the compute nodes through an LAN (e.g., an Ethernet). For instance, the user can terminate all tasks by pressing control-C from the home node keyboard. The printouts from printf statements are displayed on the screen of the home node.

The MPL and MPI Message-passing communication is realized by executing special MPL functions, through either the HPS or an Ethernet. This library provides 33 functions for process management, grouping, point-to-point communication, and collective communication. The IBM SP also supports different versions of MPI, including one that is natively implemented.

High-Performance Services The cluster architecture of the IBM SP enables the immediate use of standard, off-the-shelf software originally developed for RS/6000 workstations and distributed systems based on TCP/IP networks. In addition, SP provides a number of high-performance services, which enable new applications and subsystems that cannot execute efficiently in conventional distributed system environments. These services include a high-performance communication subsystem, a high-performance file system, parallel libraries, a parallel database, and high-performance I/O. We will just discuss a few of these services

Table 10.6 Performance of Two Protocols on the SP

Protocol	Latency (ms)	Bandwidth (MB/s)
UDP/IP	277	10.8
US	39	35.5

The SP supports two communication protocols: An IP-based one (e.g., UDP/IP or TCP/IP) that executes in the kernel space and a user space protocol called US. Both protocols can be used on either the HPS or a conventional network (e.g., Ethernet). The performance of the IP and the US protocols over the HPS is shown in Table 10.6.

The US protocol has much better performance. However, SP allows only one task per node to use the US protocol, while multiple tasks per node can use the IP to share the HPS. For applications that do not require high-performance communication, using the IP may lead to better overall system utilization.

Parallel I/O File System The SP high-performance file system is called PIOFS (for *Parallel I/O File System*). It is POSIX-compliant for most applications and system utilities. Unix operations and commands such as read, write, open, close, ls, cp, and mv behave the same as in a sequential Unix system. It supports large file size up to 2^{64} B, instead of 2^{32} B in AIX. Besides allowing traditional Unix file system interface, PIOFS provides a parallel interface to enable parallel distribution and operations on files.

IBM developed a parallel database software program, called *DB2 Parallel Edition*, to run on SP as well as other clustered platforms. It is based on the shared-nothing architecture, and it uses a *function shipping* technique. The database is distributed across multiple nodes, and a database function is shipped to the node where the data reside. DB2 Parallel Edition is scalable in both machine size and problem size. It can run on hundreds of nodes and handle large databases containing Terabytes of data.

Availability Services The SP system provides a software availability infrastructure via a set of daemons running on the nodes. The *heartbeat daemons* exchange periodic heartbeat messages to indicate which nodes are alive. *Membership services* allow nodes and processes to be identified as belonging to a certain group.

In the event of a membership change caused by node failure, shutdown, or restart, the *notification services* are used to notify the active members the event, and subsequently, *recovery services* are invoked to coordinate recovery procedures to enable the active members to continue work.

Global Services A number of global services provide selected types of single system image. An external *system data repository* (SDR) maintains systemwide information about nodes, switches, and current jobs in the system. The SDR is useful to reconfigure the system when part of the system fails, without affecting the other parts. The SDR contents can bring the system back to the state before the failure.

Global network access is realized by supporting TCP/IP and UDP/IP through the HPS. A *single file system* can be provided through the *Network File System* (NFS) or the *Andrew File System* (AFS). Besides network file systems (NFS or AFS), SP provides the *virtual shared disk* (VSD) technique for global disk access, which has an order of magnitude better performance than NFS.

VSD is a device driver layer that sits on top of the AIX *Logical Volume Manager* (LVM). When a node process wants to access a shared disk that is locally connected, the VSD directly passes the request to the LVM of the node. When a process wants to access a remote shared disk, the VSD passes the request to the VSD of the remote disk through the HPS, which then passes it to the LVM of the remote node.

System Management The SP system console is a control workstation (Fig. 10.6). The SP system administrator manages the entire SP system from this single point of control. The management functions include:

- System installation, monitoring, and configuration
- System operation
- User management
- File management
- Job accounting
- Print and mail services

In addition, each node, switch, and frame in the SP2 hardware has a supervisor card that senses environmental conditions and controls the hardware components. The administrator can use this facility to power on/off, monitor, and reset the individual node and switch components.

Load Management SP supports two types of user jobs: *interactive mode* and *batch mode*. These can be either sequential or parallel programs. IBM offers a LoadLeveler software for batch processing. LSF is also available for SP to manage load for both interactive and batch jobs.

10.3.5 The SP2 and Beyond

Since the first shipment in 1994, the SP2 system architecture has been enhanced in many ways. Table 10.7 shows samples of the architecture attributes of the 1996 version of

Table 10.7 Architecture Attributes of the IBM SP2 in 1997

Attribute	604 High Node	Wide Node
Processors	2-8 PowerPC 604	One P2SC
Per-processor peak speed	112 MHz, 224 Mflop/s	135 MHz, 540 Mflop/s
Per-processor cache	16 KB code & 16 KB data L1 cache; 1 MB L2 cache	32 KB code, 128 KB data L1 cache; no L2 cache
Main RAM memory	64MB - 2 GB	64 MB - 2 GB
Memory bus width	256 b	256 b
Internal disk capacity	2.2-6.6 GB	2.2-36.4 GB
Micro channel controllers	14	7
Maximal number of nodes in a standard system	16	128
Peak one-way point-to-pint communication bandwidth	150 MB/s	150 MB/s

the SP2 systems. Two more node types, the new wide node and the SMP node, are added to the original thin, thin2, and wide nodes in Fig. 10.7.

New Node Types A new wide node consists of a single *POWER2 superchip* (P2SC) processor, which integrates the original eight-chip POWER2 processor into a single chip, containing 160-KB on-chip cache and 15 million transistors. The processor clock frequency is also increased from 66.7 MHz to 135 MHz.

An SMP node (also known as a *high node*) consists of from two to eight PowerPC 604 processors clocked at 112 MHz. Each 604 processor has a 32-KB on-chip level-1 cache and a 1-MB off-chip cache.

Special SP Configurations In commercial configuration, an SP system has up to 128 nodes, including up to 16 PowerPC SMP nodes. In custum configuration, the machine size can increase to 512 nodes. The HPS uses the same multiswitch network architecture, but the link bandwidth is improved from 40 MB/s to 150 MB/s. To match this, the bandwidth of node I/O interfaces to the HPS, is improved from 80 MB/s to 160 MB/s.

Software Upgrades   The improvement in software includes the addition of two pieces of system software. The *High Availability Cluster Multi-Processing* (HACMP) software provides backup, fail-over, and other availability functions. The *Parallel Systems Support Programs* (PSSP) software facilitates system management, such as user and password management, job accounting, system monitoring, and system partitioning.

The IBM/LLNL Blue Pacific   The U.S. Department of Energy has contracted with IBM and Lawrance Livermore National Laboratory to build a Teraflops supercomputer, called Option *Blue Pacific*. This machine will be discussed in Section 11.3. Obviously, it will extend the SP (*Scalable Parallel*) technology further. It is speculated that the system will push the cluster concept further by having more support for single-system image and distributed-shared memory operations, beyond the current SP architecture.

10.4   The Digital TruCluster

The DEC TruCluster is a family of Unix clusters. A node can be a uniprocessor or multiprocessor system running Digital Unix. The interconnect network can be a Gigaswitch or a Memory Channel. In this section, we discuss the TruCluster using AlphaServer 8400 nodes and the Memory Channel interconnect.

10.4.1   The TruCluster Architecture

The TruCluster architecture is shown in Fig. 10.13. This configuration combines both the shared-nothing and the shared-disk approaches.

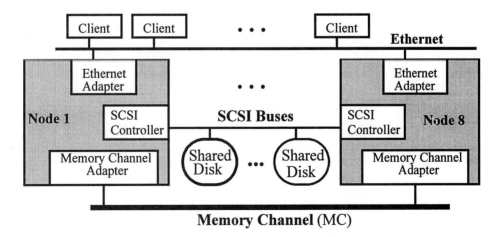

Figure 10.13 The TruCluster networking architecture. (Courtesy of Cardoza et al. *Digital Technical Journal* [127], 1996)

The TruCluster is designed as a compute or database server. It features SMP nodes, an innovative Memory Channel interconnect, the *reflective memory* mechanism for efficient communication, selected single system image, and software availability and scalability support. Different configurations are possible for different user needs. For instance, the shared disks may be distributed to individual nodes to create a shared-nothing structure.

Memory Channel   The cluster system in Fig. 10.13 has three separate clusterwide inter-connects. The Ethernet is used to connect the client workstations, PCs, or terminals to the TruCluster. Users' logins, commands and applications access the TruCluster from this standard network. A number of SCSI buses are used to enable the nodes to access files in the shared disks. Each node can have its own local disks.

The *Memory Channel* (MC) interconnect, jointly developed by Encore Computer Systems and Digital Equipment, has high speed and is responsible for supporting message passing and synchronization. This configuration is useful to describe several issues that we previously have not addressed, namely:

- How to use the MC and associated software to realize a limited form of shared memory (*reflective memory*) for synchronization and message passing
- How to use the distributed lock manager to coordinate disk sharing
- How to program a cluster of SMP nodes

The current configurations of TruCluster can have up to eight nodes interconnected by an eight-port Memory Channel. The architecture parameters of this TruCluster system are shown in Table 10.8. The parameters are all best values showing the scalability limits. They may not be simultaneously achieved.

The Node Structure   Each node is a DEC AlphaServer 8400 SMP system; the architectural characteristics are as shown in Table 8.1. The block diagram of this AlphaServer is shown in Fig. 10.14. The AlphaServer 8400 node has a nine-slot memory bus. Each slot can hold a CPU module, a memory module, or an I/O module. A node must be configured with at least one CPU module, one memory module, and one I/O module. There can be up to six CPU modules, up to seven memory modules, and up to three I/O modules.

A CPU module contains two 437-MHz Alpha 21164 processors. Each processor has a 16-KB level-1 cache and a 96-KB level-2 cache, both on chip. There is an off-chip, 4-MB level-3 cache per processor installed on the CPU module. A memory module can be populated with up to 4 GB of four-way interleaved SIMM memory.

An I/O module controls four I/O channels, each a 12-slot standard PCI bus or Digital's XMI bus. A combination of PCI and XMI buses is allowed. Up to 144 I/O slots connect various standard I/O devices, such as SCSI disks, CD-ROM drives, Ethernet adapters. The Memory Channel adapter is connected to a PCI slot.

Table 10.8 Architecture Parameters of the DEC TruCluster

Architecture Parameter	One Node	One TruCluster
No. of processors	12	96
Peak speed	10 Gflop/s	80 Gflop/s
Off-chip cache	8 MB	64 MB
Maximum memory	28 GB	224 GB
Aggregated memory bus bandwidth	2.1 GB/s peak 1.8 GB/s sustained	16.8 GB/s peak 14.4 GB/s sustained
Internal disk	192 GB	1.5 TB
I/O channels	12 PCI buses	96 PCI buses
I/O slots	144 PCI slots	1152
Aggregated I/O bandwidth	1.2 GB/s	9.6 GB/s
Node-to-node peak bandwidth	N/A	100 MB/s

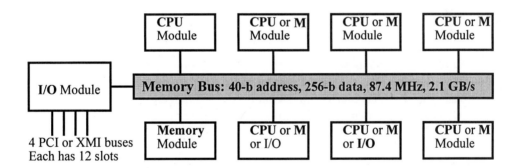

Figure 10.14 The AlphaServer 8400 block diagram. (Courtesy of Digital Equipment Co. [198], 1994)

The memory bus is a synchronous bus clocked at 87.4 MHz, one-fifth of the CPU clock frequency. The bus data width is 256 b or 32 B. A peak bandwidth is upper bounded by $32 \times 87.4 = 2796$ MB/s. Considering inherent hardware overheads, the peak bus bandwidth, i.e., the bandwidth that can never be exceeded, is 2.1 GB/s. The memory bus employs techniques such as split transactions and separate address/command bus and data bus to bring the sustained bandwidth to 1.8 GB/s.

10.4.2 The Memory Channel Interconnect

The Memory Channel is designed to provide a reliable, powerful and efficient clustering interconnect. More specifically, it has the following design requirements:

- *High Speed*: one-way message bandwidth of 100 MB/s, latency of less than 5 μs, and processor overhead of less than 0.5 μs.
- *Error Handling*: hot-swap capability and supporting error detection and recovery from network communication errors.
- *PCI Connectivity*: interface to the industry-standard PCI bus.

To achieve message latency of less than 5 μs, especially processor overhead of 500 ns, the communication software must execute entirely at user level, the communication protocol should be a zero-copy one, and software should not do checksum calculation. This is because on current uniprocessor or SMP nodes, a system call alone can consume more than 5 μs, and memory copying and checksum are expensive, especially for short messages.

This calls for user-level communication, mapping the memory channel network into the virtual address space of the individual nodes, and letting the hardware take care of error handling. DEC extended Encore's reflective memory technology as the foundation for Memory Channel.

Reflective Memory The reflective memory concept is illustrated in Fig. 10.15. Reflective memory works as follows: A *connection* is established from a virtual address page in one node to a virtual page in another node, through a page in the Memory Channel address space. For instance, connection 2 connects virtual page E in node 3 to virtual page B in node 1, virtual page B in node 1, virtual page D in node 2, virtual page F in node 3, and virtual page I in node 4.

Thereafter, when node 2 writes to page E, the Memory Channel hardware will automatically transmit the written contents to all the virtual pages connected to page E. In other words, a write to a local virtual page is reflected in all the connected local or remote virtual pages. When node 1 later reads page B, it will get the value written to page E by node 2. The reflective memory has the following features:

- First, it is virtual-address-based. Data communicate directly from one virtual page to another. This realizes zero-copy and requires the Memory Channel adapters to hold page tables and to perform address translations.
- Second, it is a user-level communication mechanism. Setting up a connection requires kernel calls. But the expensive connection establishment only needs to be performed once. The same connection can be used to perform many communications without needing any further system call.

- Third, communication granularity is at 32-B data block level. Although connections are established at page level, data communications are performed at 32-B blocks. The hardware does not have to transmit an entire page at a time.

- Fourth, communication is by ordinary *store* and *load* instructions. Although the user can communicate by message passing software, the user does not have to.

- Fifth, connections can be point-to-point (connection 1), broadcast (connection 2), or multicast (connection 3).

- Sixth, connections are essentially unidirectional. A virtual page cannot be both sending and receiving. If a node wants to send to and receive from the same connection, it has to use two separate virtual pages.

- Finally, it is allowed for more than one node to send to the same connection (e.g., connection 3). The Memory Channel software provides synchronization primitives to ensure data consistency.

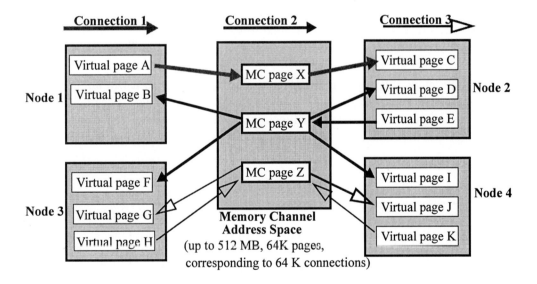

Figure 10.15 Illustrating the reflective memory concept in using the Memory Channels. (Courtesy Gillett and Kaufmann, *IEEE Micro*[268], 1997)

Communication over Memory Channel We now give more details of how a connection is established, and how a 32-B data block is communicated. The Memory Channel supports a 512-MB global address space (the MC address space in Fig. 10.15), into which the node processes can map pages of their virtual address spaces.

This MC address space is directly mapped to a 512-MB Memory Channel adapter's PCI I/O address space window in each node. The TruCluster has an 8-KB page size, implying that the Memory Channel can provide 64K pages to establish 64K different connections. We explain below how connection 1 in Fig. 10.15 is set up. Referring to Fig. 10.16, the following steps are executed to establish connection 1:

- The application calls a cluster service library function, which *allocates* a page of the MC address space (page X) for the connection.

- The process in node 1 (the sending node) calls a library function to *attach* a virtual page A in its address space to page X. This function sets up an entry in the node's virtual-to-physical map to point virtual page A to a physical page Q in the I/O space, which is directly mapped to page X in the MC address space by a *page control table* (PCT) in the MC adapter. This function also indicates in the corresponding PCT entry that the connection is enabled for point-to-point transmission by node 1.

- The process in node 2 (the receiving node) calls a library function to *attach* a virtual page (page E) in its address space to page X. This function sets up an entry in the node's virtual-to-physical map to point virtual page A to a page S in the local physical memory, which is directly mapped to page X in the MC address space by a PCT in the MC adapter.

- Furthermore, this physical page S is *pinned* (wired) so that it will not be swapped out, and the local DMA map in the MC adapter is set up to scatter data from connection X to page S. This function also indicate in the corresponding PCT entry that the connection is enabled for point-to-point receive by node 2.

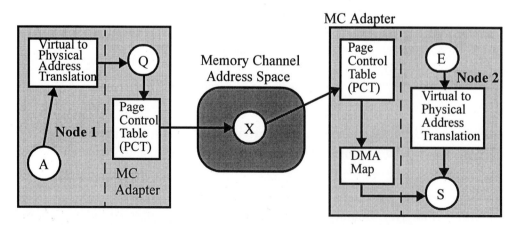

Figure 10.16 The Memory Channel communication mechanism.
(Courtesy Gillett and Kaufmann, *IEEE Micro* [268], 1997)

Page Control Table The PCT entry of a connection (i.e., an MC page) contains a number of control bits that can be used to specify the following properties and actions:

- Enabling transmit or receive
- Broadcast, multicast, or point-to-point
- Making local copy (loopback)
- Acknowledging by hardware the arrival of a write at all receive nodes
- Interrupting the destination node after a write reaches it

Block Data Transfer The following steps are executed to transmit a 32-B data block from node 1 to node 2 after the connection X is established:

- The process in node 1 performs consecutive *writes* (*stores*) to the offset in virtual page A, which translates to physical page Q. Since Q is in the I/O space, the *writes* bypass all caches, but the *writes* merge in one of 32-B write buffers.
- When the processor flushes the write buffer, a full 32-B write transfers to the PCI bus and selects the MC adapter.
- The adapter uses information from the PCT to encapsulate the PCI write into an MC packet, which consists of the 32-B data, a header identifying the destination and a trailer containing a 32-b CRC. The adapter then sends the packet out to the Memory Channel network.
- The adapter on node 2 receives the packet, strips off the MC header and trailer, and sends it to the PCI bus. The PCT and the DMA map determine that the incoming data should go to page S in the physical memory.
- When the process on node 2 subsequently reads virtual page E (which points to physical page S), it will get the new data.

The Memory Channel hardware provides much of the error handling and flow control capabilities. The hardware not only achieves very low transmission error rates, it can also do automatic error detection on all message writes. It provides hot-swap support and full hardware-based flow control. It enforces strict message-write ordering, even under errors. The hardware also provide a fast lock primitive.

10.4.3 Programming the TruCluster

The software structure for programming the TruCluster is illustrated in Fig. 10.17. End-user parallel applications have been developed using standard C, Fortran, or *High-Performance Fortran* (HPF) languages, calling PVM or MPI library functions to specify parallelism and message passing. All these libraries are implemented on top of the *Universal Message Passing* (UMP) layer. A main function of UMP is to provide a unified message-passing API that can be used on the Memory Channel, on the shared-memory architecture of an SMP node, and on other future transports and interconnects.

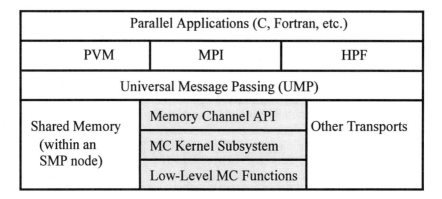

Figure 10.17 Software structure for programming the TruCluster.
(Courtesy Lawton et al.*Digital Technical Journal* [397], 1996)

Memory Channel Software Support   The shaded box is the Memory Channel software. It contains a user-level MC API and two layers of kernel-level software. End users should only use PVM, MPI, or HPF, not the lower layers, although they can do so if they choose. System programmers may use UMP as well. The MC API is too low-level even for third-party system software developers.

The MC API consists of 12 functions. Four functions are used for creating/destroying MC connections. Another four are used for managing spin locks in the global MC space. Two are used to check Memory Channel errors. A kill function sends a Unix signal to a specific process in a remote node of the cluster. Finally, there is a function to get the number of nodes in the cluster and their host names. Note that there are no *send/receive* functions, because communication is realized by ordinary *store/load* instructions.

The UMP layer contains 12 functions, as listed in Table 10.9. From the programmer's viewpoint, UMP is similar to TCP. UMP operations are based on duplex point-to-point links called *channels* (called *connections* in TCP). A channel is a pair of unidirectional buffers used to provide two-way communication between a pair of processes. These processes are called the *endpoints* of the channel, and the two processes can reside on the same node or on different nodes of the cluster.

Channels and UMP operations can be implemented either in the shared-memory space of a single node or the MC address space. The implementation is transparent to the users of UMP, although the shared-memory implementation has better performance.

A process calls *ump_open* to open a channel between a pair of tasks. However, the channel is not established until the second process also opens the same channel. The first process can use *ump_listen* and *ump_wait* to wait for the event that the channel connection is completed by the second process.

Table 10.9 Universal Message Passing (UMP) Functions.

(Courtesy Lawton et al. *Digital Technical Journal* [397], 1996)

Function	Meaning
ump_init	Initializes UMP and allocates resources.
ump_exit	Shuts down UMP and deallocates resources
ump_open	Opens a duplex UMP channel between two processes endpoints
ump_close	Closes a UMP channel
ump_listen	Registers an endpoint for a channel.
ump_wait	Waits for a UMP event to occur
ump_read	Reads a message from a channel.
ump_write	Writes a message to a channel.
ump_nbread	Reads a specified amount of message from a channel
ump_nbwrite	A nonblocking version of ump_write
ump_mcast	Multicast: writes a message to a list of channels.
ump_info	Returns UMP configuration and status information.

This is like the listen function in TCP. The *ump_wait* function is flexible, and it can wait on other events, such as arrival of a message, deletion of a channel, etc. The functions *ump_read*, *ump_nbread*, *ump_write*, *ump_nbwrite*, and *ump_mcast* are used to send and receive messages on an established channel.

The performance of various messaging layers on the TruCluster is shown in Table 10.10, including the portable MPI implementation MPICH, and the DEC-implemented PVM and MPI. Here SM stands for shared memory within a node.

Table 10.10 Performance of Point-to-Point Communication in Various Messaging Layers of DEC TruCluster. (Courtesy Lawton et al. [397])

Messaging Layer	Latency t_0 (μs)		Bandwidth r_∞ (MB/s)		Half-Peak Length $m_{1/2}$ (Bytes)	
	SM	MC	SM	MC	SM	MC
MC API	N/A	2.9	N/A	64	N/A	186
UMP	2	5.8	75	61	150	354
DEC MPI	5.2	6.9	64	61	333	421
DEC PVM	3	8	66	43	198	344
MPICH	30	N/A	24	N/A	720	421

Compared to other clusters and MPP machines, the TruCluster achieved small communication latencies of 8 μs or less, over the Memory Channel interconnect. The bandwidth numbers show that MPI and PVM can be implemented close to what the lowest-level software (the MC API) can provide. Processes on the same node can send messages more efficiently by taking advantage of the shared-memory architecture.

10.4.4 The TruCluster System Software

A TruCluster system contains both local node disks and global disks connected through shared SCSI buses (see Fig. 10.13). The TruCluster system software (see Cardoza et al. [127]) supports high availability by monitoring nodes and by automatically initiating recovery procedures in the event of component failures. It supports parallel database, allowing a database application to concurrently access raw disk and other I/O devices, regardless of whether the devices are local or remote. It supports explicit parallel programs written in C and Fortran with PVM, MPI, and HPF extensions.

The TruCluster system software includes the following component subsystems:

- *Distributed Lock Manager* (DLM): Provides software library functions to synchronize cluster-wide access to shared resources, such as disk files.

- *Connection Manager*: Keeps track of cluster membership, namely, how many and what nodes constitute the cluster. It also establishes and maintains communication paths between every pair of cluster nodes.

- *Distributed Raw Disk* (DRD): Allows a raw disk-based, user-level application (such as a distributed database management system or a transaction processing monitor) to run in a cluster, regardless of where in the cluster the physical storage is located. It allows applications to parallelly access storage media, including RAID disks, from multiple nodes.

- *Global Error Logger:* Allows the TruCluster software to log messages about events that occur in the TruCluster environment to one or more systems, so that TruCluster administrators can receive notification of critical problems when they occur.

- *Cluster Monitor:* A graphic user interface to the system administrator. It can be used to display the cluster configuration, as well as the current state of availability and connectivity in the cluster. The administrator can invoke management tools to manage the entire cluster system from a single location.

- *POLYCENTER Advanced File System* (AdvFS): A journaled, local file system to support availability. Using transaction journaling, AdvFS can recover file domains in seconds rather than hours after an unexpected restart such as a power failure.

10.5   The Berkeley NOW Project

The NOW (*Network of Workstations*) project at the University of California at Berkeley aims to develop techniques for enterprise clusters, which are also applicable to dedicated clusters. These techniques leverage small, mass-produced commercial systems to build a large-scale computer. The motivation and objective are that the clustered system should provide the individual user with fast and predictable response time of a dedicated workstation, while allowing tasks that are too large for the desktop to recruit resources throughout the cluster.

The NOW project addresses most clustering issues discussed in Chapter 9. Efficient communication is supported by utilizing commercial gigabit networks and the active message communication protocol. Single system image, resource management, and availability are provided through a user-level, clusterwide software called GLUnix. A new *serverless* network file system, called xFS, is developed to support scalable and highly available single file hierarchy. Recently, the NOW team is constructing a software framework, called WebOS, for building highly available, incrementally scalable, geographically aware Web services.

10.5.1   Active Messages for Fast Communication

Active message is an asynchronous communication mechanism to realize low-overhead communication. Its objective is to expose to the user the raw capability of the underlying communication hardware. The basic idea is to use the control information at the header of a message as a pointer to a user-level subroutine called a *message handler*. When the message header arrives at the destination node, the message handler is invoked to extract the remaining message from the network and integrate it into the ongoing computation.

The Active Message Mechanism   The active message mechanism has a generic specification [179] and different implementations. There are also suggestions to extend the original active message to provide additional functionality. See the Web Resource for pointers. We will discuss only the basic mechanism in the generic active message version 1 [179], as it is the specification followed by most implementations. Version 2 [426] has come out recently and has not been widely implemented. The version 1 specification makes use of a small number of functions (primitives) as shown in Fig. 10.18, based on the following assumptions:

- Active message is a software layer used to support communications between the processes of an SPMD application.
- The program consists of n processes, each ranked by a *virtual node number* (VNN) from 0 to n - 1. Process creation and management are provided by some system facility (e.g., GLUnix) external to the active message layer.

We call the active message layer that complies with the generic active message specification the *GAM layer*. This layer provides five basic primitives (operations), as shown in Fig. 10.18, where vnn_t and handler_t are data types defined in GAM for a virtual node number and a message handler subroutine, respectively. A number of utility functions are also provided by GAM. There are two types of messages in GAM: *request* and *reply* messages as defined below:

Request Message A request message W is sent by calling a request function:

$$am\_request\_2(Destination, request\_handler, x, y) ;$$

where the request message W consists of the *request_handler*, two integers x and y, and implicitly, the requesting process's virtual node number. A request primitive am_request_M(...) has five versions, for M = 0 to 4, each having M integer arguments. The above request has two arguments, x and y. This request sends W to the *Destination* process. The am_request is *blocking* in that it does not return until W is sent. Upon the arrival of W, the *Destination* process invokes the *request_handler* subroutine which takes as arguments x, y, and the requesting process's VNN.

Basic Functions:

int am_request_M(vnn_t dest, request_handler, int arg_0, ..., int arg_{M-1})

int am_reply_M(vnn_t dest, reply_ handler, int arg_0, ..., int arg_{M-1})

int am_get(vnn_t source, void *lva, void *rva, int nbytes,
          handler_t reply_handler, void *handler_arg)

int am_store(vnn_t dest, void *rva, void *lva, int nbytes,
          handler_t request_handler, void *handler_arg)

void am_poll(void)

Utility Functions:

int am_enable(...)   // initializes the active message layer

int am_disable(void)   // exits the active message layer

int am_procs(void)   // returns the total number of processes of program

int am_my_proc(void)   // returns the calling process's virtual node number

int am_max_size(void) // returns the max. No. of bytes for a get/store

---

Figure 10.18 Some functions in a generic active message. (Courtesy Culler et al. from http://now.cs.berkeley.edu/ [179], 1994)

Reply Message   A reply operation is very similar to the request message. A reply message W is sent by calling a reply function defined by

   am_reply_2(*Destination, reply_handler, x, y*) ;

where the reply message W consists of the *reply_handler*, two integers x and y, and implicitly the replying process's virtual node number. A reply primitive am_reply_M(...) has five versions, for M = 0 to 4, each having M integer arguments. The above reply has two arguments x and y. This reply sends W to the *Destination* process.

   The am_reply is *blocking* in that it does not return until W is sent. Upon the arrival of W, the *Destination* process invokes the *reply_handler* subroutine that takes as arguments x, y, and the replying process's VNN.

   The message handlers are meant to be quick and within some predetermined time bounds. In other words, a handler function should never suspend. The following restrictions are specified by GAM to achieve this goal:

- A reply function can be called only from a request handler subroutine.
- The *Destination* of the reply function can be only the requesting process.
- A request handler subroutine can reply at most once.
- A reply handler subroutine cannot call a request or reply function.

   The am_store and am_get functions are used to make bulk data transfers between two processes. An am_store defined by:

   am_store (*Dest, lva, rva, N, request_handler, handler_arg*) ;

transfers an N-Byte consecutive memory region with starting address *lva* in the caller (the requesting process) to the memory region starting from *rva* in the *Dest* process.

   Upon receiving all the N B of data, the *Dest* process invokes the request_handler and passes it the following parameters: the requesting process's VNN, *rva*, N, and *handler_arg*. An am_get is similar to an am_store, only in the opposite direction. The following am_get call

   am_get (*Source, rva, lva, N, reply_handler, handler_arg*) ;

fetches an N-Byte consecutive memory region with starting address *rva* in the caller (the reply process) to the memory region starting from *lva* in the *Source* process.

   Upon receiving all the N bytes of data, the caller (not the *Source*) invokes the reply_handler and passes it the following parameters: the *Source* process's VNN, *lva*, N, and *handler_arg*.

Example 10.4 Generation of active messages in the NOW

Consider a two-process program that uses GAM for communication of short messages, as shown in Fig. 10.19. Process Q computes an array A, while process P computes a scalar value x. The final result is a sum of x and $A[7]$. It is straightforward for P to remotely fetch $A[7]$ from Q and accumulate with x locally.

To implement this remote fetch, process P executes an am_request, which sends the index value 7 to process Q. Upon receiving the request message, Q invokes the request handler function request_h, whose return sends the value of $A[7]$ back to P by calling a reply function.

Process P

```
compute x
...
am_enable(...) ;
am_request_1(Q, request_h, 7);
am_poll();
sum = x+y ;
...
am_disable();



int reply_h(vnn_t Q, int z)
{ y = z; }
```

Process Q

```
compute A
...
am_enable(...) ;
...
am_poll();
...
am_disable();


int request_h(vnn_t P, int k)
{ am_reply_1(P, reply_h, A[k]); }
```

Figure 10.19 Implementing remote fetching by GAM in the NOW.

Upon receipt of the reply message, P calls the reply handler function reply_h, which passes $A[7]$ to the variable y in P. The function am_poll will poll the communication network for any arriving messages.

■

The following example shows how to use GAM to communicate with long messages, such as for large block data transfer between two workstation nodes.

Example 10.5 Long message communication using the GAM layer

As shown in Fig. 10.20, process P computes an N-byte array A and then stores its value to the array B in process Q. To avoid unnecessary buffering and copying overheads, a handshaking is performed first via a request-reply to make sure the memory area B in process Q is ready to receive data.

There can be more than one request (or reply) handler. Which handler should be used is specified in a request (or reply) function call.

- The first request message sent by P (am_request_0) triggers the request1 handler, which returns to P the starting address of memory area B via a reply message.
- The second request message sent by P (am_store) invokes request2, which is executed after the bulk data are transferred. In this example request2 is a null function.

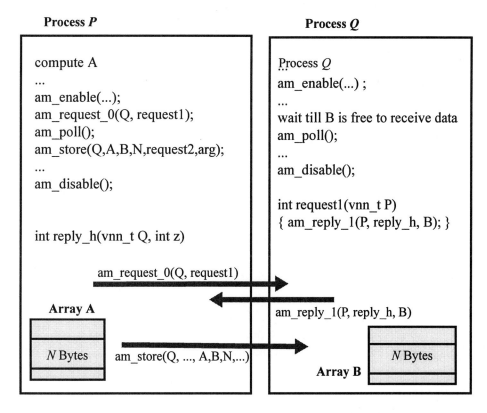

Figure 10.20 Bulk transfer of a large data block by GAM

Implementation of Active Messages Active message is a general mechanism, not required to be tied to any hardware or software platform. It has been implemented on MPPs, cluster of workstations, and even on PVM. However, active message is meant to be used as a low-level communication layer that can deliver a large percentage of the raw performance of the communication hardware.

This goal has been met with respect to latency and bandwidth. For instance, on a prototype of a cluster of HP9000/735 workstations connected by an FDDI, active message achieves full link bandwidth for bulk data transfers, and half of the peak bandwidth on 175-B messages. The start-up time is as little as 25 μs, an order of magnitude better than the TCP performance. The good performance of active message is attributed to the following reasons:

- *User-level*: Active message communications are often realized entirely at user space to eliminate the overhead associated with context switching and crossing protection boundaries. There is no need for issuing any system call. A user application can directly access the network interface hardware. The message handler is an ordinary user-level subroutine.

- *Simplicity*: Generic active messages have only five primitives, each having very simple functionality and protocol. For instance, they do not do any buffer management or error checking, which are the major sources of overhead in the protocols such as TCP.

- *Overlapping computation and communication*: Messaging software (such as PVM and MPI) supports overlapping through nonblocking send/receive operations, which requires message buffering at both the send and the receive sides. Active message deals with short and long messages differently. A short data block of 4 or fewer words can be sent by a blocking am_request. Bulk data transfers use am_store or am_get. Generic active message provides non-blocking version of am_store to facilitate overlapping of communication with computation.

GAM-2 Version 2 of Generic Active Message specification has recently been published [426], with much added functionality. GAM is meant to provide for low-overhead communications in parallel computing programs. GAM-2 generalizes active message to the support of not only parallel computing, but also networked and distributed computing. The enhanced features include the following:

- GAM supports only SPMD parallel computing programs. GAM-2 additionally supports MPMD, networked, distributed, and client-server applications.

- GAM supports only single-threaded processes. GAM-2 allows multithreaded processes.

- GAM supports only fail-safe semantics: When a process aborts, the entire parallel program will abort or hang. GAM-2 enables fault-tolerant and high-availability solutions.

10.5.2 GLUnix for Global Resource Management

Clusters need new operating system functionality to provide availability and single system image, which are absent in traditional workstation OS. Two questions must be answered: Exactly what functionality should be added? How should the needed functionality be added? We address the second question first.

Extending Operating Systems Vahdat et al. [622] noted that although there have been many operating system innovations, few have been incorporated into commercial operating systems. They pointed out that an operating system is purchased not for its functionality, but rather for the supported applications, cost/performance of the hardware/ OS combination, and system robustness. They evaluated current approaches and concluded that all have limitations. We discuss two approaches below.

Microkernel: Two well-known examples are Mach and Windows NT. A monolithic kernel that provides all services is replaced by a number of modules, and the most essential services are provided by a small module called the *microkernel*. Some services are even provided at user mode. The modularity offers flexibility and portability. For instance, Windows NT enables user-level emulation of different operating systems such as Windows 3, OS/2, and Unix. However, the microkernel approach has a number of disadvantages:

- The initial version of the OS modules (including the kernel) needs to be rewritten, possibly from scratch.
- The required effort for porting among different hardware architectures is still substantial.
- A microkernel system has higher overhead than a monolithic system, because it has more activities for context switching, crossing protection boundaries, and interprocess communication. The user-level emulation further increases overhead.

User-Level Daemon and Library: The new functions can be realized as user-level servers and libraries on top of a commercial operating system, as is done in Condor, PVM, and LSF. Here "on top" means that the user-level servers and libraries call the underlying operating system.

This approach has the advantage of not requiring kernel modification and being easy to implement. However, it is doubtful whether this approach can do general resource allocation. It also incurs the overhead for context switching and interprocess communication found in the microkernel approach.

The GLUnix Approach GLUnix stands for *Global Layer Unix*. Its main idea is that the operating system (or a virtual operating system) of a cluster should consist of two layers:

- The lower layer is the commercial operating system of the nodes, which executes in kernel mode (kernel level).
- The upper layer is a *user-level* operating system which provides the new features needed by the cluster.

In particular, this global layer provides a single system image of the nodes in a cluster, so that all of the processor, memory, network capacity, and disk bandwidth can be allocated for sequential and parallel applications. The global layer is realized as a protected, user-level, operating system library that is dynamically linked to every application.

The library intercepts all system calls, and realizes them as procedure calls in the application's address space whenever possible. User-level protection is enabled by the *software fault isolation* (SFI) scheme [630], which can be implemented in a language-independent fashion by inserting a check before every store and branch instruction in the object code.

This approach enables many cluster features to be implemented at the user level. The GLUnix approach has several advantages:

- *Ease of Implementation:* GLUnix is realized completely in user level and does not need kernel modification. The first GLUnix prototype was implemented in only 3 months!
- *Portability*: It relies on a minimal set of standard features from the underlying system, which are present in most commercial operating systems. GLUnix is portable to any operating system that supports interprocess communication, process signaling, and access to load information.
- *Efficiency*: The new features needed for clusters are invoked by procedure calls within the application's address space. There is no need to cross hardware protection boundaries, no need for kernel trap or context switching. The overhead for making system calls is eliminated in GLUnix. Coordinating of the multiple copies of GLUnix on multiple nodes can be realized by using shared-memory segments or interprocess communication primitives.
- *Robustness*: Since GLUnix is at user level, it can be tested thoroughly using conventional debugging tools. Errors can be detected, diagnosed, and eliminated.

GLUnix Features A prototype of GLUnix has been implemented that provides most of the following features, deemed important for enterprise clusters:

- Coscheduling of parallel programs
- Idle resource detection, process migration, and load balancing
- Fast user-level communication
- Remote paging
- Availability support

The GLUnix prototype provides a set of utility tools that are similar to those found in LSF. The GLUnix idea looks very appealing. Only time will tell whether the user-level idea is sufficiently flexible to provide all functionality needed by clusters, and whether the SFI overhead can be compensated by faster context switching.

10.5.3 The xFS Serverless Network File System

A serverless file system distributes functions of a file server among all the nodes of a cluster. The traditional *central file server* paradigm is compared with the *serverless file paradigm* in Fig. 10.21. A traditional centralized file server [Fig. 10.21(a)] performs the following major functions:

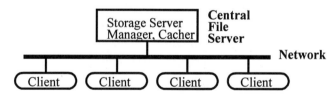

(a) A network file system with a central file server

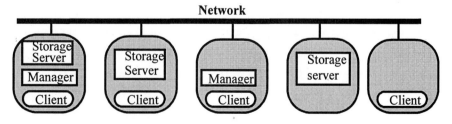

(b) The serverless network file system, xFS, in NOW

Figure 10.21 Comparison of two types of file system paradigms.

- *Central Storage*: The data files and the metadata are stored in one or more stable disks attached to the file sever. The *metadata* of a file consist of the set of *file attributes*, such as file type (a regular file, a directory, a FIFO, a link, a device, etc.), file size, device ID (where the file is located), inode number, file owner ID, file access permissions, etc.

- *Central Caching*: To improve performance, each client may cache some file blocks locally. In addition, the file server caches frequently used file blocks in its main memory to satisfy some client misses.

- *Central Management*: The server performs all file system management functions, including management for metadata and cache consistency In the xFS [Fig. 10.21(b)], all server and client functions are performed by all nodes in a distributed manner. For instance, a node may act as a storage server, a client, a manager, all at the same time (shown by the leftmost node). It may also act as a subset of these, as shown by the remaining four nodes.

The only restriction in xFS is that a node acting as a manager must also be a client, as the manager uses the client interface. There are still storage servers in xFS, but they are distributed among the cluster nodes, not centralized in a single file server.

Redundant Array of Inexpensive Disks (RAID)   The serverless file scheme in Fig. 10.21(b) can be used to create a software RAID that offers high performance and high availability without using expensive hardware RAID. Currently, xFS uses single-parity disk striping. Data blocks of a file are striped among multiple storage server nodes, plus a parity block on another node. When a node fails, the contents of the failed disk can be reconstructed by taking the exclusive-OR of the remaining disks and the parity disk.

A disadvantage of RAID is the *small-write problem*: If a write modifies only a portion of a stripe instead of the whole stripe, the system must first read the old parity and some old data to compute the new parity. This could incur a large overhead if many small writes are performed.

The xFS attacks this problem by using *log-based striping*. Each client first coalesces the *writes* into a private per-client *log*, which is a memory buffer logging all the writes. The log is then committed to disks using *log segments*. Each segment is comprised of $k - 1$ *log fragments* that, together with a parity fragment, are sent to k storage servers.

Example 10.6 Log-based striping in the xFS using software RAID

A cluster has four storage servers and employs xFS. Suppose the client coalesces the writes into a log that is committed to the disks as two log segments, as shown in Fig. 10.22.

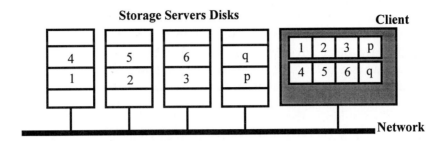

Figure 10.22 Log-based striping using the xFS over software RAID.
(Courtesy T.E. Anderson et al. *ACM Trans. Computer Systems*
[34], 1996)

The first segment consists of three fragments 1, 2, 3, and the second 4, 5, 6. The parity fragment p (q) is computed as the exclusive-OR of fragments 1, 2, and 3 (4, 5, and 6). These two segments are stored in the four storage servers in a striped fashion. ∎

The above log-based striping has two problems for large multiserver clusters. First, a client needs to send many small fragments to all the storage servers. To write large fragments, the log memory needs to be very large. Second, if several clients need to write their logs, they must compete for the servers.

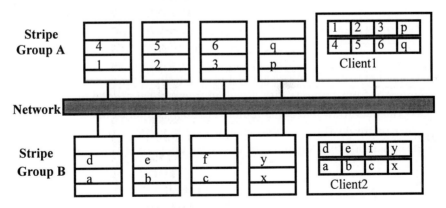

Figure 10.23 Stripe groups in using the xFS over distributed servers.
(Courtesy T.E. Anderson et al. *ACM Trans. Computer Systems*
[34], 1996)

These problems are solved in xFS by dividing the stotage servers into subsets called *stripe groups*. In Fig. 10.23, the eight servers are divided into two groups, each having 4 servers. The fragment size is 2.33 times as large as that of using a single, 8-server group. Client1 and Client2 can simultaneously write their logs to the two groups without conflict. Stripe groups have other advantages, such as offering higher availability and lower *cleaning* overhead (see below). Its disadvantage is that there is now one parity disk per group, instead of one parity disk per cluster.

Cooperative File Caching The idea of cooperative file caching is simple [182]: Each client node of a cluster allocates a portion of its memory as a file cache. A cooperative caching algorithm utilizes all these memories to create a large, clusterwide file cache. When a client encounters a local file cache miss, it does not go to the disk as is done in a conventional workstation, but goes to another client's memory to get the data. The use of remote memory is illustrated in Table 10.11.

Table 10.11 Time to Access 8-KB Block from Remote Memory or Disk
(Courtesy T.E. Anderson et al. *ACM Trans. Computer Systems* [34], 1996)

Network Media	10- Mbps Ethernet		155- Mbps ATM	
Storage	Remote Memory	Remote Disk	Remote Memory	Remote Disk
Memory copy	250 μs	250 μs	250 μs	250 μs
Star-tup time	400 μs	400 μs	400 μs	400 μs
Data transfer	6250 μs	6250 μs	400 μs	400 μs
Disk	0	14,800 μs	0	14,800 μs
Total time	6900 μs	21,700 μs	1050 μs	15,850 μs

This table shows the estimated times to serve a local cache miss by reading from a remote memory or from a remote disk. Reading from a local disk takes the same disk time (14,800 μs). The data show that using remote memory is 3 times faster on an Ethernet, but 14 times faster on an ATM. The speed lead of remote memory over disk will continue to grow, because memory copying and network speed are getting faster at much greater rates than the disk speed.

As shown in Fig. 10.24, files are stably stored in the server's disk. Traditionally, files have been cached in a client's memory (most file systems), a client's local disk (e.g., AFS), and the server's memory (most file systems). xFS allows a file to be cached in fellow clients' (remote) memories as well. A number of cooperative caching algorithms have been proposed [182]. We discuss below two cooperative file-caching algorithms: the *greedy forwarding* or *N-channce forwarding*.

Greedy Forwarding This scheme works as follows: A client accessing a file first tries its local cache. If the data block is not there, it forwards the request to the server. The server first searches its local cache with the following two results:

- Upon a hit, it sends the data back to the client. The server also maintains a cache directory which lists the contents of all clients caches at per-block level.
- With a miss, the server consults its cache directory and forwards the request to a client that holds the requested data in its cache. This latter client then forwards the data directly back to the requesting client. If the data is in none of the clients caches, the server gets the data from its disk and transmits to the target client.

A shortcoming of this greedy scheme is that the same block may be duplicated among many caches. When a client reads a block for the first time, the block will be fetched from the server disk and cached in two places: the server cache and the client cache. Another client reading the same block will cache it in its local file cache.

This duplication reduces the effective size of the cooperative cache, thus increases the miss rate. Furthermore, it makes cache consistency management more difficult. However, Greedy Forwarding is simple to implement. The cache directory in the server will take up some memory, about 24 bytes per block. For 8-KB cache blocks, that is only 0.3%. A 6-MB directory is enough for a 2-GB cooperative cache.

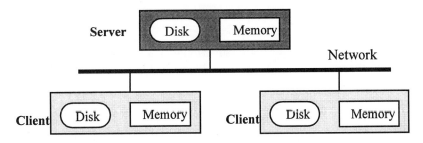

Figure 10.24 Different ways to cache files in a client-server cluster.

N-Chance Forwarding The second file caching algorithm is a generalization of Greedy Forwarding called *N-Chance Forwarding*. It avoids the duplication problem by caching a block in only one client cache. Such a block is called a *singlet*. When a client fetches a block from another client's cache, the block is discarded in the second cache, and a message is sent to the server to tell it that the block has moved. A problem with singlets is that if a singlet is discarded to leave room in the cache for a more recent data block, the singlet will be missing from the entire cooperative cache.

N-Chance Forwarding alleviates this problem as follows. The scheme works exactly as the greedy scheme as long as the client cache is not full. When any client encounters a full local cache and needs to discard a block, the client first checks if the block is a singlet. If not, the client simply discards the block. If the block is a singlet, the client sets the *recirculation count* of the block to N and sends the block to a random client to be cached there.

If the second client has to discard the block later, it decrements the recirculation count and sends the block to another client. This process continues until the recirculation count becomes zero. Then the block is simply discarded. If the block is referenced by a client, the block's recirculation count is reset to N and the recirculation process repeats.

The corresponding cache directory entry must be updated to point to the new client. It has been shown that increasing the recirculation count N from 0 (equivalent to Greedy Forwarding) to 1 improves the performance by 25% percent. Going from $N = 1$ to $N = 2$ only improves it by 5%. Further increasing N yields little improvement.

Distributed Management xFS fully distributes the file system management functions, including metadata and cache consistency management. Most importantly, xFS employs multiple managers that are located in multiple nodes. To see how to achieve distributed management, let us look at a simplified file read operation in xFS, as shown in Fig. 10.25.

When a client tries to read a file block, it issues a request with the file's path name and an in-file offset. Each directory in xFS is itself a file. Using the file's name, the client can find from the file's parent directory an *index number* of the file.

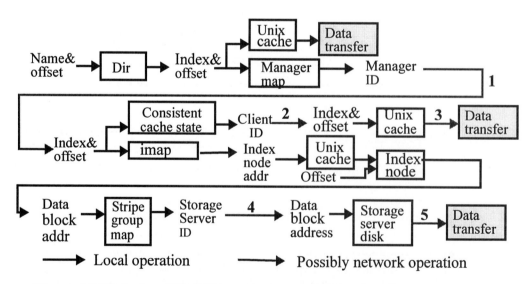

Figure 10.25 A simplified file access operation in the xFS. (Courtesy Anderson et al. *ACM Trans. Computer Systems* [34], 1996)

The index number uniquely identifies a file. The client uses the index number to search in a local cache (the Unix cache) that stores the frequently accessed file blocks. If the data block is in the cache (cache hit), the block is fetched. So, in such ideal cases, all operations are local and carried out without accessing the disk.

Upon a cache miss, the client uses the index number to look into a *manager map*, a data structure indicating which physical node manages which group of index numbers, i.e., where the correct manager is. This map is updated when a file is created. Again, looking up the manager map is a local operation, because the map is replicated in every manager and client. The index and offset information is then sent to the correct manager over the network, if it is not in the same node.

For each file for which it is responsible, the manager keeps track of information such as whether a block is cached or in disk, the exact location of the block, and whether a cached block is consistent, etc. If the block is cached in some client's local cache and consistent, the manager forwards the read request to the client, which fetches the data block from its local cache and directly sends it to the original client.

If the data block is not in the cooperative cache, the manager looks into a data structure called *imap* to find the *index node* (or *inode*), which contains the addresses of all data blocks of the file. The desired data block address is used to find the correct storage server, where the data block is fetched and forwarded to the requesting client.

Cache coherency is realized with a per-block, invalidation, and write-back protocol. A client wanting to modify a block must first send a message to the manager of the file to acquire the write ownership for that block. The manager invalidates any other cached copies, updates the cache state to a new owner, and gives the ownership to the client.

Once the client has the write ownership, it can repeatedly write the block multiple times without having to acquire the ownership each time. The client maintains write ownership until another client reads or writes the same data block. Then the manager revokes ownership, flushes any changes to stable storage, and forwards the data block to its new owner.

NFS versus xFS Performance   To improve write performance, xFS uses the *logged file system* (LFS) scheme. LFS requires *cleaning*, a process to rearrange the fragmented disk space to consolidate logs into files. The xFS implements a distributed parallel cleaning scheme so that cleaning will not become the bottleneck.

An xFS prototype has been implemented over a 32-node cluster system, which shows good scalability. The aggregated bandwidth for xFS keeps climbing as more and more clients are added to read and write the file system simultaneously. The xFS bandwidth reaches 13.8 MB/s for read and 13.9 MB/s for write, as shown in Table 10.12.

The NFS, however, peaks at five clients, with only small improvement factors. The single-client performance of xFS is not as good as that of NFS. Anderson et al. [34] believe that this is due to the prototype implementation (e.g., using user-level daemons instead of kernel service), rather than a limitation of the xFS techniques.

Table 10.12 Comparison of NFS and xFS over a 32-Node Cluster
(Courtesy Anderson et al. *ACM Trans. on Computer Systems* [34], 1996)

File Operation	NFS			xFS		
	1 Client	5 Clients	Improve	1 Client	32 Clients	Improve
Large Read	1.2 MB/s	2.7 MB/s	2.25	0.9 MB/s	13.8 MB/s	15.33
Large Write	1.2 MB/s	1.4 MB/s	1.17	0.6 MB/s	13.9 MB/s	23.17
Small Write	22 files/s	86 files/s	3.91	40 files/s	1122 files/s	28.05

To measure the performance of writing small files, execution times are measured for each client creating 2048 1-KB files. The xFS shows better performance even for the single-client case. Its scalability is much better than that of NFS, which again peaks at five clients with an aggregated throughput of creating 86 files per second.

10.6   TreadMarks: A Software-Implemented DSM Cluster

We now turn to the last aspect of single system image, *single memory space*, by examining *TreadMarks*, a DSM runtime library developed by Zwaenepoel's group at Rice University. This system provides distributed shared memory over a cluster of workstations. We first elaborate the key issues that must be considered in providing a single memory space on a cluster. Then we discuss how TreadMarks addresses these issues. Software implementation of the TreadMarks ideas is then discussed.

In evaluating how well a cluster supports single memory space, the following issues must be considered: What are the boundary conditions? What is the user interface? How can one keep the memory consistent? What performance features and techniques are provided? These questions are answered below.

10.6.1 Boundary Conditions

Boundary conditions are restrictions assumed by the DSM software. The following is a partial checklist helpful for understanding the restrictions:

- Does the operating system kernel need to be modified? A less strict question is: Does the DSM implementation need to know the OS source code?
- Does the DSM need special support from the cluster communication hardware?
- Does the DSM need special support from the memory management hardware?

If the answer to any of above is yes, implementation may be difficult and portability will become a real problem. For instance, if the operating system must be modified, not many people are in the position to make production systems. In commercial operating systems, source codes are expensive to license. A free Unix such as Linux lacks supported applications, and it is not trusted yet in the corporate world. Even if a free Unix can be used, what if we need to port the system to a Windows NT cluster?

Fortunately, TreadMarks answers no to all three questions. TreadMarks runs at the user level on a cluster of Unix workstations. It does not need kernel modification or special privileges. It uses standard Unix interface, compilers, and linkers. As a consequence, TreadMarks is quite portable, and has been ported to a number of Unix platforms from IBM, DEC, SGI, HP, and Sun Microsystems.

TreadMarks is implemented as a runtime library. All communications needed by DSM (e.g., page duplication, memory consistency operations) are implemented by message passing using standard UDP/IP through the socket interface. It can take advantage of high-performance hardware provided by a specific platform (e.g., the Memory Channel in DEC TruCluster, see Kontothanassis et al. [373]), but does not require special hardware support. It can run on a Ethernet cluster of Unix workstations.

10.6.2 User Interface for DSM

A cluster with a single memory space should look to the users like an SMP, only the central shared memory in the SMP is replaced by multiple node memories. In other words, the single memory space in a cluster should support the following:

(1) Existing binary code, probably with dynamic linking to a DSM library. In particular, a sequential binary program should be able to use the memories in all cluster nodes. This is a problem not yet effectively solved in existing systems.

(2) Unmodified source code, which is recompiled and relinked to a DSM library.

(3) The source code must be modified by inserting specific shared-memory constructs (library routines, compiler directives, etc.). Most software DSMs (TreadMarks included) follow this approach.

Since the third approach requires one to rewrite the source code, the DSM software must provide a simple API. The TreadMarks API is indeed very simple, containing only seven library functions. A TreadMarks program for Jacobi relaxation is shown Fig. 10.26, where the TreadMarks constructs are shown in boldface at their first appearance.

When a TreadMarks program starts execution, only process 0 is running. The Tmk_startup library function initializes TreadMarks and starts remote processes. The Tmk_nprocs variable is the number of processes in the parallel program. The Tmk_proc_id variable contains the process ID, ranging from 0 to Tmk_nprocs-1.

```
#define    M        1024
#define    N        1024
float      **grid;          /* shared array */
float      scratch[M][N]; /* private array */

main ( )
{
  Tmk_startup( );
  if ( Tmk_proc_id == 0 ) {
      grid = Tmk_malloc( M*N*sizeof(float));
      initialize grid ;
  }
  Tmk_barrier(0);
  length = M / Tmk_nrpocs;
  begin = length * Tmk_proc_id;
  end = length * (Tmk_proc_id+1);

  for ( number of iterations ) {
      for ( i = begin; i < end; i++ )
          for ( j = 0; j < N; j++ )
              scratch[i][j] = (grid[i-1][j] + grid[i+1][j] +
                              grid[i][j-1] + grid[i][j+1])/4;
      Tmk_barrier(1);
      for ( i = begin; i < end; i++ )
          for ( j = 0; j < N; j++ )
              grid[i][j] = scratch[i][j] ;
      Tmk_barrier(1) ;
  }
}
```

Figure 10.26 Jacobi relaxation in TreadMarks C code.
(Courtesy Amza et al. *IEEE Computer Magazine*
[30], 1996)

Shared memory space is allocated by the Tmk_malloc function. The shared space is freed by the function Tmk_free. Memory allocated statically or by a call to malloc is private. TreadMarks provides three synchronization functions. Tmk_barrier(k) is for *barrier synchronization,* where k is just a label and can be ignored here. The two functions Tmk_lock_acquire and Tmk_lock_release *acquire* and *release* a lock, respectively.

10.6.3 Implementation Issues

TreadMarks treats the shared memory as a linear array of bytes. It uses page-level data sharing to take advantage of the existing virtual memory hardware on workstation nodes. The local memory of a node is like a cache of the globally shared memory. When a node encounters a read or write miss (page fault) of a shared variable, the page containing this variable is duplicated in the local memory by TreadMarks. Replication of read-only pages does not cause any problem. However, when a page is modified by a node, the TreadMarks runtime system must invalidate the replicated pages in other nodes.

Two problems can severely degrade DSM efficiency. The first is *communication overhead* caused by enforcing consistency. The second problem is *false sharing,* when two unrelated variables (say x and y) happen to locate in the same page and each is written by a different process. The two processes do not share x or y, but they do share the page.

TreadMarks uses three techniques to alleviate these problems to make DSM efficient: *release consistency* with replication and invalidation, *lazy processing* of consistency events, and *multiwriter protocol*. The main objective of all these techniques is to reduce the number of messages that need to be communicated.

Release Consistency Release consistency is a form of weak consistency. The idea of release consistency is based on the following observation: Normal parallel programs should be determinate, i.e., they should not have data races. If a variable must be shared by multiple processes which write to the variable, synchronization operations must be used to ensure that no conflicting accesses are possible.

An access to private or read-only variable does not have to invalidate other copies of the page. A synchronization must be present between two conflicting accesses of a shared variable (recall that two accesses are conflicting if they access the same shared variable and at least one of them is a write).

Therefore, a process does not need to be informed of modification to a shared variable until the synchronization operation completes. This is the essence of release consistency: Memory consistency operations are performed after a lock is released, not after a shared variable is modified.

Lazy Processing Release consistency reduces communication needed in a sequential consistency model, but it may still perform unnecessary communications. For instance, let us consider the last three lines of code in Fig. 10.27. Suppose process 0 is the first one to finish execution of the Tmk_lock_release. Since SUM is replicated in all processes, a message must be sent to all processes to notify them that SUM is changed.

Example 10.7 Three types of variables in a shared-memory program

The code in Fig. 10.27 illustrates a TreadMarks program that computes the sum of array A. There are three types of variables: *private variables* (e.g., LocalSum) that are local to each process, read-only shared variables (e.g., array A), and read/write shared variables (e.g., Sum). The last must be protected by locks to avoid data races.

```
#define   M        1024
float     *A, *Sum;      /* shared variables */
float     LocalSum = 0.0 ; /* private variable */
int       lock_id = 1;     /* lock variable */

main ( )
{
   Tmk_startup( );
   if ( Tmk_proc_id == 0 ) {
       A = Tmk_malloc( (M+1)*sizeof(float));
       Sum = Tmk_malloc(sizeof(float));
       initialize array A; *Sum = 0.0; /* initialize
                                    shared variables */
   }
   Tmk_barrier(0);
   length = M / Tmk_nrpocs;
   begin = length * Tmk_proc_id;
   end = length * (Tmk_proc_id+1);

   for ( i = begin; i < end; i++ )
       LocalSum += A[i];

   Tmk_lock_acquire(lock_id); /* acquire lock */
       *Sum += LocalSum; /* critical section */
   Tmk_lock_release(lock_id); /* release lock */
}
```

Figure 10.27 A TreadMarks program with locking.

However, since this code fragment is a critical region, only one process (say, process 1) will be able to execute it after process 0. Therefore, only process 1 needs to be notified of the modification of SUM made by process 0. Sending a message to all processes is unnecessary and increases communication traffic.

TreadMarks uses the *lazy* release consistency protocol to solve this problem. The notification is sent not just after release, but when process 1 acquires the lock. TreadMarks uses this information to sends notification to process 1 only. This notification information can be piggybacked with the lock grant message, thus completely eliminating message traffic for notification.

Thus, when process 1 acquires the lock, TreadMarks notifies it that shared variable Sum is modified in the previous execution of the critical region, and process 1's copy of the page containing Sum is invalidated. Later on, when process 1 accesses Sum, it will encounter a page fault, which causes TreadMarks to bring in the updated page.

Multiple Writer Lazy release consistency can significantly reduce message traffic. However, it does not solve the false sharing problem, which is attacked in TreadMarks with a multiwriter technique. Simply put, a multiwriter protocol allows multiple processes to have a writable copy of the same page at the same time, so that they can write to their copies of the same pages simultaneously. Of course, there must be a way to reconcile these multiple writes. The TreadMarks multiwriter protocol is described below.

Suppose there are two processes S1 and S2, and they need to write to different locations of the same shared page P simultaneously. We will focus on the steps taken by S1, since those by S2 are similar. Initially, the shared page P is write-protected. When S1 writes to P, TreadMarks creates a copy P' of P (called the *twin* of P), saves P' on S1, and lifts the write protection on P.

Then S1 can write to P many times as ordinary local memory, until a synchronization (e.g., a *barrier*) occurs. At the barrier, the modified P is compared (word by word) to the original value saved in its twin P' to create a run-length encoding of the difference, called *diff*. Then the twin P' is discarded and P is invalidated. When S1 later accesses P, a page fault occurs. The *diff* made by S2 is sent to S1, where the *diff* is applied to page P.

10.7 Bibliographic Notes and Problems

Pfister [497], [498], Anderson et al. [33], Xu and Hwang [657] have all argued the case for clusters among other advocates. The POWER CHALLENGEarray is described in SGI [548]. The SPARCCenter is described in Cekleov et al. [134] and the SPARCcluster in [599]. The Sun Network FileSystem is given in Sandberg et al. [527]. Compaq Recovery Server Option Kit is described in [164]. Smith describes the Microsoft Wolfpack in [446]. The Tandem Himalaya is described in [605]. References for additional commercial cluster systems are cited in Table 10.1.

The NOW project is described in [469] and Anderson et al. [33], the serverless xFS system in [34], the GLUnix operating system in [36], workload evaluation in Arpaci et al. [44]. The SHRIMP project is described in [237]. The Wind Tunnel is described in [516]. The Syracuse WWVM project is described in [207]. The HKU Pearl is partially described in [328]. Additional references on cluster research projects are cited in Table 10.2.

How to support coherent shared memory on clusters is discussed in Reihardt et al. [515]. The concept of metacomputing was first discussed in Catlett and Smarr [133]. The Stardust is described in Cabillic and Puaut [119] for cluster programming. Wide-area computing issues, techniques, and systems are discussed by Foster et al. [248, 249], Grimshaw et al. [282, 410], Dincer and Fox [207], and Fox et al. [251, 252].

The IBM SP2 architecture is described in a special issue of *IBM System Journal* (Vol. 34, No. 2, 1995). POWER2 processor is assessed in Agarwal, et al [13] and the SP2 architecture in Agerwala, et al [14]. The Cornell SP2 configuration is specified in [168]. The MHPCC SP2 is described in [445]. The LoadLeveler used in SP2 is described in [336]. The communication software in SP2 is given in Snir et al. [574]. The SP2 high-performance switch is described in Stunkel et al. [592].

Digital's AlphaServer 8400 is described in Fenwick et al. [239]. An overview of Digital Unix cluster systems is given in Cardoza et al. [126]. The VAXcluster is described in Shah [551]. The TruCluster is described in Cardoza et al. [127] and in Lawton et al. [397]. The Advantage Unix cluster is described in [194] and NT cluster in [195].

TPC-C performance of the TruCluster is given in Piantenosi et al. [501]. The best place to look for information about cluster computer systems is the respective Web sites of manufacturers and research centers listed in Appendix A. TreadMarks is described in Amza et al. [30] and Kontothanassis et al. [373], where it is also compared with other software DSM systems.

Homework Problems

Problem 10.1 Describe three unique features of Wolfpack, namely, features that are not found in other clusters discussed in this chapter.

Problem 10.2 Explain the architectural and functional differences among three availability cluster configurations: the *hot standby, active takeover,* and *fault-tolerant clusters.* Give two example commercial cluster systems in each availability cluster configuration. Comment on their relative strength and weakness in commercial applications.

Problem 10.3 Refer to Section 10.3 and answer the following questions on the IBM SP cluster/MPP system:

(a)   List five decisions that the SP design team made to reduce the time to market.

(b)   List three features of the SP which make it a general-purpose system.

(c)   Explain how the SP supports the four SSI features: single entry point, single file hierarchy, single control point, and single job management system.

(d)   Explain five architectural features of SP2 that help enhance performance.

(e)   Briefly explain the main techniques used in the SP2 communication subsystem to increase bandwidth.

Problem 10.4 Refer to Section 10.4 and answer the following questions on the Digital TruCluster system:

(a)   Is the DEC TruCluster a shared nothing or shared disk system?

(b)   Of the following aspects of single system image, which does the TruCluster support and explain how: single entry point, single file hierarchy, single control point, single memory space, single process space?

(c)   Explain four architectural features of the TruCluster that are used to enhance scalable performance.

(d)   Briefly explain three techniques used in the TruCluster communication subsystem to reduce communication overhead.

(e)   Why does the TruCluster employs the Universal Message Passing layer?

Problem 10.5 Refer to Section 10.5 and answer the following questions on the Berkeley NOW project:

(a)   Of the following aspects of single system image, which does the Berkeley NOW support and explain how: single entry point, single file hierarchy, single control point, single memory space, single process space?

(b)   Explain four architectural features of the Berkeley NOW that are used to enhance performance.

(c)   Explain four architectural differences between Berkeley NOW and the SP cluster, and discuss the respective advantages.

Problem 10.6 Consider the xFS developed in the Berkeley NOW.

(a)   Briefly explain what is cooperative file caching, and what its advantages are.

(b)   Under what conditions is the xFS effective for cooperative file caching?

(c)   Under what conditions that the xFS cannot yield a good performance.

Problem 10.7 Consider the Active Messages protocol used in the Berkeley NOW.

(a)   List three differences between Active Messages and the TruCluster communication protocol, and explain their respective advantages.

(b)   In Fig. 10.20, why does Process P need to execute the am_poll function?

Problem 10.8 Consider the xFS and answer the following questions.

(a)   Explain two differences between xFS and a centralized file server, and discuss the respective advantages.

(b)   Explain the main technique used by xFS to enhance availability.

(c)   Explain the main technique used by xFS to alleviate the small-write problem.

Problem 10.9 Assume two processes are used to execute the Jacobi relaxation program in Fig. 10.26, where a float is 4 bytes, and the page size is 4 KB.

(a)   Explain why the three barriers are there. What will happen if each of them is deleted?

(b)   Derive the necessary and sufficient condition under which there will be false sharing.

(c)   Show how the multi-writer protocol can be used to reduce false sharing effects.

Problem 10.10 Assume two processes P0 and P1 are used to execute the TreadMarks program in Fig. 10.27. Explain in detail how the critical region code in the last three lines are executed, employing the lazy release consistency protocol.

Chapter 11

MPP Architecture
and Performance

After addressing basic MPP issues, this chapter presents current MPP systems representing two different approaches toward massive parallelism. These systems are the Cray
T3E representing the NCC-NUMA approach and the Intel/Sandia ASCI Option Red
adopting the NORMA architecture. We present scalable design strategies applied in three
ASCI/MPP platforms contracted with the US Department of Energy. This coverage
outlines the next generation of MPPs to be used around the turn of the century.

Besides system architectures, we study MPP benchmark results reported by several
independent research groups in recent years. The NAS results are based on MPP
benchmark experiments done at the NASA Ames Research Center and at the National
Research Center for Intelligent Computing Systems in China. The MPI and STAP results
are based on the MIT/STAP benchmark experiments performed jointly at the University of
Southern California and at the University of Hong Kong.

11.1 An Overview of MPP Technology

The term *massively parallel processors* (MPP) was used vaguely in the past, because
its meaning varied over time. Based on today's technology, it refers to a large-scale
computer system consisting of hundreds or thousands of processors. By 1997, the largest
MPP configuration ever built has 9216 processors in the Intel/Sandia Option Red.

11.1.1 MPP Characteristics and Issues

A common architecture for current MPP systems is shown in Fig. 11.1. All MPPs use
physically distributed memory, and more and more MPPs are using distributed I/O. Each

node has one or more processors and caches (P/C), a local memory, and zero or more disks. There is a local interconnect within a node that connects processors, memories, and I/O devices. In early MPPs, this local interconnect is usually a bus, while recent MPPs use higher bandwidth crossbar switching networks. Each node is connected to the network through a *network interface circuitry* (NIC).

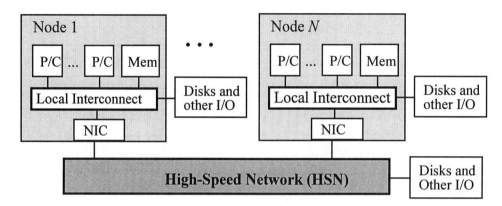

Figure 11.1   A common architecture for massively parallel processors

Scalability A distinguishing feature of MPPs is that they are designed to be scalable to thousands of processors, with proportional increase in memory and I/O capacity and bandwidth. MPPs have adopted the following techniques to increase their scalability:

- Using a physically distributed memory architecture, which offers higher aggregated memory bandwidth than a central memory architecture, thus potentially higher scalability.

- Balancing processing with memory and I/O capabilities. Without proportionally fast memory and I/O subsystems, fast processors are of little value, since data cannot be moved to the processors at sufficient speed.

- Balancing computing capability with parallelism and interaction capabilities. Otherwise, overhead due to process/thread management, communication and synchronization may dominate the execution time.

System Cost Because many node and interconnect components are needed in an MPP, it is imperative to keep the per-component cost low. A number of techniques are used to lower cost:

- Using existing, commodity CMOS microprocessors. These processors are originally developed for workstations and servers. Their commodity nature implies low price and attracts huge investment, which doubles the performance every 18 to 24 months (Moore's law).

- Using a stable architecture that is generation scalable. One such technique is the shell architecture discussed in Section 1.1.3.
- Using a physically distributed memory architecture, which is less costly than a central memory architecture of the same machine size.
- Using SMP nodes, which cuts down the size of the interconnect.

Example 11.1 A lesson learned from the custom chip design in TMC C-2

A practice in the past for building MPPs was to interconnect hundreds to thousands of custom-designed, "simple" nodes. An example is the Thinking Machine CM-2, where each node is essentially an augmented ALU plus registers and a small local memory.

Because of the simplicity, multiple nodes can be integrated on a single chip, thus the per-node cost is low. However, individual nodes gave very low performance. Dozens of them are needed to match the performance of a single commodity micro-processor. Being custom-designed, they do not benefit from Moore's law.

■

Example 11.2 The shell approach to providing generation scalability

As an example of stable architecture, all the MPPs discussed in this chapter more or less follow the shell approach (Fig. 1.2). As the microprocessor scales to the next generation, the rest of the system does not have to be changed. In addition, the MPPs all have constant node degree. For instance, the degree is 6 for Cray T3D/T3E (3-D torus), 4 for the Intel Paragon (2-D mesh), and 1 for the IBM SP2 (multistage Omega network).

■

Current commodity microprocessors are designed for small systems such as PCs, workstations, and SMP servers, not for MPPs. While using microprocessors brings many scalability and cost-effectiveness benefits, it creates several problems for MPPs, especially those based on a DSM architecture. The following is a partial list of problems that an MPP designer must address:

- Microprocessors may not have a sufficiently large physical address space. For instance, the Alpha 21064 microprocessor is used in the Cray T3D MPP. This processor provides only 33-b (8-GB) physical address space, while the T3D has a maximum physical memory of 128 GB. The Cray T3D designers had to add a special hardware called DTB Annex to extend the physical address space size.

- Microprocessor may not have a sufficiently large TLB, and a TLB miss is far more costly than a cache miss. For applications with large data sets and irregular memory access patterns, TLB misses can significantly degrade performance.

- Microprocessors access memory one cache line at a time. This makes single-word stride accesses inefficient. Current microprocessors support non-blocking cache only in a small scale, allowing one or two outstanding memory references. This limits the latency tolerance capability needed in an MPP.

- Compared to their computing capability, microprocessors have inadequate operating system support. Exception handling and crossing protection boundaries are expensive. This makes it difficult to efficiently support process management, communication, and synchronization. This issue has been addressed in Chapter 7.

Generality and Usability   Current-generation MPPs have learned from past experience: For an MPP to succeed, it must be a general-purpose system that supports different applications (e.g., technical and commercial), different algorithmic paradigms, and different operation modes. It should not just support small niche applications, while forcing the use of an eccentric environment on a peculiar architecture. More specifically, current MPPs provide the following features:

- MPPs support asynchronous MIMD modes. SIMD has been faded away in general-purpose MPPs.

- MPPs support prevalent, standard programming models such as message passing (PVM and MPI) and data parallel (HPF).

- The nodes are configured into several *pools* (or *partitions*), to support small and large jobs in interactive and batch modes.

- The interconnect topology is hidden from the users, who see a set of fully connected nodes.

- MPPs support single-system image at different levels. Tightly coupled MPPs often use a distributed operating system, providing single system image at hardware and OS levels.

- It has been estimated that at least one processor of a 1000-processor MPP will fail every day [320]. MPPs must utilize high-availability techniques.

Communication Demand   A key difference between MPPs and current clusters of workstations (COWs) is in internode communication. In a COW, the nodes are often connected through a standard LAN. In an MPP, the nodes are interconnected by a high-speed, proprietary network with higher bandwidth and low latency. Furthermore, proprietary communication software is provided to realize the high performance.

All this makes the current MPPs superior in communication performance to COWs. However, it is foreseeable that in the next decade, there will be rapid advances in standard network technologies. It is not sure how much longer the edge of having better communication support in MPPs can hold over that applied in COWs in the future.

Memory and I/O Capability Because they are scalable to large sizes, MPPs can provide very large aggregated memory and disk capacities not found in other architectures. In addition, commercial MPPs pay special attention to a high-speed I/O system. In many systems, not only the memory, but also the I/O subsystem is physically distributed. Still, the advances in I/O have lagged behind the rest of the system, and how to provide a scalable I/O subsystem is an active research area. The Bibliographic Notes and the Web Resource contain many pointers to high-performance I/O research projects.

11.1.2 MPP Systems – An Overview

Table 11.1 lists the architectural features of three current MPPs, representing three different approaches to construct large systems. The IBM SP2 offers a clustered approach to building MPPs, as discussed in Section 10.3.

The Intel ASCI system follows a more traditional MPP approach, with small nodes, tightly coupled network interconnection, and microkernel operating system on compute nodes. It is a successor of the Intel Paragon MPP system. Both SP2 and Intel ASCI are message-passing multicomputers with the NORMA model. Internode communication relies on explicit message passing in these NORMA machines.

Table 11.1 Comparison of Three Massively Parallel Processors

MPP Models	Intel/Sandia ASCI Option Red	IBM SP2	SGI/Cray Origin2000
A Large sample configuration	9072 processors, 1.8 Tflop/s at SNL	400 processors,100 Gflop/s at MHPCC	128 processors, 51 Gflop/s at NCSA
Available date	December 1996	September 1994	October 1996
Processor type	200 MHz, 200 Mflop/s Pentium Pro	67 MHz, 267 Mflop/s POWER2	200 MHz, 400 Mflop/s MIPS R10000
Node architecture and data storage	2 processors, 32 to 256 MB of memory, shared disk	1 processor, 64 MB to 2 GB local memory, 1-4.5GB Local disk	2 processors, 64 MB to 256 GB of DSM and shared disk
Interconnect and memory model	Split 2D mesh, NORMA	Multistage network, NORMA	Fat hypercube, CC-NUMA
Node operating system	Light-weighted kernel (LWK)	Complete AIX (IBM Unix)	Microkernel Cellular IRIX
Native programming mechanism	MPI based on PUMA Portals	MPI and PVM	Power C Power Fortran
Other programming models	Nx, PVM, HPF	HPF, Linda	MPI, PVM

The SGI/Cray Origin 2000 represents a different approach to constructing MPPs, characterized by a globally accessible, physically distributed memory system with hardware support for cache coherency. Another MPP that adopts a similar CC-NUMA architecture is the HP/Convex Exemplar X-Class. Cray's T3E system is also a DSM machine, but without hardware support for cache coherency. It is therefore an NCC-NUMA machine. The native programming environment of such DSM machines provides a shared-variable model. At the application programming level, all MPPs now support standard languages and libraries, such as C, Fortran, HPF, PVM, and MPI.

The Past MPPs   In the past, MPPs were primarily used for scientific supercomputing. Notable systems include the Thinking Machine's CM2/CM5, NASA/Goodyear's MPP, nCUBE, Cray T3D/T3E, Intel Paragon, MasPar MP1, Fujitsu VPP500, and KSR 1, etc. Some of these MPPs have vector hardware or explore only SIMD fine-grain data parallelism. Some of these earlier MPPs are treated in Hwang's *Advanced Computer Architecture* [327], published in 1993.

Today, most of these machines are gone, and many thought MPPs were dead with the down sizing operations of Thinking Machines Corporation, Cray Research Inc., Intel Scalable Systems Division, among many other supercomputer companies. The reality is that parallel processing is reborn with an increasing demand from industrial, commercial, and business applications in recent years.

Commercial MPP Applications   Most technical and research literatures on MPPs focus on technical computing. Many papers discuss how parallel computing on MPPs can solve grand challenge problems. This could give the wrong impression that MPPs are only good for extremely large, parallel scientific computing applications.

In reality, many MPPs have been successfully used in commercial and network applications. For instance, of the 3000 SP2 systems sold by 1997, about half of which are used in commercial applications. Of the remaining half, a large portion is used for LAN consolidation. Only a small percentage are used for scientific supercomputing.

One particularly hot area of commercial MPP applications is data warehousing, decision support systems, and digital libraries. Scalability, availability, and manageability are more important in the high-performance commercial application market. After studying the Cray T3E /MPP in Section 11.2, we shall assess a whole new generation of MPPs in Section 11.3 and some details of the ASCI Option Red system in Section 11.5.

11.2 The Cray T3E System

The Cray T3E shipped in 1995 is a successor of the Cray T3D system produced in 1993. It uses faster components and has mede several architectural changes to improve the performance over the T3D. This is resulted from a sequence of benchmark and applicational evaluations of the T3D architecture [40,361,540].

11.2.1 The System Architecture of T3E

As shown in Fig. 11.2, the Cray T3E is a distributed shared-memory multiprocessor. The system is comprised of a number of *processing elements* (PEs) interconnected by a 3D bidirectional torus network for fast communication, and by a number of GigaRing channels which provide connectivity to I/O devices.

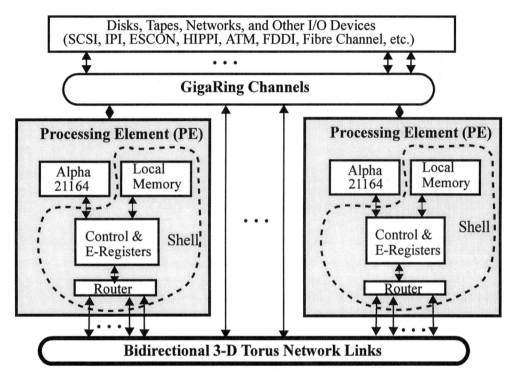

Figure 11.2   The Cray T3E architecture diagram. (Courtesy of Scott and Thorson, *Proc. of Hot Interconnects IV* [541], 1996, see also http://www.cray.com/products/systems/crayt3e/ [175], 1997)

The architecture attributes of the T3E are listed in Table 11.2. The T3E-900 is a T3E enhancement announced in late 1996. The main differences are that the T3E-900 uses faster processor clock (450 MHz versus 300 MHz in T3E) and offers lower price (price starts at less than $500K versus $1M for T3E).

Processing Element   Each PE in the T3E consists of a DEC Alpha 21164 microprocessor [196] surrounded by a *shell* circuitry, which includes a local memory, a control chip, and a router chip. The system (shell) logic runs at 75 MHz. The 21164 processor is clocked at 300 MHz with a peak speed of 600 Mflop/s.

Table 11.2   Performance Attributes of Cray T3E and T3E-900

Attribute	T3E	T3E-900
Processor clock frequency (MHz)	300	450
Peak processor speed (Mflop/s)	600	900
Number of processors	6-2048	6-2048
Peak system speed (Gflop/s)	3.6-1228	5.4-1843
Physical memory capacity (GB)	1-4096	1-4096
Aggregated peak memory bandwidth (GB/s)	7.2-2450	7.2-2450
Maximal number of I/O channels	1-128	1-128
Aggregated peak I/O bandwidth (GB/s)	1-128	1-128
Peak 3-D torus link bandwidth (MB/s)	600	600

The local memory provides 64-MB to 2-GB capacity and a 1.2 GB/s peak bandwidth (The STREAM copy benchmark sustained 470 MB/s). The router chip has 7 bidirectional ports, one for connection to the PE and the remaining six for the six links ended in the 3D torus network.

The custom-made control chip realizes distributed shared memory, which consists of all the local memories in the PEs. Every processor can access the memory in any PE, and every PE can access any I/O device through the GigaRing channels. This chip is also responsible for supporting latency hiding and efficient synchronization.

The T3E PEs do not have board-level caches, but use the caches in the 21164 processors. There are two levels of on-chip caches: The first level consists of an 8-KB instruction cache and an 8-KB data cache; the second level is a three-way set-associative, 96-KB cache unified for both instructions and data. Not using a board level cache can improve the main memory bandwidth.

Interconnect   The T3E supports low-latency, high-bandwidth communication through a 3D torus network [541], which is capable of delivering a 64-bit word every system clock (13.3 ns) in each of all six directions. The bisection bandwidth for a 512-PE system exceeds 122 GB/s. The network implements an adaptive, minimum-path routing algorithm to allow messages to go around hot spots.

The I/O Subsystem   At the heart of the T3E I/O subsystem are the GigaRing channels [539], which are connected to the torus network and to the PEs. Each GigaRing channel is a pair of counter-rotating, 32-bit rings, with data in the two rings traveling in opposite directions, for high bandwidth and reliability.

Up to 16 PEs can be connected to a GigaRing channel, and each channel can deliver a peak bandwidth of 1 GB/s. In the maximal configuration, the I/O subsystem with multiple

GigaRing channels can provide an I/O bandwidth of 128 GB/s. Besides PEs, other types of I/O nodes are connected to a GigaRing channel, including the following:

- Multipurpose node, which accepts Sbus controller cards to support SCSI, FDDI, Ethernet, and ATM
- Disk node, which supports disks and RAID over Fibre Channel and IPI (a Cray disk technology); and tape node that supports Block Mux and ESCON tape drives
- HIPPI node, which supports 100-MB/s or 200-MB/s HIPPI channels

11.2.2 The System Software in T3E

Unlike the T3D, which requires a Cray C90 as a front end, the T3E is a *self-host* system. The T3E runs a derivation of the Cray 64-bit Unix system (UNICOS) called UNICOS/mk, which is a distributed operating system that provides single system image at the kernel level. The T3E provides an integrated environment that supports shared-variable, message-passing, and data-parallel programming.

The Operating System The UNICOS/mkUNICOS/mk system is divided into a number of local and global servers. The PEs are divided into *user* and *system* PEs. User PEs run user applications and commands. System PEs are dedicated to providing global operating system services. Each user PE contains local servers and a Unix microkernel derived from the CHORUSCHORUS system [156].

All process-specific requests are handled by the microkernel and local servers, including memory allocation and message/data passing. Global servers provide system-wide services, such as process management, file space allocation, scheduling, security, and I/O management.

The UNICOS/mk system supports availability by automatic job recovery via kernel-supported checkpoint/restart, shared file system, guaranteed file transfer, etc. It includes a set of tools to provide resource management, administration, system monitoring, accounting, job scheduling, and security services.

To achieve scalable I/O, the UNICOS/mk system implements distributed file system management. Local file servers in a user PE service unbuffered read/write requests. Global file servers need to be involved only for less frequent requests, such as file open and close. Multiple file servers enable parallel I/O transfers.

The Programming Environment The T3E provides optimizing compilers for Fortran 90, C, and C++, and a suite of optimized and parallelized scientific and mathematical libraries. The T3E supports data-parallel programming by using the Fortran 90 and HPF languages, message-passing programming by PVM and MPI libraries, and shared-variable programming by using the Cray Shared Memory Library SHMEM and the CRAFT compiler directives and library routines. Any combination of these tools can be used.

In addition, the T3E provides a set of environment tools to aid the development of efficient parallel programs, including the following:

- A symbolic, source-level debugger for parallel applications called *TotalView*, which enables a user to control and display the progress of individual processes or process groups

- A parallel performance analysis tool called *MPP Apprentice*, which features an expert system that interprets performance information and makes performance-enhancing recommendations

11.3   New Generation of ASCI/MPPs

There are two major efforts in the U.S. to develop high-performance supercomputers: The *Petaflops project* and the *ASCI program*. The Petaflops project aims at a long term goal to achieve a supercomputer speed of Pflop/s (10^{15} flop/s) speed. Recent studies revealed that using conventional technology and *commodity off-the-shelf* (COTS) components, such an MPP could be built by year 2015 at a cost of US $1 billion.

Using exotic technologies, a 1-Pflop/s supercomputer could probably be built by 2007 with unknown cost. It is unclear whether noval technology can win over COTS, since the latter is desired by the mainstream of computer industry. Either way, a 1-Pflop/s system must effectively exploit massive parallelism in the range of thousands of processors to over a million of threads concurrently. Readers are Referred to the Web Resources for links to more information on the Petaflops project.

In 1994, the U.S. *Department of Energy* (DOE) launched the *Accelerated Strategic Computing Initiative* (ASCI) program, a 10-year $1-billion program to build Tflop/s supercomputer systems for simulating aging effects on the nuclear weapons stockpile, biotechnology, medical and pharmaceutical research, weather prediction, aircraft and automobile design, industrial process improvement, environment protection, etc.

11.3.1   ASCI Scalable Design Strategy

The ASCI program will develop supercomputer systems as a reliable substitute to underground nuclear testing. The goal of the ASCI program is to deploy a 1-Tflop/s system by 1996, a 10- to 30-Tflop/s system around year 2000, and a 100-Tflop/s system by 2004. These systems should be available at similar costs.

Going from 1 Gflop/s in 1994 to 100 Tflop/s in 2004 calls for a 10^5 times performance increase in 10 years. This cannot be achieved by following the current trend in computer industry, where Moore's law holds. According to Moore's law, a 10^5-fold performance increase while keeping the cost constant needs 26 to 32 years, which implies that 100 Tflop/s cannot be achieved until year 2025.

The strategies that the ASCI program adopted have two notable features: the *accelerated development* and the *balanced scalable design*. The ASCI program aims at not just the peak speed, but a total system solution that can deliver sustained application performance that is five orders of magnitude better than that in 1994. This calls for a balanced scalable design approach as characterized by the following accelerated steps:

- Focus on high-end platforms for scientific computing applications, not on mass market platforms and applications (the market sweet spot).
- Use as many commodity off-the-shelf (COTS) hardware and software components as possible, and focus on developing those key enabling technologies that are not effectively offered by the mainstream computer industry.
- Use massively parallel architectures. Focus on scaling and integrating technologies that can glue thousands of COTS nodes into an efficient platform with a single-system image.

Sclable Design Roadmap Table 11.3 shows the ASCI roadmap for delivering a balanced scalable computing environment. This is in reponse to the supercomputing need of the U.S. nuclear energy and weapon research programs, up to year 2003.

Table 11.3 ASCI Strategy for Balanced Design of Future MPPs
(Courtesy: D. Crawford, Sandia National Laboratories, [172],1996)

Attribute	1996	1997	2000	2003
Application performance (times)	1		1000	100,000
Peak computing speed (Gflop/s)	100	1000	10,000	100,000
Memory capacity (TB)	0.05	0.5	5	50
Disk capacity (TB)	0.1-1	1-10	10-100	100-1000
Archival storage capacity (PB)	0.13	1.3	13	130
I/O speed (GB/s)	5	50	500	5000
Network speed (GB/s)	0.13	1.3	13	130

The balanced design strategy includes the following engineering considerations :

- *End-to-end performance*. Impressive peak speed or benchmark performance are not enough. A scientist sitting in front of a desktop should see a 10^5-fold performance increase of the target application. This includes the entire process of developing codes, submitting and running the jobs, visualizing the results, etc.
- *Balanced scalable hardware*. The hardware must be scalable and balanced. This implies the scaling from desktop and networking hardware all the way to a supercomputer platform. A balanced design rule is set: A 1-Gflop/s peak speed should

be matched with 1 GB memory, 50 GB disk, 10 TB achival storage, 16 GB/s cache bandwidth, 3 GB/s memory bandwidth, 0.1 GB/s I/O bandwidth, 10 MB/s disk bandwidth, and 1 MB/s archival storage bandwidth.

* *Balanced scalable software.* ASCI believes that 10- to 100-fold performance improvement will come from developing new software. Such software tools must enable not only scientists to efficiently use the supercomputer, but also improve their productivity. ASCI estimates that without such tools, it would take over 30 years to develop production-quality applications

11.3.2 Hardware and Software Requirements

The ASCI funding agency has given complete sets of hardware and software requirements to three major computer companies in the US to build the ASCI prototype platforms. These requirements are interpreted below.

Hawrdware Requirements Requirements in processors, memory hierarchy, and I/O subsystems have been all specified for building the ASCI supercomputers. For example, the ASCI memory requirements are shown in Table 11.4.

There are two sets of requirement entries: Most entries are in demand for 1998. Those entries within parentheses are to be met by year 2000. The boldface entries show where the industry lags behind the requirements as of 1997 time frame. The lightface entries are attainable by current industrial capabilities..

Consider the per-processor local memory requirements. Assume a processor with a 200-MHz clock (5 ns CPU cycle time) and a 400-Mflop/s speed. Then the memory latency requirements in Table 11.3 correspond to 150 to 400 ns by year 1998 and 75 to 150 ns by year 2000. The memory bandwidth for random read/write should be 800 to 3200 MB/s peak (the same sustained by year 2000), and the local memory capacity should be at least 400 GB.

Software Requirements The ASCI software requirements are specified in Table 11.5. In software areas, the industry is lagging far behind. Therefore, a main thrust of the effort is geared to software development. Listed below are ASCI requirements in human/computer interface, application and programming environments, distributed operating systems, etc.

(1) *Human/computer interface*: Visualization and Internet technology

(2) *Application Environment*: Mathematical algorithms, mesh generation, domain decomposition, and scientific data management

(3) *Programming Envirnment*: Programming model, libraries, compilers, debuggers, performance tools, and object technologies

Table 11.4 ASCI Memory Requirements by The US Dept. of Energy
(Courtesy: D. Crawford, Sandia National Laboratories, [172],1996)

Memory Level	Effective Latency (CPU cycles)	Read/Write Bandwidth*	Storage Capacity**
On-chip cache, L1	2 to 3	16 to 32 B/cycle	10^{-4} B/flop/s
Off-chip cache, L2	5 to 6	16 B/cycle	10^{-2} B/flop/s
Local main memory	**30 to 80** **(15 to 30)**	**2 to 8 B/flop peak** **(2 to 8 B/flop sustained)**	1 B/flop/s
Nearby nodes	**300 to 500** **(30 to 50)**	**1 to 8 B/flop** **(8 B/flop)**	1 B/flop/s
Far away nodes	**1000** **(100 to 200)**	**1 B/flop**	1 B/flop/s
I/O speed (Memory-disk)	10 ms	0.01 to 0.1 B/flop	10 to 100 B/flop/s
Archive (disk-tape)	In seconds	**0.001 B/flop** **(0.01 to 0.1 B/flop)**	**100 B/flop/s** **(10^4 B/flop/s)**
User access time	0.1 s (1/60 s)	**OC3/desktop** **(OC12-48/desktop)**	100 users
Multiple sites	0.1 s	Unknown	Unknown

Note : Boldface entries refer to where the industry is incapable of meeting the requirement as of 1997. Lightface entries correspond to otherwise. Most entries are required for 1998 and those in parentheses are set for year 2000.
* Bandwidth per unit workload or per CPU cycle. ** Capacity per unit speed (flop/s).

(4) *Distributed Operating Software*: I/O, file and storage systems, reliability, communication, system administration, distributed resource management

(5) *Diagonostic performance monitor*: System health and monitoring aperatus..

11.3.3 Contracted ASCI/MPP Platforms

Three projects have started under ASCI, shown in Table 11.6. Intel was awarded $55 million to build a machine (code-named *Option Red*) with sustained Linpack performance of more than 1 Tflop/s. The Option Red machine will be discussed in Section 11.4.

IBM was awarded $93 million to build a 3-Tflop/s supercomputer (code-named *Blue Pacific*) to be on-line at Lawrence Livermore National Laboratory in 1998. Obviously,

Table 11.5 ASCI Software Requirements and Industrial Readiness
(Courtesy: D. Crawford, Sandia National Laboratories, 1996 [172])

Software Requirement	Security	Scalability	Functionality	Portability
Human/computer interface	↑ △	↓ △	Visualization ↓ △ Internet ↑ △	↑ ●
Application environment	↑ ●	↓ △	↓ △	↑ △
Programming environment	↓ △	↓ △	↓ △	↓ △
Distributed operating software	↓ △	↓ △	↓ △	↓ △
Diagnostics performance monitor	↑ ●	↓ △	↑ ●	↓ ●

↑ Industry meeting requirements; ↓ Industry not meeting requirements;
● Requirements stay the same; △ Requirements increase with time.
Double entries refer to joint requirements at the same time frame.

IBM Deep Blue project has some common interest. Details of this Blue Pacific option will not be released until late 1998 or at an even later time.

Table 11.6 Summary of the Three ASCI MPP Platforms

Feature	Intel/SNL Option Red	IBM/LLNL Blue Pacific	SGI/LANL Blue Mountain
Processor selection	200-MHz Pentium Pro with 200 Mflop/s	200-MHz POWER3 with 800 Mflop/s	1-Gflop/s MIPS processor
System architecture	NORMA-MPP	Cluster	CC-NUMA
No. of processors	9216	4096	3072
Peak speed	1.8 Tflop/s	3.2 Tflop/s	over 3 Tflop/s
Memory capacity	594 GB	N/A	N/A
Disk capacity	1 TB	75 TB	N/A
Available date	June 1997	December 1998	December 1998

SGI/Cray won a $110.5 million contract to build two supercomputers with combined peak performance of more than 4 Tflop/s. One system (code-named *Blue Mountain*) consisting of 3072 processors targeted at 3-Tflop/s speed is supposed to be delivered to the Los Alamos National Laboratory by December 1998. The development and applications of these ASCI/MPP options are worth of watching in the years to come.

11.4 Intel/Sandia ASCI Option Red

The Option Red is an MPP system jointly developed by Intel Scalable System Division and Sandia National Laboratories. The system was delivered to Sandia National Laboratories in December 1996. The full configuration was completed in June 1997.

11.4.1 The Option Red Architecture

The Option Red is a distributed-memory MPP shown in Fig. 11.3. The system uses a total of 4608 nodes (each having two 200-MHz Pentium Pro processors), 594 GB of memory, with a peak speed of 1.8 Tflop/s and a peak cross-section bandwidth of 51 GB/s. Of these nodes, 4536 are *compute nodes*, 32 *service nodes*, 24 *I/O nodes*, 2 *system nodes*, and the rest are hot spares. The system has 1540 power supplies, 616 interconnect backplanes, and 640 disks (more than 1-TB capacity).

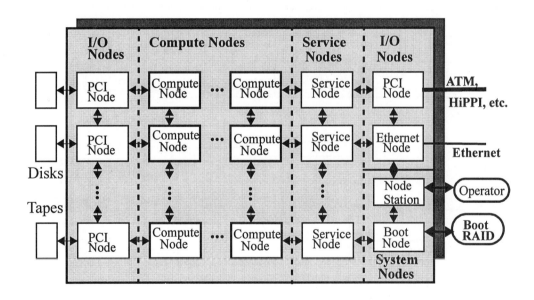

Figure 11.3   The ASCI Option Red system block diagram.
(Courtesy of Mattson et al. *Proc. of the 10th Int'l Parallel Processing Symp.* [433], 1996)

Node Architecture   The compute nodes are used to run parallel computing applications. The service nodes are used to support log-in, software development, and other interactive operations. The I/O nodes are used to access disks, tapes, networks (Ethernet, FDDI, ATM, etc.), and other I/O devices.

There are two system nodes: The *boot node i*s responsible for initial system booting and provides the service and the *I/O nodes* for single system image support. The *system node* is used to support the system's RAS capabilities.

The compute nodes and the service nodes are implemented identically. Two nodes are implemented on a single board, as shown in Fig. 11.4(a). The two SMP nodes are chained together by connecting their NICs (*Network Interface Component*). Only one NIC goes to the interconnect backplane.

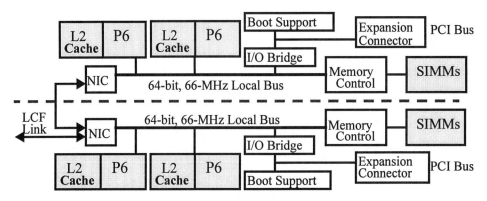

(a) Dual-node board for compute and service nodes

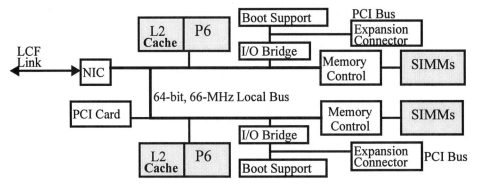

(b) Single-node board for I/O and system nodes

Figure 11.4   Two Option Red node boards for compute, I/O, and services. (Courtesy of Mattson et al. *Proc. of the 10th Int'l Parallel Processing Symp.* [433], 1996)

Each node's local I/O includes the following parts. A serial port, called *Node Mainte-nance Port*, is connected to the system's internal Ethernet and is used for system bootstrap, diagnosis, and RAS. The expansion connector is provided for node testing. The boot support hardware includes a flash ROM which contains the Node Confidence Tests, BIOS, and other code needed to diagnose node failures and to load operating systems.

The board for I/O and system nodes [Fig. 11.4(b)] is similar to the dual-node board [Fig. 11.4(a)]. However, there are only two processors (1 node), a single local bus, and a single NIC. The per-node memory capacity is increased from 32 to 256 MB to 64 MB to 1 GB. The number of 133-MB/s PCI cards is increased from 2 to 3. Every I/O node board also has on-board basic I/O facilities accessible through the front panel, such as RS232, Ethernet (10 Mb/s), and Fast-Wide SCSI.

System Interconnect The nodes are connected by an *Interconnection Facility* (ICF), which employs a two-plane mesh network topology as shown in Fig. 11.5. Each node board is connected to a custom ASIC called *Mesh Routing Component* (MRC) via the NIC on the node board. The MRC has six bidirectional ports, each is capable of transmitting data at a peak unidirectional speed of 400 MB/s, and 800 MB/s at full duplex.

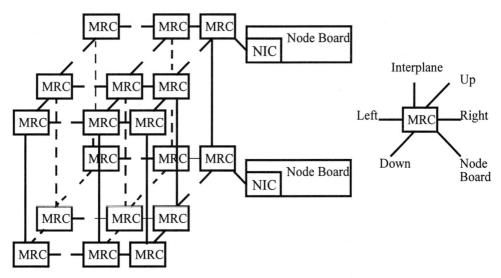

Figure 11.5   The Option Red Interconnection architecture
(Courtesy of Mattson et al. *Proc. of the 10th Int'l Parallel Processing Symp.* [433],1996)

Four ports are used for intraplane mesh connections as Left, Right, Up, and Down in Fig. 11.5. There is also a port for interplane connection. Messages from any node may be wormhole-routed to another node via either plane. This reduces latency and increases the system availability.

11.4.2 Option Red System Software

The ASCI Option Red system software is an evolution of the Paragon environment. The system, the service, and the I/O nodes run the Paragon operating system, which is a distributed Unix system based on OSF. The compute nodes run a *light-weight kernel* (LWK) called *Cougar*. Support is provided to interface these two systems, including fast communication, Unix programming interface, and a parallel file system.

The Light Weight Kernel   The light-weight kernel (LWK) operating system is derived from the PUMA system [642], with the following design features:

- The LWK design emphasizes performance over functionality, efficiently supporting MPPs with thousands of nodes. It provides only those functionalities needed by parallel computation, not general OS services.

- There are thousands of compute nodes in the TFLOPS system. Cougar is designed to occupy less than 0.5 MB memory to prevent the aggregate memory consumed by the LWK from growing rapidly.

- The design assumes that the communication network is trusted and controlled by the kernel. There is no need for protection checking and message authentication.

- LWK provides an open architecture, allowing the development of efficient, user-level library routines.

As shown in Fig. 11.6, LWK consists of two layers: the *process control thread* (PCT) and the *quintessential kernel* (Q-Kernel). Each node has a number of user processes, a PCT, and a Q-Kernel. The Q-Kernel is the only software that can directly access the address mapping and the communication hardware. It provides the basic computing, communication, and address space protection functionalities. The PCT provides process management, naming services, and group protection functionalities.

The LWK environment assumes a partial order of trust: Every component trusts the communication hardware to provide correct and secure communication. That is, the hardware will route a message reliably to the correct physical node. The message will not be corrupted during transmission or routed to a wrong node. The Q-Kernel trusts the hardware and the Q-Kernels in other nodes, but not PCTs or application processes.

The PCT trusts the Q-Kernel and other PCTs, but not the processes. An application process trusts the Q-Kernel and the PCT, but not the other processes. The LWK architecture ensures that data structure in one level can only be corrupted by the same or a more trusted level. This is achieved by constructing a series of *protection domains*.

The Q-Kernel's domain includes all physical resources in a node. The PCT's address space forms a subdomain, which again includes several subdomains, each for a process. Each entity has direct access to only its domain. Thus a process cannot corrupt the PCT, which cannot corrupt the Q-Kernel.

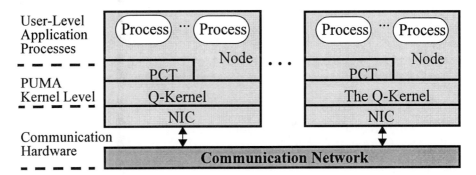

Figure 11.6 The layer structure of the LWK (Cougar) environment.
(Courtesy: Wheat, et al., *Proc. 27th Hawaii Int'l Conference on Systems Sciences,* [642], 1994)

Dividing the kernel into two layers has several advantages. First, it increases portability. While most of the Q-Kernel code needs to be rewritten for a different MPP, most of the PCT code is portable. Second, layering separates functionality concerns. More specifically, the Q-Kernel is responsible for controlling access to physical resources, while the PCT is responsible for *managing* such accesses.

The designers of PUMA did not formally define controlling and managing. Roughly speaking, management dictates what and how to accomplish a task, while controlling refers to the actual execution and enforcement of the what/how decisions. For instance, selecting the time slice (quanta) size, determining which process can be a communication's destination, and selecting a process scheduling policy are examples of management operations, while enforcing the quanta, checking if a message's destination is valid, and enforcing a scheduling policy are control operations.

The designers of PUMA believe that control operations occur far more frequently than management operations, and management policies change more frequently than control mechanisms. Dividing the LWK into two levels facilitates optimizing control operations without impacting the management policies, and enables the running of different PCTs (e.g., single-tasking or multitasking) on the same Q-Kernel.

Message Passing The ASCI Option Red system supports MPI, NX, and portals for message passing. MPI is the standard library in the system. NX is implemented to provide backward compatibility with Paragon, on which many applications use the NX message-passing library. Portals provide the most efficient, low-level message passing library. The portal concept was first introduced in the PUMA operating system. It can be used to eliminate memory copy overhead in message passing.

However, message-passing using portals is not a user-level communication mechanism like those in Section 7.4.3; kernel crossing is still needed. A *portal* is a portion of the address space of the destination process that is open to other processes for sending messages. To send a message, a sender process executes a kernel routine with a prototype like the following:

```
send_user_msg (
    void *buf,              // start of the send message buffer
    size_t len,             // size of the send message
    int tag,                // message tag
    proc_id dest,           // destination process ID
    portal_id portal,       // index of the destination portal
    int *flag               // flag to be incremented when the msg is sent
)
```

This is a nonblocking call, which returns as soon as the kernel has recorded the information necessary for message transmission. The source kernel sends to the destination kernel a message header containing the sender process ID, the destination process ID, the message length and tag, the destination portal, but not the message buffer address or the flag. When the header arrives, the destination kernel translates it and verifies that the message can be stored in the portal. Then it arranges to have the message body sent to the destination portal.

After all the message has been stored in the destination portal, the destination kernel sets a bit in the portal descriptor to indicate that a new message has arrived. Once the message is sent, the source kernel increments the flag value. The sender process can poll the flag to determine when it is safe to reuse the send message buffer. No explicit routine needs to be called to receive a message. The receiver process can poll the portal descriptors to find out if a new message has arrived. The process can also arrange to have a signal generated when a message arrives at any of its portals.

11.5 Parallel NAS Benchmark Results

When designing an efficient parallel computer, the system architect faces many issues. A main question is: What are the architectural characteristics supporting the majority of the applications? The following parameters are a partial list of important application characteristics:

- Main memory requirement (in MB)
- I/O rate (in MB/s) and I/O data amount (in MB)
- Communication latency (t_0 in μs) and bandwidth (r_∞ in MB/s)
- Average parallelism, sequential bottleneck (Amdahl's law), granularity, and load imbalance, etc.

The values of these parameters depend on the application, the problem size, the machine size, and the processor speed, among other things. In this section, we discuss the architectural implications of three NAS benchmarks for scientific simulation, computational fluid dynamics, and signal processing applications. These research results do give partial answers to three frequently asked questions:

- How large should the memory be ?
- How fast should the communication subsystem be?
- How fast should the I/O system be?

11.5.1 The NAS Parallel Benchmarks

The *NAS Parallel Benchmarks* (or NPB) were already discussed in Section 3.1.2. Table 11.7 shows the problem sizes of the eight NPB benchmarks in total number of Gflop. In [593], Sun et al. characterized the NPB MG and FT parallel benchmarks and discussed the implications in designing parallel computer architectures.

Table 11.7 NAS Parallel Benchmark Problem Sizes

Benchmark Program Name	Total Workload (Gflop)		
	Class A	Class B	Class C
Embarrassingly Parallel (EP)	26.68	100.9	N/A
Multigrid (MG)	3.905	18.81	155.5
Conjugate Gradient (CG)	1.508	54.89	N/A
3D FFT PDE (FT)	5.631	71.37	N/A
Integer Sort (IS)	0.7812	3.150	N/A
LU Solver (LU)	64.57	319.6	2039
Pentadiagonal Solver (SP)	102.0	447.1	1450
Block Tridiagonal Solver (BT)	181.3	721.5	2866

The characterization is based on the *phase-parallel* model in Section 1.3.3. We partitioned each benchmark into a sequence of supersteps and identified the memory, I/O, and communication requirements for each superstep. The following values are then found for each benchmark with different machine size n and problem size N.

- The *workload structure*, including the average parallelism, the percentage of the sequential portion (Amdahl's law), the granularity w, the load imbalance σ, etc.
- The memory (in MB) required by the benchmark per node
- I/O rate (in MB/s) and the I/O data amount (in MB)

- Communication requirements, including communication pattern, message length, latency (t_0 in μs) and bandwidth (r_∞ in MB/s) requirements

We present the results for the FT and the MG benchmarks for the class C problem sizes. Some of the results use the processor speed as a parameter. We always mean the *sustained speed* (in Mflop/s). Current- and next-generation microprocessors have a peak speed range from 200-1000 Mflop/s. However, often the sustained speed is only 5% to 10% of the peak. Thus we present results for three cases: the sustained processor speed is 10, 50, or 100 Mflop/s. The methodology is valid for other values as well.

11.5.2 Superstep Structure and Granularity

The superstep structures of the FT and the MG benchmarks are shown in Fig. 11.7. Each oval represents a computation phase, while a rectangle a communication phase. The arrows represent iterations.

For instance, the MG benchmark is comprised of four supersteps, with superstep 3 iterates 680 times. The communication phase of superstep 1 consists of six point-to-point communications, and that of superstep 2 has two reductions. According to the phase parallel model in Section 1.3.3, the granularity of a superstep is the workload (e.g., Mflop) performed by one node in one superstep.

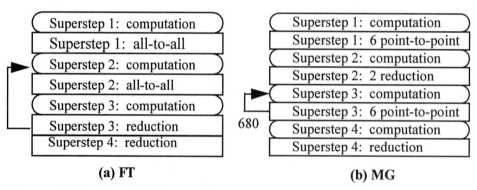

(a) FT   (b) MG

Figure 11.7   Simplified superstep structures for FT and MG.

The granularity w of the dominant supersteps for FT and MG are shown in Table 11.8.The granularity varies greatly among different benchmarks, and decreases as the machine size increases. The superstep models revealed the following two factual points:

- The phase-parallel model (Section 1.3.3) is a valid approach to the study of all NAS benchmarks. The superstep structure abstracts away many details and greatly simplifies performance analysis.

Table 11.8 Granularities (Mflop) of the Dominant Supersteps in FT and MG

Nodes	2	8	32	128	512
FT	10036	1221	305	76	19
MG	285.2	71.3	17.8	4.5	1.1

- Each benchmark program has one to a few *dominant supersteps*, where most of the computation workload is executed. As a first round approximation, one can concentrate on those dominant supersteps. For instance, superstep 3 is the dominant superstep in MG, as it iterates 680 times for the class C problem size.

11.5.3 Memory, I/O, and Communications

A key parameter that must be determined in designing any parallel computer system is the physical memory capacity per node. This parameter may also affect processor design by demanding a certain minimum size for the address space.

Memory Requirements The memory requirement can be defined as the total amount of memory needed per node, as a function of the machine size. This includes all types of memory usage: data storage input/output, temporary storage, communication message buffers, and so on. The per-node memory requirements for FT and MG are shown in Fig. 11.8, which leads to three implications:

- First, class C benchmarks have large problem sizes, and need more memory than what can be accommodated by a single node on current MPPs. In fact, the FT and the MG benchmarks need 3.5 to 4.3 GB main memory, which is larger than some processors' physical address space size.

- Second, the required memory per node decreases almost linearly when more nodes are used, and the memory used for message passing is just a small percentage of the total memory. For example, if 8 nodes are used to execute MG, the total memory required per node and the message buffer per node are 457 MB and 4.26 MB respectively. But they become 59.2 MB and 1.08 MB for 64 nodes.

- Third, Fig. 11.8 indicates how many nodes are needed for an MPP to execute the NAS class C benchmarks. For instance, most MPPs have a per-node memory less than 128 MB that can be used by applications. To be able to execute the FT and the MG benchmarks, at least 64 nodes must be used.

I/O Requirements The I/O amount is first calculated for each benchmark. Then the I/O rate required is derived. The I/O rate depends on the I/O `mount, the number of processors, and the processor speed. We derive the I/O rate with the following balancing assumption: the total I/O time should be equal to the total computation time while ignoring the message-passing overhead.

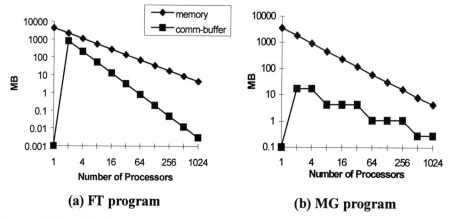

(a) FT program   (b) MG program

Figure 11.8   Node memory requirements for two NAS programs

Table 11.9 shows the I/O rates required of FT and MG over 2 to 2048 nodes. For each benchmark, the three rows represent the three cases in which the per-processor sustained speed is 10 Mflop/s, 50 Mflop/s and 100 Mflop/s, respectively. The required I/O rates increase significantly when more nodes are used or when the processors become faster.

For example, if the sustained processor speed equals 10 Mflop/s, the required I/O rates for MG are 0.14 MB/s and 2.29 MB/s when the machine size equals 2 nodes and 32 nodes, respectively. But when the sustained processor speed increases to 100 Mflop/s, the required I/O rates jump to 1.4 MB/s and 22.9 MB/s.

Table 11.9 I/O Demand in FT and MG Programs

NAS Code	Processor Speed	Number of Nodes					
		2	8	32	128	512	2048
FT	10 Mflop/s	0.05	0.21	0.85	3.4	14	56
	50 Mflop/s	0.27	1.06	4.25	17	68	272
	100 Mflop/s	0.53	2.12	8.5	34	136	544
MG	10 Mflop/s	0.14	0.57	2.29	9.39	38.7	164
	50 Mflop/s	0.7	2.82	11.47	46.94	193.6	822
	100 Mflop/s	1.4	5.65	22.9	93.9	387	1643

With current disk and RAID technology, sequential I/O (single-disk or single RAID) is probably enough for small to medium machine sizes (e.g., no more than 64 nodes) or slow processors. But if the machine size is greater than 64 nodes and the processor is fast, parallel I/O is needed to maintain a balanced system. The good news is that all NAS benchmarks have regular I/O access patterns and parallel I/O where a file is strided across multiple disks is not difficult to realize.

Communications Requirement Another important factor is how fast the communication subsystem should be. Two popular parameters are the communication *latency* (or start-up time) t_0 and the *asymptotic bandwidth* r_∞ for sending a point-to-point message between two nodes. We determine these parameters as follows: We first set the communication time equal to the computation time per superstep, which can be calculated from the superstep structure and the processor speed. Then we derive the latency and the bandwidth requirements from the communication time, using Eq.(1.1).

The bandwidth requirements are shown in Fig. 11.9. It happens that the latency effect is small. It is obvious that the required bandwidth increases proportionally to the processor speed. However, the bandwidth required does not necessarily increase as more processors are used, as the FT curves show a flat trend. For the current and the next generations of processors, a bandwidth of 40-50 MB/s is enough for up to 2000 processors..

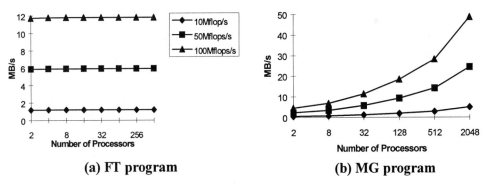

(a) FT program   (b) MG program

Figure 11.9   Communication bandwidths required in NAS programs.

Architectural Implications To sum up, the NAS FT and MG benchmarks lead to the following architectural implications: The granularity is from a few Mflop to a few hundred Mflop. To execute class C NAS benchmarks, the per-processor memory should be at least 64 to 128 MB per node. The I/O rate requirement increases as the processor speed and the machine size increase, ranging to hundreds of MB/s.

NAS results suggest parallel I/O. Furthermore, the latency is less important than bandwidth, contrary to Berkeley conclusion. A 50-MB/s per-node bandwidth is enough for up to 2000 processors, where each processor sustains 100 Mflop/s. We should caution the reader that these results are based on one case study, using NPB on message-passing machines. For different benchmarks on different platforms with different programming models, the results and conclusion may be very different.

11.6 MPI and STAP Benchmark Results

In this final section, we present some of the MPI and STAP benchmark results obtained by Hwang's research group at the University of Southern California (USC) and at the University of Hong Kong (HKU) from 1994 to 1997.

11.6.1 MPI Performance Measurements

Message Passing Interface (MPI) has become a commonly accepted communication standard for specifying message-passing functions in programming NUMA, multicomputers, or clusters of workstations. Details of MPI functionality will be treated in Chapter 14. The MPI benchmark results are based on performance measuremnts on three MPPs: T3D, SP2, and Paragon as orginally reported in [332].

A collective messaging operation involves a group of processes, residing in the same or different nodes that call the same collective communication routine, with matching arguments. Typical collective operations are *broadcast*, *gather*, *scatter*, *total exchange*, *barrier*, *reduce*, *scan* (*prefix*), etc. These operations provide a common interface for users to design their application codes.

MPI Implementations   Several implementations of MPI are available in the public domain, such as the CHIMP/MPI [19], LAM [462], mpi++ [357], MPICH [423]. Many MPP venders have created their own implementations, optimized with respect to their own machine architecture. For examples, the CRI/EPCC MPI [122] is available on the T3D. The MPICH has been ported on SP2 and Paragon, among other machines.

These MPI implementations offer a rich set of collective operations summarized in Table 11.10. In our experiments, the data structures used are made small enough to fit in each node memory to avoid page faults. The test program is written in standard C and MPI primitives. No machine-specific library functions or any assembly codes are used. The best compiler option avaialable to each machine is always applied. .

Table 11.10 MPI Collective Operations

Operation	Functional description
MPI_Bcast	Broadcasts a message to all processes in the same group.
MPI_Barrier	Blocks until all process have reached this routine.
MPI_Alltoall	Sends data from all to all processes.
MPI_Gather	Gathers together values from a group of processes.
MPI_Reduce	Reduces values on all processes to a single value.
MPI_Scan	Compute the prefix (partial reductions) of a collection of processes.
MPI_Scatter	Sends data from one task to all other tasks in a group.

Testing Conditions We measure the wall clock time using `MPI_Wtime()` function. The test program is executed repeatedly for more than 22 times, with timing starting on the third iteration to exclude the warm-up effect.

The minimal time, the maximal time, and the mean time from all processes are collected. To interpret the results, we focus on the maximum time, because we feel it reflects the condition that all processes involved in the machine have finished the operation.

Aggregate Message Length Let $f(m, n)$ be the *aggregate message length* in a collective operation involving n nodes. This is equal to the sum of all messages being transmitted among all pairs of nodes in a collective communication operation.

For example, in a *broadcast* operation, m is the length (in B) of the message broadcast from the source node, thus $f(m, p) = m(n-1)$, because $n-1$ destinations need to receive the same message.

Similarly, $f(m, n) = m(n-1)$ for the *scatter, gather, reduce,* and *scan* operations. We have $f(m, n) = mp(n-1)$ for a *total exchange* operation. Four performance metrics are summarized in Table 11.11 for evaluation of collective communication operations.

We have defined the *total messaging time* $T(m, n)$ in Eq. 3.4 as the *communication overhead*. This time can be divided into two terms : the *start-up time* $t_0(n)$ and the *transmission delay* $d(m, n)$. Based on today's standard, all time parameters are measured in μs.

Table 11.11 Collective Communication Metrics

Total messaging time (μs)	$T(m, p) = t_0(n) + d(m, n)$
Start-up latency (μs)	$t_0(n)$
Transmission delay (μs)	$d(m, n)$
Aggregate bandwidth (MByte/s)	$R_\infty(n)$

An *asymptotic bandwidth* $r_\infty(n)$ was defined in Eq. 3.3 We define below an *aggregate bandwidth* $R_\infty(n)$, that is related to $r_\infty(n)$ by the general formula :

$$R_\infty(n) = r_\infty(n) \times f(m, p) \tag{11.1}$$

For a point-to-point communication, Eq. 11.1 is written as

$$R_\infty(n) = r_\infty(n) \tag{11.2}$$

For a *total exchange* operation, we have

$$R_\infty(n) = n(n-1) r_\infty(n) \qquad (11.3)$$

For other collective operations (*broadcast, gather, scatter, scan etc.*), we have

$$R_\infty(n) = (n-1) r_\infty(n) \qquad (11.4)$$

11.6.2 MPI Latency and Aggregate Bandwidth

The *start-up latency* $t_0(n)$ is plotted in Fig.11.10 for broadcast and total exchange operations on the three machines. This latency increases monotonically with the machine size n. The start-up latency increases linearly with increase in machine size for the *gather, scatter,* and *total exchange* operations. The latency increases logarithmically for the *broadcast, scan, reduce,* and *barrier* operations, as larger machine are used.

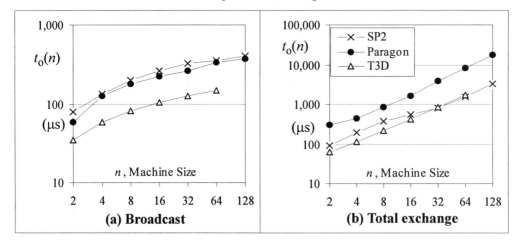

Figure 11.10 Start-up latencies of two MPI collective operations over machines with 2 to 128 nodes.

Latencies on T3D Except for the *scan* operation, the T3D has demonstrated the lowest *start-up latency* in all collective operations. The T3D has a lower latency with special hardware support for fast messaging, lower network latency (20 ns per hop as opposed to 125 ns for the SP2, and 40 ns for Paragon), and the use of *prefetch queue* and *remote processor store* to hide remote memory access latencies.

Latencies on Paragon and SP2 For *broadcast* operation, the SP2 has slightly higher latency than Paragon [Fig. 11.10(a)]. The Paragon has the longest latency in *total* *exchange, scatter, gather,* and *reduce* operations. However, it performs the *scan* operation with even shorter latency than the T3D. Collectively messaging over a large machine, Paragon has much greater latency than others. For collective operations over small machines, the SP2 ranks a second place in latency for the *total exchange, scatter,* and *gather* operations.

Breakdown of Timing Results   The relative performance of the six collective functions is illustrated in Fig. 11.11 for three MPPs with n = 32 nodes and *m* = 1 KB per message.

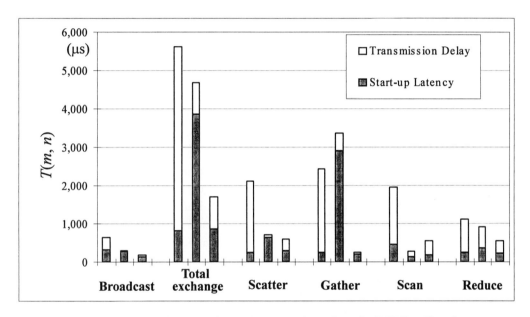

Figure 11.11 Breakdown of total messaging time in MPI collective operations (SP2: left bar, Paragon: middle bar, T3D: right bar within each operationblock).

The dark section of each bar shows the start-up latency and the white bar section the transmission delay. For the same collective operation, the performance ranking of the three machines is inversely proportional to the height of the bars. Obviously, the *total exchange* demands the longest time to complete.

The T3D shows the lowest start-up latency in *broadcast, gather,* and *reduce* operations. The latency in Paragon varies dramatically with different collective functions performed. For example, in the *total exchange* and *gather* operations, Paragon experienced 3857 μs and 2918 μs latencies, about 4 to 15 times greater than the SP2 and T3D counterparts. In the case of *scan,* Paragon even shows a lower latency. As the message length increases, the transmission delays (the white bars) increase linearly.

Aggregate Bandwidth   The aggregate bandwidth can set the limit on the relative performance of machines with different sizes. For 64 nodes, the aggregated bandwidths of *total exchange* for T3D, Paragon, and SP2 are 1.745, 0.879, and 0.818 GB/s, respectively.

The aggregate bandwidth offers a better metric to quantify the data transfer rate in a collective messaging operation. The asymptotic bandwidth by Hockney [314] can be applied only to point-to-point communications. The aggregate bandwidths defined for various MPI collective functions (Eq. 11.1 to 11. 4) can serve this purpose.

Upgrading MPI with Active Messages: The MPI implementations in the three MPPs are still very slow. The collective operations are the major bottlenecks in MPP communication performance. The use of *active messages* in MPI have been practiced in MPI-AM [177] and MPI-FM [424]. Data prefetching can be also employed to reduce the communication latency [124, 150, 522].

Should MPP be pushed for fine-grain parallelism, reducing the CCR to a lower value or hiding long latency within computations is important [101]. The application codes should be implemented independently of the hardware platforms or network topologies.

Architectural Attributes   Considering message length, the SP2 outperforms Paragon in any short messages less than 1 KB. The Paragon performs better than the SP2 in long messages, except the *reduce* operation. For short messages, the SP2 and Paragon perform about the same in the *broadcast* and *barrier* operations. For long messages, the T3D and SP2 have similar performance in the *broadcast, scatter*, and *reduce* operations.

The T3D is well supported by some special hardware features, which helped lower the start-up latency as well as the transmission delay of large number of messages in a collective communication. The hardwired barriers are effective to reduce the synchronization time on the T3D significantly.

The Paragon is weak in handling short messages for its long latency from NX overhead in collective messaging passing. This surge in latency is especially true in performing the *total exchange* and *gather* operations on Paragon. The SP2 is weak in handling long messages for its limited network bandwidth.

The above MPI evaluation results should be useful to those who are developing parallel applications on message-passing multicomputers. The collective latency and messaging time expressions and reported numerical data can be applied in tradeoff studies of SPMD or MPMD computations using collective message passing. The latency and messaging delays can be used to predict MPP performance as reported in [656]. Details of the prediction scheme are left in a homework problem.

11.6.3 STAP Benchmark Evaluation of MPPs

The performance of existing MPPs has been evaluated by the STAP benchmark reported in [334, 634, 656]. The bulk of material presented below is based on the reported results. Lessons learned from STAP benchmark experiments are also provided for other users to consider in developing their own pallel applications on MPPs.

Parallelization of STAP Programs   We parallelized the STAP benchmarks using the *phase-parallel* paradigm [656]. This approach enables the portability of the parallel STAP programs on different MPP platforms, we chose the MPI to implement all interprocessor communication functions

Node Program   Figure 11.12 shows the data flow and control flow of a typical parallel STAP benchmark program on any of the three MPPs. Following the phase-parallel model, each processor executes the same SPMD program over different data subsets.

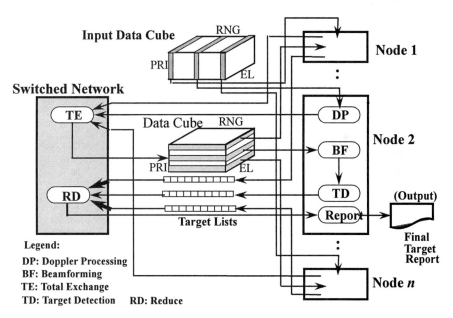

Figure 11.12 Data/control flow in each node program and internode collective communication through the network.

Each node program consists of essentially three major steps: *Doppler processing* (DP), *beamforming* (BF), and *target detection* (TD). The input data cube is partitioned into *p* equal chunks (shaded vertical slices) along the *RNG* dimension. Each node is required to process only a slice of the data cube as shown in Fig. 11. 12.

All subdivided slices are then processed by all nodes simultaneously. When all subprograms within a phase are done, the processors are allowed to interact or exchange data via message passing through the switched network. A multistage Omega network is used in SP2, the 2D mesh network in Paragon, and the 3D torus in T3D.

Sustained Speed   The sustained speed varies with machine platforms and node numbers allocated, as shown in Fig. 11.13. Among all three machines, we observed that no machine has exceeded 10 Gflop/s speed in executing any of the STAP programs up to 128 nodes. In general, the SP2 achieved the highest sustained speed of all machine sizes, followed by the T3D, and the trailed by Paragon. This is true for all 24 machine-program combinations. Only the speed results of four programs are shown.

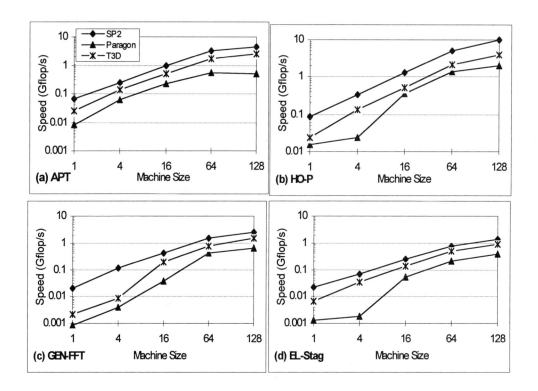

Figure 11.13 Sustained speed of four STAP benchmark programs.
(Courtesy C. J. Wang, et al. *Proc. 10th Int'l Conf. Parallel and Distributed Computing Systems*, [634], 1997)

Figure 11.13(a) shows the speed results of the APT program. The APT program scored 66 Mflop/s on one SP2 node and 4.5 Gflop/s on 128 SP2 nodes. The same APT achieved 8 Mflop/s sequential speed and 490 Mflop/s on 128 Paragon nodes.

In Fig. 11.13(b), the speed results of the HO-PD program are shown. In the past, we have measured the 23 Gflop/s speed on 256 processors in the Maui SP2. The HO-PD is a computationally intensive program with a workload of 10.99 Gflop and a nominal input data cube of 61.44 MB.

Its communication cost is relatively smaller, compared with its computational workload. On a 64-processor configuration, the SP2, T3D, and Paragon achieved the speeds of 5.2, 2.1, and 1.4 Gflop/s, respectively. With 128 nodes, the SP2 achieved the highest speed of 9.8 Gflop/s in Fig. 11.13(b).

A 128-node Paragon achieved 2.0 Gflop/s speed in executing the HO-PD program. The speed results of the GEN-FFT program are shown in Fig. 11.13(c). The ranking of the three MPPs did not change. The SP2 reached 2.5 Gflop/s, and Paragon reached 640 Mflop/s for a machine size of 128 nodes.

The FFT program uses a large data cube of 98 MB. Both Paragon and T3D experienced a speed dip on a small configuration of 16 or fewer nodes. The speed results of the EL-Stag program are shown in Fig. 11.13(d). This program has experienced a large amount of communication overload, because it needs to handle collective messages exceeding 110.6 MB.

Breakdown of Parallel Execution Time To illustrate the bottlenecks, the breakdown of the HO-PD execution time on all three MPPs is plotted in Fig. 11.14(a) for the smaller machines of no more than 8 nodes, and in Fig. 11.14(b) for machines from32 to 128 nodes

In each machine size group, the left bar refers to the SP2, the middle bar to Paragon and the right bar to the T3D. The disk access time appears only in small configurations of Paragon. The bottleneck for a small configuration is from the disk access penalties. For Paragon, the disk access time is 117 s, about 73% of the overall execution time.

The bottleneck is beamforming for all machines with medium or large number of nodes. Obviously, there is no communication cost on a single-node machine. The disk access time for collective communication is larger than that in computational tasks.

The disk access is a bottleneck in performance for machines with limited local memory. Apart from the disk access time, the beamforming consumes the most time in running all STAP programs on all MPPs with 32 or more nodes.

Even though we do not have measured data on very large machines with more than 256 nodes, we believe the situation will not change. The computing power of the node processors determines the time to perform the beamforming operations.

Communication-to-Computation Ratio The *communication-to-computation ratio* (CCR) is defined by dividing the total communication time by the total computation time measured in the STAP benchmark experiments. The CCR equals zero for a one-node application partition, where there is no need of communication.

The lower value of the CCR means that the communication mechanism is relatively more effective than the raw computation power of the MPP. In Fig. 11.15, the variation of CCR with the machine size are plotted for the EL-Stag program. The T3D has the lowest stable CCR among the three MPPs. The T3D is equipped with the most impressive communication mechanisms among the three machines. The SP2 has higher CCR than either Paragon or T3D.

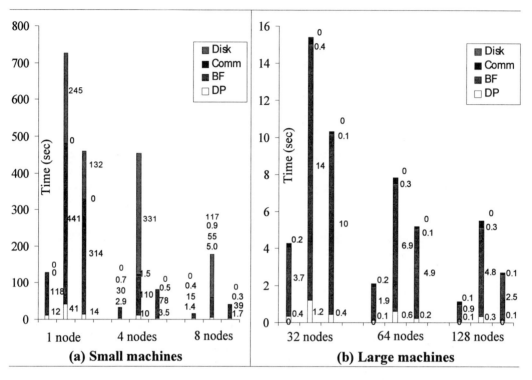

Figure 11.14 Breakdown of the HO-PD execution time on three MPP systems. (For each machine size, left bar: SP2, middle bar: Paragon, right bar: T3D. Courtesy Wang, et al. *Proc. 10th Int'l Conf. Parallel and Distributed Computing Systems*, [634], 1997)

Both SP2 and T3D have comparable communication times, whereas the computation time on SP2 is less than half of that on T3D. The high CCR on SP2 [Fig. 11.15(a)] is mainly attributed to the fast processors in SP2 node. Although the overall performance of SP2 is the best among the three MPPs, a better communication subsystem for SP2 would improve performance more significantly than one for T3D or Paragon.

In Fig. 11.15(b), the speed variation with the CCR value is shown for the EL-Stag program. The speed curve starts from the lower left corner and increases to the upper right direction. Further increase in CCR will result in a speed reduction.

In general, if the CCR value is small, higher speedup can be achieved with some performance tuning. With large CCR, the parallel algorithm should be redesigned. The lower CCR is attributed to the lower software overhead in initiating a message passing function, the speed of a message coprocessor in assembling and disassembling the messages, the routing algorithm used in the interconnection network, etc.

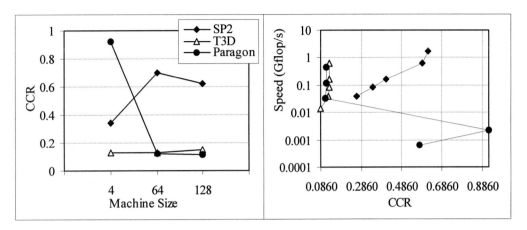

Figure 11.15 Effects of CCR value on the MPP speed performance.

Large-scale signal processing applications, such as STAP, can only exploit coarse-grain parallelism for much higher communication delay than the computation time consumed in each node. In our experiments, the T3D has the lowest CCR among the three MPPs as seen in Fig. 11.15(a). The 3D torus topology in T3D network ensures short connection paths as well as a high bisection bandwidth. Three latency hiding mechanisms in T3D, *prefetch queue* (PFQ), *read ahead, and Block Transfer Engine* (BLT), help improve the overall T3D performance.

System Utilization Rate   The ranges of the system utilization rates of all three MPPs are shown in Fig. 11.16 for all STAP programs. The SP2 achieved the highest overall utilization rate of around 30%.

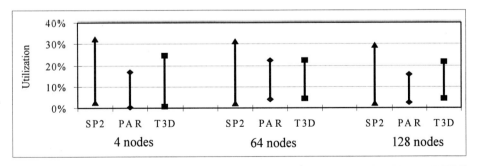

Figure 11.16 System utilization rates of SP2, Paragon (PAR), and T3D for executing all STAP programs on machines with 4, 64, and 128 nodes, respectively.

The Paragon (PAR in Fig. 11.16) has the lowest utilization range among three machines. For machine size 128, it only achieved 16% utilization at best. Generally speaking, the system utilization rate decreases with increasing machine size in most parallel applications.

11.6.4 MPP Architectural Implications

The performance of MPPs can be attributed to many hardware mechanisms and software support systems. We identify below the architectural strength and weakness of each MPP with respect to STAP applications. We examine node architectures, switched interconnection networks, parallel I/O subsystems, software environments, and MPI collective communications. Some MPP operational parameters are given in Table 11.12.

Table 11.12   Operational Parameters of Three MPPs Evaluated

Architectural Features	IBM SP2	Cray T3D	Intel Paragon
Configurations used and machine sites	256 nodes used out of 400 nodes used at MHPCC	64 PEs used at Cray Eagan Center	128 nodes used at San Diego Super-computer Center
Point-to-point latency and bandwidth	39 μs, 35 MB/s	2 μs, 150 MB/s	30 μs, 175 MB/s
Aggregate bandwidth of total exchange over 64 nodes	0.818 GB/s	1.745 GB/s	0.879 GB/s
MPI implementation	MPICH	CRI/EPCC MPI	MPICH

Node Architecture The node architecture greatly affects the total system performance. Among the three MPPs, the SP2 performs the best in speed and utilization for almost all STAP benchmark programs. The highest speed was measured as 19.7 Gflop/s in executing the HO-PD program over 256 SP2 nodes. This is primarily attributed to the peak 266 Mflop/s speed and a good optimizing compiler designed for the POWER2 microprocessor.

The Alpha 21064 has higher clock rate but lower ILP exploited. Alpha was outper-formed by the POWER2 node for this reason. For 64 nodes, the relative speeds of SP2, T3D, and Paragon are 5: 2: 1 Gflop/s, respectively. This ratio is very close to the relative peak speed, 5: 3: 2 for the three node processors, respectively.

Another strength of SP2 lies in large node memory allowed, compared to those allowed in each T3D or Paragon node. For example, many of Paragon nodes in the San Diego Supercomputer Center have only 16 MB of local memory. Loading a large data set into this node causes many disk swaps, which significantly degraded the performance.

The STAP experience told us that a decent MPP node should have at least 128 MB of memory, in order to accommodate large data sets often encountered in scientific and signal processing applications. In real-time applications such as radar signal processing, we may want to avoid using local disks completely for their intolerable access latencies. On the other hand, all three MPPs allow up to 50 GB of shared disk array in their I/O or gateway nodes, making them suitable for use in large database applications.

Scalability of Switched Networks Communication is rather expensive on the MPPs. The communication performance is heavily tied to the switched network latency and bandwidth, as well as to the CCR, which depends on both network traffic and node granularity. The point-to-point message passing on the T3D offers the lowest latency of 2 μs. The SP2 and Paragon have comparable latency of less than 40 μs.

Among the three switched network topologies, Paragon 2D mesh has demonstrated the highest size scalability, as evidenced by the prototype development of the 4,536-node Intel/Sandia ASCI Option Red machine [347]. The next is the 3D torus network scalable to 1024 nodes as installed at the U.S. National Security Agency [212]. The SP2 high-performance switch [592] has been scaled to 512 nodes in a custom-ordered configuration at the Cornell Theory Center [168].

Parallel I/O Three different I/O architectures are employed in our MPP systems. In a large Paragon, the nodes are partitioned into 512 *compute nodes* and 16 *I/O nodes.* The file I/O is provided through one or more I/O nodes. These nodes usually reside on the outer columns of the 2D mesh. The operating system on the machine was Paragon/OSF R1.3.3. Each I/O node is connected to a 4.8-GB RAID-3 disk array.

Intel's *Parallel File System* (PFS) provided parallel access to files. Every disk access from a compute node requires a message exchange with the I/O node. We found that I/O performance is more affected by the network traffic, since the network is also used to deliver regular messages among compute nodes. Hot-spot problems can arise if several compute nodes access the same disk concurrently.

The SP2 has different I/O architecture, in which every node is connected to a local disk. There is no need to distinguish between I/O node or compute node in this case. In SP2, each compute node runs a full IBM/AIX operating system. The disk is attached directly to each node. The I/O node is defined by software dynamically. The PFS allows the user to create files which span many SP2 nodes. This eliminates the inconvenience and administrative overhead in maintaining multiple data files.

In T3D, disks are only attached to the host Cray C90 or Cray Y-MP. The I/O nodes are attached to the host through the I/O gateway. Each I/O gateway consists of two nodes, each of which contains a single Alpha processor with 4-M words of memory (half of that in the compute processor) and special communication hardware. One node handles I/O in each direction, and they are used for system calls as well as file access.

Extravagant Memory/Disk Events A typical STAP program requires one to process more than 100 MB of radar signal data cube within each pulse repetition interval. It is impossible to load the entire data cube into the local memory of a single node. Among the three MPPs we have used, a single node of SP2, T3D, and Paragon has at most 64 MB of local memory. Paragon has only 16 MB per node. Extravagant memory or disk accesses are experienced.

For example, Paragon supports microkernel operating system on each compute node. The kernel and server will take over 6.5 MB of memory. The NX message buffer occupies another 1 MB of memory. This leaves less than 8 MB for data memory, besides the space needed for the node programs.

The *disk swap delay* occurs often in using a small MPP configuration to solve a large-scale problem. The problem comes from the fact that the local memory is not sufficient to hold the entire node program and data subcube.

I/O is potentially the major bottleneck in MPP applications [103]. The extrvagant memory/disk events clearly demonstrated the damaging effects on performance. Enlarging the node memory to 256 B or higher is a must, if we expect a small MPP to become truly useful. Optimization of collective communications and use of larger communication buffers will improve the MPP in I/O performance.

OS and Programming Environments Cray T3D is coupled with the UNICOS MAX operating system. UNICOS is a distributed OS, where the functions are divided between the microkernel OS in PE nodes and the UNICOS on the Y-MP or C90 host. The microkernel design minimizes the amount of OS software running on a PE. Thus the software is reduced, releasing local memory for application data storage and processing.

The software environment affects the performance measurement. The T3D has the smallest execution time variance, and little warmup effect, which are desirable properties for high-performance signal processing applications. High-speed synchronization and gang scheduling are the key factors to achieve the stable computation. In T3D, the SHMEM library provides low-latency *remote read/write* operations. In addition, the *atomic swap* capability is provided to automatically read the content of the word at a remote target address.

Paragon supports a microkernel operating system on each compute node. The kernel and server will take over 6.5 MB main memory on each node. The NX message buffer will consume another 1 MB of main memory space. On SP2, complete AIX makes each SP2 node work as a stand-alone workstation. The local disk can be used to accommodate large data sets or programs. It can be used for message buffer if the message is large.

Essentially, the distributed OS design affects the parallel programming environment. In our STAP experiments, the SP2 is the most user-friendly. These include the ease of parallel code development, job submission, debugging, and the supports in handling large input data and working memory. The lack of virtual address support in T3D also raises a problem in program compilation. Some of our codes couldn't even be compiled due to their large memory requirement.

All three MPPs runs with Unix-based operating systems. None of them is supported by a truly real-time OS. This has created some problem in getting accurate measurement in real-time applications, be it signal processing or transaction processing. The problem comes from interferences from other users in competing the resources.

11.7 Bibliographic Notes and Problems

Dongarra et al.[212] have ranked the top 500 supercomputers. The U.S. High Performance Computing initiative is specified in [233]. Many supercomputers and past MPPs are surveyed below, even though they are not detailed in this chapter. Hwang assessed parallel processing on supercomputers in [324]. Carey edited the book on parallel supercomputing [128]. Parallel processing for supercomputers and AI machines is covered in Hwang and DeGroot [330]. Bell [67] has assessed supercomputers.

Cray T3D is described in Adams [3], Koeninger et al. [371], and Cray announcement [174]. Empirical evaluation of T3D can be found in Arpaci et al. [40]. Dunigan released the Beta test data on Intel Paragon [222]. Hwang et al. [332] evaluated MPI performance. Wang et al. [634] reported the STAP performance on T3D, SP2, and Paragon.The Intel/ Sandia ASCI Option Red machine is described in [347] and by Mattson et al. [433]. The T3E is described in Scott [540], [541] and in Cray's Web page [175].The GigaRing is described in [539].

The Cedar project is reported in Kuck et al. [382]. The NYU Ultracomputer is described in Gottlieb et al. [276]. The Mosaic is given in Seitz [546]. The Tera, an exploratory multithreaded MPP, is described in Alverson, et al [27]. The TC 2000 MPP is described in BBN technical summary [65]. The MIT J machine is described in Dally et al. [186]. The Connection machine is described in Hillis [309], CM5 in [612], and fat tree network of CM5 in Leiserson et al. [404].The HEP is described in Kowalik [376] and Smith [570]. The Titan supercomputer is given in Siewiorek and Koopman [557].

Hockney studied benchmark and performance metrics for MPPs in [313] to [318]. The SPMD model for parallel computing is characterized in Daniel et al. [187]. Anderson and Lam [35] have considered global optimization techniques for scalable systems [35]. Metacomputing is described in Foster, et al [249]. Architectural support for messaging in CM5, and T3D is given in Karamchetti and Chien [360].

Frye et al [256] considered external user interface in using MPPs. Scalable algorithms for STAP benchmarks have been suggested in Bhat and Prasanna [79]. Gunther has studied OLTP on MPP platforms [284]. Horst assessed MPPs in [320]. Blevins, et al have suggested the BLITZEN architecture for MPPs [90].

We have also evaluated MPP performance in [655] to [657].The problem of parallel I/O is addressed in Miller and Katz [447], Henderson, et al [304], Bordawekar et al [103], and Del Rosario and Choudhary [201]. Collective communication is assessed in McKinley, et al [439], and Nupairoj and Ni [472]. The Petaflops computing is elaborated in Sterling et al. [587].

Homework Problems

Problem 11.1   An MPP is scalable to thousands of processors with Tflop/s speed, hundreds of GB of memory, and many TB of disk storage.

    (a)   Explain with reasoning two features of required processor architecture to meet the above MPP system goals. Use examples to support your claims if needed.

    (b)   Identify four system architectural features that are necessary to achieve high performance in modern MPP systems.

Problem 11.2   For the six features identified in Proble. 11.1, discuss how

    (a)   the Intel ASCI Option Red system incorporates the six features.

    (b)   the Cray T3E system incorporates the six features.

Problem 11.3   Study the early performance prediction scheme for MPPs proposed by the authors in [656]. Choose a benchmark program from the NAS suite and predict its performance on a real MPP with known overhead figures. Verify the predicted performance with the measured performance reported by the NAS group.

Problem 11.4   Compare the Intel ASCI Option Red and Cray T3E in the following areas. Discuss their relative strengths and weaknesses in each case.

    (a)   Node architecture and processing power

    (b)   System interconnect and bandwidth

    (c)   Memory model and latency hiding

    (d)   Native programming environment

Problem 11.5   Consider the data reported in Fig. 11.16.

    (a)   Why does Paragon utilization increases as machine size goes from 8 to 64?

    (b)   Why does SP2 have higher utilization than the other two machines?

Problem 11.6   Use the phase parallel model to characterize the architecture implications of the following simplified parallel 2D Jacobi relaxation problem:

    The problem data domain consists of a square of $N \times N$ grid points, which is evenly decomposed into n subdomains. Each subdomain consists of a square of $(N/\sqrt{n}) \times (N/\sqrt{n})$ grid points and is assigned to a node to process. The parallel computation consists of 10000 iterations.

In each iteration, a node first gets boundary grid points from four neighbors, and then computes the new values of the $(N/\sqrt{n}) \times (N/\sqrt{n})$ grid points in its own subdomain. The following SPMD code illustrate the parallel algorithm. The notation [.] indicates that all array elements along that dimension may be accessed.

```
M = N/sqrt(n) ;
for ( k= 0; k< 10000; i++) {     /* iterate 10000 times */

    /* Communication Phase: get the boundary values from the 4 neighbors */
    Exchange boundary values with the North neighbor to get Grid [0][.] ;
    Exchange boundary values with the South neighbor to get Grid [M+1][.] ;
    Exchange boundary values with the West neighbor to get Grid [.][0] ;
    Exchange boundary values with the East neighbor to get Grid [M+1][0] ;
    /* Computation Phase: Update the M × M inner elements of Grid */
    for ( i= 1; i < M+1; i++)
        for ( j= 1; j< M+1; j++)
            Grid[i][j]=(Grid[i-1][j] + Grid[i+1][j] +
                        Grid[i][j-1] + Grid[i][j+1])/4;
}
```

Assume a problem size of $N=16{,}384$ grid points in each dimension, the number of processing nodes $n = 4, 16, 64, 256, 1024$, and the sustained speed is 100 Mflop/s per each processing node.

(a) What is the total flop count in the workload? What is the granularity per node?

(b) Assume Grid is a float (32-b) array. What is the per-node memory requirement, and how much is the portion for communication buffer?

(c) What is the communication requirement (start-up time and bandwidth)?

(d) What are the I/O requirements in term of I/O rate and I/O data amount?

Problem 11.7 Consider the entries in Table 11.12. Compare Paragon with the SP2

(a) Peak point-to-point hardware latency and bandwidth

(b) User level (e.g., MPI) point-to-point hardware latency and bandwidth

(c) MPI total exchange performance

(d) Draw some conclusions on their relative performance.

Problem 11.8 Redo Problem 11.7 with $N = 1024$ and $N = 131{,}072$ grid point in each dimension. Compare the results for three different problem sizes over different machines.

The purpose of the comparison study is twofold as asked below:

(a)   Perform a scalability analysis on the machine size.

(b)   Perform a scalability analysis on the problem size.

Problem 11.9 Study the ASCI/MPP material in Section 11.3, in particular, the entries in Tables 11.3 and the 11.6 and answer the following questions:

(a)   Why COTS hardware and software were suggested for the construction of future generation of supercomputers or MPPs?

(b)   Justify all the numerical values specified in these two tables. Are the three ASCI platform options satisfying the requirements for the time frame between 1997 and 2000?

Problem 11.10 Repeat a similar study of as in Problem 11.9. But now focus your study on the memory and software requirements stated in Tables 11.4 and 11.5.

(a)   Repeat question Part (b) in problem 11.9 by comparing the entries in Table 11.4 and Table 11.6. Are the three ASCI platform options satisfying the memory requirements between 1997 and 2000?

(b)   For software support and programming environments, why today's computer industry, including the manufacturers of the ASCI platforms, is lagging behind?

(c)   Per each of the entries in Table 11.5 where the industry cannot meet the ASCI requirement, explain why based on your knowledge on state of the art in parallel programming software environments.

Part IV

Parallel Programming

A misconception in parallel programming is that it is much more difficult to program a parallel computer than to program a sequential workstation. This is not necessarily true. Consider the following counter examples:

- A parallel machine can be used to provide greater throughput by running multiple sequential applications independently or in a sequence. The user sees a faster sequential machine, even without parallelizing the application codes. It may be surprising to many readers that quite a few parallel machines are used in this way, although with poor efficiency.

- The user runs a database software (e.g., Oracle) on top of a parallel machine. The user sees familiar Oracle functions as in a workstation environment. The Oracle system exploits parallelism automatically which is transparent to the user. Or the user writes a sequential C program and leaves to an intelligent compiler to exploit the available parallelism automatically.

IV. 1 Objectives

Realizing the above facts, we should take parallel programming more seriously. Researchers in academia and industry have devised many ways to perform parallel programming. They have probably caused more confusions than clarifications. Fortunately, these paradigms have converged to essentially three: shared-variable, message-passing, and data parallel programming to be studied in this final part of the book.

Chapter 12   This chapter discusses various *parallel programming paradigms* (i.e. ways to design a parallel program). We study the details of shared-variable programming here, including software multithreading. The material is useful not only to parallel platforms but also to sequential machines.

Chapter 13   This chapter is devoted to message passing programming, using the MPI and PVM libraries as major tools. MPI is now a well accepted Standard that is widely used. Although not formally a standard, the PVM is very stable and also widely used. The two libraries are mutually supportive in nature.

Chapter 14   This chapter studies the data parallel approach, emphasizing Fortran 90 and HPF. These two language approaches are not limited to MPPs. In fact, today's SMPs, CC-NUMAs, clusters, or MPPs are all supported with MPI, PVM, Fortran 90, and HPF.

IV.2 Notes to Readers

Chapter 12 should be read ahead of Chapters 13 and 14. Chapter 13 is most important for distributed computing. If you are not a Fortran fan or not hungry for supercomputing power, Chapter 14 can be skipped.

For those who have limited interest or no direct access of a parallel computer, reading this part may motivate you to obtain access to those machines from remote connections such as through the Internet.

Chapter 12

Parallel Paradigms and Programming Models

This chapter discusses methodologies to perform parallel programming. We first discuss frequently used parallel algorithmic paradigms, and then characterize four parallel programming models: implicit, shared-variable, data-parallel and message-passing.

We will discuss the shared-variable model in detail in this chapter. Five approaches are presented for shared-variable programming: the ANSI X3H5, the IEEE POSIX threads (Pthreads), the SGI Power C, the OpenMP standard, and a new parallel C language, called C//, developed by the authors.

12.1 Paradigms and Programmability

A parallel computer system should be flexible and easy to use. It should exhibit good *programmability* in supporting various parallel *algorithmic paradigms*. Here *parallel algorithmic paradigms* (also known as parallel programming paradigms, or simply paradigms) are ways to structure algorithms to run on a parallel system.

12.1.1 Algorithmic Paradigms

A number of parallel algorithmic paradigms have emerged, as summarized in Fig. 12.1. These paradigms are often combined in various ways in a real parallel program. The Bibliographic Notes contain pointers to references where more detailed treatment and examples can be found.

Phase Parallel   The phase-parallel model discussed in Section 1.3.3 offers a paradigm that is widely used in parallel programming. This paradigm is illustrated in Fig. 12.1(a). The parallel program consists of a number of *supersteps*, and each superstep has two *phases*. In a computation phase, multiple processes each perform an independent computation C. In the subsequent interaction phase, the processes perform one or more synchronous interaction operations, such as a barrier or a blocking communication. Then the next superstep is executed.

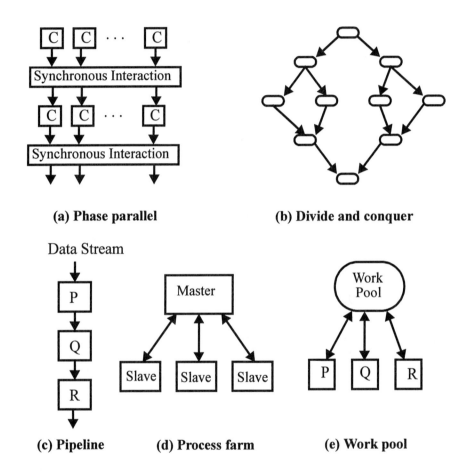

(a) Phase parallel   (b) Divide and conquer

(c) Pipeline   (d) Process farm   (e) Work pool

Figure 12.1   Five parallel algorithmic paradigms.

This paradigm is also known as the *loosely synchronous paradigm* and the *agenda paradigm*. The loose synchrony facilitates debugging and performance analysis, but it has two main shortcomings: Interaction is not overlapped with computation, and it is difficult to maintain balanced workload among the processes.

Synchronous and Asynchronous Iteration A special case of the phase-parallel paradigm is the *synchronous iteration paradigm*, where the supersteps are a sequence of iterations in a loop, as illustrated by the following code to compute iteratively the vector function $\mathbf{x} = \mathbf{f}(\mathbf{x})$, where $\mathbf{x}$ is an n-dimensional vector:

```
parfor ( i = 0; i < n; i++)   // create n processes, each executing a for loop
{
    for ( j = 0; j < N; j++) {
        x[i] = fᵢ(x) ;
        barrier;
    }
}
```

For n = 9, Fig. 12.2 shows how the code works. With the barrier synchronization, none of the n processes can start the $(j+1)$th iteration of the **for** loop until all of them have finished the jth iteration.

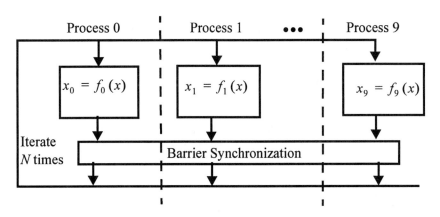

Figure 12.2   Barrier synchronization in synchronous iteration.

In contrast, the *asynchronous iteration paradigm* allows a process to proceed to the next iteration, without waiting for the remaining processes to catch up. This is illustrated in the following code:

```
parfor ( i := 0; i < n; i++) {
    for ( j := 0; j < N; j++)
        x[i] = f(x);
}
```

The above code could be indeterminate, because when a process is computing x[i] in the jth iteration, the x[i-1] value used could be computed by another process in iteration $j, j-1$, *j*+1, etc. However, under certain conditions, an asynchronous iteration algorithm will converge to the correct results and is faster than a synchronous iteration algorithm.

Divide and Conquer   The *parallel divide-and-conquer paradigm* is very similar to its sequential counterpart, as shown in Fig. 12.1(b). A parent process divides its workload into several smaller pieces and assigns them to a number of child processes. The child processes then compute their workload in parallel and the results are merged by the parent. The dividing and the merging procedures are done recursively. This paradigm is very natural for computations such as quick sort. Its disadvantage is the difficulty in achieving load balance.

Pipeline   With the *pipeline paradigm*, a number of processes form a virtual pipeline, as shown in Fig. 12.1(c). A continuous data stream is fed into the pipeline, and the processes execute at different pipeline stages simultaneously in an overlapped fashion.

Process Farm   This paradigm is also known as the *master-slave* paradigm, as shown in Fig. 12.1(d). A master process executes the essentially sequential part of the parallel program and spawns a number of slave processes to execute the parallel workload. When a slave finishes its workload, it informs the master which assigns a new workload to the slave. This is a very simple paradigm, where the coordination is done by the master. A disadvantage is that the master could become the bottleneck.

Work Pool   This paradigm, shown in Fig. 12.1(e), is often used in a shared-variable model. A pool of works is realized in a global data structure. A number of processes are created. Initially, there may be just one piece of work in the pool. Any free process fetches a piece of work from the pool and executes it, producing zero, one, or more new work pieces put into the pool. The parallel program ends when the work pool becomes empty.

This paradigm facilitates load balancing, as the workload is dynamically allocated to free processes. However, implementing the work pool to allow efficient accesses by multiple processes is not easy, especially in a message-passing model. The work pool may be an unordered set, a queue, or a priority queue.

12.1.2 Programmability Issues

We define *programmability* as the combination of generality, portability, and structuredness. These concepts are explained below:

Structuredness   A program is *structured* if it is comprised of *structured constructs*, where each structured construct has the following properties:

- It can be considered as a *single-entry single-exit* construct
- Different semantic entities are clearly identified in the syntax of the construct

- Related operations are enclosed in one construct, instead of being scattered among different places in the program.

We have learned in sequential programming that structured constructs are preferred to unstructured ones [448]. The same is true for parallel programming. For instance, it is harder to give formal semantics for unstructured constructs than structured ones [41]. Structured parallel programs tend to be easier to understand and to have other "nice" properties.

For instance, data flow analysis of structured sequential programs can be conducted in a purely syntax-directed manner, which implies more efficient compile time checking and optimization. However, if unstructured constructs are allowed, we need more complex methods, such as interval analysis (if the underlying flow graph is reducible) or iterative analysis (for general flow graphs) [17].

Structured parallel programming is needed more than sequential programming, because the former is more difficult. However, structured methodology is lacking for parallel programming. The structured programming principle is not just applicable to designing language constructs, but also for designing programs. One can write highly structured programs in Fortran, which is not a structured language. On the other hand, one can write unstructured, difficult-to-understand programs in C, which is considered a more structured language.

Generality Given a parallel programming model, we would be interested in knowing how flexible its constructs are in specifying various programs. For sequential programming, a well known result is the *structured theorem* [448], which says that any Turing computable function can be expressed using just structured sequential constructs: assignment, sequential composition, conditional, and while loop.

If we are interested only in Turing computability, all parallel programming models are equally general, since they contain the four structured sequential constructs. But the main motivation for using parallel programs is to speed up computations. Therefore, when considering the generality, we need to find out not only what functions can be represented, but how fast they can be computed.

Intuitively, generality should convey the following meaning: A program class C is as general as or more general than program class D if for any program Q in D, we can write a program P in class C, such that not only both programs have the same semantics, but also P performs as well as or better than Q. In other words, the generality concept covers efficiency, and a general parallel model should efficiently support various parallel algorithmic paradigms discussed in Section 12.1.1.

Portability A program is *portable* across a set of computer systems if it can be transferred from one machine to another with little effort. Portability is a relative measure, which depends on both the language used and the target machines. For instance, portability increases from (1) to (4):

(1) The algorithm of the program needs to be changed.

(2) The algorithm remains the same, but the source code needs to be changed.

(3) The source code remains the same, but it has to be recompiled and relinked.

(4) The executable can be directly used.

Also, a program has a high portability if it can be transferred among machines of different architecture classes (sequential computers, shared-memory machines, or message-passing machines).

12.1.3 Parallel Programming Examples

We specify below four example programs used throughout Part IV of this book. These examples will explain different paradigms and parallel programming models.

Example 12.1 The π computation problem

The π computation problem is to calculate an approximate value of π through the following integration:

$$\pi = \int_0^1 \frac{4}{(1+x^2)} dx \approx \sum_{0 \le i < N} \frac{4}{1 + \left(\frac{i+0.5}{N}\right)^2} \times \frac{1}{N} \qquad (12.1)$$

A sequential C code follows to compute π by numerical integration:

```
#define    N   1000000
main() {
    double local, pi=0.0, w ;
    long i ;
    w = 1.0 / N ;
    for ( i = 0; i < N; i++) {
        local = ( i + 0.5 ) * w ;
        pi = pi +  4.0 / ( 1.0 + local * local ) ;
    }
    printf("pi is    %f \n", pi*w) ;
} /* main() */
```

The constant N is the problem size. A bigger N value should lead to better precision but requires longer time to compute.

■

Example 12.2 The target detection problem in radar signal processing

In this problem, one is given a searching space with $M \times N$ points, where some points are *targets,* such as airplanes in an aviation system. A numeric function IsTarget is used to test if a point is a target. Each point has a direction and a distance.

The problem is to find at most *MaxT* targets with the smallest distances. A sequential program follows, where a point A[i][j] has a direction i and a distance j:

```
#define     MaxT        32
#define     M           2048
#define     N           512
int     i, j, k = 0 ;
float   A[M][N] ;
generate the points A


for ( j = 0 ; j < N; j ++ )
   for ( i = 0 ; i < M; i ++ )
   {
       if ( IsTarget(i,j) ) {
            Target[k].direction = i ;
            Target[k].distance = j ;
            Target[k].feature = A[i][j] ;
            k = k + 1 ;
            if ( k > MaxTarget ) goto finished ;
       }
   }
finished:
```

The target detection problem can be solved by many paradigms. A parallel program should reduce both the worst-case and the average execution times. A parallel solution might first execute IsTarget $M \times N$ times to find all targets, and then sort them to get the first MaxT targets.

Such an algorithm may be optimal in the worst case, but not in the case when the first MaxT points happen to be the targets.

∎

Example 12.3 The Gaussian elimination problem

Gaussian elimination is a popular direct method for solving a dense linear system of equations such as Eq.(12.2). A sequential Fortran code for Gaussian elimination follows:

```
real A(n,n+1), x(n)
integer i, j, k, pivot_location(1)

do i = 1, n-1
    ! pivoting
    k = i
    do j = i, n
        if (ABS(A(i, j) .LT. ABS(A(i,k)) then k = j
    end do
    swap( A(i, i:n+1), A(i-1+pivot_location(1), i : n+1) )

    ! triangulization
    A(i, i:n+1) = A(i, i:n+1) / A(i,i)
    A(i+1:n, i+1:n+1) = A(i+1:n,i+1:n+1) -
&       SPREAD(A(i, i+1:n+1), 1, n-i) * SPREAD(A(i+1:n, i), 2, n-i+1)
end do

!Back substitution follows:
do i = n, 1, -1
    x(i) = A(i, n+1)
    A(1:i-1, n+1) = A(1:i-1, n+1) - A(1:i-1, i)* x(i)
end do
```

The code consists of two do loops. The first **do** loop performs triangulization. The second loop performs back-substitution. In each iteration *i* of the first loop, the location of the "pivot" element is first found, which has the largest absolute value. The swap subroutine exchanges the pivot row with the current row *i*. Then all elements of *A* in the *i*th column below the *i*th row are zeroed out.

■

Example 12.4 The Jacobi relaxation problem

The *Jacobi relaxation* method is often realized in a synchronous iteration paradigm. It is a simple iterative method to solve a system of linear equations $Ax = b$:

$$\begin{bmatrix} a_{11} & a_{12} & \cdots & a_{1n} \\ a_{21} & a_{22} & \cdots & a_{2n} \\ \cdots & \cdots & \cdots & \cdots \\ a_{n1} & a_{n2} & \cdots & a_{nn} \end{bmatrix} \times \begin{bmatrix} x_1 \\ x_2 \\ \cdots \\ x_n \end{bmatrix} = \begin{bmatrix} b_1 \\ b_2 \\ \cdots \\ b_n \end{bmatrix} \qquad (12.2)$$

We assume that the A matrix is *dense*, i.e., most of its elements are not zero. The Jacobi method starts with an initial estimation of the unknown vector x. Then in a number of iterative steps, a new (and better) estimation of x is computed from the current estimation. The equation for each iterative step is

$$\text{New } x_i = \left[b_i - \sum_{j \neq i} a_{ij} \times x_j\right] / a_{ii} = \left[b_i - \sum_{1 \leq j \leq n} a_{ij} \times x_j\right] / a_{ii} + x_i \qquad (12.3)$$

The iteration continues until the new x value is sufficiently close to the current x value, i.e., $\sum_{1 \leq i \leq n} (\text{new } x_i - x_i)^2 < ErrorBound$, where *ErrorBound* is a given positive number.

∎

12.2 Parallel Programming Models

In this section, we introduce the implicit parallel programming model and three explicit parallel programming models (data-parallel, shared-variable, message-passing).

12.2.1 Implicit Parallelism

Explicit parallelism means that parallelism is explicitly specified in the source code by the programmer using special language constructs, compiler directives, or library function calls. If the programmer does not explicitly specify parallelism, but lets the compiler and the run-time support system automatically exploit it, we have the *implicit parallelism* as discussed below.

Parallelizing Compilers A most popular approach of implicit parallelism is the automatic parallelization of sequential programs. The compiler performs *dependence analysis* on a sequential program's source code and then uses a suite of *program transformation* techniques to convert the sequential code into native parallel Fortran code. Such a compiler is called a *parallelizing compiler* or a *restructuring compiler*.

A key in parallelizing a sequential code is *dependence analysis*, which identifies *data dependence* and *control dependence* in the code. If operation A depends on operation B, then A must be executed after B. Two operations can be executed in parallel if they are *independent* (i.e., there is no direct or indirect dependence between them).

When dependencies do exist, *program transformation* (also known as *restructuring*, or *optimizing*) techniques are used to either eliminate them or otherwise make the code parallelizable, if at all possible.

Most existing analyzing and restructuring techniques focus on loop level, i.e., how to exploit parallelism in Fortran **do** loops or C **for** loops.

Effectiveness of Parallelizing Compilers   A crucial question to ask is, *Are parallelizing compilers effective in generating efficient code from real sequential programs?* Some performance studies indicate that they may not be effective. Blume et al. [91] at University of Illinois measured the performance of Kap, a leading commercial parallelizing compiler, on the Cedar computer using 12 Perfect benchmark programs.

The results are shown in the Automatic column of Table 12.1. Cedar is a research multiprocessor consisting of 32 processors, each having a 4-stage vector pipeline. When both parallelization and vectorization are fully utilized, the Cedar system can achieve a peak speedup of 128. However, as shown in Table 12.1, the Kap compiler achieved very limited speedup. In half of the cases the speedup is less than 1.

Table 12.1   Effects of Compiler Transformations for the Perfect Benchmark
(Courtesy: Blume et al. *IEEE Trans. Parallel and Distributed Systems,* [91], 1992)

Program	Automatic	Manual	Privatization	Reduction	Induction
ADM	0.6	10.1	9.6		
ARC2D	13.5	20.8	1.2		
BDNA	1.8	8.5	1.4	3.3	
DYFESM	2.2	11.4	2.2	2.1	
FLO52	5.5	15.3	1.0	1.1	
MDG	1.0	20.6	21.0	21.0	
MG3D	0.9	48.8	18.0	15.2	
OCEAN	0.7	16.7	3.8		8.3
QCD	0.5	20.8	8.2		
SPEC77	2.4	15.7	6.8	3.4	
TRACK	0.4	5.2	6.0		
TRFD	0.8	43.2	13.3		12.7

These researchers analyzed how the Kap compiler parallelizes each individual code and concluded that the main reason for the disappointing performance is the compiler's inability to exploit parallelism. This prompted them to investigate if automatic parallelization is a feasible approach. They manually transformed the 12 programs into parallel Cedar codes, and studied whether the applied transformations could be automated.

The speedup performance of these manually transformed codes is shown in the Manual column of Table 12.1, which shows significant improvement over the Kap generated codes. Most encouragingly, they found out that the majority of the transformations applied in the manual experiments *can* be automated. Furthermore, these transformations could be derived from the program text, not needing knowledge of the application.

Three of the most important transformations are *privatization, parallel reduction,* and the elimination of *induction variables.* Blume et al. studied the effects of various individual restructuring techniques. The three rightmost columns of Table 12.1 shows their results in terms of the slowdown factor when the labeled transformation is not performed in the manual experiments. For instance, the speedup of the ADM program will be only 10.1/9.6 if privatization is not used.

Example 12.5 Optimizing techniques for eliminating data dependency

In the following sequential do loop (on the left), statement Q depends on statement P, since Q needs the value of variable A computed in P. The N iterations of the loop can not be executed in parallel due to this dependency. The *privatization* technique eliminates this dependency by making A into an array such that each iteration i has its own private copy $A(i)$. The parallelized loop is shown on the right.

do i = 1, N	**pardo i = 1, N**
P: A = ...	P: A(i) = ...
Q: X(i) = A + ...	Q: X(i) = A(i) + ...
...	...
end do	**end pardo**

In following sequential do loop (on the left), statement Q depends on itself, since computing Sum in the ith iteration needs the value of Sum from the previous iteration. variable A computed in P. The *reduction* technique eliminates this dependency by turning the loop-carried sequence of N sequential additions into an explicit reduction function.

do i = 1, N	**pardo i = 1, N**
P: X(i) = ...	P: X(i) = ...
Q: Sum = Sum + X(i)	Q: Sum = **sum_reduce**(A(i))
...	...
end do	**end pardo**

An induction is a statement in a loop such as $X(i) = X(i-1)+Y(i)$, e.g., computing $X(i)$ in the ith iteration needs $X(i-1)$ from the previous iteration. Techniques such as that used in designing a carry-lookahead adder can be used to eliminate dependencies caused by induction.

Beyond Automatic Parallelization The Illinois research has shown that automatic parallelization is feasible. However, from Table 12.1, we can see that the efficiency of even the manually optimized codes is not high, ranging from 5.2/128 = 4% for TRACK to 48.8/128 = 38% for MG3D.

These results are obtained on Cedar, a machine with hierarchical shared-memory. The performance could be further reduced for distributed-memory MPPs and COWs, as the communication latency is higher. Other techniques have been developed to complement the compiler optimization approach. We will discuss several important ones.

User Direction The user can help the compiler in parallelizing by providing it with more information. For instance, the *interactive parallelization* approach allows a parallelizing compiler (or the run-time system) to ask questions, and the programmer can provide additional information to guide the parallelization process. But the most popular method in this direction allows the programmer to insert *compiler directives* (also called *pragmas* in C) in the source code. These are formatted comments carrying additional information to help the compiler do a better job in restructuring. For instance, consider the following code in Convex Exemplar C:

```
#pragma _CNX loop_parallel
for (i = 0; i < 1000; i++){
    A[i] = foo(B[i], C[i]);
}
```

The **loop_parallel** pragma forces the compiler to parallelize the immediately following loop, no matter what the loop body is. The compiler does not have to analyze the loop body to check for dependence. It is the user's responsibility to ensure that correctness of the code after parallelization. The above code works because the user knows that the function foo is reentrant and side-effect free. Without this application knowledge provided by the pragma, the compiler would not parallelize the loop, because it cannot determine that the function foo is reentrant and side-effect free.

Run-Time Parallelization Even with the help of compiler directives, not all parallelism can be recognized at compile time. Some can only be revealed at run time. Techniques have been proposed to perform *run-time parallelization*. For instance, in the Jade language [517] developed at Stanford University, parallelization involves both the compiler and the run-time system.

A programmer starts with an existing sequential program. Additional language constructs are used to decompose the sequential program into multiple tasks, and to specify how each task will access data. The compiler and the run-time system will automatically recognize and exploit parallelism at both compile time and runtime. The Jade approach provides two additional benefits. First, more parallelism can be recognized. Second, Jade allows the automatic exploitation of irregular and dynamic parallelism.

Explicit Parallelism The implicit parallelism approach has many advantages (see Section 12.2.3 for details) and parallelizing compiler technology is an active research topic hotly pursued by both academia and industry alike. However, there are many people who believe that explicit parallelism is needed. Their belief is influenced partly by the currently disappointing performance of automatic tools, and partly by a theoretical result obtained more than twenty years ago by A. J. Bernstein [74]:

Bernstein's Theorem: It is undecidable whether two operations in an imperative sequential program can be executed in parallel.

∎

An implication of this theorem is that there is no automatic technique, compiler time or run time, that can fully exploit all potential parallelism in a sequential program. To overcome this theoretical limitation, two solutions have been suggested.

The first is to use explicit parallelism, as discussed in Section 12.2.2. The second solution is to abolish the imperative style altogether (note that Bernstein's theorem applies only to imperative programs), and to use a programming language which makes parallelism recognition easier. See Section 12.2.4 for further discussion.

12.2.2 Explicit Parallel Models

Although many explicit parallel programming models have been proposed, three models have become the dominant ones. They are *data-parallel*, *message-passing*, and *shared-variable* models. In this section, we briefly illustrate these three models using a simple example and summarize their main features in Table 12.2.

Table 12.2 Main Features of Data-Parallel, Message-Passing, and Shared-Variable Parallel Programming Models

Feature	Data-Parallel	Message-Passing	Shared-Variable
Control flow (threading)	Single	Multiple	Multiple
Synchrony	Loosely synchronous	Asynchronous	Asynchronous
Address space	Single	Multiple	Multiple
Interaction	Implicit	Explicit	Explicit
Data allocation	Implicit or semiexplicit	Explicit	Implicit or semiexplicit

The Data-Parallel Model   This model applies to either SIMD or SPMD modes. The idea is to execute the same instruction or program segment over different data sets simultaneously on multiple computing nodes. Essentially, data-parallel programming has a single thread of control and massive parallelism is exploited at data set level. Furthermore it has a global naming space and parallel operations on aggregate data structures, and it applies loosely synchronous operations. We shall study this approach in detail in Chapter 14.

Example 12.6 An example data-parallel program to compute the π function.

Following is a simple example to explain the idea of data parallelism. This data-parallel code computes π. Before we formally introduce HPF in Chapter 14, we use a C notation here similar to HPF.

```
main() {
        double local[N], tmp[N], pi, w ;
        long i, j, t, N=1000000 ;
A:      w = 1.0 / N ;
B:      forall ( i = 0; i < N; i++) {
  P:        local[i] = ( i - 0.5 ) * w ;
  Q:        tmp[i] = 4.0 / ( 1.0 + local[i] * local[i] ) ;
          }
C:      pi = sum(tmp) ;
D:      printf("pi is    %f \n", pi*w) ;
} /* main() */
```

This program has four statements A, B (which has two substatements P and Q), C, and D. The programmer can assume that these four statements are executed by a single process one after another. However, each statement can perform the same operation to multiple data items simultaneously.

Statements A and D are just ordinary sequential statements. Statement C performs a reduction to sum all N elements of the tmp array and assigns the result to variable pi. Statement P simultaneously evaluates N right-hand side expressions and updates all N elements of the local array. All the N elements of the local array have to be updated in statement P before any operations in statement Q can be executed. Similarly, all tmp elements are assigned in statement Q before the reduction in statement C takes place.

■

Message-Passing Model   Message passing has the following characteristics:

- *Multithreading*: A message-passing program consists of multiple processes, each of which has its own thread of control and may execute different code. Both control parallelism (MPMD) and data parallelism (SPMD) are supported.

- *Asynchronous Parallelism*: The processes of a message-passing program execute asynchronously. Special operations, such as barrier and blocking communication (e.g., the MPI_Reduce in Example 12.7), are used to synchronize processes. Note that fine-grain MIMD does not benefit today's message-passing systems.

- *Separate Address Spaces:* The processes of a parallel program reside in different address spaces. Data variables in one process are not visible to other processes. Thus, a process cannot read from or write to another process's variables. The processes interact by executing special message-passing operations, such as the five MPI library function calls in Example 12.7.

- *Explicit Interactions:* The programmer must resolve all the interaction issues, including data mapping, communication, synchronization, and aggregation. The workload allocation is usually done through the *owner-compute* rule; i.e., the process that owns a piece of data performs the computations associated with it.

- *Explicit Allocation*: Both workload and data are explicitly allocated to the processes by the user. To reduce designing and coding complexity, the user often realizes applications by using the single-code approach to write SPMD programs.

Example 12.7 A message-passing code for computing the π function

We use a C notation with MPI in the following code. Readers need not fully understand the code. Details of message-passing operations will be treated in Chapter 13.

```
      #define   N   1000000
        main() {
            double local, pi, w;
            long i, taskid, numtask;
A:      w = 1.0 / N;
        MPI_Init( &argc, &argv );
        MPI_Comm rank( MPI_COMM_WORLD, &taskid);
        MPI_Comm_size( MPI_COMM_WORLD, &numtask );
B:      for ( i = taskid; i < N; i = i+numtask) {
    P:      local = ( i + 0.5 ) * w ;
    Q:      local = 4.0 / (1.0 + local * local) ;
        }
C:      MPI_Reduce (&local, &pi, 1, MPI_Double,
            MPI_MAX, 0, MPI_COMM_WORLD);
D:      if (taskid == 0) printf("pi is    %f \n", pi*w) ;
        MPI_Finalize();
    } /* main() */
```

■

Shared-Variable Model   The shared-memory model is similar to the data-parallel model in that it has a single address (global naming) space. It is similar to the message-passing model in that it is multithreading and asynchronous. However, data reside in a single, shared address space, thus does not have to be explicitly allocated. Workload can be either explicitly or implicitly allocated. Communication is done implicitly through shared reads and writes of variables. However, synchronization is explicit.

Example 12.8 A shared-variable parallel code for computing the π function

We use a C-like notation to illustrate the ideas of the shared-variable parallel programming model.

```
        #define    N   1000000
        main() {
                double local, pi = 0.0, w ;
                long i ;
A:              w = 1.0 / N ;
B:              #pragma parallel
                #pragma shared ( pi, w)
                #pragma local ( i, local )
                {
                        #pragma pfor iterate (i = 0; N ; 1)
                        for ( i = 0; i < N; i++) {
P:                      local = ( i +  0.5 ) * w ;
Q:                      local = 4.0 / ( 1.0 + local * local ) ;
                        }
C:                      #pragma critical
                                pi = pi + local ;
                }
D:      printf("pi is    %f \n", pi*w) ;
        } /* main() */
```

■

12.2.3 Comparison of Four Models

Table 12.3 compares the four models from a user's viewpoint. We use stars to indicate the relative merits. A four-star (★★★★) entry indicates that the model is most advantageous with respect to that particular issue, while one star (★) indicates the weakest. Table 12.3 only provides a snapshot of the state of the art. The relative merits are not necessarily inherent to any model. In fact, the relative ranking of these models may change in the future.

Table 12.3 Comparison of Four Parallel Programming Models

Issues		Implicit	Data Parallel	Message Passing	Shared Variable
Platform-independent examples		Kap, Forge	Fortran 90, HPF,	PVM, MPI	X3H5
Platform-dependent examples			CM C*	SP2 MPL, Paragon Nx	Cray Craft, SGI Power C
Parallelism issues		★★★★	★★★	★	★★
Allocation issues		★★★★	★★	★	★★★
Interaction issues	Communication	★★★★	★★★	★	★★★
	Synchronization	★★★★	★★★★	★★	★
	Aggregation	★★★★	★★★	★★★★	★
	Irregularity	★★★★	★★	★★	★★★
Semantic issues	Termination	★★★★	★★★★	★★	★
	Determinacy	★★★★	★★★★	★★	★
	Correctness	★★★★	★★★	★★	★
Program-ability issues	Generality	★	★★	★★★	★★★
	Potability	★★★★	★★★	★★	★
	Structuredness	★★★★	★★	★	★

Implicit Parallelism The implicit approach has many advantages. The huge repository of existing sequential software can be reused for parallel computers. The programmer only needs to use the familiar sequential language, without having to know about parallel programming or the underlying parallel architecture.

The developed programs are portable among different parallel computers. It is much easier to understand the semantics of implicit programs than explicit ones. The user has no need to worry about any of the interaction and semantic issues.

Data Parallelism The data-parallel model assumes a single address space, and data allocation is not required. However, to achieve high performance in distributed memory machines, data-parallel languages such as HPF use explicit data allocation directives. A data-parallel program is single-threaded and loosely synchronous.

There is no need for explicit synchronization. Single-threading ensures that data-parallel programs are always determinate and free of deadlocks and livelocks. The other two models do not have these advantages. The loose synchrony property makes it difficult to realize some data-parallel paradigms, such as asynchronous iteration and work pool.

There are two influential standard data-parallel languages: Fortran 90 (now Fortran 95) and High-Performance Fortran (HPF). Programs written in such languages are portable on a wide range of machines (MPPs, DSM, SMP, COWs, and PVPs). However, there are also some machine-specific languages such as C* for CM-5 that are not standard.

Message Passing   Users must explicitly allocate data and workload to processes in message-passing programs. These programs are multithreading and asynchronous, requiring explicit synchronizations (e.g., barrier, blocking messaging) to maintain correct execution order. However, these processes have their own separate address spaces. Subtle synchronization mistakes from reading/writing of shared variables will not occur.

The message-passing model is more flexible than the data-parallel model, but it still lacks support for the work pool paradigm and applications that need to manage a global data structure. There are two widely used standard libraries: PVM and MPI, which are implemented in virtually all types of parallel computers and greatly help the portability of message-passing programs.

Furthermore, message-passing programs usually exploit large-grain parallelism. They can be executed on machines that have a native shared-variable model (multiprocessor: DSMs, PVPs, and SMPs). On the other hand, shared-variable programs normally cannot be directly executed on a multicomputer.

Shared Variable   The shared-variable model assumes the existence of a single, shared address space where all shared data reside. So there is no need for data allocation. Shared-variable programs are multi-threading and asynchronous, requiring explicit synchronizations to maintain correct execution order among the processes.

Furthermore, the processes share a single address space. Subtle synchronization mistakes from reading/writing of shared variables are easy to make. There is no widely accepted standard yet. A program written for SGI Power Challenge cannot run directly on a Convex Exemplar. A shared-variable program developed for SMPs or DSMs cannot run on multicomputers such as MPPs or clusters.

A widespread misconception is that the shared-variable (shared-memory) model supports fine-grain parallelism better than the message-passing model. The shared-variable model is a parallel *programming* model and can be implemented on any parallel platform, be it PVP, SMP, DSM, MPP, or cluster. Fine-grain parallelism can be supported if the underlying platform has efficient communication/synchronization mechanism. A shared-variable program could incur higher interaction overhead and run more slowly than a message-passing one on a cluster, an MPP, or even an SMP.

Another popular statement is that shared-variable programming is easier than message-passing programming. This statement is not wrong, but it is not an established fact backed by scientific experiments, either. The fact is that there are many more *existing* applications developed for SMPs and PVPs than MPPs and clusters. However, for developing *new*, efficient parallel programs that are loosely synchronous and have regular communication patterns, the shared-variable approach is not necessarily easier than the message-passing one. The shared-variable model does have an edge for irregular parallel programs, where it is difficult to specify the needed interactions using message-passing primitives. It also allows global pointer operations, a capability lacking in the message-passing model. Shared-variable programming does not have to explicitly partition and allocate data, but this may hurt performance.

As far as debugging is concerned, shared-variable programs may be more difficult than message-passing ones. All processes in a shared-variable program reside in a single address space, and access to shared data must be protected by synchronization constructs such as locks and critical regions. Subtle synchronization errors can easily occur that are difficult to detect. These problems occur less frequently in a message-passing program, as the processes do not share a single address space.

12.2.4 Other Parallel Programming Models

All the parallel programming models discussed above share a common computational style: they are all *imperative*. That is, the program contains commands that order the system to perform a sequence of actions. Other computational styles exist, including the following:

- *Functional programming*: The traditional von Neumann model specifies a computation by describing *how* to do it in an imperative program. In other words, it must specify the control flow of the program. A functional programming model (also known as a *dataflow* style) describes the functional relationship between the input data and the result of the computation. The compiler and the run-time system will generate the desired execution sequence from the functional specification.

- *Logic programming*: The logic programming style goes one step further. It specifies neither the control flow nor the function. Instead, it just specifies *what* is to be computed, i.e., the logic relation between the input and the output. The compiler and the run-time system must derive an algorithm from the logic specification.

- *Computing-by-learning*: In some applications, the user not only does not have an algorithm, but also cannot specify the functional or logic relation. For instance, how does the user tell the computer to recognize a smiling face? The compute-by-learning style first constructs a learning program for the application. Then a number of examples are presented to the learning algorithm, which eventually develops an algorithm to accomplish the task.

- *Object-oriented programming*: With this style, a program is comprised of a number of *objects*, where an object is an *abstract data type*, i.e., a set of data structures (data types) and a set of operations (known as *methods*) that operate on the data structures. Furthermore, objects are grouped in an *inheritance hierarchy*: A child object inherits the data structures and methods of a parent object.

Example 12.9 Comparison of parallel programming styles

Suppose we want to compute $y = (a + b) / c$. With the imperative style, we can write the following code:

```
t1 = a + b;
t2 = 1 / c;
y = t1 * t2 ;
```

Note that this code overspecifies by forcing the addition before the reciprocation. The functional style is more relaxed:

```
y = mul( +(a, b), rec(c) ) ;        /* rec denotes reciprocation */
```

The logic programming style specifies even less:

```
find y such that y * c = a + b
```

■

These nonimperative styles have many advantages over the imperative style. Take the functional style. It has been shown that a purely functional program is

- *Coherent*: Afunctional program is guaranteed to be determinate, deadlock-free, and compositional.
- *Abstract*: there is no need for users to explicitly specify parallelism, communication, synchronization, scheduling, etc. In fact, many issues will not arise in a functional program.

However, the imperative style has its advantages, too. First, it is a mature, time-proven style familiar to most users. Second, it shares the von Neumann model with the computer hardware (e.g., commodity processors), thus can be efficiently implemented. Third, even if functional programming has good performance, many people would still use the imperative style, which is a fundamental style that can be used to naturally represent many problems solutions.

The imperative style dominates programming of today's sequential or parallel computers. For this and other practical considerations, this book concentrates on the imperative programming style only

12.3 Shared-Memory Programming

Parallel programming based on the shared-memory model has not progressed as much as message-passing parallel programming. An indicator is the lack of a widely accepted standard such as MPI or PVM for message-passing. The current situation is that shared-memory programs are written in a platform-specific language for multiprocessors (mostly SMPs and PVPs). Such programs are not portable even among multiprocessors, not to mention multicomputers (MPPs and clusters).

On the other hand, message-passing programs using PVM or MPI can run on all parallel computers, including SMPs and PVPs. Consequently, most modern high-performance computers explicitly support PVM, MPI, and HPF. An SMP or PVP system also supports its native shared-memory parallel language. Establishing a practical standard is the number one task for the research and development of shared-memory parallel programming. The new OpenMP effort is extremely important.

We first discuss three platform-independent shared-memory programming models: X3H5, Pthreads, and OpenMP. We next discuss two commercial languages. The SGI Power C uses a small set of structured constructs to extend C to a shared-memory parallel language.

The Cray MPP model integrates the supports for data-parallel, message-passing, and shared-memory. We finally discuss a new language called C//, which adds a small number of orthogonal constructs to facilitate the development of coherent programs, i.e., parallel programs that are structured, terminative, determinate, and compositional.

Shared-Memory Standards There are three shared-memory programming standards. The X3H5 standard has not gained wide acceptance, but has influenced the design of several commercial shared-memory languages.

The X3H5 has evolved into the OpenMP standard. Pthreads is an influential standard for multithreading at the Unix operating system layer. We use Solaris Threads to illustrate multithread programming, as it has many common features with Pthreads and is a mature commercial product.

12.3.1 The ANSI X3H5 Shared-Memory Model

The X3H5 shared-memory model is a de jure, ANSI standard established in 1993. But it has not received as wide acceptance as the de facto MPI standard for message passing. In fact, no commercial shared-memory systems adhere to X3H5, although they have adopted many of its ideas.

We will see the influence of X3H5 in Section 12.3.1. The X3H5 defines one conceptual standard programming model, and three bindings of the model for C, Fortran 77, and Fortran 90. We will just discuss the main ideas. For more details, see the complete specification [39].

Parallelism Constructs The X3H5 uses a number of constructs to specify parallelism, as shown in Fig. 12.3 and Fig. 12.4. An X3H5 program does not explicitly specify how many threads are used to execute the program. A program starts in sequential mode with a single initial thread, called the *base thread* or *master thread*.

program main	! The program begins in sequential mode
A	! *A* is executed by only the base thread
parallel	! Switch to parallel mode
B	! *B* is replicated by every team member
psections	! Starts a parallel block
section	
C	! One team member executes *C*
section	
D	! Another team member executes *D*
end psections	! Wait till both *C* and *D* are completed
psingle	! Temporarily switch to sequential mode
E	! *E* is executed by one team member
end psingle	! Switch back to parallel mode
pdo i=1,6	! Starts a pdo construct
F(i)	! The team members share the 6 iterations of *F*
end pdo no wait	! No implicit barrier
G	! More replicate code
end parallel	! Switch back to sequential mode
H	! *H* is executed by only the initial process
...	! There could be more parallel constructs
end	

Figure 12.3   Parallel constructs in the ANSI X3H5 standard.

A *parallel construct* (also known as *parallel region*) is a pair of **parallel** and **end parallel** with the enclosed code. When the program encounters a **parallel**, it switches to parallel mode by spawning zero or more children threads. The base thread and its children form a team. All the team members proceed in parallel to execute the subsequent code till an **end parallel**. Then the program switches back to sequential mode. Only the base thread continues execution.

Parallel Blocks Inside a parallel construct, there can be a number of *work-sharing* constructs, which are either a *parallel block* (**psections**... **end psections**), a *parallel loop* (**pdo** .. **end pdo**), or a *single process* (**psingle** ... **end psingle**) construct.

Any code that is not a work-sharing construct is duplicatedly executed by every team members. A parallel block is used to specify MPMD parallelism. It consists of several component *sections* (units of work), each to be executed by a team member.

Parallel Loop A parallel loop is used to specify SPMD parallelism. It can be considered a special form of a parallel block, where all the sections (iterations) have the same code. A single process construct dictates that the code shall be executed by only one team member sequentially.

The sequencing of the code in Fig. 12.3 is depicted in Fig. 12.4, assuming three threads are used. Initially, there is just the base thread P executing code A. When encountering **parallel**, it spawns two child threads Q and R, so the team now has three members. The code B is not in a work-sharing construct, thus is executed by all three threads.

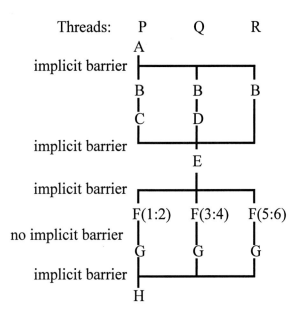

Figure 12.4   Illustration of the X3H5 program in Fig. 12.3.

The parallel block (**psections**) has only two sections C and D, which are executed by any two members of the team. The single-process code E is executed by any one of the team members (e.g., Q). The **pdo** has six iterations which are executed by the team in an arbitrary fashion. The only constraint is that each iteration is to be executed by one team member (thread).

Example 12.10   Work-sharing construct

A work-sharing construct allows the work to be load-balanced among the member threads. The load distribution in Fig. 12.4 for the **pdo** in Fig. 12.3 is just one possible scenario. When a thread finishes an iteration, it can go to execute another of the remaining iterations.

Suppose iterations $F(1)$ and $F(2)$ take a long time to execute, while the remaining iterations take short times. Then threads P and Q will execute $F(1)$ and $F(2)$, respectively, while thread R will execute iterations $F(3..6)$.

■

Implicit Barrier There is an implicit barrier at **parallel, end parallel, end psections, end pdo,** or **end psingle** . This barrier is also an implicit *fence operation* that forces all memory accesses up to this point to become consistent. If there is no need for the implicit barrier, a **no wait** should be added. This is shown in the **end pdo no wait** statement of Fig. 12.3. Now thread P, after finishing F(1:2), can proceed to execute G without having to wait for the completion of the entire parallel loop (see Fig. 12.4).

Parallel and work-sharing constructs can be nested. Then a team member that encounters an inner parallel or work-sharing construct becomes the base thread, which can spawn additional threads to form a new team.

Thread Interaction The X3H5 model includes many interesting features for threads interaction. We will briefly discuss several that have been widely adopted.

- A variable has a *shared/private* attribute within a parallel construct. A private variable belongs to one team member and is invisible to other team members. A shared variable belongs to (can be read/modified by) all team members.
- The X3H5 pays attention to the memory consistency issue, and provides ways to guarantee memory consistency. An example is the implicit barrier mentioned above. The X3H5 provides also explicit barrier and fence operations.
- The X3H5 model introduces four types of synchronization objects (variables) *latch*, *lock*, *event*, and *ordinal*, and associated initialization and synchronization operations. Table 12.4 shows the possible states and operations applicable to these synchronization objects.

Table 12.4 States and Operations of Synchronization Objects

Object Type	Possible States	Representative Operations
Latch	Uninitialized, unlatched, latched	Critical region
Lock	Uninitialized, unlocked, locked	Test, lock, unlock
Event	Uninitialized, clear, posted	Wait, post, clear
Ordinal	Uninitialized, integer	Sequence

Each type of object has an initialize operation and a destroy operation. Any object must first be initialized before it can be used by any other operation. A latch, lock, or event object is initialized by assigning to the unlatched, unlocked, or clear state, respectively. An ordinal object is initialized to either an integer value or a two-element integer array. Any object can be destroyed by assigning it an uninitialized state.

Synchronization Lock and event synchronizations are similar to those in Section 7.2.2. An ordinal object is used to synchronize threads according to their ranks. For instance, we can use an ordinal object to specify that thread T_i cannot enter a critical section until after $T_{i-1}, ..., T_1$ have all executed the critical section. The latch object is used with a critical region, which has the following syntax:

> **critical region** [(latch_variable)]
>   critical_section_code
> **end critical region**

This construct is similar to the generic critical region construct in Section 7.2.2, except that an optional latch variable may be included. All critical regions without a latch are regarded as having a common, implicit, system-supplied latch. Using latches can reduce contentions and increase parallelism, as shown in the following example.

Example 12.11 Mutual exclusion using latches

Suppose a sequential program needs 10,000 s to execute on one processor. By partitioning it into 100 threads and running on a 100-processor multiprocessor, we want to reduce the execution time to 100 s. The problem is that there is a critical section that must be executed in a mutually exclusive manner.

Thread T1	T2	...	T100
...	...	...	...
Critical region	Critical region	...	Critical region
...	...	...	...

A trivial solution follows that uses critical regions to enforce mutual exclusion. Suppose each critical region needs 10 s. This seems to be small compared to the original 10000 s time. However, due to mutual exclusion, the 100 threads must execute their critical regions one by one, for a total execution time of $100*10+100 = 1100$ s. The speedup is reduced to only 10000/1100=9.

On closer inspection, we find that the 100 threads interact in 50 pairs: T1 needs to be mutually exclusive only with T2, T3 with T4,... , T99 with T100. A better solution follows that uses 50 different latch variables.

Thread T1	T2	...	T100
...	...	...	...
critical region(L1)	critical region(L1)	...	critical region(L50)
...	...	...	...

Only critical regions with the same latch need to be mutually exclusive. The total execution time is reduced to 2*10+100 = 120, and the speedup is improved to 83. ∎

12.3.2 The POSIX Threads (Pthreads) Model

The POSIX Threads (or *Pthreads*) stands for official IEEE POSIX 1003.1c-1995 thread standard, which was established by the IEEE standards committee. Its functionality and interface are similar to those of Solaris Threads. In this section, we focus on features that are common to both, selected from IEEE standard and commercial product.

Thread Management   The thread library routines used to manage threads are shown in Table 12.5. The *pthread_create*() routine creates a new thread within a process. The new thread executes *myroutine* with an argument *arg*.

The attributes of the new thread are specified with *attr*, or assume default attributes if *attr* is NULL. If the *pthread_create*() succeeds, it will return 0 and put the new thread ID into *thread_id*; otherwise, an error code will be returned, indicating the error type.

Table 12.5 Basic Thread Management Primitives in Pthreads

Function Prototype	Meaning
int *pthread_create*(pthread_t * *thread_id*, pthread_attr_t * *attr*, void * (**myroutine*)(void *), void * *arg*)	Create a thread
void *pthread_exit*(void * *status*)	A thread exits
int *pthread_join*(pthread_t thread, void ** *status*)	Join a thread
pthread_t *pthread_self*(void)	Returns the calling thread ID

The *pthread_exit*() routine terminates the calling thread and makes *status* available to the thread that successfully joins with the terminating thread. In addition, *pthread_exit*() executes any remaining cleanup handlers in the reverse order they were pushed, after which all appropriate thread-specific destructors are called.

An implicit call to *pthread_exit()* is made if any thread, other than the root thread where main() was first called, returns from *myroutine* specified in *pthread_create()*. The return value takes the place of *status*. Ff the terminating thread is the last thread in the process, the process exits as if exit() were called with a status of 0. The *pthread_self()* routine returns the thread ID of the calling thread.

Pthreads Synchronization Examples of Pthreads synchronization primitives are listed in Table 12.5. We focus on mutual-exclusion (*mutex*) variables and conditional (*cond*) variables. The former are similar to the semaphore construct and the latter the event construct. Note that before a synchronization variable can be used, it needs to be created (initialized). After the user is done with using a synchronization variable, it should be destroyed to clean up (release memory, etc.).

Table 12.6 Example Thread Interaction Primitives in Pthreads

Function	Meaning
pthread_mutex_init(...)	Creates a new mutex variable
pthread_mutex_destroy(...)	Destroy a mutex variable
pthread_mutex_lock(...)	Lock (acquire) a mutex variable
pthread_mutex_trylock(...)	Try to acquire a mutex variable
pthread_mutex_unlock(...)	Unlock (release) a mutex variable
pthread_cond_init(...)	Creates a new conditional variable
pthread_cond_destroy(...)	Destroy a conditional variable
pthread_cond_wait(...)	Wait (block) on a conditional variable
pthread_cond_timedwait(...)	Wait on a conditional variable up to a time limit
pthread_cond_signal(...)	Post an event, unlock one waiting process
pthread_cond_broadcast(...)	Post an event, unlock all waiting process

The *pthread_mutex_lock()* routine locks a mutex variable, if it is not already locked. Then the routine returns and the calling thread acquires the mutex and becomes its owner. If the mutex is already locked, the calling thread blocks until the mutex becomes available (unlocked). The *pthread_mutex_trylock()* routine is similar to a test-and-set. It locks the mutex and returns immediately. The returned value indicates whether the mutex was already locked or not. This routine never blocks.

The *pthread_mutex_unlock()* routine releases a previously acquired mutex. In other words, the calling thread must be the owner of the mutex for the unlock operation to be successful. When the mutex is released, it can be acquired by another thread as follows: If there are threads waiting for the mutex, the scheduler will determine which thread obtains the lock. If no thread is blocked waiting for the mutex, the mutex will be available to the next thread that calls *pthread_mutex_lock*() or *pthread_mutex_trylock*().

The *pthread_cond_wait*() routine atomically blocks the current thread waiting on a condition variable and unlocks a mutex variable. The waiting thread unblocks only after the same condition variable is signaled (posted) by another thread calling *pthread_cond_signal*() or *pthread_cond_broadcast*() *and* the current thread aquires the lock on the mutex. The *pthread_cond_timedwait*() is similar to *pthread_cond_wait*(), except that it will unblock when the waiting time reaches a limit.

The *pthread_cond_signal*() routine unblocks one thread that has been blocked waiting for the condition variable. The scheduler will determine which thread will be unblocked. The *pthread_cond_broadcast*() routine unblocks *all* threads that have been blocked waiting for the condition variable.

12.3.3 The OpenMP Standard

The OpenMP [475] is an shared-memory standard supported by a group of hardware and software vendors, such as DEC, Intel, IBM, Kuck & Associates, SGI, Portland Group, Numerical Algorithms Group, U.S. DOE ASCI program, etc. It is a set of compiler directives, library routines, and environment variables that collectively provides an shared-memory API for Unix and Windows NT platforms. The first standard specification with a Fortran binding came out in October 1997 [476]. Supports for C and C++ are planned.

OpenMP in Perspective Table 12.7 compares OpenMP with existing parallel programming standards, based on OpenMP's perspective..

Table 12.7 Comparison of Five Parallel Programming Standards
(Courtesy: OpenMP Standards Board [475], 1997)

Attribute	X3H5	MPI	Pthreads	HPF	OpenMP
Scalable	no	yes	sometimes	yes	yes
Fortran binding	yes	yes	no	yes	yes
C binding	yes	yes	yes	no	planned
High level	yes	no	no	yes	yes
Performance oriented	no	yes	no	yes	yes
Supports data parallelism	yes	no	no	yes	yes
Potable	yes	yes	yes	yes	yes
Vendors support	no	widely	Unix SMP	widely	starting
Incremental parallelization	yes	no	no	no	yes

We use the Pthreads column to explain this table. The OpenMP group believes that Pthreads is not scalable, since it is targeted at low-end Unix SMPs, not technical high-performance computing. It does not even have a Fortran binding. Pthreads is low level, because it uses the library approach, not compiler directives.

The library approach precludes compiler optimization. Pthreads supports only threads parallelism, not fine-grain data parallelism. Pthreads is portable and supported among major Unix SMP platforms (including single-processor workstations).

OpenMP inherits many X3H5 concepts. We only discuss several important features that differentiate OpenMP from X3H5, as shown in Table 12.8.

Table 12.8 Comparison of Main Features of X3H5 and OpenMP
(Courtesy: OpenMP Standards Board, [475], 1997)

Attribute		X3H5	OpenMP
Overview	Orphan directives	None	Yes
	Library functions and environment variables	None	Standard
	Throughput mode	Undefined	Yes
Parallelism construct	Parallel region	PARALLEL	PARALLEL
	Iterative	PDO	DO
	Noniterative	PSECTION	SECTION
	Single thread	PSINGLE	SINGLE, MASTER
Data environment	Autoscope	None	DEFAULT
	Reduction	None	REDUCTION
	Private Initialization	None	FIRSTPRIVATE, COPYIN
	Private persistence	None	LASTPRIVATE
Synchronization constructs	Barrier	BARRIER	BARRIER
	Fence	SYNCHRONIZE	FLUSH
	Critical region	CRITICAL SECTION	CRITICAL
	Atomic region	None	ATOMIC
	Locks	None	Full functionality

This concept makes OpenMP flexible to support coarse-grain parallelism. Many sequential codes can be parallelized in the simple way illustrated in Fig. 12.5.

```fortran
        parameter(N = 512, NZ = 16)
        common /setup/ npoints, nzone
        dimension field(N), ispectrum(NZ)
        data npoints, nzone / N, NZ /

!$OMP PARALLEL DEFAULT(PRIVATE) SHARED(field, ispectrum)
        call initialize_field(field, ispectrum)
        call compute_field(field, ispectrum)
        call compute_spectrum(field, ispectrum)
!$OMP END PARALLEL
        call display(ispectrum)
        stop
        end


   subroutines initialize_field and compute_field are ignored


        subroutine compute_spectrum(field, ispectrum)
        common /setup/ npoints, nzone
        dimension field(npoints), ispectrum(nzone)

!$OMP DO
        do i= 1, npoints
             index = field(i)*nzone + 1
!$OMP ATOMIC
             ispectrum(index) = ispectrum(index) + i
        enddo
!$OMP END DO NOWAIT
        return
        end
```

Figure 12.5   OpenMP code to illustrate orphan directives (Courtesy: OpenMP Standards Board, 1997 [475])

Pthreads does not support incremental parallelization well. Given a sequential computing program, it is difficult for the user to parallelize it using Pthreads. The user has to worry about many low-level details, and Pthreads do not naturally support loop-level data parallelism. Concequently, the user needs to modify much of the sequential code. As will be seen in Problem 12.6, the Pthreads code for π-computation is much more complex than the sequential one in Example 12.1.

OpenMP is designed to alleviate the shortcomings discussed above. We think OpenMP is a promising standard, much improved from PCF and X3H5. However, for its wide acceptance, the key is to develop good compilers and run-time environments.

Key Features of OpenMP OpenMP incorporates the concept of *orphan directives*, do not appear in the lexical extent of a parallel construct but lie in its dynamic (execution) extent. As illustrated in Fig. 12.5, the three directives in the subroutine compute_spectrum are orphaned directives. They are not lexically enclosed in the parallel construct of the main program, but they are in its dynamic execution path.

We parallelize the main program using one or more parallel directives, and use other directives to control execution in the called subroutines. This way, we could enable parallel execution of major portions of the code with small code modification. This concept also facilitates the development and reuse of modular parallel programs. We suggest the reader rewrite Fig. 12.5 in X3H5 (which does not support orphan directives) to fully appreciate the usefulness of the concept of orphan directives (see Problem 12.8).

Beisdes compiler directives, OpenMP provides a set of run-time library routines with associated environment variables. They are used to control and query the parallel execution environment, provide general-purpose lock functions, set execution mode, etc. For instance, OpenMP allows a *throughput mode*.

The system then dynamically sets the number of threads used to execute parallel regions. This can maximize the throughput performance of the system, probably at the expense of prolonging the elapsed wall-clock time of one application. X3H5 and OpenMP have similar parallelism directives. Only the notations are slightly different. OpenMP includes a new MASTER directive. Only the master thread should execute the code.

OpenMP provides more flexible functionality to control the data environment than X3H5. For instance, OpenMP supports reduction by a REDUCTION($+$, *sum*) clause in a PARALLEL DO directive, demonstrated in Fig. 12.6. A private copy of the reduction variable *sum* is created for each thread. The private copy is initialized to 0 according to the reduction operator $+$. Each thread computes a private result.

At the end of PARALLEL DO, the reduction variable *sum* is updated to equal the result of combining the original value of the reduction variable *sum* with the the private results using the operator $+$. Reduction operators other than $+$ can be specified. The autoscope feature refers to the DEFAULT clause in Fig. 12.5.

The clasue DEFAULT(PRIVATE) directs all variables in a parallel region to be private, unless overwritten by other explicit SHARED clauses. There is also a DEFAULT(SHARED) clause to direct all variables as shared. Autoscoping makes it unnecessary to explicitly enumerate all variables. This can save programmers' time and reduce errors.

```
            program compute_pi
            integer n, i
            real*8 w, x, sum, pi, f, a
c function to integrate
            f(a) = 4.d0 / (1.d0 + a*a)
            print *, 'Enter number of intervals: '
            read *, n
c calculate the interval size
            w = 1.0d0/n
            sum = 0.0d0
!$OMP PARALLEL DO PRIVATE(x), SHARED(w), REDUCTION(+: sum)
            do i = 1, n
                x = w * (i - 0.5d0)
                sum = sum + f(x)
            enddo
!$OMP END PARALLEL DO
            pi = w * sum
            print *, 'computed pi = ', pi
            stop
            end
```

Figure 12.6   Computing π in OpenMP. (Courtesy: OpenMP Standards Board, [475], 1997)

OpenMP introduces an ATOMIC directive (see Fig. 12.5) which allows the compiler to take advantage of the most efficient scheme to implement atomic updates to a variable. This is superior to mutually exclusive constructs such as critical regions and locks, as discussed in Section 2.4.1 and Section 7.2.1

12.3.4 The SGI Power C Model

The SGI Power C language is an extension to the sequential C language with compiler directives (pragmas) and library functions, to support shared-variable parallel programming. Similar extended constructs are also provided for Fortran.

The Power C design is elegant in that it is structured and very simple. There are only a few pragmas and a small number of library functions. We have found it easy to learn and use, yet flexible enough for many types of technical computing problems. The main constructs for thread management are illustrated in Example 12.12.

Example 12.12 The Power C threads management primitives

Consider the following Power C code, the structure of which is depicted in Fig. 12.7. The **parallel** pragma indicates that the next block (e.g., code enclosed in "{" and "}") is a parallel block, which can be executed by one or more threads. The **shared** pragma specifies that variables x and y are shared among the threads.

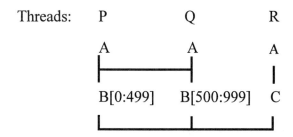

Figure 12.7   Illustration of Power C threads management primitives.

In other words, there is only one copy of x and y throughout the entire code. When one thread writes x, other threads will see the modified value. The **local** pragma indicates that variables a and b are local to each thread. In other words, there is one copy of a and b per thread. When one thread writes its a, other threads cannot see the change.

```
#pragma  parallel     shared (x,y)    local (a,b)  // a parallel block
{
    code A                   // A is duplicated and executed by every thread
    #pragma pfor iterate (i=0; 1000; 1)    // a work sharing construct
    {    for (i=0; i<1000; i++) code B }
    #pragma independent { code C }
}
```

Within a parallel block, additional pragmas can be used to denote specific parallel structures. Any code that is not annotated with a directive is called a *local code*, such as *A*. Local code is always duplicated in all threads. The **pfor** pragma denotes that the next block is a *work-sharing* one. It happens that the block is a loop. So the workload of 1000 iterations is shared among threads.

Assume two threads are used. Then each thread could get 500 iterations to process. Note that the work distribution does not have to be so even. If one thread happens to be faster, it can perform more work.

The **independent** pragma says that the next statement (code *C*) should be executed by a thread independently of the other work in the parallel block. Derived from the sequential code "*A*; **for** (i=0; i<1000; i++) *B*; *C*;", this parallel code has the same semantics as the sequential code. This is a main feature of the compiler directive approach: when all directives are deleted, we have a valid sequential program that computes the same function as the parallel one.

■

Besides pragmas, the Power C language provides a few library functions for thread management, which can be used to inquire about the number of threads, the thread ID, etc.

Synchronization Primitives A small number of primitives (pragmas and library routines) are used for thread synchronization, including the following:

- **Critical region**. Power C uses the structured *critical region* construct to realize mutual exclusion in a parallel code. This is a better construct than unstructured ones such as locks or semaphores.
- **One processor**. This pragma is used to indicate that some code should be executed by only one thread. Note that the one-processor primitive is different from critical region, which must be executed by all threads involved. This pragma is similar to that in the X3H5 standard.
- **Synchronize**. This is the standard barrier synchronization primitive.
- **Fuzzy barrier**. In the standard barrier, all threads must synchronize at the barrier point defined by the synchronize operations. If one thread has not arrived at the barrier point, all other threads must wait and cannot execute any useful operations. A fuzzy barrier can alleviate this problem, as shown in Fig. 12.8.

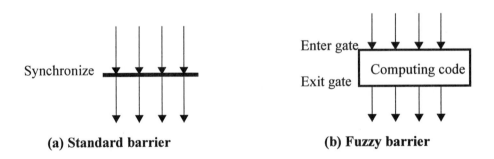

(a) Standard barrier   (b) Fuzzy barrier

Figure 12.8   Comparison of a standard barrier and a fuzzy barrier.

The barrier is not a point any more, but a *zone* enclosed by **enter gate** and **exit gate** primitives. When a thread reaches an enter gate, it can proceed to execute the useful computing code in the barrier zone, without having to wait for other threads. However, no threads can exit a barrier zone until all threads have entered the zone. With proper insertion of useful computation codes in the barrier zone, fuzzy barrier can significantly reduce the barrier-waiting overhead.

12.3.5 C//: A Structured Parallel C Language

We have proposed in [658] a new parallel language C// (pronounced *C Parallel*), which is based on the standard C language with a small set of extended constructs for parallelism and process interaction. At the core of C// is a construct called *coherent region*, which facilitates the development of structured, determinate, terminative, and compositional parallel programs. We will discuss only the salient features of C//.

Coherent Parallel Programs   We believe a key to overcoming the parallel software crisis is a parallel language that facilitates the development of *coherent parallel programs*, where coherent means the following properties.

(1) *Terminative*: The program always finishes execution and enters a final state, starting from any initial state. Recall that when a parallel program is not terminative, it can be in an infinite loop, livelock or deadlock.

(2) *Determinate*: A program is *determinate* if for any initial data state, there is only one final data state. That is, the program is a one-to-one function over the set of data states. Determinacy guarantees that the results of a program are reproducible, which facilitates debugging, among other things.

(3) *Compositional*: A program is *compositional* if its semantics can be uniquely determined by the program's structure and the semantics of its components. For instance, the semantics of **if** (*C*) *R* **else** *T* are completely determined by its structure "**if** (...)... **else**..." and the semantics of its components *C*, *R* and *T*. Compositionality facilitates program development and ease of understanding.

(4) *Structured*: A program is *structured* if it is composed of *structured constructs*, where each structured construct has the following properties: (a) It is a single-entry single-exit construct. (b) Different semantic entities are clearly identified in the syntax of the construct. (c) Related operations are enclosed in one construct, instead of being scattered among different places in the program.

Parallelism Constructs   We assume the usual C definitions of variables and expressions, and a *sequential block* is a sequence of statements enclosed in a pair of curly brackets: $\{S_1 S_2 ... S_n\}$. MPMD parallelism is specified using a *parallel block*:

$$\textbf{par } [\text{region\_name\_declarations}] \ \ \{S_1 S_2 ... S_n\}$$

where each component process S_i can be any process, such as another parallel block. The optional region name declaration will be discussed shortly. C// has also a shorthand notation, the *parallel loop* construct for specifying SPMD parallelism:

> **parfor** (i=e1; e2; e3)   /* loop head with loop index variable i */
> [region_name_declarations]
> { S }   /* loop body */

The expressions e1, e2, and e3 in the loop head are used to generate a set $I = \{I_1, I_2, ..., I_n\}$ of loop index values, in a way similar to the sequential for loop:

> $I = \varnothing; j = e1;$ **while** (e2) { $I = I \cup \{i\}$; e3; }

Let $n(I)$ (or simply n) be the number of elements of I. The parallel loop construct creates $n(I)$ processes, where each process is a copy of the identical code S and each uses a different loop index value I_j. Thus a parallel for loop can be viewed as a shorthand notation for an equivalent parallel block. In general, **parfor** (e1; e2; e3) { S } is equivalent to

> **par** { $\{S(I_1)\}$ $\{S(I_2)\}$ ... $\{S(I_n)\}$ }

For instance, parallel block **par** {Process(1) ... Process(n)} is equivalent to **parfor** (i=1; i<=n; i++) { Process(i) }, where $I = \{ 1, 2, ..., n \}$.

The loop head is evaluated to generate the *complete* index value set I before any loop body statement is executed. I should always be finite, otherwise the parallel loop will not terminate, as in the case of **parfor** (i=1; ; i++) { ... }. If I is empty, the parallel loop is equivalent to an empty statement ";". I should be evaluated from the loop head alone. To avoid confusion, C// requires that e1 and e2 be side-effect free, and that e3 can only affect the loop index variable.

Coherent Region The component processes of a parallel block (or a parallel for loop) interact through calls on *coherent regions*. Such a region call starts with a reserved keyword **core**, followed by an optional *region name*, an optional *region condition*, and an optional *region body* which is enclosed in two barriers:

> **core** region_name (region_condition)
> {   /* entry barrier */
>   region_body
> }   /* exit barrier */

A region may or may not have a name. Any region with a name must have the region name declared at the beginning of a **par** block (or parallel loop) as follows:

> **par**
>   **region** region_name [participant_list] ;
>   { ... }

where the optional participant list enumerates all processes (called the *participants* of the region) that use the region to interact. When this list is not explicitly specified, all component processes of the parallel block are its participants.

A region condition is a logic expression without side effect. When the region condition is missing, it is considered as if the condition expression were always TRUE. When a named region is called by its participants, the conditions in all the region calls of its participants must be identical. Thus we are justified to say *the* condition of a region.

A region body consists of a sequence of *parallel segments* or/and *sequential segments*. A sequential segment contains a sequence of (conditional) assignment statements. A parallel segment contains a sequence of the following statements:

- Assignment;
- *Aggregation assignment* of the form "var = op (expression);", where op is a reduction or scan operation, such as "sum_reduction", "minimum_scan", etc.
- Conditional assignment of the form "**if** (condition) assignment" or "**if** (condition) aggregation assignment". The condition should not have any side effect.

Each parallel segment must satisfy the *single-assignment rule*, which says that no variable can be assigned more than once in a segment.

The Semantics of Coherent region   When a process reaches a **core** R, it has to wait until all participants of the region R also reach their **core** R. Then the region condition for this region is evaluated. If the condition evaluates to FALSE, all participant processes wait until the condition becomes TRUE. Then all participants enter the region in parallel to execute the region body. After the body is executed, all participants leave the region to execute their respective subsequent computations. Note that the only participant of a nameless region is the process that calls it.

The evaluation of the region condition and execution of the region body is viewed as an *atomic* action. The body of a region (we assume all bodies of a region are combined into a single region) is executed according to the following rules:

- If there are multiple segments in a region body, they should be executed one after another in the order specified.
- A sequential segment is executed as a sequential block.

- All assignments in a parallel segment can be viewed to conceptually form a parallel assignment statement with the following syntax:

    Variable-1, ..., Variable-N = Expression-1, ..., Expression-N

    The semantics is that all N expressions are first evaluated to N values. Then the N values are assigned to the N corresponding variables.

Support of Process Interactions   The coherent region is a versatile construct. Communications and aggregations are demonstrated in Example 12.13 and Example 12.13. How to realize synchronizations is shown in Fig. 12.9.

In general, an atomicity or a critical region synchronization is realized by a unconditional, nameless coherent region; a control synchronization is realized by a named region; and a data synchronization is realized by a conditional coherent region.

We say a *competitive interaction* is involved when processes interact by competing for shared resource. All others interactions are called *cooperative*. Usually, cooperative interactions are realized by named unconditional regions, while competitive interactions are realized by nameless conditional regions.

Example 12.13 How communications are realized in C//

The following parallel block shows how permutation, broadcasting, multicasting, and conditional assignment can be realized:

```
parfor (i=0;i<n;i++)
    region Region1;
{   ...
    core Region1
    { a[i] = d ;          // 1-to-n broadcast: d is broadcast to a[0], ..., a[n-1]
      b[i] = e[f[i]] ;  // m-to-n multicast if f[i] evaluate to m different values
      c[i] = c[i+1 mod n]; // a permutation
      if (i % 2 == 0) left[i+1]=right[i-1]=u[i];}  //a conditional assignment
      }   ...
}
```

The conditional assignment is useful for some processes to communicate. For instance, only even-numbered processes send out their u[i] values in the above region.

The entire segment is a parallel assignment. In particular, the portion of the parallel assignment for "c[i] = c[i+1 **mod** n]" is equivalent to

```
        LOCK(S)                    core (S) { S=S-1; }
         Critical section             Critical section
        UNLOCK(S)                  core { S=S+1; }
```

(a) Mutual exclusion using locks   **(b) C// code for mutual exclusion**

```
        await(S);                  core (S){}
        signal(S);                 core {S = 1;}
        clear(S);                  core {S = 0;}
```

(c) Three event operations   **(d) C// codes for the three event operations**

```
        await (condition)          core (condition)
          do critical_section        { seq critical_section }
```

(e) Conditional critical region   **(f) C// code for conditional critical region**

```
        result = F&A(x, v)         core { result = x; x = x + v; }
```

(g) A fetch-and-add operation   **(h) C// code for fetch-and-add**

```
        C&S(w, old, new)           core {
        int * w, old, new;             seq
        { if (*w == old)                   result = (w ==old)? w=new,1:0;
          {*w = new; return 1; }      }
          else  return 0;
        }
```

(i) Compare and swap   **(j) C// code for "result = C&S(&w,old,new)"**

```
        core R {}                  core R {s = Sum_reduction(a[i]) ; }
```

(k) Barrier synchronization in C//   **(l) Sum reduction in C//**

Figure 12.9   Realization of various synchronizations using the coherent region construct in C//.

c[0], c[1], ..., c[n-1] = c[1], c[2], ..., c[n-1], c[0]

At the end of the execution of Region1, a circular left-shifting of array c is performed, i.e., new c[0] = old c[1], new c[1] = old c[2], ..., new c[n-1] = old c[0].

■

Example 12.14 How aggregations are realized in C//

Another form of cooperative interaction is when processes want to aggregate some of their values. This type of interaction is supported in C// through the aggregation operators (e.g., reduction and scan). C// provides mechanisms to allow users to define their own aggregation operators. The following code computes the histogram, hist[i] (for i = 0, 1, ..., binnumber-1), of an *n*-dimensional array A. It demonstrates the use of parallel and sequential segments, as well as the use of reduction operators.

```
parfor (i = 0; i < n; i++)
    region R;
{   core R
        amax = max_reduction (A[i]);   /* assign the maximum of A to amax*/
        amin = min_reduction (A[i]);   /* assign the minimum of A to amin */
    seq
        binsize = (amax - amin) / binnumber;
    parallel
        bin[i] = 1 + (a[i] - amin) / binsize;
    parallel
        hist[bin[i]] = count_reduction(bin[i]);
    }
}
```

The count_reduction operator counts the number of occurrences of each data value. Suppose $n = 4$, bin[0] = b[1] = b[2] = 2, and bin[3] = 1. Then hist[0] = hist[3] = 0, his[1] = 1, and his[2] = 3.

■

Status of The C// Language   The C// programming language is an on-going research project at the National Center for Intelligent Computing, Beijing, China. A number of example C// programs have been written and hand compiled to execute on some SMP and clustered parallel computers. No compiler has been developed for the C// yet.

The full C// language specification and its compilers are very much needed to prove its usefulness. We encourage the parallel programming community to develop compiler, preprocessor, and runtime libraries to implement C// on scalable parallel computer systems.

12.4 Bibliographic Notes and Problems

Several books have discussed various parallel algorithmic paradigms. The most recent is by Brinch-Hansen [111]. The book by Bertsekas and Tsitsiklis [78] contains in-depth discussions of synchronous and asynchronous iteration algorithms. Carriero and Gelernter [130] discussed three paradigms: the agenda, the result, and the specialists paradigms. Barrett et al [63] discussed templates used for linear systems.

Banerjee [60], Kennedy [367], Wolfe [647], and Zima and Chapman [669] are good introductions to parallelizing compiler technology. The Illinois work on parallelizing compilers can be found in Blume et al [91, 92, 93]. Hall et al [292] showed that the Stanford SUIF compiler effectively parallelize widely used sequential benchmark programs such as SPEC92 and SPEC95.

The biggest problem in shared-variable programming is the lack of a widely accepted standard. The ANSI X3H5 standard committee is inactive now, and the X3H5 shared-memory standard is not well accepted by the industry. A good treatment of concurrent programming can be found in Andrews [37].

In practice, most programmers are using either a proprietary parallel programming language (e.g., SGI Power C or Cray Craft) or a threads library (e.g., Solaris Threads or Ptheads). The OpenMP standard may unify shared-memory multiprocessing paradigm.

Pthreads is an influential IEEE/ANSI standard and is described in [345]. Many commercial operating systems have implemented a threads model similar or compatible to Pthreads. These specific threads systems are described in [132, 193, 340, 601].

The X3H5 standard is described in [39]. This standard is much influenced by an earlier standardization effort by a consortium of computer vendors and users, which resulted in a shared-memory parallel Fortran called PCF [488].

The OpenMP standard is described in [475, 476]. The Cray Craft model is described by Pase et al [490]. Convex [166] discussed the shared-memory programming model on the Exemplar machines. SGI [547] developed the Power C programming model. There is also a similar model for Fortran. The coherent region construct and the C// language are presented by the authors in [658].

Parallel programming using functional languages are discussed in Allan and Oldehoeft [24]. Object-oriented approaches are given in Agha [15]. Parallel programming using C++ is given in Wilson and Lu [645]. Programming languages for distributed computing systems are surveyed in Bal. et al [58].

Homework Problems

Problem 12.1   Use a concrete example and a counter-example of parallel language constructs to illustrate the following programmability concepts.

(a)   Structuredness

(b)   Generality

(c)   Portability

Problem 12.2   Compare the implicit and explicit parallel programming models.

(a)   Explain why the implicit model is superior to the explicit models in almost all issues related to parallelism handling, data allocation, interaction, semantics, and programmability, except generality.

(b)   Indicate the main limitations that prevented adoption of the implicit compiler approach in many parallel and distributed computer platforms.

Problem 12.3   Write a parallel code to compute the inner-product of two integer arrays using each of the following paradigms. You can use any pseudocode that is clearly defined. Assume there are n processors. The sequential code is given as follows:

$$\text{sum} = 0; \textbf{for } (i{=}0;\ i{<}N;\ i{+}{+})\ \text{sum} = \text{sum} + A[i]{*}B[i];$$

(a)   Phase parallel

(b)   Divide-and-conquer

(c)   Pipeline

(d)   Process farm

(e)   Process pool

Problem 12.4   Answer the following questions on parallel programming models:

(a)   Why is the message-passing model inferior to the shared-variable model with respect to allocation issues?

(b)   Why is the message-passing model superior to the shared-variable model with respect to the synchronization, semantic, and portability issues?

(c)   Why is the implicit model inferior to the shared-variable model with respect to the generality issues?

Problem 12.5 Write an SPMD parallel program to compute the π value with two problem sizes: $N = 100,000$ and $N = 10,000,000$. You should develop and test your code on any shared-memory language and platform on your site. The program should be optimized and executed on different machine sizes using 1, 2, 3, 4, 5, 6, 7, 8 processors. The deliverable of this exercise includes:

(a) Complete, well-documented sequential and parallel programs.

(b) A description of the test environment and procedures, including platforms, languages, compiler options, libraries, environment parameters, execution modes (batch/interactive, dedicated/shared), how to measure time, etc.

(c) Tabulated and plotted performance numbers measured, including sequential execution time, parallel execution time, speed (in Mflop/s), speedup, efficiency, and utilization.

(d) An analysis of your measured results. Pay attention to observations and explanations of the utilization and scalability over machine size and problem size.

Problem 12.6 Use the Pthreads notation to write parallel programs for

(a) The π computation problem

(b) The Jacobi relaxation problem

(c) The Gaussian elimination problem

(d) The target detection problem

You can define the detailed notations that are not explicitly given in this chapter.

Problem 12.7 Use the X3H5 notation to write parallel programs for

(a) The π computation problem

(b) The Jacobi relaxation problem

(c) The Gaussian elimination problem

(d) The target detection problem

You can define the detailed notations that are not explicitly given in this chapter.

Problem 12.8 Compare X3H5 and OpenMP in terms of ease to use and code efficiency.

(a) Rewrite Fig. 12.5 in X3H5 and compare it with the code in Fig. 12.5.

(b) Compare X3H5 code and OpenMP code (Fig. 12.6) for π computation.

Problem 12.9 Use the SGI Power C notations to write parallel programs for

    (a)   The π computation problem

    (b)   The Jacobi relaxation problem

    (c)   The Gaussian elimination problem

    (d)   The target detection problem

You can define the detailed notations that are not explicitly given in this chapter.

Problem 12.10 Compare the notations used in Problem 12.6, Problem 12.7, and Problem 12.9 in terms of ease of use and how well they support parallelism and collective communications.

Problem 12.11 Give a concrete example to show why event synchronization constructs are needed. Follow the style of Example 12.11 in illustrating the use of latches.

Chapter 13

Message-Passing Programming

This chapter deals with the principles of parallel programming by passing messages among processing nodes. We show how to achieve parallel processing by using two public domain message-passing systems: the PVM and MPI. These two widely accepted systems result from many years' work by researchers and users in academia, industry, and government laboratories. They both have incorporated many useful features from previous programming systems.

13.1 The Message-Passing Paradigm

The message-passing approach to exploiting distributed parallelism is characterized in this section. Basic concepts have already been explained in Section 12.2.2; more details are presented here.

13.1.1 Message-Passing Libraries

Many message-passing software packages have been created in recent years. Some of the most influential ones are listed in Table 13.1. Pointers to these systems can be found in the Bibliographic Notes and the Web Resources list. Some good ideas in the early message-passing systems have been merged into the PVM and MPI standards.

An end user is suggested to use PVM or MPI for developing new message-passing applications. The other systems are not as widely available or portable, and most have become obsolete. In fact, most parallel computer vendors now use PVM or MPI as their native support for message passing, replacing previous proprietary software. A brief review of these packages are gven below:

Table 13.1 Samples of Message-Passing Software

Name	Original Creator	Distinct Features
CMMD	Thinking Machines	Use Active Messages for low latency
Express	Parasoft	Collective communication and I/O
Fortran-M	Argonne National Lab	Modularity and Determinacy
MPI	MPI Forum	A widely adopted standard
Nx	Intel	Originated from the Intel hypercube MPPs
P4	Argonne National Lab	Integrate shared memory and message passing
PARMACS	ANL/GMD	Mainly used in Europe
PVM	Oak Ridge National Lab	A widely-used, stand-alone system
UNIFY	Mississippi State	A system allowing both MPI and PVM calls
Zipcode	Livermore National Lab	Contributed to the context concept

Proprietary Software The CMMD [618] is the message-passing library used in Thinking Machines CM-5 system. It features user-space communication based on the active message mechanism to reduce communication latency.

The Express software [372] from Parasoft Corporation is a programming environment that supports both point-to-point and collective message communications, as well as parallel I/O.

Nx [502] is the microkernel system developed for Intel MPPs (e.g., hypercubes and Paragon). It has been replaced by a new microkernel called PUMA in Intel/Sandia ASCI TFLOP system.

Public-Domain Software Fortran-M [247] is an extension to Fortran 77, designed to support both shared memory and message passing. Only message-passing support is currently implemented. The language provides a number of mechanisms to facilitate the development of modular parallel programs which exhibit determinate behavior.

P4 stands for *Parallel Programs for Parallel Processors* [118]. It is a set of macros and subroutines used for programming both shared-memory and message-passing systems. P4 is meant to be portable among many architectures. The PARMACS [303] is a message-passing package derived from P4 and mainly used in Europe.

PVM and MPI The most important and popular message-passing software packages are MPI and PVM, which will be studied in detail shortly. MPI is a standard specification for a library of functions developed by the *MPI Forum*, a broadly based consortium of parallel computer vendors, library writers, and application specialists. It has been adopted by all major parallel computer vendors.

PVM is a self-contained, public-domain software system to run parallel applications on a network of heterogeneous Unix computers. As its popularity grows, it has been ported to SMPs, PVPs, MPPs, clusters of workstations, and PCs.

13.1.2 Message-Passing Modes

It is customary in message-passing systems to use the term *communication* to refer to all interaction operations, i.e., *communication, synchronization,* and *aggregation.* Consequently, the interaction modes in Section 2.4.2 are often called *communication modes.* Communications usually occur within processes of the same group. However, inter-group communications are also supported by some systems (e.g., MPI). There are three aspects of a communication mode that a user should understand:

- How many processes are involved?
- How are the processes synchronized?
- How are communication buffers managed?

Three communication modes are used in today's message-passing systems. We describe these communication modes below from the user's viewpoint, using a pair of *send* and *receive,* in three different ways. We use the following code example to demonstrate the ideas.

Example 13.1 Send and receive buffers in message passing

In the following code, process P sends a message contained in variable M to process Q, which receives the message into its variable S.

Process P:		Process Q:	
	M = 10;		S = −100
L1:	**send** M **to** Q;	L1:	**receive** S **from** P;
L2:	M = 20;;	L2:	X = S + 1;
	goto L1;		

The variable M is often called the *send message buffer* (or *send buffer*), and S is called the *receive message buffer* (or *receive buffer*).   ■

Synchronous Message Passing When process P executes a *synchronous send* M **to** Q, it has to wait until process Q executes a corresponding *synchronous receive* S **from** P. Both processes will not return from the send or the receive until the message *M* is both sent and received. This means in the above code that variable *X* should evaluate to 11.

When the *send* and *receive* return, *M* can be immediately overwritten by *P* and *S* can be immediately read by *Q*, in the subsequent statements L2. Note that no additional buffer needs to be provided in synchronous message passing. The receive message buffer *S* is available to hold the arriving message.

Blocking Send/Receive   A *blocking send* is executed when a process reaches it, without waiting for a corresponding receive. A *blocking send* does not return until the message is sent, meaning the message variable *M* can be safely rewritten. Note that when the send returns, a corresponding receive is not necessarily finished, or even started. All we know is that the message is out of *M*. It may have been received. But it may be temporarily buffered in the sending node, somewhere in the network, or it may have arrived at a buffer in the receiving node.

A *blocking receive* is executed when a process reaches it, without waiting for a corresponding send. However, it cannot return until the message is received. In the above code, *X* should evaluate to 11 with *blocking send/receive*. Note that the system may have to provide a temporary buffer for blocking-mode message passing.

NonBlocking Send/Receive   A *nonblocking send* is executed when a process reaches it, without waiting for a corresponding receive. A *nonblocking send* can return immediately after it notifies the system that the message *M* is to be sent. The message data are not necessarily out of *M*. Therefore, it is unsafe to overwrite *M*.

A *nonblocking receive* is executed when a process reaches it, without waiting for a corresponding send. It can return immediately after it notifies the system that there is a message to be received. The message may have already arrived, may be still in transient, or may have not even been sent yet. With the nonblocking mode, *X* could evaluate to 11, 21, or -99 in the above code, depending on the relative speeds of the two processes. The system may have to provide a temporary buffer for nonblocking message passing.

Comparison of Three Modes   These three modes are compared in Table 13.2. The synchronous mode has been used in Hoare's theoretical message-passing language CSP [312] and a similar mode is used in Ada rendezvous. This mode has two main advantages:

- First, the synchronous semantics is clean and easy to use. When a process returns from a send subroutine, it knows for certain that the message is sent and received.
- Second, a separate data buffer is not required to be set up by the user or by the system. The message *M* can be directly copied into *S* in the receiver process' address space. A disadvantage is that the sender must wait for the receiver, which leads to some wasted cycles.

In real parallel systems, there are variants on this definition of synchronous (or the other two) mode. For example, in some systems, a synchronous send could return when

Table 13.2 Comparison of Three Communication Modes

Communication Event	Synchronous	Blocking	Nonblocking
Send Start Condition	Both send and receive reached	Send reached	Send reached
Return of send indicates	Message received	Message sent	Message send initiated
Semantics	Clean	In-Between	Error-Prone
Buffering Message	Not needed	Needed	Needed
Status Checking	Not needed	Not needed	Needed
Wait Overhead	Highest	In-Between	Lowest
Overlapping in Communications and Computations	No	Yes	Yes

the corresponding receive is started but not finished. A different term may be used to refer to the synchronous mode. The term *asynchronous* is used to refer to a mode that is not synchronous, such as blocking and nonblocking modes.

Blocking and nonblocking modes are used in almost all existing message-passing systems. They both reduce the wait time of a synchronous send. However, sufficient temporary buffer space must be available for an asynchronous send, because the corresponding receive may not be even started; thus the memory space to put the received message may not be known.

The nonblocking mode reduces the wait time to a minimum. However, it causes an additional problem. Assume the code in Example 13.1 uses nonblocking send/receive. The statement $M = 20$ may be executed before message M is sent. Thus the new value 20 will be transmitted to Q, instead of the intended value 10.

On the receiver side, the statement $X = S + 1$ may be executed before the message (either 10 or 20) from P arrives at S. Thus the old value -100 of S may be used. To rectify such undesirable behaviors, message-passing systems provide a *status checking* or *wait-for* function which forces the process to wait until it is safe to continue. With such a function, the code in Example 13.1 would look like the following:

	Process P:	Process Q:
	$M = 10$;	$S = -100$
L:	**send M to Q;**	**receive S from P;**
	do some computation	*do some computation*
	which does not change M	*which does not use S*
	wait for M to be sent	**wait for S to be received**
	$M = 20$;	$X = S + 1$;

The buffer space may be automatically provided by the system (e.g., in SP2) or explicitly allocated by the user (e.g., in MPI). In either case, the user must be concerned with this issue when developing code. If sufficient buffer space is not available, the system may fail, or the parallel program may hang (deadlock).

The main motivation for using the nonblocking mode is to overlap communication and computation. That is, while the message passing is in progress, the processes can proceed to carry out subsequent computations, which do not interfere with the message passing. (This is shown in the above code in italic.) Such an overlapping idea seems to be attractive, especially on machines with separate message processors.

However, nonblocking introduces its own inefficiencies, such as extra memory space for the temporary buffers, allocation of the buffer, copying message into and out of the buffer, and the execution of an extra wait-for function. These overheads may significantly offset any gains from overlapping communication with computation.

Furthermore, there are no sufficient quantitative data from real applications suggesting the effectiveness of such overlapping. More research is needed to hide communication latency within computation more effectively.

13.2 Message-Passing Interface (MPI)

MPI is a standard specification for a library of message-passing functions. MPI was developed by the *MPI Forum*, a broadly based consortium of parallel computer vendors, library writers, and application specialists.

MPI achieves portability by providing a public-domain, platform-independent standard of message-passing library. MPI specifies this library in a language-independent form, and provides Fortran and C bindings. This specification does not contain any feature that is specific to any particular vendor, operating system, or hardware.

Due to these reasons, MPI has gained wide acceptance in the parallel computing community. MPI has been implemented on IBM PCs on Windows, all main Unix workstations, and all major parallel computers. This means that a parallel program written in standard C or Fortran, using MPI for message passing, could run without change on a single PC, a workstation, a network of workstations, an MPP, from any vendor, on any operating system.

A nice feature of the MPI design is that MPI provides powerful functionality based on four orthogonal concepts. One can learn these concepts individually, knowing their combinations follow well-defined semantic patterns. MPI provides more than 200 functions, much more than PVM. However, the authors experience is that the orthogonal design make MPI easier to learn and use than PVM.

These four concepts are *message data types, communicators, communication operations*, and *virtual topology*. We will focus on the first three concepts, since they are most fundamental in message passing and frequently used.

Parallelism Issues in MPI MPI is silent on how processes come into being at the beginning, leaving it to implementation. Static processes are assumed: All processes are created when a parallel program is loaded. They stay alive until the entire program terminates. There is a default process group consisting of all such processes, identified by MPI_COMM_WORLD. No process can be created or terminated in the middle of a program execution.

We use the MPI program in Fig. 13.1 to explain the issues involved. This is an SPMD program computing $\sum foo\,(i)$. The routine MPI_Init must be called in every process to initialize MPI, before any other MPI routine is called. Then the size of the default group and the rank of each process are found by calling the MPI_Comm_size and the MPI_Comm_rank routines.

The processes communicate with one another by calling MPI_Send and MPI_Recv routines to send and receive messages. Finally, when the MPI functions are not needed any more, every process calls the MPI_Finalize to terminate the MPI environment. These six routines form a minimal set to write complete message-passing programs in MPI.

.Suppose the program in Fig. 13.1 is stored in a "myprog.c" file. On the IBM SP2 multicomputer system, this program is compiled using the parallel C compiler *mpcc*:

    mpcc myprog.c -o myprog

The executable myprog is loaded on n nodes by executing the following command:

    MPIRUN myprog -np n

The same executable myprog will be loaded as n processes in n nodes. These node programs are specified in a host.file, which will be looked up when loading the code. These n processes stay alive until the end of the program. They form the default process group MPI_COMM_WORLD. For process k (k = 0, 1, ..., $n-1$), the MPI_Comm_size routine will return n in group_size, and the MPI_Comm_rank will return k in my_rank.

Note that the number of nodes (i.e., the size of the default process group) n is a load-time parameter. The same program can run on different n nodes. Process 0 performs all the I/O operations in the above code. It first reads a value of N from the user, and then sends this value to all the other processes in the **for** loop. It then computes its share of the work, sum = foo(0)+foo(n)+foo($2n$)+... Afterwards, it receives $n-1$ partial sums from the other processes and accumulate them into a total sum, which is then printed.

For each of the other $n-1$ processes, the following operations are performed: Process k ($0 < k < n$) receives N sent from process 0, computes its share of the work to generate a partial sum, sum = foo(k)+foo($n+k$)+foo($2n+k$)+. . ., and then sends this partial sum to process 0. Note that the send S1 in process 0 matches the receive S3 in process k, and S4 matches S2.

```
#include "mpi.h"
int foo(i) int i ; { ... }
main(argc, argv)
int argc;
char * argv[] ;
{    int i, tmp, sum = 0, group_size, my_rank, N;
     MPI_Init (&argc, &argv) ;
     MPI_Comm_size (MPI_COMM_WORLD, &group_size) ;
     MPI_Comm_rank (MPI_COMM_WORLD, &my_rank) ;
     if ( my_rank == 0 ) {
          printf("Enter N: "); scanf("%d", &N);
          for (i=1; i<group_size; i++)
S1:            MPI_Send(&N, 1, MPI_INT, i, i, MPI_COMM_WORLD) ;
          for ( i = my_rank ; i<N ; i = i + group_size )
               sum = sum + foo(i) ;
          for (i=1; i<group_size; i++) {
S2:            MPI_Recv(&tmp, 1, MPI_INT, i, i, MPI_COMM_WORLD, &status) ;
               sum = sum + tmp ;
          }
          printf ("\n The result = %d", sum) ;
     }
     else {          /* if my_rank != 0 */
S3:       MPI_Recv(&N, 1, MPI_INT, 0, i, MPI_COMM_WORLD, &status) ;
          for ( i = my_rank ; i<N ; i = i + group_size )
               sum = sum + foo(i) ;
S4:       MPI_Send(&sum, 1, MPI_INT, 0, i, MPI_COMM_WORLD) ;
     }
     MPI_Finalize() ;
}
```

Figure 13.1   An MPI message-passing program.

MPI Implementations   MPI is not a stand-alone, self-contained software system. It serves as a message-passing communication layer on top of the native parallel programming environment, which takes care of necessities such as process management and I/O. For instance, MPI can be built on top of POE/MPL on the IBM SP2, and on top of OSF/Nx on the Intel Paragon.

Besides these proprietary environments, there are several public-domain MPI environments. Examples include the *CHIMP* implementation developed at Edinburg University, and the *LAM* (Local Area Multicomputer) developed at the Ohio Supercomputer Center, which is an MPI programming environment for heterogenous Unix clusters.

The MPICH   The most popular public-domain implementation is *MPICH*, developed jointly by Argonne National Laboratory and Mississippi State University. MPICH is a *portable* implementation of MPI on a wide range of machines, from IBM PC's, networks of workstations, to SMPs and MPPs.

The portability of MPICH means that one can simply retrieve the same MPICH package and install it on almost any platform. MPICH also has good performance on many parallel machines, because it often runs in the more efficient native mode rather than over the common TCP/IP sockets.

13.2.1 MPI Messages

MPI processes are heavy-weighted, single-threaded processes discussed in Section 2.2.7. They have separate address spaces. Thus one process cannot directly access variables in another process's address space. Interprocess communication is realized by message passing.

The subroutines MPI_Send and MPI_Recv in Fig. 13.1 are point-to-point communication operations, which pass a message between a pair of processes. These subroutines

> MPI_Send(&N, 1, MPI_INT, i, i, MPI_COMM_WORLD) ;
> MPI_Recv(&n, 1, MPI_INT, 0, i, MPI_COMM_WORLD, &status) ;

are much more complex than the send and the receive operations in Example 13.1:

> **send** M **to** Q ;   and   **receive** S **from** P ;

Why all these complexities in MPI? To answer this question, we need to answer another question first: What is a message?

What Is a Message?   As an analogy, a message is like a letter. We need to specify the contents of the message (the letter itself), and the intended recipient of the message (what is on the envelope). The former is often called a *message buffer* and the latter a *message envelope*.

Example 13.2 Sending an array of data via MPI

Consider an array of N complex numbers declared in C notation **double** A[100]. Now suppose process P wants to send to process Q:

(1) The entire array A;

(2) The first two elements of array A (i.e., A[0} and A[1]); or

(3) All the even-indexed components of array A (i.e., A[0], A[2], A[4], ...).

Case (1) can be specified by process P executing a simple **send** A **to** Q. The message buffer is identified by the array name A, and the envelope is just a process name Q. Concise notations similar to this have been used in theoretical message-passing languages (e.g. CSP [312]). However, it is difficult to use such simple notation to handle cases (2) and (3), where the message is only part of array A.

■

Send in Commercial Messaging Systems In commercial message-passing systems (e.g., Nx in Intel Paragon and MPL in IBM SP2), a message send routine is more flexible, including four arguments:

send(*address, length, destination, tag*)

where the message buffer is specified by the pair (*address, length*) and the message envelope by (*destination, tag*). The *address* field specifies the starting address of the memory area where the message is held. The *length* field specifies how many bytes need to be sent. Assume a double-precision number is 64-bit, or 8 bytes, on a computer. Then case (2) can be specified by **send**(A, 16, Q, tag). In C and Fortran, the starting address of an array is identified by the array name. The message length is 16 bytes since it contains two double precision numbers.This scheme still does not handle case (3) well, though.

Send in MPI The MPI send subroutine in Fig. 13.1 is redrawn in Fig. 13.2. The three italicized attributes (message *address*, message *count*, message *datatype*) specify the message buffer, and the other four arguments form the message envelope.

Note that the address field does not have to be the starting address of a data structure. It could be any memory address in the application's address space. For instance, to send the third, the fourth, and the fifth elements of A, we could use **send**(A+2, 24, Q, tag).

Why Message Data Type? The pair (*address, length*) is used to specify a message buffer by most message-passing systems before MPI. So why MPI introduces the additional message *data type* attribute? There are two reasons: to support *heterogeneous computing* and to allow messages from non-contiguous, non-uniform memory sections.

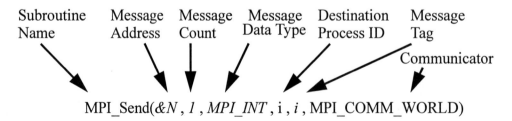

Figure 13.2 Anatomy of MPI components in sending a message.

The term *heterogeneous computing* refers to running applications on a system consisting of different computers, such as an network of workstations. Each computer in the system could come from a different vendor, using different processors and operating systems. An important issue is how to ensure inter-operability even if these computers use different data representations. Let us consider a somewhat extreme case:

Example 13.3 Sending data of different types

A workstation in Los Angeles is to send a string of 100 English characters to a workstation in Beijing. A naive method is for the USA workstation to execute a **send**(String, 100, China), and for the Chinese workstation to execute a corresponding **recv**(String, 100, USA). The data structure String is declared on both machines by **char** String[1000].

Unfortunately, this method may fail because the **char** data type is defined differently on the USA and the Chinese workstations. The meaning of **send**(String, 100, China) is to send out 100 *bytes* (each byte is 8 bits), with the starting address String. Similarly, the meaning of **recv**(String, 100, USA) is to receive 100 *bytes*.

This is fine for the USA machine, since a character is represented as a byte. But the **char** data type on the Chinese machine must accommodate all Chinese characters (more than 50,000), plus all ASCII characters. Thus each character on the Chinese machine is a 16-bit, or 2-byte data item.

■

MPI solves this problem by offering a set of predefined basic data types, as shown in Table 13.3. Now communicating 100 characters can be realized by

USA: MPI_Send(String, 100, MPI_CHAR, China, ...)

China: MPI_Recv(String, 100, MPI_CHAR, USA, ...)

The second argument 100 is no longer the message length in bytes, but the number of data items of type MPI_CHAR. This quantity is called message *count*, to differ from message *length*. The data type MPI_CHAR can be implemented as 8-bit on the USA machine, but 16-bit on the Chinese machine. The implementation is responsible for ensuring that the character string is properly converted from the 8-bit format to the 16-bit format.

The predefined data types of MPI include all basic types in C and Fortran, plus two additional types: MPI_BYTE and MPI_PACKED. MPI_BYTE indicates a byte of 8 bits. The reason for including MPI_PACKED will be discussed shortly.

Table 13.3 Predefined Data Types in MPI
(Nonexisting cases are left blank)

MPI (C Binding)	C	MPI (Fortran Binding)	Fortran
MPI_BYTE		MPI_BYTE	
MPI_CHAR	Signed char	MPI_CHARACTER	CHARACTER(1)
		MPI_COMPLEX	COMPLEX
MPI_DOUBLE	Double	MPI_DOUBLE_PRECISION	DOUBLE_PRECISION
MPI_FLOAT	Float	MPI_REAL	REAL
MPI_INT	Int	MPI_INTEGER	INTEGER
		MPI_LOGICAL	LOGICAL
MPI_LONG	Long		
MPI_LONG_DOUBLE	Long double		
MPI_PACKED		MPI_PACKED	
MPI_SHORT	Short		
MPI_UNSIGNED _CHAR	Unsigned char		
MPI_UNSIGNED	Unsignedint		
MPI_UNSIGNED _LONG	Unsigned long		
MPI_UNSIGNED _SHORT	Unsigned short		

Derived Data Types   Even when empowered with the three attributes (address, count, datatype), it is still difficult to solve the third problem in Example 13.2. That is, sending all even-indexed elements of array A.

Example 13.4 Sending noncontiguous data items

One way to send non-contiguous data items is to first pack these items into a contiguous temporary buffer and then send it. The following code outlines how to use MPI to pack and send all even-indexed elements of an array A.

```
double A[100] ;
MPI_Pack_size( 50, MPI_DOUBLE, comm, &BufferSize ) ;
TempBuffer = malloc( BufferSize ) ;
j = sizeof( MPI_DOUBLE ) ;
Position = 0 ;
for ( i = 0 ; i <50 ; i++ )
    MPI_Pack (A+i*j, 1, MPI_DOUBLE, TempBuffer, BufferSize,
                &Position, comm) ;
MPI_Send ( TempBuffer, Position, MPI_PACKED, destination, tag, comm ) ;
```

The MPI_Pack_size routine is used to determine how big a temporary buffer is needed to hold 50 MPI_DOUBLE data items. The buffer size is next used to dynamically allocate memory for this TempBuffer. The 50 even-indexed elements of array A are then packed into the TempBuffer in a **for** loop.

The first argument in the MPI_Pack routine is the address of an array element to be packed. The second argument is its data type. The variable Position is used by the MPI_Pack routine to keep track of how many items have been packed. The final value of Position is used in the following MPI_Send as the message count. Note that in MPI all packed messages use the data type MPI_PACKED.

∎

This packing method is tedious and error-prone. MPI enables a user to specify a message with the following flexible format: A message consists of a number of *data items*, each of which has an *address* and a *data type* (and of course, a data value).

Example 13.5 Sending mixed data types in message passing

Assume each double-precision number is 8 bytes long, a character is 1 byte, and an integer is 4 bytes in the following messages.

(1) All elements of a double-precision, 100-element array A. This message consists of 100 items. Each item has a **double** data type, which also determines the size of each item (8 bytes). The (starting) address of the ith item is $A+8(i-1)$.

(2) The third and the fourth elements of array A. This message consists of two items A[2] and A[3]. Each item has a **double** data type. The first item starts at A+16, and the second at A+24.

(3) The even-indexed elements of array A. This message consists of 50 items A[0], A[2], A[4], ..., A[98]. Each item has a **double** data type. The address of the ith item is $A+16(i-1)$.

(4) The third element of array A, followed by a character c, followed by an integer k. This message consists of three items with different data types.

Suppose they are part of a structure:

struct { **double** A[100]; **char** b, c; **int** j, k; } S ;

Then the address is S+16 for the first item (A[2]), S+801 for the second (*c*), and S+806 for the third element (*k*).

■

Messages in (1) and (2) have two properties: The data items are stored consecutively, and all data items have the same data types. Such messages are conveniently specified by a triple (address, count, datatype). For instance, the message in case (1) are specified by (A, 100, MPI_DOUBLE), and that in (2) by (A+16, 2, MPI_DOUBLE). However, such a simple scheme cannot handle cases (3) and (4). The data items in (3) do not reside at contiguous locations. The data items in (4) are noncontiguous and have mixed data types.

MPI introduces the concept of *derived data types* to allow the specification of any message consisting of noncontiguous data items of possibly mixed data types. A derived data type is constructed by the user application at run time, from basic data types. MPI has very powerful and versatile subroutines to build complex data types. We will just use an example to illustrate the basic idea.

Example 13.6 Send all even-indexed elements of an array

This example explains how to handle part of an indexed array.

```
double A[100] ;
MPI_Data_type    EvenElements ;
... ...
MPI_Type_vector ( 50, 1, 2, MPI_DOUBLE, &EvenElements ) ;
MPI_Type_commit ( &EvenElements ) ;
MPI_Send ( A, 1, EvenElements, destination, ...) ;
```

The programmer uses MPI_Data_type to declare a new data type EvenElements. MPI_Type_vector(*count, blocklength, stride, oldtype, &newtype*) is one of MPI routines for constructing derived data types. The derived type *newtype* consists of *count* copies of *blocks*. Each block in turn consists of *blocklength* copies of consecutive items of existing data type *oldtype*. The *stride* specifies the number of *oldtype* elements between every two consecutive blocks. Thus (*stride − blocklength*) is the gap between two blocks.

The MPI_Type_vector(50, 1, 2, MPI_DOUBLE, &EvenElements) routine thus creates a derived type EvenElements, which consists of 50 blocks. Each block consists of 1 double-precision number, followed by an 8-byte gap, followed by the next block. The stride is the size of two double-precision numbers, or 16 bytes. The 8-byte gap is used to skip the odd-indexed elements of array A.

This new type must be committed before being used in the send routine. Note that one element of EvenElements contains all 50 even-indexed elements of A. Thus the count field has value 1 in MPI_Send.

∎

Message Buffer The term *message buffer* (or simply, *buffer*) has been used with different meanings by the message passing community.

Some examples are given below:

- A buffer refers to an application memory area specified by the programmer, where the data values of a message are stored. For instance, in **send**(A, 16, Q, tag), the buffer A is a variable declared in the user application. The starting address of the buffer is used in a message-passing routine.
- A buffer could also mean some memory area created and managed by the message-passing system (not the user), which temporarily stores a message while it is being sent. Such a buffer does not appear in the user's application program and is sometimes called a *system message buffer* (or system buffer).
- MPI allows a third possibility. The user may set aside a memory area of a certain size, to be used as an intermediate buffer to hold arbitrary messages that could appear in her application.

Example 13.7 Sending messages between a pair of processes

Consider the following code to pass a message M stored in array A of process P to array B of process Q:

Process P: Process Q:

double A[2000000] ;   **double** B[32] ;
send (A, 32, Q, tag);   **recv**(B, 32, P, tag);

The use of the three types of buffers is illustrated in Fig. 13.3, where M indicates the 32-byte message, consisting of the first four elements of array A.

In case (a), the message is directly transferred between user-level buffers A and B, both of which are declared in the user application. The receive buffer must be large enough.

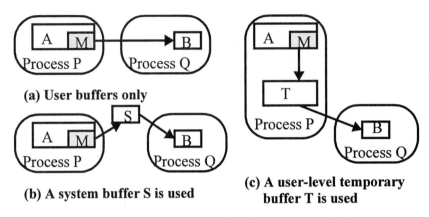

Figure 13.3   Three types of message buffers used in MPI.

For instance, an error would occur if **send** (A, 64, Q, tag) were executed by P. Such an error is often not detected by current message-passing systems, creating much difficulty in debugging.

With asynchronous message passing, it is difficult to perform the above direct transfer [case (a)]. A number of alternatives have been suggested. In case (b), the message is first temporarily copied into a dynamically created system buffer S, before being deposited in the receive user buffer B. There are two problems. First, an extra copying introduces additional overhead. Second, a system buffer with sufficient size is not guaranteed. The buffer needed may be too big to fit in the physical memory, resulting in excessive paging. The buffer may even be too big to fit into the virtual memory, causing the program to abort or hang.

Case (c) is another alternative used in MPI. The user first declares a temporary buffer T which is large enough to hold any message that needs to be buffered. During a message-passing operation, the message is first copied into the buffer T, before being deposited in the receive user buffer B. If the system cannot accommodate buffer T, it will inform the application program to generate an error message and gracefully terminate. Otherwise, the application is guaranteed to have enough buffer space.

■

13.2.2 Message Envelope in MPI

How does one specify the intended recipient of a message? Listed below is an MPI send routine where the message envelope consists of three entities, shown in italic. The *destination* field is an integer identifying the process to which the message is sent.

MPI_Send(address, count, datatype, *destination, tag, communicator*)

Why the Tag? A *message tag*, also known as a *message type*, is an integer used by the programmer to label different types of message and to restrict message reception. To see why tags are needed, let us first consider the following code where tags are absent:

Process P:	Process Q:
send(A,32,Q)	**recv**(X, P, 32)
send(B,16,Q)	**recv**(Y, P, 16)

The intent is to transfer the first 32 bytes of A into X and the first 16 bytes of B into Y. But what if message B, although sent later, arrives at process Q earlier, thus being received into X by the first **recv**(). This error can be avoided by using tags:

Process P:	Process Q:
send(A,32,Q, tag1)	**recv**(X, P, tag1, 32)
send(B,16,Q, tag2)	**recv**(Y, P, tag2, 16)

Now message A (with tag1) is guaranteed to be received first. If message B arrives at Q first, it will be queued (buffered) until **recv**(Y, P, tag2, 16) is executed.

Example 13.8 Tags used in message passing

Another reason for tags is exemplified by the following scenario. Suppose two client processes P and R each send a service request message to a server process Q:

Process P:	Process R:
send(request1,32,Q)	**send**(request2,32,Q)

It is unknown which send will be executed first. The server process Q is required to process the clients' requests in a first-come first-served order:

```
Process Q:
while (true ) {
    recv(received_request, Any_Proccess, 32);
    process received_request ;
}
```

This code is not very flexible, since all requests are processed the same way. By using tags, different messages can be processed differently:

Process P: Process R:
send(request1,32,Q, tag1)   **send**(request2,32,Q, tag2)

Process Q:
while (true) {
    recv(received_request, Any_Proccess, 32, Any_Tag, Status)
    if (Status.Tag==tag1) *process received_request in one way*;
    if (Status.Tag==tag2) *process received_request in another way*;
}

The recv() statement says that it will receive a message of 32 bytes with any tag (Any_Tag is known as a *wild-card tag*) from any process (Any_Process is known as a *wild-card process ID*) into received_request. The actual tag of the message received is stored in the Tag field of Status to guide the branching in program.

■

What Is a Communicator? A *communicator* is a *process group* plus a *context*. A process group is a *finite* and *ordered* set of processes. The finiteness implies that a group has a finite number n of processes, where n is called the group size. The ordering means that the n processes are ranked by integers 0, 1, ..., $n-1$. A process is identified by its rank in a communicator (group). As illustrated in Fig. 13.1, the group size and the rank of a process are obtained by calling the two MPI routines:

MPI_Comm_size(communicator, &group_size)

MPI_Comm_rank (communicator, &my_rank)

Most MPI users only need to use routines for communication within a group (called *intra-communicators* in MPI). All discussions of MPI in this book fall into this category, including all point-to-point and collective communication routines.

This implies that all MPI communication routines have a communicator argument. MPI allows inter-group communication through *inter-communicators*, which we will not discuss, but we refer the reader to Snir et al. [575] and the MPI home pages.

Contexts in MPI are like system-designated supertags that safely separates different communications from adversely interfering with one another. Each communicator has a distinct context. A message sent in one context cannot be received in another context. To see why the concept of context is needed, let us look at the following example.

Example 13.9 **The use of communicator**

Consider the following code fragment:

Process 0:
MPI_Send (msg1, count1, MPI_INT, 1, tag1, comm1) ;
parallel_fft (...) ;

Process 1:
MPI_Recv (msg1, count1, MPI_INT, 0, tag1, comm1) ;
parallel_fft (...) ;

The intent is that process 0 will send msg1 to process 1, and then both execute a subroutine parallel_fft(). Now suppose parallel_fft() contains another send routine:

if (my_rank==0) MPI_Send (msg2, count1, MPI_INT,1,tag2, comm2);

If there were no communicators, the MPI_Recv in process 1 could mistakenly receive the msg2 sent by the MPI_Send in parallel_fft(), when tag2 happens to have the same integer value as tag1. So why cannot we use different tag values? It is impossible to guarantee that tag1 and tag2 are distinct for the following three reasons:

- Tags are integer values specified by users, and users make mistakes.
- Even if a user does not make mistakes, it is difficult or impossible to ensure that the value of tag1 is different from that of tag2. The function parallel_fft() could be written by another user, or it could be a library routine. So the user cannot know the value of tag2.
- Even if the user could always know the value of tag2, errors could still occur, because the MPI_Recv routine may decide to use a wildcard tag MPI_Any_tag.

■

MPI solves these problems by using communicators. Communications in parallel_fft use different communicators, which may contain the same group of processes (e.g., processes 0 and 1), but each communicator will have a distinct, system-assigned context that is different from that of comm1. Therefore, there is no danger that MPI_Recv accidentally receives msg2 from the MPI_Send in parallel_fft.

MPI is designed so that communications within different communicators are separated and any collective communication is separate from any point-to-point communication, even if they are within the same communicator. This communicator concept facilitates the development of parallel libraries, among other things.

Managing Communicators MPI includes several predefined communicators. For instance, MPI_COMM_WORLD contains the set of all processes and is defined after the routine MPI_Init is executed. MPI_COMM_SELF contains only the process which uses it. MPI also provides several routines for building user-defined communicators. We illustrate only two of them through an example.

Example 13.10 New communicator in MPI

Consider the following code, which is executed by 10 processes:

```
MPI_Comm        MyWorld, SplitWorld ;
int             my_rank, group_size, Color, Key;

MPI_Init (&argc, &argv) ;
MPI_Comm_dup (MPI_COMM_WORLD, &MyWorld) ;
MPI_Comm_rank (MyWorld, &my_rank) ;
MPI_Comm_size (MyWorld, &group_size) ;
Color = my_rank % 3;
Key = my_rank / 3;
MPI_Comm_split (MyWorld, Color, Key, &SplitWorld) ;
```

Executing the routine MPI_Comm_dup(MPI_COMM_WORLD, MyWorld) will create a new communicator MyWorld, which contains the same group of 10 processes as the original MPI_COMM_WORLD but has a different context. How MPI_Comm_split works is illustrated in Table 13.4.

The 10 processes of the communicator MyWorld are split into three new communicators. All processes with the same color form a new communicator named SplitWorld. They are ranked in the new communicator in the order given by Key. Note that since there are three different values of Color, three new communicators are formed. Although all are named SplitWorld, they are distinct communicators.

Table 13.4 Splitting a Communicator Myworld

Rank in MyWorld	0	1	2	3	4	5	6	7	8	9
Color	0	1	2	0	1	2	0	1	2	0
Key	0	0	0	1	1	1	2	2	2	3
Rank in SplitWorld (Color=0)	0			1			2			3
Rank in SplitWorld (Color=1)		0			1			2		
Rank in SplitWorld (Color=2)			0			1			2	

Summary of Message Features   Let us summarize the features of a message in the send routine in the general form:

MPI_Send(buffer, count, datatype, destination, tag, communicator) ;

and as a specific example in Fig. 13.1 and Fig. 13.2:

MPI_Send(&N, 1, MPI_INT, i, i, MPI_COMM_WORLD) ;

which says to label 1 integer stored at location &N with tag i and send it to process i.

Both the sender and the destination processes are in the communicator of all processes. This is summarized below:

- The first argument indicates the starting address of the message buffer, i.e., the memory area holding the data to be sent.
- The second argument specifies how many items of a certain data type, which is given in the third argument, are contained in the message.
- The data type could be either a basic type or a derived type, the latter created by the user to specify a message with possibly noncontiguous data items of mixed data types.
- The fourth argument is the destination process ID (rank), while the fifth is the message tag.
- The sixth argument identifies a process group and a context, i.e., a *communicator*. Usually, messages are passed only among processes of the same group. But MPI allows intergroup communications through *intercommunicators*.

The corresponding receive routine is very similar to the send routine:

MPI_Recv(&tmp, 1, MPI_INT, i, i, MPI_COMM_WORLD, &Status)

or more generally,

MPI_Recv(address, count, datatype, source, tag, communicator, status)

The first argument indicates the starting address of the receive message buffer, i.e., the memory area to hold the data to be received. The second argument specifies the *maximum* number of items of a certain data type, which is given in the third argument, that can be received. The actual number of items received may be less.

The fourth argument is the source process ID (rank), while the fifth is the message tag. These two fields can be wild-card MPI_Any_source and MPI_Any_tag. The sixth argument identifies a communicator. The seventh argument is a pointer to a structure

MPI_Status   Status

which holds various information about the message received. For instance, the actual source process rank and the actual message tag can be found in the two fields Status.MPI_SOURCE and Status.MPI_TAG. The actual number of data items received can be found through another MPI routine

MPI_Get_count (&Status, MPI_INT, &C) ;

This routine uses the information in Status to determine the actual number of items of a certain data type (MPI_INT in this case) and puts this number in variable C.

Example 13.11 Status word in message passing

The status information is useful when a receiver could receive messages of different sizes and tags from different processes. Let us revisit the client-server program in Example 13.8. A number of client processes request services by sending messages to the server. There are three types of service, identified by the message tag. The input parameters associated with a request constitute the actual data items in the message. Thus different messages could have different sizes. Two messages requesting the same service use the same tag.

```
while (true ) {
    MPI_Recv(received_request, 100, MPI_BYTE, MPI_Any_source,
            MPI_Any_tag, comm, &Status ) ;;
    switch ( Status.MPI_Tag ) {
        case    tag_0 :    perform service type 0 ;
        case    tag_1 :    perform service type 1 ;
        case    tag_2 :    perform service type 2 ;
    }
}
```

■

13.2.3 Point-to-Point Communications

MPI provides both blocking and nonblocking send/receive operations. That is, a blocking send (receive) cannot return until the message buffer can be safely written (read), while a nonblocking send (receive) can immediately return. MPI offers quite a number of send and receive routines, based on different combinations of the blocking/nonblocking attribute and different ways to use message buffers.

Communication Modes   The term *communication mode* is used in MPI to refer to one of the following ways of buffer management and synchronization between a send and a receive. The terminology is different from that in Section 13.1.2. These modes apply to both blocking and nonblocking *send* operations. MPI has the usual blocking and nonblocking receives discussed in Section 13.1.2.

- *Synchronous*: The send cannot return until a corresponding receive has started, thus an application buffer is available in the receiver side to hold the arriving message. Note that it is possible to have a *nonblocking*, synchronous send in MPI, whose return does not mean the message has been sent! The implication is that there is no need for additional buffering at the receive side. A system buffer may be still needed at the send side. To eliminate extra copying of message, a blocking, synchronous send should be used.
- *Buffered*: A buffered send can assume the availability of a certain amount of buffer space, which must be previously specified by the user program through a subroutine call MPI_Buffer_attach (*buffer*, *size*) that allocates a user buffer of *size* bytes. This buffer can be released by MPI_Buffer_detach (*buffer*, *size*). This buffered mode was shown in Fig. 13.3b.
- *Standard*: The send can be either synchronous or buffered, depending on the implementation.
- *Ready*: The send is certain that the corresponding receive has already started. It does not have to wait as in the synchronous mode. This knowledge allows implementation to use a more efficient communication protocol in some cases.

These modes, together with the blocking attribute, generate eight send routines in MPI, as shown in Table 13.5. There are only two receive routines.

Table 13.5 Different Send/Receive Operations in MPI

MPI Primitive	Blocking	Nonblocking
Standard Send	MPI_Send	MPI_Isend
Synchronous Send	MPI_Ssend	MPI_Issend
Buffered Send	MPI_Bsend	MPI_Ibsend
Ready Send	MPI_Rsend	MPI_Irsend
Receive	MPI_Recv	MPI_Irecv
Completion Check	MPI_Wait	MPI_Test

Nonblocking Communication   There are four nonblocking send routines and one nonblocking receive routine in MPI. They are used to *initiate* (start) a send or receive (hence the letter I in their names). MPI provides other routines to check the completion of a send or receive. We discuss only two basic ones through examples.

Example 13.12 Process pipeline using message passing

Shown in Fig. 13.4 is a three-process pipeline, where a process continuously receives an input data stream from the left, computes a new data stream, and sends it to the right. The following code fragment shows the basic idea:

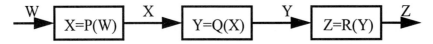

Figure 13.4   The flow of data stream in a process pipeline.

```
while ( Not_Done ) {
    MPI_Irecv (NextX, ...) ;
    MPI_Isend (PreviousY, ...) ;
    CurrentY = Q( CurrentX ) ;}
```

Process Q executes a nonblocking receive to initiate receiving the next data item X from process P, and a nonblocking send to initiate sending the previous data item Y to process R. Then the computation of the current X is executed while the send/receive are in progress. That is, computation is overlapped with communication.

∎

We need to take care of several details. First, we need two separate buffers for X and Y, so that while the next X is put into a buffer by a receive, the computation can use the current X in the other buffer. Such a double-buffering scheme is a well-known technique and is illustrated in Fig. 13.5.

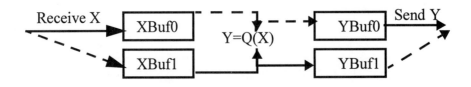

Figure 13.5   Double buffering in a process pipeline.

The solid lines show how the buffers are used in an iteration of the while loop. They will switch to the dashed lines in the next iteration. Second, we need to make sure that data in a buffer are used before the buffer is updated. This results in the code below:

```
while ( Not_Done ) {
    if (X==XBuf0) {X=XBuf1; Y=YBuf1; Xin=XBuf0; Yout=YBuf0;}
    else { X=XBuf0; Y=YBuf0; Xin=XBuf1; Yout=YBuf1; }
    MPI_Irecv (Xin, ..., recv_handle) ;
    MPI_Isend (Yout, ..., send_handle) ;
    Y = Q( X ) ;                          /* overlapping computation */
    MPI_Wait ( recv_handle, recv_status ) ;
    MPI_Wait ( send_handle, send_status ) ;
}
```

The **if-else** statement is used to switch the double buffer. The nonblocking send routine

MPI_Isend (buf, count, datatype, dest, tag, comm, send_handle)

has the same format as a blocking send, except for an additional parameter send_handle, which is used to check the completion of the send. Similarly, the nonblocking receive routine MPI_Irecv has one more parameter receive_handle. Checking the completion of a send/receive can be done through a routine

MPI_Wait (Handle, Status)

which will not return until the send or the receive indicated by Handle is completed, and some status information is passed to parameter Status. Thus the above code will not proceed to the next iteration until the two nonblocking send and receive are completed. MPI also provides another routine

MPI_Test (Handle, Flag, Status)

which merely tests whether the send or receive indicated by Handle is completed. If so, Flag is assigned a True value. This routine, unlike MPI_Wait, does not block.

Example 13.13 Deadlock in sending and receiving messages

What will happen when the following code is executed? The code is purposely written to have some bugs.

Process P: Process Q:

... ...

X = 0 ;   **if** (Y == 5) MPI_Irecv (&Y, ...,P, ...) ;

MPI_Issend (&X, ...,Q, ...);   printf("Y is %d ", Y) ;

This is an erroneous code where the following outcomes are possible:

- *Deadlock.* The routine MPI_Issend is a nonblocking, synchronous send. It will not return until the corresponding MPI_Irecv has started to execute, which will not happen if the condition Y==5 is not true.
- *Y is 0.* Suppose Y==5 is true when the **if** statement is executed. Then MPI_Irecv could receive the X value (0) from process P into its variable Y, and print it out.
- *Y is 5.* Another possible scenario when Y==5 is true is that the printf is executed before the X value is received into Y (recall that MPI_Irecv is a nonblocking receive). Thus the old value of Y is printed out.

■

13.2.4 Collective MPI Communications

Collective communication involves all processes of a group (communicator) to participate in a *global communication* operation. MPI provides a rich set of collective communication routines, as listed in Table 13.6. We will illustrate the basic collective communication principles by discussing several of the routines, assuming a communicator *Comm* containing *n* processes.

Broadcast   In the following *broadcast* operation, process ranked Root sends the same message to all processes (including itself) in the communicator labeled Comm:

MPI_Bcast (Address, Count, Datatype, Root, Comm)

The content of the message is identified by the triple (Address, Count, Datatype), just as in a point-to-point communication. For the Root process, this triple specifies both the send buffer and the receive buffer. For other processes, this triple specifies the receive buffer.

Gather   A *gather* routine is given below:

MPI_Gather (SendAddress, SendCount, SendDatatype,
RecvAddress, RecvCount, RecvDatatype, Root, Comm)

The Root process receives a personalized message from each of the *n* processes (including itself). These *n* messages are concatenated in rank order, and stored in the

Table 13.6 Collective Communications in MPI

Type	Routine	Functionality
Data movement	MPI_Bcast	One-to-all, identical message
	MPI_Gather	All-to-one, personalized messages
	MPI_Gatherv	A generalization of MPI_Gather
	MPI_Allgather	A generalization of MPI_Gather
	MPI_Allgatherv	A generalization of MPI_Allgather
	MPI_Scatter	One-to-all, personalized massages
	MPI_Scatterv	A generalization of MPI_Scatter
	MPI_Alltoall	All-to-all, personalized messages
	MPI_Alltoallv	A generalization of MPI_Alltoall
Aggregation	MPI_Reduce	All-to-one reduction
	MPI_Allreduce	A generalization of MPI_Reduce
	MPI_Reduce_scatter	A generalization of MPI_Reduce
	MPI_Scan	All-to-all parallel prefix
Synchronization	MPI_Barrier	Barrier synchronization

receive buffer of the Root process. Each send buffer is identified by the triple (SendAddress, SendCount, SendDatatype). The receive buffer is ignored for all non-Root processes. For the Root process, it is identified by the triple (RecvAddress, RecvCount, RecvDatatype).

Scatter   A *scatter* routine is given below:

MPI_Scatter (SendAddress, SendCount, SendDatatype,
            RecvAddress, RecvCount, RecvDatatype, Root, Comm)

A scatter performs just the opposite operation of a gather. The Root process sends out a personalized message to each of the n processes, itself included. These n messages are originally stored in rank order in the send buffer of the Root process. Each receive buffer is identified by the triple (RecvAddress, RecvCount, RecvDatatype). The send buffer is ignored for all non-Root processes. For the Root process, it is identified by the triple (SendAddress, SendCount, SendDatatype).

Total Exchange   In an all-to-all, total exchange routine:

MPI_Alltoall (SendAddress, SendCount, SendDatatype,
            RecvAddress, RecvCount, RecvDatatype, Comm)

every process sends a personalized message to each of the *n* processes, including itself. These *n* messages are originally stored in rank order in its send buffer. Looking at the communication from another way, every process receives a message from each of the *n* processes.

These *n* messages are concatenated in rank order, and stored in the receive buffer. Note that a total exchange is equivalent to *n* gathers, each by a different process. Therefore, the Root argument is not needed any more. All in all, n^2 messages are communicated in a total exchange.

Aggregation   MPI provides two types of aggregation (see Section 2.4.1): reduction and scan. An MPI reduction routine has the following syntax:

$$\text{MPI\_Reduce(SendAddress, RecvAddress, Count, Datatype, Op, Root, Comm)}$$

Here each process holds a partial value stored in SendAddress. All processes reduce these partial values into a final result and store it in RecvAddress of the Root process. The data type of the data items are specified in the Datatype filed. The reduction operator is specified by the Op field.

The scan operation has very similar syntax to reduction:

$$\text{MPI\_Scan (SendAddress, RecvAddress, Count, Datatype, Op, Comm)}$$

The Root field is absent, since a scan combines partial values into *n* final results and stores them in RecvAddress of the *n* processes. The scan operator is specified by the Op field.

The MPI reduction and scan routines allow each process to contribute a vector, not just a scalar value. The length of the vector is specified in Count. MPI supports user-defined reduction or scan operations.

Barrier   In a *barrier operation* MPI_Barrier(Comm), all processes in the communicator Comm synchronize with one another; i.e., they wait until all processes execute their respective MPI_Barrier function.

Common Features of Collective Routines   The following features are common to every collective communication operation in MPI:

- All processes of the communicator *must* call the collective communication routine. A code in which some members of a communicator call a collective routine while others may not call it is erroneous. The behavior of an erroneous code is *undefined*, meaning it could be anything, including deadlock or wrong results.

- Every collective routine, except MPI_Barrier, employs a communication mode analogous to the standard, blocking mode in point-to-point communication. That is, a process returns from a collective routine as soon as it finishes its participation in the collective operation.

  For instance, when the Root process returns from an MPI_Bcast, it simply means that its send buffer Address can be safely reused. The other processes may not have even started their corresponding MPI_Bcast yet!

- A collective routine may or may not be synchronous, depending on implementation. MPI requires that the user take responsibility to ensure that his code is correct whether a synchronous implementation is chosen or not.

- The Count and Datatype must be compatible on all processes involved.

- There is no tag argument in a collective routine. The message envelope is specified by the communicator argument and the source/destination processes. For instance, in MPI_Bcast, the source of the message is Root, and the destinations are all processes (including Root).

Example 13.14 Array handling in an MPI program

Consider an MPI program executed by three processes. They all execute the following code initially:

```
int     i, j, my_rank, group_size, A[3], B[3], tag=1, root=0;
MPI_Comm comm;
MPI_Init (&argc, &argv);
comm = MPI_COMM_WORLD;
MPI_Comm_rank (comm, &my_rank);
MPI_Comm_size (comm, &group_size);
for (i = 0; i < 3 ; i++) { A[i] = B[i] = my_rank * group_size + i; }
```

What are the values A[1] and B[0] in process 2, after the execution of each of the following code segment?

```
(1) if (my_rank==0){
        MPI_Bcast (A, 3, MPI_INT, root, comm);
        MPI_Send (B, 3, MPI_INT, 2, tag, comm);
    } else if (my_rank == 1) {
        MPI_Bcast (A, 3, MPI_INT, root, comm);
    else {
        MPI_Recv (B, 3, MPI_INT, 0, tag, comm);
        MPI_Bcast (A, 3, MPI_INT, root, comm);
    }
```

(2) MPI_Bcast (A, 3, MPI_INT, root, comm);

(3) MPI_Scan (A, B, 1, MPI_INT, MPI_SUM, comm);

(4) MPI_Scatter(A,1,MPI_INT,B,1,MPI_INT,root,comm);

Solution After the initial code is executed, arrays A and B have the following values:

Table 13.7 Array Values in Three Processes

	Process 0	Process 1	Process 2
A[0]=B[0]	0	3	6
A[1]=B[1]	1	4	7
A[2]=B[2]	2	5	8

(1) The code attempts to send array B of process 0 to process 2, and to broadcast array A of process 0 to all processes. Thus the answer should have been $A[1] = 1$ and $B[0] = 0$ in process 2. However, this code is erroneous, because it will deadlock when MPI_Bcast is implemented synchronously. Thus this code is not MPI-compatible, and the result is undefined.

(2) In process 2, we have $A[1] = 1$ and $B[0] = 6$.

(3) This operation scans the three A[0]'s (0, 3, 6) and generates $B[0] = 0$ in process 0, $B[0] = 0+3 = 3$ in process 1, and $B[0] = 0+3+6 = 9$ in process 2.

(1) This routine scatters the three elements of array A of process 0 to all three processes, one element per process. The answer is $A[1] = 7$ and $B[0] = 2$ in process 2.

∎

13.2.5 The MPI-2 Extensions

MPI has been implemented on most parallel systems since the standard was published in 1994. The implementation platforms range from MPP supercomputer, SMP servers, to clusters of workstations and PCs. However, users soon asked for more functionality to be added.

In 1997, the MPI Forum announced a revised standard, called MPI-2. The original MPI is renamed MPI-1. The MPI-2 has added many new features, and the user/vendor community may have difficulty to embrace all of them. We use *dynamic process* and *one-sided communication* to get to the point. We also briefly outline other MPI-2 features.

Dynamic Processes Processes in an MPI-1 parallel program are static: The program starts execution with a given number of processes, and no process can be added or deleted while the program is running. MPI-2 decides to support dynamic processes, which provide the following benefits:

- MPI-1 does not specify how processes are created or how they establish communication. Therefore, MPI-1 needs the underlying platform to provide such capability, such as POE in SP2 and rsh on a network of workstation. The dynamic process mechanism in MPI-2 provides this capability in a portable (platform-independent) way.
- Dynamic processes facilitate porting PVM codes to MPI. They enable important classes of applications, such as client-server and process farm.
- Dynamic processes allow more efficient use of resources and load balancing. For instance, the number of nodes used can shrink or grow as needed.
- Fault tolerance can be supported. When a node fails, the process running on it can failover to a new process created on another node.

MPI-2 introduces a number of new functions for dynamic process management. We will just discuss MPI_Spawn, which has the following prototype:

```
int MPI_Spawn(
    char* command_line,       /* executable and arguments */
    int minprocs,             /* minimum number of processes to spawn */
    int maxprocs,             /* maximum number of processes to spawn */
    char* info,               /* where and how to start the processes */
    int root,                 /* rank of the root (parent) process */
    MPI_Comm comm,            /* the root process's communicator */
    MPI_Comm* intercomm,      /* intercommunicator between comm and
                                  the newly spawned group of processes */
    int* array_of_errcodes    /* one error code per spawned process */
)
```

This function tries to spawn *maxprocs* children processes, each executing the identical code specified in *command_line*. If MPI is unable to spawn *maxprocs* processes, it may spawn as few as *minprocs* processes. MPI_Spawn returns an error if it cannot spawn even *minprocs* processes. The *info* argument should be a null string, allowing the runtime system to determine where and how to start the processes.

The *array_of_errcodes* is an array of length *maxprocs*. When *k* processes are success-fully started, the first *k* elements of *array_of_errcodes* will each contain the value MPI_SUCCESS, while each of the remaining elements will contain an error code indicating the reason why a process was not successfully started.

MPI_Spawn is a collective operation, meaning all members of the communicator *comm* must call it to spawn processes. However, only the arguments *command_line*, *minprocs*, *maxprocs*, and *info* in the *root* process are significant, and the value of these arguments on other processes is ignored. So, although the spawned processes can be considered children to all the processes in *comm*, only the root process is the true parent.

The children are MPI processes, meaning they must call MPI_Initialize, which is collective with MPI_Spawn in the parent processes. The children and the parent processes can communicate through the intercommunicator *intercomm*, which is returned by MPI_Spawn in the parent processes, and the a child process can obtain the handle of the intercommunicator by calling an MPI_Parent function, which has a prototype

int MPI_Parent(MPI_Comm* intercomm)

One-Sided Communication   The MPI-2 includes a new point-to-point communication model called *remote memory access* (RMA), which allows a process to performs *one-sided communication*, whereby a process can send a data value to a destination or retrieves a data value from a source, without the other party's participation. This is in contrast to MPI-1, where all point-to-point communications are two-sided, in that both sender and receiver processes must participate.

MPI-2 includes a rich set of functions to support RMA. We will only describe three RMA functions. Before remote operations can be performed, a *memory window* must be set up. This is accomplished by the MPI_RMA_init function as prototyped below:

```
int MPI_RMA_init(
    void* window_base,          /* base address of memory window */
    int window_size,            /* size of memory window in bytes */
    int disp_unit,              /* unit size for displacements */
    MPI_Comm comm,              /* current communicator */
    MPI_Comm* newcomm           /* new communicator for RMA */
)
```

All processes in the communicator *comm* must call the same MPI_RMA_init collec-tively. This collective call returns a new communicator *newcomm*. Each calling process specifies a window in its address space that can be remotely accessed by any process in *newcomm*. The window starts at *window_base* and is *window_size*-byte large. Any RMA operation must access the window with a unit of *disp_unit*. For instance, when *disp_unit* = 8, an RMA can access a data buffer starting at *window_base* + 8, *window_base* + 16, etc., but not at *window_base* + 10.

A one-sided retrieve is called a *get*. The MPI_Get has the following prototype:

> int MPI_Get(void* *dest_addr*, MPI_Init *dest_count*,
>   MPI_Datatype *dest_datatype*, int *source_rank*, int *source_disp*,
>   int *source_count*, MPI_Datatype *source_datatype*,
>   MPI_Comm *comm*,
>)

This function transfers the contents of a data buffer in a memory window of a source process to a data buffer in the destination process (the calling process). The source process is uniquely specified by the *comm* and *source_rank* pair. The source memory window is determined by *comm*, which is actually the *newcomm* in a previous MPI_RMA_init call.

The source data buffer starts at $window\_base + source\_disp \times disp\_unit$. It contains *source_count* data items of type *source_datatype*. The destination data buffer starts at *dest_addr*, and contains *dest_count* data items of type *dest_datatype*. MPI_Get is a blocking function. It returns after the data arrive at the destination data buffer.

A one-sided send is called a *put*. The MPI_Put is very similar to the MPI_Get:

> int MPI_Get(void* *source_addr*, MPI_Init *source_count*,
>   MPI_Datatype *source_datatype*, int *dest_rank*, int *dest_disp*,
>   int dest_count, MPI_Datatype *dest_datatype*,
>   MPI_Comm *comm*,
>)

The only difference is that the direction of data transfer is reversed. Data are copied from the source data buffered the memory window of the destination. MPI_Get is a blocking function. It returns after the data leave the source data buffer (but have not necessarily arrived at the destination memory window).

Other MPI Extensions Besides dynamic process and one-side communication, MPI-2 extends MPI-1 in other aspects. The following is a partial list:

- Collective communications are extended to include nonblocking mode and inter-communicator ones. In MPI-1, only blocking and intra-communicator collective communications are supported.
- MPI-2 adds scalable I/O support, called MPI-IO. In MPI-1, the I/O issue is totally ignored.

- MPI-1 only defines language bindings for Fortran 77 and C. MPI-2 extends the binding to Fortran 90 and C++.
- MPI-2 adds supports for real-time processing.
- MPI-2 extends the MPI-1 external interface to give environment tool writers more accessibility to MPI objects. This could aid in the development of profiling, monitoring, and debugging tools.
- MPI-2 extends MPI-1 to allow nonblocking collective communications, inter-communicator collective communications, inter-communicator topology, and other additional features to enhance functionality or reduce memory requirement.

13.3 Parallel Virtual Machine (PVM)

PVM is a self-contained, public-domain software system that was originally designed to enable a network of heterogeneous Unix computers to be used as a large-scale, message-passing parallel computer. As its popularity grows, it has been ported to SMPs, PVPs, MPPs, clusters of workstations and PCs. PVM has been implemented for non-Unix platforms such as Windows NT and Windows 95. The programming languages supported include C, Fortran, and Java.

Development History The development of PVM was started in 1989 at Oak Ridge National Laboratory. Its research and development are still an on-going project involving researchers from universities and research institutions. Although more and more people are using MPI, PVM is still the most popular software platform for parallel processing.

With PVM, a user can construct a *virtual machine*, a set of fully connected nodes. Each node can be any Unix computer, such as any sequential, vector, or parallel computer. The user can then dynamically create and manage a number of processes to run on this virtual machine. PVM provides library routines to support interprocess communication and other functions.

Comparison with MPI The main difference between PVM and MPI is that PVM is a self-contained system while MPI (or more specifically, MPI-1) is not. MPI relies on the underlying platform to provide process management and I/O functions. These functions are included in PVM.

On the other hand, MPI has more powerful support for message passing. MPI and PVM are evolving towards each other. For instance, MPI-2 added process management functions, while PVM now has more collective communication functions. It will be beneficial to the parallel processing community if PVM and MPI eventually merge into a single, standard library.

Unlike MPI, PVM is not a standard, which implies that PVM can undergo version changes more easily and frequently than MPI. Also, PVM can have quite different implementations that are not necessarily portable. However, PVM has become quite stable in the past two years. By *generic PVM*, we refer to the PVM that is portable among a network of Unix workstations. We will focus on generic PVM from now on.

13.3.1 Virtual Machine Construction

The PVM system consists of two parts: a PVM daemon (called *pvmd*) that resides on every computer of the virtual machine, and a user-callable library (called *libpvm3.a*) that is linked to the user application for process management, message passing, and virtual machine management.

We will use a number of simple examples to illustrate the basic ideas. PVM is much more powerful and flexible than these examples demonstrate. The details of how to install and use PVM can be found in the *PVM Users Guide* available from the PVM software package. See the Web Resources for how to get PVM from several Web sites. Note that in PVM calls a node a *host*. A process is also called a *task*.

PVM Console   After installing PVM, a user types the following command from any host to create a *PVM console*:

    pvm host_file

The successful execution of this command will start a pvmd daemon on the invoking host and on every host listed in an optional host_file, and will display the following prompt at the invoking host (called the *master host*):

    pvm>

which indicates that the host is in the PVM console mode. The PVM console is a stand-alone, interactive PVM process, like a shell. The user types a number of commands to manage the virtual machine, to invoke a PVM application job, and to monitor the execution of the job. Table 13.8 lists several frequently used PVM console commands.

Dynamic Configuration   The virtual machine can be dynamically configured by a user application calling PVM library functions. The pvm_addhosts and the pvm_delhosts functions are used to add and delete one or more hosts to the virtual machine, respectively:

```
int     info, nhost=2, infos[2];
char    *hosts[] = { "apple", "orange.usc.edu"}
info = pvm_addhosts( hosts, nhost, infos);
info = pvm_delhosts( hosts, nhost, infos);
```

Table 13.8 Major PVM Console Commands

Command	Operational Meaning
pvm> add apple	Add the host "apple" to the virtual machine
pvm> delete apple	Delete the host "apple" from the virtual machine
pvm> conf	List the configuration of the virtual machine
pvm> spawn -count 4 app	Start 4 tasks to run "app" on the virtual machine
pvm> jobs	List jobs running in the virtual machine
pvm>halt	Kill all PVM processes and shut down PVM

These functions add and delete two hosts "apple" and "orange", respectively. Note that the host name can be the full or abbreviated name. The variable *hosts* holds the names of the hosts, and *nhost* the number of hosts to be added or deleted. The integer array *infos* of length *nhost* holds a status code returned for each host. A negative value indicates an error condition. These two functions return the number of hosts successfully added or deleted. When completely successful, *info* should be equal to *nhost*.

Mechanisms and Implementation   The first pvmd started by hand is called the *master* pvmd. Other pvm daemons are subsequently started by the master and are called *slaves*. The set of the master and the slave daemons (together with their hosts) forms the virtual machine. Any virtual machine has exactly one master at all times.

The minimal virtual machine consists of just one member: the master. Only the master can add or delete a slave from the virtual machine. However, requests to add or delete a slave come from outside of the master daemon. Such a request is always passed to the master daemon which then starts the slave on a specific host. PVM allows the master to start up a slave via **rsh**, **rexec()**, and other means.

A data structure called the *host table* resides on every host. The host table has an entry called a *host descriptor* for each host of the virtual machine. The host descriptor holds the host configuration information, as well as packet queues and message buffers for communication. Initially, the host table has only one entry for the master host.

As a new slave is added to the virtual machine, the host table of the master host is updated to include a new entry for the newly added slave. Information of the updated host table is then broadcast to the entire virtual machine, including the newly added slave hosts. This way, the host tables of all hosts of the virtual machine are synchronized and consistent, barring any host crash or network failure.

To see how the virtual machine is reconfigured, we discuss how a host is dynamically added in Fig. 13.6. Suppose a virtual machine consists of two hosts H1 and H2. A PVM task T in host H2 calls a PVM library function pvm_addhosts() to add a new slave host H3 to the virtual machine. This message is passed to the PVM daemon pvmd2 on host H2, triggering it to call a function dm_addhost(), which then passes to the master pvmd1.

The master daemon uses **rsh** (or other means) to start pvmd3 and gets the slave configuration information back. Then the master daemon parses the configuration information and updates its host table. The results are broadcast to the slaves to update their host tables. Upon completion, the slaves each send an acknowledgment back to the master.

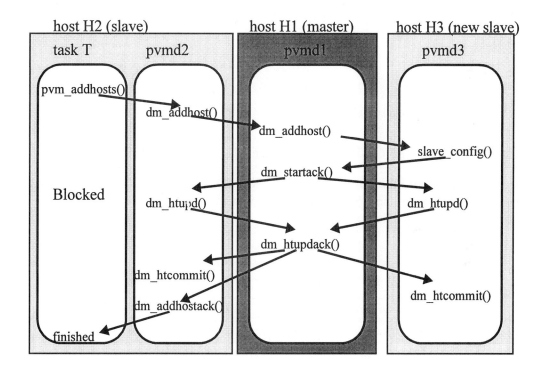

Figure 13.6   Dynamically add a host H3. (Courtesy of Beguilin et al., *A User's Guide To PVM*, [66]. 1991)

After the master decides that all host tables are updated, it broadcasts a message to order all slaves to commit their host tables. At the same time, the master informs the requesting slave that the add-host operation is successful. Note that task T blocks (waits) until it receives a notification from its local daemon that pvm_addhosts has completed.

13.3.2 Process Management in PVM

PVM enables both static and dynamic parallelism. To start a static parallel program, a user may execute a "spawn -count 4 *foo*" at the PVM console. This will create a parallel application of four tasks (processes) to run on the virtual machine. Each task executes the same code *foo*. The application will have a static DOP of 4 if foo does not call the dynamic task creation function pvm_spawn().

The command "spawn -count 4 *foo*" only says that an SPMD program of four tasks needs to be created. It does not say how the tasks are mapped to the hosts, i.e., which task should be executed on which host. PVM provides an allocation algorithm to map tasks to hosts. Multiple tasks are mapped to each host. PVM allows the user to explicitly specify a specific host or architecture type for each task. For instance, the console command:

> pvm> spawn -(*apple*) *foo*

will start a task on host *apple* to execute the code *foo*. The command

> pvm> spawn -(*RS6K*) *foo*

will start a task on any RS/6000 node that uses the AIX operating system.

Process Management Functions   PVM allows MPMD parallelism and dynamic process management. The most important function is pvm_spawn(). This and some of the other process management functions are listed in Table 13.8.

Table 13.9 PVM Functions for Process Management

PVM Function Call	Meaning
tid=pvm_mytid();	Get the task ID of the calling task
tid=pvm_parent();	Get the task ID of the parent task
info=pvm_exit();	The calling task exits PVM. The task continues running as a Unix process
numt =pvm_spawn(...);	Spawn a PVM task
info=pvm_kill(tid);	Terminate a PVM task
tstat=pvm_pstat(tid);	Get the status of a PVM task
info=pvm_tasks(...);	Get the information of all tasks running on the virtual machine
mstat=pvm_mstat(host);	Get the status of a host
info=pvm_config(...);	Get the configuration information of the entire virtual machine

Let us look at the pvm_spawn function in detail. Its function prototype and synopsis in C are given as follows:

```
int numt                    // the actual number of tasks started
= pvm_spawn (
char *progm,                // the executable file name
char **argv,                // pointer to the array of augment
int flag,                   // option to specify on which hosts to spawn, etc.
char *where,                // work with flag
int ntask,                  // number of copies of the executable to start up
int *tid                    // Holds the tid's of the tasks spawned
)
```

This function spawns *ntask* copies of the executable named *progm*. The arguments to progm are contained in an array pointed to by *argv*. The *flag* should be set to the sum of the option values shown in Table 13.10.

Table 13.10 The *flag* Parameter in pvm_spawn

Option Symbol	Option Value	Meaning
PvmTask Default	0	The hosts are chosen by PVM
PvmTaskHost	1	A host is specified by *where*
PvmTaskArch	2	A architecture is specified by *where*
PvmTaskDebug	4	Start up processes under debugger
PvmTaskTrace	8	PVM trace data will be generated
PvmMppFront	16	Start up process on MPP frontend
PvmHost-Compl	32	Use complement host set

Grouping To support collective operations, PVM uses a separate library called libpvm3.a, which is built on top of the core PVM library libpvm3.a. The PVM daemon pvmd does not handle grouping functions, which are performed by a separate daemon called the *group server*. This daemon is automatically started when the first group function is invoked.

PVM supports *dynamic grouping*: Any task can join or leave a group at any time. Some of the group functions are listed in Table 13.8The PVM dynamic grouping concept is very flexible. There can be multiple groups, and a task can belong to different groups at the same time. A task can join or leave a group at any time without having to notify other members of the group. Unlike MPI, where a task always has a unique rank within a group, a PVM task can be assigned different instance numbers when it leaves and rejoins a group. A task can broadcast a message to a group even though it is not a member of the group..

Table 13.11 PVM Group Functions

PVM Function Call	Meaning
inum=pvm_joingroup("World");	The calling task joins group World and is assigned an instance number *inum*. The instance number is like rank in MPI
info=pvm_lvgroup("World");	The calling task leaves group World
tid=pvm_gettid("World", inum);	Get task ID from instance number
inum=pvm_getinst("World", tid);	Get instance number from task ID
gsize=pvm_gsize("World");	Get group size
info=pvm_barrier("World", 10);	The calling task blocks (waits) until 10 members of World have called pvm_barrier.
info=pvm_bcast("World", tag);	Broadcast a message identified by tag to all members of World (excluding self)
info=pvm_reduce(...);	Similar to reduction in MPI

Dynamic grouping increases the nondeterministic behavior of a program. For instance, a broadcast operation can have different results if a task is joining or leaving a group. The task may or may not get the broadcast message. A barrier operation may deadlock if member tasks are leaving a group.

Task Identifier   PVM uses a 32-bit integer to address any task, group, or pvmd within a virtual machine. This integer is called a *task identifier* (or TID), although it can refer to a group or a daemon. The generic TID format is shown in Fig. 13.7. The 32 bits are divided into four fields: a *server* bit S (a daemon is often called a server), a *group* bit G, a 12-bit *host* field H, and an 18-bit *local* field L.

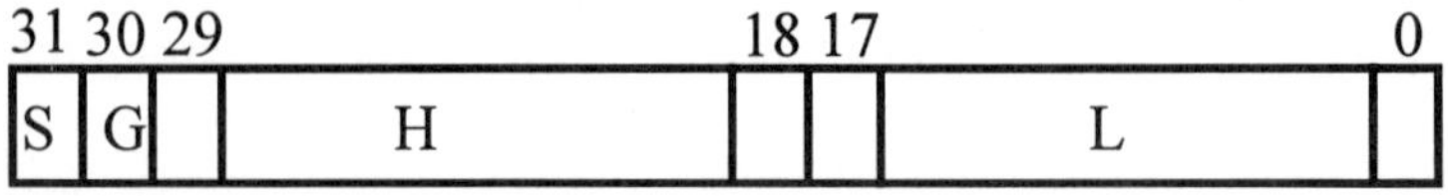

```
31 30 29                    18 17                        0
┌─┬─┬──────────────┬──┬─┬──────────────────┬─┐
│S│G│      H       │  │ │        L         │ │
└─┴─┴──────────────┴──┴─┴──────────────────┴─┘
```

Figure 13.7   PVM generic TID format.

The interpretation of various combinations of these fields are listed in Table 13.12, where $4095 = 2^{12} - 1$ is the maximum number of hosts in a virtual machine. The maximum number of tasks per host is $262143 = 2^{18} - 1$. In real implementations, constraints of the operating system (e.g., number of processes allowed per user, number of open files allowed) will limit the number of tasks per host to a much smaller number.

Table 13.12 Interpretation of the Task Identifier

S	G	H	L	Meaning
0	0	1..4095	1..262143	A task ID
0	1	1..4095	Don't care	Multicast address
1	0	0	0	The local pvmd
1	0	1..4095	0	A pvmd ID
1	1	Small negative number		An error code

When $S = G = 0$ (i.e., not a server or group), the TID identifies a task. The task resides on the host as indicated by the H filed, and the local process ID is contained in the L field. Note that with this encoding scheme, tasks can be assigned unique TIDs by their local pvmd's without inter-host communication.

When $S = 0$ and $G = 1$, the TID identifies a group of tasks that allow a message to be multicast to multiple hosts. This is different from the task grouping concept which supports collective (or global) operations.

When $S = 1$ and $G = H = L = 0$, the TID identifies the local pvmd. If the H field is positive, the TID identifies either a local or a remote pvmd, with the host number indicated in the H field. When $S = G = 1$, the TID is used to indicate an error condition.

13.3.3 Communication with PVM

A PVM program running on a virtual machine involves daemons and tasks. Several types of communications are possible, including communications between a daemon and a task, between two daemons, and between two tasks. Besides passing data for parallel computation, communication is needed to pass heartbeat messages among pvmd's. In addition, PVM allows Unix-like signals to be sent to tasks.

Communications Protocols The generic PVM assumes each host in the virtual machine can connect directly to every other host using TCP and UDP. These protocols are chosen because they are standard and widely available. PVM protocol drivers are run as user-level processes. This has the advantage of not needing to modify the kernel, but introduces high overhead. The communication protocols are used in PVM as shown in Fig. 13.8. There are three types of communications: between two pvmd's, between a pvmd and its tasks, and between two tasks.

Communications between two pvmd's use UDP. There are three reasons for using UDP instead of TCP. The first is scalability. Each TCP connection consumes a file descriptor in the pvmd. The number of opened files is restricted by any operating system. On the other hand, a single UDP socket can communicate with many number of remote UDP sockets.

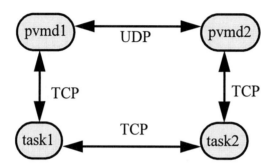

Figure 13.8   Communication protocols used in generic PVM.

The second reason is that using TCP would require establishing up to $n(n-1)/2$ connections for an n-node virtual machine, which is expensive to maintain. The UDP can be initialized to bypass the connection phase: connections do not have to be explicitly established between pvmd's. The third reason is that it is easier to set timeouts in UDP to detect host, pvmd, and network failures.

Communications between a pvmd and its tasks are through the TCP protocol, mainly because TCP offers reliable delivery while UDP is unreliable. A communication between two tasks can be accomplished in two ways. By default, any message sent from one task to another is routed through the path: source task to source pvmd to destination pvmd to destination task.

Alternatively, a message can be directly sent from the source task to the destination task over a TCP connection. The second method reduces message routing overhead, but there are the drawbacks of TCP connections discussed above.

PVM Communication Functions   Users of PVM do not need to know the three different types of communications discussed above. They see only a set of high-level communication functions. PVM provides many such functions. We will discuss only a few of them to illustrate the ideas. Sample PVM communication functions are shown in Fig. 13.9.

```
int bufid = pvm_initsend( int encoding )
int info = pvm_pkint(int *p, int nitem, int stride)
int info = pvm_send(int tid, int tag)
int info = pvm_mcast(int *tids, int ntasks, int tag)
int bufid = pvm_recv(int tid, int tag)
int bufid = pvm_nrecv(int tid, int tag)
int bufid = pvm_trecv(int tid, int tag, struct timeval *tmout)
int bufid = pvm_probe(int tid, int tag)
int bufid = pvm_bufinfo(int bufid, int *bytes, int *tag, int *tid)
int info = pvm_upkint(int *p, int nitem, int stride)
```

Figure 13.9   A sample of PVM communication functions.

At any given time, a PVM task has one *active send buffer* and one *active receive buffer*. To send data from one task to another, the source task first packs the data into a message in the active send buffer, and then calls a library function to send the message to the destination task's active receive buffer.

The destination task executes a receive function to get the message into the active receive buffer, and then unpacks the message to get the data. Note that these send/receive functions do not have to specify any message buffers, because they always use the active ones. This is different from MPI.

The function *pvm_initsend* creates a new send buffer and makes it the current active buffer. The *encoding* parameter specifies the data format that should be used in packing data into messages, and it can be one of the following PVM constants:

- PvmDataDefault denotes that data are encoded by the XDR (*eXternal Data Representation*) format, which is a platform-independent format allowing data exchange among heterogeneous nodes that use different data formats.
- PvmDataRaw denotes that the raw data will be used without any encoding. This is appropriate for communication among homogeneous nodes, and can avoid the overhead of encoding/decoding.
- PvmDataInPlace denotes a third scheme that can further reduce overhead. It says that the real data will not be copied to the active buffer, only the address and the length of the data.

The actual packing of data is done by a number of PVM functions pvm_pk*, one for each data type. For instance, to pack the even-indexed elements of a 100-element integer array A (i.e., A[0], A[2], ..., A[98]) into the active send buffer, we can call the PVM function pvm_pkint(A, 50, 2), where A is the address of the data, 50 is the number of data elements, and 2 is the stride.

As another example, we use pvm_pkfloat(...) to pack data of float type. PVM also provides a pvm_packf(...) function to pack data consisting of elements of different types. The function *pvm_send*(*tid*, *tag*) send the message in the active send buffer to the task denoted by *tid* and tags the message with *tag*.

To broadcast a message to several tasks, we can use the function *pvm_mcast*(*tids*, *ntasks*, *tag*), where *tids* lists the destination tasks, *ntasks* is the number of tasks. The functions pvm_recv and pvm_nrecv are blocking and nonblocking receives, respectively. Both *tid* and *tag* could have a wildcard value –1, indicating any task and any tag.

Example 13.15 PVM program to compute π in Example 12.1

A PVM program to compute π is shown in Fig. 13.10. This SPMD program divides the interval [0, 1] into N regions and uses n tasks. Each task is responsible for computing a subtotal *mypi* from N/n regions. The n subtotals are then aggregated into a final sum by a reduction operation.

Suppose the source code is contained in the file pi.c. The executable file *pi* is obtained by using the following command:

cc -I/pvm3/include pi.c libgpvm3.a libpvm3.a -o pi

The user invokes the first task, which spawns the other $n-1$ tasks. Every task first executes the following statement:

me = *pvm_joingroup*("PI");

to join a group PI, and the variable *me* will contain the rank (instance number) of the task within the group. The initial task has a rank of 0. The next statement

parent = *pvm_parent*();

finds the ID of the parent task. For the initial task which has no parent, this function call returns 0. For the other tasks, the function call returns the initial task's ID. The initial task executes

pvm_spawn("pi", (char**)0, 0, "", n–1, tids);

to spawn $n-1$ children tasks, which should execute the code in file "pi" and their tasks' IDs are returned in array *tids*. The other three arguments have null (0) values, which are typical of many PVM programs. The initial task then prompts the user to enter the number of regions in variable N by executing

```
    printf("Enter the number of regions: ");
    scanf("%d",&N);
```

```
#define n16              /* number of tasks */
#include "pvm3.h"
main (int argc, char ** argv)
{
    int mytid, tids[n], me, i, N, rc, parent;
    double mypi, h, sum=0.0, x ;
    me = pvm_joingroup( "PI" );
    parent = pvm_parent();
    if (me == 0) {
        pvm_spawn("pi", (char**)0, 0, "", n-1, tids);
        printf("Enter the number of regions: ");
        scanf("%d",&N);
        pvm_initsend( PvmDataRaw );
        pvm_pkint(&N,1,1);
        pvm_mcast(tids,n-1,5);
    } else {
        pvm_recv(parent, 5);
        pvm_upkint(&N, 1, 1);
    }
    pvm_barrier("PI", n);   /* optional */
    h  = 1.0 / (double) N;
    for (i = me + 1; i <= N; i += n)  {
        x = h * ((double)i - 0.5);
        sum += 4.0 / (1.0 + x*x);
    }
    mypi = h * sum;
    pvm_reduce( PvmSum, &mypi, 1, PVM_DOUBLE, 6, "PI", 0);
    if ( me == 0 ) printf("pi is approximately  %.16f \n", mypi);
    pvm_lvgroup("PI");
    pvm_exit();
}
```

Figure 13.10 A π program written in PVM.

It then broadcasts this value to all children tasks by executing

```
pvm_initsend( PvmDataRaw );
pvm_pkint(&N,1,1);
pvm_mcast(tids,n-1,5);
```

where 5 is an arbitrary message tag. Meanwhile, every child task will match the broadcast by executing

```
pvm_recv(parent, 5);        /* 5 is a tag matching that in pvm_mcast */
pvm_upkint(&N, 1, 1);
```

All tasks then synchronize by executing a barrier

```
pvm_barrier("PI", n);  /* n tasks in group PI should synchronize here */
```

The next few lines are just normal computation:

```
h  = 1.0 / (double) N;
for (i = me + 1; i <= N; i += n)  {
   x = h * ((double)i - 0.5);
   sum += 4.0 / (1.0 + x*x);
}
mypi = h * sum;
```

The next reduction operation

```
pvm_reduce( PvmSum, &mypi, 1, PVM_DOUBLE, 6, "PI", 0);
```

says that this is a sum reduction of all tasks in group PI. Each task contributes one data item of type PVM_DOUBLE from location *mypi*. The final result should be stored in *mypi* of the initial task (which has a rank 0). The number 6 is again an arbitrary message tag.

In general, a PVM reduction has the function prototype

```
pvm_reduce(void (*func)(), void *buffer, int nitem, int datatype,
              int tag, char *group, int root);
```

After printing the results, the program cleans up by calling

```
pvm_lvgroup("PI") and pvm_exit().
```

■

13.4 Bibliographic Notes and Problems

Turcotte [620] gave a comprehensive survey of cluster-supporting software, including many message-passing software packages, that appeared before 1993. For in-depth discussions of representative message-passing software, see [118, 303, 372, 502, 574] and [618]. Chandy and Foster [141, 247] discussed the importance of developing modular and determinate parallel programs, and the Fortran-M approach.

A good tutorial of PVM is the book by Geist et al. [261] and the paper by Sanderram [597]. Konuru et al. [375] discuss how to efficiently implement PVM. An in-depth discussion of MPI is given in Gropp et al. [283] and Snir et al. [575] .

Bruck et al. [112] and Lauria and Chien [394] offer two examples of efficient imple-mentations of MPI. The authors have given a systematic evaluation of MPI performance. on MPPs in [332] and [655]. The most updated information and complete references are found in the PVM and MPI home pages listed in the Web Resources.

Homework Problems

Problem 13.1 Briefly define and use the first MPI_Send statement in Example 13.9 to elaborate the following MPI terms:

(a) Message buffer

(b) Message envelope

(c) Message context

Problem 13.2 Refer to Example 13.14. What are the values A[1] and B[0] in process 2 after each of the following code fragments are executed?

(a) MPI_Alltoall (A,1,MPI_INT,B,1,MPI_INT, comm);

(b)
```
if (my_rank == 0){
        MPI_Bcast (A, 3, MPI_INT, root, comm) ;
        MPI_Send (B, 3, MPI_INT, 2, tag, comm) ;
} else if (my_rank == 1) {
        MPI_Send (B, 3, MPI_INT, 2, tag, comm) ;
        MPI_Bcast (A, 3, MPI_INT, root, comm) ;
} else {
        MPI_Recv (B, 3, MPI_INT, MPI_ANY_SOURCE, tag, comm) ;
        MPI_Bcast (A, 3, MPI_INT, root, comm) ;
        MPI_Recv (B, 3, MPI_INT, MPI_ANY_SOURCE, tag, comm) ;
}
```

(c)   **if** (my_rank == 0){
       MPI_Send (A, 3, MPI_INT, 1, tag, comm) ;
       MPI_Recv (B, 3, MPI_INT, 2, tag, comm) ;
  } **else if** (my_rank == 1) {
       MPI_Send (A, 3, MPI_INT, 2, tag, comm) ;
       MPI_Recv (B, 3, MPI_INT, 0, tag, comm) ;
  } **else** {
       MPI_Send (A, 3, MPI_INT, 0, tag, comm) ;
       MPI_Recv (B, 3, MPI_INT, 1, tag, comm) ;
  }

(d)   **if** (my_rank == 0){
       MPI_Isend (A, 3, MPI_INT, 1, tag, comm) ;
       MPI_Recv (B, 3, MPI_INT, 2, tag, comm) ;
  } **else if** (my_rank == 1) {
       MPI_Isend (A, 3, MPI_INT, 2, tag, comm) ;
       MPI_Recv (B, 3, MPI_INT, 0, tag, comm) ;
  } **else** {
       MPI_Isend (A, 3, MPI_INT, 0, tag, comm) ;
       MPI_Recv (B, 3, MPI_INT, 1, tag, comm) ;
  }

Problem 13.3 Write an MPI program to solve the banking problem in Section 7.2.1. Specifically, assume there is only one account which allows withdraw and deposit transactions. The program includes a *banker task* amd two *customer tasks*. The banker manages the account and each of the customers initiates withdraw or deposit transactions.

(a)   Write both the banker and the customer codes.

(b)   Show that your program will not deadlock.

(c)   Show that your program will not starve any customer. If your program can not guarantee starvation freedom, give the necessary and sufficient conditions under which the program is starvation free.

(d)   Show that your program will generate correct results. For instance, when a customer withdraws $100 from the account, the banker should deduct $100 from the account and send the same ammount to the customer. The program should not mess up even when the other customer is requesting her transaction.

Problem 13.4 Use the MPI notation to write a complete (ready-to-run) parallel program for the Jacobi relaxation problem (Example 12.4). A skeleton code is shown in Fig. 13.11.

You need to perform the following tasks:

(a)   Write a complete, well-documented parallel program.

(b)   Explain the step-by-step logic of the code, including why a variable or statement is there.

(c)   Show under what conditions your code will not deadlock.

(d)   Summarize your coding experience by explaining what you like and what you do not like about parallel programming in MPI. You must give concrete examples from this exercise. For those you do not like, suggest alternatives.

```
float Arow[N], X[N], NewX, B, error, Temp ;
MPI_Comm comm ;
error = SomeLargeValue ;
initialize A, X, and B
MPI_Init (&argc, &argv) ;
comm = MPI_COMM_WORLD ;
MPI_Comm_rank (comm, &my_rank) ;
while ( error > SomeErrorBound ) {
    Temp = B;
    for ( i=0; i<N; i++) Temp -= Arow[i] * X[i] ;
    Temp /= A[my_rank] ;
    NewX = Temp + X[my_rank] ;
    MPI_Allgather (&NewX, 1, MPI_FLOAT, &X, 1, MPI_FLOAT, comm) ;
    Temp = Temp * Temp ;
    MPI_Allreduce (&Temp, &error, 1, MPI_FLOAT, MPI_SUM, comm) ;
}
```

Figure 13.11 Jacobi relaxation in MPI code.

Problem 13.5   Repeat Problem 13.4 using the PVM. Comment on the programmability and software productivity between the corresponding MPI and PVM codes.

Problem 13.6   Write an SPMD program in MPI to compute π with two problem sizes: $N = 100,000$ and $N = 10,000,000$. Develop and test your code on any parallel computer on your site. The program should be optimized and executed using 1, 2, 3, 4, 5, 6, 7, 8 processors. The deliverable of this exercise includes:

(a)   Complete, well-documented sequential and parallel programs.

(b)   A description of the test environment and procedures, including platforms, languages, compiler options, libraries, environment parameters, execution modes (batch/interactive, dedicated/shared), how to measure time, etc.

(c)   Tabulated and plotted performance numbers measured, including sequential execution time, parallel execution time, speed (in Mflop/s), speedup, efficiency, and utilization.

(d)   An analysis of your measured results. Pay attention to observations and explanations of the utilization and scalability over machine size and problem size.

Problem 13.7 Repeat Problem 13.6 using the PVM. Comment on the programmability and software productivity between the corresponding MPI and PVM codes.

Problem 13.8 Use the MPI notation to write a complete (ready-to-run) parallel program for the target detection problem (Example 12.2). A skeleton code is shown in Fig. 13.12.

```
#define        MaxTarget        10
#define        M                1000
#define        N                256
int   i, j, k = 0, group_size, my_rank;
float   A[M][N] ;
MPI_Init (&argc, &argv) ;
comm = MPI_COMM_WORLD ;
MPI_Comm_rank (comm, &my_rank) ;
compute A[*][my_rank]
j = my_rank ;
for ( i = 0 ; i < m ; i ++ )
   if ( IsTarget( A(i,j) ) ) {
      LocalTarget[k].direction = j ;
      LocalTarget[k].distance = i;
      k = k + 1 ;
      if ( k > MaxTarget ) break ;
   }
MPI_Op_create ( & target_reduce, Commute, & TargetReduce ) ;
MPI_Reduce(LocalTargets, GlobalTargets,
              TargetsType, TargetReduce, 0, comm);
MPI_Finalize() ;
```

Figure 13.12 A target detection code using MPI.

You need to perform the following tasks:

(a)   Write a complete, well-documented parallel program with the MPI commands.

(b)   Explain the step-by-step logic of the code, including why a variable or statement is there.

(c)   Show under what conditions your code will not end up with a deadlock.

(d)   Summarize your coding experience by explaining what you like and do not like in parallel programming in MPI. Give ive concrete examples from this exercise. For those you do not like in using the MPI, suggest alternatives and explain.

Problem 13.9 Repeat Problem 13.8 using the PVM. Comment on the programmability and software productivity between the corresponding MPI and PVM codes.

Problem 13.10 Compile a large table to show the relative strength and weakness of MPI and of PVM. The table must be able to cover important communication functions and major performance attributes.

Justify with reasoning the table entries. Your study could be handled at various application levels, ranging from high- to lower levels. The program comparison results you have obtained in Problem 13.4 to Problem 13.9 for individutal programs can be generalized to sort out some commonalities as general entries in the table.

Chapter 14

Data-Parallel Programming

This final chapter is devoted to exploiting data parallelism in user applications. We discuss in detail two data-parallel programming languages, Fortran 90 and *High-Performance Fortran* (HPF). These programming environments are the result of extensive research and development over the years by the parallel Fortran computing communities.

Then we discuss some recent efforts to extend Fortran, such as Fortran 95, Fortran 2001, and HPF-2. We also assess non-Fortran efforts for exploiting data parallelism, paying special attention on a new functional language called Nesl.

14.1 The Data-Parallel Model

The data-parallel programming model is characterized by several features. Some of them are quite different from those in shared-variable and message-passing models. We briefly list these features below and elaborate on them later using program examples from Fortran 90, HPF, and other data-parallel languages.

- *Single-threading*: From the programmer's viewpoint, a data-parallel program is executed by exactly one process, with a single thread of control. In other words, as far as control flow is concerned, a data-parallel program is just like a sequential program. There is no control parallelism.

- *Parallel operations on aggregate data structure*: A single step (statement) of a data-parallel program can specify multiple operations that are simultaneously applied to different elements of an array or other aggregate data structure.

- *Loosely synchronous*: There is an implicit synchronization after every statement. This statement-level synchrony is *loose*, compared with the *tight* synchrony in an SIMD system that synchronizes after every instruction.

- *Global naming space*: All variables reside in a single address space. All statements can access any variable, subject to the usual variable scope rules. This is in contrast to the message-passing approach to be discussed shortly, where variables may reside in different address spaces.

- *Implicit interaction:* Because there is an implicit barrier after every statement, there is no need for explicit synchronization in a data-parallel program. Communication is implicitly done by variable assignment.

- *Implicit data allocation:* The programmer does not have to explicitly specify how to allocate data. She may give the compiler hints about data allocation to improve data locality and reduce communication.

14.2   The Fortran 90 Approach

A popular data-parallel language is Fortran 90, which improves Fortran 77 in many aspects. We will discuss only the two important features that support parallelism: element-wise parallel array operations and array intrinsic functions.

Intrinsic Operations   Every language has some *intrinsic operators* (e.g., +, −, *, /) and *intrinsic functions* (e.g., sin, log), which are considered part of the language. Therefore, any implementation of the language must support them with the same semantics. The compiler knows the syntax and semantics of an intrinsic function, without further definition or specification by the user.

In contrast, *extrinsic* functions are those that may be implemented by some systems but not by others. The same extrinsic function may have different syntax and semantics in different implementations. For instance, the print() function in C is considered intrinsic, because it is implemented in all systems with the same syntax and semantics. However, a C function for FFT is extrinsic.

14.2.1 Parallel Array Operations

In keeping with the tradition of strong support for scientific computation in Fortran 77, the new Fortran 90 standard has significantly extended the array capability of Fortran 77. Now an entire array or a portion of an array (called an *array section*) may appear as an operand.

Array Sections   All intrinsic operators and functions may be applied to whole arrays or array sections, operating element-wise. It is also possible to perform parallel array operations that are not element-wise, through intrinsic functions. This will be discussed in Section 14.2.2.

Example 14.1 Array handling in a Fortran 90 program

Let us consider the following Fortran 90 code fragment:

real, dimension (3, 3):: A
real, dimension (2, 2):: B, C

S1:   C = 0
S2:   B = A(1:3:2, 2:3)
S3:   **where** ((B-6) > 0) C = sin(A(1:2, 1:2))

For a given array A, the results of the above code are

$$A = \begin{bmatrix} 1 & 2 & 3 \\ 4 & 5 & 6 \\ 7 & 8 & 9 \end{bmatrix} \quad B = \begin{bmatrix} 2 & 3 \\ 8 & 9 \end{bmatrix} \quad C = \begin{bmatrix} 0 & 0 \\ -0.7568 & -0.9589 \end{bmatrix} \qquad (14.1)$$

Interpretations   The four features of the data-parallel programming style can be elaborated with this example code:

- *Single threading* and *loosely synchronous*: The code consists of three statements S1, S2, S3, which are executed as a single thread, one after another. Statement S2 (S3) does not start until all operations of S1 (S2) finish.
- *Parallel array operations*: A statement may contain multiple operations on array elements, which are executed simultaneously. Statement S1 assigns 0 to all four elements of array C. Statement S2 assigns four elements of array A to array B. The **where** statement S3 assigns array C according to the condition $(B-6)>0$. Only those elements of C where the condition is true are modified. The two elements in the first row of C are not changed, because the condition is false.
- *Global naming space*: All array variables and elements are visible to all statements. There is no notion of processor private data.

Array Shaping   The number of dimensions of an array is called its *rank*. The variables A, B, and C in Example 14.1 are all rank-2 arrays. An array section are selected by using a three-tuple $L{:}U{:}S$ (called *subscript triplet*) for each array dimension, where L, U, and S are the *lower bound*, the *upper bound*, and the *stride* of the array section. Each of L, U, and $S can be any scalar expression, but the stride expression must not be zero. When the stride part is missing, a default stride of 1 is assumed. Thus, $A(1{:}3{:}2, 2{:}3)$ is the array section formed by the first and the third rows and the second and the third columns of an array A.

When several arrays or array sections appear in a statement, they must be *conformable*. For instance, in an array addition $X + Y$, the two arrays must have the same *shape*. That is, for each dimension, they must have the same *size* (i.e., number of elements). Thus two $2{\times}3$ arrays have the same shape, but a $2{\times}3$ array and a $3{\times}2$ array do not. When an operator is applied to a scalar s and an array X, the scalar s is first expanded to an array of the same shape as X, with all elements being s. This is shown in Example 14.1, where $B{-}6$ is computed in the following way:

$$B - 6 = \begin{bmatrix} 2 & 3 \\ 8 & 9 \end{bmatrix} - 6 = \begin{bmatrix} 2 & 3 \\ 8 & 9 \end{bmatrix} - \begin{bmatrix} 6 & 6 \\ 6 & 6 \end{bmatrix} = \begin{bmatrix} -4 & -3 \\ 2 & 3 \end{bmatrix} \quad (14.2)$$

14.2.2 Intrinsic Functions in Fortran 90

Besides element-wise array operations discussed above, Fortran 90 provides additional intrinsic functions to manipulate arrays as a whole. We discuss several below.

Matrix Operations   Fortran 90 provides several operators that operate on entire arrays. For instance, a matrix X can be transposed by calling *transpose*(X). The matrix multiplication function *matmul* multiplies an $m \times n$ array X by an $n \times p$ array Y and returns an $m \times p$ array. An example is the statement $Z = matmul\,(X,Y)$, where Z must be an $m \times p$ array.

Array Construction   A rank-1 array can be created through an *array constructor*, which is a sequence of values separated by commas and delimited by "(/" and ")/". For instance, a three-element vector X can be constructed by assignment $X = (/\ 1,3,5\ /)$ or, equivalently, by $X = (/\ 1{:}5{:}2\ /)$. Another way to create an array is to use the intrinsic function *spread*, which constructs a new array by adding an extra dimension and copying the operand.

For instance, the function spread((/ 1,3,5 /), 1, 3) makes 3 copies of (/ 1,3,5 /), and spread them along dimension 1 (the row dimension). The function spread((/ 1,3,5 /), 2, 2) makes 2 copies of (/ 1,3,5 /), and spread them along dimension 2 (the column dimension). The results are shown as follows:

$$\mathrm{spread}(\ (/\ 1,3,5\ /),\ 1,\ 3) = \begin{bmatrix} 1 & 3 & 5 \\ 1 & 3 & 5 \\ 1 & 3 & 5 \end{bmatrix}, \qquad \mathrm{spread}(\ (/\ 1,3,5\ /),\ 2,\ 2) = \begin{bmatrix} 1 & 1 \\ 3 & 3 \\ 5 & 5 \end{bmatrix} \quad (14.3)$$

Vector Reductions   Fortran 90 supports several intrinsic reduction functions, as listed in Table 14.1, where the arguments DIM and MASK are optional..

Table 14.1 Reduction Functions in Fortran 90

Function	Meaning
SUM(X, DIM, MASK)	The arithmetic sum of the elements of array X
PRODUCT(X, DIM, MASK)	The arithmetic product of the elements of array X
MAXVAL(X, DIM, MASK)	The arithmetic maximum of the elements of array X
MINVAL(X, DIM, MASK)	The arithmetic minimum of the elements of array X
ALL(X, DIM, MASK)	The logic product (AND) of the elements of array X
ANY(X, DIM, MASK)	The logic sum (OR) of the elements of array X
COUNT(X, DIM, MASK)	The number of the elements of array X
MAXLOC(X, MASK)	The location of the maximal element of array X
MINLOC(X, MASK)	The location of the minimal element of array X

We use several examples below to illustrate how to apply these functions.

Let $A = \begin{bmatrix} 1 & 2 & 3 \\ 4 & 5 & 6 \end{bmatrix}$. Then SUM(A) = 21, MAXVAL(A)=6, and COUNT(A)=6.

The DIM argument is used to restricted the reduction along the given dimension. The MASK argument restricts the reduction to those elements where the condition set by MASK is true. Thus SUM(A,1) = [5 7 9], because A is sum-reduced along the columns. COUNT(A,A>2)=4 since there are four elements that are greater than 2. MAXLOC(A,A<5)=[2 1] since the location (subscripts) of the maximum element that is less than 5 is row 2 and column 1.

Array conformability has a different meaning when parallel intrinsic functions are used. For instance, when two matrices are multiplied using the intrinsic function $Z=MALMUL(X,Y)$, conformability implies that the second dimension of X must have the same size as the first dimension of Y. Thus X and Y are conformable if X is a 2×3 array and Y a 3×2 array. They are not conformable if both are 2×3 arrays.

Next, we give two examples to show that Fortran 90 facilitates the writing of simple and clean programs for regularly structured parallel computing applications. With Fortran 90, programmers need not worry about process management, communication, data allocation, and synchronization issues, which are all left to the compiler. Fortran 90 provides more features and functionalities than Fortran 77. For some applications, the Fortran 90 programs can be shorter and simpler than corresponding Fortran 77 codes.

Example 14.2 Fortran 90 program for Jacobi relaxation

A Fortran 90 program for Jacobi relaxation (see Example 12.4) is shown in Fig. 14.1. The main computation is contained in the **do while** loop. The sequential nature of the loop simply says that the iteration steps should be performed one by one.

Parallel computation is possible within each iteration. Because of the array capability of Fortran 90, only three statements are needed. Note that the statement labels S1, S2, and S3 are not part of the code. They are used for discussing the code.

```
parameter n = 1024
real A(n,n), x(n), b(n), error, Temp(n), DiagA(n)
integer i
error = SomeLargeValue
initialize A, x, and b
do i = 1, n
        DiagA(i) = A(i,i)
end do
do while ( error > ErrorBound )
S1:     Temp = ( b - MATMUL(A, x) ) / DiagA
S2:     x = Temp + x
S3:     error = SUM ( Temp * Temp )
end do
```

Figure 14.1   Jacobi relaxation in Fortran 90

We can make several observations:

- The code is simple and concise, very close to the algorithm.
- Using intrinsic operator SUM and MATMUL simplifies coding.
- Fortran 90 is not very flexible, compared to HPF, to be discussed shortly. For instance, the Fortran 90 code in Fig. 14.1 needs a **do** loop to get the array DiagA for the diagonal elements of A.

■

Example 14.3 Gaussian elimination written in Fortran 90

A Fortran 90 code for Gaussian elimination (see Example 12.3) is given in Fig. 14.2. The code consists of two **do** loops. The first iteration performs triangularization. The second loop performs back-substitution. :

```
real A(n,n+1), x(n)
integer i, pivot_location(1)
do i = 1, n-1
    ! pivoting
    pivot_location = MAXLOC( ABS( A(i : n, i) ) )
    swap( A(i, i:n+1), A(i-1+pivot_location(1), i : n+1) )

    ! triangularization
    A(i, i:n+1) = A(i, i:n+1) / A(i,i)
    A(i+1:n, i+1:n+1) = A(i+1:n,i+1:n+1) - &
    SPREAD(A(i, i+1:n+1), 1, n-i) * SPREAD(A(i+1:n, i), 2, n-i+1)
end do
! back substitution
do i = n, 1, -1
    x(i) = A(i, n+1)
    A(1:i-1, n+1) = A(1:i-1, n+1) - A(1:i-1, i)* x(i)
end do
```

Figure 14.2   Gaussian elimination in Fortran 90.

In each iteration i of the first loop, the location of the "pivot" element is first found which has the largest absolute value. The swap subroutine exchanges the pivot row with the current row i. Then all elements of A in the ith column below the ith row are zeroed out.

14.3 High-Performance Fortran

High-Performance Fortran (HPF) is a language standard, designed as a superset (extension) of Fortran 90 to meet the following goals [370, 419]:

- Support for data-parallel programming
- Top performance on MIMD and SIMD computers with nonuniform memory access costs
- Capability of tuning code for various architectures

To achieve the first goal, HPF introduces a FORALL construct, an INDEPENDENT directive, and some additional intrinsic functions. These new language features allow HPF to be more flexible in specifying more general array sections and parallel computation patterns than Fortran 90.

To meet the second goal, HPF offers directives such as ALIGN and DISTRIBUTE, which enable the user to tell the compiler how data should be allocated among processors, so that communication overhead is minimized and workload is evenly divided.

HPF also has features targeted toward the third goal, such as EXTRINSIC procedures, which allow a user to take advantage of low-level features of a specific architecture. In this section, we discuss the HPF features for the first two goals, which are the most important. We will use a number of simple examples to illustrate the fundamental ideas.

14.3.1 Support for Data Parallelism

HPF provides four mechanisms for the user to specify data parallelism. They are array expressions and assignments, array intrinsic functions, the FORALL statement, and the INDPENDENT directive. They first two already appear in Fortran 90, but HPF adds many more intrinsic. The FORALL and the INDEPENDENT are new features.

The FORALL Construct   A FORALL statement of HPF is similar to a Fortran 90 array assignment statement, but more flexible. Let us consider the following example code:

$$\text{FORALL } (i=2:5, X(i) > 0) \; X(i) = X(i-1) + X(i+1)$$

The "$i = 2:5$" part is called a *forall triplet spec* in HPF, where i is the index variable. The subscript triplet 2:5 is equivalent to 2:5:1, with lower bound 2, upper bound 5, and a default stride of 1. The forall triplet spec determines a set of *valid* index values, $\{2,3,4,5\}$. The $X(i) > 0$ part is a scalar valued expression, called the *mask*. The set of *active* index values is a subset of the valid index values, such that the mask is true.

Assume initially $X = [1, -1, 2, -2, 3, -3]$ in the above FORALL statement. Then the active set is $\{3,5\}$, since $X(2) = -1 < 0$ and $X(4) = -2 < 0$. After the active index value set is determined, all the expressions in the assignment are computed simultaneously, in any order, for all active index values 3 and 5:

$$X(3-1) + X(3+1) \text{ evaluates to } -1+(-2) = -3$$
$$X(5-1) + X(5+1) \text{ evaluates to } -2+(-3) = -5$$

Then for all active index values (3 and 5), the left-hand side array elements are assigned with the corresponding right-hand side values simultaneously. The other elements of the left-hand side array are not changed. Thus the resulting array X after the above FORALL is executed is $X[1, -1, -3, -2, -5, -3]$.

There may be more than one forall-triplet specs in a FORALL statement. Then *combinations* of index values are used. For instance, the FORALL statement

FORALL (i=1:2, j=1:3, Y(i, j)>0) Z(i, j) = 1 / Y(i, j)

is equivalent to the Fortran 90 statement

where (Y(1:2, 1:3) > 0) Z(1:2, 1:3) = 1 / Y(1:2, 1:3)

The set of valid combinations of index values is the set {(1,1), (1,2), (1,3), (2,1), (2,2), (2,3)}, and the set of active combinations is the subset for which Y(i,j) is greater than zero.

Many parallel computations can be easily specified using FORALL but not in Fortran 90. The FORALL statement

FORALL(i=1:4, j=1:5, k=1:6) X(i, j, k) = i+j−k

can be expressed equivalently by the following Fortran 90 array assignment:

```
        X = SPREAD( SPREAD ( (/1:4 /), 2, 5), 3, 6)
 &          + SPREAD( SPREAD ( (/1:5 /), 1, 4), 3, 6)
 &          − SPREAD( SPREAD ( (/1:6 /), 1, 4), 2, 5)
```

Note that the Fortran 90 code is much more difficult to understand and implement efficiently. As another example, the simple FORALL statement

FORALL(i=1:n) X(i, j(i)) = Y(i)

cannot be expressed with a single array assignment in Fortran 90.

Sometimes a user may want to include several assignments in a FORALL statement. This can be accomplished with a more general form of the FORALL statement, called *FORALL construct*, or *multi-statement FORALL*, as illustrated by the following code:

```
FORALL (i=1:n)
   A(i) = sin(B(i))
   C(i) = sqrt(A(i) * A(i))
   D(i) = B(i) +2
END FORALL
```

The second assignment begins only after all the computations (there are *n* evaluations of the sine function) in the first assignment finish. Thus array A used in the second assignment has the new values computed in the first assignment.

Similarly, the third statement starts only after the second finishes, even though it does not seem to use array C computed in the second assignment. It is possible that C is an alias of B (through EQUIVALENT). Then array B uses the new value of C computed in the second assignment.

We note that functions and procedures can be called in a FORALL statement. The only requirement is that they be *pure*, that is, side-effect-free. HPF offers a list of syntactical restrictions to help the compiler determine whether a function or procedure is pure or not. For instance, one such restriction is that no global variable appear as the left-side of an assignment.

Independent Directives   The programmer can use this directive to assert to the compiler that there is no *loop-carried dependence* among the iterations of a loop. In other words, all iterations can be executed independently.

Example 14.4 The INDEPENDENT directive

The effects of the INDEPENDENT directive in the the following codes are illustrated in Fig. 14.3. The arrows indicate execution order restrictions.

```
    !HPF$ INDEPENDENT                    !HPF$ INDEPENDENT
        FORALL (i=1:2)                       do i = 1, 2
            A(i) = foo(B(i))                     A(i) = foo(B(i))
            C(D(i)) = log(E(i))                  C(D(i)) = log(E(i))
        END FORALL                           end do
```

Consider Fig. 14.3(a), where the INDEPENDENT directive is absent from the FORALL statement. Because of a loose synchrony, both instances of the first statement A(i) = foo(B(i)) must be finished before any of the second statement C(D(i)) = log(E(i)) can start. However, the FORALL postulates that both instances of each statement can be executed in parallel. We can visualize that there is an implicit barrier between the two statements.

When the INDEPENDENT directive is added, the compiler knows then that the barrier synchronization can be removed. Thus we get Fig. 14.3(b). Note that the compiler cannot perform this optimization without the help of the INDEPENDENT directive, because it cannot tell whether the subroutine foo() is pure, and whether there is an alias.

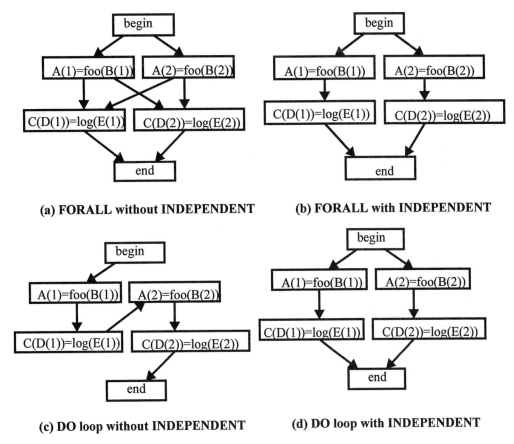

(a) FORALL without INDEPENDENT   **(b) FORALL with INDEPENDENT**

(c) DO loop without INDEPENDENT   **(d) DO loop with INDEPENDENT**

Figure 14.3   Visualization of the INDEPENDENT directive.

The DO loop [Fig. 14.3(c)] follows a different order, executing the two iterations sequentially. But when the INDEPENDENT directive is added [Fig. 14.3(d)], the compiler can optimize it to behave exactly like the FORALL statement with INDEPENDENT.

∎

14.3.2 Data Mapping in HPF

Data mapping refers to the allocation of data to processors, which should be done to achieve two goals:

- The interprocessor communication is minimized;
- The workload is evenly distributed among available processors.

An HPF compiler can use the *owner-compute rule* to distribute workload: computation associated with a data item is performed by the processor which owns that data item. Therefore, data mapping indirectly determines workload distribution.

HPF provides a number of directives which can be used by the programmer to recommend to the compiler how to best map data to the nodes. We use several examples to explain three important directives, which also illustrate the HPF data mapping principle.

Example 14.5 Data mapping in HPF program segments

Consider the following HPF code fragment:

```
integer A(100), B(100), C(101), i
!HPF$ ALIGN A(i) WITH B(i-1)
!HPF$ PROCESSOR N(4)
!HPF$ DISTRIBUTE A(BLOCK) ONTO N
!HPF$ DISTRIBUTE C(CYCLIC) ONTO N
      FORALL ( i = 2 : 100 )
          A(i) = A(i) + B(i-1)
          C(i) = C(i-1) + C(i) + C(i+1)
      END FORALL
```

The data mapping realized by the above code is shown in Fig. 14.4. Data mapping in HPF is realized in two phases: *logical mapping* and *physical mapping*. Logical mapping is specified by the user through compiler directives, which map data to a number of virtual nodes (abstract processors).

Physical mapping maps these virtual nodes to physical nodes (processors) in a real computer. Physical mapping is done by the system and is transparent to the user. A virtual node is mapped to only one physical node. In other words, all data allocated to the same virtual node must be mapped to the same physical node.

Logical mapping consists of two steps: *alignment* and *distribution*. Referring to the above code, the PROCESSOR directive

!HPF$ PROCESSOR N(4)

informs the compiler that this application wishes to use a linear array of four *virtual* nodes, shown as N1, N2, N3, and N4 in Fig. 14.4.

The name of this node array is N. HPF allows the specification of the set of virtual nodes as any rectilinear arrangement. For instance, the compiler directive

!HPF$ PROCESSOR N(4,5)

specifies 20 nodes arranged as a 4×5 mesh, and the directive

!HPF$ PROCESSOR N(4,5,6)

specifies 120 nodes arranged as a $4 \times 5 \times 6$ mesh.

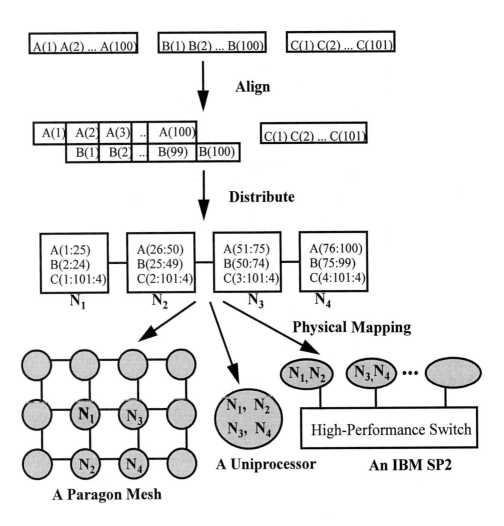

Figure 14.4 Data mapping in programming both sequential and parallel processors in using HPF data arrays.

Example 14.5 uses three data arrays: A, B, and C. The alignment step consists of only one ALIGN directive

!HPF$ ALIGN A(i) WITH B(i–1)

which suggests to the compiler that A(i) and B(i–1) should be mapped to the same node. But it does not say which node. Aligned data elements are stored in the same processor to reduce interprocessor communication. The scope of the dummy variable i is limited to the single directive. It can be replaced by any variable. Note that no directive to indicate how array C should be aligned. Consequently, it is not aligned with any other arrays.

In the distribution step, the programmer uses two DISTRIBUTE directives

!HPF$ DISTRIBUTE A(BLOCK) ONTO N
!HPF$ DISTRIBUTE C(CYCLIC) ONTO N

to recommend that both A and B be distributed in a block fashion, and C in a cyclic fashion, to the node array. Note that although only array A appears in the first DISTRIBUTE directive, it implicitly distributes array B as well, since B is aligned with A.

A block distribution evenly divides an array into a number of chunks (or *blocks*) of consecutive array elements, and allocates a block to a node. A cyclic distribution evenly divides an array so that every ith element is allocated to the ith node. For instance, in Example 14.5, the array elements are allocated to the four nodes in the following way:

N(1): A(1), A(2), ..., A(25), B(1), B(2), ..., B(24), C(1), C(5), C(9), ..., C(97), C(101)

N(2): A(26), A(27), ..., A(50), B(25), B(26), ..., B(49), C(2), C(6), C(10), ..., C(98)

N(3): A(51), A(52), ..., A(75), B(50), B(51), ..., B(74), C(3), C(7), C(11), ..., C(99)

N(4): A(76), A(77), ..., A(100), B(75), B(76), ..., B(99), C(4), C(8), C(12), ..., C(100)

The same logical mapping can be used in different physical mappings. Three such mappings are shown in Fig. 14.4. The four virtual nodes can be mapped to four physical nodes in an Intel Paragon computer, which has a mesh topology. The same virtual nodes can be mapped to two physical nodes in an IBM SP2, which uses a multistage interconnection network.

In fact, the HPF code can be executed by a uniprocessor, where all virtual nodes map to the same physical processor. A nice feature of HPF is that the same code (with all the directives) can be used on all three computers without change. Each system takes care of its physical mapping.

Now let us see how much communication is needed in the code of Example 14.5. The assignment A(i) = A(i) + B(i–1) does not need any communication, thanks to aligning A(i) with B(i-1). However, the assignment C(i) = C(i–1) + C(i) + C(i+1) needs to communicate two array elements, because the cyclic distribution allocates C(i–1), C(i), C(i+1) in different nodes. This gives about 200 elements in total to be communicated. Thus cyclic distribution is a bad choice. We should use a block distribution to reduce communication:

!HPF$ DISTRIBUTE C(BLOCK) ONTO N

Then, array C is distributed as follows:

N1: C(1), C(2),..., C(26)
N2: C(27), C(28),..., C(52)
N3: C(53), C(54),..., C(78)
N4: C(79), C(80),..., C(101)

Node N4 is allocated only 23 elements. Now only the boundary elements of each block need to be communicated, i.e., a total of six elements: C(26), C(27), C(52), C(53), C(78), and C(79).

This demonstrates another advantage of HPF: Neither the executable statements nor the declarations need to be changed. All we need to do to enhance performance is to change a compiler directive.

Example 14.6 HPF program for Gaussian elimination

An HPF program for Gaussian elimination is shown in Fig. 14.5. The DISTRIBUTE directive says that the first dimension of A should be block distributed. The * in the second dimension of A states that the second dimension is *collapsed*, meaning array section A(i,1:n) is mapped to the same node.

Assume there are 4 virtual nodes N1, N2, N3, and N4, and n divides 4. Then the data distribution is as follows:

N1: A(1:n/4, 1:n), b(1:n/4), Temp(1: n/4), x(1:n)
N2: A(n/4+1:2*n/4, 1:n), b(n/4+1:2*n/4), Temp(n/4+1:2*n/4), x(1:n)
N3: A(2*n/4+1:3*n/4, 1:n), b(2*n/4+1:3*n/4), Temp(2*n/4+1:3*n/4), x(1:n)
N4: A(3*n/4+1:n, 1:n), b(3*n/4+1:n), Temp(3*n/4+1:n), x(1:n)

```
parameter n = 32
real A(n,n+1), x(n)
integer i, pivot_location(1)
!HPF$ PROCESSOR Nodes(4)
!HPF$ ALIGN x(i) WITH A(i,j)
!HPF$ DISTRIBUTE A(BLOCK,*) ONTO Nodes
do i = 1, n-1
  ! pivoting
  pivot_location = MAXLOC( ABS( A(i : n, i) ) )
  swap( A(i, i:n+1), A(i-1+pivot_location(1), i : n+1) )
  ! triangularization
  A(i, i:n+1) = A(i, i:n+1) / A(i,i)
  FORALL ( j = i+1 : n, k = i+1 : n+1 ) A(j, k) = A(j, k) - A(j, i) * A(i, k)
end do
! back substitution
do i = n, 1, -1
  x(i) = A(i, n+1)
  A(1:i-1, n+1)  = A(1:i-1, n+1) - A(1:i-1, i)* x(i)
end do
```

Figure 14.5   Gaussian elimination in HPF.

The corresponding data distribution is given in Table 14.2. A problem with this distribution is *load imbalance*. Consider the triangularization step. After the 8th iteration, node 1 will be idle. After the 16th iteration, both nodes 1 and 2 will be idle. The work can be balanced by distributing the data cyclically, which is left as an exercise.

Table 14.2 Data Distribution among Four Nodes

Node 1	A(1:8, 1:33), x(1:8)
Node 2	A(9:16, 1:33), x(9:16)
Node 3	A(17:24, 1:33), x(17:24)
Node 4	A(25:32, 1:33), x(25:32)

14.3.3 Summary of Fortran 90 and HPF

We summarize below a number of practical issues in using either Fortran 90 or HPF to exploit data parallelism in user applications. Users must learn how to deal with these issues before developing efficient applications on parallel computers.

Parallelism Issues Because of single threading, a data-parallel program logically has only one process, although multiple nodes can execute the same program on different data subdomains. Most parallelism issues are taken care of by the system. The user does not have to worry about creating, terminating, and grouping processes, or how many processes are running the program.

The statement-level data parallelism and the where construct allow different nodes to execute different instructions at the same time. Thus both SIMD and SPMD computations are supported, but not MPMD. The degree of parallelism in a program varies dynamically when different array assignments are used or differnt FORALL statements are executed.

Interaction Issues Because of loose synchrony, most interaction issues are sidestepped in data-parallel programming. There is no interaction operation in Fortran 90 or HPF, except the intrinsic functions for reduction.

Fortran 90 provides nine intrinsic functions to support reduction, but no support for prefix, descend, or sorting. It does not support user-defined reduction operation, either. HPF extends Fortran 90's capability through a set of (nonintrinsic) library routines. These include additional reduction functions, scan functions, and sorting functions, among others.

Interactions in data-parallel programs are cooperative. There is only one interaction mode: statement-level loose synchrony. All operations of a statement must be finished before any operation of the next statement can start. Communication is implicit, through array assignment. Both regular and irregular communication patterns can be specified by different array index expressions, as shown below:

```
!HPF$ ALIGN A(i) WITH B(i)
A(i) = B(i–1)          ! a regular left-shift
A(V(i)) = B(i)         ! an irregular communication, the pattern is
                       ! specified through an index array V
```

HPF provides a set of array-mapping directives to minimize communication overhead by exploiting data locality. The most important directives are PROCESSOR, ALIGN, and DISTRIBUTE. HPF also provides dynamic REALIGN and REDIS-TRIBUTE statements to accommodate changing communication patterns. These data mapping constructs are more effective if the communication patterns are regular.

Semantic Issues Due to single threading and loose synchrony, data-parallel programs have clean semantics. It is impossible to have deadlock or livelock. The only possible non-termination is due to infinite looping, just as in a sequential program. Fortran 90 and HPF data-parallel programs are similar to sequential Fortran 90 programs with respect to the structuredness, compositionality, and correctness issues. There is no additional complexity introduced by parallelism.

Data-parallel programs are determinate as long as the *single assignment rule* is satisfied: Any array element is assigned at most once in a statement. Any program that does not satisfy the single-assignment rule is deemed illegal in Fortran 90 and HPF.

Example 14.7 Single assignment in data parallelism

Consider the following code fragment in HPF notation:

```
Integer    A(5), I(5), J(5)
S1:    I = (/ 1, 2, 3, 4, 5 /)
S2:    J = (/ 1, 2, 2, 4, 5 /)
S3:    A(I) = I + 2
S4:    A(2:4) = A(1:3) + A(3:5)
S5:    A(J) = J
```

Because all the elements of array I are different, single assignment is satisfied by statement S3, which generates A=(/ 3, 4, 5, 6, 7/). In contrast, statement S5 is illegal, since J(2)=J(3)=2, implying that A(2) is assigned twice. Note that S4 is equivalent to FORALL (k=2:4) A(k) = A(k–1)+A(k+1). It is not equivalent to the sequential loop

```
do k=2,4
    A(k) = A(k-1)+A(k+1)
end do
```

In S4, All three right-hand expressions are first evaluated, in any order, using existing values of array A: A(1)+A(3)=3+5=8, A(2)+A(4)=10, A(3)+A(5)=12. Then these three values are assigned to A(2), A(3), A(4), again in any order. Thus after S4 is executed, A=(/3, 8, 10, 12, 7 /).

∎

Programmability Because Fortran 90 and HPF are a high-level language standard, any conforming program should have good portability. Because most parallelism, interaction, and semantic issues are taken care of by the system, and because of their simple semantics, Fortran 90 and HPF are easy to use.

So why has it not dominated practical parallel programming? Why are still people investigating and using other approaches? The main reason is that the current Fortran 90 and HPF language specifications have severe limitations on generality and efficiency.

Fortran 90 and HPF do not support the exploitation of control parallelism. They are not well suited in supporting algorithmic paradigms such as asynchronous iteration, work queue, and pipelining, or applications such as database.

For instance, it is difficult to write efficient data parallel codes for the banking problem in Section 7.2.1 or the target detection problem of Section 12.1.3. The HPF data-mapping directives seem to only support arrays with regular communication patterns. It is doubtful whether it can efficiently support parallel algorithms with general data structures and irregular communication patterns.

Example 14.8 Target detection in HPF

The following HPF target detection code skeleton is based on a solution suggested by Michael Kumbera of Maui High-Performance Computing Center:

```
complex A(N,M)
integer temp1(N,M) temp2(N,M),
integer direction(MaxTargets), distance(MaxTargets)
integer i, j
L1:    FORALL (i=1:N, j=1:M) temp1(i,j) = IsTarget(A(i,j))
L2:    temp2 = SUM_PREFIX ( temp1, , MASK=(temp1>0) )
L3:    FORALL (i=1:N, j=1:M; temp2(i,j)>0 .and. temp2(i,j)<=MaxTargets )
           distance(temp2(i,j)) = i
           direction(temp2(i,j)) = j
       END FORALL
```

The FORALL statement L1 simultaneously evaluates every element of array A, and assign temp1(i,j)=1 if A(i,j) is a target. Suppose there are four targets A(1,3), A(1,4), A(2,1), and A(4,4). Then temp1 has the value shown below:

$$
temp1 = \begin{bmatrix} 0 & 0 & 1 & 1 \\ 1 & 0 & 0 & 0 \\ 0 & 0 & 0 & 1 \end{bmatrix}, \quad temp2 = sum\_prefix\,(temp1) = \begin{bmatrix} 0 & 0 & 2 & 3 \\ 1 & 0 & 0 & 0 \\ 0 & 0 & 0 & 4 \end{bmatrix} \quad (14.4)
$$

L2 assigns to array temp2 the prefix sum of all positive elements of temp1. Another FORALL updates the target list (represented by two arrays distance and direction). Note that since all non-zero elements of temp2 are distinct, the single assignment rule is satisfied. The target list is correctly computed as follows:

distance(1)=distance(temp2(2,1))=2, direction(1) = direction(temp2(2,1)) = 1
distance(2)=distance(temp2(1,3))=1, direction(2) = direction(temp2(1,3)) = 3
distance(3)=distance(temp2(1,4))=1, direction(3) = direction(temp2(1,4)) = 4
distance(4)=distance(temp2(4,4))=4, direction(4) = direction(temp2(4,4)) = 4

The main problem of this solution is that the IsTarget function must be evaluated N*M times, no matter how the targets are distributed.

                                                                      ■

HPF Performance   Although not enough performance data are available to draw general conclusions, there is some evidence suggesting that Fortran 90 and HPF compilers can achieve good performance for computational programs with static parallelism and regular communication patterns.

For instance, Table 14.3 shows the speedup performance of several machines running a shallow-water numerical computation program written in HPF. Good speedup is achieved on MPPs (the Intel Paragon and the IBM SP2), an SMP (the DEC AlphaServer 8400), and a DEC cluster.

The speedup of the HPF programs is measured over the sequential Fortran 77 programs on one processor. The DEC HPF compiler even achieves superlinear speedup in some cases, due to caching effects and additional optimizations performed by the HPF compiler.

Table 14.3   Speedup Performance of HPF Shallow-Water Codes
(Courtesy: Harris et al, *Digital Technical Journal,* Vol. 7, No. 3, 1995 [298])

System	Number of Processors			
	1	2	4	8
Intel Paragon	1.00	1.95	3.84	7.38
IBM SP2	1.00	1.97	3.81	7.50
DEC AlphaServer 8400	1.12	1.97	5.30	10.6
DEC Advantage Cluster	1.00	1.59	3.13	8.57

For some applications, the data-parallel approach can achieve performance comparable to that of the message passing approach. Table 14.4 shows the speedup performance of a DEC Advantage Cluster, running two programs of the same red-black relaxation algorithm. One program is written using Fortran 77 and PVM, while the other program is written in HPF.

**Table 14.4 Speedup Comparison of HPF and PVM Red-Black Codes
Running on a DEC Advantage Cluster with GIGAswitch**

(Courtesy: Harris et al, *Digital Technical Journal*, Vol. 7, No. 3, 1995 [298])

Data-Parallel Language	Number of Processors			
	1	**2**	**4**	**8**
DEC Fortran77/PVM	N/A	1.79	3.73	7.01
DEC HPF	1.05	1.95	4.10	8.04

Again, the speedup numbers are measured against sequential Fortran 77 code running on one processor. As can be seen, the data-parallel code outperforms the explicit message-passing code in all cases.

14.4 Other Data-Parallel Approaches

There are several on-going research efforts that alleviate the generality and efficiency deficiency of Fortran 90 and HPF. We group these efforts into two categories: those that are similar to HPF but offer more flexibility, and those that are based on non-Fortran languages. We briefly introduce these approaches below, along with a review of the Connection Machine approaches to data parallelism.

14.4.1 Fortran 95 and Fortran 2001

In the first category, we mention four efforts: **Fortran 95, Fortran 2001, Fortran D,** and **Vienna Fortran.** The first two extend Fortran 90, while the others extend HPF.

At the writing of this book, two extensions to Fortran 90 are under development by a joint committee of two international standards bodies: the *International Organization for Standardization* (ISO) and the *International Electrotechnical Commission* (IEC). These extensions are informally known as *Fortran 95* and *Fortran 2001.*

Fortran 95, published in 1997, is a relatively minor enhancement and correction of Fortran 90. The major new features in Fortran 95 are the FORALL statement and the FORALL construct, PURE and ELEMENTAL procedures, and structure and pointer default initialization.

In contrast, Fortran 2001 will be a major revision, with a target publication date of year 2001. Some of the new features being considered for Fortran 2001 are exception handling; interoperability with the C language; enhancements to derived data types; additional supports for high performance numerical computing, such as interval arithmetic; new I/O features such as asynchronous I/O and derived-type I/O; object orientation; and support of internationalization features provided through the operating system.

Extending HPF   Fortran D (later Fortran 90 D) and Vienna Fortran provide more general data mapping functionalities, such that dynamic and irregular data distributions are supported. We will use an example in Fortran D to illustrate the basic ideas. Vienna Fortran has similar philosophy and language features.

Example 14.9 The *N*-body problem solved in HPF code

In the *N-body* problem, the behavior of a system of N particles x(1), x(2), ..., x(N) is simulated in a sequence of time steps as follows: In each step, the force(i) imposed on each particle x(i) is computed by

$$force\,(i) \;=\; \sum_{x\,(j)\ \text{is a neighbor of}\ x\,(i)} f(i,j) \tag{14.5}$$

where particle x(j) is called a *neighbor* of particle x(i) if the distance between them is less than some threshold value. Then every particle's position and velocity are updated, using the newly computed force(i).

For instance, a 6-body problem is shown in Fig. 14.6, where each vertex represents a particle. An edge is drawn between a pair of particles if they are neighbors. Without a threshold, we need to consider all 6*5=30 pairs. But in Fig. 14.6 we only need to consider 7 pairs, which are represented in the above code by two arrays

PairA = (/ 1, 1, 2, 2, 3, 4, 3,... /) and PairB = (/ 2, 5, 5, 3, 4, 6, 6,... /)

The two vertices of edge i are labeled in PairA(i) and PairB(i); e.g., the two vertices of edge 4 are PairA(4) = 2 and PairB(4) = 3.

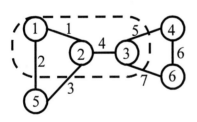

(a) A regular block distribution

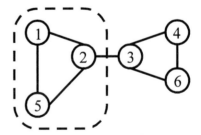

(b) An irregular distribution

Figure 14.6   Data distribution in a six-body problem.
(Adapted from Ponnusamy et al., *IEEE Parallel and Distributed Technology,* Spring 1995, [507])

The force acted on x(2) is from its three neighbors x(1), x(3), and x(5). Thus force(2) = f(2,1) + f(2,3) + f(2,5). This is a sum reduction over all neighbors. This functionality is realized in Fortran D by a new REDUCE intrinsic, which can be used in a FORALL. Data distribution onto two nodes can be realized as follows:

```
!HPF$ PROCESSOR Nodes(2)
!HPF$ DISTRIBUTE x(BLOCK) ONTO Nodes
    B = (/ 1, 1, 2, 2, 1, 2 /)
!HPF$ REDISTRIBUTE x(INDIRECT(B))
```

This block distribution is shown in Fig. 14.6(a). The three particles inside the dashed circle are allocated to Nodes(1), while the other three to Nodes(2). This distribution, together with the owner-compute rule, causes six particles to be communicated.

One can improve locality through irregular distribution through an indirect array B. The programmer first specifies the value of the indirect array. Then a dynamic REDISTRIBUTE statement will reallocate the particle array x to the form shown in Fig. 14.6(b).

Unlike BLOCK or CYCLIC for regular distributions, the x(INDIRECT(B)) says that x(i) should be allocated to Nodes(B(i)). Thus x(5) should be mapped to Nodes(B(5)), which is Nodes(1). This irregular distribution needs only two particles to be communicated.

To summarize, the above simulation is written as a HPF skeleton as follows:

```
do time_step = 1, number_of_time_steps
! compute the set of pairs

   ......
! compute the force on each particle
   FORALL ( i = 1 : Number_of_Pairs )
       REDUCE ( SUM, force(PairA(i)), f(PairA(i), PairB(i)) )
       REDUCE ( SUM, force(PairB(i)), f(PairA(i), PairB(i)) )
   END FORALL

   FORALL ( i = 1 : N ) update_position_and_velocity(i)
end do          ! end of time step iteration
```

■

An effort is under way to define a successor of HPF, called HPF-2. A nice feature of the HPF-2 effort is that the rationale and justifications for the new features are documented, together with a set of motivating applications. These are real applications which demonstrate concretely why a certain capability should be added to HPF. The above discussed irregular mapping of arrays is but one issue addressed in the HPF-2 effort. Other new capabilities considered by the HPF Forum for HPF-2 include:

- Mapping of more general data structures, not just arrays. These include pointer-based data structures such as linked lists, and derived data types.
- Distributing to processor subsets, not just the entire grid of virtual nodes.
- More flexible workload allocation schemes, not just owner-compute.
- Support for control parallelism, including dynamic creation and termination of multiple processes (not just single threading).
- Support for atomic operations.
- Parallel I/O and checkpoints.
- A Kernel HPF, which is an official subset of HPF designed for realizing high performance across a broad spectrum of computers.

14.4.2 The pC++ and Nesl Approaches

There are several data-parallel efforts that use a different language from Fortran. For instance, the **pC++ project** at Indiana University [98] adopts an object-oriented approach. Data domains are specified by more flexible *objects*, not just arrays. So irregular data-parallel computing is supported. We discuss below in some detail the **Nesl** language developed by Blelloch's group at CMU. Both pC++ and Nesl have been implemented and are available in the public domain.

The Nesl Language Nesl is a simple functional, data-parallel language, aimed at specifying and teaching parallel algorithms. Nesl has two important features. First, data are specified as *sequences*, which are an ordered set of data elements. Each item itself can be a sequence. Unlike arrays, which are a regular data structure, sequences can represent both regular and irregular data structures, including recursive ones.

Second, Nesl supports *nested parallelism*, by allowing parallel functions to be simultaneously applied to all elements of a sequence, which can be themselves sequences. This is in contrast to Fortran 90 or HPF, which only support *flat* data parallelism.

The main parallel construct in Nesl is *apply-to-each*, denoted by a pair of braces {}. This construct is illustrated by the expression { a*a : a **in** [1, 2, 3, 4] | a > 2 }. This expression says: "In parallel, for each element a in sequence [1, 2, 3, 4] such that $a > 2$, apply the square operator". It returns a sequence [9, 16]. The filter a > 2 is optional.

The operator in an apply-to-each construct can be a built-in operator or a user-defined function. For instance, with the definition

function factorial(n) = **if** (n==1) **then** 1 **else** n*factorial(n-1);

the expression { factorial(b) : b in [3, 2, 1] } returns a sequence [6, 2, 1].

Nesl provides functions that operate on a sequence as a whole, instead of element-wise. Several such functions are listed in Table 14.5. In addition, the **write**(*s, p*) function takes two arguments: the first is a sequence *s*. The second argument *p* is a sequence of integer-value pairs. Each pair (*i, v*) specifies that the *i*th element of *s* is assigned with *v*. Thus after executing

Result = **write**([1, 3, 5, 7], [(0,2),(3,3)])

we have Result = [2, 3, 5, 3]. Note that the sequence index starts at 0.

It is possible to repeat an index in *p*. Then one of the values is written in a nondeterministic way. Thus

write([1, 3, 5, 7], [(0,2), (0,4), (3,3)])

returns either [2,3,5,3] or [4,3,5,3]. Nesl provides another exclusive-write function **e_write**(s,p), which disallows such concurrent writes.

Table 14.5 Several Nesl Sequence Functions

Function	Meaning	Example
dist(*a, n*)	Returns a sequence of *n* copies of *a*	**dist**(t, 3) returns [t, t, t]
#s	Returns length of sequence s	#[t,t,t] returns 3
[*s:e:n*]	Returns an integer sequence from *s* to *e* by *n*	[2:9:3] returns [2, 5, 8]
drop(*s, n*)	Drops the first *n* elements of sequence *s*	**drop**([1 ,2, 3, 4], 2) returns [3, 4]
sum(*s*)	Sum-reduction of sequence *s*	**sum**([1, 2, 3]) returns 6
flatten(*s*)	Flatten a nested sequence *s*	**flatten**([[1,2],[[3],[4]]]) returns [1,2,3,4]

Example 14.10 Simultaneous sum of a sparse matrix

Suppose we want to simultaneously sum each row of a sparse matrix:

$$rowsum\ (A)\ =\ rowsum \left(\begin{bmatrix} 1.0 & 0 & 0 & 2.0 & 0 \\ 0 & 3.0 & 0 & 0 & 0 \\ 5.0 & 0 & 0 & 0 & 0 \\ 6.0 & 0 & 0 & 7.0 & 0 \end{bmatrix} \right) = \begin{bmatrix} 3.0 \\ 3.0 \\ 5.0 \\ 13.0 \end{bmatrix} = S \qquad (14.6)$$

Since most elements are zero, we should represent this sparse matrix in some special form to conserve memory. Then it would be difficult to realize this parallel operation in Fortran 90 or HPF. In Nesl, the sparse matrix A can be represented as a sequence of four rows, each of which is a sequence of (column, value) pairs:

$A = [\ [(0, 1.0), (1, 2.0)], [(1, 3.0)], [(0, 5.0)], [(0, 6.0), (3, 7.0)]\ ]$

Now the rowsum operation can be specified in Nesl very concisely as follows:

$S = \{\ \textbf{sum}\ (\ \{v : (i, v)\ \textbf{in}\ row\}\ ) : row\ \textbf{in}\ A\ \}$

This example demonstrates the two types of nested parallelism: the data sequence A is nested, and the four sum reductions are applied in parallel.

■

Example 14.11 The Eratosthenes method in data-parallel code

The set of all primes less than or equal to N can be found by the sieve of Eratosthenes method using the following Nesl procedure:

```
function primes(N)
if (N==2) then ([] int)
else
   let  sqr_primes = primes(isqrt(N)) ;
        composites = { [2*p:N:p] : p in srq_primes } ;
        flat_comps = flatten(composites) ;
        flags = write( dist(t,N), { (i, f): i in flat_comps } ) ;
        indices = { i in [0:N]; fl in flags | fl }
   in drop(indices, 2) ;
```

The expression [] **int** denotes an empty sequence, which is the result when $N = 2$. The function isqrt(N) returns the integer value of the square root of N. The two constants t and f indicate logical true and false, respectively.

Suppose N=20. The above procedure will recursively call itself to find that primes(isqrt(20))=primes(4), which eventually returns [2,3]. The other variables have the following values:

sqr_primes = [2, 3]
composites = [[4, 6, 8, 10, 12, 14, 16, 18], [6, 9, 12, 15, 18]]
flat_comps = [4, 6, 8, 10, 12, 14, 16, 18, 6, 9, 12, 15, 18]
flags = [t, t, t, t, f, t, f, t, f, f, f, f, t, f, t, f, f, f, t, f, t]
indices = [0, 1, 2, 3, 5, 7, 11, 13, 17, 19]

And the final result returned by primes(20) is [2, 3, 5, 7, 11, 13, 17, 19].

Blelloch advocated measuring the performance of parallel algorithms using two metrics *work* and *depth* [87]. For work-efficient parallel algorithms (i.e., where the parallel algorithm performs approximately the same amount of work as the best sequential algorithm), work is analogous to the sequential workload T_1 in Chapter 3 and depth is similar to the critical path T_∞.

Furthermore, he suggested that performance metrics should be based on a parallel programming language, instead of on a machine model. The work and the depth of a Nesl program are obtained from those of the primitive operations (shown in Table 14.5) and rules for composing the metrics across expressions, as discussed below. .

Table 14.6 Work and Depth of Nesl Primitive Functions

Operation	Work	Depth	Notes
dist(*a,n*)	1	1	
#s	1	1	
[*s:e:n*]	(e-s)/n	1	
sum(s)	L(s)	log(L(s))	L(s) is the length of sequences
drop(s,n)	L(*result*)	1	*result* is the result sequence
flatten(s)	L(result)	1	
write(s,p)	L(result)	1	

Denote by $W(e)$ and $D(e)$ the work and the depth of expression e, respectively. Then the two metrics of a composite expression can be computed from those of the subexpressions as follows:

(1) In most cases, the work and depth of an expression is the sum of the work and depth of its subexpressions.

(2) For an apply-to-each, we have

$$W(\{e_1(a) : a \text{ in } e_2\}) = 1 + W(e_2) + \sum_{a \text{ in } e_2} W(e_1(a))$$

$$D(\{e_1(a) : a \text{ in } e_2\}) = 1 + D(e_2) + \sum_{a \text{ in } e_2} D(e_1(a))$$

$$(14.7)$$

(3) For conditional expression e, that is, "if e_1 then e_2 else e_3", we have

$$W(e) = 1 + W(e_1) + \begin{cases} W(e_2) & e_1 = true \\ W(e_3) & e_1 = false \end{cases}$$

$$D(e) = 1 + D(e_1) + \begin{cases} D(e_2) & e_1 = true \\ D(e_3) & e_1 = false \end{cases}$$

$$(14.8)$$

Example 14.12 Parallel algorithm primes (N)

What are the work and the depth of the parallel algorithm primes(N) in Example 14.11? What is the average parallelism?

Solution: Assume primes(N) has a work of $f(N)$. The first recursive call in sqr_primes = primes(isqrt(N)) has a work of $f(\sqrt{N})$, the second recursive call has a work of $f(\sqrt[4]{N})$, etc. That is, the problem size is reduced from N to $\sqrt{N}$, $\sqrt[4]{N}$, until it reaches the last recursion level d, when the problem size becomes 2.

Thus the dominant work is done at recursion level 0. The total work in this level is $O(N\log\log N)$ and the depth at any recursion level is constant (see Problem 14.6). We have $d = \log\log N$ recursion levels.

Thus the total work is $O(N\log\log N)$, the total depth is $O(\log\log N)$, and the average parallelism is $O(N\log\log N)/O(\log\log N) = O(N)$. ■

Nesl is a simple and elegant language. It demonstrated that with a data-parallel, functional style, a small set of well-designed constructs can be used to concisely and clearly specify many parallel computing algorithms. For such algorithms, Nesl programs are often much simpler than corresponding programs specified in mainstream languages (C and Fortran with parallel extensions).

The Nesl programs are very close to informal algorithm descriptions. All these and the public-domain nature make Nesl an ideal language for teaching data-parallel algorithm design. The disadvantages of Nesl are those of data-parallel and functional languages. It is not in the mainstream and not flexible enough to specify MIMD computations.

For instance, it is difficult to specify parallel algorithms in Nesl for solving the banking problems and the target problem (see Section 12.1.3). Another problem with Nesl is that it did not sufficiently take into consideration the parallelism and communication overeheads or data locality issues for reducing such overheads.

Blelloch et al. [89] have tested their Nesl implementations on DECstation 5000, CM-2, CM-5, and Cray C90 with some simple benchmark applications. The performance of Nesl programs is shown competitive with that of machine-specific Fortran or C codes for regular dense data, and is often superior for irregular data.

14.5 Bibliographic Notes and Problems

The data-parallel approach was discussed by Hill and Steel [310], Albert et al. [20], Blelloch et al. [89], Chandy et al. [142], with [142] being a critique of the data-parallel style. Fortran 90 was described in Adams et al. [4] and Fortran 95 was described in Adams et al. [5].

High-Performance Fortran was described in Koelbel et al. [370] and Loveman [419]. Critiques and extensions of HPF can be found in Chapman et al. [149] and Ponnusamy et al. [506,507]. Example 14.9 is adapted from Ponnusamy et al. [507].

Nested data parallelism and the Nesl language were discussed by Blelloch and colleagues [87, 89]. The discussion of Nesl in this chapter draws much material from these two papers.

Thinking Machines Corporation developed several influential data-parallel languages, such as C*, CM Fortran, and *Lisp, for its Connection Machines: CM-2 and CM-5, which are no longer in production. Interested readers may refer to Hwang's earlier book [327] for some coverage of the architecture and programming of the TMC Connection Machines.

There exists a large body of Internet information about the data-parallel programming style. The Web Resources contain many pointers. A good starting point is the Fortran resources maintained at Northeast Parallel Research Laboratory at Syracuse University.

Homework Problems

Problem 14.1   Use the Jacobi relaxation code in Fig. 14.1 to explain the following features of data-parallel programming:

   (a)   Single thread of control

   (b)   Global naming space

   (c)   Parallel operations on aggregate data structure

   (d)   Loosely synchrony

   (e)   Implicit interaction

   (f)   Implicit data allocation

Problem 14.2   Write a Gaussian elimination program in HPF such that the workload in the triangularization step is more balanced than Fig. 14.5.

   (a)   Write a program for matrix size N and machine size n. You can make simplifying assumptions such as N is much larger than n, and N evenly divides n, etc.

   (b)   Assume a two-node system. What is the flop workload for each of the two nodes, with the program in (a)?

   (c)   Assume a two-node system. What is the flop workload for each of the two nodes, with the program in Fig. 14.5?

Problem 14.3   Write an HPF program for Jacobi relaxation (see Example 12.4 and Fig. 14.1). Explain how the HPF code is simpler and faster than Fig. 14.1. (Hint: The HPF code does not need a sequential do loop to compute the array DiagA).

Problem 14.4   Compare the Fortran 90 and HPF approaches to data parallelism and answer the following questions.

   (a)   What are the similarities and differences in intrinsic functions in Fortran 90 and HPF approaches?

   (b)   Repeat the same question in Part (a) for parallel array operations.

   (c)   In handling loop-carried dependences, explain how it is supported in HPF.

   (d)   Compare the strengths and weaknesses in Fortran 90 and HPF in terms of compiler support for data parallelism. What are the limits of today's Fortran 90 and HPF compilers?

Problem 14.5 Write a program to find the set of all primes less than or equal to N by the sieve of Eratosthenes method (Example 14.11):

(a)   In Fortran 90

(b)   In HPF

(c)   Compare these programs with Example 14.11 and comment on the advantages and disadvantages of Fortran 90, HPF, and Nesl.

Problem 14.6 Show that the prime algorithm in Example 14.11 has $O(N \log \log N)$ work and $O(1)$ depth at recursion level 0.

Problem 14.7 Regarding research efforts to improve Fortran 90 and HPF, answer the following questions by looking more deeply into the latest developments on functional and feature extensions beyond what you have learned in this chapter:

(a)   What are the major improvements of Fortran 95 from those features in today's Fortran 90?

(b)   Repeat part (a) for Fortran 2001.

Problem 14.8 Answer the following regarding non-Fortran approaches to exploiting data parallelism:

(a)   Study the pC++ project at Indiana University and comment on the object-oriented approach to data parallelism.

(b)   Comment on the advantages and the new idea of *sequences* developed by the Nesl language project at Carnegie-Mellon University, as opposed to using arrays in handling regular and irregular data structures.

(c)   Identify the weaknesses of using object-oriented and functional approaches to handling data parallelism in real-world applications.

Problem 14.9 Now, you have studied all three parallel programming models for explicit parallelism. Revisit Table 12.3 and answer the following with full justifications:

(a)   Compare and justify the relative ranking of the three models with respect to those issues that the data-parallel approach scores the best, such as the handling of parallelism, communication, synchronization, termination, determinacy, correctness, compositionality, and portability.

(b)   Repeat part (a) with respect to those issues that the data-parallel model shows the poorest in ranking, such as the handling of irregularity and application generality and efficiency.

(c)   Repeat part (a) for the message-passing model with respect to the issues of aggregation and generality and efficiency.

(d)   Repeat part (b) for the message-passing model with respect to the issues of parallelism, allocation, communication, and structuredness.

(e)   Repeat part (a) for the shared-memory model with respect to the issues of allocation, communication, irregularity, and generality.

(f)   Repeat part (b) for the shared-memory model with respect to the issues of synchronization, aggregation, all four semantics issues, and program portability and structuredness.

Problem 14.10 Fully justify the column entries in Table 12.3 that the implicit compiler approach shows the best rating in all issues except the generality with respect to programmability. Your justifications here should be much deeper than those given in Problem 12.2. Use platform-dependent examples and quote benchmark results to justify those claims.

Bibliography

[1]   B. Abeli, C. B. Stunkel, and C. Benveniste, "Clock Synchronization on a Multicomputer," *Journal of Parallel and Distributed Computing*, Vol. 40, No. 1, 1997, pp. 118-130.

[2]   ACM, *Resources in Parallel and Concurrent Systems,* with an Introduction by Charles Seitz, ACM Press, New York, 1991.

[3]   D. Adams, *Cray T3D System Architecture Overview Manual,* Cray Research, Inc., Sept. 1993.

[4]   J. Adams, W. Brainerd, J. Martin, B. Smith, and J. Wagener, *The Fortran 90 Handbook*, McGraw-Hill, New York, 1992.

[5]   J. Adams, W. Brainerd, J. Martin, B. Smith, and J. Wagener, *The Fortran 95 Handbook,* MIT Press, 1997.

[6]   S. V. Adve and M. D. Hill, "A Unified Formalization of Four Shared-Memory Models," *IEEE Trans. on Parallel and Distributed Systems,* June 1993, Vol. 4, No. 6, pp. 13-24.

[7]   S. V. Adve and M. D. Hill, "Weak Ordering: A New Definition," *Proc. 17th Ann. Int'l. Symp. Computer Arch.,* 1990.

[8]   S. V. Adve and K. Gharachorloo, "Shared Memory Consistency Models: A Tutorial," *IEEE Computer*, Vol. 29, No. 12, Dec. 1996, pp. 66-76.

[9]   A. Agarwal, "Performance Tradeoffs in Multithreaded Processors," *IEEE Trans. Parallel and Distributed Systems*, May 1992, pp. 525-539.

[10]   A. Agarwal, R. Bianchini, D. Chaiken, K.L. Johnson, D. Krantz, J. Kubiatowicz, B.-H. Lim, K. Mackenzie, and D. Yeung, "The MIT Alewife Machine: Architecture and Performance," *Proc. of the 22nd Annual International Symposium on Computer Architecture*, pp. 2-13, 1995.

[11]   A. Agarwal, R. Simoni, J. Hennessy, and M. Horowitz, "An Evaluation of Directory Schemes for Cache Coherence," *Proc. 15th Int'l. Symp. on Computer Architectue*, pp. 280-289, June 1988.

[12]   R. C. Agarwal et al., "High-Performance Implementations of the NAS Kernel Benchmarks on the IBM SP2," *IBM System Journal,* Vol. 34, No. 2, 1995, pp. 263-272.

[13]   R. C. Agarwal, F. G. Gustavson, and M. Zubair, "Exploiting Functional Parallelism of POWER2 to Design High-Performance Numerical Algorithms," *IBM J. Res. Develop.*, Vol. 38, No. 5, 1994, pp. 563-576.

[14]   T. Agerwala, J. L. Martin, J. H. Mirza, D. C. Sadler, D. M. Dias, and M. Snir, "SP2 System Architecture," *IBM Systems Journal,* Vol. 34, No. 2, 1995, pp. 152-184.

[15]   G. Agha, "Concurrent Object-Oriented Programming," *Comm. of the ACM*, Vol. 33, No.9, Sept, 1990, pp. 125-141.

[16]   A. Aho, J. Hopcroft, and J. Ullman, *The Design and Analysis of Computer Algorithms*, Addison-Wesley, Reading, Mass., 1974.

[17]   A. V. Aho, R. Sethi, and J. D. Ullman, *Compilers: Principles, Techniques, and Tools*, Addison-Wesley, Reading, Mass., 1988.

[18]   M. Ajtai, J. Komlos, and E. Szemeredi, "An $O(n \log n)$ Sorting Network," *Proc. 15th ACM Symp. on Theory of Computing*, 1983, pp. 1-9.

[19]   R. Alasdair, A. Bruce, J. G. Mills, and A. G. Smith; "CHIMP Version 2.0 User Guide," University of Edinburgh, March 1994.

[20]   E. Albert, J. Lukas, and G. Steele, "Data Parallel Computers and the FORALL Statement" *J. Parallel and Distributed Computing*, 13(2):185-192, 1991.

[21]   T. B. Alexander et al., "Corporate Business Servers: An Alternative to Mainfrmaes for Business Computing," *Hewlett-Packard Journal*, June 1994, pp. 8-33.

[22]   A. Alexandrov, M. Ionescu, K. Schauser, and C. Scheiman, "LogGP: Incorporating Long Messages into the LogP Model," *Proc. of the 7th Annual ACM Symp. on Parallel Algorithms and Architectures*, 1995, pp. 95-105.

[23]   M. S. Allan and M. C. Becker, "Multiprocessing Aspects of the PowerPC 601," *COMPCON,* Spring 1993.

[24]   S. J. Allan and R. Oldehoeft, "HEP SISAL: Parallel Functional Programming", in Kowalik (Ed.), *Parallel MIMD Computation: HEP Supercomputers and Applications*, MIT Press, Cambridge, Mass., 1985.

[25]   V. H. Allan, R. B. Jones, R. M. Lee, S. J. Allan, "Software Pipelining," *ACM Computing Survey,* 27(3):367-432, 1995.

[26]   G. S. Almasi and A. J. Gottlieb, *Highly Parallel Computing*, 2d ed., Benjamin/Cummings, New York, 1993.

[27]   R. Alverson, D. Callahan, D. Cummings, B. Koblenz, A. Porterfield, and B. Smith, "The Tera Computer System," in *1990 Int'l. Conf. on Supercomputing*, pp. 1-6, September 1990.

[28]   S. P. Amarasingbe, J. M. Anderson, C. S. Wilson, S.-W. Liao, B. R. Murphy, R. S. French, M. S. Lam, and M. W. Hall, "Multiprocessors from a Software Perspective," *IEEE Micro*, Vol. 16, No. 3, 1996, pp. 52-61.

[29]   G. M. Amdahl, "Validity of Single-Processor Approach to Achieving Large-Scale Computing Capability," *Proc. AFIPS Conf.,* Reston, Va., 1967, pp. 483-485.

[30]   C. Amza, A.L. Cox, S.D. Warkadas, P. Keleher, H. Lu, R. Rajamony, W. Yu, and W. Zwaenepoel, "TreadMarks: Shared Memory Computing on Networks of Workstations," *IEEE Computer*, 29(2):18-28, 1996.

[31]   D. Anderson and T. Shanley, *Pentium Processor System Architecture*, 2d ed., Addison-Wesley, Reading, Mass., 1995.

[32]   J. Anderson and M. Moir, "Using Local-Spin *k*-Exclusion Algorithms to Improve Wait-Free Object Implementations ," *Distributed Computing*, to appear 1997.

[33]   T.E. Anderson, D.E. Culler, D.A. Patterson et al., "A Case for NOW (Networks of Workstations),"   *IEEE Micro,* February 1995, pp. 54-64.

[34]   T.E. Anderson, M.D. Dahlin, J.M. Neefe, D.A. Patterson, D.S. Roseli, and R.Y. Wang, "Serverless Network File Systems," *ACM Transactions on Computer Systems*, Vol. 14, No. 1, 1996, pp. 41-79.

[35]   J. Anderson and M. Lam, "Global optimizations for Parallelism and Locality on Scalable Parallel Machines," *Proc. of the SIGPLAN'93 Conf. on Programming Language Design and Implementation*, 1993.

[36]   T. E. Anderson, H. M. Levy, B. N. Bershad, and E .D. Lazowska, "The Interaction of Architecture and Operating System Design," *Proc. 4th International Conference on Architectural Support for Programming Languages and Operating Systems*, 1991, pp. 108-120.

[37]   G.R. Andrews, *Concurrent Programming: Principles and Practice*, Benjaming/Cummings, Redwood City, California, 1991.

[38]   I. Angus, G. Fox, J. Kim, and D. Walker, *Solving Problems on Concurrent Processors,* Vol. II, Prentice-Hall, Englewood Cliffs, N.J., 1990.

[39]   ANSI Technical Committee X3H5, *Parallel Processing Model for High-Level Programming Languages*, 1993.

[40]   R. H. Arpaci, D.E. Culler, A. Krishnamurthy, S.G. Steinberg, and K. Yelick, "Empirical Evaluation of the Cray T3D: A Compiler Perspective," *Proc. the 22d Annual Int'l. Symp. on Computer Architecture*, 1995, pp. 320-331.

[41]   K.R. Apt and E.-R. Olderog, *Verification of Sequential and Concurrent Programs,* Springer-Verlag, New York, 1991.

[42]   J. Archibald and J.L. Baer, "Cache Coherence Protocols: Evaluation Using a Multiprocessor Simulation Model," *ACM Trans. Computer Systems,* Vol. 4, March 1986, pp. 273-296.

[43]   B.W. Arden and H. Lee, "Analysis of Chordal Ring Network," *IEEE Trans. on Computers*, Vol. 30, No.4, April 1981, pp. 291-295.

[44]   R. H. Arpaci, A. C. Dusseau, A. M. Vahdat, L. T. Liu, T. E. Anderson, and D. A. Patterson, "The Interaction of Parallel and Sequential Workloads on a Network of Workstations," *Proc. of the 1995 ACM SIGMETRICS Conf. on Measurement and Modeling of Computer Systems*, 1995, pp. 267-278.

[45]   W. C. Athas and C. L. Seitz, "Multicomputers: Message-Passing Concurrent Computers," *IEEE Computer,* August 1988, pp. 9-24.

[46]   ATM Forum, *ATM User Network Interface: Version UNI 3.1 Specification*, March 1994.

[47]   A. Azagury, D. Dolev, G. Goft, J. Marberg, and J. Satran, "Highly Available Cluster: A Case Study," *Proc. 24th Symp. on Fault-Tolerant Computing*, 1994, pp. 404-413.

[48]   R.G. Babb II (Ed.), *Programming Parallel Processors*, Addison-Wesley, Reading, Mass., 1988.

[49]   M. J. Bach and S. J. Buroff, "Multiporcessor UNIX Operating Systems", *AT&T Bell Lab. Tech. Journal*, 63(8), Oct. 1984.

[50]   J. Bacon, *Concurrent Systems: An Integrated Approach to Operating Systems, Database, and Distributed Systems*, Addison-Wesley, Reading, Mass., 1993.

[51]   D. A. Bader, D. R. Helman, and J. JaJa, "Practical Parallel Algorithms for Personalized Communication and Integer Sorting," *UMIACS Technical Report*, University of Maryland, Dec. 1995.

[52]   J. L. Baer and W .H. Wang, "Multi-Level Cache Hierarchies: Organizations, Protocols, and Performance," *J. of Parallel and Distributed Computing,* Vol. 6, 1989, pp. 451-476.

[53]   J.C.M. Baeten and W.P. Weijland, *Process Algebra*, Cambridge University Press, 1990.

[54]   D.H. Bailey, T. Harris, W. Saphir, R. van der Wijngaart, A. Woo, and M. Yarrow, "The NAS Parallel Benchmarks 2.0," NASA Ames Research Center Report NAS-95-020, Dec. 1995.

[55]   W. E. Baker, R. W. Horst, D. P. Sonnier, and W. J. Watson, "A Flexible ServerNet-Based Fault-Tolerant Architecture," *25th Annual Symp. on Fault-Tolerant Computing*, June 1995.

[56]   M. A. Baker, G. C. Fox and H. W. Yau, "Cluster Computing Review," Northeast Parallel Architectures Center, Syracuse University, Nov. 1995.

[57]   H. E. Bal, M.F. Kaashoek, and A.S. Tanenbaum, "Orca: A Language for Parallel Programming of Distributed Systems," *IEEE Trans. Soft. Eng.,* March 1992, pp. 190-205.

[58]   H. E. Bal, J. G. Steiner, and A. S. Tanenbaum, "Programming Languages for Distributed Computing Systems," *ACM Computing Surveys*, 21(3):261-322, 1989.

[59]   V. Bala, J. Bruck, R. Cypher, P. Elustondo, A. Ho, C.-T. Ho, S. Kipnis, and M. Snir, "CCL: A Portable and Turnable Collective Communication Library for Scalable Parallel Computers," *IEEE Trans. on Parallel and Distributed Systems*, 1995, 6(2):154-164.

[60]   U. Banerjee, *Dependence Analysis*, Kluwer Academic Publishers, Boston, 1996.

[61]   J. Barlett et al., "Fault Tolerance in Tandem Computer Systems," in *Reliable Compute Systems: Design and Evaluation*, D. Siewiorek and R. Swarz(Eds.), Digital Press, Maynard, Mass., 1992.

[62]   R. Baron et al., "Mach-1: An Operating Environment for Large-Scale Multiprocessor Applications," *IEEE Software*, July 1985, pp. 65-67.

[63]   R. Barrett, M. Berry, T. F. Chan, J. Demmel, J. Donato, J. Dongarra, V. Eijkhout, C. Pozo, C. Romine, and H. van der Vorst, *Templates for the Solution of Linear Systems: Building Blocks for Iterative Methods,* SIAM Press, Philadelphia, 1992.

[64]   B. E. Bauer, *Practical Parallel Programming*, Academic Press, San Diego, Calif., 1992.

[65]   BBN, *TC2000 Technical Product Summary*, BBN Advanced Computers Inc., Cambridge, Mass., Nov. 1989.

[66]   A. Beguilin, J. Dongarra, G. A. Geist, R. Manchek, and V. S. Sunderam, *A Users' Guide to PVM*, Technical Report ORNL TM-11286, Oak Ridge National Lab., July 1991.

[67]   G. Bell, "Ultracomputers: A Teraplop before Its Time," *Communications of the ACM*, August 1992, pp. 27-47.

[68]   G. Bell, "An Insider's Views on the Technology and Evolution of Parallel Computing," in *Software for Parallel Computers*, R. H. Perrott (Ed), Chapman & Hall, London, 1992, pp 11-26.

[69]   G. Bell, "Why There Won't Be Apps: The Problem with MPPs," *IEEE Parallel and Distributed Technology*, Fall 1994, pp. 5-6.

[70]   G. Bell, "1995 Observations on Supercomputing Alternatives: Did the MPP Bandwagon Lead to a Cul-de-Sac?" *Communications of the ACM.*, Vol. 39, No. 3, 1996, pp. 11-15.

[71]   G. Bell and J. N. Gray, "The Revolution Yet to Happen," in P.J. Denning and R.M. Metcalfe (Eds.), *Beyond Calculation: The Next Fifty Years of Computing*, Springer-Verlag New York, Inc., 1997.

[72]   M. Ben-Ari, *Principles of Concurrent and Distributed Programming*, Prentice-Hall, Englewood Cliffs, N.J., 1990.

[73]   R. J. Bergeron, "The Performance of the NAS HSPs in 1st Half of 1994," Report NAS-95-008, NASA Ames Research Center, February 1995.

[74]   A. J. Bernstein, "Analysis of Programs for Parallel Processing," *IEEE Trans. Elect. Comput.*, Vol. EC-15, No. 5, 1966, pp. 757-763.

[75]   P.A. Bernstein and E. Newcomer, *Principles of Transaction Processing*, Morgan Kaufmann Publishers, San Francisco, 1997.

[76]   M. Berry et al. "The Perfect Club Benchmarks: Effective Performance Evaluation of Supercomputers". *Technical Report* CSRD No. 827, Center for Supercomputing Research and Development, University of Illinois, Urbana, Ill., May 1989.

[77]   J. Berstin, "Sybase for HACMP/6000: An Architectured Approach to Clustered Systems," IBM *AIXpert*, May 1994, pp. 46-52.

[78]   D.P. Bertsekas and J.N. Tsitsiklis, *Parallel and Distributed Computation: Numerical Methods*. Prentice-Hall, New Jersey, 1989.

[79]   P. Bhat, Y. Lim, and V. Prasanna, "Scalable Portable Parallel Algorithms for STAP," *Proc. of the Adaptive Sensor Array Processing Workshop,* MIT Lincoln Lab., March 1996.

[80]   L. N. Bhuyan, R. Iyer, T. Askar, A. Nanda, and M. Kumar, "Performance of Multistage Bus Networks for a Distributed Shared Memory Multiprocessor," *IEEE Trans. on Parallel and Distributed System*, Jan. 1997, pp. 82-95.

[81]   L. N. Bhuyan and X. Zhang, *Multiprocessor Performance Measurement and Evaluation*, IEEE Computer Society Press, 1995.

[82]   A. Bilas and E.W. Felten, "Fast RPC on the SHRIMP Virtual Memory Mapped Network Interface," *Journal of Parallel and Distributed Computing*, Vol. 40, No. 1, 1997, pp. 138-146.

[83]   R. Bisiani and M. Ravishankar, "PLUS: A Distributed Shared-Memory System," *Proc. 17th Int'l. Symp. on Computer Architecture*, May 1990, pp. 115-124.

[84]   P. Bitar and A. M. Despain, "Multiprocessor Cache Synchronization Issues, Innovations, and Evolution," *Proc. 13th Annu. Int'l. Symp. Computer Arch.*, 1986.

[85]   D. L. Black, "Scheduling Support for Concurrency and Parallelism in the Mach Operating System," *IEEE Computer*, Vol. 23, No. 5, May 1990, pp. 35-42.

[86]   G. E. Blelloch, "Scans as Primitive Parallel Operations and Their Use in Algorithm Design," *IEEE Trans. on Computers*, TC-38(11), 1989, pp. 1526-1538.

[87]   G. E. Blelloch, "Programming Parallel Algorithms," *Comm. of ACM*, Vol. 39, 1996, pp. 85-97.

[88]   G. E. Blelloch, K.M. Chandy, and S. Jagannathan (Eds.), *Specification of Parallel Algorithms*, American Mathematical Society, 1994.

[89]   G. E. Blelloch, S. Chatterjee, J.C. Hardwick, J. Sipelstein, and M. Zagha, "Implementation of a Portable Nested Data Parallel Language," *Journal of Parallel and Distributed Computing*, Vol. 21, No. 1, 1994, pp. 4-14.

[90]   D. W. Blevins, E. W. Davies, R. Heaton, and J. H. Reif, "BLITZEN: A Highly Integrated Massively Parallel Machine," *Journal Parallel and Distributed Computing*, 1990, pp. 150-160.

[91]   W. Blume and R. Eigenmann, "Performance Analysis of Parallelizing Compilers on the Perfect Benchmarks Programs," *IEEE Trans. on Parallel and Distributed Systems*, 1992, Vol. 3, No. 6, pp. 643-656.

[92]   W. Blume, R. Eigenmann, J. Hoeflinger, D. Padua, P. Petersen, L. Rauchwerger, and P. Tu, "Automatic Detection of Parallelism: A Grand Challenge for High-Performance Computing," *IEEE Parallel and Distributed Technology*, 2(3):37-47, 1994.

[93]   W. Blume, R. Doallo, R. Eigenmann, J. Grout, J. Hoeflinger, T. Lawrence, J. Lee, D. Padua, Y. Paek, B. Pottenger, L. Rauchwerger, and P. Tu, "Parallel Programming with Polaris," *IEEE Computer*, Vol. 29, No. 12, Dec. 1996, pp. 78-82.

[94]   R. D. Blumofe, C. F. Joerg, B. C. Kuszmaul, C. E. Leiserson, K. H. Randall, and Y. Zhou, "Cilk: An Efficient Multithreaded Runtime System," *Journal of Parallel and Distributed Computing*, Vol. 37, No. 1, 1996, pp. 55-69.

[95]   M. A. Blumrich, C. Dubnicki, E.W. Felten, and K. Li, "Protected, User-Level DMA for the SHRIMP Network Interface," *Proc. 2nd Int'l. Symp. on High-Performance Computer Architecture*, Feb. 1996.

[96]   M. A. Blumrich, C. Dubnicki, E.W. Felten, K. Li, and M.R. Mesarina, "Virtual-Memory-Mapped Network Interfaces," *IEEE Micro*, Feb. 1995, pp. 21-27.

[97]   N. J. Boden, D. Cohen, R. E. Felderman, A. E. Kulawik, C. L. Seitz, J. N. Seizovic, and W.-K. Su, "Myrinet: A Gigabit-per-Second Local Area Network," *IEEE Micro*, Feb. 1995, pp. 29-36.

[98]   F. Bodin, P. Beckman, D. B. Gannon, S. Narayana, and S. Yang, "Distributed pC++: Basic Ideas for an Object Parallel Language," *Proc. Supercomputing '91*, pages 273-282, 1991.

[99]   S. H. Bokhari, "Multiphase Complete Exchange: A Theoretical Analysis," *IEEE Trans. on Computers,* Vol. 45, No. 2, 1996, pp. 220-229.

[100]   S. H. Bokhari, "Multiphase Complete Exchange on Paragon, SP2, annd CS-2," *IEEE Parallel and Distributed Technology*, Vol. 4, No. 3, 1996, pp. 45-59.

[101]   R. Bond, "Measuring Performance and Scalability Using Extended Versions of the STAP Processor Benchmarks," *Technical Report*, MIT Lincoln Laboratories, Lexington, MA. December 1994.

[102]   B. Boothe and A. Ranade, "Improved Multithreading Techniques for Hiding Communication Latency in Multiprocessor," *Proc. 19th Annu. Int'l. Symp. Computer Arch.*, Australia, May 1992.

[103]   R. Bordawekar, J. del Rosario, and A. Choudhary, "Design and Evaluation of Primitives for Parallel I/O," *Proc. Supercomputing '93*, pp. 452-461, 1993.

[104]   R. Bordawekar, A. Choudhary, K. Kennedy, C. Koelbel, and M. Paleczny, "A Model and Compilation Strategy for Out-of-Core Data Parallel Programs," *Proc. of the Fifth ACM SIGPLAN Symposium on Principles and Practice of Parallel Programming*, August 1995, pp. 1-17.

[105]   W. J. Bouknight, S. A. Denenberg, D. E. McIntyre, J. M. Randall, A. H. Sameh, and D. L. Slotnick, "The Illiac IV System," *Proceedings of the IEEE*, Vol. 60, No. 4, April 1972, pp. 369-388.

[106]   N. S. Bowen, C. A. Polyzois, and R. D. Regan, "Restart Services for Highly Available Systems," *Proc. 7th IEEE Symp. on Parallel and Distributed Processing*, Oct. 1995, pp. 596-601.

[107]   S. Brawer, *Introduction to Parallel Programming*, Academic Press, New York, 1989.

[108]   M. Brehob, T. Doom, R. Enbody, W.H. Moore, S. Q. Moore, R. Sass, and C. Severance, "Beyond RISC - The Post-RISC Architecture," *Technical Report CPS-96-11*, Dept. of Computer Science, Michigan State University, 1996.

[109]   R. P. Brent, "The Parallel Evaluation of General Arithmetic Expressions," *Journal of the ACM*, Vol. 21, No. 2, 1972, pp. 201-206.

[110]   T. Brewer and G. Astfalk, "The Evolution of the HP/Convex Examplar," *Proc. of COMPCON Spring '97*, IEEE Computer Society, Feb.1997, pp. 81-86.

[111]   P. Brinch Hansen, *Studies in Computational Science: Parallel Programming Paradigms*, Prentice-Hall, Englewood Cliffs, N.J., 1995.

[112]   J. Bruck, D. Dolev, C.-H. Ho, M.C. Rosu, and R. Strong, "Efficient Message Passing Interface (MPI) for Parallel Computing on Clusters of Workstations," *Journal of Parallel and Distributed Computing*, Vol. 40, No. 1, 1997, pp. 19-34.

[113]   J. Bruck, L. de Coster, N. Dewulf, C.-T. Ho, and R. Lauwereins, "On the Design and Implementation of Broadcast and Global Combine Operations Using the Postal Model," *IEEE Trans. on Parallel and Distributed Systems*, Vol. 7, No. 3, March 1996, pp. 256-265.

[114]   H. Bryhni and B. Wu, "Initial Studies of SCI LAN Topologies for Local Area Clustering," the *First International Worksthop on SCI-Based Low-Cost/High-Performance Computing*, Santa Clara University, August 1994.

[115]   D. Burger and J.R. Goodman, "Guest Editors' Introduction: Billion-Transistor Architectures," *IEEE Computer Magazine*, Vol.30, No. 9, Sept. 1997, pp. 46-48.

[116]   H. Burkhart III et al., *Overview of the KSR1 Computer System*, Technical Report KSP-TR-9202001, Kendall Square Research, Boston, February 1992.

[117]   G. Burns, R. Daoud, and J. Vaigl, "LAM: An Open Cluster Environment for MPI," Ohio Supercomputer Center, (http://www.osc.edu/lam.html), May 1994.

[118]   R. Butler and E. Lusk, "Monitors, Message, and Clusters: The p4 Parallel Programming System," *Parallel Computing*, Vol. 20, April 1994, pp. 547-564.

[119]   G. Cabillic and I. Puaut, "Stardust: An Environment for Parallel Programming on Networks of Heterogeneous Workstations," *J. Parallel and Distributed Computing*, Vol. 40, No. 1, 1997, pp. 65-80.

[120]   B. Callaghan, "WebNFS: The Filesystem for the World Wide Web," *SunSoft White Paper*, 1996.

[121]   D. Callahan, K. Cooper, R. Hood, K. Kennedy, and L. Torczon, "ParaScope: A Parallel Programmign Environment", *Int. Journal of Supercomputer Applications*, 2(4), 1988.

[122]   K. Cameron, L. J. Clarke, and A. G. Smith, "CRI/EPCC MPI for Cray T3D," http://www.epcc.ed.ac.uk/t3dmpi/Product/, August 1996.

[123]   D. C. Cann, J. T. Feo, and T. M. DeBoni, "SISAL 1.2: High-Performance Applicative Computing," *Proc. Symp. Parallel and Distributed Processing*, 1990, pp. 612-616.

[124]   P. Cao, E. W. Felten, A. R. Karlin, and K. Li, "A Study of Integrated Prefetching and Caching Strategies," *Proceedings of the 1995 ACM SIGMETRICS*, 1995.

[125]   J. Cao, W. Jia, X. Jia, and T.Y. Cheung, "Design and Analysis of an Efficient Algorithm for Coordinated Checkpointing in Distributed Systems," *Proc. of 1997 IEEE Symp. Reliable Distributed Systems*, March 1997, pp. 261-268.

[126]   W. M. Cardoza, F. S. Glover, and W. E. Snaman, "Overview of Digital UNIX Cluster System Architecture," *Proc. of COMPCON'96*, Feb. 1996, pp. 254-259.

[127]   W. M. Cardoza, F.S. Glover, W.E. Snaman, Jr., "Design of the TruCluster Multicomputer System for the Digital Unix Environment," *Digital Technical Journal*, Vol. 8, No. 1, 1996, pp. 5-17.

[128]   G. F. Carey (Ed.), *Parallel Supercomputing: Methods, Algorithms and Applications*, Wiley, New York, 1989.

[129]   N. Carriero and D. Gelernter, *How to Write Parallel Programs*, MIT Press, Cambridge, Mass., 1990.

[130]   N. Carriero and D. Gelernter, "Linda in Context," *Comm. ACM*, 32(4):444-458, 1989.

[131]   J. B. Carter, J. K. Bennett, and W. Zwaenepoel, "Techniques for Reducing Consistency-Related Communication in Distributed Shared Memory Systems," *ACM Trans. on Computer Systems*, 13(3):205-243, 1995.

[132]   B. Catenzaro, *Multiprocessor System Architectures: Multithreaded Systems Using SPARC, Multilevel Bus Architectures, and Solaris (SunOS)*, SunSoft Press, Mountain View, Calif., 1994.

[133]   C. Catlett and L. Smarr, "Metacomputing," *Comm. of ACM*, Vol. 35, No. 6, June 1992, pp. 44-52.

[134]   M. Cekleov et al., "SPARCcenter 2000: Multiprocessing for the 90's," *Digest of Papers. COMPCON Spring '93*, San Francisco, IEEE Comput. Soc. Press, 1993, pp. 345-53.

[135]   L. Censier and P. Feautrier, "A New Solution to Coherence Problems in Multicache Systems," *IEEE Trans. on Computers*, C-27(12): 1112-1118, December 1978.

[136]   Y. Censor and S.A. Zenios, *Parallel Optimization: Theory, Algorithms, and Applications*, Oxford University Press, New York, 1997.

[137]   D. Chaiken, C. Fields, K. Kurihara, and A. Agarwal, "Directory-Based Cache Coherence in Large-Scale Multiprocessors", *IEEE Computer* 23(6): 49-58, June 1990.

[138]   R. Chandra, S. Devine, B. Verghese, A. Gupta, and M. Rosenblum, "Scheduling and Page Migration for Multiprocessor Compute Servers," *Proc. 6th Int'l. Conf. on Architectural Support for Programming Languages and Operating Systems*, pp. 12-24, 1994.

[139]   R. Chandra, A. Gupta, and J. Hennessy, "COOL: An Object-Based Language for Parallel Programming," *Computer*, 27(8):14-26, 1994.

[140]   S. Chandra, J. R. Larus, and A. Rogers, "Where Is Time Spent in Message-Passing and Shared-Memmory Programs?" *Proc. 6th Int'l. Conf. on Architectural Support for Programming Languages and Operating Systems*, pp. 61-73, 1994.

[141]   K. M. Chandy and I. Foster, "A Deterministic Notation for Cooperating Processes," *IEEE Trans. Parallel and Distributed Systems*, 1995.

[142]   K. M. Chandy, I. Foster, K. Kennedy, C. Koelbel, and C.-W. Tseng, "Integrated Support for Task and Data Parallelism," *Int'l. J. Supercomputer Applications*, 8(2): 80-98, 1994.

[143]   K. M. Chandy and C. Kesselman, "Parallel Programming in 2001," *IEEE Software*, 1991, 8(6):1-20.

[144]   K. M. Chandy and C. Kesselman, "CC++: A Declarative Concurrent Object-Oriented Programming Notation," *Research Directions in Concurrent Object-Oriented Programming*, MIT Press, 1993.

[145]   K. M. Chandy and S. Taylor, *An Introduction to Parallel Programming*, Jones and Bartlett, 1992.

[146]   A. Chang and M. F. Mergen, "801 Storage: Architecture and Programming," *ACM Trans. Computer Systems,* 6(1): 28-50, 1988.

[147]   P. P. Chang, S. A. Mahlke, W. Y. Chen, N. J. Warter, and W.W. Hwu, "IMPAXR: An Architectural Framework for Multiple-Instruction Issue Processors," *Proc. of Int'l. Symp. Computer Architecture,* 1991, pp. 226-232.

[148]   J. Chapin, S.A. Herrod, M. Rosenblum, and A. Gupta, "Memory System Performance of Unix on CC-NUMA Multiprocessors," *Proc. of the 1995 ACM SIGMETRICS Conf. on Measurement and Modeling of Computer Systems*, 1995, pp. 1-13.

[149]   B. Chapman, P. Mehrotra, and H. Zima, "Extending HPF for Advanced Data-Parallel Applications," *IEEE Parallel and Distributed Technology*, Vol. 2, No. 3, 1994, pp. 15-27.

[150]   T. F. Chen and J. L. Baer, "A Performance Study of Software and Hardware Data Prefetching Schemes," *Proc. of the 21st Ann. Int'l. Symp. on Computer Architecture*, April 1994, pp. 223-232.

[151]   P. M. Chen, E. Lee, G. Gibson, R. Katz, and D. Patterson, "RAID: High-Performance, Reliable Secondary Storage," *ACM Computing Surveys*, Vol. 26, June 1994, pp. 145-186.

[152]   D. Y. Cheng, "A Survey of Parallel Programming Languages and Tools," *Technical Report RND-93-005*, NASA Ames Research Center, Moffett Field, Calif., 1993.

[153]   C. Y. Chin and K. Hwang, "Packet Switching Networks for Multiprocesssors and Dataflow Computers," *IEEE Trans. Computers,* Nov. 1984, pp. 991-1003.

[154]   F. T. Chong, B.-H. Lim, R. Bianchini, J. Kubiatowicz, and A. Agarwal, "Application Performance on the MIT Alewife Machine," *IEEE Computer*, Vol. 29, No. 12, Dec. 1996, pp. 57-64.

[155]   Y. K. Chong and K. Hwang, "Evaluation of Four Consistency Models for Shared-Memory Multiprocessors," *IEEE Trans. Parallel and Distributed Systems*, Oct. 1996, pp. 1085-1099.

[156]   Chorus Systems, *CHORUS Kernel V3 R4.2 Specification and Interface*, CS/TR-91-69.1, 1993.

[157]   C. K. Chow, "On Optimization for Storage Hierarchies," *IBM Journal of Research and Development,* 1974, pp. 194-203.

[158]   A. Church and J.B. Rosser, "Some Properties of Conversion," *Trans. of the American Mathematical Society*, Vol. 39, 1936, pp. 472-482.

[159]   R. Clark and K. Alnes, "An SCI Interconnect Chipset and Adaptor," *Symposium Records of Hot Interconnect IV*, 1996, pp. 221-235.

[160]   J. Cocke and V. Markstein, "The Evolution of RISC Technology at IBM," *IBM Journal of Research and Development*, Vol. 34, No. 1, 1990, pp. 4-11.

[161]   W. W. Collier, *Reasoning about Parallel Architecture*, Prentice-Hall, Englewood Cliffs, N.J., 1992.

[162]   R. Colwell and R. Steck, "A Technical Tour of the Intel Pentium Pro Processor: A 0.6um BiCMOS Processor Employing Dynamic Execution," *ISSCC*, Feb. 1995.

[163]   D. E. Comer, *Internetworking with TCP/IP*, 3d ed., Prentice Hall, Englewood Cliffs, N.J., 1995.

[164]   Compaq , *Recovery Server Option Kit*, Compaq Computer Corp., http://www.compaq.com., 1997.

[165]   Convex, , *CONVEX Exemplar Architecture*, CONVEX Press, Richardson, TX. 1994.

[166]   Convex, *CONVEX Exemplar Programming Guide,* Order No. DSW-067, CONVEX Press, Richardson, TX. 1994.

[167]   S.A. Cook, "A Taxonomy of Problems with Fast Parallel Algorithms," *Information and Control*, Vol. 64, pp. 2-22, 1985.

[168]   Cornell Theory Center, "IBM RS/6000 Scalable POWERparallel System (SP)," 1995. http://www.tc.cornell.edu/UserDoc/Hardware/SP/.

[169]   A. Cox, S. Dwarkadas, P. Keleher, B. Lu, R. Rajamony, and W. Zwaenepoel, "Software versus Hardware Shared-Memory Implementation: A Case Study," in *Proc. 21st Int'l. Symp. on Computer Architecture*, pp. 106-1117, April 1994.

[170]   H. G Cragon, *Branch Strategy Taxonomy and Performance Models,* IEEE Computer Society Press, Los Alamitos, Calif., 1992.

[171]   H. G. Cragon, *Memory Systems and Pipeline Processors*, Jones and Bartlett Publishers, Sudbury, Mass., 1996.

[172]   D. Crawford, "ASCI Academia Strategic Alliances Program: Research Interests in Computer Systems and Computational and Computer Science Infrastructure," Sandia National Laboratories, December 1996.

[173]   J.H. Crawford, "The i486 CPU: Executing Instructions in One Clock Cycle," *IEEE Micro,* 10(1):27-36, Feb. 1990.

[174]   Cray Research, Inc. *Cray/MPP Announcement*, Eagan, Minn., 1992.

[175]   Cray Research, Inc. Cray T3E Information, (http://www.cray.com/products/systems/crayt3e/), 1997.

[176]   CSRD, "Perfect Club Benchmark Evaluation Package," Technical Report, Center for Supercomputer Research and Development, University of Illinois, Urbana, 1991.

[177]   D. E. Culler et al. (NOW Project), "Efficient and Portable Implementation of MPI using Active Messages," University of California, Berkeley, (http://now.cs.berkeley.edu/Fastcomm/mpi.html), July 10, 1996.

[178]   D. E. Culler, R. Karp, D. Patterson, A. Sahay, K.E. Schauser, E. Santos, R. Subramonian, T. von Eicken, "LogP: Towards a Realistic Model of Parallel Computation," *Proc. ACM Symp. on Principles and Practice of Parallel Programming,* 1993, pp. 1-12.

[179]   D. E. Culler, K. Keeton, L.T. Liu, A. Mainwaring, R.P. Martin, S. Rodrigues, and K. Wright, *Generic Active Message Specification*, Version 1.1, Nov. 1994. Available at http://now.cs.berkeley.edu/.

[180]   D. E. Culler, L.T. Liu, R.P. Martin, and C.O. Yoshikawa, "Assessing Fast Network Interfaces," *IEEE Micro*, Feb. 1996, pp. 35-43.

[181]   D. E. Culler, J. P. Singh, and A. Gupta, *Parallel Computer Architecture: A Hardware/Software Approach*, Morgan Kaufmann Publishers, San Francisco, 1998.

[182]   M. D. Dahlin, R. Y. Wang, T. E. Anderson, and D.A. Patterson, "Cooperative Caching: Using Remote Client Memory to Improve File System Performance," *Proc. of the 1st Symp. on Operating Systems Design and Implementation*, 1994, pp. 267-280.

[183]   W. J. Dally, "Performance Analysis of *k*-ary *n*-Cube Interconnection Networks ," *IEEE Trans. Computers,* June 1990, pp. 775-785.

[184]   W. J. Dally, "Network and Processor Architecture for Message-Driven Computers", in Suaya and Birtwistle(Eds.), *VLSI and Parallel Computation*, Chapter 3, Morgan Kaufmann, San Mateo, CA, 1990.

[185]   W. J. Dally and C. L. Seitz, "Deadlock-Free Message Routing in Multiprocessor Interconnection Networks," *IEEE Trans.on Computers*, C-36(5): 547-553, 1987.

[186]   W. J. Dally, J. Fiske, J. Keen, R. Lethin, M. Noakes, P. Nuth, R. Davison, and G. Fyler, "The Message-Driven Processor: A Multicomputer Processing Node with Efficient Mechanism," *IEEE Micro,* 12(2): 23-39, Apr. 1992.

[187]   R. G. Daniels, "A Participant's Perspective," *IEEE Micro*, Vol. 16, No. 6, 1996, pp. 20-31.

[188]   F. Darema, D. A. George, V.A. Norton, and G.F. Pfister, "A Single-Program-Multiple-Data Computational Model for EPEX Fortran," *Parallel Computing,* Vol. 7, 1988, pp. 11-24.

[189]   H. Davis, S.R. Goldschmidt, and J. Hennessy, "Multiprocessor Simulation and Tracing Using Tango," *Proc. of Int'l. Conf. on Parallel Processing*, Vol. II, Aug. 1991, pp. II 99-107.

[190]   J. R. Davy and P.M. Dew (Eds.), *Abstract Machine Models for Highly Parallel Computers*, Oxford Science Pub., Oxford, 1995.

[191] DEC, *Alpha Architecture Handbook,* Digital Equipment Corporation, Boxboro, MA., 1992.

[192] DEC, *Open VMS Clusters Handbook*, Document No. EC-H2207-93, Maynard, MA., 1993.

[193] DEC, *DEC OSF/1—Guide to DECthreads*, Part No. AA-Q2DPB-TK, Boxboro, MA., 1994.

[194] DEC, *AdvantageCluster: Digital's UNIX Cluster*, Sept. Boxboro, MA, 1994.

[195] DEC, *Digital Clusters for Windows NT,* http:/www.windowsnt.digital.com/clusters/default.htm, 1997.

[196] DEC, *Alpha 21164 Microprocessor Hardware Reference Manual*, Boxboro, MA., 1995.

[197] DEC, *TruCluster: Digital's UNIX Cluster*, Digital Part # EC-Z6310-43, Feb. 1996.

[198] DEC, *Digital's Unix Clusters Lead Industry in High Availability Commercial Solutions*, Maynard, MA., October, 1994.

[199] G. Deconinck, J. Vounckx, R. Cuyvers, and R. Lauwereins, "Survey of Checkpointing and Rollback Techniques," *Tech. Reports O3.1.8 and O3.1.12,* Katholieke Universiteit Leuven, June 1993.

[200] T. DeFanti, I. Foster, M. Papka, R. Stevens, and T. Kuhfuss, "Overview of the I-WAY: Wide Area Visual Supercomputing," *Int'l. Journal of Supercomputer Applications*, Vol. 10, No. 2, 1996.

[201] J. Del Rosario and A. Choudhary, "High-Performance I/O for Parallel Computers: Problems and Prospects," *IEEE Computer,* 27(3):59-68, 1994.

[202] M. Denman, P. Anderson, and M. Snyder, "Design of the PowerPC 604e Microprocessor," *Proc. of IEEE Compcon'96*, 1996, pp. 126-131.

[203] P. J. Denning, "Working Set Model for Program Behavior," *Comm. of ACM*, 11(6):323-333, 1968.

[204] S. Dickey, A. Gottlieb, and Y.-S. Liu, "Interconnection Network Switch Architectures and Combining Strategies," Ultracomputer Note #187, New York University, Sept. 1993.

[205] K. Diefendorff and P. K. Dubey, "How Multimedia Workloads Will Change Processor Design," *IEEE Computer,*Vol.30, No. 9, Sept. 1997, pp. 43-45.

[206] E.W. Dijkstra, "Solution of a Problem in Concurrent Programming Control", *Comm. ACM,* Vol. 8, Sept. 1965, pp.569-578.

[207] K. Dincer and G.C. Fox, "Building a World-Wide Virtual Machine Based on Web and HPCC Technologies," *Supercomputing'96 Conference Proceedings*, 1996.

[208] J. J. Dongarra, J. Martin, and J. Worlton, "Computer Benchmarking: Paths and Pitfalls," *IEEE Spectrum*, July 1986, p. 38.

[209] J. J. Dongarra, "The Linpack Benchmark-Parallel Report," http://performance.netlib.org/performance/html/linpack-parallel.data.co10.html, 1996.

[210] J. J. Dongarra and D. Walker, "Software Libraries for Linear Algebra Computations on High Performance Computers," *SIAM Review*, 1995.

[211] J. J. Dongarra, "The Performance Database Server (PDS): Reports: Linpack Benchmark - Parallel," http://performance.netlib.org/ performance/html/linpack-parallel.data.co10.html.

[212] J. J. Dongarra et al., "TOP 500 Supercomputers", http://www.netlib.org/benchmark/top500.html, June 20, 1996.

[213] X. Du and X. Zhang, "Coordinating Parallel Processes on Networks of Workstations," to appear in *Journal of Parallel and Distributed Computing.*

[214] J. Duato, S. Yalamanchili, and L. Ni, *Interconnection Networks: An Engineering Approach,* IEEE Computer Society Press, 1997.

[215] C. Dubnicki, A. Bilas, K. Li, and J.F. Philbin, "Design and Implementation of Virtual Memory-Mapped Communication on Myrinet," *Proc. of the 11th Int'l. Parallel Processing Symp.*, 1997.

[216] C. Dubnicki, L. Iftode, E.W. Felton, and K. Li, "Software Support for Virtual Memory-Mapped Communication," *Proc. of the 10th Int'l. Parallel Processing Symp.*, 1996.

[217] M. Dubois and C. Scheurich, "Memory Access Depndencies in Shared-Memory Multiprocessors," *IEEE Trans. Computers*, Vol. 16, No. 6, 1990. pp. 660-673.

[218] M. Dubois, C. Scheurich, and F.A. Briggs, "Memory Access Buffering in Multiprocessors," *Proc. 13th Int'l. Symp. on Computer Architecture*, 1986, pp. 434-442.

[219]   M. Dubois, C. Scheurich and F. A. Briggs, "Synchronization, Coherence and Event Ordering in Multiprocessors," *IEEE Computer*, 21(2), 1988.

[220]   M. Dubois and S. Thakkar (Eds.), *Scalable Shared-Memory Multiprocessors*, Kluwer Academic Publishers, Boston, Mass., 1992.

[221]   M. Dubois and S. Thakkar (Eds.), *Cache and Interconnect Architecture in Multiprocessors*, Kluwer Academic Publishers, Boston, Mass., 1990.

[222]   T. H. Dunigan; "Beta Testing the Intel Paragon MP," *Technical Report*, ORNL/TM-12830, Oak Ridge National Lab., June 1995.

[223]   G. Eddon. *RRC for NT: Building Remote Procedure Calls for Windows NT Networks*, Prentice Hall, Englewood Cliffs, N.J., 1994.

[224]   R. Edenfield, M. Gallup, W. Ledbetter, R. McGarity, E. Quintana, and R. Reininger, "The 68040 Processor: Part I. Design and Implementation," *IEEE Micro, Vol.* 10, No. 1, Feb. 1990, pp. 66-78.

[225]   J. Edmondson, P. Rubinfeld, R. Preston, and V. Rajagopalan, "Superscalar Instruction Execution in the 21164 Alpha Microprocessor," *IEEE Micro*, April 1995, pp. 33-44.

[226]   S. Eggers and R. Katz, "Evaluting the Performance of Four Snooping Cache Coherency Protocols," *Proc. of 16th Annual Int'l. Symp. on Computer Architecute*, May 1989, pp. 2-15.

[227]   S. Eggers, J. S. Emer, H. Levy. J. Lo, R. Stamm, and D. Tullsen, "Simultaneous Multithreading: A Platform for Next-Generation Processors," *IEEE Micro*, September/October 1997.

[228]   E. N. Elnozahy, D. B. Johnson, and Y. M. Wang, "A Survey of Rollback-Recovery Protocols in Message-Passing Systems," *Technical Report CMU-CS-96-181*, Department of Computer Science, Carnegie Mellon University, Sept. 1996.

[229]   E. N. Elnozahy and W. Zwaenepoel, "Manetho: Transparent Rollback-Recovery with Low Overhead, Limited Rollback and Fast Output Commit," *IEEE Transactions on Computers*, Special Issue on Fault-Tolerant Computing, May 1992, pp. 526-531.

[230]   H. El-Rewini, T. Lewis, and H.H. Ali, *Task Scheduling in Parallel and Distributed Systems*, Prentice-Hall, Englewood Cliffs, N.J., 1994.

[231]   D.E. Engerbresten, D.M. Kuchta, R.C. Booth, J.D. Crow, and W.G. Nation, "Parallel Fiber-Optic SCI Links," *IEEE Micro*, Vol. 16, No. 1, 1996, pp. 20-26.

[232]   P. H. Enslow (ed.), *Multiprocessors adn Parallel Processing*, Wiley, New York, 1974.

[233]   Executive Office of the President, Office of Science and Technololgy Policy, *A Research and Development Strategy for High Performance Computing*, November 1987.

[234]   F. Faggin, M.E. Hoff, Jr., S. Mazor, and M. Shima, "The History of the 4004," *IEEE Micro*, Vol. 16, No. 6, 1996, pp. 10-20.

[235]   R. Fatoohi, "Performance Evaluation of Communication Networks for Distributed Computing," NASA Ames Research Center Report NAS-95-009, 1995.

[236]   D. G. Feitelson and L. Rudolph (Eds.), *Job Scheduling Strategies for Parallel Processing*, LNCS 949, Springer-Verlag, Berlin, 1995.

[237]   E. W. Felten, R.D. Alpert, A. Bilas, M.A. Blumrich, D.W. Clark, S.N. Damianakis, C. Dubnicki, L. Iftode, and K. Li, "Early Experience with Message-Passing on the SHRIMP Multicomputer," *Proc. the 23rd Int'l. Symp. on Computer Architecture*, 1996.

[238]   T. Y. Feng, "A Survey of Interconnection Networks," *IEEE Computer*, Vol. 14. No. 12, 1981, pp. 12-27.

[239]   D. M. Fenwick, D.J. Foley, W.B. Gist, S. R. VanDoren, and D. Wissel, "The AlphaServer 8000 Series: High-End Server Platform Development," *Digital Technical Journal*, Vol. 7, No. 1, 1995, pp. 43-65.

[240]   J. A. Fisher, "Trace Scheduling: A Technique for Global Microcode Compaction," *IEEE Trans. Computers*, Vol. 30, No. 7, 1981, pp. 478-490.

[241]   J. A. Fisher, "Very Long Instruction Word Architectures and the ELI-512," *Proc. 10th Annual Symp. Computer Arch.*, ACM Press, New York, 1983, pp. 140-150.

[242] J. A. Fisher, "Walk-Time Techniques: Catalyst fro Architectural Change," *IEEE Computer Magazine*, Vol.30, No. 9, Sept. 1997, pp. 40-42.

[243] J. L. Flanagan, "Technologies for Mutlimedia Communications," *Proceedings of the IEEE*, Vol. 82, No. 4, April 1994, pp. 590-603.

[244] M. J. Flynn, "Some Computer Organizations and Their Effectiveness," *IEEE Trans. on Computers*, Vol. C-21, 1972, pp. 948-960.

[245] M. J. Flynn, *Computer Architecture: Pipelined and Parallel Processor Design*, Jones and Bartlett, Boston, Mass., 1995.

[246] S. Fortune and J. Wyllie, "Parallelism in Random Access Machines," *Proc. ACM Symp. on Theory of Computing*, 1978, pp. 114-118.

[247] I. Foster and K. M. Chandy, "Fortran M: A Language for Modular Parallel Programming," *J. Parallel and Distributed Computing*, Vol. 25, No. 1, Jan. 1995.

[248] I. Foster, J. Geisler, W. Nickless, W. Smith, and S. Tuecke, "Software Infrastructure for the I-WAY High-Performance Distributed Computing Experiment," *Proc. of the 5th IEEE Symp. on High Performance Distributed Computing*, 1996.

[249] I. Foster, J. Geisler, C. Kesselman, and S. Tuecke, "Multimethod Communication for High-Performance Metacomputing Applications," *Supercomputing'96 Conference Proceedings*, 1996.

[250] I. Foster, C. Kessekman, and S. Tuecke, "The Nexus Approach to Integrating Multithreading and Communication," *Journal of Parallel and Distributed Computing*, Vol. 37, No. 1, 1996, pp. 70-82.

[251] G. Fox and W. Furmanski, "Towards Web/Java Based High Performance Distributed Computing - An Evolving Virtual Machine," *Proc. of the 5th IEEE Symp. on High Performance Distributed Computing*, August 1996.

[252] G. Fox, W. Furmanski, M. Chen, C. Rebbi and J. Cowie, "WebWork: Integrated Programming Environment Tools for National and Grand Challenges," *NPAC Tech. Report SCCS-0715*, Syracuse University, 1995.

[253] G. C. Fox, M.A. Johnson, G.A. Lyzenga, S.W. Otto, J.K. Salmon, and D.W. Walker, *Solving Problems on Concurrent Processors,* Vol. I, Prentice-Hall, Englewood Cliffs, N.J., 1988.

[254] G. C. Fox, R.D. Williams, and P.C. Messina, *Parallel Computing Works!*, Morgan Kaufmann Publishers, San Francisco, 1994.

[255] A. G. Fraser, "Future WAN Telecommunications," *IEEE Micro*, Vol. 16, No. 1, 1996, pp. 53-57.

[256] D. Frye, Ray Bryant, H. Ho, R. Lawrence, and M. Snir. "An External User Interface for Scalable Parallel Systems," *Technical Report*, International Businesss Machines, Armonk, N.Y., May 1992.

[257] E. M. Frymoyer, "Fibre Channel Fusion: Low Latency, High Speed," *Data Communications*, http://www.data.com/, Feb. 1995.

[258] D. D. Gajski and J. K. Peir, "Essential Issues in Multiprocessor Systems," *IEEE Computer*, Vol. 18, June, 1985.

[259] J. Gary, and D.P. Siewiorek, "High-Availability Computer Systems," *IEEE Computer*, Sept. 1991, pp. 39-48.

[260] G. A. Geist and V. S. Sunderam, "The Evolution of the PVM Concurrent Computing System," *Proceeding of the 26th IEEE Compcon Symposium*, pages 471-478, San Francisco, February 1993.

[261] A. Geist, A. Beguelin, J. Dongarra, W. Jiang, R. Mancheck, and V. Sunderam, *PVM: Parallel Virtual Machine - A User's Guide and Tutorial for Networked Parallel Computing*, MIT Press, Cambridge, Mass., 1994.

[262] A.V. Gerbessiotis and L.G. Valiant, "Direct Bulk-Synchronous Parallel Algorithms," *J. Parallel and Distributed Computing*, Vol. 22, No. 2, Feb. 1994, pp. 251-267.

[263] D. Gelernter, "Generative Communication in Linda," *ACM Trans. Program Lang. and Syst.*, 1985, 7(1): 80-112.

[264] K. Gharachorloo et al., "Memory Consistency and Event Ordering in Scalable Shared-Memory Multiprocessors," *Proc. of the 17th Annual Symp. on Computer Architecture*, June 1990, pp. 15-25.

[265]   K. Gharachorloo, S. Adve, A. Gupta, J. L. Hennessy, and M. Hill, "Programming for Different Memory Consistency Models," *Journal of Parallel and Distributed Computing,* Aug. 1992.

[266]   G. A. Gibson, *Redundant Disk Arrays: Reliable, Parallel Secondary Storage*, MIT Press, Cambridge, Mass., 1992.

[267]   Gigabit Ethernet Alliance, "White Paper of Gigabit Ethernet", March 1997. Also avaialble from web site http ://www.gigabit-ethernet.org/.

[268]   R. B. Gillett and R. Kaufmann, "Using the Memory Channel Networks," *IEEE Micro,* Jan./Feb., 1997, pp. 19-25.

[269]   L. M. Goldschlager, "A Universal Interconnection Pattern for Parallel Computers," *Journal of ACM,* Vol. 29, 1982, pp. 1073-1086.

[270]   S. C. Goldstein, K.E. Schauser, and D.E. Culler, "Lazy Threads: Implementing a Fast Parallel Call," *Journal of Parallel and Distributed Computing*, Vol. 37, No. 1, 1996, pp. 5-20.

[271]   H. Goldstine and J. von Neumann, "On the Principles of Large-Scale Computing Machines," *Collected Works of John von Neumann*, Vol. 5. Pergamon, New York, 1963.

[272]   J. R. Goodman, M. Vernon, and P. Woest, "Efficient Synchronization Primitives for Large-Scale Cache-Coherent Multiprocessors," *Proc. Third Int'l. Conf. on Architectural Support for Programming Languages and Operating systems*, April 1989, pp. 64-73.

[273]   J. R. Goodman, "Cache Consistency and Sequential Consistency," *Technical Report 61,* IEEE SCI Committee, 1990.

[274]   B. Groop, R. Lusk, T. Skjellum, and N. Doss; "Portable MPI Model Implementation," Argonne National Laboratory, *Technical Report*, July 1994.

[275]   J. C. Gomez, V. Rego, and V.S. Sunderam, "Efficient Multithreaded User-Space Transport for Network Computing: Design and Test of the TRAP Protocol," *Journal of Parallel and Distributed Computing*, Vol. 40, No. 1, 1997, pp. 103-117.

[276]   A. Gottlieb, R. Grishman, C. P. Kruskal, K. P. McAuliffe, L. Rudolph, and M. Snir, "The NYU Ultracomputer: Designing a MIMD, Shared Memory Parallel Computer," *IEEE Trans.on Computers*, C-32(2):175-189, 1983.

[277]   H. Grahn and P. Stenstrom, "Efficient Strategies for Software-Only Directory Protocols in Shared Memory Multiprocessors," *Proc. of the 22nd Annual Int'l. Symp. on Computer Architecture*, 1995.

[278]   A.Y. Grama, A. Gupta, and V. Kumar, "Isoefficiency: Measuring the Scalability of Parallel Algorithms and Architectures," *IEEE Parallel & Distributed Technology*, 1(3):12-21, 1993.

[279]   J. Gray (Ed.), *The Benchmark Handbook for Database and Transaction Processing Systems*, Morgan Kaufmann, San Mateo, Calif., 1991.

[280]   B. Grayson, A.P. Shah, and R.A. van de Geijn, "A High Performance Parallel Stranssen Implementation," *Tech. Report*, Dept. of Computer Science, University of Texas at Austin, 1995.

[281]   D. Greenley et al., "UltraSPARC: The Next Generation Superscalar 64-bit SPARC," *Digest of Papers, Compcon*, Spring 1995, pp. 442-451.

[282]   A. Grimshaw, W. Wulf, J. French, A. Weaver, and P. Reynolds, Jr., "Legion: The Next Logical Step towards a Nationwide Virtual Computer," *Tech. Report CS-94-21*, Department of Computer Science, University of Virginia, 1994.

[283]   W. Gropp, E. Lusk, and A. Skjellum, *Using MPI: Portable Parallel Programming with the Message Passing Interface,* MIT Press, Cambridge, Mass., 1994.

[284]   N. J. Gunther, "Issues Facing Commercial OLTP Applications on MPP Platforms," *Digest of Papers, IEEE Compcon'94*, pp. 242-247.

[285]   A. Gupta, J. Hennessy, K. Gharachorloo, T. Mowry, and W.-D. Weber. "Comparative Evaluaton of Latency Reducing and Tolerating Techniques," *Proc. 18th Int'l. Symp. on Computer Architecture*, June 1991, pp. 254-263.

[286]   J. L. Gustafson, "Reevaluating Amdahl's Law," *Comm. of ACM,* Vol. 31, No. 5, 1988, pp. 532-533.

[287]   J. L. Gustafson and Q.O. Snell, "HINT: A New Way to Measure Computer Performance," *Proc. of Supercomputing'96*, 1996.

[288]   D. B. Gustavson, "The Scalable Coherent Interface and Related Standards Projects," *IEEE Micro*, Vol. 12, No. 2, 1992, pp. 10-12.

[289]   D. B. Gustavson and Q. Li, "Low Latency, High-Bandwidth, and Low Cost for Local-Area Multi-Processor," Dept. of Computer Engineering, Santa Clara University, 1996.

[290]   L. Gwennap, "Intel's P6 Uses Decoupled Superscalar Design," *Microprocessor Report,* February 1995, pp. 5-15.

[291]   L. Gwennap, "MIPS R10000 Uses Decoupled Architecture," *Microprocessor Report*, October 1994, pp. 18-22.

[292]   M. W. Hall, J.M. Anderson, S.P. Amarasinghe, B.R. Murphy, S.W. Liao, E. Bugnnion, and M.S. Lam, "Maximizing Multiprocessor Performance with the SUIF Compiler," *IEEE Computer*, Vol. 29, No. 12, Dec. 1996, pp. 84-89.

[293]   E. Hagersten, A. Landin, and S. Haridi, "Multiprocessor Consistency and Synchronization Through Transient Cache States," in Dubois and Thakkar (Eds.), *Scalable Shared-Memory Multiprocessors,* Kluwer Academic Published, Boston, MA, 1992.

[294]   E. Hagersten, A. Landin, and S. Haridi, "DDM - A Cache-Only Memory Architecture," *IEEE Computer,* 25(9):44-54, Septemeber 1992.

[295]   F. Halsall, *Data Communications, Computer Networks and Open Systems* (Fourth Edition), Addison-Wesley, Reading, Mass., 1996.

[296]   L. Hammond, B. Nayfeh, and K. Olukotun, "A Single-Chip Multiprocessor," *IEEE Computer Magazine*, Vol.30, No. 9, Sept. 1997, pp. 79-85.

[297]   T. J. Harris and N.P. Topham, "The Scalability of Decoupled Multiprocessors," *Proc. of Scalable High-Performance Computing Conference*, Knoxville, Tenn., May 1994, pp. 17-22.

[298]   J. Harris, J.A. Bircsak, M.R. Bolduc, J.A. Diewald, I. Gale, N.W. Johnson, S. Lee, C.A. Nelson, and C.D. Offner, "Compiling High Performance Fortran for Distributed-Memory Systems," *Digital Technical Journal*, Vol. 7, No. 3, 1995, pp. 5-23.

[299]   P. Hatcher and M. Quinn, *Data-Parallel Programming on MIMD Computers*, MIT Press, 1991.

[300]   A. H. Hayes, M.L. Simmons, J.S. Brown, and D.A. Reed (Eds.), *Debugging and Performance Tuning for Parallel Computing Systems*, IEEE Computer Society Press, July 1996.

[301]   M. Heath and J. Etheridge, "Visualizing the Performance of Parallel Programs," *IEEE Software*, 8(5):29-39, 1991.

[302]   J. Heinrich, *MIPS R10000 Microprocessor User's Manual,* MIPS Technologies, Inc., 1994.

[303]   R. Hempel, "The ANL/GMD Macros (PARMACS) in Fortran for Portable Parallel Programming Using the Message Passing Model: Users Guide and Reference Manual," Gesellschaft fur Mathematik und Datenverarbeitung (GMD) mbH, 1991.

[304]   M. Henderson, B. Nickless, and R. Stevens, "A Scalable High-Performance I/O System," *Proc. 1994 Scalable High-Performance Computing Conf*, pp. 79-86 IEEE Computer Society, 1994.

[305]   J. L. Hennessy and D. A. Patterson, *Computer Architecture: A Quantitative Approach,* Morgan Kaufmann, San Mateo, CA, 1995.

[306]   M.P. Herlihy, "Wait-Free Synchronization," *ACM Trans. on Program Lang. and Syst*, 1990, 12(3):463-492.

[307]   M.P. Herlihy and J.E.B. Moss, "Transactional Memory: Architectural Support for Lock-Free Data Structures," *Proc. of the 20th Int'l. Symp. Computer Architecture*, pp. 289-300, 1993.

[308]   M. Hill, "What Is Scalability?," ACM *Computer Architecture News*, Dec. 1990, pp 18-21.

[309]   W. D. Hillis, "The Connection Machine," *Scientific American*, Vol. 256, June 1987, pp. 108-115.

[310]   W.D. Hillis and G. L. Steele, "Data Parallel Algorithms," *Comm. ACM*, 29(12):1170-1183, 1986.

[311]   S. Hiraanandani, K. Kennedy, and C.W. Tseng, "Compiling Fotran D for MIMD Disytributed Machines," *Comm.of the ACM*, Aug.1992, pp. 66-80.

[312]   C. A. Hoare, *Communicating Sequential Processes,* Prentice-Hall, London, U.K. , 1984.

[313]   R. W. Hockney, "Performance Parameters and Benchmarking of Supercomputers," *Parallel Computing*, Vol. 17, pp. 1111-1130, 1991.

[314]   R. W. Hockney, "The Communication Challenge for MPP: Intel Paragon and Meiko CS-2," *Parallel Computing, Vol. 20*, 1994, pp. 389-398.

[315]   R. W. Hockney, "A Framework for Benchmark Performance Analysis," in J.J. Dongarra and W. Gentzsch (Eds.), *Computer Benchmarks, Advances in Parallel Computing*, Vol. 8, Elsevier Science Pub., 1993, pp. 65-76.

[316]   R. W. Hockney, *The Science of Computer Benchmarking*, The Society for Industrial and Applied Mathematics, Philadelphia, 1996.

[317]   R. W. Hockney and M. Berry, "Public International Benchmarks for Parallel Computers: PARKBENCH Committee Report No. 1," *Scientific Computing*, 3(2):101-146, 1994.

[318]   R. W. Hockney and C. R. Jesshope, *Parallel Computers: Architecture, Programming, and Algorithms*, Adam Hilger, Philadelphia, 1988.

[319]   C. Holt, J. P. Singh, and J. Hennessy, "Application and Architectural Bottlenecks in Large Scale Distributed Shared Memory Machines," *Proc. of the 23rd Int'l. Symp. on Computer Architecture*, 1996, pp. 134-145.

[320]   R. W. Horst, "Massively Parallel Systems You Can Trust," *Digest of Papers, IEEE Compcon '94*, pp. 236-241.

[321]   Hotchips, *Proc. Hot Chips III Symp. on High-Performance Chips,* Stanford University, Palo Alto, CA, 1991.

[322]   D. Hunt, "Advanced Features of the 64-bit PA-8000," *Proceedings 1995 IEEE COMPCON*, 1995, pp. 123-128.

[323]   K. Hwang, W. Croft, G Goble, B. W. Wah, F. A. Briggs, W. Simmons, and C. L. Coates, "A UNIX-Based Local Computer Network with Load Balancing," *IEEE Computer*, April 1982, pp. 55-66.

[324]   K. Hwang, "Advanced Parallel Processing with Supercomputer Architectures," *Proc. of the IEEE*, 1987, pp. 1348-1379.

[325]   K. Hwang, P. S. Tseng, and D. Kim, "An Orthogonal Multiprocessor for Parallel Scientific Computations," *IEEE Trans. Computers*, Vol. C-38, No. 1, Jan. 1989, pp. 47-61.

[326]   K. Hwang, "Exploiting Parallelism in Multiprocessors and Multicomputers," in *Parallel Processing for Supercomputers and Artificial Intelligence*, K. Hwang and D. DeGroot (Eds.), McGraw-Hill, New York, NY, 1989, pp. 1-68.

[327]   K. Hwang, *Advanced Computer Architecture: Parallelism, Scalability, and Programmability*, McGraw-Hill, New York, 1993.

[328]   K. Hwang, "Gigabit Networks for Building Scalable Multiprocessors and Multicomputer Clusters," *HKIE Transactions* , Hong Kong Institute of Engineers, Hong Kong, 1997.

[329]   K. Hwang and F.A. Briggs, *Computer Architecture and Parallel Processing*, McGraw-Hill, New York, 1983.

[330]   K. Hwang and D. DeGroot (Eds.), *Parallel Processing for Supercomputers and Artificial Intelligence,* McGraw-Hill, New York, 1989.

[331]   K. Hwang and J. Ghosh, "Hypernet: A Communication-Efficient Architecture for Multicomputers," *IEEE Trans. on Computers,* Nov. 1989, pp. 1450-1466.

[332]   K. Hwang, C. J. Wang, and C.-L. Wang, "Evaluating MPI Collective Communication on the SP2, T3D, and Paragon Multicomputers," *IEEE High-Performance Computer Architecture* (HPCA-3), San Antonio, Texas, Feb. 1-5, 1997.

[333]   K. Hwang and Z. Xu, "Scalable Parallel Computers for Real-Time Signal Processing," *IEEE Signal Processing,* Vol. 13, No. 4, 1996, pp. 50-66.

[334]   K. Hwang, Z. Xu, and M. Arakawa, "Benchmark Evaluation of the IBM SP2 for Parallel Signal Processing," *IEEE Transactions on Parallel and Distributed Systems*, Vol. 7, No. 5, 1996, pp. 522-536.

[335]   IBM Corp., *RISC System/6000 Technology,* IBM Advanced Workstations Division, Austin, TX, 1990.

[336] IBM Corp., *LoadLeveller General Information Manual*. Document GH26-7227, Armonk, NY.

[337] IBM Corp., *AIX Parallel Environment: Programming Primer*, Release 2.0, Pub. No. SH26-7223, June 1994.

[338] IBM Corp. "*High Availability Cluster Multiprocessing/6000 (HACMP) System Overview,*" Doc. No. SC23-2408-02, Armonk, N.Y., 1993.

[339] IBM Corp., *Sysplex Hardware and Software Migration*, Doc. No. GC 28-1210-00, Armonk, N.Y., 1994.

[340] IBM Corp., *AIX 4.1 Threads*, Armonk, N.Y., 1995.

[341] IBM Corp., *IBM System/370 Principles of Operation*, IBM Pub. GA22-7000-9, File S370-01, 1983.

[342] L. Iftode, C. Dubnicki, E.W. Felten, and K. Li, "Improving Release-Consistent Shared Virtual Memory Using Automatic Update," *Proc. 2nd Int'l. Symp. on High-Performance Computer Architecture*, 1996.

[343] IEEE, *ANSI/IEEE Standard 1596-1992: Scalable Coherent Interface*, IEEE, Piscataway, N.J., 1993; and *Standard 1596.3-1996*, 1996.

[344] IEEE, *Futurebus+: Logical Layer Specifications 896.1-1991*, Microprocessor Standards Subcommittee, IEEE Computer Society, 1991.

[345] IEEE, *POSIX P1003.4a: Threads Extension for Portable Operating Systems*, IEEE, Piscataway, N.J., 1994.

[346] Intel Corp., Material for Standard High Volume (SHV) servers can be found on Intel's Web sites, such as http://www.intel.com/intel/june297/foils/grove/index.htm, http://www.intel.com/intel/june297/foils/miner/, and http://www.intel.com/procs/SERVERS/feature/shv/

[347] Intel/Sandia , *The ASCI Teraflop Machine*, http://abacus.ssd.intel.com/tflop1.html, 1996.

[348] D. V. James et al., "Distributed-Directory Scheme: Scalable Coherence Interface," *IEEE Computer*, June 1990, pp. 74-77.

[349] E. H. Jensen, G.W. Hagensen, and J.M. Broughton, "A New Approach to Exclusive Data Access in Shared-Memory Multiprocessors," *Technical Report UCRL-97663*, Lawrence Livermore National Laboratory, 1987.

[350] T. Jermoluk, *Multiprocessor UNIX*, Silicon Grpahics Inc., Santa Clara, CA, 1990.

[351] D. Johnson, D. Lilja, J. Riedl, and J. Anderson, "Low-Cost, High-Performance Barrier Synchronization on Networks of Workstations," *Journal of Parallel and Distributed Computing*, Vol. 40, No. 1, 1997, pp. 118-130.

[352] M. Johnson, *Superscalar Microprocessor Design*, Prentice-Hall, Englewood Cliffs, N.J., 1991.

[353] L. Johnsson, "Communication in Network Architectures," in *VLSI and Parallel Computation*, Morgan Kaufmann Publishers, San Mateo, CA. 1990.

[354] L. Johnsson and C.-T. Ho, "Matrix Transposition on Boolean n-Cube Configured Ensemble Architectures," *SIAM J. Matrix Analysis and Applications*, 9(3):419-454, 1988.

[355] L. Johnsson and C.-T. Ho, "Optimum Broadcasting and Personalized Communication in Hypercubes," *IEEE Trans. Computs.*, C-38(9):1249-1268, 1989.

[356] N. P. Jouppi and D. W. Wall, "Available Instruction-Level Parallelism for Superscalar and Superpipelined Machines," *Proc. Third Int'l. Conf. Arch. Support for Prog. Lang. and OS*, pp. 272-282, ACM Press, New York, 1989.

[357] D. Kafura and L. Huang, "MPI++: A C++ Language Binding for MPI," *MPI Developers Conference*, University of Notre Dame, June 1995.

[358] A. Kagi, D. Burger, and J.R. Goodman, "Efficient Synchronization: Let Them Eat QOLB," *Proc. of the 24th Annual Int'l. Symp. on Computer Architecture*, 1997, pp. 170-180.

[359] G. Kane, *MIPS R2000 RISC Architecture*, Prentice-Hall, Englewood Cliffs, NJ, 1988.

[360] V. Karamcheti and A. A. Chien, "Software Overhead in Messaging Layers: Where Does the Time Go?" *Proc. 6th Inl. Conf. on Architectural Support for Programming Languages and Operating Systems*, pp. 51-60, 1994.

[361] V. Karamcheti and A.A. Chien, "A Comparison of Architectural Support for Messaging in the TMC CM-5 and the Cray T3D," *Proc. 22nd Int'l. Symp. on Computer Architecture*, pp. 298-307, 1995.

[362] A. H. Karp and R. G. Babb II, "A Comparison of 12 Parallel Fortran Dialects," *IEEE Software*, Vol.5, No.5, 1988, pp. 52-66.

[363] A. H. Karp, "Programming for Parallelism," *IEEE Computer,* Vol. 20, No. 5, 1987, pp. 43-57.

[364] A. H. Karp and H. P. Flatt, "Measuring Parallel Processor Performance," *Comm. of the ACM,* May, 1990, pp. 539-543.

[365] W. J. Kaufmann and L. L. Smarr, *Supercomputing and the Transformation of Science*, Scientific American Library, 1993.

[366] R. M. Keller, "Formal Verification of Parallel Programs," *Comm. of ACM*, Vol.19, July 1976, pp. 371-384.

[367] K. Kennedy, "Compiler Technology for Machine-Independent Parallel Programming," *Int'l. Journal of Parallel Programming*, Vol. 22, No. 1, 1994, pp. 79-98.

[368] B. Kernighan and D. Ritchie, *The C Programming Language,* 2d ed., Prentice Hall, 1988.

[369] Y. A. Khalidi, J. M. Bernabeu, V. Matena, K. Shirriff, and M. Thadani, "Solaris MC: A Multicomputer OS," Sun Microsystems Lab. SMLI TR-95-48, Nov. 1995. Also appeared in *Proc. of the 1996 USENIX Conference*.

[370] C. Koelbel, D. Loveman, R. Schreiber, G. Steele, and M. Zosel. *The High Performance Fortran Handbook*, MIT Press, 1994.

[371] R. K. Koeninger, M. Furtney, and M. Walker, "A Shared Memory MPP from Cray Reseach," *Digital Technical Journal*, Vol. 6, No. 2, Spring 1994, pp. 8-21.

[372] A. Kolawa, "Parasoft: A Comprehensive Approach to Parallel and Distributed Computing," *Proc. Workshop on Cluster Computing*, 1992.

[373] L. Kontothanassis et al., "VM-Based Shared Memory on Low-Latency, Remote-Memory-Access Networks", *Proc. of the 24th Int'l. Symp. Computer Architecture,* June 1997, pp. 157-169.

[374] L. I. Kontothanassis, R.W. Wisniewski, and M.L. Scott, "Scheduler-Conscious Synchronization," *ACM Trans. on Computer Systems*, Vol. 15, No. 1, 1997, pp. 3-40.

[375] R. B. Konuru, S.W. Otto, and J. Walpole, "A Migratable User-Level Process Package for PVM," *Journal of Parallel and Distributed Computing*, Vol. 40, No. 1, 1997, pp. 81-102.

[376] J. S. Kowalik (Ed.), *Parallel MIMD Computation: HEP Supercomputer and Applications*, MIT Press, Cambridge, MA, 1985.

[377] C. E. Kozyrakis, S. Perissakis, D. Patterson, T. Anderson, K. Asanovic, N. Cardwell, R. Fromm, J. Golbus, B. Gribstad, K. Keeton, R. Thomas, N. Treuhaft, and K. Yelick, "Scalable Processors in the Billion-Transistors Era: IRAM," *IEEE Computer*, Vol.30, No. 9, Sept. 1997, pp. 75-78.

[378] N. P. Kronenberg, H. M. Levy, W. D. Strecker, and R. J. Merewood. "The VaxCluster Concept: An overview of a Distributed System". *Digital Technical Journal*, Vol. 4, No. 7, Sept. 1987, pp. 15-21.

[379] C. P. Kruskal and A. Weiss, "Allocating Independent Subtasks on Parallel Processors," *IEEE Trans. Software Engineering,*" Vol. SE-11, pp. 1001-1016, 1985.

[380] C.P. Kruskal, L. Rudolph, and M. Snir, "Efficient Synchronization of Multiprocessors with Shared Memory," *ACM Trans. Program. Lang. Systems*, April, 1988, pp. 579-601.

[381] Kuck and Associates, *The KAP Preprocessor,* http://www.kai.com/kap/kap_what_is.html.

[382] D.J. Kuck, E.S. Davidson, D.H. Lawrie, and A.H. Sameh, "Parallel Supercomputing Today - The Cedar Approach," *Science*, 231(2), Feb. 1986.

[383] V. Kumar, A. Grama, A. Gupta, and G. Karypis, *Introduction to Parallel Computing,* Benjamin/Cummings, New York, 1993.

[384] V. Kumar, A. Grama, and V. Rao, "Scalable Load Balancing Techniques for Parallel Computers," *J. Parallel and Distributed Computing*, 22(1):60-79, 1994.

[385] H. T. Kung and C. E. Leiserson, "Systolic Arrays for VLSI " in *SIAM Sparse Matrix Proceedings*, edited by Duff and Stewart, Knoxville, Tenn., 1978.

[386] H. T. Kung el.al., "Network-Based Multicomputers: An Emerging Parallel Architecture," *Proc. of Supercomputing*, IEEE Computer Society Press, 1991, pp. 664-673.

[387] H. T. Kung, "Gigabit Local Area Networks: A Systems Perspective," *IEEE Communicatins Magazine*, April, pp. 79-89.

[388] J. Kuskin et al., "The Stanford FLASH Multiprocessor," *Proc. 21st Int. Symp. on Computer Architecture*, April 1994, pp. 302-313.

[389] M. S. Lam, "Software Pipelining: An Effective Scheduling Technique for VLIW Machines," *Proc. ACM SIGPLAN Conf. Prog. Lang, Design and Implementation*, 1988, pp. 318-328.

[390] L. Lamport, "Solved Problems, Unsolved Problems and Non-Problems in Concurrency," *Proc. of Int'l. Conference on Distributed Computing*, 1983, pp. 1-11.

[391] L. Lamport, "How to Make a Multiprocessor Computer That Correctly Executes Multiprocess Programs," *IEEE Trans. Computers*, Vol. C-28, Sept. 1979, pp. 690-691.

[392] J. R. Larus, "Loop-Level Parallelism in Numeric and Symbolic Programs," *IEEE Trans. on Parallel and Distributed Systems*, Vol. 4, No. 7, 1993, pp. 812-826.

[393] J. Laudon and D. Lenoski, "The SGI Origin: A ccNUMA Highly Scalable Server," *Proc. of the 24th Int'l. Symp. Computer Architecture,* June 1997, pp. 241-251.

[394] M. Lauria and A. Chien, "MPI-FM: High Performance MPI on Workstation Clusters," *Journal of Parallel and Distributed Computing*, Vol. 40, No. 1, 1997, pp. 4-18.

[395] D. H. Lawrie, "Access and Alignment of Data in an Array Processor," *IEEE Trans. on Computers*, Dec. 1975.

[396] H. Lawson, *Parallel Processing in Industrial Real-time Applications*, Prentice Hall, N.J., 1992.

[397] J. V. Lawton, J.J. Brosnan, M.P. Doyle, and S.D. Riodain, "Building a High-Performance Message-Passing System for MEMORY CHANNELClusters," *Digital Technical Journal*, Vol. 8, No. 2, 1996, pp. 96-116.

[398] R. B. Lee, "Precision Architecture," IEEE *Computer*, Vol. 22, No. 1, Jan. 1989, pp. 78-91.

[399] R. Lee, P. C. Yew, and D. Lawrie, "Data Prefetching in Shared Memory Multiprocessors", *Proc. Int'l. Conf. on Parallel Processing*, August 1987, pp. 28-31.

[400] R. B. Lee, "Subword Parallelism with MAX-2," *IEEE Micro*, Vol. 16, No. 4, 1996, pp. 51-59.

[401] R. B. Lee and M. D. Smith, "Media Processing: A New Design Target," *IEEE Micro*, August, 1996, pp. 6-9.

[402] F. T. Leighton, *Introduction to Parallel Algorithms and Architectures*, Morgan Kaufmann, 1992.

[403] C. Leiserson, "Fat Tree: Universal Networks for Hardware-Efficient Supercomputing," *IEEE Trans. on Computers*, Oct. 1985, pp. 892-901.

[404] C. E. Leiserson, Z. S. Abuhamdeh, D. C. Douglas, C. R. Feynman, M. N. Ganmukhi, J. V. Hill, W. D. Hills, B. C. Kuszmaul, M. A. St. Pierre, D. S. Wells, M. C. Wong, S. Yang, and R. Zak, "The Network Architecture of the CM-5," *Int'l. Symp. on Parallel and Distributed Algorithms '92*, June 1992, pp. 272-285.

[405] D. E. Lenoski et al., "The DASH Prototype: Logic and Performance," *IEEE Trans. on Parallel and Distributed Systems,* Jan. 1993, pp. 41-61.

[406] D. E. Lenoski and W.-D. Weber, *Scalable Shared-Memory Multiprocessing,* Morgan Kaufmann, San Francisco, CA, 1995.

[407] E. Levin, "Grand Challenges in Computational Science," *Comm. of the ACM*, Vol. 32, No. 12, Dec. 1989, pp. 1456-1457.

[408] D. Levitan, T. Thomas and P. Tu, "The PowerPC 620 Microprocessor: A High Performance Superscalar RISC Microprocessor," *Digest of Papers, Compcon95,* Spring 1995, pp. 285-291.

[409] E. Levy and A. Silberschartz, "Distibuted File Systems: Concepts and Examples," *ACM Computing Surveys*, Vol. 22, No. 4, 1990, pp. 321-374.

[410] M. J. Lewis and A. Grimshaw, "The Core Legion Object Model," *Proc of the Fifth IEEE Int'l. Symp. on High Performance Distributed Computing*, IEEE Computer Society Press, August 1996.

[411]   T. G. Lewis and H. El-Rewini, *Introduction to Parallel Computing,* Prentice-Hall, N.J., 1992.

[412]   T. G. Lewis, "The Nethead Gang," *IEEE Computer*, 1995, 28(12): 8-10.

[413]   K. Li, "IVY: A Shared-Virtual Memory System for Parallel Computing," *Proc. Int'l. Conf. Parallel Processing*, August 1988, pp. 94-101

[414]   K. Li and P. Hudak, "Memory Coherence in Shared Virtual Memory Systems," *ACM Trans. Computer Systems*, Nov. 1989, pp. 321-359.

[415]   K. Li, J.F. Naughton, and J.S. Plank, "Low-Latency, Concurrent Checkpointing for Parallel Programs," *IEEE Trans. on Parallel and Distributed Systems*, Vol. 5, No. 8, 1994, pp. 874-879.

[416]   D. J. Lilja, *Architectural Alternatives for Exploiting Parallelism*, IEEE Computer Society Press, Los Alamitos, CA, 1992.

[417]   X. Lin and L. M. Ni, "Deadlock-Free Multicast Wormhole Routing in Multicomputer Networks," *Proc. 18th Annu. Int'l. Symp. Computer Arch.*, pp. 116-125, 1991.

[418]   M. H. Lipasti and J. P. Shen, "Superspeculative Microarchitecture for Beyond AD 2000," *IEEE Computer*, Vol.30, No. 9, Sept. 1997, pp. 59-66.

[419]   D. Loveman, "High Performance Fortran," *IEEE Parallel and Distributed Technology*, 1(1):25-42, 1993.

[420]   T. Lovett and R. Clapp, "STiNG: A CC-NUMA Computer System for the Commercial Marketplace," *Proc. of the 23rd Annual Int'l. Symp. on Computer Architecture*, 1996, pp. 308-317.

[421]   T. D. Lovett, R. M. Clapp and R. J. Safranek, *"NUMA-Q: An SCI-Based Enterprise Server,"* Sequent Computer White Paper, http://www.sequent.com/, 1996.

[422]   E. Lusk et al., *Portable Programs for Parallel Processors*, Holt, Rinehardt, and Winston, New York, 1987.

[423]   R. Lusk, N. Doss, A. Skjellum, and W. Gropp; "MPICH: A High-Performance Portable Implementation of the MPI Standard," *MPI Developers Conference,* University of Notre Dame, June 1995.

[424]   S. Mahlke, R. Hank, J. McCormick, D. August, and W.M. Hwu, "A Comparison of Full and Partial Predicated Execution Support for ILP Processors," *Proc. of the 22nd Annual Int'l. Symp. on Computer Architecture,* June 1995, Italy, pp. 138-149.

[425]   Marathon Corp., *MIAL1/2 Fault-Tolerant Cluster*, Web: http://www.mial.com, 1997.

[426]   A. Mainwaring and D.E. Culler, "Generic Active Message Applications Programming Inteface and Communication Subsystems Organization," Division of Computer Science, University of California at Berkeley, Nov. 1996. Available at http://now.cs.berkeley.edu/.

[427]   J. MarcFrailong et al., "The Next Generation SPARC Multiprcoessing System Architecutre," *Proceedings of COMPCON*, Spring 1993, pp. 475-480.

[428]   N. Margulis, *i860 Microprocessor Architecture,* Intel Osborne/McGraw-Hill, Berkeley, CA, 1990.

[429]   J. Markoff, "The Microprocessor's Impact on Society," *IEEE Micro*, Vol. 16, No. 6, 1996, pp. 54-59.

[430]   M. A. Marson, G. Balbo, and G. Conte, *Performance Modules of Multiprocessor Systems*, MIT Press, Cambridge, MA, 1988.

[431]   R. P. Martin et al., "Effects of Communication Latency, Overhead, and Bandwidth in a Cluster Architecture," *Proc. of the 24th Int'l. Symp. Computer Architecture,* June 1997, pp. 85-96.

[432]   J. R. Mashey, "From Buses to Modular, Distributed Crossbars: Not All Shared-Memory Systems Are Equal," Silicon Graphics, 1997. Available at http://www.sgi.com/Prooducts/hardware/servers/techtalk.html.

[433]   T. G. Mattson, D. Scott, and S. Wheat, "A TeraFLOPS Supercomputer in 1996: The ASCI TFLOPS System," *Proc. of the 6th Int'l. Parallel Processing Symp.*, 1996, pp. 84-93.

[434]   C. May et al. (Eds.), *The PowerPC Architecture: A Specification for a New Family of RISC Processors,* Morgan Kaufmann, San Francisco, 1994.

[435]   D. Matzke, "Will Physical Scalability Sabotage Performance Gains?" *IEEE Computer Magazine*, Sept. 1997, Vol.30, No. 9, pp. 37-39.

[436] O. McBryan, "An Overview of Message Passing Environments," *Parallel Computing*, Vol. 20, No. 4, 1994, pp. 417-444.

[437] J. D. McCalpin,"STREAM: Sustainable Memory Bandwidth in Recent and Current High Performance Computers". An on-line technical report, available at http://www.cs.virginia.edu/stream/ or http://reality.sgi.com/mccalpin/papers/bandwidth/bandwidth.html. 1995.

[438] J. D. McCalpin, "Memory Bandwidth and Machine Balance in Current High Performance Computers," *IEEE Technical Committee on Computer Architecture Newsletter*, December 1995.

[439] P. K. McKinley, Y.-J. Tsai, and D.F. Robinson, "Collective Communication in Wormhole-Routed Massively Parallel Computers," *IEEE Computer*, 1995, 28(12): 39-50.

[440] L. McVoy and C. Staelin, "lmbench: Portable Tools for Performance Analysis," *Proc. of the 1996 USENIX Technical Conference*, San Diego, CA, January 1996, pp. 279-295.

[441] J. M. Mellor-Crummey and M. L. Scott, "Algorithms for Scalable Synchronization on Shared Memory Multiprocessors", *ACM Trans. Computer Systems*, 1991, 9(1): 21-65.

[442] Message Passing Interface Forum, *MPI: A Message-Passing Interface Standard*, Version 1.1, June 12, 1995, ftp://ftp.mcs.anl.gov/pub/mpi/mpi-1.jun95/mpi-report.ps, August 1996.

[443] P. Messina and T. Sterling (Eds.), *System Software and Tools for High Performance Computing Environment*, SIAM, Philadelphia, 1993.

[444] R. Metcalfe and D. Boggs, "Ethernet: Distributed Packet Switching for Local Area Networks," *Comm. of ACM*, Vol. 19, No. 7, 1976, pp. 711-719.

[445] MHPCC, "MHPCC 400-Node SP2 Environment", Maui High-Performance Computing Center, Maui, HI, August 1995 (http://www.mhpcc.edu).

[446] Microsoft, *Wolfpack for Windows NT Servers,* Web: http://www.microsoft.com, 1997.

[447] E. Miller and R. Katz, "Input/Output Behavior of Supercomputing Applications," *Proc. Supercomputing '91*, pages 567-576. ACM, 1991.

[448] H.D. Mills, "Structured Programming: Retrospect and Prospect," *IEEE Software,* 1986, 3(6):58-66.

[449] R. Milner, *Communication and Concurrency,* Prentice-Hall, Englewood Cliffs, N.J., 1989.

[450] S. Mirapuri, M. Woodacre, and N. Vasseghi, "The MIPS R4000 Processor," *IEEE Micro,* Vol. 12, No. 2, April 1992, pp. 10-22.

[451] G. E. Moore, "Can Moore's Law Continue Indefinitely?," *Computerworld*, July 1996.

[452] C. Morin, A. Gefflaut, M. Banatre, and A.-M. Kermarrec, "COMA: An Opportunity for Building Fault-Tolerant Scalable Shared Memory Multiprocessors," *Proc. of the 23rd Int'l. Symp. on Computer Architecture*, 1996, pp. 56-65.

[453] J. H. Morris, M. Satyanarayanan, M.H. Conner, J.H. Howard, and D.S.H. Rosenthal, "Andrew: A Distributed Personal Environment," *Comm. ACM*, Vol. 29, No. 3, 1986, pp. 184-201.

[454] T. Mowry and A. Gupta, "Tolerating Latency through Software-Controlled Prefetching in Shared-Memory Multiprocessors," *Journal of Parallel and Distributed Computing*, June 1991, pp. 87-106.

[455] MPI Forum, "MPI: A Message-Passing Interface Standard," *Int'l. Journal of Supercomputer Applications*, Vol.8, March 1994.

[456] S. S. Muchnick, "Optimizing Compilers for SPARC," *Sun Technology,* Summer, 1988, pp. 64-77.

[457] T. N. Mudge, J. P. Hayes, and D. D. Winsor, "Multtiple Bus Architectures", *IEEE Computer*, Vol. 20, No. 6, 1987, pp. 42-49.

[458] S. S. Mukherjee, B. Falsafi, M.D. Hill, and D.A. Wood, "Coherent Network Interfaces for Fine-Grain Communication," *Proc. of the 23rd Int'l. Symp. on Computer Architecture*, 1996, pp. 247-257.

[459] H. S. Murayama et al., "A Study of High-Performance Communication Mechanism for Multicomputer Systems," *10th Int'l. Parallel Processing Symp.*, Honolulu, Hawaii, April, 1996, pp. 76-83.

[460] R. Nelson, D. Towsley, and A.N. Tantawi, "Performance Analysis of Parallel Processing Systems," *IEEE Trans. Software Engineering*, April 1988, pp. 532-540.

[461] R. Netzer and J. Xu, "Necessary and Sufficient Conditions for Consistent Global Snapshot," *IEEE Trans. on Parallel and Distributed Systems*, Vol. 6, No. 2, 1994, pp. 165-169.

[462]   N. Nevin, "The Performance of LAM 6.0 and MPICH 1.0.12 on a Workstation Cluster," *Technical Report,* OSC-TR-1996-4, Ohio Supercomputer Center, 1996.

[463]   L. M. Ni and K. Hwang, "Optimal Load Balancing in a Multiple Processor System with Many Job Classes," *IEEE Trans. on Software Engineering,* May 1985, pp. 491-496.

[464]   L. M. Ni, "A Layered Classification of Parallel Computers," *Proc. 1991 Int'l. Conf. for Young Computer Scientists*, Beijing, China, May 1991, pp. 28-33.

[465]   L. M. Ni and P. K. McKinley, "A Survey of Wormhole Routing Techniques in Direct Networks," *IEEE Computer,* Vol. 26, No. 2, Feb. 1993, pp. 62-76.

[466]   A. Nicolau and J. A. Fisher, "Measuring the Parallelism Available for Very Long Instruction Word Architectures," *IEEE Trans. on Computers*, Nov. 1984, pp. 968-976.

[467]   R. S. Nikhil, *Tutorial Notes on Multithreaded Architectutre,* presented at *the 19th Annual Int'l. Symp. Computer Architecture (ISCA)*, Austrialia, 1992. Contact DEC Cambridge Laboratory, Bldg. 700, 1 Kendall Square, Cambridge, MA.

[468]   B. Nitzberg and V. Lo, "Distributed Shared Memory: A Survey of Issues and Algorithms," *IEEE Computer,* Vol. 24, No. 8, August 1991, pp. 52-60.

[469]   NOW Project, "Efficient and Portable Implementation of MPI using Active Messages," University of California, Berkeley (http://now.cs.berkeley.edu/Fastcomm/mpi.html), July 10, 1996.

[470]   A. G. Nowatzyk, M.C. Browne, E. J. Kelly, and M. Parking, "S-Connect: From Networks of Workstations to Supercomputer Performance," *Proc. of the 22nd Annual Int'l. Symp.on Computer Architecture*, 1995, pp. 71-82.

[471]   NSF, "Grand Challenges: High Performance Computing and Communications," A Report by the Committee on Physical, Mathematical and Engineering Sciences, NSF/CISE, 1800 G Street NW, Washington, DC 20550, 1991.

[472]   N. Nupairoj and L. M, Ni, "Performance Evaluation of Some MPI Implementations on Workstation Clusters," *Proc. of the Scalable Parallel Libraries Conf.,* Oct. 1994.

[473]   D. Nussbaum and A. Agarwal, "Scalability of Parallel Machines," *Comm. of the ACM,* 34(3): 56-61, March 1991.

[474]   Open Software Foundation, *OSF DCE Application Environment Specification*, Prentice-Hall, Englewood Cliffs, N.J., 1992.

[475]   OpenMP Standards Board, *OpenMP: A Proposed Industry Standard API for Shared Memory Programming*, October 1997, http://www.openmp.org/openmp/mp-documents/paper/paper.html

[476]   OpenMP Standards Board, *OpenMP Fortran Application Program Interface Version 1.0*, October 1997, http://www.openmp.org/openmp/mp-documents/fspec.pdf

[477]   Oracle Corporation, *Oracle Parallel Server in the Digital Environment*, Oracle Corporation White Paper, Part A19242, June 1994.

[478]   J. K. Ousterhourt et al., "The Sprite Network Operating System," *IEEE Computer*, Vol. 21, No. 2, Feb. 1988, pp. 23-26.

[479]   D. A. Padua, D. J. Kuck, and D. H. Lawrie, "High-Speed Multiprocessors and Compilation Techniques," *IEEE Trans. on Computers*, pp. 763-776, Sept. 1980.

[480]   S. Pakin, V. Karamcheti, and A. Chien, "Fast Messages: Efficient Portable Communication for Workstation Clusters and Massively-Parallel Processors," *IEEE Concurrency*, Vol. 5, No. 2, 1997, pp. 60-73.

[481]   C. M. Pancake, "Software Support for Parallel Computing: Where Are We Headed?" *Comm. of the ACM*, Vol. 34, No. 11, 1991, pp. 53-64.

[482]   C. M. Pancake and D. Bergmark, "Do Parallel Languages Respond to the Needs of Scientific Programmers?" *IEEE Computer*, Vol. 23, Dec. 1990, pp. 13-23.

[483]   C. M. Pancake, M.L. Simmns, and J.C. Yan, "Performance Evaluation Tools for Parallel and Distributed Systems," *IEEE Computer*, Vol. 28, Nov. 1995, pp. 16-19.

[484]   D. K. Panda and K. Hwang, "Fast Data Multipulation in Multiprocessors Using Parallel Pipelined Memories," *J. Parallel and Distributed. Computing*, Vol. 12, June 1991, pp. 130-145.

[485]   D. K. Panda and C. B. Stunkel (Eds.) *Communication and Architectural Support for Network-Based Parallel Computing*, Spinger-Verlag, Heidelberg, Germany, 1997.

[486]   M. S. Papamaroos and J.H. Patel, "A Low-Overhead Coherence Solution for Multiprocessor with Private Cache Memories", *Proc. of the 11th Annual Symp. Computer Achitecure*, June 1984, pp. 348-354.

[487]   D. B. Papworth, "Tuning the Pentium Pro Microarchitecture," *IEEE Micro*, Vol. 16, No. 2, 1996, pp. 8-15.

[488]   Parallel Computing Forum, "PCF: Parallel Fortran Extensions," *Fortran Forum*, Vol. 10, No. 3, September 1991.

[489]   C. Partridge, *Gigabit Networking*, Addison-Wesley, Reading, Mass., 1994.

[490]   D. M. Pase, T. MacDonald, and A. Meltzer, "The CRAFT Fortran Programming Model," *Scientific Programming*, Vol. 3, 1994, pp. 227-253.

[491]   Y. N. Patt, S. J. Patel, M. Evers, D. H. Friendly, and J. Stark, "One Billion Transistors, One Uniprocessor, One Chip," *IEEE Computer Magazine*, Vol.30, No. 9, Sept. 1997, pp. 51-57

[492]   D. Patterson and J.L. Hennessy, *Computer Organization and Design: The Hardware/Software Interface*, Morgan Kaufmann, San Francisco, 1994.

[493]   D. Patterson and C. Sequin, "A VLSI RISC," *IEEE Computer,* Vol. 15, Sept. 1982.

[494]   B. Pawlowski, C. Juszczak, P. Staubach, C. Smith, D. Lebel, and D. Hitz, "NFS Version 3 Design and Implementation," *Proc. of the 1994 Summer USENIX Conference*, USENIX Association, Berkeley, CA, June 1994.

[495]   A. Peleg, "MMX Technology Extension to the Intel Architecture," *IEEE Micro*, Vol. 16, No. 4, 1996, pp. 42-50.

[496]   S. Petri and H. Langendorfer, "Load Balancing and Fault Tolerance in Workstation Clusters Migrating Groups of Communicating Processes," *Operating Systems Review*, Vol. 29, No. 4, Oct. 1995, pp. 25-36.

[497]   G. F. Pfister, *In Search of Clusters,* Prentice-Hall PTR, Upper Saddle River, NJ, 1995.

[498]   G. F. Pfister, "Clusters of Computers: Characteristics of an Invisible Architecture," Keynote address presented at *IEEE Int'l. Parallel Processing Symp.*, Honolulu, April 1996.

[499]   G. F. Pfister and V.A. Norton, "Hot Spot Contention and Combining in Multistage Interconnection Networks," *Proc. Int''l. Conf. Parallel Processing*, 1995, pp. 790-797.

[500]   G. F. Pfister, W. C. Brantley, D. A. George, S. L. Harey, W. J. Kleinfelder, K. P. McAuliffe, E. A. Melton, V. A. Norlton, and J. Weiss, "The IBM Research Parallel Processor Prototype (RP3): Introduction and Architecture," *Proc. Intl Conf. on Parallel Processing*, 1985, pp. 764-771.

[501]   J.A. Piantenosi, A. S. Sathaye, and D. J. Shakshober, "Performance Measurement of TruCluster Systems under the TPC-C Benchmark," *Digital Technical Journal*, Vol. 8, No. 3, 1996, pp. 46-57.

[502]   P. Pierce and G. Regnier, "The Paragon Implementation of the NX Message Passing Interface," *Proc. of the Scalable High-Performance Computing Conference*, May 1994, pp. 184-190.

[503]   J. S. Plank, M. Beck, G. Kingsley, and K. Li, "Libckpt: Transparent Checkpointing under Unix," *Proc. of 1995 Winter USENIX Technical Conference*, pp. 213-224.

[504]   J. S. Plank and K. Li, "Faster Checkpointing with $N+1$ Parity," *Proc. of the 24th Int'l. Symp. on Fault Tolerant Computing*, 1994, pp. 288-297.

[505]   J. S. Plank, J. Xu, and R. Netzer, "Compressed Differences: An Algorithm for Fast Incremental Checkpointing," *Tech. Report CS-95-302*, University of Tennessee, 1995.

[506]   R. Ponnusamy, J. Saltz, A. Choudhary, Y.-S. Hwang, and G. Fox, "Runtime Support and Compilation Methods for User-Specified Data Distributions," *IEEE Trans. on Parallel and Distributed Systems*, August, 1995.

[507]   R. Ponnusamy, J. Saltz, A. Choudhary, Y.-S. Hwang, and G. Fox, "Supporting Irregular Data Distributions in FORTRAN 90D/HPF Compilers," *IEEE Parallel and Distributed Technology*, Spring 1995.

[508] F. P. Preparata and J. Vuillemin, "The Cube-Connected Cycles: A Versatile Network for Parallel Computation," *Comm. of the ACM*, Vol. 24, 1981, pp. 300-309.

[509] J. Protic, M. Tomaševic and V. Milutinovic (Eds.), *Distributed Shared Memory: Concepts and Systems*, IEEE Computer Society Press, August 1997.

[510] S. Przybylski, *Cache and Memory Hierarchy Design,* Morgan Kaufmann, San Mateo, CA, 1990.

[511] M. Quinn, *Parallel Computing: Theory and Practice*, McGraw-Hill, New York, 1994.

[512] M. J. Quinn and P. J. Hatcher, "Data-Parallel Programming on Multicomputers," *IEEE Software*, 7(5): 69-76, Sept. 1990.

[513] J. Rattner, "Desktops and TeraFLOP: A New Mainstream for Scalable Computing," *IEEE Parallel and Distributed Technology,* August 1993, pp. 5-6.

[514] D. A. Reed, R. A. Aydt, R. J. Noe, P. C. Roth, K. A. Shields, B. W. Schwartz, and L. F. Tavera, "Scalable Performance Analysis: The Pablo Performance Analysis Environment," *Proc. Scalable Parallel Libraries Conf.*, 1993, pp. 104-113.

[515] S. K. Reinhardt, M.D. Hill, J.R. Larus, A.R. Lebeck, J.C. Lewis, and D.A. Wood,"The Wisconsin Wind Tunnel: Virtual Prototyping of Parallel Computers," *Proc. of the 1993 ACM SIGMETRICS Conf.* 1993, pp. 48-60.

[516] S. K. Reinhardt, R.W. Pfile, and D.A. Wood, "Decoupled Hardware Support for Distributed Shared Memory," *Proc. of the 23rd Int'l. Symp. on Computer Architecture*, 1996.

[517] M.C. Rinard, D.J. Scales, and M.S. Lam, "Jade: A High-Level, Machine-Independent Language for Parallel Programming," *IEEE Computer,* Vol. 26, June 1993, pp. 28-38.

[518] E. Roberts, *"Gigabit Ethernet: Weighed Down by Doubts,"* *Data Communications*, November 1996.

[519] M. Rosenblum, E. Bugnion, S.A. Herrod, E. Witchel, and A. Gupta, "The Impact of Architectural Trends on Operating System Performance," *Proc. of the 15th ACM Symp. on Operating Systems Principles*, December 1995, pp. 285-298.

[520] M. Rosenblum, J. Chapin, D. Teosdosiu, S. Devine, T. Lahiri, and A. Gupta, "Implementing Efficient Fault Containment for Multiprocessors". *Comm. of the ACM*, Vol. 39, Sept. 1996, pp. 52-61.

[521] R. H. Saavedra, D. E. Culler, and T. von Eicken, "Analysis of Multithreaded Architecture for Parallel Computing," *Proc. of ACM Symp. Parallel Algorithms and Architecture*, Greece, July 1990.

[522] R .H. Saavedra, W. Mao and K. Hwang, "Performance and Optimization of Data Prefetching Strategies in Multiprocessors," *Journal of Parallel and Distributed Computing,* Sept. 1993, pp. 427-448.

[523] R. H. Saavedra and A.J. Smith, "Analysis of Benchmark Characteristics and Benchmark Performance Prediction," *ACM Trans. on Computer Systems*, Vol. 14, No. 4, 1996, pp. 344-384.

[524] S. Sahni and V. Thanvantri, "Performance Metrics: Keeping the Focus on Runtime," *IEEE Parallel and Distributed Technology,* Spring 1996, pp. 43-56.

[525] S. Saini and D.H. Bailey, "NAS Parallel Benchmark Results 12-95," *Technical Report* NAS-95-021, NASA Ames Research Center, Dec. 1995.

[526] M. Santayaraman, "Scalable, Secure, and Highly Aavailable Distributed File Access," *IEEE Computer*, Vol. 23(5), May 1990, pp. 9-21.

[527] R. Sandberg, D. Goldberg, S.R. Kleiman, D. Walsh, and B. Lyon, "Design and Implementation of the Sun Network Filesystem," *Proc. of the Summer 1985 USENIX Technical Conference*, 1985, pp. 119-130.

[528] W. Saphire, L.A. Tanner, and B. Traversat, "Job Management Requirements for NAS Parallel Systems and Clusters," *Technical Report NAS-95-006*, NASA Ames Research Center, 1995.

[529] W. Saphire, A. Woo, and M. Yarrow; "The NAS Parallel Benchmarks 2.1 Results," NASA Ames Research Center, *Technical Report NAS-96-010*, NASA Ames Research Center, August, 1996.

[530] S. Saunders (Ed.), *The McGraw-Hill High-Speed LANs Handbook,* McGraw-Hill, New York, 1996.

[531] J. Singh, W.-D. Weber, and A. Gupta, "SPLASH: Stanford Parallel Applications for Shared Memroy," ACM SIGARCH *Computer Architecture News* Vol. 20(1), March 1992, pp. 5-44.

[532] R. Sites, "Alpha AXP Architecture," *Comm. of the ACM,* Vol. 36, No. 2, Feb. 1993, pp. 33-44.

[533] J. E. Smith and S. Vajapeyam, "Trace Processors Moving to Fourth Generation Microarchitectures," *IEEE Computer*, Vol.30, No. 9, Sept. 1997, pp. 68-74.

[534] M. Syanarayanan, "Commercial Multiprocessign Systems," *IEEE Computer*, May 1980, pp. 75-96.

[535] K .E. Schauser and C.J. Scheiman, "Active Messages Implementations for Meiko CS-2," Department of Computer Science, University of California, Santa Barbara, 1994.

[536] C. Scheurich, *Access Ordering and Coherence in Shared-Memory Multiprocessors*, Ph.D. Thesis, University of Southern California, 1989.

[537] C. Schimmel, *UNIX Systems for Modern Architectures: Symmetric Multiprocessing and Caching for Kernel Programmers*, Addison-Wesley Pub. Co., Menlo Park, CA. 1994.

[538] J. T. Schwartz, "Ultra-Computers," *ACM Trans. Prog. Lang. and Systems*, Vol. 2, April, 1980, pp. 484-521.

[539] S. L. Scott, "The GigaRing Channel," *IEEE Micro*, Feb. 1996, pp. 27-34.

[540] S. L. Scott, "Synchronization and Communication in the T3E Multiprocessor," *Proc. of ASPLOS 7*, Oct. 1996, pp. 26-36.

[541] S. L. Scott and G. Thorson, "The Cray T3E Network: Adaptive Routing in a High Performance 3D Torus," *Hot Interconnects IV*, August, 1996.

[542] SDSC, "SDSC's Intel Paragon," San Diego Supercomputer Center, http://www.sdsc.edu/Services/Consult/Paragon/paragon.html.

[543] C. L. Seitz, "The Cosmic Cube," *Comm. of the ACM*, Vol. 28, Jan. 1985, pp. 22-33.

[544] C. L. Seitz, J. Seizovic, and W. K. Su, "The C-Programmer's Guide to Multicomputer Programming," *Technical Report CS-TR-88-1,* California Institute of Technology, Pasadena, CA, 1989.

[545] C. L. Seitz, Concurrent Architectures," in Suaya and Birtwistle (eds.), *VLSI and Parallel Computation*, Chapter 3, Morgan Kaufmann, San Mateo, CA, 1990.

[546] C. L. Seitz, "Mosaic C: An Experimental Fine-Grain Multicomputer," *Technical Report,* California Institute of Technology, Pasadena, CA, 1992.

[547] Silicon Graphics, *IRIS Power C User's Guide*, Silicon Graphics Computer Systems, Mounbtain View, CA. 1989.

[548] Silicon Graphics, *POWER CHALLENGEarray Technical Report*, Silicon Graphics Computer Systems, Mounbtain View, CA. 1996.

[549] Silicon Graphics, *POWER CHALLENGE Technical Report*, Silicon Graphics Computer Systems, Mountain View, CA. 1996.

[550] Silicon Graphics, *Origin 200 and Origin 2000 Technical Report*, Silicon Graphics Computer Systems, Mountain View, CA. 1997.

[551] J. Shah, *VAXclusters*, McGraw-Hill, New York, 1991.

[552] S. S. Shang and K. Hwang, "Distributed Hardwired Barrier Synchronization for Scalable Multiprocessor Clusters," *IEEE Trans. Parallel and Distributed Computing Systems*, June 1996, pp. 591-605.

[553] N. Shavit and D. Touitou, "Software Transactional Memory," *Distributed Computing*, Vol. 10, No. 2, 1997, pp. 99-116.

[554] M. S. Shur, S.M. Sze, and J.M. Xu (Eds.), "Special Issue on Present and Future Trends in Device Science and Technology," *IEEE Trans. on Electronic Devices*, Vol. 43, No. 10, 1996.

[555] H. J. Siegel, *Interconnection Networks for Large-Scale Parallel Processing: Theory and Case Studies*, McGraw-Hill, New York, 1989.

[556] H. J. Siegel et al., "Report of the Purdue Workshop on Grand Challenges in Computer Architecture for the Support of High Performance Computing," *J. Parallel and Distributed Computing*, Vol. 16, No. 3, 1992, pp. 199-211.

[557] D. P. Siewiorek and P. J. Koopman, *The Architecture for Supercomputers, TI-TAN: A Case Study*, Academic Press, New York, 1991.

[558] P. S. Sindhu, J. M. Frailong, and M. Cekleov, "Formal Specification of Memory Modules," in *Scalable Shared-Memory Multiprocessors,* Kluwer Academic Publishes, Boston, MA, 1992.

[559]   J. Singh, J. L. Hennessy, and A. Gupta, "Scaling Parallel Programs for Multiprocessors: Methodology and Examples," *IEEE Computer*, Vol. 26, No. 7, 1993, pp. 42-50.

[560]   J. Singh, J. L. Hennessy, and A. Gupta, "Implication of Hierachical N-Body Methods for Multiprocessor Architectures," *ACM Trans. on Computer Systems*, Vol. 13, Feb. 1995, pp. 141-202.

[561]   A. Singhal, D. Broniarczyk, F. Cerauskis, J. Price, L. Yuan, C. Cheng, D. Doblar, S. Fosth, N. Agarwal, K. Harvey and E. Hagersten, "Gigaplane: A High Performance Bus for Large SMPs," *Symposium Record of Hot Interconnects IV*, August 1996.

[562]   M. Singhal, "Deadlock Detection in Distributed Systems," *IEEE Computer*, Nov. 1989, pp. 37-48.

[563]   R L. Sites and R. T. Witek (Eds.), *Alpha AXP Architecture Reference Manual*, 2d ed., Digital Press, Boston, MA. 1995.

[564]   A. Skjellum, "The Multicomputer Toolbox: Current and Future Directions," *Proc. 1993 IEEE Scalable Parallel Libraries Conf.*, pp. 94-103.

[565]   A. Skjellum, P. Vaughan, and C. Roberts, "UNIFY: Interoperable MPI and PVM Programming in a Workstation-Network Environment," *MPI Developers Conference,* University of Notre Dame, June 1995.

[566]   M. Slater, "The Microprocessor Today," *IEEE Micro*, Vol. 16, No. 6, 1996, pp. 32-45.

[567]   M. D. Smith, M. Johnson, and M. A. Horowitz, "Limits on Multiple Instruction Issue," *Proc. of the Third Int'l. Conf. on Architectural Support for Programming Languages and Operating Systems*, April 1989, pp. 290-302.

[568]   A. J. Smith, "Cache Memories," *ACM Computing Survey,* pp. 473-530, Sept. 1982.

[569]   B. J. Smith, "The Quest for General-Purpose Parallel Computing," Tera Computer Company, 1990, http://www.tera.com.

[570]   B. J. Smith, "The Architecture of HEP," *Parallel MIMD Computation: The HEP Supercomputer and its Applications*, edited by J. S. Kowalik, MIT Press, 1985, pp. 41-55.

[571]   J. E. Smith, "Using Standard Microprocessors in MPPs," presented at the *Int'l. Symp. on Computer Architecture*, 1992.

[572]   J. E. Smith and S. Vajapeyam, "Trace Processor: Moving to Fourth-Generation Microarchitectures," *IEEE Computer*, Sept. 1997, pp. 68-74.

[573]   M. Smith, "Closing on Clusters: Will Wolfpack Dominate the High-Volume Windows NT Cluster Market?" *Windows NT Magazine*, Aug. 1996, also from http://www.winntmag.com/issues/Aug96/ wolfpack.htm, July 14, 1997.

[574]   M. Snir, P. Hochschild, D.D. Frye, and K.J. Gildea, "The Communication Software and Parallel Environment of the IBM SP2," *IBM Systems Journal*, Vol. 34, No. 2, 1995, pp. 205-221.

[575]   M. Snir, S.W. Otto, S. Huss-Lederman, D.W. Walker, and J. Dongarra, *MPI: The Compelete Reference*, The MIT Press, Combridge, Mass., 1996.

[576]   L. Snyder. "Type Architectures, Shared Memory, and the Corollary of Modest Potential," *Annual . Review of . Comput. Science*, Vol. 1, 1986, pp. 289-317.

[577]   G. Sohi, S. Breach, and T. N. Vijaykumar, "Multiscalar Processors," *Proc. of the 22nd Annual International Symposium on Computer Architecture*, June 1995, pp. 414-425.

[578]   S. Song, M. Denman, and J. Chang, "The PowerPC 604 RISC Microprocessor," *IEEE Micro,* Oct. 1994, pp. 8-17.

[579]   W. Stallings (Ed.), *Advances in Local and Metropolitan Networks*, IEEE Computer Society Press, 1994.

[580]   W. Stallings (Ed.), *Advances in ISDN and Broadband ISDN*, IEEE Computer Society Press, 1992.

[581]   W. Stallings, *Operating Systems* (2nd Ed.), Prentice-Hall, Englewood Cliffs, NJ., 1995.

[582]   D.A. Stamper, *FDDI Handbook: High-Speed Networking Using Fiber and Other Media,* Addison Wesley, Reading Mass., 1994.

[583]   D. Stein and D. Shah, "Implementing Lightweight Threads," *Proc. of the USENIX Summer Conference*, 1992.

[584]   P. Stenstrom, "A Survey of Cache Coherence Schemes for Multiprocessors," *IEEE Computer* 23(6):12-24, June 1990.

[585]   P. Stenstrom et al., "Comparative Performance Evaluation of Cache Coherent NUMA and COMA Architectures," *Proc. of the 19th Annual Int'l. Symp. on Computetr Architecture*, 1992, pp. 80-91.

[586]   T. Sterling, P. Merkey, and D. Savarese, "Improving Application Performance on the HP/Convex Exemplar," *IEEE Computer*, Vol. 29, No. 12, Dec. 1996, pp. 50-55.

[587]   T. Sterling, P. Messina, and P. H. Smith, *Enabling Technologies for Petaflops Computing*, MIT Press, Cambridge, Mass., 1995.

[588]   H. S. Stone, "Parallel Processing with the Perfect Shuffle," *IEEE Trans. Computers*, Vol. 20, 1971.

[589]   J. M. Stone, H. S. Stone, P. Heidelberger, and J. Turek, "Multiple Reservations and the Oklahoma Update," *IEEE Parallel and Distributed Technology,* Spring 1993, 1(4): 58-71.

[590]   H. S. Stone, *High-Performance Computer Architectures (Third Edition)*, Addison-Wesley, 1993.

[591]   T. Stricker and T. Gross, "Optimizing Memory System Performance for Communication in Parallel Computers," *Proc. of the 22nd Annual Int'l. Symp. on Computer Architecture*, 1995, pp. 308-319.

[592]   C. B. Stunkel, D. G. Shea, B. Abali, M. G. Atkins, C. A. Bender, D.G. Grice, P. Hochschild. D. J. Joseph, B. J. Nathanson, R. A. Swetz, R. F. Stucke, M. Tsao, and P. R. Varker, "The SP2 High-Performance Switch," *IBM Systems Journal,* 34(2):185-204, 1995.

[593]   Y. Sun, J. Wang, and Z. Xu, "Architectural Implications of the NAS MG and FT Parallel Benchmarks," *Proc. of the Int'l. Conference on Advances in Parallel and Distributed Computing*, March 1997, pp. 235-240.

[594]   X. H. Sun and D. Rover, "Scalability of Parallel Algorithm-Machine Combinations," *IEEE Trans. on Parallel and Distributed Systems,* May, 1994.

[595]   X. H. Sun and J. Zhu, "Performance Considerations of Shared Virtual Memory Machines," *IEEE Trans. on Parallel and Distributed Systems,* Nov. 1995.

[596]   X. H. Sun, and L. Ni, "Scalable Problems and Memory-Bounded Speedup," *Journal of Parallel and Distributed Computing*, Vol. 19, Sept. 1993, pp. 27-37.

[597]   V. S. Sunderram, "PVM: A Framework for Parallel Distributed Computing," *Concurrency: Practices and Experience*, Dec. 1990, pp. 315-339.

[598]   Sun, *SPARC Architecture Reference Manual V8*, Sun Microsystems, Inc., CA. Dec. 1990.

[599]   Sun, *SPARC-Cluster 1 Product Overview*, Sun Microsystems, Mountain View, CA. 1994.

[600]   Sun, *The Ultra Enterprise 10000 Server: Technical White Paper*, Sun Microsystems, Mountain View, CA. 1997. Available from http://www.sun.com/.

[601]   SunSoft, *Multithreaded Programming Guide*, Sun Microsystems, 1994.

[602]   D. Tabak, *Advanced Microprocessors,* McGraw-Hill, New York, 1995.

[603]   D. Tabak, *RISC Systems and Applications,* Research Studies Press, Wiley, New York, 1996.

[604]   R. Take, "A Routing Method for the All-to-All Burst on Hypercube Network," *Proc. 35th National Conf. of Information Processing Society of Japan*, 1987, pp. 151-152.

[605]   Tandem Computers Incorporated. *NonStop Himalaya Range*. Document # CD0194-0993.

[606]   C. K. Tang, "Cache Design in the Tightly Coupled Multiprocessor System", *Proc. AFIPS National Computer Conf.*, New York, June 1976, pp. 749-753.

[607]   A. S. Tannenbaum, *Distributed Operating Systems*, Prentice-Hall, N.J., 1995.

[608]   T. Tannenbaum and M. Litzkow, "The Condor Distributed Processing System," *Dr. Dobb's Journal,* Feb. 1995, pp. 40-48.

[609]   D. Teodosiu, J. Baxter, K. Govil, J. Chapin, M. Rosenblum, and M. Horowitz, "Hardware Fault Containment in Scalable Shared-Memory Multiprocessors," *Proc. of the 24th Annual Int'l. Symp. on Computer Architecture*, 1997, pp. 73-84.

[610]   S. S. Thakkar, M. Dubois, A. T. Laundrie, G. S. Sohi, D. V. James, S. Gjessing, M. Thapar, B. Delagi, M. Carlton, and A. Despain, "New Directions in Scalable Shared-Memory Multiprocessor Architectures," *IEEE Computer*, Vol. 23, No. 6, 1990, pp. 71-83.

[611] R. Thekkath and S.J. Eggers, "The Effectiveness of Multiple Hardware Contexts," *Proc. 6th Intl. Conf. on Architectural Support for Programming Languages and Operating Systems*, 1994, pp. 328-337.

[612] TMC, *The CM-5 Technical Summary*, Thinking Machines Corporation, Cambridge, MA, 1995.

[613] D. Tolmie and D. Flanagan, "HIPPI: It's Not Just for Supercomptuers Anymore," *Data Communciatiuons*, http://www.data.com/

[614] R. M. Tomasulo, "An Efficient Algorithm for Exploiting Multiple Arithmetic Units," *IBM J. Res. and Develop.*, 11(1):25-33, 1997.

[615] H. C. Torng and S. Vassiliadis, *Instruction-Level Parallel Processors,* IEEE Computer Society Press, Los Alamitos, CA. 1995.

[616] J. Torrellas, M. S. Lam, and J. L. Hennessy, "False Sharing and Spatial Locality in Multiprocessor Caches," *IEEE Trans. on Computers*, June 1994, Vol. 43, No. 6, pp. 651-663.

[617] M. Tremblay and J.M. O'Connor, "UltraSparc I: A Four-Issue Processor Supporting Multimedia," *IEEE Micro*, Vol. 16, No. 2, 1996, pp. 42-50.

[618] L. W. Tucker and A. Mainwaring, "CMMD: Active Messages on the CM-5," *Parallel Computing*, Vol. 20, 1994, pp. 481-496.

[619] D. M. Tullsen and S. J. Eggers, "Effective Cache Prefetching on Bus Based Multiprocessors," *ACM Trans. on Computer Systems*, Vol.3, Jan. 1995, pp. 57-88.

[620] L. H. Turcotte, "A Survey of Software Environments for Exploiting Networked Computing Resources," *Technical Report* MSU-EIRS-ERC-93-2, Mississippi State University, 1993.

[621] U. Vahalia, *Unix Internals: The New Frontiers*, Prentice-Hall, Englewood Cliffs, NJ., 1996.

[622] A. Vahdat, D. Ghormley, and T. Anderson, "Efficient, Portable, and Robust Extension of Operating System Functionality," *Technical Report,* Computer Science Division, UC Berkeley, Dec. 1994.

[623] L. G. Valiant, "A Bridging Model for Parallel Computation," *Comm. of ACM,* 33(8):103-111, 1990.

[624] A. Varma and C. Raghavendra, *Interconnection Networks for Multiprocessors and Multicomputers: Theory and Practice.* IEEE Computer Society Press, Los Alamitos, CA, 1994.

[625] E. F. Van De Velde, *Concurrent Scientific Computing*, Springer-Verlag, 1994.

[626] T. von Eiken, A. Basu, and V. Buch, "Low-Latency Communication over ATM Network Using Active Messages," *IEEE Micro*, Feb. 1995, pp. 46-53.

[627] T. von Eiken, A. Basu, V. Buch, and W. Vogels, "U-Net: A User-Level Network Interface for Parallel and Distributed Computing," *Proc. of 15th ACM Symp. on Operating Systems Principles*, 1995.

[628] T. von Eiken, D.E. Culler, S.G. Goldstein, and K.E. Schauser, "Active Messages, A Mechanism for Integrated Communication and Computation," *Proc. of the 19th Annual Intl. Symp. on Computer Architecture*, 1992, pp. 256-266.

[629] B. W. Wah and C. V. Ramamoorthy (Eds.), *Computers for Artificial Intelligence Processing*, Wiley, New York, 1990.

[630] R. Wahbe, S. Lucco, and T. Anderson, "Efficient Software-Based Fault Isolation," *Proc. 14th ACM Symp. on Operating Systems Principles*, Dec. 1993, pp. 203-216.

[631] E. Waingold, M. Taylor, D. Srikrishna, and V. Sarkar, "Baring It All to Software: Raw Machines," *IEEE Computer Magazine*, Vol.30, No. 9, Sept. 1997, pp. 86-93.

[632] D. Wall, "Limits in Instruction-Level Parallelism," *Proc. Arch. Support for Prog. Languages and Operating Systems*, April 1991, pp. 176-188.

[633] B. J. Walker, J. Lilienkamp, and J. Hopfield, *Open Single System Image Software for Multicomputer or MPP (Massively Parallel Processor)*, Locus Computing Corporation, 9800 La Cienega Blvd., Ingelwood CA 90301-4400, 1993.

[634] C. J. Wang, C.-L. Wang, and K. Hwang, "STAP Benchmark Evaluation of the T3D, SP2, and Paragon," *Proc. of 10th Int'l. Conf. on Parallel and Distributed Computing Systems,* New Orleans, Oct. 1-3, 1997.

[635] H. C. Wang and K. Hwang, "Multicoloring for Solving Multigrid PDE Problems on Shared-Memory Multiprocessors," *IEEE Trans. Parallel and Distributed Systems*, Nov. 1996, pp. 1195-1205.

[636]   T. M. Warschko, J. M. Blum, and W. F. Tichy, "The ParaStation Project: Using Workstations as Building Blocks for Parallel Computing," *Proc. of Int'l. Conf. on Parallel and Distributed Processing, Techniques and Applications (PDPTA'96)*, August 1996, Sunnyvale, CA, Vol I, pp. 375-386.

[637]   D. L. Weaver and T. Germond (Eds.), *SPARC Architecture Manual,* Prentice-Hall, Englewood Cliffs, N.J., 1994.

[638]   W. -D. Weber, *Scalable Directories for Cache-Coherent Shared-Memroy Multiprocessors,* Ph.D. Thesis, Stanford University, January 1993.

[639]   W. -D. Weber et al., "The Mercury Interconnect Architecture: A Cost-Effective Infrastructure for High-Performance Servers," *Proc. of the 24th Annual Intl. Symp. on Computer Architecture*, 1997, pp. 98-107.

[640]   U.Weiser, "Intel MMX Technology - An Overview," *Proc. Hot Chips Symp.*, Aug. 1996, p.142.

[641]   M. Welsh, A. Basu, and T. von Eicken, "ATM and Fast Ethernet Network Interfaces for User-level Communication," *Proc. of the Third International Symposium on High Performance Computer Architecture (HPCA)*, San Antonio, Texas, 1997.

[642]   S. R. Wheat, R. Riesen, A.B. Maccabe, D.W. van Dresser, and T.M. Stallcup, "PUMA: An Operating System for Massively Parallel Systems," *Proc. 27th Hawaii Int'l. Conf. on Systems Sciences*, Vol. II, 1994, pp. 56-64.

[643]   S. Whitney et al., "The SGI Origin Software Environment and Application Performance," *Proc. of COMPCON Spring 97,*IEEE Computer Society, Feb, 1997, pp. 165-170.

[644]   A. W. Wilson, "Hierarchical Cache/Bus Architecture for Shared-Memory Multiprocessors," *Proc. 14th Annu. Int. Symp. Computer Arch.*, pp. 244-252, 1987.

[645]   G. V. Wilson and P. Lu (Eds.), *Parallel Programming Using C++*, The MIT Press, Cambridge, Massachusetts, 1996.

[646]   I. Wladawsky-Berger, "The Power and Promise of Parallel Computing," *IBM System Journal,* 1995, 34(2):146-151.

[647]   M. Wolfe, *High-Performance Compilers for Parallel Computing*, Addison-Wesley, Redwood City, CA, 1996.

[648]   K. F. Wong and M. Franklin, "Checkpointing in Distributed Computing Systems," *Journal of Parallel and Distributed Computing*, Vol. 35, 1996, pp. 67-75.

[649]   S. C. Woo, M. Ohara, E. Torrie, J.P. Singh, and A. Gupta, "The SPLASH-2 Programs: Characterization and Methodological Considerations," *Proc. of the 22nd Annual Intl. Symp. on Computer Architecture*, 1995, pp. 24-36.

[650]   P. H. Worley, "Limits on Parallelism in the Numerical Solution of Linear PDEs," *SIAM J. Sci. and Stat. Computing*, Vol. 12, No. 1, 1991, pp. 1-35.

[651]   J. Worlton, "Characteristics of High-Performance Computers," *in Supercomputers: Directions in Technology and its Applications, National* Academy Press, 1989, pp. 21-50.

[652]   J. Xu and K. Hwang, "Heuristic Methods for Dynamic Load balancing in a Message-Passing Multicomputer," *Journal of Parallel and Distributed Computing*, Vol. 18, No. 1, May 1993, pp. 1-13.

[653]   Z. Xu and K. Hwang, "Molecule: A Language Construct for Layered Development of Parallel Programs," *IEEE Trans. on Software Engineering,* Vol. SE-15, No. 5, 1989, pp. 587-599.

[654]   Z. Xu and K. Hwang, "Language Constructs for Structured Parallel Programming," *Proc. of the Sixth Int'l. Parallel Processing Symp.*, 1992.

[655]   Z. Xu and K. Hwang, "Modeling Communication Overhead: MPI and MPL Performance on the IBM SP2 Multicomputer," *IEEE Parallel and Distributed Technology,* Spring 1996, pp. 9-23.

[656]   Z. Xu and K. Hwang, "Early Prediction of MPP Performance: SP2, T3D, Paragon Experiences," *Parallel Computing*, Oct. 1996.

[657]   Z. Xu and K. Hwang, "MPP versus Clusters for Scalable Computing," *Proc. of the 2nd Int'l. Symp. on Parallel Architectures, Algorithms, and Networks,* IEEE Computer Society Press, June, 1996, pp. 117-123.

[658]   Z. Xu and K. Hwang, "Coherent Parallel Programming in C//," *Proc. of Int'l. Conf. on Advances in Parallel and Distributed Computing*, IEEE Computer Society Press, March 1997, pp. 116-122.

[659] J.-H. Yang and J. Anderson, "A Fast, Scalable Mutual Exclusion Algorithm," *Distributed Computing*, Vol. 9, No. 1, August 1995, pp. 51-60.

[660] K.C. Yeager, "The MIPS R10000 Superscalar Microprocessor," *IEEE Micro*, Vol. 16, No. 2, 1996, pp. 28-41.

[661] P. C. Yew, N. F. Tseng, and D. Lawrie, "Distributing Hot-Spot Addressing in Large-Scale Multiprocessors," *IEEE Trans. on Computers*, pp. 388-395, April 1987.

[662] P. C. Yew and B. W. Wah (Eds.), Special Issue on Shared-Memory Multiprocessors, *J. Parallel and Distributed Computing*, June 1991.

[663] A. Yu, "The Future of Microprocessors," *IEEE Micro*, Vol. 16, No. 6, 1996, pp. 46-53.

[664] E. W. Zegura, "Architecture for ATM Switched Systems," *IEEE Communications,* Feb. 1993, 28-37.

[665] X. Zhang, R. Castaneda, and E. W. Chan, "Spin-Lock Synchronization on the Butterfly and KSR1," *IEEE Parallel and Distributed Technology,* Spring 1994, pp. 51-63.

[666] X. Zhang, Y. Yan, and R. Castaneda, "Evaluating and Designing Software Mutual Exlusion Algorithms on Shared-Memory Multiprocessors," *IEEE Parallel and Distributed Technology,* Spring 1996, pp. 25-42.

[667] S. Zhou, *LSF: Load Sharing and Batch Queueing Software*, Platform Computing Corporation, North York, Canada, 1996.

[668] S. Zhou, X. Zheng, J. Wang, and P. Delisle, "Utopia: a Load Sharing Facility for Large, Heterogeneous Distributed Computer Systems," *Software – Practice and Experience,* 23(12):1305-1336, December 1993.

[669] H. Zima and B. Chapman, *Supercompilers for Parallel and Vector Computers*, Addison-Wesley, Reading, Mass., 1991.

[670] G. Zorpetta, "The Power of Parallelism," *IEEE Spectrum*, Vol. 29, Sept.1992, pp. 28-33.

[671] R. Zucker and J. L. Baer, "A Performance Study of Memeory Consistency Models," *Proc. 19th Int'l. Symp. on Computer Architecture*, May 1992, pp. 2-12.

Web Resources List

This list of Web links is created based on public information available as of November 1997. Due to the very nature of constant changes in Web resources, some of the URLs may have been updated when you read this book.

To keep you up to date, an on-line version of this list of WWW Resources is maintained at the following Web site: **http://www.cs.hku.hk/~kaihwang/book98.html**. You can access the file directly to read more recently updated link information.

Table of Contents:

Web1. Computing Trends and Benchmarks   766
1.1. Parallel Computing Trends   766
1.2. Parallel Computer Models   766
1.3. Parallel Programming Fundamentals   766
1.4. Benchmarks and Performance   767

Web2. Enabling Technologies   768
2.1. Commodity Microprocessors   768
2.2. Systems Interconnects   770
2.3. Distributed Memory Technologies 772
2.4. Threads, Synchronization and Communication 773

Web3. Systems Architecture   774
3.1. Symmetric Multiprocessor (SMP)   774
3.2. NUMA (CC-NUMA, NCC-NUMA, COMA)   774
3.3. Clustering Technology   774
3.4. Cluster Computer Systems   776
3.5. Massively Parallel Processors (MPPs)   777

Web4. Parallel Programming   778
4.1. Shared-Variable Programming   778
4.2. Message-Passing Programming   778
4.3. Data-Parallel Programming   779
4.4. Environments and Tools   779

Web5. Other Relevant Links   781
5.1. Parallel Computing Bibliography on WWW   781
5.2. Glossary, News Groups, and Home Pages   782
5.3. Research Centers at Academic Institutions   783
5.4. Industrial Companies and Software Vendors   784
5.5. Agencies, Laboratories, Consortiums, and Organizations   785

Web1. Computing Trends and Benchmarks

1.1. Parallel Computing Trends

- TOP 500 (http://www.netlib.org/benchmark/top500.html), World's TOP 500 most powerful computing sites (at Netlib, University of Tennessee)
- HPC Modernization Program (http://www.hpcm.dren.net/), at US Department of Defense
- PetaFLOPS web site (http://www.aero.hq.nasa.gov/hpcc/petaflops/) at NASA HPCC Program (http://www.aero.hq.nasa.gov/hpcc/)
- The USA National Coordinate Office web site (http://www.hpcc.gov/) of Federal Computing, Information, and Communications Program on future technologies such as next-generation Internet, high-end computing, equitable access, and new approaches to key applications.
- Examples of parallel applications at USA HPCC Bluebook (http://www.ccic.gov/pubs/blue97/) and Syracuse NPAC (http://www.npac.syr.edu/crpc/), as well as other research centers
- Parallel Databases Introductions (http://www.npac.syr.edu//users/gcf/cps616rdbms1/), done in 1995, contains basic introduction and a comprehensive survey.
- Commercial parallel databases can be found in IBM DB2 Parallel Edition: (http://www.software.ibm.com/data/db2/db2pev12.html), in Informix Parallel Server: (http://www.informix.com/informix/products/techbrfs/servconn/21384/21384.htm), Sybase/MPP: (http://www.sybase.com/products/system11/mppdata.html), in Oracle Parallel Server: (http://www.oracle.com/products/oracle7/server/whitepapers/parallel/html/pararchit.html),
- The Data Warehousing Information Center (http://pwp.starnetinc.com/larryg/). Data warehousing and decision support technology.
- Jim Gray's home page (http://www.research.microsoft.com/research/barc/Gray) has many interesting articles on clusters, databases, and scalable servers
- Kai Hwang's home page for this book (http://www.cs.hku.hk/~kaihwang/book98.html).

1.2. Parallel Computer Models

- **Bulk Synchronous Parallel (BSP)** has several sites:
  - The Oxford BSP Library (http://www.comlab.ox.ac.uk/oucl/oxpara/bsplib.html)
  - h-relation personalized communication (http://www.umiacs.umd.edu/~dbader/3548.html) paper and code in Split-C
  - Bulk Synchronous Parallel (BSP) Repository (http://www.scs.carleton.ca/~palepu/BSP.html)
- Workshop on models of parallel computation (http://www.cs.ruu.nl/docs/wmpc/#16), Utrecht Univ. Jan. 27, 1995. PRAM FORK compiler, coarse-grained machines models
- The Concurrent Systems site (http://www.comlab.ox.ac.uk/archive/concurrent.html) at Oxford Univ. links to theoretical and formal models for parallel and distributed systems
- NHSE catalog of commercial HW/SW vendors (http://www.netlib.org/nhse/catalog.html)
- Free C/C++ sources for numerical computation. (http://gauge.phys.uva.nl:2001/c-sources.html).
- Parallel Monte Carlo simulation at Berkeley (http://dnclab.berkeley.edu/codef/may/prj.html)
- Partitioning unstructured meshes (http://www.gre.ac.uk/~wc06/papers/papers.html) at London
- Numerical recipes (http://cfata2.harvard.edu/nr/nrhome.html) has on-line book and code
- PLAPACK, InterCom (iCC), SUMMA (http://www.cs.utexas.edu/users/rvdg) at Austin

1.3. Parallel Programming Fundamentals

- Bibliography on multithreading (http://liinwww.ira.uka.de/bibliography/Os/threads.html)
- Software for MIMD computers (http://www.ccsf.caltech.edu/software.html) a list at Caltech

- National HPCC Software Exchange (**http://www.netlib.org/nse/home.html**)
- **Parallel Object Oriented Programming**
  - CC++ (**http://www.compbio.caltech.edu**) at Caltech
  - CHARM++ (**http://charm.cs.uiuc.edu**) at UIUC
  - LPARX ftp site (**ftp://cs.ucsd.edu/pub/skohn/lparx/**) at UCSD
  - Object Oriented Parallel Programming (**http://www.arc.unm.edu/workshop/oop/oop.html**) a survey at U. of New Mexico
  - pC++, Sage++ (**http://www.cica.indiana.edu/sage**) at Indiana University
- **Parallel software**
  - PIPS (**http://www.cri.ensmp.fr/~pips/**)
  - Adaptor (**http://www.gmd.de/SCAI/lab/adaptor/adaptor_home.html**)
  - D System (**http://www.cs.rice.edu/fortran-tools/DSystem/DSystem.html**)
  - EPPP (**http://www.crim.ca/Domaines_Services/APAR/Anglais/eppp.html**)
  - Omega (**http://www.cs.umd.edu/projects/omega/index.html**) (Maryland)
  - PAF (**http://www.prism.uvsq.fr/english/parallel/paf/autom_us.html**)
  - Parafrase-2 (**http://www.csrd.uiuc.edu/parafrase2/index.html**)
  - Pandore (**http://www.irisa.fr/EXTERNE/projet/pampa/PANDORE/pandore.html**)
  - Prepare (**http://www.irisa.fr/EXTERNE/projet/pampa/PREPARE/prepare.html**)
  - Polaris (**http://www.csrd.uiuc.edu/polaris/polaris.html**)
  - SUIF (**http://suif.stanford.edu**) (**Stanford**)
  - Nascent (**http://cse.ogi.edu/Sparse/nascent.html**) (OGI)
  - Vienna Fortran Compilation System (**http://www.par.univie.ac.at/project/vfcs.html**)
- **Parallel Programming Issues**
  - AIMS (**http://www.nas.nasa.gov/NAS/Tools/Projects/AIMS/**) (NASA Ames)
  - Falcon(**http://www.cc.gatech.edu/systems/projects/FALCON/**), (Georgia Tech)
  - Pablo (**http://www-pablo.cs.uiuc.edu/**) (Univ. of Illinois)
  - Paradyn (**http://www.cs.wisc.edu/~paradyn**) (Univ. of Wisconsin)
  - Poirot (**http://www.cs.uoregon.edu/~bhelm/poirot/index.html**) (Univ. of Oregon)
  - TASS (**http://www.cc.gatech.edu/computing/Architecture/projects/tass.html**) (Georgia Tech)
  - TAU(**http://www.cs.uoregon.edu:80/paracomp/tau/**) (Univ. of Oregon)
  - The Lost Cycles Toolkit for Performance Prediction (**http://www.cs.rochester.edu/u/leblanc/prediction.html**), at Univ. of Rochester

1.4. Benchmarks and Performance

- **Benchmarks Family and Repository**
  - IDEAS International Pty. Ltd at Melbourne, Australia (**http://www.ideas.com.au**), has benchmark results for TPC, SPEC, AIM, GPC, etc. Updated monthly
  - comp.benchmarks news group (**http://hpwww.epfl.ch/bench/bench.html**) has info on SPEC, molecule dynamics, and several other benchmarks
  - Netlib Repository (**http://www.netlib.org/liblist.html**), has performance data on many benchmarks: LINPACK, LAPACK, BLAS, BLACS, Livermore loops, Dhrystone, Whetstone, NAS, SPEC, Sim, etc. It contains sources for some benchmarks.
  - PDS: The Performance Database Server at Netlib with 13 benchmarks on machines from 44 companies (**http://netlib2.cs.utk.edu/performance/html/PDStop.html**).
- **Micro Benchmarks**
  - HBench-OS performance (**http://www.eecs.harvard.edu/~vino/perf/hbench/**).
  - The HINT Benchmark (**http://www.scl.ameslab.gov/scl/HINT/HINT.html**), for testing system

scalability and emphasizing memory access effects

- LMBENCH **(http://reality.sgi.com/lm/lmbench/)**, free, portable software evaluating Unix systems performance
- The Stream benchmark home page **(http://www.cs.virginia.edu/stream/)**
- Netperf Homepage **(http://www.cup.hp.com/netperf/NetperfPage.html)**, a networking benchmark by Hewlett-Packard. This page allows one to seach for networking performance numbers for different system/card/network combinations. Code can be downloaded.
- An SGI ftp site **(ftp://ftp.sgi.com/sgi/src/)**, contains Cray Research TCP benchmark NETTEST and a TCP/UDP benchmark TTCP by US Army Ballistics Research Lab
- Benchmarking Methodology Working Group **(http://www.ietf.cnri.reston.va.us/html.charters/bmwg-charter.html)** Internet Engineering Task Force **(http://www.ietf.cnri.reston.va.us/home.html)** Benchmarking Methodology for Network Interconnect Devices **(ftp://ds.internic.net/internet-drafts/draft-ietf-bmwg-methodology-01.txt)** Benchmarking Terminology for Network Interconnection Devices **(ftp://ds.internic.net/rfc/rfc1242.txt)**
- PC Benchmarks (DOS, Windows, OS/2, Unix, systems and components) **(http://www.dfw.net/~sdw/bench.html)**
- AIM Technology Homepage **(http://www.aim.com/aim.html)**, measures Unix performance
- The WPI Benchmark Suite **(http://wpi.wpi.edu/benchmarks/)** testing OS performance
- The TraceBase **(http://tracebase.nmsu.edu/tracebase.html)** for memory access traces.

- **Numerical Computing Benchmarks**
  - The NAS Parallel Benchmarks **(http://www.nas.nasa.gov/NAS/NPB/)**
  - Sequential C version of the NAS LU benchmark **(ftp://a.cs.uiuc.edu/pub/CHARM/LU.tar.gz)**
  - Perfect Benchmarks **(http://www.psc.edu/general/software/packages/perfect/perfect.html)**
  - PARKBENCH (PARallel Kernels and BENCHmarks) **(http://netlib2.cs.utk.edu/parkbench/)**
  - The SPLASH Benchmarks **(http://www-flash.stanford.edu/apps/SPLASH/)**

- **Commercial Applications Benchmarks**
  - Business Applications Performance Corp (BAPCo) Home Page, has data of SYSmark95, which rates (PC) systems in six major Windows business application areas: Word Processing, Spreadsheets, Database, Desktop Graphics, Desktop Presentation and Desktop Publishing
  - Graphics Performance Characterization committee (GPC)
  - Transaction Processing Performance Council (TPC) home page **(http://www.tpc.org/)**
  - TPC Benchmarks in a Nutshell from SGI **(http://www.sgi.com/Technology/tpc.html)**

- Parallel System Performance Projects **(http://www.cs.wm.edu/hpcs/WWW/HTML/research.html)** **(http://rabbit.cs.utsa.edu/performance.html)** at the College of William and Mary

Web2. Enabling Technologies

2.1. Commodity Microprocessors

- **Microprocessors Technology Surveys**
  - Chip directory **(http://www.hitex.com/chipdir/chipdir.htm)** and **(http://infopad.eecs.berkeley.edu/CIC/)**
  - CPU Info Center at Berkeley **(http://infopad.eecs.berkeley.edu/CIC/)**
  - Offermann's chip list **(http://einstein.et.tudelft.nl/~offerman/chiplist.html)**.
  - Great microprocessors of the past and present **(http://www.cs.uregina.ca/~bayko/cpu.html)**
  - Microcomputer History **(http://www.islandnet.com/~kpolsson/comphist.htm)**

- **(http://www.cs.umd.edu/users/fms/comp)** A review of CPU history
- Microprocessor instruction set cards **(http://www.comlab.ox.ac.uk/archive/cards.html)**
- Silicon technology roadmap **(http://www.sematech.org/public/roadmap/index.htm)**, by Semiconductor Industry Association (SIA)
- Sematech Semiconductor Dictionary **(http://www.sematech.org/member/division/its/dict/home.htm)**
- CHIPS Semiconductor Desk Reference **(http://www.marshall.com/pub/chips/chipmenu.htm)**, from Marshall Industries
- Semiconductor subway **(http://www-mtl.mit.edu/semisubway.html)**, at MIT
- Comparison table of the major embedded microprocessors **(http://infopad.eecs.berkeley.edu/CIC/embed/esum.html)**, at UC-Berkeley
- The Transputer Archival **(http://www.comlab.ox.ac.uk/archive/transputer.html)**
- **Commodity Processors (Sorted by architecture)**
  - **HaL** Computer Systems, Inc. **(http://www.hal.com/)**
  - **680x0**
    - Motorola M680x0 Processors site **(http://www.mot.com/SPS/HPESD/prod/0X0/)**
  - **80x86 and compatibles**
    - AMD PC processors site **(http://www.amd.com/products/cpg/cpg.html)**, has information on K6, K5, Am5x86, Am486, etc.
    - Centaur Technology's Products page **(http://www.centtech.com/prodinfo/welcome.html)**, has information on the IDT-C6 processor
    - Cyrix On-Line Products page **(http://www.cyrix.com/)**, has 6x86MX, 6x86, and MediaGX processors information
    - IBM x86 Microprocessors page **(http://www.chips.ibm.com/products/x86/)**, has 6x86MX, 6x86L, and 6x86 processors information
    - IMS products site **(http://www.imes.com/)**, has information on its Meta 6000 processors
    - Intel products site **(http://www.intel.com/procs/index.htm)**, has links to 80486, Pentium, Pentium Pro, and Pentium II
  - **80960**
    - Intel i960 processors home page **(http://developer.intel.com/design/i960/INDEX.HTM)**
  - **Alpha** DEC documentations and catalogs **(http://www.digital.com/catmag.html)**
    - DEC Alpha home page **(http://www.digital.com/semiconductor/alpha/alpha.htm)**
  - **ARM**
    - Advanced RISC Machines, Ltd., (ARM) home page **(http://www.arm.com/)**
    - DEC's StrongARM **(http://www.digital.com/semiconductor/strongarm/strongar.htm)**
  - **ColdFire**
    - Motorola ColdFire **(http://www.mot.com/SPS/HPESD/prod/coldfire/)** is a variable-length RISC processor designed for the embedded consumer market.
  - **Java Chips**
    - Sun JavaChips page **(http://www.sun.com/sparc/java/)**, covering picoJava, microJava, and UltraJava processors
    - IMS products site **(http://www.imes.com/)**, has information on its Meta Expresso Java chip
  - **MIPS**
    - SGI MIPS Group site **(http://www.sgi.com/MIPS/)**
    - QED products site **(http://www.qedinc.com/prodinfo.html)**
    - IDT RISC microprocessors site **(http://www.idt.com/risc/)**

- **PA (Precision Architecture)** HP E business magazine **(http://www.hp.com/Ebusiness)**
  - HP microprocessors technology page **(http://hpcc920.external.hp.com/computing/framed/technology/micropro/)**
  - HP Computing Technology page **(http://www.wsg.hp.com/wsg/strategies/strategy.html)**
- **PowerPC**
  - IBM's PowerPC microprocessors site **(http://www.chips.ibm.com/products/ppc/)**
  - Motorola PowerPC home page **(http://www.mot.com/PowerPC/)**
  - Motorola PowerPC Library **(http://www.mot.com/SPS/PowerPC/library/library.html)**
- **SPARC**
  - Sun SPARC page **(http://www.sun.com/sparc/)**
  - HAL **(http://www.hal.com/)** HAL Scalable Servers (http://www.hal.com/ssd.htm)
  - Fujitsu SPARClite Embedded Microprocessors **(http://www.fujitsumicro.com/products/embctrl/embctrl.html)**
  - Fujitsu TurboSPARC Microprocessors **(http://www.fujitsumicro.com/sparcupgrade/sparcmicro.html)**
- **Processor Research Projects**
  - **Asynchronous Logic**
    - The Amulet group **(http://www.cs.man.ac.uk/amulet/index.html)** implemented the ARM processors using asynchronous logic
    - Asynchronous Logic home page **(http://www.cs.man.ac.uk/amulet/async/index.html)**
  - **Advanced processor architecture**
    - M-Machine at MIT Kemal Ebcioglu (IBM), Joseph Fisher (Hewlett-Packard), Wen-Mei Hwu (Univ of Illinois, Urbana), Monica Lam (Stanford), Alex Nicolau (Univ of Calif, Irvine), and Bob Rau (Hewlett-Packard)
    - The IMPACT group **(http://www.crhc.uiuc.edu/Impact/index.html)** at UIUC is conducting many projects in advanced processors architecture
    - Chromatic's MpactVLIW architectures media engine **(http://www.mpact.com/Technical/specs.html)** VLIW SIMD
    - MOST project at McGill **(http://www-acaps.cs.mcgill.ca/~jwang/MOST-Web/)**
  - **VLIW architecture**
    - The IMPACT Compiler Group at Illinois **(http://www.crhc.uiuc.edu/Impact/index.html)**
    - Chromatic's MpactVLIW architectures media engine **(http://www.mpact.com/Technical/specs.html)**
    - MOST Project **(http://www-acaps.cs.mcgill.ca/~jwang/MOST-Web/)**
  - **Multithreaded Processor Architecture**
    - MIT M-Machine **(http://www.ai.mit.edu/projects/cva/cva_m_machine.html)**
    - Earth project **(http://www-acaps.cs.mcgill.ca/info/EARTH/index.html)**
    - Tera Computer Company Homepage **(http://www.tera.com/tera.html)**
    - Rishiyur S. Nikhil's Home Page on multithreading architecture **(http://www.research.digital.com/CRL/personal/nikhil/home.html)**
    - Simultaneous Multithreading **(http://www.cs.washington.edu/research/smt/)**
    - Tera Computer Company **(http://www.tera.com/tera.html)**

2.2. Systems Interconnects
- Computer and communication entry Page **(http://www.cmpcmm.com/cc/).**

- CERN high-speed interconnect Project **(http://www.cern.ch/HSI/)** links to ATM, DS/HS link, Fibre Channel, Gigabit Ethernet, HIPPI, SCI, etc.
- Essential Communications' home page **(http://www.esscom.com/)** on gigabit networking, especially HiPPI and gigabit Ethernet.
- Univ. of New Hampshire, InterOperability Lab **(http://www.iol.unh.edu/sitemap/index.html)** standard network technology, including tutorials, consortiums, and test results.
- Doorn's homepage **(http://www.wco.com/~schelto/)** Fibre Channel, HiPPI, Gigabit Ethernet.
- The advanced telecommunications program at Lawrence Livermore National Laboratory **(http://www.llnl.gov/atp/)** has pointers to leading edge high-speed networks.
- **Switch**
  - Why a switch **(http://www.austin.ibm.com/tech/whyswitch/whyswit.html)** a light-hearted discussion of IBM SMP, worth a read just for the fun!
  - DEC Gigaswitch **(http://www.networks.digital.com/)**
  - Giga Labs Ethernet and HiPPI switches **(http://www.gigalabs.com/)**
  - Myrinet **(http://www.myri.com)**
- **Bus**
  - PCI SIG Home Page **(http://www.teleport.com/~pc2/pcisigindex.html)**
  - VITA (VMEbus International Trade Association) **(http://www.vita.com/)**
- **ATM**
  - The ATM Forum **(http://atmforum.com/)**
  - The ATM Consortium **(http://www.iol.unh.edu/consortiums/atm/index.html)**
  - The ATM Web Knowledgebase **(http://www.npac.syr.edu/users/dpk/ATM_Knowledgebase/ ATM-technology.html)**
- **Ethernet, Fast Ethernet, and Gigabit Ethernet**
  - The INTERNET COMPUTER XCHANGE's Ethernet contents **(http://www.planet.net/icxc/ ethernet.html).** The site also has links to Frame Relay , ISDN, Videoconferencing, glossary and acronyms, companies, and electronic magazines.
  - Charles Spurgeon's Ethernet Web site provides 10-Mbps Ethernet (IEEE 802.3), 100-Mbps Fast Ethernet (802.3u), and the Gigabit Ethernet (802.3z) **(http://wwwhost.ots.utexas.edu/ethernet/ethernet-home.html)**
  - The Fast Ethernet Consortium **(http://www.iol.unh.edu/consortiums/fe/index.html)**
  - Gigabit Ethernet Alliance **(http://www.gigabit-ethernet.org/)**
  - IEEE 802.3z Task Force **(http://www.wco.com/~schelto/Ethernet/gigabit.htm)**
- **FDDI**
  - The Fiber Distributed Data Interface (FDDI) Consortium **(http://www.iol.unh.edu/consortiums/fddi/index.html)**
  - ANSI X3T12 (FDDI) Home Page **(http://sholeh.nswc.navy.mil/x3t12/)**
- **Fibre Channel**
  - Fibre Channel Association **(http://www.fibrechannel.com/)**
  - Fibre Channel Ass. Membership Listing **(http://www.fibrechannel.com/members.html)**
- **HiPPI**
  - HIPPI Networking Forum (HNF) **(http://www.esscom.com/hnf/index.html)**
  - HIPPI standards ctivities home page **(http://www.cic-5.lanl.gov/~det/)**
- **Scalable Coherent Interface (SCI)**
  - RD24 Project at CERN **(http://www1.cern.ch/RD24/)**
  - SCIzzL home page **(http://sunrise.scu.edu/)**
  - SCI for Dept. of Physics, Univ. of Oslo **(http://www.fys.uio.no/internat/sci/sci.html)**

- SCI home page **(http://baugi.ifi.uio.no/~sci/endex.html)**
- SCI News **(http://www.dolphinics.no/SCI_News/SCI_News.html)** at Dolphin Interconnect Solutions **(http://www.dolphinics.no/Dolphin.html)**

2.3. Distributed Memory Technology

- Compiler for DSM, Wei Li at Rochester **(http://www.cs.rochester.edu/u/wei/dsm.html)**
- Memory System Project at DEC **(http://www.research.digital.com/wrl/projects/memorySystems/MemorySys.html)**
- DSM Home Page **(http://www.cs.umd.edu/~keleher/dsm.html)** has links to dozens of pages
- DSM bibliography **(http://www.cs.ualberta.ca/~rasit/dsmbiblio.html)**
- RPM Project **(http://www.usc.edu/dept/ceng/dubois/RPM.html)**
- **Hardware DSM -- CC-NUMA**
  - Alewife **(http://cag-www.lcs.mit.edu:80/alewife/)**
  - Avalanche **(http://www.cs.utah.edu/projects/avalanche/)**
  - Cashmere **(http://www.cs.rochester.edu/u/kthanasi/cashmere.html)**
  - DICE **(http://www-mount.ee.umn.edu/~dice/)**
  - FLASH **(http://www-flash.stanford.edu/)**
  - NUMAchine at the Univ. of Toronto **(http://www.eecg.toronto.edu/EECG/RESEARCH/ParallelSys/numachine.html)**
  - S3.mp **(http://playground.sun.com/pub/S3.mp/s3mp.html)**
- **Hardware DSM -- COMA**
  - COMA-F **(http://powderkeg.Stanford.EDU:80/~ubetcha/)**
  - Data Diffusion Machine **(http://www.pact.srf.ac.uk/DDM/)**
  - I-ACOMA **(http://iacoma.cs.uiuc.edu/iacoma)**
  - Simple COMA **(http://playground.Sun.COM:80/pub/S3.mp/simple-coma/)**
  - Another Simple COMA **(http://www.sics.se/~ans/simple-coma/)**
- **Software DSM**
  - Calypso **(file://ftp.cs.nyu.edu/pub/calypso/bdk95a.ps)**
  - Cilk **(http://theory.lcs.mit.edu/~cilk)**
  - The C Region Library (CRL) **(http://www.pdos.lcs.mit.edu/crl/)** an all-software DSM system at MIT Lab. for Computer Science.
  - The Coherent Virtual Machine (CVM) **(http://www.cs.umd.edu/projects/cvm/)** (Maryland), a software DSM supporting multiple protocols, multithreading, and limited fault tolerance.
  - DiSOM **(http://cretina.inesc.pt/documents/)** (Distributed Shared Object Memory), memory consistency models, data replication, security, and optimized checkpointing algorithms.
  - Distributed Filaments **(http://www.cs.arizona.edu/people/filament/)**
  - Emerald **(http://www.diku.dk/research-groups/distlab/emerald/)**
  - GMS **(http://www.cs.washington.edu/homes/levy/gms/index.html)**
  - KOAN/Fortran-S **(http://www.irisa.fr/EXTERNE/projet/caps/HTML/Fortran-S_US.html)**
  - Larchant **(http://www-sor.inria.fr/SOR/projects/larchant.html)**
  - Locust **(http://www.cs.sunysb.edu:80/~manish/locust/)**
  - Midway **(http://www-cgi.cs.cmu.edu/afs/cs.cmu.edu/project/midway)**
  - Millipede **(http://www.cs.technion.ac.il/Labs/Millipede)**
  - Mirage **(http://olympia.ucr.edu/)**
  - ORCA **(http://www.cs.vu.nl//vakgroepen/cs/orca.html)**
  - PAMS **(http://www.myrias.ab.ca/)**

- Phosphorus (**http://www-inf.enst.fr/~demeure/hosphorus/contents.html**)
- Quarks (**http://www.cs.utah.edu:80/projects/flexmach/quarks.html**) (Utah), user-level library that supports DSM on a Unix workstation cluster.
- SAM (**http://suif.stanford.edu/~scales/sam.html**)
- SHRIMP (**http://www.cs.Princeton.EDU:80/shrimp/html/shared_memory.html**)
- Treadmarks (**http://softlib.rice.edu/CITI/comp~sys/software/treadmarks.html**)
- Vote (**http://www.first.gmd.de/vote**) (GMD First) provides runtime support for both message-passing and shared memory communication within a single address space.
- Wind Tunnel (**http://www.cs.wisc.edu/~wwt**) at Univ. of Wisconsin
- **Latency Tolerance**
  - Locality Management in Large-Scale Multiprocessors (**http://www.cs.rochester.edu/users/faculty/scott/locality.html**), at University of Rochester, has papers and codes

2.4. Threads, Synchronization and Communication

Threads

- pthreads (**file://ftp.cs.fsu.edu/pub/PART**) POSIX Threads from Florida State University
- Some documents of pthreads at MIT (**ftp://rtfm.mit.edu/pub/pthreads**)
- Pthread at MIT (**http://www.mit.edu:8001/people/proven/pthreads.html**).
- OSF ftp site (**file://riftp.osf.org/pub/**) has papers on ANDF, task migration, real time, and threads, including migrating and distributed threads, and source codes
- Quick Threads (**http://www.cs.washington.edu/homes/pardo/papers.d/thread.html**) at Washington
- Pthread in IBM AIX 4.1 (**http://www.austin.ibm.com:80/powerteam/aix41/diff/df4threa.html**)
- Threads at Sun (**http://www.sun.com/sunsoft/Developer-products/sig/threads/index.html**)
- Documentation on Solaris2.x (**http://www.ece.uc.edu/sun-tips.html**)
- Georgia Tech Cthreads (**http://www.cc.gatech.edu/systems/projects/Cthreads/**)
- MIT Cilk (**http://theory.lcs.mit.edu:80/~cilk/**), a user-level thread library

Synchronization

- PAPERS, Purdue's Adapter for Parallel Execution and Rapid Synchronization (**http://garage.ecn.purdue.edu/~papers/Index.html**)
- Univ. of Florida Technical Reports (**http://www.cis.ufl.edu/cis/tech-reports/tr94-abstracts.html**).
- Portable Thread Synchronization Using C++ (**http://world.std.com/~jimf/c++sync.html**) classes for portable multithreaded applications on Win32 and Solaris
- High-performance synchronization papers and codes. (**http://www.cs.rochester.edu/users/faculty/scott/synchronization.html**).

Communication

- **Active Message**
  - UTK/HTML active message paper (**http://www.netlib.org/utk/papers/am/am_report.html**)
  - Berkeley Active Message page (**http://now.cs.berkeley.edu/AM/active_messages.html**)
  - Cornell Active Message page (**http://www.cs.cornell.edu/Info/Projects/CAM/**)
  - Cornell/NYNET ATM Cluster Project (**http://www.cs.cornell.edu/Info/Projects/ATM/**), has an active message implementation
  - Active Messages for SP-2 (**http://www.cs.cornell.edu/Info/Projects/CAM/sp2.html**) at Cornell
  - Active Messages, J-Machine (**http://www.ai.mit.edu/people/ellens/Papers/ppopp95-abs.html**)
  - Active Messages for the Meiko CS-2 (**http://www.cs.ucsb.edu/TRs/TRCS94-25.html**)

- Illinois Fast Messages **(http://www-csag.cs.uiuc.edu/projects/communication/sw-messaging.html)**
- PUMA/SUNMOS **(http://www.cs.sandia.gov/~rolf/puma/puma.html)**
- SHRIMP Base Library **(http://www.cs.princeton.edu/shrimp/htMan/SBLmodel.html)**
- SHRIMP VMMC **(http://www.cs.Princeton.EDU:80/shrimp/html/communication.html)**
- U-Net **(http://www2.cs.cornell.edu/U-Net/Default.html)**

Web3. Systems Architecture

3.1. Symmetric Multiprocessor (SMP)

- Intel's Multiprocessor Specification **(http://www.intel.com/IAL/processr/mpovr.html)**
- Digital AlphaServer Family **(http://www.digital.com/alphaserver/)**
- Digital Technical Journal **(http://www.digital.com/info/DTJ/)** has .ps and .txt papers
- HP Enterprise Server page **(http://www.hp.com/gsy/products.html)**
- Integrated Computing Engines, Inc. **(http://www.iced.com/)** maker of the Desktop RealTime Engine
- IBM Server page **(http://www.rs6000.ibm.com/hardware/#eservers)**
- SGI Power Challenge white paper **(ftp://ftp.sgi.com/sgi/whitepaper/challenge_paper.ps.Z)**
- SGI Tech Center **(http://www.sgi.com/Technology/tech_center.html)**
- Sun SMP Architectures **(http://www.sun.com/smi/ssoftpress/books/Catanzaro/Catanzaro.html)**
- Sun Server page **(http://www.sun.com/servers/)**
- The Cray Research system page **(http://www.cray.com/products/systems/)**

3.2. NUMA Systems

CC-NUMA Systems

- HP/Convex Exemplar Servers **(http://www.hp.com/wsg/products/servers/servhome.html)**
- Data General's Numaliine servers **(http://www.dg.com/numaliine/)**
- Sequent NUMA-Q **(http://www.sequent.com/numaq/technology/)**
- SGI/Cray Origin 2000 **(http://www.sgi.com/Products/hardware/servers/index.html)**

NCC-NUMA System

- Cray T3E **(http://www.cray.com/products/systems/crayt3e/)**

3.3. Clustering Technology

Checkpoint, Recovery, Process Migration

- Kai Li **(http://www.cs.princeton.edu/~li/)**
- Rob Netzer **(http://www.cs.brown.edu/people/rn/home.html)**
- James S. Plank (libckpt, etc.) **(http://www.cs.utk.edu/~plank)**
- The Warp system **(http://warp.dcs.st-and.ac.uk/warp/systems/checkpoint/)**
- The MIST PVM checkpointer **(http://www.cse.ogi.edu/DISC/projects/mist/)**
- Consistent Checkpoints (CoCheck)
  (http://wwwbode.informatik.tu-muenchen.de/~stellner/CoCheck.html)
- Condor **(http://www.cs.wisc.edu/condor/)**

- Checkpointing and recovery references (**http://alphamor.zdv.uni-tuebingen.de/PARALLEL/cpft/**)
- Manetho (**http://www.cs.cmu.edu/afs/cs.cmu.edu/user/mootaz/ftp/html/manetho.html**), has a 1996 survey on rollback recovery
- OSF ftp site (**file://riftp.osf.org/pub/**) has papers on ANDF, task migration, real time, and threads, including migrating and distributed threads, and source codes
- OSF Mobile Objects and Agents (MOA) Project (**http://www.osf.org/RI/java/moa/**)
- Survey of Checkpointing and Rollback Techniques (**ftp://ftp.esat.kuleuven.ac.be/pub/ACCA/FTMPS/REPORTS/f93-04.ps.Z**)
- Overview of Checkpoint (**http://warp.dcs.st-and.ac.uk/warp/systems/checkpoint/**)
- OGI MIST Project (**http://www.cse.ogi.edu/DISC/projects/mist/**)

Single System Image and Metacomputing

- The AppLes project (**http://www-cse.ucsd.edu/groups/hpcl/apples/apples.html**)
- Berkeley WebOS (**http://now.CS.Berkeley.EDU/WebOS/**)
- Broadway (**http://www.camb.opengroup.org/tech/desktop/x/**)
- The GLOBE Project (**http://www.cs.vu.nl/~steen/globe/**)
- The Globus project (**http://www.globus.org/**)
- GLUnix (**http://now.cs.berkeley.edu/Glunix/glunix.html**)
- GOST: Global Operating System Technologies (**http://gost.isi.edu/gost-group/projects/gost/**)
- The Legion Project (**http://www.cs.virginia.edu/~legion/**)
- The NetSolve Project (**http://www.cs.utk.edu/~casanova/NetSolve/**)
- Solaris MC (**http://www.sun.com/960201/cover/solaris-mc.html**)
- Another Solaris MC Page (Research at Sunlabs) (**http://www.sunlabs.com/research/solaris-mc/**)

File Systems

- IBM SP's GPFS (**http://www.rs6000.ibm.com/software/sp_products/gpfs.html**) and PIOFS (**http://www.rs6000.ibm.com/software/sp_products/piofs.html**)
- The HA-NFS project (**http://www.cs.cmu.edu/afs/cs.cmu.edu/user/mootaz/ftp/html/hanfs.html**)
- The Alex FTP Filesystem (**http://www.ludd.luth.se/~kavli/alex.html**)
- The Berkeley xFS (**http://now.cs.berkeley.edu/Xfs/xfs.html**)
- The Bayou Project (**http://www.parc.xerox.com/csl/projects/bayou/**)
- CIFS (Common Internet File System) (**http://www.microsoft.com/intdev/cifs/**)
- The Coda project (**http://www.cs.cmu.edu/afs/cs/project/coda/Web/coda.html**)
- The File Mobility Group (**http://ficus-www.cs.ucla.edu/**)
- VERITAS (**http://www.veritas.com/**)
- WebNFS (**http://www.sun.com/webnfs/index.html**)

Workload Management (Load sharing, load balancing, batching, scheduling)

- NHSE Review: Cluster Management Software (**http://www.crpc.rice.edu/NHSEreview/CMS/**), an HTML article by Baker, Fox and Yau, reviews 20 job management software packages
- The AppLes project (**http://www-cse.ucsd.edu/groups/hpcl/apples/apples.html**)
- Prospero Resource Manager (PRM) (**http://nii-server.isi.edu/gost-group/products/prm/**) at USC ISI
- DQS (**http://www.scri.fsu.edu/~pasko/dqs.html**) or (**file://ftp.scri.fsu.edu/pub/dqs**) Distributed Queueing System, Florida State University
- NQS (**file://minnie.zdv.uni-mainz.de/pub/batch/nqs/**) NQS mirror at CERN

(ftp://shift.cern.ch/pub/NQS/)
- GENIAS Software's CODINE **(http://www.genias.de/genias_welcome.html)**
- CONDOR **(file://ftp.cs.wisc.edu/)**
- DJM **(http://s1.arc.umn.edu/html/cm-software/djm.html)**, Distributed Job Manager at Minnesota.
- IBM LoadLeveler **(http://www.rs6000.ibm.com/software/sp_products/loadlev.html)**
- Platform Computing' Load Sharing Facility (LSF) suite **(http://www.platform.com/Products/)**
- Oregon Resource Management group **(http://www.cs.uoregon.edu/research/Distributed Computing/)**, software tools for allocation, mapping, placement, scheduling and migration, and fault tolerance in heterogeneous environments.
- NASA Langley Research Center **(file://techreports.larc.nasa.gov/pub/techreports/larc/94/tm109\025.ps.Z)**
- OSF Mobile Objects and Agents (MOA) Project **(http://www.osf.org/RI/java/moa/)**
- Scheduling, batch processing, load balancing **(http://www.epm.ornl.gov/~zhou/lds/lds.html)** at Oak Ridge National Laboratory

3.4. Cluster Computer Systems

Academic Research Clusters

- Cluster Computing Conference - CCC '97 online proceedings **(http://www.mathcs.emory.edu/~ccc97/sessions.html)**
- Beowulf clusters based on commodity PC-class hardware and the Linux OS. The Stone SouperComputer page **(http://www.esd.ornl.gov/facilities/beowulf/)**
- The Berkeley NOW project **(http://now.cs.berkeley.edu/)**
- The PeralCluster for distributed multimedia and Internet applications at Univ. of Hong Kong **(http://www.cs.hku.hk/~jmma/mmcentre/background/index.html)**
- Buffalo Cluster-based Computing via ATM **(http://piranha.eng.buffalo.edu/)**
- National Scalable Cluster Project (NSCP) **(http://www_lac.eecs.uic.edu/NSCP2.html)**
- Toronto Cluster computing **(http://www.utirc.utoronto.ca/HPC/cluster.html)**

Commercial Clusters

- **Bull** offers Escala Powercluster and High-Availability Solutions (HAS) **(http://www-frec.bull.fr/OSBU/hacmp_eg.htm)** based on PowerPC nodes and AIX/OS.
- **DEC** offers clusters on three platforms
  - Digital Open VMS Clusters **(http://www.openvms.digital.com/openvms/products/clusters/)**
  - Digital UNIX clusters (TruClusters) **(http://www.unix.digital.com/cluster/)**
  - Digital Clusters for Windows NT **(http://www.windowsnt.digital.com/clusters/)**
- **HP offers**
  - MC/ServiceGuard **(http://hpcc920.external.hp.com/gsy/high_availability/)** High-Availability Cluster, including ClusterView and Lock Manager
  - Enterprise Parallel Server **(http://www.hp.com/gsy/products/eps_brief.html)**
- **IBM offers clusters for Unix and mainframe platforms**
  - S/390 Parallel Sysplex cluster **(http://www.s390.ibm.com/products/pso/psohp.html)**
  - S/390 Sysplex **(http://www.almaden.ibm.com/journal/sj36-2.html)** in IBM *Systems Journal.*
  - IBM SP **(http://www.rs6000.ibm.com/hardware/largescale/)** (formerly known as SP1 and SP2), RS6000/AIX clusters
  - SP2 user group home page at Cornell University **(http://spud-web.tc.cornell.edu/HyperNews/get/SPUserGroup.html)**

- SP2 at MHPCC (**http://www.mhpcc.edu/doc/SP2.general/SP2.general.html**)
- SP2 Special Issue in *IBM Systems Journal*, 1995
  (**http://www.almaden.ibm.com/journal/sj34-2.html**)
- HACMP (**http://www.rs6000.ibm.com/software/Apps/hacmp/**) (High Availability Cluster Multi-Processing), high availability cluster software for AIX Version 4
- IBM SP system and RS/6000 web site (**http://www.rs6000.ibm.com/**) and AIX Redbooks (**http://www.rs6000.ibm.com/resource/aix_resource/Pubs/redbooks/**).
- The IBM Scalable Parallel Systems group
  (**http://www.research.ibm.com/parallel_systems/**)
- **Microsoft develops clusters software and systems on NT platforms**
  - The Microsoft NT Server Strategy and the Wolfpack
    (**http://www.microsoft.com/syspro/technet/boes/bo/winntas/prodfact/clustrwp.htm**)
  - The SNAP (Scalable Networks and Platforms) Project in Microsoft
    (**http://www.research.microsoft.com/**)
  - Jim Gray's home page has many interesting articles on clusters, database, and scalable servers
    (**http://www.research.microsoft.com/research/barc/Gray**)
- Sequent Symmetry 5000 Cluster (**http://www.sequent.com/public/solution/server/s5000cl.html**)
- **Sun Microsystems**
  - Solaris MC (**http://www.sun.com/960201/cover/solaris-mc.html**)
  - Another Solaris MC Page (Research at Sunlabs)
    (**http://www.sunlabs.com/research/solaris-mc/**)
  - Sun Microsystems SPARCcluster
    (**http://www.sun.com/products-n-solutions/hw/servers/cluster.html**)
  - Ultra Enterprise Cluster PDB
    (**http://www.sun.com/products-n-solutions/hw/servers/product/PDB/**)
    (**http://www.sun.com/servers/ultra_enterprise/PDB/**), and Parallel Database Server
    (**http://www.sun.com/products-n-solutions/hw/servers/product/PDB/**)
  - Sun Microsystems Ultra HPC Cluster
    (**http://www.sun.com/products-n-solutions/hw/servers/product/PDB/**)
  - Sun Microsystems Enterprise Cluster (**http://www.sun.com/clusters/products.html**)
  - Sun Microsystems Full Moon Cluster Roadmap
    (**http://www.sun.com/webtone/roadmaps/fullmoonclusters.html**)
- Tandem Himalaya (**http://www.tandem.com/MENU_PGS/PROD_PGS/SERV&STR.HTM**)

3.5. Massively Parallel Processors (MPPs and HPC)

- DEC HPC site (**http://www.digital.com/info/hpc/welcome.html**)
- TOP 500 (**http://www.netlib.org/benchmark/top500.html**).World's TOP 500 most powerful computing sites (at Netlib, University of Tennessee)
- HPC Modernization Program (**http://www.aero.hq.nasa.gov/hpcc/petaflops/**)
- PetaFLOPS web site (**http://www.aero.hq.nasa.gov/hpcc/**)
- NASA HPCC Program (**http://www.aero.hq.nasa.gov/hpcc/**)
- Cray T3E (**http://www.cray.com/products/systems/crayt3e/**)
- IBM SP (**http://www.rs6000.ibm.com/hardware/largescale/**)
- Intel Paragon (**http://www.sdsc.edu/Services/ Consult/Paragon/paragon.html**)
- Tera Computer Company (**http://www.tera.com**)
- Supercomputer (**http://www.cs.cmu.edu/afs/cs.cmu.edu/project/scandal/public/www/gifs.html**)
- The Stone SouperComputer (Beowulf) (**http://www.esd.ornl.gov/facilities/beowulf/**)
- Dept. of Energy ASCI Platforms: Options Red (Sandia/Intel), Blue Pacific (LLNL/IBM), and Blue

Mountain (LANI/SGI-Cray).**(http://www.llnl.giv/asci/platforms)**

Web4. Parallel Programming

4.1. Shared-Variable Programming

- ANSI X3H5 (shared-memory model) drafts **(ftp://ftp.cs.orst.edu/standards/)**, at Oregon.
- X3H5 - Parallel Processing Constructs for High Level Programming Languages **(http://www.x3.org/tc_home/x3h5.html)**, at ANSI (inactive)
- The Orca Parallel Programming **(http://www.cs.vu.nl/vakgroepen/cs/orca.html)**
- The SR Programming Language **(http://www.cs.arizona.edu/sr/www/index.html)**
- High-Performance Synchronization for Shared-Memory Parallel Programs **(http://www.cs.rochester.edu/u/scott/synchronization.html)**, at Univ. of Rochester
- ANL Shared-Variable Macros on Solaris **(ftp://dit.lth.se/pub/sun_thread_ANL_macros)**
- PROTEUS simulator **(http://www.cs.berkeley.edu/~brewer)**
- uC++ **(http://plg.uwaterloo.ca/~pabuhr/uC++.html)**
- Ultracomputer Project **(http://cs.nyu.edu/cs/projects/ultra/)** at NYU
- Convex Exemplar Technical Information **(http://www.convex.com/tech_cache/technical.html)**
- Cray MPP Fortran Model **(ftp://ftp.cray.com/product-info/program_env/program_model.html)**
- Linda Group **(http://www.cs.yale.edu/HTML/YALE/CS/Linda/linda.html)**
- The Fortran M Programming Language **(http://www.mcs.anl.gov/fortran-m/index.html)**
- **Split-C**
  - FTP site for Split-C at UCB **(ftp://ftp.cs.berkeley.edu/ucb/CASTLE/Split-C)**
  - Split-C ERRATA **(http://www.umiacs.umd.edu/~dbader/Split-C_ERRATA.txt)**
- PAMS (Parallel Applications Management System) **(http://www.myrias.ab.ca/)**, by Myrias Computer Technologies

4.2. Message-Passing Programming

- Collective Communication Tutorial **(http://www.cs.utexas.edu/users/rvdg/tutorial.html)** at Austin
- **PVM at UTK (http://netlib2.cs.utk.edu/pvm)**
  - Another PVM Home Page at ORNL **(http://www.epm.ornl.gov/pvm)**
  - PVM Workshop **(http://www.arc.unm.edu/workshop/pvm/pvm.html)** from UNM
  - Introduction to PVM **(http://www.mhpcc.edu/training/workshop/html/pvm/PvmIntro.html)** from Maui workshop
  - PVM introduction **(http://www.eece.ksu.edu/pvm3/pvm3.html)** from KSU
- **MPI Standard** site **(http://www.mcs.anl.gov/mpi/index.html)**
  - An Introduction to the MPI Standard **(http://www.netlib.org/utk/papers/intro-mpi/intro-mpi.html)**,Oak Ridge National Laboratory
  - Introduction to MPI **(http://www.mhpcc.edu/training/workshop/html/mpi/MPIIntro.html)**
  - EPCC **(http://www.epcc.ed.ac.uk/epcc-tec/documents.html)**, on MPI, HPF, Parallel Tools Consortium, and Performance Analysis Tools for Parallel Programs
  - MPI Workshop **(http://www.arc.unm.edu/workshop/mpi/mpi.html)** from UNM
  - IBM MPL **(http://www.mhpcc.edu/training/workshop/html/mpl/MPLIntro.html)**
- **Other Message Passing Software**
  - Kent Retargetable Occam Compiler **(http://www.hensa.ac.uk/parallel/occam-for-all/kroc)**
  - Southampton Portable Occam Compiler

(**http://www.hensa.ac.uk/parallel/occam/compilers/spoc/index.html**)
- Express (**http://www.parasoft.com/express**)
- A Comparison of CPS, Linda, PVM POSYBL, and TCGMSG
  (**file://netlib2.cs.utk.edu/nse/docs/mattson_hicss.ps**)
- p4 (**file://info.mcs.anl.gov/pub/p4/**) at Argonne National Laboratory
- LAM (**http://www.osc.edu/lam.html**) at Ohio Supercomputer Center.
- CHIMP (**file://ftp.epcc.ed.ac.uk/pub/chimp**) at University of Edinburg
- Chameleon (**file://info.mcs.anl.gov/pub/pdetools**) at Agonne National Lab
- PICL (**http://netlib2.cs.utk.edu/picl**) at Oak Ridge National Lab
- TCGMSG (**file://ftp.tcg.anl.gov/pub/tcgmsg**) at Argonne National Lab
- CPS (**http://www-fermitools.fnal.gov/abstracts/cps.html**) at Fermi Nat'l Accelerator Lab
- PARA++ (**http://www.loria.fr/~coulaud/parapp.html**), C++ Bindings for Message Passing
- Interprocessor Collective Communications Library (iCC)
  (**http://www.cs.utexas.edu/users/rvdg/intercom/**) for Intel Paragon

4.3. Data-Parallel Programming

- **Fortran 90 and HPF**
  - Fortran Market (**http://www.fortran.com/fortran/market.html**) has Fortran 90 and links to a lot of Fortran stuff
  - Fortran 95 draft standard (**file://ftp.ncsa.uiuc.edu/sc22wg5/N1122/ps**)
  - Fortran 90 Frequently Asked about News (**http://lenti.med.umn.edu/~mwd/f90-faq.html**)
  - DEC Fortran 90 and HPF (**http://www.digital.com/info/hpc/f90/**) has performance comparison (PVM vs HPF, and Gigaswitch network vs AlphaServer)
  - HPFA (**http://www.npac.syr.edu/hpfa/**), High Performance Fortran Applications, Syracuse
  - High Performance Fortran Forum Home Page (**http://www.erc.msstate.edu/hpff/home.html**)
  - FTP site at Rice for HPFF (**ftp://cs.rice.edu/public/HPFF**)
  - FORGE XHPF Introduction and Exercise
    (**http://www.mhpcc.edu/training/workshop/html/xhpf/XhpfExercise.html**) at Maui
  - Applied Parallel Research, Inc. (**http://www.infomall.org/apri/**), maker of Forge, xHPF, and Shared-memory parallelizer, has HPF source code for NAS benchmarks
  - Good MPI and HPF reviews from Edinburg
    (**http://www.epcc.ed.ac.uk/epcc-tec/documents.html**)
  - Welcome to The Portland Group, Inc. (PGI) (**http://www.pgroup.com/**):compilers for C, C++, F77, HPF
  - Some benchmarks in Germany
    (**http://www.mpa-garching.mpg.de/~tomek/htmls/refs/ppm_bench.html**)
- **Non-Fortran**
  - Scandal Supercomputing Project Home Page
    (**http://www.cs.cmu.edu/afs/cs.cmu.edu/project/scandal/public/mosaic/mosaic.html**)
  - pC++/Sage++ (**http://www.extreme.indiana.edu/sage/docs.html**) at Indianna University
  - Extreme! Computing (**http://www.extreme.indiana.edu/**), at Indianna University

4.4. Parallel Programming Environments and Tools

- Software for MIMD Computers (**http://www.ccsf.caltech.edu/software.html**) a list at Caltech
- Parallel Tools Consortium (**http://www.ptools.org/**) and
  (**http://www.nhse.org/rib/repositories/ptlib/catalog/**)
- Documentation of Parallel Tools (**http://www.tc.cornell.edu/UserDoc/Software/PTools/**)
- Compilers, Analysis, Transformations

- Free Compilers/Interpreters, Berkeley (**http://remarque.berkeley.edu/~muir/free-compilers/**)
- UIUC IMPACT Research Group (**http://www.crhc.uiuc.edu/Impact/**)
- CODE Visual Parallel Programming System at Austin (**http://www.cs.utexas.edu/users/code**)
- NIST Parallel Applications Development Environment (PADE) (**http://math.nist.gov/pade/pade.html**) has doc and code
- Kuck & Associates, Inc., Home of the KAP Optimizer (**http://www.kai.com/index.html**)
- KAP for DEC Fortran and DEC C programs (**http://www.digital.com/info/hpc/kap.html**)
- CMU Fx project: (**ftp://warp.cs.cmu.edu/afs/cs.cmu.edu/project/iwarp/member/fx/public/mosaic/fx.html**)
- Adaptor (**file://ftp.gmd.de/gmd/adaptor/**) Interactive analysis and transformation tool for data parallel FORTRAN 90 programs
- Stanford SUIF Compiler (**http://suif.stanford.edu/**), Rob French (rfrench@cs.stanford.edu)
- Parallaxis (**file://ftp.informatik.uni-stuttgart.de/pub/parallaxis**)
- PCN (**file://info.mcs.anl.gov/pub/pcn/**) Parallel programming system
- Forge (**ftp://ftp.netcom.com/pub/forge/home.html**) , Applied Parallel Research Inc.
- PARADIGM (**http://www.crhc.uiuc.edu/Paradigm/**), Parallelizing Compiler for Distributed-memory General-purpose Multicomputers
- Polaris (**http://www.csrd.uiuc.edu/polaris/polaris.html**), Source Restructurer for Fortran
- Berkeley Titanium Project (**http://www.cs.berkeley.edu/~yelick/titanium/**) with a link to Bacon,Graham and Sharp's survey paper

- **Scalable I/O**
  - Parallel I/O Bibliography (**http://www.cs.dartmouth.edu/cs_archive/pario/bib.html**)
  - Parallel I/O Archive (**http://www.cs.dartmouth.edu/pario.html**) at Dartmouth
  - High Performance Storage System (HPSS) (**http://www.ccs.ornl.gov/HPSS/HPSS.html**), provides a scalable parallel storage system, based on IEEE Mass Storage Reference Model
  - MPI-IO report (**http://lovelace.nas.nasa.gov/MPI-IO/mpi-io.html**)
  - IEEE P1285 Scalable Storage Interface (**http://sunrise.scu.edu/P1285Home.html**)
  - PASSION (Parallel And Scalable Software for I/O) (**http://www.cat.syr.edu/passion.html**)

- **Debugging**
  - High Performance Debugging Forum (**http://www.ptools.org/hpdf/**)
  - Ariadne (**http://www.cs.uoregon.edu/~sameer/ariadne/**)
  - IBM SP's pdbx and xpdbx (**http://www.tc.cornell.edu/UserDoc/Software/PTools/pdbx/**)
  - Lightweight Corefile Browser (LCB) (**http://www.cs.orst.edu/~pancake/ptools/lcb/**)
  - Mantis Parallel Debugger (**http://HTTP.CS.Berkeley.EDU/projects/parallel/castle/mantis/**)
  - Rob Netzer (**http://www.cs.brown.edu/people/rn/home.html**)
  - The p2d2 project (**http://science.nas.nasa.gov/Groups/Tools/Projects/P2D2/**)
  - Panorama (**http://www-cse.ucsd.edu/users/berman/panorama.html**)
  - PRISM (**http://www.nrl.navy.mil/CCS/help/GWS/Doc/Welcome2.html**)
  - The Sneezy Project (**http://www.cs.uoregon.edu/research/paracomp/tau/sneezy/index.html**)
  - TotalView (**http://www.dolphinics.com/tw/tv37ann.htm**), see also the PTools TotalView page (**http://www.tc.cornell.edu/UserDoc/Software/PTools/totalview/**)
  - Xmdb (**http://www-c8.lanl.gov/dist_comp2/mdb/mdb.html**) A parallel debugging tool for PVM at Los Alamos

- **Performance Monitoring, Profiling, and Tuning**
  - AIMS (**http://www.nas.nasa.gov/NAS/Tools/Projects/AIMS/**)
  - Falcon (**http://www.cc.gatech.edu/systems/projects/FALCON/falcon_home.html**)
  - The Lost Cycles Toolkit for Performance Prediction (**http://www.cs.rochester.edu/u/leblanc/prediction.html**)

- Pablo (**http://www-pablo.cs.uiuc.edu/**)
- Paradyn (**http://www.cs.wisc.edu/~paradyn**)
- Parallel Performance Project in Michigan :
  (**http://www.eecs.umich.edu/~boyd/PPP/publist.html**) and
  (**http://www.eecs.umich.edu/PPP/PPP.html**)
- Poirot (**http://www.cs.uoregon.edu/~bhelm/poirot/index.html**)
- TASS (**http://www.cc.gatech.edu/computing/Architecture/projects/tass.html**)
- TAU (Tuning and Analysis Utilities) (**http://www.cs.uoregon.edu/research/paracomp/tau/**)
- **Other Software Tools**
  - Scalable Unix Commands for Parallel Computers (**http://www.mcs.anl.gov/home/lusk/ptools/**)
  - Douglas Jensen (**http://www.realtime-os.com/dresour.html**), on real-time and distributed OS
  - newlib, a C-library for embedded systems (**file://ftp.cygnus.com//pub/newlib**)
  - Public Domain RT Exec for M68K, F68KANS
    (**ftp://taygeta.oc.nps.navy.mil/pub/Forth/ANS/f68kans.zip**)
  - Run-time software tools :
    (**http://wwwbode.informatik.tu-muenchen.de/parallelrechner/tools/runtime_tools.html**)
  - The Nexus Multithreaded Runtime System (**http://www.mcs.anl.gov/nexus/**)

Web5. Other Relevant Links

5.1. Parallel Computing Bibliography on WWW

- Distributed Algorithms and/or Distributed Systems
  (**http://www.cwi.nl/cwi/departments/AA1/distcom/distcom.html**)
- The national CS TR library project at Cornell (**http://cs-tr.cs.cornell.edu/Info/cstr.html**)
- University of Rochester CS Technical Reports (**http://www.cs.rochester.edu/trs/systems-trs.html**)
- Sun Microsystems Lab Technical Reports
  (**http://www.sun.com/smli/technical-reports/index.html**)
- The CSP archive at Oxford (**http://www.comlab.ox.ac.uk/archive/csp.html**)
- Digital Technical Reports (**http://www.research.digital.com/wrl/techreports/**)
- Computer Science Bibliography at Arizona
  (**http://donkey.CS.Arizona.EDU:1994/bib/Parallel/**)
- *Parallel Computing Works* (**http://www.infomall.org/npac/pcw/**), by Geoffrey, et al.
- SGI Technical Library (**http://techpubs.sgi.com/library/**)
- Concurrent Systems links (**http://www.comlab.ox.ac.uk/archive/concurrent.html**) at Oxford
- Templates for the Solution of Linear Systems: Building Blocks for Iterative Methods
  (**http://www.netlib.org/linalg/html_templates/Templates.html**), by Barrett, et al.
- Computational Science Electronic Textbook (**http://csep1f.phy.ornl.gov:80/toc.html**)
- (Caltech) Archetypes "Electronic Textbook" Table of Contents (**http://www.etext.caltech.edu/**)
- *Designing and Building Parallel Programs* by Ian Foster
  (**http://www.mcs.anl.gov/home/toonen/book/book.html**).
- *PVM: Parallel Virtual Machine: A Users' Guide and Tutorial for Networked Parallel Computing*
  (**http://www.netlib.org/pvm3/book/pvm-book.html**) a book by Geist, et al.
- Bibliographies on Parallel Processing
  (**http://liinwww.ira.uka.de/bibliography/Parallel/index.html**)
- Parallel Computing Archive at HENSA Unix (**http://unix.hensa.ac.uk/parallel/index.html**)

- SEL-HPC Article Archive **(http://www.lpac.qmw.ac.uk/SEL-HPC/Articles/index.html)**
- The World-Wide Web Virtual Library: Distributed / MetaComputing **(http://www.dataspace.com:84/vlib/comp-distributed.html)**
- High Performance FORTRAN Educational Materials **(http://www.npac.syr.edu/EDUCATION/PUB/hpfe/index.html)**
- Parallel Tools Projects Around the World **(http://www.llnl.gov/ptools/projects.world.html)**.
- Parallelism Bibliographies **(file://unix.hensa.ac.uk/parallel/bibliographies)**
- Transputer, occam and Parallel Computing Archive **(http://unix.hensa.ac.uk/parallel/index.html)**
- Distributed Algorithms & Systems at Centre for Math. and Computer science, Amsterdam. **(http://www.cwi.nl/cwi/departments/AA1/distcom/distcom.html)**
- Reports, Journals, and Societies Related to HPCC **(http://www.netlib.org/nse/pubs.html)**
- NHSE URL list **(http://www.netlib.org/nse/nse_list.html)** contains a total of 720 links
- Parallel Simulation **(http://www.cs.wm.edu/nicol/nicol.html)**
- Documentation for Parallelists **(http://www.ccsf.caltech.edu/documentation.html)**
- Parallel supercomputing **(http://www.cs.cmu.edu/Web/Groups/scandal/www/resources.html)**
- Parallel and Heterogeneous Publications **(http://www-cse.ucsd.edu/users/jenny/parpubs.html)**

5.2. Glossary, News Groups, and Homepage Lists

Glossary

- High Performance Computing and Communications Glossary (1.3) Index Page **(http://www.npac.syr.edu:80/nse/hpccgloss/)**
- Sun Glossary **(http://www.sun.com/smcc/Products/glossary.html)**
- Fortran Glossary **(http://www.npac.syr.edu/hpfa/fortgloss/fortgloss.html)**, at Syracuse Univ.
- Denis Howe's On-line Dictionary of Computing **(http://wombat.doc.ic.ac.uk/)**
- Harris Semiconductor Terms **(http://rel.semi.harris.com/docs/lexicon/preface.html)**.
- (The New Hacker's Dictionary) **(http://www.ccil.org/jargon/jargon.html)** by Eric Raymond
- Cybernetics Glossary **(http://pespmc1.vub.ac.be/ASC/indexASC.html)**
- Tom Van Vleck's page **(http://www.best.com/~thvv/tvv.html)** articles on software
- Philips Semiconductors Glossary **(http://server1.pa.hodes.com/ps/philips3.html)**

News Groups

- Architecture **(news://comp.arch)**
- Benchmarks **(news://comp.benchmarks)**
- MPI (news://**comp.parallel.mpi**)
- OS Research (news://comp.os.research)
- Parallel **(news://comp.parallel)**
- PVM **(news://comp.parallel.pvm)**
- Supercomputer **(news//:comp.sys.super)**

Sites List of Home Pages

- IEEE/CS ParaScope **(http://computer.org/parascope/),** world-wide parallel computing sites
- A list of major HPC institutions from Caltech

(http://www.ccsf.caltech.edu/other_sites.html)
- CMU high-performance computing research groups:
  (http://www.cs.cmu.edu:8001/afs/cs.cmu.edu/project/scandal/public/www/research-groups.html)
- High Performance Computing List by Oliver McBryan
  (http://www.cs.colorado.edu/homes/mcbryan/public_html/bb/2/summary.html) .
- World Wide Computer Architecture Page at Univ. of Wisconsin:
  (http://www.cs.wisc.edu/p/arch/www/www_architecture.html) .
- Comp.Parallel (http://dragon.acadiau.ca:1667/~rmuise/parallel.html)
- High Performance Computing Lists
  (http://www.cs.colorado.edu/homes/mcbryan/public_html/bb/2/summary.html)
- Transputer, Occam and parallel computing archive
  (http://www.hensa.ac.uk/parallel/index.html) (Dave Beckett)
- Parallell computing list (http://argo.cis.temple.edu/parallel.html) (Jonathan Wang)
- Nan's Parallel Computing Page (http://www.cs.rit.edu/~ncs/parallel.html)

5.3. Research Centers at Academic Institutions
- Berkeley Software Warehouse (http://http.cs.berkeley.edu/csdiv/SWW/)
- Caltech Concurrent Supercomputing Facilities (http://www.ccsf.caltech.edu/)
- CMU Parallel Systems Group (http://parsys.cs.cmu.edu/)
- Center for Applied Parallel Processing, University of Colorado (CAPP)
  (http://www.cs.colorado.edu/home/capp/Home.html)
- Center for Discrete Mathematics and Theoretical Computer Science (http://dimacs.rutgers.edu/)
- Center for Parallel Computing, Michigan (CPC) (http://www.engin.umich.edu/labs/cpc/cpc.html)
- Center for Research in Parallel Computation, Rice University (CRPC)
  (http://softlib.cs.rice.edu/CRPC.html)
- Center for Distributed Multimedia and Metacomputing Appolications at Univ. of Hong Kong
  (http://www.cs.hku.hk/~jmma/mmcentre/background/index.html)
- Center for Reliable an HPC, Univ. of Illinois (CRHC) (http://www.crhc.uiuc.edu/)
- Center for Innovative Computer Application, Indiana University (CICA)
  (http://www.cica.indiana.edu/)
- Coalition of Academic Supercomputing Centres (http://www.osc.edu/casc.html)
- Cornell Theory Center (CTC) (http://www.tc.cornel.edu/ctc.html)
- Mississipi Center for Supercomputing (http://www.mcsr.olemiss.edu/)
- MIT Parallel and Distributed Operating Systems Group (http://www.pdos.lcs.mit.edu/).
- Nat'l Center for Atmospheric Res. Colorado (NCAR) (http://http.ucar.edu/metapage.html)
- National Center for Supercomputer Applications at UIUC (NCSA) (http://www.ncsa.uiuc.edu/)
- National Supercomputing Center for Energy and the Environment at Univ. of Nevada (NSCEE)
  (http://www.nscee.edu/nscee/index.html)
- Supercomputing Research Institute at Florida University (SCRI) (http://www.scri.fsu.edu)
- Univ. of Minnesota Supercomputer Institute (http://web.msi.umn.edu/)
- Univ. of New Mexico High Performance Computing Center (http://www.arc.unm.edu/)
- Univ. of Texas System Center for HPC (gopher://almach.chpc.utexas.edu/)
- Univ. of Washington HPCC Group (http://www-hpcc.astro.washington.edu/)
- Yale Center for Parallel Supercomputing (http://www.yale.edu/HTML/YaleYCPS-Info.html)
- Nara Institute of Science and Technology, NAIST (http://www.aist-nara.ac.jp/)

- NCSA Virtual Reality Faclities **(http://www.ncsa.uiuc.edu/Viz/VR/)** and CAVE. **(http://evlweb.eecs.uic.edu/EVL/VR/systems.html#CAVE)**
- Ohio State PAC Group **(http://www.cis.ohio-state.edu/~panda/pac.html)**
- Cecil Project **(http://www.cs.washington.edu/research/projects/cecil/www/cecil-home.html)**
- Large program traces site **(ftp://tracebase.nmsu.edu/pub/)**
- Another large program traces site **(ftp://demon.eecs.umich.edu)**
- Universal Stub Compiler, Arizona **(http://www.cs.arizona.edu/xkernel/www/software-distr.html)**
- The Language List **(http://cuiwww.unige.ch/langlist)** enumerate programming languages

5.4. Commercial and Industrial Companies

- Ada Home page **(http://www.adahome.com/)**
- Alta Technology **(http://www.xmission.com:80/~altatech/)**
- Amdahl Corporation **(http://www.amdahl.com/)**
- Apple **(http://www.apple.com/)**
- Applied Parallel Research **(ftp://ftp.netcom.com/pub/forge/home.html)**
- Bolt Beranek and Newman Inc. (BBN) **(http://www.bbn.com/)**
- Computational Engineering International, Inc. **(http://www.ceintl.com/)**
- Convex Computer Corporation **(http://www.convex.com/)**
- Cray Research **(http://www.cray.com/)**
- Digital Equipment Corporation **(http://www.dec.com/info.html)**
- Elegant Mathematics, Inc. **(http://elegant-math.com/)**
- Fortran Market **(http://www.fortran.com/fortran/market.html)**
- Fujitsu **(http://www.fujitsu.co.jp/)**
- Fujitsu, USA **(http://www.fai.com/)**
- Hewlett-Packard **(http://www.hp.com/)**
- Hitachi **(http://www.hitachi.co.jp/)**
- IBM **(http://www.ibm.com/)**
- IBM Almaden Research Center **(http://www.watson.ibm.com)**
- IBM High-Performance Computing **(http://ibm.tc.cornell.edu/)**
- IBM Kingston, Large Scale Computing **(http://lscftp.kgn.ibm.com)**
- IBM SP **(http://www.rs6000.ibm.com/hardware/largescale/)**
- Intel **(http://www.intel.com/)**
- Lateiner Dataspace **(http://www.dataspace.com/WWW/welcome.html)**
- Legent Corporation **(http://www.legent.com/)**
- The MathWorks,Inc. **(http://www.mathworks.com/)**
- Matsushita Electric **(http://mei.co.jp/)**
- Meiko Scientific **(http://www.meiko.com/)**
- MIPS Corporation **(http://www.mips.com/)**
- Myrias Computer Technologies, Inc. **(http://www.myrias.com/)**
- Myricom **(http://www.myri.com/)**
- nCUBE **(http://www.ncube.com/)**
- NEC **(http://www.nec.com/)**

- Network Buyer's Guide: **(http://www.sresearch.com/)**, Strategic Research Corporation
- NTT Transmission Systems Labs, Japan **(http://www.ntt.jp)**
- The Numerical Algorithms Group **(http://www.nag.co.uk:70/)**
- Pacific-Sierra Research Corporation **(http://www.psrv.com/)**
- ParaSoft Corporation **(http://www.parasoft.com/)**
- Parsytec Computer GmbH **(http://www.parsytec.de/)**
- The Portland Group, Inc. (PGI) **(http://www.pgroup.com/)**
- Silicon Graphics **(http://www.sgi.com/)**
- Software.net **(https://software.net/index.htm/SK:annajickobcppdao)**
- Sun Microsystems **(http://www.sun.com/)**
- Tandem Computers **(http://www.tandem.com/)**
- Tera Computer Company **(http://www.tera.com/tera.html)**
- Thinking Machines Corp. **(http://www.think.com/)**
- Visual Numerics **(http://www.vni.com/)**

5.5. Agencies, Laboratories, Consotiums and Organizations

- Program (ATIP) Reports **(http://www.cs.arizona.edu/japan/www/atip_reports.html)**
- Arctic Region Supercomputing Center **(http://www.arsc.edu/)**
- Argonne Nat'l Laboratory, Mathematics & Computer Science Div. **(http://www.mcs.anl.gov/)**
- Army HPC Research Center (AHPCRC) **(http://www.arc.umn.edu/html/ahpcrc.html)**
- Army Research Lab **(http://www.arl.army.mil/)**
- Alabama Supercomputer Network (ASN)**(http://sgisrvr.asc.edu/)**
- Dept of Energy Labs and Programs **(http://www.esd.ornl.gov/doe-labs/doe-labs.html)**
- Fermilab **(http://www.fnal.gov/)**
- Lawrence Livermore National Laboratory **(http://www.llnl.gov/comp/comp.html)**
- Los Alamos Nat'l Laboratory (LANL) Advanced Computing Laboratory **(http://www.acl.lanl.gov/).**
- Maui High Performance Computing Center (MHPCC) **(http://www.mhpcc.edu/mhpcc.html)**
- North Carolina Supercomuting Center (NCSC) **(http://www.mcnc.org/HTML/ITD/NCSC/ncschome.html)**
- NASA Ames Research Center, Numerical Aerodynamic Simulation Systems Division, Applied Research Branch **(http://www.nas.nasa.gov/RNR/rnr.html)**
- NASA Ames NAS **(http://www.nas.nasa/HPCC/general.html)**
- NASA Jet Propulsion Laboratory (JPL)**(http://www.jpl.nasa.gov/)**
- National Center for Atmospheric Research (NCAR) **(http://http.ucar.edu/metapage.html)**
- National Center for Supercomputer Applications (NCSA) **(http://www.ncsa.uiuc.edu/)**
- Nat'l Supercomputing Center for Energy and Environment (NSCEE) **(http://www.nscee.edu/nscee/index.html)**
- National Energy Research Supercomputer Center (NERSC) **(http://www.nersc.gov/)**
- National Aeronautics and Space Admin (NASA) **(http://hypatia.gsfc.nasa.gov/)**
- Nat'l Oceanic and Atmospheric Administration (NOAA) HPCC **(http://hpcc.noaa.gov/hpcc.html)**
- National Consortium for High Performance Computing Page **(http://www.nchpc.lcs.mit.edu/)**
- Nat'l Coordination Office for HPCC **(http://www.hpcc.gov/)**

- Naval Research Laboratory **(http://www.nrl.navy.mil/)**
- Oak Ridge National Laboratory: Center for Computing **(http://www.ccs.ornl.gov/)**
- Ohio Supercomputer Center (OSC) **(http://www.osc.edu/pcrm.html)**
- Parallel Tools Consortium **(http://www.llnl.gov/ptools/)**
- Pittsburgh Supercomputing Center (PSC) **(http://pcsinfo.psc.edu/)**
- San Diego Supercomputer Center **(http://www.sdsc.edu/SDSCHome.html)**
- Sandia National Laboratories **(http://www.cs.sandia.gov/)** Massively Parallel Comp. Res. Lab.
- Utah Supercomputing Institute **(http://ute.usi.utah.edu)**
- German Scientific Computing **(http://www5.informatik.tu-muenchen.de/sci-comp/home.html)**
- Institute for New Generation Computer Technology **(http://www.icot.or.jp/)**
- Japan Advanced Institute of Science and Technology (JAIST), Ishikawa, Japan **(http://www.jaist.ac.jp/jaist/iscenter/home.html)**
- Agency of Industrial Science & Technology, AIST **(http://www.aist.go.jp/)**
- Parallel Processing in Japan **(http://fuji.stanford.edu/papers/ppij.html)**
- Japan Information Center of Science and Technology (JICST) **(http://www.jicst.go.jp/)**
- Ministry of International Trade and Industry (MITI), Japan **(http://www.rwcp.or.jp/organization_scheme.html)**
- National Center for Science and Information Systems, NACSIS, Japan **(http://www.nacsis.ac.jp/nacsis.f-index.html)**
- Project Mentifex, in AI **(http://www.newciv.org/Mentifex/)**
- Good OO On-line Documentation, Object Agency, Inc **(http://www.toa.com/online/html/html.html)**
- DoD High Performance Computing Modernization Program **(http://www.hpcm.dren.net/)**
- JNNIE Project Home Page **(http://www.tc.cornell.edu/JNNIE/JNNIEHome.html)** Joint NSF-NASA Initiative on Evaluation of scalable parallel computing systems
- MetaCenter **(http://www.ncsa.uiuc.edu/)** Nat'l MetaCenter for Compu. Sci. and Eng.
- National Metacenter for Computational Science and Engineering **(http://pscinfo.psc.edu/MetaCenter/)**
- NHSE catalog of commercial HW/SW vendors **(http://www.netlib.org/nhse/catalog.html)**
- Computational Bioscience and Engineering Laboratory **(http://hpccwww.dcrt.nih.gov/)**
- National Cancer Institute Laboratory of Mathematical Biology **(http://www-lmmb.ncifcrf.gov/)**
- Nat'l Cancer Institute, Frederick Supercomputer Center **(http://www.ncifcrf.gov:1994/FBSC)**
- National Cancer Institute MPsrch - massively parallel sequence comparison **(http://www-ips.ncifcrf.gov/~bshapiro/mpsrch-doc.html)**
- The Energy Science Network **(http://www.es.net/)**
- National Energy Research Supercomputer Center **(http://www.nersc.gov/)**
- Adventures in Supercomputing **(http://www.ornl.gov/ocnm/AiS/AiS.html)**
- MCNC North Carolina Supercomputing Center **(http://www.mcnc.org/HTML/ITD/NCSC/biz/seed.html)**
- IEEE/ACM SC97:High performance Networking and Computing, San Jose, Nov. 15-21, 1997 **(http://www.spercomp.org./sc97)**

Subject Index

A

Accelerated Strategic Computing Initiative (ASCI) 574
   Blue Mountain 578
   Blue Pacific 577
   Option Red 93, 113, 569, 570, 578, 581, 601
Access frequency 217
Access time 212, 214
Acquire (lock) 233
Active availability cluster 514
Active message 541, 546, 594
Adapter processor 379
Adaptive Processing Testbed (APT) 97, 596
Address bus 167
Address space 15, 44, 63
Advantage Cluster 724
Affinity scheduling 412
Agenda paradigm 610
Aggregate bandwidth 281, 305, 306, 591, 594
Aggregate message length 591
Aggregation 79, 116, 655, 679, 680
Aggregation assignment 645
ALCACHE 240
Alpha 21164 179, 180, 182, 183, 532, 571
AMD K5/6 177
Amdahl Corp. 513
   EnVista Server 515
   Scaling multiserver cluster 516
Amdahl's law 40, 134
Amdahl's rule 40
American National Standards Institute (ANSI) 39
   Standard 1596-1992 327
   Standard X3 307, 315, 334
   Standard X3H5 62, 609, 629, 630, 636, 637
Andrew file system (AFS) 476, 479, 529
ANSI X3H5 62, 609, 629, 630, 636, 637
Apollo 9000 Cluster 507
Application
   Characteristic 163
   Environment 576
Application programming interface (API) 375
Arbitrated loop 309
Architectural support 250
Argonne National Laboratory
   Fortran-M 654
   Globus 510, 511
   P4 654
   PARMACS 654
Array conformability 709
Array handling 681, 707
Array shaping 707
ASCI software requirement 576, 578
Asymmetric network 281, 282
Asymptotic bandwidth 17, 18, 23, 123, 124, 589, 591
Asynchronous
   Link 276
   Network 282
Asynchronous iteration paradigm 611
Asynchronous message passing 657
Asynchronous transfer mode (ATM) 315, 318, 319, 336, 552
   Cell format 319
   Cell switch 279, 319
   Cell switching 320
   Forum 334
   Layers 323
   Network 279, 322, 326
   Protocol 307, 308
   Speed 320
   Technology 335
Atomic operation 16, 60, 351
Atomic region 352
Atomic swap 602
ATOMIC/LAN 313
Atomicity problem 349, 352
Attraction memory 241
Automatic parallelization 620
Availability 34, 459, 460, 461, 500
Availability cluster 456, 508
Availability support 457
Available parallelism 131
Average CPI 168
Average granularity 127, 128
Average overhead 127, 128
Average parallelism 127, 128, 133

B

Backward compatibility 44
Backward recovery 466
Balanced scalable design 575
Balanced scalable hardware 575
Balanced scalable software 576
Bandwidth 212, 214
Bandwidth management 278
Banking problem 352
Barrel shifter 287, 288
Barrier synchronization 78, 124, 349, 590, 611, 680
Base communication layer (BCL) 381
Batch job 487
Batch processing 68
Bay Networks Centillion 322
BBN Butterfly 253
Beam Space PRI-Staggered Post Doppler (BM-Stag) 97
Beamforming (BF) 97, 111, 595
Benchmark 91
   Class C NAS 589
   Embarrassingly parallel 96
   GPC 100
   LINPACK 92, 93, 149
   LMBENCH 92, 93
   Macro 92
   Micro 92
   MP3D 265
   NPB 96, 585
   Parallel NAS 92, 149, 584
   PARKBENCH 92, 96
   Perfect 132, 618
   SDM 100
   SFS 100
   SPEC 92, 100
   SPEC95 100
   SPECfp 101
   SPECfp95 180
   SPEChpc96 100
   SPECint 101
   SPECint95 180
   SPECweb96 100

Splash 92, 447
STAP 92, 97, 111, 594
STREAM 92, 94, 95, 572
TPC 92, 98, 433
TPC-B 431, 432
TPC-C 432
TPC-D 431, 432
Benefits of caching 254
Benefits of prefetching 255
Berkeley
   Active message 381, 386, 399
   Fast socket 381
   IRAM processor 204, 206
   NOW 505, 510
   RISC 175
   Unix (BSD 4.2) 375
Bernstein's Theorem 621
Binary cube 286
Binary semaphore 355
Binary tree network 287, 289,
293
Bisection bandwidth 281
Block transfer engine (BLT) 599
Block tridiagonal solver
  (BT) 585
Blocking network 282
Blocking receive 656
Blocking send 656
Bound thread 345
Branch history table (BHT) 186
Branch prediction 167
Broadcast 128, 286, 590, 591,
678
Buffered communication
 mode 675
Buffered send 675
Bulk gap 385
Bulk synchronous parallel
 (BSP) 15, 20
Bus cycle 165, 166
Business application 162
Butterfly network 301
Bytes per second (B/s) 3

C

C++ 573
C// (C Parallel) 609, 643, 647,
648, 649
Cabling distance 311, 335
Cache
   Coherence protocol 236, 450
   Directory 243, 244
   Line 215

Cache coherency 220, 265
   Problem 220
   Property 215
Cache coherent nonuniform mem-
ory access (CC-NUMA) 211,
238, 241, 242, 246, 297, 399,
407, 410, 415, 448, 578
Cache directory 243
Cache only memory access
(COMA) 211, 238, 239, 240,
241
Calendar scheduling 488
CDC
   1604 5
   6600 5
Cell switching 279
Central file server 549
Central processing unit
 (CPU) 168
Chained cache directory 245
Channel
   Bandwidth 280
   Delay 280
   Width 281
Checkpoint 467, 468, 500
   Compression 470
   Coordinated 472
   Forked 469
   Incremental 469
   Independent 472
   Interval 469
   Overhead 469
   Transparent 471
   User directed 470
CHIMP/MPI 590, 660
Chordal ring 287, 288
CHORUS system, Inc.
 CHORUS 573
Cilk thread library 398
Circuit switched network 278
Circular shift 124
Cisco LightStream 322
Client side caching 478
Clock
   Cycle 165
   Period 165
   Rate 17, 168
   Synchronization 303
Cluster 578
   Cache (CC) 297
   Compact 454
   Monitor 540
Cluster of computers 29, 453
Cluster of workstations
 (COW) 26, 453, 568

CM5 290, 291, 570, 654
Coarse granularity 161
Coherent region 352, 644, 645
COMA cache 239, 240
Commodity off the shelf
 (COTS) 574
Communication 116, 655
   Collective 18, 121, 123
   Latency 280
   Mode 675
   Operation 77
   Overhead 123, 124, 591
   Subsystem 157
Communication-to-computation
 ratio (CCR) 25, 241, 509, 597
Communicator 658, 662, 670,
680
Compaq Corp. 513, 514
   NT cluster 506
   NT server 516
   On line recovery server 516
Compare and swap 359, 361, 364
Competitive interaction 84, 646
Compiler directive 56, 236, 620
Completely connected
 network 287, 288, 293
Completion check 675
Complex instruction set comput-
ing (CISC) 169, 170, 171, 172,
175
Complex operation 164
Computation phase 23
Computational granularity 163
Compute node 579
Concurrent read concurrent write
 (CRCW) 16
Concurrent read exclusive write
 (CREW) 16
Conjugate gradient (CG) 585
Consistency problem 228
Consistency rule 16
Consistent checkpointing 472
Contention delay 280
Context 670
Context switch 65, 117
   Overhead 258
   Policy 260
Context switch on
   Block of instruction 261
   Cache miss 260
   Every instruction 261
   Every load 260
Control state 60
Control synchronization 78, 349
Control variable 60

Convoying blocking 357
Cooperative file caching 552
Cooperative interaction 84, 646
Coprocessor 168
Cornell U-Net 386, 399
Cost effectiveness 102
costScale 132
Coverage factor 256
CP-PACS 93
CPU bound problem 139
Cray Research, Inc.
  64-bit Unix system
    (UNICOS) 573
  C90 36, 106, 113, 316, 573
  CRAFT 58, 573
  Cray 1 113
  CRI/EPCC MPI 590
  GigaRing 266, 572
  MPP Apprentice 574
  Shared memory library
    (SHMEM) 573
  Supercomputers 158, 159
  T3D 64, 93, 106, 235, 253,
    316, 378, 379, 382, 386, 567,
    570, 597, 600, 602
  T3E 316, 382, 386, 567, 570,
    571, 572
  T3E-900 571, 572
  T90 113
  TotalView 574
  UNICOS MAX 602
  UNICOS/mk 573
  X/MP 5, 113
  Y/MP 113, 304, 316
Cray T3E 316, 382, 386, 567,
  570, 571, 572
Critical path 133
Critical region 351, 359, 633,
  642
Critical section 351
Crossbar network 301
Crossbar switch 298, 305
CSMA/CD access protocol 311,
  312
Cube connected cycle
  (CCC) 291, 293
  k-cube connected cycle 292
Current microprocessor 204
Cut-through routing 283, 284
Cycles per instruction
  (CPI) 164, 168
Cyrix Corp.
  6x86 177
  M2 177

D

Data
  Bus 167
  Cache 170
  Dependency 619
  Distribution 720, 726
  Encapsulation 368
  Locality 16, 251
  Mapping 715, 717
  Movement 679
  Parallel construct 74
  Parallelism 625
  Path 167
  Prefetching strategy 255
  Routing function 284
  Synchronization 78, 349
  Transfer rate 212
  Transfer unit 215
Data diffusion machine
  (DDM) 240
Data General
  CC-NUMA cluster 331
  NUMA server 253
dcgc 132
Deadlock prevention 85, 356,
  357, 358, 677
Decision support system
  (DSS) 431
Decoupled CISC/RISC 173,
  180, 187
Dedicated cluster 453, 455
Degree of parallelism (DOP) 76,
  163
Dependence analysis 617
Diagonostic performance
  monitor 577
Digital Equipments Corp. 513
  Advantage Cluster 724
  Alpha 64, 117, 174, 266
  Alpha 21064 175, 567
  Alpha 21164 179, 180, 182,
    183, 532, 571
  Alpha NTFS 516
  AlphaServer 235
  AlphaServer 8400 413, 532,
    533, 724
  AXP 506
  CVAX 344
  Fortran 77 725
  GIGAswitch/ATM 322
  GIGAswitch/FDDI 298
  HPF 725
  Memory channel 378, 386,
    394, 531, 532, 534, 535, 536,
    537, 538, 540, 557
  NT Cluster 506, 507, 511,
    512, 516
  Open VMS Cluster 506, 507
  PDP-11 5
  Projection 202
  PVM 725
  TruCluster 5, 114, 378, 386,
    394, 505, 507, 508, 531, 532,
    533, 536, 537, 538, 539, 540,
    557, 562, 563, 564
  VAX 506
  VAX 9000 5
Digital signal processor
  (DSP) 168, 181
Dining philosophers 349
Direct exchange algorithm 396
Direct memory access
  (DMA) 378, 388, 390
Direct network 278
Directory architecture for shared
  memory (Dash). See Stanford
  Dash
Directory based protocol 224,
  243, 248, 249
Disk swap delay 602
Distributed coherent cache 252,
  253
Distributed directory 243
Distributed multimedia system
  (DMS) 326
Distributed operating
  software 577
Distributed queueing system
  (DQS) 492
Distributed shared memory
  (DSM) 16, 26, 29, 238, 241,
  253, 442, 447, 626
DMA buffer 379
Domain name server
  (DNS) 375, 377, 466
Domino effect 472
Doppler processing (DP) 111,
  595
Double buffering 676
Double word 3
DSM software 556
Dynamic branch prediction 186
Dynamic configuration 687
Dynamic execution 191
Dynamic load balancing 408
Dynamic network 278
Dynamic program execution 167
Dynamic random access memory
  (DRAM) 6, 213

Dynamic system domain   420

E

Effective access time   217, 218
Effective latency   577
Efficient communication   457
Elapsed time   109
Element Space PRI-Staggered
  Post Doppler (EL-Stag)   97
Embarrassingly parallel (EP)   585
Embedded microprocessor   180,
  181
ENIAC   5
Enterprise cluster   453, 455
Environment tool   56
Equivalent directive   714
Eratosthenes method   730
E-register   239
Escon (enterprise systems
  connection)   308
espresso   132
Ethernet   311, 552
    Fast   278, 311
    Frame   369
    Generations   310
    Gigabit   311, 334, 376
Event scheduling   488
Examplar X-Class   106, 240, 253,
  421, 449
Exclusive read exclusive write
  (EREW)   16
Execution mode   62
Execution time   102
Explicit allocation   77
Explicit parallelism   617, 621
Exponential backoff   362
Extravagant memory/disk
  event   601

F

Failover   465
Failure diagnosis   466
Failure notification   466
Failure recovery   466
Fast Message   386, 390, 392, 399
Fat hypercube topology   440,
  441, 450
Fat tree network   290
Fault handling   383
Fault tolerant multiserver
  cluster   465, 516

FC standard   308
FDDI full duplex technology
  (FFDT)   298
Fetch and add   359, 363, 366
Fetch and increment   366
Fetch-and-phi   364
Fiber channel (FC)   278, 307,
  315, 335, 336
Fiber Channel Physical Standard
  (FCPH)   308
Fiber Channel Systems Initiative
  (FCSI)   307
Fiber distributed data interface
  (FDDI)   278, 308, 309, 336
FIFO queue   392
File caching   483
File transfer protocol (FTP)   369
Financial digital library
  (FDL)   326
First in first out (FIFO)   216
Floating point count   110
Floating point operation (flop)   3
Floating point operations per sec-
  ond (flop/s)   3
Floppy disk   213
Flow control   383
Flow control digits (flit)   283, 302
FORALL construct   712, 714,
  721
FORE ASX-1000   322
Fortran 2001   725
Fortran 77   629, 686
Fortran 90   58, 573, 629, 705,
  706, 707, 708
Fortran 90 D   726
Fortran 95   725
Forward recovery   467
Fourth generation language
  (4GL)   52
Frame scheduling   446
Fujitsu Corp.
    VPP500   93, 570
Full-map cache directory   244
Functional programming   627
Futurebus   294, 297
Fuzzy barrier   642

G

GAM layer   542, 544, 545
GAM-2   546
Gang scheduling   446, 491
Gather   124, 590, 591, 678
Gaussian elimination   615, 711,

  719
Gbps (Giga bits per second)   276
gcc   132
General (GEN) program   97, 597
General purpose register
  (GPR)   170, 212
Generic flow control (GFC)   321
Gigabit ethernet   311, 334, 376
Gigaplane bus   295, 296, 299
Gigaplane-XB crossbar
  centerplane   416, 419
GIGAswitch/ATM   322
GIGAswitch/FDDI   298
Global coherent memory
  (GCM)   332
Global communication   678
Global layer Unix
  (GLUnix)   541, 547, 548, 549
Global memory order   230
Global naming space   706, 707
GMD Corp.
    PARMACS   654
Granularity   76, 587
Graphic user interface (GUI)   459
Greedy forwarding   552, 553
Group server   691
Guaranteed delivery   383
Gustafson's Law   136

H

h relation   21, 398
Half peak length   123
Hard disk   213
Hardware multithreading   343
HARNET (Hong Kong Academic
  Research Network)   324
Heartbeat daemon   466
Heterogeneous cluster   454
Heterogeneous computing   663
Hewlett Packard   513
    9000   413
    9000 Unix   516
    Apollo 9000 Cluster   507
    MC/Service Guard   507, 515,
      516
    P7   203
    PA8000   179, 180, 191, 195
    PA-RISC   174
    T600   413
    Wolfpack   516
Hierarchical bus   297
Hierarchical memory   211
High level language (HLL)   6

High performance computing and communication (HPCC) 511
High performance Fortran (HPF) 58, 520, 537, 538, 540, 626, 636, 705, 711, 724
High performance parallel interface (HiPPI) 307, 308, 314, 316, 317, 335, 336, 509
High-Order Post-Doppler (HO-PD) 97, 596, 597
HiPPI-FC (Fiber Channel Protocols) 318
HiPPI-IPI (IPI-3 Command Sets) 318
HiPPI-LE Link Encapsulation 318
Hit ratio 217
Home node 240
Homogeneous cluster 454
Hong Kong University 324
Horizontal microprogramming 173
Host interface 277
Hot standby cluster 465, 513
Hot-spot 299
Householder Transform (HT) 97, 111
HP/Convex
 Examplar X-Class 106, 240, 253, 421, 449
 SPP 235
Human/computer interface 576
Hybrid CISC/RISC 173
Hyper text transfer protocol (HTTP) 369
Hypercube network 290, 293
Hypercube routing function 285

I

I/O bound problem 139
I/O bus 275
I/O node 579
I/O processor (IOP) 294
IBM 160, 307, 308, 514
 AIX 690
 ASCI Blue Pacific 577
 CICS/6000 526
 DB2 526
 DB2 Parallel Edition 520, 528
 HACMP 464, 507
 High performance switch 13, 280, 302, 464, 508, 521
 Mathematical and engineering

 libraries (ESSL) 526
 Message passing library (MPL) 387, 520, 527, 660
 Parallel debugger (pdbx) 527
 Parallel I/O file system (PIOFS) 528
 Parallel Operating Environment (POE) 13, 46, 527, 660, 683
Personal computer 5, 157
POWER 117
POWER2 117, 522, 600
POWER2 superchip (P2SC) 530
PowerPC 174, 234, 266
PowerPC 604 191, 194
PowerPC 604e 179, 180
PowerPC 620 179
R40 413
RP3 253, 363
RS6000 413
Sysplex 506, 507
System 360 5
System 370 5, 266
System 390 316
System 701 5
System 7030 5
Visualization tool (VT) 527
Vulcan chip 300
IBM SP 5, 38, 93, 104, 106, 118, 280, 303, 378, 379, 381, 383, 386, 396, 399, 457, 464, 505, 507, 508, 518, 521, 522, 525, 530, 567, 569, 597, 600, 660, 683, 724
IBM SP architecure 521
IBM SP communication adapter 524
IBM SP I/O subsystem 523
IBM Sysplex 506, 507
IEEE Standard
 1014-1987 297
 1296-1987 297
 1496-1993 418
 1596-1992 327, 334
 802 307, 311, 334
 896.1-1991 297
 POSIX Threads (Pthreads) 344, 609, 629, 634, 636
IETF 334
Illiac mesh network 289, 293
Implicit barrier 632
Implicit data allocation 706
Implicit interaction 706

Implicit parallelism 617, 625
Inclusion property 215
Independent directive 714
Independent scheduling 490
Indirect network 278
Induction variable 619
Infinite looping 85
Informix 160, 430
In-order
 Decode/translation 190
 Execution 167
 Front end 188, 189
 Retirement 189, 190
 Translation engine 186
Instruction
 Cache 170
 Count 108
 Cycle 165
 Execution rate 168
 Issue latency 164
 Issue rate 164
Instruction level parallelism (ILP) 173, 175
Integrated services digital network (ISDN) 326
Intel Corp. 516
 486DX 46
 486SX 46
 80x86 170, 174, 177, 343
 ASCI Option Red 93, 113, 386, 569, 570, 578, 581, 582, 601
 Cougar 582
 i860 113, 175
 iPSC 113
 iPSC/1 291
 iPSC/2 291
 Klamath 177, 197
 MPP family 158, 159
 nCUBE 291, 570
 NX 654, 660
 Orion PCI bridge (OPB) 426
 OSF 660
 P7 203
 Paragon 36, 93, 106, 113, 378, 379, 386, 396, 570, 598, 600, 660, 724
 Parallel file system (PFS) 601
 Pentium 117, 166, 171, 224, 266, 343
 Pentium Pro 177, 179, 180, 186, 189
 Pentium-2 177, 197
 Projection 201
 Standard high volume

(SHV) 332, 409, 413, 414, 426
Intelligent peripheral interface (IPI) 308
Interaction
  Blocking 82
  Nonblocking 82
  Operation 17, 77
  Overhead 163
  Phase 23
  Synchronous 81
Intercluster bus 297
Interconnection facility (ICF) 581
Interconnection network 9, 273
Interleaved memory modules 213
Internal cycle 165
International Electrotechnical Commission (IEC) 725
International Standards Organization (ISO) 39, 322, 367, 725
Internet Activities Board (IAB) 368
Internet protocol (IP) 308, 373, 380
Internetworking 315
Interprocess communication (IPC) 67
Inter-processor-memory access 298
Interrupt handler 65
Intracluster communication
  Enclosed 455
  Exposed 455
Intrinsic function 706
Intrinsic operator 706
Invalidate instruction 226
Inverse perfect shuffle 285
IP datagram 369
IRAM processor 204, 206
Irregular communication 397
ISA 170, 171, 192
Isoefficiency 145
Isolated redundancy 463
Isoperformance model 144
Isospeed 146
Isoutilization 146, 147
Issue rate 175

J

Jacobi relaxation 616, 710
Job management system

(JMS) 457, 486
Job scheduling 488
Journaling (Logging) 447

K

Karsruhe Parastaion 510
k-ary n-cube network 292, 293
Kendell Square, Inc.
  KSR-1 240, 253, 570
  KSR-2 240
Kernel
  Overhead 280
  Process 62
  Space 64, 380
  Thread 345

L

LAM (Local Area Multicomputer) 590, 660
Lamport's pets problem 351, 354, 399
LAN backbone 312
Language construct 56
Large job problem 490
Latency avoidance 250
Latency hiding 252, 265
Latency reduction 252
Lawrence Livermore National Latoratory
  ASCI Blue Pacific 578
Least frequently used (LFU) 216
Least recently used (LRU) 216
Legion 510
Library subroutine 56
Light weight kernel (LWK) 582
Lightweight process (LWP) 71, 116, 345
Limited cache directory 245
Linear array 287, 288, 293
Link level protocol (LLP) 439
LINPACK benchmark 92, 93, 149
Livelock 85
Livermore National Laboratory Zipcode 654
Load index 497
Load information manager (LIM) 495
Load sharing facility (LSF) 56, 492, 494
Load sharing policy 497

Load/store architecture 171
Local area network (LAN) 30, 244, 274, 275, 374
Local bus 275
Local memory 16, 238
Locality of reference 216
Lock(S) 355, 356, 359
Lock-free 356
Locking overhead 356
LogGP model 385, 399
Logic programming 627
LogP model 384, 399
Loop interchange 251
Loop parallelism 132
Loosely synchronous paradigm 610, 707
Los Alamos National Laboratory
  ASCI Blue Mountain 578
Low level construct 349
LSF architecture 495
LSF utilities 495
LU solver (LU) 585

M

Machine balance 94, 95
Machine cycle 165
Machine size 17
Macroarchitecture 9
Main memory 213
Many-to-many communication 83
Many-to-one communication 83
Marathon 513
  Fault-tolerant cluster 505
  MIAL2 507, 517
MasPar MP1 570
Massively parallel processor (MPP) 26, 28, 33, 398, 565, 626
Master device 294
Master election scheme 486, 501
Master LIM 498
Maui High Performance Computing Center (MHPCC) 10, 525
MAX extension 197
Maximum parallelism 127, 133
Maximum performance 127
Mbps (Million bits per second) 276
Mean time to failure (MTTF) 460, 463, 467, 469
Mean time to repair (MTTR) 460, 467
Media access control (MAC) 312

Media coprocessor 168
Meiko CS-2 386, 394, 396
Memory
  Bandwidth 213
  Bank 213
  Bound problem 139
  Bus 275
  Capacity 212
  Consistency model 229
  Event ordering 230
  Hierarchy 114, 219
  Locality domain (MLD) 444
  Migration 444
  Order 229
  Placement 444
  Replication 444
  Subsystem 220
  Window 684
Mesh network 289
Mesh routing component
  (MRC) 581
MESI protocol 224, 225, 228,
  266, 408
Message
  Buffer 667
  Envelop 668
  Length 23, 280
  Logging 473
  Tag 669
  Type 669
Message buffer 668
Message passing interface
  (MPI) 38, 58, 326, 380, 387,
  394, 489, 520, 527, 537, 538,
  539, 540, 546, 590, 626, 636,
  653, 654, 658, 683
Message tag 669
Metropolitan area network
  (MAN) 274, 275
Microarchitecture 9
MicroChannel 294
Microinstruction 177
Microkernel OS 547, 602
Microoperation (uop) 186
Microsoft Wolfpack 505, 507,
  512, 516
Middleware 160
Migration overhead 492
Minimum memory latency 410
MIPS 174
  R10000 179, 180, 191, 195,
    435
  R2000 344
  R3000 245
  R3010 245

R5000 179
MIT Alewife 253, 266
Monte Carlo simulation 86
Moore's Law 156, 566
Motorola Corp.
  M6800 170
  M680x0 174
  M68x0 174
  MC68060 171
MPI Forum 654, 658, 683
MPI implementation 660
mpi++ 590
MPI-2 682, 684, 685, 686
MPICH 590, 661
MPI-FM 390
MPI-IO 685
MPI-Java Interface (MJI) 326
MSI protocol 227, 228
Multiaccess network 278
Multibus 294, 297
Multicast 286
Multicomputer 237
Multigrid (MG) 585
Multilevel caches 213
Multimedia extension
  (MMX) 177, 178, 196
Multimedia register 197
Multimode fiber 336
Multiphase complete exchange
  algorithm 396
Multiple context 252, 260, 265
Multiple instruction multiple data
  (MIMD) 14, 301, 568
Multiple program multiple data
  (MPMD) 72, 382, 594, 690
Multiprocessing support 191
Multiprocessors 15, 237, 294
Multiprogramming 68
Multiscalar processor 204, 205
Multistage interconnection net-
  work (MIN) 301, 305
Multitasking 68
Multithreaded architecture 264
Multithreaded processor 258,
  261, 263
Multithreading 257
  Linear region 263
  Saturated region 262
Multiway shuffle network 301
Multiway superscalar
  processor 204, 205
Mutual exclusion 349, 350, 351,
  633
Mutual takeover cluster 465
Myrinet 313, 314, 335, 336, 393,

399
Myrinet control program
  (MCP) 313
Myrinet host interface 313
Myrinet switch 313

N

N version programming
  (NVP) 464
NASA/Goodyear's MPP 570
National Scalable Cluster Project
  (NSCP) 511
N-body problem 726
N-chance forwarding 553
NCR LifeKeeper 515, 516
NEC SX-4/32 93
Nesl 728
Nesl primitive function 731
Nesl sequence 729
Nested parallelism 728
Network
  Application 162
  Cable 276
  Channel 276, 307
  Diameter 282
  Latency 280
  Link 276
  Switch 276
Network file system (NFS) 476,
  479, 529, 556
Network interface circuitry
  (NIC) 9, 277, 378, 379, 566
Network interface driver
  (NID) 387
Network of workstations
  (NOW) 36, 453, 541
Network queueing system
  (NQS) 56, 492
Network-network interface
  (NNI) 320
No remote memory access
  (NORMA) 237, 242
Node availability 491
Node complexity 32
Node degree 282
Non cache coherent nonuniform
  memory access (NCC-
  NUMA) 238, 239, 242
Nonblocking network 282
Nonblocking receive 656
Nonblocking send 656
Nonserializability 356, 357
Nonuniform memory access

(NUMA) 15, 237
Non-write-allocate policy 225
NORMA-MPP 578
NOW project 505, 510
NP hard problem 230
NT Cluster 506, 507, 511, 512, 516
NUMA-Q 2000 235, 240, 253, 425, 427, 431, 449, 450
Numerical aerodynamic simulation (NAS) 486
NX message buffer 602
NYU Ultracomputer 363

O

Object oriented programming 627
Octopus Technologies 514
Oklahoma Update 362
Omega network 301, 567
On line transaction processing (OLTP) 99, 458, 506
100 BaseT 311, 336
100VG-AnyLAN 311, 336
One-to-many communication 82
One-to-one communication 82
Open system interconnect (OSI) 321, 367
Open VMS Cluster 506, 507
OpenMP standard 609, 629, 636, 637, 638, 639, 640
Oracle 160, 430
Ordered delivery 384
Ordinary differential equation (ODE) 36
Out of order core 188, 189
Out of order execution 167, 190, 192
Overdesign 44, 45
Overhead 384
    Communication 18
    Load imbalance 18
    Parallelism 18, 22
    Synchronization 18

P

PA8000 179, 180, 191, 195
Packet buffer 283
Packet switched network 278
Packet switching 279, 414
Page control table (PCT) 536

Paging rate 497
Paragon 36, 93, 106, 113, 378, 379, 386, 396, 570, 598, 600, 660, 724
Parallel algorithm 160
Parallel array operation 707
Parallel block 72, 630
Parallel divide and conquer paradigm 610, 612
Parallel execution time 17
Parallel loop 73, 631
Parallel NAS benchmark 92, 149, 584
Parallel random access machine (PRAM) 14, 18
Parallel reduction 619
Parallel region 630
Parallel vector processor (PVP) 26, 28
Parallel virtual machine (PVM) 13, 58, 380, 387, 394, 489, 520, 537, 538, 539, 540, 546, 547, 626, 653, 654, 683, 686, 690, 692
Parallelizing compiler 617, 618
Parasoft Express 654
Parbegin 73
Parent process 61
Parfor 73
PARKBENCH 92, 96
Partial differential equation (PDE) 41
Partial failure 461
Partition Manager 527
pC++ 728
PCI bus 294
Pearl cluster 325, 326, 510, 511
Pentadiagonal solver (SP) 585
Pentium Pro 177, 179, 180, 186, 189
Per byte messaging time 24
Perfect shuffle network 285, 301
Performance attribute 14, 17
Performance debugging 56
Performance/cost ratio 33, 35, 36, 102
Permutation 284
Per-port bandwidth 280
Per-processor bandwidth 297, 305, 306
PetaFLOPS Project 40, 574
Phase parallel model 23, 126, 586, 595, 609
Pi function 614, 623
Ping pong scheme 120

Pipe 389
Pipe bandwidth 117
Pipe latency 117
Pipeline cycle 164, 165
Pipeline paradigm 610, 612
Pipelining 103, 176
Plaintree Systems WaveSwitch 313
Planned shutdown 461
Point-to-point communication 18, 120, 122, 394, 539
Polling 389
POLYCENTER Advanced File System (AdvFS) 540
Portability 578
POSIX Threads (Pthreads) 344, 609, 629, 634, 636
Post-RISC 191, 192, 193, 194
POWERnode 509
POWERpath-2 bus 295, 296
PowerPC 604 191, 194
PowerPC 604e 179, 180
Pragmas in C 620
Preemption 347
Preemptive scheduler 69
Prefetch 255, 265
    Binding 255
    Hardware-controlled 255
    Nonbinding 255
    Queue 592
    Software-controlled 255
Principle
    Balanced design 37, 39
    Design for scalability 37
    Independence 37
    Latency hiding 37
Private network-network interface (PNNI) 321
Privatization 619
Privileged mode 62
Process 59
    Context 65
    Control block 66
    Creation 61, 117
    Descriptor 66
    Farm 610
    Group 670
    ID 673
    Interaction 646
    Migration 221
    Pipeline 676
    Switching 61
Process control thread (PCT) 582
Processor 168

Affinity 447
  Bus 275
  Consistency 231, 233
  Cycle 164, 166
  Efficiency 259, 261, 263
Producer consumer
  synchronization 349
Program
  Locality 16
  Order 229
  State 60
  Transformation 617
Program determinacy 85
Programmable protocol
  engine 428
Programming environment 237
Programming model
  Data parallel 622
  Message passing 622, 626
  Shared variable 624, 626
Protection domain 582
Proxy file system (PXFS) 482
Pthreads synchronization 635
PUMA operating system 583
PVM console 687

Q

Quad (see also Intel SHV) 332,
  409, 413, 414, 426, 427, 428,
  431, 432
Quadword 3, 196

R

R10000 179, 180, 191, 195, 435
Random access machine
  (RAM) 14
Raw (configurable)
  processor 204, 206
Read miss 227
Read/write bandwidth 577
Read-modify-write 233, 236, 364
Ready communication mode 675
Ready send 675
Real time system 69
Receive overhead 384
Receive-side congestion
  problem 383
Recorder buffer (ROB) 186
Recruitment threshold 492
Recurive doubling 80
Red black code 725

Reduce 124, 128, 590, 591
Reduced instruction set computing
  (RISC) 169, 170, 171, 172, 175
Redundant array of inexpensive
  disks (RAID) 31, 317, 428, 462,
  477, 508, 573, 588
Reflective memory 534
Register file 212
Register width 167
Regular communication 396
Regularity in algorithm 163
Relaxed memory
  consistency 252, 257
Relaxed memory model 235
  Alpha 235
  IBM370 235
  PC 235
  PowerPC 235
  PSO 235
  RCpc 235
  RCsc 235
  RMO 235
  SC 235
  TSO 235
  WO 235
Relaxed memory order 234
Release (unlock) 233
Release consistency 231, 233,
  234, 257, 265, 559
Reliability 459
Reliability, availability, and ser-
  viceability (RAS) 459, 460, 581
Reliable delivery 383
Remote cache 239
Remote caching 412
Remote memory 16, 238
Remote memory access
  (RMA) 684
Remote processor store 592
Remote read/write 602
Reply mesh 245
Request mesh 245
Reservation station (RS) 186
Restructuring compiler 617
Ring network 287, 288, 293
Routing delay 280
Routing distance 280
Run time parallelization 620

S

Sandia National Laboratory
  ASCI Option Red 578
Saturation arithmetic 196

SBus 294, 418
Scalability 3, 6, 147, 566, 578
  Generation 12, 52, 567
  Heterogeneity 13, 52
  Machine size 11
  Network 283
  Problem size 11
  Resource 9
  Size 9
  Software 11
  Space 13
  Technology 12
  Time 12
Scalable coherence interface
  (SCI) 327, 335, 336
  Coherence protocol 332
  Interconnect 327
  Interface 426
  Ring 328, 330, 450
  Standard 329
Scalable network 282
Scalable POWERparallel. See
  IBM SP
Scalable shared memory multipro-
  cessing (S2MP) 434
Scale down 6
Scale up 6
Scan 124, 590, 591
Scatter 124, 590, 591, 679
Scheduler 67
Security 578
Semantic issue 722
Semaphore 355, 359
Semiconductor Industry Associa-
  tion (SIA) 201
Send overhead 384
Sequent Computers
  DYNIX/ptx 430
  IQ-Link 426, 428, 429
  NUMA-Q 2000 240, 253,
    425, 427, 431, 449, 450
  ptx/CLUSTERS 430
  ptx/Debug 430
  STiNG 456
Sequent Computers.
  NUMA-Q 2000 235
Sequential code 242
Sequential consistency 231, 233,
  257
Sequential execution time 17
Sequential locality 216
Serial program parallel system
  (SPPS) 54
Service access point (SAP) 486
Serviceability 459

sgefa 132
SGI Power C 609, 629, 640
SGI Power Challenge Array 506
SGI/Cray Origin 2000 5, 240,
  253, 434, 436, 437, 438, 449,
  450, 569, 570
Shallow water code 724
Shared disk architecture 9
Shared media network 278, 334
Shared memory architecture 9
Shared nothing architecture 9
Shared SCSI bus 540
Shared virtual memory
  (SVM) 241
Sharing list
  Creation 333
  Structure 332
  Update 333
Shell 62
Shielded twisted pair (STP) 276,
  335
SHRIMP VMMC 241, 386, 394,
  395, 399, 510
SI protocol 227, 228
Silicon Graphics, Inc.
  4D/340 Powerstation 245
  ASCI Blue Mountain 578
  Cellular IRIX 410, 443
  Origin 2000 5, 240, 253, 434,
    436, 437, 438, 449, 450, 569,
    570
  Power C 609, 629, 640
  Power Challenge 106, 113,
    295, 296, 413, 505, 626
  Power Challenge Array 506
  POWERpath-2 bus 295, 296
  SPIDER router 438, 439, 440
Simple mail transfer protocol
  (SMTP) 369
Simple network management pro-
  tocol (SNMP) 369, 480
Simple operation 164
Simultaneous multithreaded
  processor 204, 205
Single address space 15, 407
Single assignment 722
Single chip multiprocessor 204,
  206
Single entry point 458, 475, 486,
  504
Single entry single exit 612
Single file hierarchy 476
Single input/output 480
Single instruction multiple data
  (SIMD) 14, 26, 72, 286, 301,

568
Single instruction single data
  (SISD) 14
Single job management
  system 481
Single memory space 481
Single mode fiber 336
Single networking 480, 485
Single point of control 480
Single point of failure 461, 462
Single process space 482
Single program multiple data
  (SPMD) 14, 72, 378, 382, 541,
  546, 594
Single system image (SSI) 31,
  32, 326, 457, 473, 474, 478, 479,
  521
Single thread of control 60
Single threaded processor 261
Single threading 705, 707
Single user interface 481
Slack cluster 454
Slave device 294
Sliding window mechanism 373
Small computer systems interface
  (SCSI) 275, 294, 307, 308, 428
Small in-line memory module
  (SIMM) 213
Snoopy bus 222, 226
Snoopy coherency protocol 223
Socket 375
Socket-FM 390
Software
  Combining 363
  Migration 420
  Multithreading 343
  Overhead 280
  Support 251
  Tool 160
Software coherent NUMA (SC-
  NUMA) 238, 241
Software-implemented
  DSM 211, 505
Solaris MC 479, 482, 484, 485,
  486, 502, 504, 507
Solaris thread 344
Space sharing 70, 490
SparcCenter 2000 235
SPARCcluster 507
Spatial locality 216
Specific performance 123
Speedup 17, 43
  Fixed memory 143
  Fixed time 136, 142

Fixed workload 135, 142
  Memory bound 140
  Scaled 137
SPIDER router 438, 439, 440
Spin locking scheme 366
Splash benchmark 92, 447
Split I/D caches 177
Split transaction 415
Split-C 385
Standard communication
  mode 675
Standard exchange
  algorithm 396, 397
Standard high volume
  (SHV) 332, 409, 413, 414, 426
Standard send 675
Stanford Dash 245, 246, 253,
  254, 265, 266, 449, 450
STAP benchmark 92, 97, 111,
  594
Star network 289, 293
Startup latency 17, 123, 124,
  591, 592
Starvation freedom 351
Static network 277
Static priority 488
Static random access memory
  (SRAM) 213
Storage capacity 577
Store-and-forward routing 283,
  284
Stratus RADIO cluster 513, 515,
  516
Strict consistency 233
Stripe group 552
Subword parallelism 196
SUN Microsystems, Inc.
  Gigaplane bus 295, 296, 299
  MicroSPARC-2 179
  SPARC 174, 266, 344
  SPARCcluster 507
  SunFire Enterprise 10000 421
  SunFire Enterprise 6000 421
  Ultra Enterprise 295, 299
  Ultra Enterprise 10000 409,
    412, 416
  Ultra Enterprise 6000 413
  Ultra Servers 235
  UltraSPARC 180, 194
  UltraSPARC-2 179, 191
  UltraSPARC-I 417
Superpipelined architecture 173,
  175
Superscalar architecture 173,
  175, 178, 198

Superspeculative processor 204, 205
Superstep 24, 609
Supervisor mode 62
Sustained speed 586, 596
Switch
  Degree 281
  Delay 280
  Module 301
  Selection 306
Switched fabric 309
Switched network 278
Sybase 160, 430
Symmetric multiprocessor (SMP) 10, 26, 28, 33, 407, 408, 626
Symmetric network 281, 282
Synchronization 78, 116, 633, 655, 679
Synchronization objects
  Process-local 348
  Process-shared 348
Synchronization operation 15
Synchronizing load 253
Synchronous clocking 276
Synchronous communication mode 675
Synchronous iteration paradigm 610
Synchronous message passing 655
Synchronous network 282
Synchronous send 675
System
  Availability 35, 520
  Bus 275, 305
  Interconnect 273
  Mode 62
  Utilization 599
System area network (SAN) 244, 274, 275
Systolic array network 289, 290

T

Tandem Computers, Inc. 513
  Himalaya 507, 515
  ISV Portfolio 516
  NonStop ServerWare server 516
  ServerNet 516
Tape unit 214
Target detection (TD) 111, 595, 723

Task descriptor 66
Task identifier 692
Tbps (Tera bits per second) 276
TCP
  Flow control 373
  IP 307, 308, 367, 369, 370, 375, 399, 455, 458, 528, 529
  Segment 369
  Socket 377
TELNET 369, 371
Temporal locality 216
10 BaseT 311, 334
Termination 85
Test and set 359, 360, 364
Thin node 521
Thin node-2 521
Thinking Machines Corp.
  *Lisp 733
  C* 733
  C-2 567
  CM Fortran 733
  CM2 291, 570
  CM5 290, 291, 570, 654
  CMMD 654
Thread 343
  Control 344
  Creation 346
  Library 71
  Local storage 347
  Management 346, 634, 641
  Scheduling 347
  Structure 346
  Synchronization 348
3-D FFT PDE (FT) 585
3-D torus network 567, 572
Throughput processing 54, 102, 103, 162
Ticket algorithm 365
Tightly coupled MPP 379
Tiling 488, 490
Time sharing 68, 70, 490
Time synchronization problem 119
Torus network 289
Total exchange 124, 396, 397, 590, 679
Total failure 461
Total messaging 591
Trace processor 204, 205
Transaction memory 359, 362
Transaction Processing Council (TPC) 326
Transactions per second (TPS) 11
Transient failure 461

Transmission control protocol (TCP) 372, 380
Transmission delay 591
Transport level interface (TLI) 369
TreadMarks DSM cluster 241, 505, 510, 556, 557, 560, 561
TruCluster 5, 114, 378, 386, 394, 505, 507, 508, 531, 532, 533, 536, 537, 538, 539, 540, 557, 562, 563, 564
2-D mesh network 289, 293, 329, 567
2-D torus network 289, 293

U

UDP socket 375
UDP/IP 367, 528, 529
Ultra Enterprise 10000 409, 412, 416
Ultra large scale integrated (ULSI) 6
Ultra port architecture (UPA) 417
UltraSPARC-2 179, 191
Unbound thread 345
Unified cache 169
Uniform device naming 485
Uniform memory access (UMA) 15, 237, 242
UNIFY 654
Uniprogramming 68
Unit of transfer 212
Univac LARC 5
Universal message passing (UMP) 394, 539
Unix (BSD 4.2) 375
Unix cluster 506
Unlock(S) 355, 356
Unplanned failure 460
Unshielded twisted pair (UTP) 276, 335
US protocol 387, 390, 464, 520, 528
User datagram protocol (UDP) 371, 380, 694
User mode 62
User network interface (UNI) 320
User process 62
User space (US) 64, 380, 383, 386
Utilization 17, 102
Utopia system 494

V

Vector reduction 709
Very long instruction word
 (VLIW) 173, 199, 200
Vinca 514
Virtual address 63
Virtual channel identifier
 (VCI) 319
Virtual file system (VFS) 483
Virtual machine 686
Virtual memory 64
Virtual memory mapped commu-
 nication (VMMC) 394
Virtual node number (VNN) 541
Virtual path identifier (VPI) 319
Virtual shared disk (VSD) 529
Visual instruction set (VIS) 197
VME bus 294, 297
von Neumann model 52
vRPC 394

W

Wait-for graph 358
Wallclock time 109
Weak consistency 231, 232, 233
Weak ordering 234
WebWindow 459
Wide area network (WAN) 274,
 276
Wide node 521
Wind Tunnel 93, 241, 510
Windows NT server cluster 505
Wolfpack API 512
Work pool paradigm 610, 612
Work sharing construct 631, 641
Working set 241
Workload 17, 126, 585
World wide virtual machine
 (WWVM) 511
World Wide Web (WWW) 162
Writable data 221
Write back (WB) 215, 220, 225,
 227
Write invalidate protocol 222,
 223
Write miss 225
Write through (WT) 215, 220,
 227
Write update protocol 222, 224

X

XFS file system 447, 541, 549,
 554, 555, 556, 562
xlisp 132

Z

Zero copy protocol 395

Author Index

A

Adams, D. 603
Adams, J. 733, 737
Adve, S.V. 235, 236, 266, 737
Agarwal, A. 263, 264, 266, 737
Agarwal, R.C. 149, 562
Agerwala, T. 562
Agha, G. 649, 737
Aho, A. 737
Albert, E. 733
Alexander, T.B. 451
Alexandrov, A. 737
Allan, M.S. 738
Allan, S.J. 649
Almasi, G.S. 47, 148, 738
Alnes, K. 451
Alverson, R. 603, 738
Amarasingbe, S.P. 451
Amdahl, G.M. 47, 148, 738
Amza, C. 558, 562
Anderson, D. 207, 738
Anderson, J. 399, 603
Anderson, T.E. 344, 555, 561, 562, 738
Andrews, G.R. 87, 649, 738
Apt, K.R. 87
Arden, B.W. 337, 738
Arpaci, R.H. 488, 501, 562, 603, 738
Astfalk, G. 424, 451
Athas, W.C. 149
Azagury, A. 502

B

Bader, D.A. 397, 398, 399
Baer, J.L. 738, 739, 743, 764
Baeten, J.C.M. 87
Bailey, D.H. 149
Baker, M.A. 492, 493, 501

Bal, H.E. 649
Bala, V. 399
Banerjee, U. 649, 739
Baron, R. 451
Barrett, R. 649
Bell, G. 47, 148, 258, 454, 603, 739
Berger, D. 204
Bernstein, A.J. 621
Berstin, J. 501
Bertsekas, D.P. 649
Bhat, P. 603
Bhuyan, L.N. 451, 740
Bisiani, R. 266
Blelloch, G.E. 731, 733
Blevins, D.W. 603
Blume, W. 132, 148, 618, 649
Blumrich, M.A. 399
Boden, N.J. 337, 741
Bokhari, S.H. 396
Bond, R. 741
Bordawekar, R. 603
Bowen, N.S. 502
Brehob, M. 207
Brent, R.P. 741
Brewer, T. 424, 451
Briggs, F.A. 232, 745, 750
Brinch-Hansen, P. 649
Bruck, J. 699
Bryhni, H. 338
Burger, D. 207, 741
Burkhardt, H. 266
Butler, R 654

C

Cabillic, G. 562
Callaghan, B. 502
Callahan, D. 742
Cameron, K. 590
Cao, J. 502
Cardoza, W.M. 540, 562, 742

Carey, G.F. 603
Carriero, N. 649
Catenzaro, B. 451
Catlett, C. 562, 742
Cekleov, M. 451, 561
Censier, L. 243, 266
Chaiken, D. 266
Chandy, K.M. 699, 733, 743
Chapman, B. 649, 733
Chien, A.A. 399, 603, 699, 751
Chin, C.Y. 743
Chong, Y.K. 266, 743
Choudhary, A. 603
Chow, C.K. 266, 743
Clapp, R.M. 427, 429, 432, 433, 451
Clark, R. 451
Clarke, L.J. 590
Colwell, R. 207
Comer, D.E. 399
Cragon, H.G. 266, 744
Crawford, D. 575, 577, 578, 744
Crawford, J.H. 206
Culler, D.E. 47, 267, 384, 399, 451, 744

D

Dally, W.J. 284, 337, 603, 744
Daniel, R.G. 603
Davy, J.R. 47
Deconinck, G. 502
DeGroot, D. 603, 750
Del Rosario, J. 603
Denman, M. 207
Denning, P.J. 266, 745
Despain, A.M. 740
Dew, P.M. 47
Diefendorff, K. 207
Dijkstra, E.W. 745

Dincer, K. 562
Dongarra, J.J. 93, 149, 603, 745
Du, X. 501
Duato, J. 337, 745
Dubey, P.K. 207
Dubois, M. 232, 266, 451, 745
Dunigan, T.H. 603

E

Edenfield, R. 206
Edmondson, J. 207
Eggers, S.J. 207, 746, 762
Elnozahy, E.N. 502
Engerbretsenh, D.E. 338
Enslow, P.H. 746

F

Feautrier, P. 243, 266
Feitelson, D.G. 501
Felten, E.W. 746
Feng, T.Y. 337, 746
Fenwick, D.M. 451, 562
Fisher, J.A. 207, 746, 756
Flanagan, D. 338, 762
Flynn, M.J. 47, 266, 747
Fortune, S. 47
Foster, I. 562, 603, 699, 743, 747
Fox, G.C. 459, 562, 747
Franklin, M. 469, 502
Frye, D. 603
Frymoyer, E.M. 338

G

Gajski, D.D. 451, 747
Gary, J. 502
Geist, A. 699
Gelernter, D. 649, 742
Germond, T. 207
Gharachorloo, K. 233, 235, 236, 255, 266, 737, 748
Ghosh, J. 337, 750
Gibson, G.A. 748

Gillett, R.B. 337
Goodman, J.R. 204, 207, 232, 266, 399, 741, 748
Gottlieb, A.J. 47, 148, 399, 603, 738, 748
Grama, A.Y. 145
Gray, J.N. 47, 740, 748
Greenley, D. 207
Grimshaw, A. 562
Gropp, W. 699
Gross, T. 379, 761
Gunther, N.J. 603
Gupta, A. 145, 254, 255, 256, 257, 265, 267, 748
Gustafson, J.L. 136, 148, 748
Gwennap, L. 207, 749

H

Hagersten, E. 241, 266, 749
Hall, M.W. 649
Hammond, L. 207
Harris, J. 724, 725
Harris, T.J. 47
Hatcher, P.J. 749, 758
Henderson, M. 603
Hennessy, J.L. 47, 206, 245, 255, 266, 737, 742, 744, 748, 749, 750, 757, 760, 762
Herlihy, M.P. 362, 399
Hill, M.D. 47, 266, 737
Hill, W.D. 733
Hillis, W.D. 603, 749
Ho, C.-T. 396, 399, 751
Hoare, C.A.R. 749
Hockney, R.W. 118, 123, 148, 594, 603, 750
Holt, C. 451
Horst, R.W. 603
Hunt, D. 207
Hwang, K. 47, 148, 149, 206, 266, 267, 337, 364, 399, 451, 501, 561, 570, 603, 699, 733, 743, 750, 756, 758, 759, 762, 763

J

Jain, R. 338
James, D.V. 332, 338
Jermoluk, T. 751
Jesshope, C.R. 750
Johnson, D. 364
Johnson, M. 207, 751
Johnsson, L. 337, 396, 399, 751
Jouppi, N.P. 207, 751

K

Kagi, A. 399
Kane, G. 751
Karamchetti, V. 399, 603
Karp, A.H. 148, 752
Katz, R. 603, 746
Kaufmann, R. 337
Kaufmann, W.J. 752
Keller, R.M. 87
Kennedy, K. 649, 741, 742, 743, 749, 752
Khalidi, Y.A. 482, 483, 484, 485, 502
Koelbel, C. 733
Koeninger, R.K. 603
Kontothanassis, L.I. 365, 399, 557, 562
Konuru, R.B. 699
Koopman, P.J. 603
Kowalik, J.S. 603, 752
Kozyrakis, C.E. 207
Kruskal, C.P. 364, 399
Kuck, D.J. 603, 752
Kumar, V. 145, 148, 501
Kung, H.T. 337, 752

L

Lam, M.S. 603, 753
Lamport, L. 231, 266, 399, 753
Langendorfer, H. 501
Larus, J.R. 132
Laudon, J. 451, 753

Lauria, M. 699
Lawrie, D.H. 337, 752, 753, 756
Lawton, H. 538, 539
Lawton, J.V. 562
Lee, H. 337
Lee, R.B. 207, 753
Leighton, F.T. 399
Leiserson, C.E. 337, 603, 752, 753
Lenoski, D.E. 47, 148, 149, 266, 451, 753
Levitan, D. 207
Levy, E. 502
Lewis, T.G. 754
Li, K. 266, 502, 742, 754, 757
Lilja, D.J. 754
Lin, X. 754
Lipasti, M.H. 207
Litzkow, M. 502
Lo, V. 233, 266, 756
Loveman, D. 733
Lovett, T.D. 427, 429, 432, 433, 451
Lu, P. 649
Lusk, E. 654

M

Mao, W. 758
Margulis, N. 207
Markoff, J. 207
Marson, M.A. 451
Martin, R.P. 385, 501
Mashey, J.R. 409
Mattson, T.G. 579, 580, 581, 603
Matzke, D. 207
McCalpin 95
McKinley, P.K. 337, 603
McVoy, L. 93, 94, 116, 117, 409
Mellor-Crummey, J.M. 366, 399
Messina, P. 755
Miller, E. 603

Milner, R. 87
Mirapuri, S. 207
Moir, M. 399
Moore, G.E. 156, 755
Morris, J.H. 502
Moss, J.E.B. 362
Mowry, T. 255, 267
Munchnick, S.S. 207

N

Netzer, R. 472, 502
Ni, L.M. 47, 139, 148, 337, 603, 745, 754, 756
Nicolau, A. 756
Nikhil, R.S. 267, 756
Nitzberg, B. 233, 266, 756
Nupairoj, N. 603

O

O'Connor, J.M. 207
Oldehoeft, R. 649
Olderog, E.-R. 87

P

Padua, D.A. 756
Pakin, S. 399
Pancake, C.M. 756
Panda, D.K. 756
Partridge, C. 337, 757
Pase, D.M. 649
Patel, J.H. 757
Patt, Y.N. 207
Patterson, D.A. 47, 206, 207, 266, 738, 743, 744, 749, 752, 757
Pawlowski, B. 502
Peir, J.K. 451
Peleg, A. 207
Petri, S. 501
Pfister, G.F. 47, 87, 501, 505, 561, 757
Piantenosi, J.A. 562
Plank, J.S. 502
Ponnusamy, R. 726, 733

Prasanna, V. 603
Preparata, F.P. 337, 758
Przybylski, S. 266, 758
Puaut, I. 562

Q

Quinn, M. 749, 758

R

Raghavendra, C. 337
Rattner, J. 47, 758
Ravishankar, M. 266
Reihardt, S.K. 562
Roberts, E. 338
Rosenblum, M. 501
Rudolph, L. 501

S

Saavedra, R.H. 258, 263, 267, 758
Sahni, S. 148, 152, 758
Sandberg, R. 561
Sanderram, V.S. 699
Santayaraman, M. 502
Saphire, W. 486, 501, 758
Saunders, S. 337, 758
Scheurich, C. 232, 451
Schwartz, J.T. 759
Scott, M.L. 366, 399
Scott, S.L. 364, 603, 759
Seitz, C.L. 149, 284, 603, 737, 739, 741, 744, 759
Sequin, C. 207
Shah, J. 562
Shang, S.S. 364, 451, 759
Shanley, T. 207
Shen, J.P. 207
Siegel, H.J. 337, 759
Siewiorek, D.P. 502, 603, 747, 759
Silberschartz, A. 502
Sindhu, P.S. 266, 759
Singh, J. 760
Singhal, A. 337, 451

Sites, R.L. 758, 760
Slater, M. 760
Smarr, L. 562, 742, 752
Smith, A.G. 590
Smith, A.J. 266, 760
Smith, B. 47, 252, 267, 738
Smith, B.J. 603, 760
Smith, J.E. 207, 759, 760
Snell, Q.O. 148
Snir, M. 399, 562, 670, 699, 760
Snyder, L. 760
Sohi, G. 760
Song, S. 207
Staelin, C. 94, 117
Stallings, W. 206, 337, 760
Steck, R. 207
Steel, G.L. 733
Stenstrom, P. 149, 241, 267, 761
Sterling, T. 451, 603
Stone, H.S. 266, 285, 337, 761
Stone, J.M. 362, 399
Stricker, T. 379
Stunkel, C.B. 562, 737, 761
Sun, X.-H. 139, 148, 761
Syanarayanan, M. 451

T

Tabak, D. 206, 761
Take, R. 396
Tananenbaum, T. 502
Tang, C.K. 243
Tannenbaum, T. 761
Thakkar, S.S. 266, 451, 746
Thanvantri, V. 148, 152
Tolmie, D. 316, 318, 338, 762
Tomasulo, R.M. 762
Topham, N.P. 47
Torng, H.C. 762
Torrellas, J. 762
Tremblay, M. 207
Tsitsiklis, J.N. 649
Turcotte, L.H. 455, 501, 699

U

Ullman, J.D. 737

V

Vahalia, U. 502
Vahdat, A. 547
Vajapeyam, S. 207
Valiant, L.G. 47, 398, 762
Varma, A. 337
Vassiliadis, S. 762
von Eiken, J. 762
Vuillemin, J. 337

W

Wah, B.W. 451, 762, 764
Waingold, E. 207
Walker, B.J. 501
Wall, D.W. 131, 148, 207, 751, 762
Wang, C.J. 603, 750, 762
Wang, C.-L. 750, 762
Wang, H.C. 149, 451, 762
Weaver, D.L. 207
Weber, W.-D. 47, 148, 255, 451
Weijland, W.P. 87
Weiser, U. 207
Wheat, S.R. 583
Whitney, S. 451
Wilson, A.W. 763
Wilson, G.V. 649
Wolfe, M. 649, 763
Wong, K.F. 469, 502
Worley, P.H. 148
Worlton, J. 763
Wu, B. 338
Wyllie, J. 47

X

Xu, J. 472, 502, 763
Xu, Z. 47, 148, 399, 501, 561, 699, 750, 763

Y

Yalamanchili, S. 337
Yeager, K.C. 207
Yew, P.C. 451, 764
Yu, A. 201, 207, 764

Z

Zegura, E.W. 338
Zhang, X. 365, 366, 367, 451, 501, 740, 764
Zhou, S. 495, 497, 498, 502, 764
Zima, H. 649, 764
Zucker, R. 764
Zwaenepoel, W. 744